Protein Deimination in Human Health and Disease

Anthony P. Nicholas • Sanjoy K. Bhattacharya
Paul R. Thompson
Editors

Protein Deimination in Human Health and Disease

Second Edition

Springer

Editors
Anthony P. Nicholas
Department of Neurology
University of Alabama at Birmingham
and The Birmingham Veterans
Administration Medical Center
Birmingham, AL, USA

Sanjoy K. Bhattacharya
Bascom Palmer Eye Institute
University of Miami
Miami, FL, USA

Paul R. Thompson
Department of Biochemistry and Molecular
Pharmacology
UMASS Medical School
Worcester, MA, USA

ISBN 978-3-319-58243-6 ISBN 978-3-319-58244-3 (eBook)
DOI 10.1007/978-3-319-58244-3

Library of Congress Control Number: 2017948284

Printed on acid-free paper

This Springer imprint is published by Springer Nature
The registered company is Springer International Publishing AG
The registered company address is: Gewerbestrasse 11, 6330 Cham, Switzerland

On behalf of the editors, I wish to dedicate this book to the memory of Kerri Anne Mowen, who sadly passed away at age 41 on February 14, 2016 of a brain aneurysm. Although I had long been familiar with her work, I first met Kerri in 2008 at the FASEB Methylation meeting where we shared our equal passion for both the protein arginine deiminases and protein arginine methyltransferases. What impressed me most about our first meeting was that Kerri was not only whip smart but also a joy to be around. We quickly became collaborators,

and most importantly friends, leading us eventually to cofound Padlock Therapeutics. Kerri's contributions to the PAD field are indelible and include developing both PAD2 and PAD4 knockout mice, helping establish the key role of PAD4 in NETosis, and establishing the importance of PAD4 activity in the initiation versus effector phases of rheumatoid arthritis. Her imprint on the PAD field will long be felt, and her future contributions sadly missed.

PRT

Contents

Chapter 1
A History of Deimination Research in Japan: The Founding Fathers

Hidenari Takahara

The research on protein deimination and the enzymes responsible for catalyzing this posttranslational modification started from investigations of the hair follicle and myelin in central nerve system. About 60 years ago, in 1958, the presence of protein-bound citrulline was first reported by Dr. George Rogers (Adelaide University, Australia) in the protein of the inner root sheath (IRS) of hair follicles (Rogers 1958). In order to obtain information about the protein composition of the IRS, he conducted a quantitative amino acid analysis of an acid hydrolysate on sufficient amounts of IRS that were dissected from the vibrissae follicle of rats. At that time, the common method for separating amino acids was paper chromatography, and when applied to the IRS hydrolysates, citrulline was discovered as a distinct ninhydrin-positive spot in an area adjacent to the basic amino acids. About 10 years after Rogers's discovery, Dr. Mario Moscarello (Toronto University, Canada) started an intensive investigation of myelin sheath proteins in the central nerve system. In 1971, he also found the presence of peptide-bound citrulline in myelin basic protein (MBP) using similar methods to Rogers (Finch et al. 1971). Moscarello continued the investigation of MBP until he passed away in 2013, publishing many papers concerning the hyper-deimination of MBP in the pathology of multiple sclerosis. His research career involving the deimination of MBP was described in a eulogy in the first volume of this book series (Nicholas and Bhattacharya 2014).

Although it was unclear as to how the citrulline was incorporated into proteins, the source was thought to be arginine. In 1977, Rogers and colleagues (1977) were the first to conclusively demonstrate that arginine residues were indeed converted to citrulline via a deimination reaction where they combined hair follicle extracts with calcium to promote this reaction. Following this report, research to identify the specific deiminating enzyme was energetically carried out in Japan. In 1979, Dr. Kiyoshi Sugawara (Fig. 1.1; Ibaraki University, Japan) reported the presence of protein-bound

H. Takahara (✉)
College of Agriculture, Ibaraki University,
3-21-1, Chuuo, Ami, Inashiki 300-0393, Ibaraki, Japan
e-mail: hidenari86ta@msn.com

A.P. Nicholas et al. (eds.), *Protein Deimination in Human Health and Disease*,
DOI 10.1007/978-3-319-58244-3_1

Fig. 1.1 The founding fathers: *left*, Professor Dr. George Rogers (Adelaide University, Australia). *Center*, Professor Dr. Hidenari Takahara (Ibaraki University, Japan). *Right*, Professor Dr. Kiyoshi Sugawara (Ibaraki University, Japan). Taken at the first International Symposium of Deimination and Skin Biology, April 2009 in Osaka, Japan

citrulline in the epidermal proteins of newborn rat (Sugawara 1979). Successively, he demonstrated the existence of the enzyme that converts arginyl to citrulline residues in the extracts from the newborn rat epidermis (Fujisaki and Sugawara 1981). To assay the enzyme activity, he introduced a colorimetric method using simple synthetic substrates of arginine blocked at the N- and C-terminals as a substrate. In this procedure, high temperatures (over 50 °C) and the presence of dithiothreitol (DTT) greatly enhanced enzyme activity. According to these procedures, he could overcome tedious and laborious work that was needed to measure enzyme activity using an amino acid analyzer. Dr. Sugawara then introduced the logical name peptidylarginine deiminase (PAD) for the enzyme, because it acts on arginine residues embedded in a peptide backbone and is distinct from deiminases that act on free arginine (Fujisaki and Sugawara 1981). In this year, PAD was registered as new enzyme to IUPAC Enzyme Committee and was classified into EC 3.5.3.15.

In 1982, I joined Dr. Sugawara's laboratory. This was just after he obtained a new finding that the extract from rabbit skeletal muscle contains very high deiminase activity, about 120-fold compared to that of the newborn rat epidermis. Since the available amounts of the newborn rat epidermis were very low and the tissue preparations were burdensome, the high abundance of a PAD in rabbit skeletal muscle was an exciting research finding that gave me a tremendous head start to further characterize this enzyme. My first research project at Ibaraki University was to purify the PAD from rabbit skeletal muscle. Very fortunately, I quickly purified the enzyme to homogeneity and determined the chemical, physiochemical, and kinetic properties toward several synthetic arginine derivatives including natural proteins (Takahara et al. 1983). This was the first and most definitive report demonstrating that the enzyme could catalyze the conversion of arginyl to citulline residues in native protein substrates in vitro. Among the protein substrates examined using this purified PAD, the reaction toward the Kunitz soybean trypsin inhibitor (STI) attracted our attention (Takahara et al. 1985). The effect of the enzyme on STI activity was remarkable as treatment with this PAD rapidly abolished the inhibitory activity of STI without altering its overall conformation; complete inactivation of STI was attained within several minutes at 37 °C. Surprisingly, only the modified arginine residue was the reactive site (or primary contact site) despite the fact that all of the remaining nine arginine residues in STI are exposed on the protein surface (Takahara et al. 1985). This study was first an indication of a biological function for the deimination and biochemical application of PAD. Furthermore, the observation that skeletal muscle PAD showed a high affinity for only the functional arginine residue in STI inspired the idea of an effective affinity adsorbent composed of immobilized STI for PAD purification. Our expectation was fully realized, as a 1800-fold purification with 50% yield was achieved by this affinity column (Takahara et al. 1986). Thereafter, we could supply a sufficient amount of purified rabbit skeletal muscle PAD to other researchers. Although recombinant enzymes from various sources superseded the rabbit skeletal muscle PAD since the latter half of the 1990s, the natural enzyme is still under requisition today. Several earlier experiments conducted with rabbit skeletal muscle PAD elicited important insights into the physiology and pathophysiology of protein deimination. For instance, our collaborative work with Dr. Masaki Inagaki (Aichi Cancer Center, Japan) provided very interesting results. In general, vimentin, an intermediated filament protein, is expressed by various cells and forms a stable, less dynamic molecular network. In 1989, we found that there was a complete loss of filament-forming ability of vimentin after PAD treatment. The enzyme could also deiminate the filaments that had been polymerized and induced filament disassembly. The deimination reduced the isoelectric point of the head domain, in which the positive charge of arginine residues are essential for maintaining the ability to form filaments, resulting in the complete loss of their intermediate filament constructs. Similar results were obtained with other intermediate filaments such as desmin and glial fibrillary acidic protein (GFAP) (Inagaki et al. 1989). Thus, we presumed that deimination of intermediated filaments controls the cytoskeletal network. This hypothesis was verified by several subsequent reports by others in the field. In particular, deiminated vimentin was found in vivo, and this modification triggers structural collapse and promotes apoptosis (Asaga et al. 1998; Hsu et al. 2014). There is also some evidence that deimination of GFAP is a characteristic feature of neurodegenerative diseases (Ishigami et al. 2005).

In parallel with research on rabbit skeletal muscle PAD, we attempted to develop a model system using the mouse/rat for investigation of the physiological function of PAD. From 1988, Dr. Tatsuo Senshu (Tokyo Metropolitan Institute of Gerontology, Japan) started his investigations into the PADs. Together, by 1995, our findings, coupled with those of Senshu's laboratory, established that PAD is widely distributed in many tissues with the notable exception of serum and the location of the enzyme was essentially in cytoplasm. Among the tissues tested thus far, the activity of PAD in the salivary glands, pancreas, and uterus far exceeded those of any other tissues.

Immunohistochemical analyses indicated that the enzyme is preferentially located in acinal cells of the salivary glands and pancreas and in the luminal and glandular epithelia of the uterus (Takahara et al. 1989; Watanabe et al. 1988). Additionally, we noted estrous cycle-dependent changes in enzyme expression in the uterus, with the level being highest and lowest at diestrus (Takahara et al. 1989). Senshu's group also found estrous cycle-dependent change of this enzyme in the rat pituitary gland (Senshu et al. 1989). The expression of PAD in the pituitary and uterus responds adequately to administration with 17β-estradiol (Senshu et al. 1989; Takahara et al. 1992). In the uterus, a remarkable series of events takes place during the estrous cycle. The luminal and glandular epithelia of the uterus at the estrous stage show hyperplasia and vigorously secrete fluid into the lumen. Therefore, PAD may be important for exocrine events, but the physiological roles of PAD in the uterus and pituitary are still unknown.

During investigations of PAD activity in the skin, we had a question: why are the substrate specificities of the skeletal muscle PAD toward several arginine derivatives different from those of the epidermal PAD reported previously (Fujisaki and Sugawara 1981)? Both PADs showed high activity toward the synthetic arginine derivatives blocked at the N- and C-termini (i.e., benzoyl-L-arginine ethyl ester), whereas the skeletal muscle PAD showed very low activity to C-terminal free arginine derivatives such as benzoyl-L-arginine and acetyl-L-arginine. On the other hand, these C-terminal free substrates were comparably processed by epidermal PAD. This question was resolved by our comprehensive work published in 1991 (Terakawa et al. 1991). We compared the elution profiles of the PAD activities of the extracts from several tissues of mouse using anion-exchange chromatography, in which PAD activity was simultaneously measured with the different substrates. As shown in Fig. 1.2, three peaks were eluted upon chromatography of the skin extract. Since each peak showed different substrate specificities, we proposed designating them as peptidylarginine deiminase type I (PAD1), II (PAD2), and III (PAD3) according to the order of elution. The extracts of the skeletal muscle, pancreas, salivary gland, and brain (spinal cord) showed a single peak that corresponded to type II enzyme. Type I enzyme is specifically located in the uterus and epidermal cells, and type III enzyme is present in the hair follicle. These three types of enzyme were not significantly different in catalytic properties, including absolute dependence on calcium ions for activity and the stimulation with DTT. Senshu's group also described the presence of three isozymes in rat tissues and called them "epidermal type, skeletal muscle type, and hair follicle type," which correspond to PAD1, PAD2, and PAD3, respectively (Watanabe et al. 1988). Thereafter, by innovative techniques such as molecular genetics and proteomics, two new PAD isozymes were found in rat epidermis (Yamakoshi et al. 1998)/a keratinocyte cell line (Ishigami et al. 1998) and in the mouse ovary (Wright et al. 2003), and they were named PAD4 and ePAD, respectively. In 1999, Dr. Michiyuki Yamada (Yokohama City University, Japan) and colleagues identified a novel PAD in human myeloid leukemia HL-60 cells, which can induce to differentiate into granulocytes by retinoic acid (Nakashima et al. 1999). By comparison of the amino acid sequence and substrate specificity of HL-60 PAD with those of the four known rat PADs, they

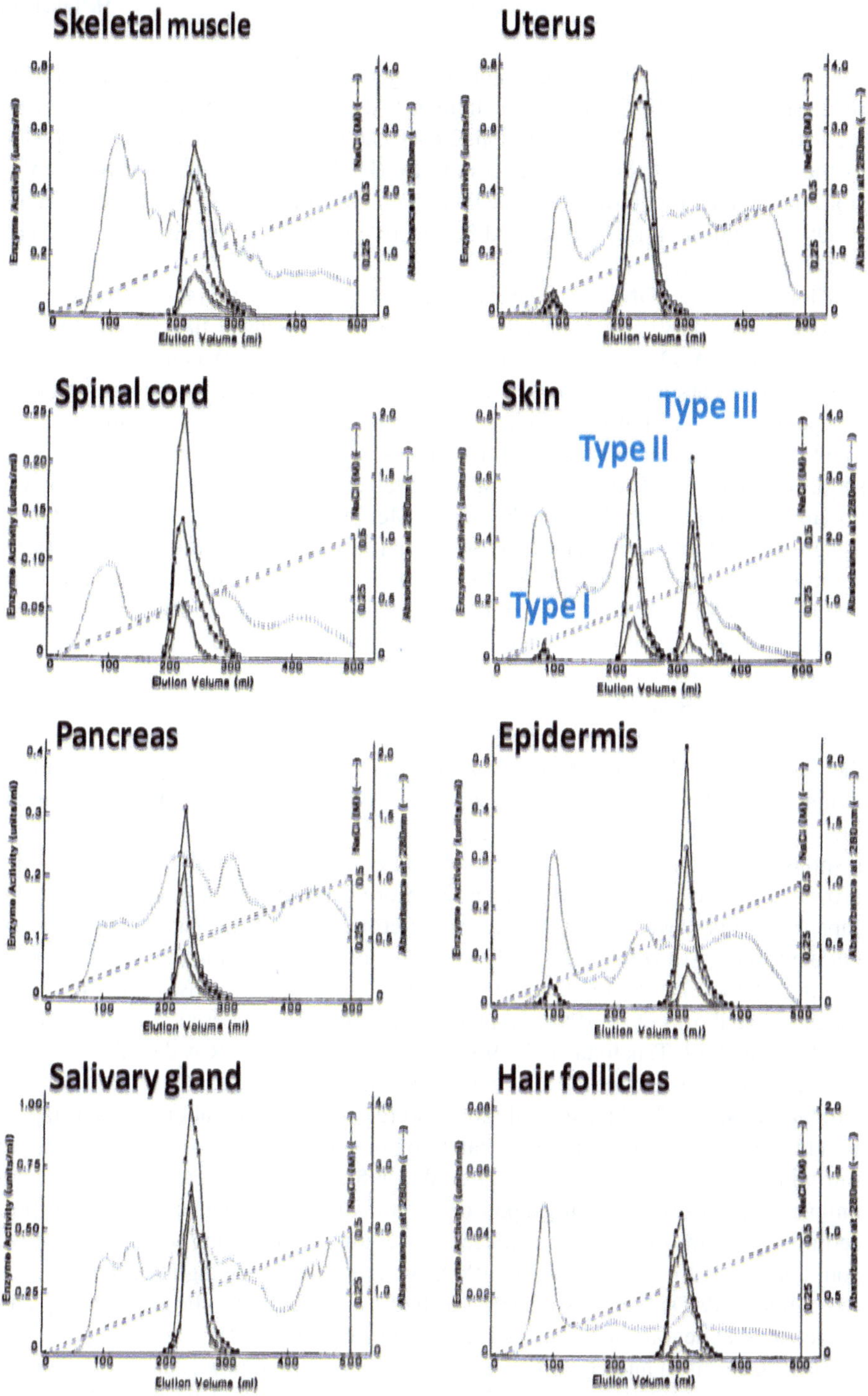

Fig. 1.2 DEAE-Sephacel ion-exchange column chromatography of the PAD activity from various tissues from mouse. The substrates used for measuring PAD activity were Benzoyl-L-Arg-*O*-ethyl ester (*open circle*), Benzoyl-L-Arg (*open triangle*), and protamine (*filled circle*)

concluded that HL-60 PAD did not belong to any PADs and named it PAD5 (Nakashima et al. 1999). However, human PAD5 proved to be the human orthologue of mouse PAD4 (Chavanas et al. 2004), and it was subsequently named PAD4 by the HUGO Gene Nomenclature committee (HGNC). In addition, to avoid confusion, the HGNC recommended that PAD5 remains unused and ePAD be renamed PAD6. In total, it is now recognized that there are five PAD isozymes, i.e., PAD1, PAD2, PAD3, PAD4, and PAD6.

Cloning of the cDNAs for the five PAD isozymes was a historic struggle that stretched from 1989 to 2004. In 1989, Watanabe and Senshu first reported on the cDNA of rat PAD2 and deduced its amino acid sequence (Watanabe and Senshu 1989). Four years later, we revealed at long last in the full nucleotide sequence of mouse the PAD2 cDNA (Tsuchida et al. 1993). Looking back on that time, we had to overcome several obstacles to reach our goal. The N-terminal amino acid sequence of mouse PAD2 was N^{α}-acetyl-Met-Gln-, a sequence which has never previously been reported among N^{α}-acetyl-Met protein. As such, it was difficult to assign the methionine codon (ATG) at the translational start of the cDNA. In 1997, we purified a small amount of PAD3 from newborn rat epidermis by a procedure that included STI-affinity chromatography and carried out peptide mapping by the in situ protease digestion method (Nishijyo et al. 1997). Subsequently, we succeeded in cloning the cDNA and sequenced the full-length cDNA encoding rat PAD3 by RT-PCR and 3′–/5′-RACE methods using synthesized nucleotide primers designed from the internal amino acid sequences (Nishijyo et al. 1997). This was the first report exhibiting the entire amino acid sequence of the isozyme, and by alignment of the PAD2 and PAD3 sequences, we found that a half of C-terminal region was highly conserved, and we predicted that the conserved region was likely responsible for the catalytic activity of this enzyme. This notion was ultimately proved by the excellent works of Dr. Mamoru Sato (Yokohama City University, Japan) and his coworkers in 2004, who determined the X-ray crystal structure of human PAD4 (Arita et al. 2004). These leading researches ultimately resulted in the cloning of the cDNAs for other PAD isozymes. Most cDNA isozymes from the rodent (Yamakoshi et al. 1998; Ishigami et al. 1998; Wright et al. 2003; Watanabe and Senshu 1989; Tsuchida et al. 1993; Nishijyo et al. 1997; Rus'd et al. 1999) and human PAD (Nakashima et al. 1999; Chavanas et al. 2004; Kanno et al. 2000; Ishigami et al. 2002; Guerrin et al. 2003) were established in laboratories in Japan. In addition, we produced the bacterial recombinant PADs of mouse and human in run-up to other laboratories (Ohsugi et al. 1995). The constructed plasmids had a unique DNA linker containing a pair of Shine-Dalgarno sequences and a short preceding cistron inserted into the adjacent 5′-region of the coding region, so that we could obtain a large quantity of the PADs without a sequence tag in bacteria. These recombinant PADs were also easily purified by STI-affinity chromatography and helped a great deal for many investigations (Ohsugi et al. 1995).

In 1992, Senshu and coworkers developed an excellent procedure for the detection of deiminated proteins on membranes or fixed tissues (Senshu et al. 1992). This method involves a three-step process. In the first step, citrulline-containing proteins immobilized on the membrane or fixed tissues are chemically modified. In the second step, immunoblotting is performed using specific antibodies that were developed against in vitro deiminated histones that were chemically modified in a similar manner. In the third step, citrullinated proteins are visualized using a secondary antibody. This method enabled detection of citrulline-containing proteins at fmol levels, regardless of backbone protein molecules. As known, histones contain a large amount of arginine residues and are highly conserved at the amino acid sequences, regardless of species, which should lead them to the idea for producing the specific antibodies against the chemically modified citrulline. Thereafter, this antibody was called the anti-modified citrulline (AMC) antibody or the "Senshu antibody" and contributed exponentially to the expansion of research on protein deimination and the PADs. Using this immunoblotting analysis, they found that filaggrin and keratin K1/K10 were citrullinated in the epidermis under physiological conditions (Senshu et al. 1995).

For more than 20 years (from 1977 to 1998), most investigations into protein deimination and PAD activity were carried out by only two research groups: ours and Senshu's group in Japan and Moscarrello's group in Canada. Unfortunately, these works were only noticed to a limited extent. There might be two major reasons why protein deimination did not receive a worldwide attention. First, the deiminated proteins were mostly structural proteins in vivo. Second, the PAD reaction is irreversible, and a "peptidyl citrulline-iminase" enzyme has not been discovered. Thus, it was not felt that this modification could reversibly regulate metabolic events and bioprocesses in the same way as phosphorylation/dephosporylation cycles.

However, in 1999, the report of Dr. Guy Serre (University Toulouse III, France) and colleagues developed a new wave that changed the image of the PADs. In this study, Serre and colleagues showed that a "mysterious" antibody present in the sera of patients with rheumatoid arthritis (RA) recognized citrullinated proteins. These antibodies were subsequently named anti-citrullinated protein antibodies (ACPA) (Girbal-Neuhauser et al. 1999). The AMC antibody and purified PAD from rabbit muscle were critical in establishing this finding. Several successive experiments conducted by Serre's laboratory finally identified the antigen recognized in the joints as deiminated α- and β-chains of fibrin (Masson-Bessière et al. 2001). Since the levels of ACPA, including anti-cyclic citrullinated peptide antibodies, are significantly higher in RA patient sera. With high specificity and positive correlation, they are the best diagnostic marker for the disease to date. The direct association between the presence of deiminated proteins and occurrence of RA enticed enormous investigations of protein deimination and PAD. Furthermore, in 2003, the paper of Dr. Akari Suzuki (RIKEN, Japan) and colleagues that described an association of a functional haplotype of the PAD4 gene (*PADI4*) in a Japanese population with RA (Suzuki et al. 2003) further drove an influx of researchers into the field. Once again, in 1999, Yamanda and his coworkers first found human PAD4 in HL-60 cells when cells were induced to differentiate into granulocytes (Nakashima et al. 1999). Their successive finding in 2002, in which PAD4 contains a canonical nuclear localization signal within N-terminal domain and is localized to nucleus, where it deiminated histones H3 and H4, suggested that PAD4 may be a new factor modulating a variety of the nuclear functions dependent on chromatin structure

(Nakashima et al. 2002). These studies also aroused enthusiasm among molecular biochemists who had showed a disinterest in the PADs until then. This is when investigations into protein deimination and PAD activity spread across the world, and the relevant reports increased over the years. To date, PADs are now known to play functional roles in key cellular processes (terminal epidermal and hair follicle differentiations, apoptosis, and gene regulation) under physiological conditions, and dysregulated PAD activity is involved in the pathogenesis of autoimmune diseases such as multiple sclerosis, rheumatoid arthritis, psoriasis, Alzheimer's disease, and cancers. So far, these findings were reviewed in many journals including this book series.

For the past decade, the field of deimination research has not so explosively accelerated, but certainly advanced. I believe that these valuable findings are founded on the studies started from their humble beginnings in Japan.

References

Arita, K., et al. (2004). Structural basis for Ca(2+)-induced activation of human PAD4. *Nature Structural & Molecular Biology, 11*(8), 777–783.

Asaga, H., Yamada, M., & Senshu, T. (1998). Selective deimination of vimentin in calcium ionophore-induced apoptosis of mouse peritoneal macrophages. *Biochemical and Biophysical Research Communications, 243*(3), 641–646.

Chavanas, S., et al. (2004). Comparative analysis of the mouse and human peptidylarginine deiminase gene clusters reveals highly conserved non-coding segments and a new human gene, *PADI6. Gene, 330*, 19–27.

Finch, P. R., Wood, D. D., & Moscarello, M. A. (1971). The presence of citrulline in a myelin protein fraction. *FEBS Letters, 15*(2), 145–148.

Fujisaki, M., & Sugawara, K. (1981). Properties of peptidylarginine deiminase from the epidermis of newborn rats. *Journal of Biochemistry, 89*(1), 257–263.

Girbal-Neuhauser, E., et al. (1999). The epitopes targeted by the rheumatoid arthritis-associated antifilaggrin autoantibodies are posttranslationally generated on various sites of (pro)filaggrin by deimination of arginine residues. *Journal of Immunology, 162*(1), 585–594.

Guerrin, M., Ishigami, A., Mechin, M. C., Nachat, R., Valmary, S., Sebbag, M., Simon, M., Senshu, T., & Serre, G. (2003). cDNA cloning, gene organization and expression analysis of human peptidylarginine deiminase type I. *Biochemical Journal, 370*(Pt 1), 167–174.

Hsu, P. C., et al. (2014). Vimentin is involved in peptidylarginine deiminase 2-induced apoptosis of activated Jurkat cells. *Molecules and Cells, 37*(5), 426–434.

Inagaki, M., et al. (1989). Ca^{2+}-dependent deimination-induced disassembly of intermediate filaments involves specific modification of the amino-terminal head domain. *The Journal of Biological Chemistry, 264*(30), 18119–18127.

Ishigami, A., et al. (1998). Molecular cloning of two novel types of peptidylarginine deiminase cDNAs from retinoic acid-treated culture of a newborn rat keratinocyte cell line. *FEBS Letters, 433*(1–2), 113–118.

Ishigami, A., et al. (2002). Human peptidylarginine deiminase type II: Molecular cloning, gene organization, and expression in human skin. *Archives of Biochemistry and Biophysics, 407*(1), 25–31.

Ishigami, A., et al. (2005). Abnormal accumulation of citrullinated proteins catalyzed by peptidylarginine deiminase in hippocampal extracts from patients with Alzheimer's disease. *Journal of Neuroscience Research, 80*(1), 120–128.

Kanno, T., et al. (2000). Human peptidylarginine deiminase type III: Molecular cloning and nucleotide sequence of the cDNA, properties of the recombinant enzyme, and immunohistochemical localization in human skin. *The Journal of Investigative Dermatology, 115*(5), 813–823.

Masson-Bessière, C., et al. (2001). The major synovial targets of the rheumatoid arthritis-specific antifilaggrin autoantibodies are deiminated forms of the alpha- and beta-chains of fibrin. *Journal of Immunology, 166*(6), 4177–4184.

Nakashima, K., et al. (1999). Molecular characterization of peptidylarginine deiminase in HL-60 cells induced by retinoic acid and 1alpha,25-dihydroxyvitamin D(3). *The Journal of Biological Chemistry, 274*(39), 27786–27792.

Nakashima, K., Hagiwara, T., & Yamada, M. (2002). Nuclear localization of peptidylarginine deiminase V and histone deimination in granulocytes. *The Journal of Biological Chemistry, 277*(51), 49562–49568.

Nicholas, A. P., & Bhattacharya, S. K. (Eds.). (2014). *Protein deimination in human health and disease*. New York: Springer.

Nishijyo, T., et al. (1997). Isolation and molecular cloning of epidermal- and hair follicle-specific peptidylarginine deiminase (type III) from rat. *Journal of Biochemistry, 121*(5), 868–875.

Ohsugi, I., et al. (1995). Expression of mouse uterine peptidylarginine deiminase in *Escherichia coli*: Construction of expression plasmid and properties of the recombinant enzyme. *Archives of Biochemistry and Biophysics, 317*(1), 62–68.

Rogers, G. E. (1958). Some observations on the proteins of the inner root sheath cells of hair follicles. *Biochimica et Biophysica Acta, 29*(1), 33–43.

Rogers, G. E., Harding, H. W., & Llewellyn-Smith, I. J. (1977). The origin of citrulline-containing proteins in the hair follicle and the chemical nature of trichohyalin, an intracellular precursor. *Biochimica et Biophysica Acta, 495*(1), 159–175.

Rus'd, A. A., et al. (1999). Molecular cloning of cDNAs of mouse peptidylarginine deiminase type I, type III and type IV, and the expression pattern of type I in mouse. *European Journal of Biochemistry, 259*(3), 660–669.

Senshu, T., et al. (1989). Peptidylarginine deiminase in rat pituitary: Sex difference, estrous cycle-related changes, and estrogen dependence. *Endocrinology, 124*(6), 2666–2670.

Senshu, T., et al. (1992). Detection of citrulline residues in deiminated proteins on polyvinylidene difluoride membrane. *Analytical Biochemistry, 203*(1), 94–100.

Senshu, T., et al. (1995). Detection of deiminated proteins in rat skin: Probing with a monospecific antibody after modification of citrulline residues. *The Journal of Investigative Dermatology, 105*(2), 163–169.

Sugawara, K. (1979). Presence of citrulline in the membranous proteins of stratum corneum of newborn rat and cow snout. *Agricultural and Biological Chemistry, 43*(12), 2215–2217.

Suzuki, A., et al. (2003). Functional haplotypes of PADI4, encoding citrullinating enzyme peptidylarginine deiminase 4, are associated with rheumatoid arthritis. *Nature Genetics, 34*(4), 395–402.

Takahara, H., Oikawa, Y., & Sugawara, K. (1983). Purification and characterization of peptidylarginine deiminase from rabbit skeletal muscle. *Journal of Biochemistry, 94*(6), 1945–1953.

Takahara, H., Okamoto, H., & Sugawara, K. (1985). Specific modification of the functional arginine residue in soybean trypsin inhibitor (Kunitz) by peptidylarginine deiminase. *The Journal of Biological Chemistry, 260*(14), 8378–8383.

Takahara, H., Okamoto, H., & Sugawara, K. (1986). Affinity chromatography of peptidylarginine deiminase from rabbit skeletal muscle on a column of soybean trypsin inhibitor (Kunitz)-Sepharose. *Journal of Biochemistry, 99*(5), 1417–1424.

Takahara, H., et al. (1989). Peptidylarginine deiminase of the mouse. Distribution, properties, and immunocytochemical localization. *The Journal of Biological Chemistry, 264*(22), 13361–13368.

Takahara, H., et al. (1992). Expression of peptidylarginine deiminase in the uterine epithelial cells of mouse is dependent on estrogen. *The Journal of Biological Chemistry, 267*(1), 520–525.

Terakawa, H., Takahara, H., & Sugawara, K. (1991). Three types of mouse peptidylarginine deiminase: Characterization and tissue distribution. *Journal of Biochemistry, 110*(4), 661–666.

Tsuchida, M., et al. (1993). cDNA nucleotide sequence and primary structure of mouse uterine peptidylarginine deiminase. Detection of a 3′-untranslated nucleotide sequence common to the mRNA of transiently expressed genes and rapid turnover of this enzyme's mRNA in the estrous cycle. *European Journal of Biochemistry, 215*(3), 677–685.

Watanabe, K., & Senshu, T. (1989). Isolation and characterization of cDNA clones encoding rat skeletal muscle peptidylarginine deiminase. *The Journal of Biological Chemistry, 264*(26), 15255–15260.

Watanabe, K., et al. (1988). Combined biochemical and immunochemical comparison of peptidylarginine deiminases present in various tissues. *Biochimica et Biophysica Acta, 966*(3), 375–383.

Wright, P. W., et al. (2003). ePAD, an oocyte and early embryo-abundant peptidylarginine deiminase-like protein that localizes to egg cytoplasmic sheets. *Developmental Biology, 256*(1), 73–88.

Yamakoshi, A., et al. (1998). Cloning of cDNA encoding a novel isoform (type IV) of peptidylarginine deiminase from rat epidermis. *Biochimica et Biophysica Acta, 1386*(1), 227–232.

Chapter 2
Aspects of Peptidylarginine Deiminase Regulation that May Predispose to Autoreactivity Against Citrullinated Proteins

Indira Neeli and Marko Radic

2.1 Summary

Autoimmune diseases represent a long-standing puzzle. In an incompletely understood series of steps, the immune system loses immune tolerance to self and acquires the ability to recognize and respond to defined and characteristic autoantigens. The involvement of an infectious agent has been suspected to trigger this transition, but a specific etiologic stimulus has not been identified. Recent years have seen an evolution in the understanding of events that lead to autoimmunity. A central role has been assigned to posttranslational modifications of autoantigens during the initial, preclinical phase of autoimmune syndromes. In response to various infections or even under sterile inflammatory conditions, the innate immune system activates a characteristic set of enzymatic reactions, including the regulated conversion of certain arginine residues to citrulline residues. The conversion, carried out by the peptidylarginine deiminases (PADs), results in the conversion of arginine residues to citrulline residues in many notable autoantigens. In turn, an important category of autoantibodies, referred to as anti-citrullinated protein antibodies (ACPA), specifically recognizes the citrullinated form of these autoantigens. Thus, the concept is gaining acceptance that diverse infections (or sterile inflammation) result in the citrullination of self-proteins, which—given genetic predisposition or a conducive infectious microenvironment—break tolerance and trigger a self-perpetuating autoimmune process. This chapter highlights aspects of PAD regulation and the development of ACPA in order to propose a unifying principle for the induction of autoimmune disorders.

I. Neeli, Ph.D. • M. Radic, Ph.D. (✉)
Department of Microbiology, Immunology and Biochemistry, University of Tennessee Health Science Center, 858 Madison Avenue, Memphis, TH 38163, USA
e-mail: mradic@uthsc.edu

A.P. Nicholas et al. (eds.), *Protein Deimination in Human Health and Disease*,
DOI 10.1007/978-3-319-58244-3_2

2.2 Tolerance and Autoimmunity

Autoimmunity was anticipated as a possible outcome that could arise from the adaptive nature of the immune system (Ehrlich P 1800s). Because the immune system is capable of recognizing almost any foreign molecule, it appeared obvious to Paul Ehrlich that autoreactivity, or "horror autotoxicus," would also predictably appear. Indeed, several dozen human autoimmune disorders are currently recognized as distinct clinical entities, and new disorders continue to be assigned to the category of maladies that arise due to a malfunction of the immune system. Many of these disorders are quite common, such that, for example, rheumatoid arthritis (RA) affects nearly 1 in 100 persons. In aggregate, several percent of the human population develop a serious health condition because of the inappropriate anti-self-reactivity of the immune system.

Conversely, it may be even more astounding that more of us do not suffer from an autoimmune disease, given the enormous rates at which cells of the immune system proliferate and the large numbers of cells that comprise the innate and adaptive immune systems. Moreover, our immune systems are constantly stimulated by microbes that survive in our environment and form part of our extensive microbiota. In view of the continued microbial challenge, it is important to consider ways in which our immune systems distinguish foreign pathogens from self. The mechanism that safeguards against the formation of autoreactive B and T lymphocytes is called immune tolerance. During development and throughout their functional responses, lymphocytes are regulated by active mechanisms that suppress immune responses aimed at molecules that form part of ourselves. The encounter of autoreactive lymphocytes with self-molecules ensures the functional inactivation or correction of the inappropriate B and T cell antigen receptors. Multiple mechanisms are engaged in central tolerance, some of these lead to anergy, deletion, or receptor editing in autoreactive B cells (Radic and Zouali 1996), and others induce cell death of autoreactive T cells in the organs that support lymphopoiesis (Palmer 2003).

The sophistication and power of immune tolerance can best be appreciated by immunization methods in experimental animal species. Standard approaches can readily yield antibodies of high affinity and exquisite specificity for the antigen of choice, provided that the immunogen contains structural motifs (epitopes) that are distinct from the endogenous antigens. For example, a protein that is isolated from the hemolymph of a marine invertebrate, keyhole limpet hemocyanin, is a versatile carrier in numerous immunization experiments. However, if the immunogen of interest is highly related or identical to the animal's own molecular components, multiple immunizations may not elicit an immune response to the intended target. A notable example is provided by the collective experience of frequent but often futile efforts to raise antibodies to various members of the histone protein family (Rubin et al. 1990).

Histones are highly conserved, basic proteins that associate into octamers of two histone H3, two H4, and two each of H2A and H2B monomers (Moudrianakis and Arents 1993). Together these eight histones form the core of a structure that serves to organize approximately 147 base pairs of nuclear DNA into a single nucleosome, the basic unit of chromatin. Histones are among the most ubiquitous, abundant, and highly conserved proteins in eukaryotes, given that histone primary amino acid sequences vary by less than 1% from humans to yeast. Thus, immunizations of mice or rabbits with histones elicit very little to no immune response due to the strong negative selection of lymphocytes against self-reactivity by immune tolerance. A similar resistance to immunization with DNA is observed, yet both DNA and histones are targets of anti-self-antibodies (Stollar 1971) in an autoimmune disease called systemic lupus erythematosus (SLE). Other conserved and abundant proteins are characteristic autoantigens in other autoimmune diseases. For example, myelin basic protein, a membrane-associated protein of neurons, is recognized by T cells in multiple sclerosis (Martin et al. 1992), insulin is a target of autoantibodies in diabetes (Michels and Nakayama 2010), and collagen or even the immunoglobulin molecule itself is attacked by the immune system in RA (Rowley et al. 2008).

But how does the immune system overcome ("break") immune tolerance and begin an autoimmune attack? Some early illustrative experiments with the multifunctional protein cytochrome c provided important clues. Immune tolerance prevents mice from responding to mouse cytochrome c. However, human cytochrome c, due to only six amino acid variations in sequence from the mouse protein, can induce a B cell response (Lin et al. 1991). Remarkably, following a subsequent immunization with a combination of mouse and human cytochrome c, the mouse B and T cells respond to the mouse cytochrome c protein. The conclusion from those experiments is that small molecular differences between autoantigens and immunogens can break tolerance and induce a self-sustaining autoimmune response, provided that B cells to the altered antigen arise which can present antigen to incompletely deleted T cells.

It was observed that small molecular differences between an immunogen and an autoantigen can also be introduced by various covalent modifications of autoantigens and that such differences can stimulate a response by the immune system. In pioneering experiments, Bill Weigle discovered that chemically modified albumin could break tolerance in rabbits tolerant of the unmodified albumin (Weigle 1962). More recent work from Mark Mamula's laboratory expanded this field of research and showed that isoaspartate racemization of self-antigens, including histones, increases their immunogenicity (Doyle et al. 2013). These experiments established the possibility that posttranslational modification (PTM) of autoantigens may contribute a possible mechanism to disrupt immune tolerance. In turn, studies in a number of laboratories identified autoantibodies that preferentially reacted against the modified form of a variety of autoantigens. This independent evidence solidified the idea that the covalent, enzymatic modifications of autoantigens play a central role in the induction of autoimmunity.

2.3 Discovery of Arginine Deimination

The first identification of citrulline residues in a protein was reported in 1958 by researchers from the Melbourne Wool Research Laboratories examining hair follicles in sheep (Rogers and Simmonds 1958). The discovery was notable for the fact that citrulline is not incorporated into proteins during translation; thus, it must be introduced into proteins by posttranslational modifications. However, formal proof for this conclusion had to await the discovery of peptidylarginine deiminases (PADs), the enzymes that convert arginine residues in proteins to citrulline residues (Fujisaki and Sugawara 1981). Subsequent research identified additional members of this small protein family, which established the expression of five isoforms of PADs in mammals (Chavanas et al. 2004). PADs are a relatively recent evolutionary adaptation, which are present in vertebrates but absent from other eukaryotic subphyla (Vossenaar et al. 2003).

Research into the structure and function of PADs intensified following the discovery that autoantigens, including myelin basic protein, keratin, and histones, are substrates for PADs (Muller and Radic 2015). These hallmark autoantigens that have high diagnostic value for the detection and prognosis of a variety of difficult to treat autoimmune disorders previously had not been considered to be linked by a common pathway. The discovery of shared PTM between autoantigens in different autoimmune disorders immediately raised expectations that a common underlying mechanism may be at the root of this category of clinical entities. Most notably, a series of elegant studies traced the RA autoreactivity against cells from the oral mucosa to yet another citrullinated antigen. Schellekens and collaborators observed that autoantibodies in RA display a high degree of specificity and sensitivity for peptides derived from filaggrin but only if the arginine residue in these peptides is replaced by citrulline (Schellekens et al. 1998). The diagnostic utility of the serologic test for anti-citrullinated protein antibodies (ACPA) was demonstrated by the observation that ACPA can be detected months or even years ahead of the appearance of clinical manifestations of RA (Johansson et al. 2016). The inclusion of ACPA as classification criteria for RA further boosted basic and applied research into PADs (Liao et al. 2008).

The five PAD isozymes differ from each other in the cell types and subcellular locations where they are expressed and, in turn, by what substrates they modify. PAD1 is expressed in the epidermis and uterus (Terakawa et al. 1991), PAD3 is active in hair follicles (Kanno et al. 2000), and PAD6 is most abundant in the oocyte and during embryonic development (Yurttas et al. 2008). PAD2 and PAD4 are of greatest relevance in cells that comprise the immune system (Vossenaar et al. 2004). Notably, PAD2 is also expressed in the central nervous system and in muscles (van Beers et al. 2013). PAD4 is highly expressed in leukocytes, the white blood cells of the innate immune system, along with other immune cells (Anzilotti et al. 2010). PAD4 is the only PAD with a distinct nuclear localization sequence, and thus, it is the probable isozyme in charge of modifying nuclear proteins, including histones (Nakashima et al. 2002). Accordingly, activation of PAD4 makes important contributions to chromatin structure modifications that regulate gene expression (Cuthbert et al. 2004; Wang et al. 2004). However, PAD2 has also been implicated in the modification of histones (Zhang et al. 2012), whereas PAD4 may also be active in the cytoplasm and even on extracellular substrates (see below).

All PADs require calcium for enzymatic activity, and maximal activity may require extracellular calcium influx to supplement intracellular calcium stores. Calcium ions organize the overall PAD structure and shape the conformation of the active site, as determined by X-ray diffraction patterns of PAD4 obtained in the presence and absence of the divalent cations (Arita et al. 2004). PADs are relatively promiscuous enzymes because they accept substrates with different amino acid sequences, provided the target arginine is preceded by a residue with a small side chain at the R-2 position and the polypeptide chain flanking the arginine can fold into a tight beta turn (Arita et al. 2006). Catalysis, which consists of the hydrolysis of the guanidino group on arginine to yield the ureido group on citrulline (Fig. 2.1), leads to the release of ammonia, the loss of arginine's positive charge, and the nearly exact gain of 1 Da in mass by the citrulline-containing product (Rohrbach et al. 2012b). Although PAD4 can accept methylated arginine as a substrate, it prefers unmodified arginine by a factor of over 100:1 in an in vitro reaction (Kearney et al. 2005; Raijmakers et al. 2007). Therefore, biochemically, one consequence of arginine citrullination is to preclude any further arginine modifications (Thompson and Fast 2006).

As the evidence for autoantibodies to citrullinated antigens accumulated, it became clear that useful applications would emerge from the development of PAD inhibitors. Insights into the conformation of the PAD4 active site were starting points for the design of different versions of PAD4 inhibitors. Work spearheaded by the Thompson lab used various chemical approaches to screen and confirm the efficacy, specificity, bioavailability, and mode of action of various compounds (Bicker and Thompson 2013). One of the earliest inactivators that were effective at inhibiting the action of several PAD enzymes was *N*-alpha-benzoyl-*N*5-(2-fluoro-1-iminoethyl)-L-ornithine amide, or F-amidine, which became available about 10 years ago (Luo et al. 2006), and was soon followed by several other irreversible and selective inhibitors (Knuckley et al. 2010). For example, a natural compound with antibiotic properties, streptonigrin, was identified as a selective PAD4 inhibitor (Dreyton et al. 2014), whereas the inorganic dye used in histology, ruthenium red,

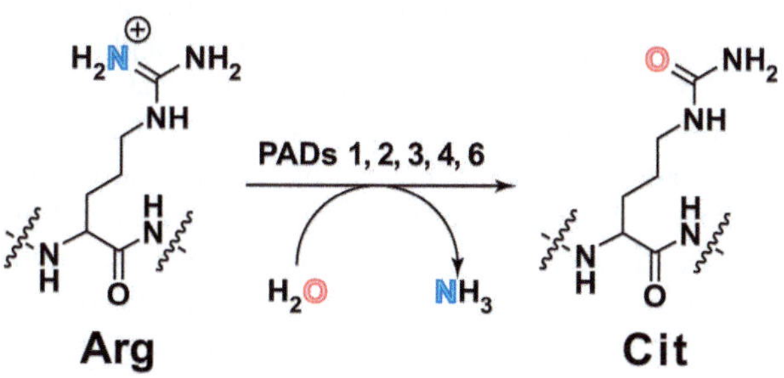

Fig. 2.1 Diagram of peptidylarginine deiminase (PAD)-mediated conversion of arginine to citrulline. One of five PADs converts arginine to citrulline by using oxygen from water and releasing nitrogen as ammonia

was a selective inhibitor of PAD2 (Lewallen et al. 2014). Ruthenium red binds to PAD2 at a site that normally accepts calcium, and thus, it prevents the activation of the enzyme. These inhibitors and inactivators are of immense value for determination of PAD activity and its consequences in vivo and in vitro.

Mounting evidence places PAD4 at the center of attention in events shaping the interactions between the innate and adaptive immunity. In particular, research indicates that PAD4 activation is intimately involved in the initial stimulation and subsequent tissue damage associated with autoimmune diseases. Nevertheless, additional open questions remain and deserve further attention. For example, although PAD4 can auto-citrullinate (Andrade et al. 2010; Mechin et al. 2010), the precise consequences of this modification are unclear, as the modified PAD shows little change in enzymatic activity on selected in vitro substrates (Slack et al. 2011). Clearly, one possibility in vivo is that auto-deimination may affect the binding with other interacting partners of the enzyme. More broadly, several unsolved questions of PAD4 regulation remain, and efforts at finding answers will drive much ongoing and future research. Some of the questions are highlighted in Box 2.1.

Box 2.1: Unsolved Questions in Cell Biology of PADs

1. Is PAD4 enzymatic activity controlled by other interacting proteins, many of which are yet to be identified?
2. Do PAD4 modifications regulate the access of calcium to its five binding sites?
3. What principles determine PAD4 recognition of other proteins and the cellular location of the enzyme?

2.4 Role of PAD4 in Innate Immune Responses

PAD4 supports fundamental functions in neutrophils, the most abundant type of granulocytes in blood. The PAD4 gene is expressed relatively late during neutrophil maturation but reaches high expression levels during a short window preceding the release of neutrophils from the bone marrow (Theilgaard-Monch et al. 2005). Even though PAD4 clearly can be transported into the nucleus and its most abundant cellular substrates are core histones, large quantities of PAD4 are also packaged into ficolin-1-rich neutrophil granules (Rorvig et al. 2013). The issue of subcellular localization can be addressed with the use of specific antibodies raised against PAD4. Figure 2.2 shows the results of probing human peripheral blood neutrophils with a monoclonal antibody to PAD4 (kind gift of Prof. Katsuhiko Nakashima). Using a preparation of unstimulated neutrophils, it can be seen that cells exhibit heterogeneous staining patterns, in which some cells have abundant nuclear PAD4, whereas others have a more cytoplasmic PAD4 distribution. In addition, particular cells exhibit a patchy nuclear distribution, perhaps indicating that PAD4 shows preference for certain chromosomal domains. The resolution of the confocal microscope does not allow the precise localization of PAD4 to cytoplasmic granules, but

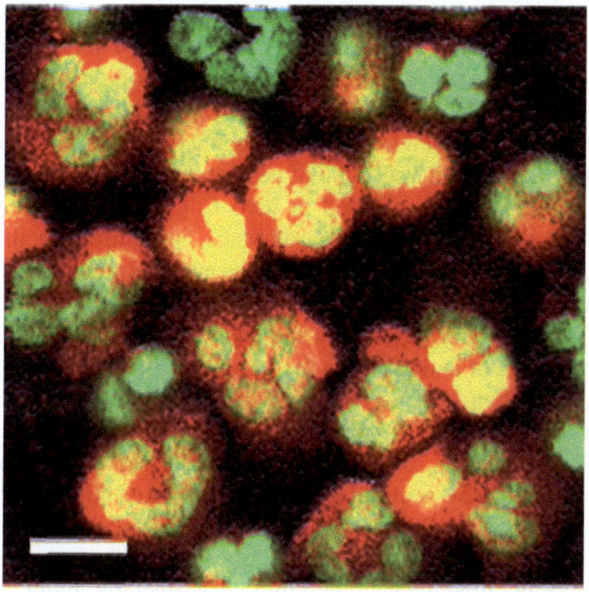

Fig. 2.2 Detection of PAD4 in purified human blood neutrophils. Human neutrophils were isolated from blood and purified using standard conditions. Fixed cells were permeabilized and incubated with a monoclonal antibody to PAD4 (kind gift from Dr. Nakashima, Japan). Antibody binding was visualized by secondary anti-mouse antibodies shown in *red*, and nuclear DNA was detected with *Sytox Green*. The overlap between the colors yields *yellow*. In this preparation, neutrophils were heterogeneous, and PAD4 was localized to both nuclei and cytoplasm. Bar is equivalent to 10 μm

that assignment is consistent with biochemical fractionation experiments carried out by Niels Borregaard's laboratory in Copenhagen.

The protective function of neutrophils in the immune system is to act as sentinels for infections or inflammation. As white blood cells, neutrophils circulate throughout the body, yet they are exquisitely sensitive to any local signs of inflammation. The presence of IL-8 (CXCL8) and expression of selectins on endothelial cells lining the blood vessels induce neutrophil attachment to the vessel wall (Riese et al. 2014) and their extravasation (migration) into tissues. Neutrophils express metalloproteases such as MMP-9 which transiently dissociate tight junctions that bind endothelial cells to each other, thus generating gaps that neutrophils exploit to squeeze through the endothelium and reach the inflamed tissues (McColl et al. 2008). Once in tissues, neutrophils convert into highly motile cells capable of chemotaxis toward increased concentrations of microbial patterns such as f-MLP and various cytokines, such as IL-1 or TNF. Neutrophils also migrate toward signals from damaged cells, such as ATP, ADP, and related molecular messengers that promote inflammation (McDonald et al. 2010). Neutrophils integrate these diverse signals by expression of multiple cell surface receptors that operate in a hierarchical system of dominant vs. supplementary signaling cascades (Heit et al. 2002; Mocsai et al. 2015).

The functional transition of a neutrophil from passive transport in the blood to an active, directional movement in tissues is mirrored in a notable increase in gene transcription. In part, the upregulation of gene expression reflects the activation of cell surface adhesion molecules, such as Mac1 integrins, which are important in the process of extravasation as well as in the subsequent migration in tissues (Kobayashi et al. 2002). The coordinated migration of neutrophils in tissues has been described as "swarming" which can be viewed in vivo following a focused tissue injury (Lammermann et al. 2013). Expression of Mac1 is required for the formation of tight clusters that comprise the center of swarms and assist in the repair of damaged extracellular matrix components. A variety of gene products is induced, most notably several pro-inflammatory cytokines and chemokines that serve to condition the tissue microenvironment and promote the migration of other cell types to the site of inflammation (Nathan 2006).

PAD4 expression is induced by several of the signals that promote inflammation, including TNF (Neeli et al. 2008) and IL-8 (Gupta et al. 2005). Moreover, PAD4 expression is stimulated by neutrophil adhesion to extracellular matrix components via Mac1 and is sustained by an active and functional cytoskeleton (Neeli et al. 2009). Most importantly, PAD4 expression is strongly induced by substances emanating from pathogens or by direct contact with microbes. Bacteria, fungi, and viruses that upregulate PAD4 activity include clinically important pathogens such as *Staphylococcus aureus* (Kolaczkowska et al. 2015) and *Shigella flexneri* (Li et al. 2010). It is therefore tempting to say that PAD4 activation begins with the stimulation of neutrophils in the blood, progresses through the stages of active migration toward a site of infection, and culminates during the precise moment that contact with an invasive pathogen is made. Upon encounter with microbes, the neutrophils may induce alternative responses. Neutrophils are active phagocytes capable of the internalization and destruction of a pathogen. Alternatively, neutrophils can release bactericidal products and reactive oxygen species (ROS). Although the precise role of PAD4 during phagocytosis is not known, it has been reported that various secreted cytokines and chemokines are modified by deimination (Proost et al. 2008; Moelants et al. 2013). It thus appears that PAD4 is intimately involved in essential functions of the innate immune system, for which it regulates and fine-tunes many complementary but independent processes.

2.5 PAD4 and NETs

An additional and unexpected innate defense mechanism was initially described in 2004 (Brinkmann et al. 2004). These authors observed that in response to various bacterial pathogens, neutrophil cell membranes rupture and release nuclear chromatin into the cellular surroundings. The released chromatin forms a matrix that was named a "neutrophil extracellular trap" (NET) because it has the ability to bind and immobilize the pathogens that it contacts. Subsequent studies have more carefully examined the steps in NET release (Fuchs et al. 2007). Morphologically, it was observed that the nuclear chromatin relaxes, and, as a result, the lobed nucleus expands until it is roughly spherical. Thereupon, the nuclear envelope distends until gaps appear, the cytoplasmic granules break open, and granule contents associate with the chromatin that breaks out from the confines of the nucleus. Ultimately, the plasma membrane ruptures and NETs that consist of nuclear chromatin and associate with several granule constituents, such as myeloperoxidase and elastase (Papayannopoulos et al. 2010), are deployed. It is assumed that the granule contents such as proteases and antibacterial peptides, which decorate the NETs, contribute to their bactericidal potential. Interestingly, certain strains of bacteria express nucleases that allow them to escape from neutrophil traps (Buchanan et al. 2006). In vivo, these bacteria are more virulent than variants that are deficient in nuclease production.

Separate experiments have provided evidence for the release of NETs in response to certain viruses and fungal infections. Infections with poxviruses identified a protective function of NETs in the microvessels of the liver (Jenne et al. 2013). In that study, NETs were visualized in vivo and provided effective containment of viruses in liver sinusoids which protected the body from viral dissemination. Interestingly, neutrophils also release NETs in response to fungi such as *Aspergillus* (Gazendam et al. 2016) and *Candida* (Byrd et al. 2013). In fact, the relative size of a pathogen may determine, in part, whether the neutrophil will release NETs or undergo an alternative innate immune defense mechanism (Branzk et al. 2014). Cell wall glucans of *Candida* are also effective inducers of neutrophil swarming in vitro (Byrd et al. 2013). These and other in vivo studies determined that there is a type of NET release that does not result in the massive rupture of the cell but, in fact, leaves the neutrophil able to continue chemotaxis (Yipp et al. 2012). It was further determined that this alternative form of NETs may be composed of mitochondrial DNA (Yousefi et al. 2009). If so, the mitochondrial NETs may lack histones and perhaps other granule-derived bactericidal components of classical NETs. Clearly, important characteristics of the mitochondrial nucleoid NETs and their ability to damage or destroy pathogens remain to be established.

In the classical form of NET release, numerous independent and consistent studies demonstrated that PAD4 carries out an essential function. In pioneering studies, our lab showed that various inflammatory stimuli induce histone deimination and identified deiminated histones as integral components of NETs (Neeli et al. 2008). Subsequent studies confirmed these results (Wang et al. 2009) and established that PAD4 activity is essential for the regulated release of NETs, as neutrophils deficient in PAD4 fail to deploy extracellular chromatin (Li et al. 2010; Rohrbach et al. 2012a). These results are consistent with evidence that PAD4 inhibitors are effective at blocking NET release (Lewis et al. 2015). However, it is not clear how PAD4 contributes to NETosis. It is plausible, albeit unproven, that PAD4 contributes an essential function for NET deployment by modifying arginine residues in histones. The amino termini of core histones extend from the nucleosome core particle in unstructured fashion. Each of the four core histones has extended amino termini of about 18–25 residues in length that contain numerous positively charged arginine and lysine side chains (Radic and Muller 2013). Through these extensions, each of the eight histones in a nucleosome can contribute to bind adjacent nucleosomes into condensed and relatively inert chromatin. A fifth histone, the linker histone H1, does not form part of the nucleosome core particle yet contributes in an important way to chromatin structure. Histone H1 binds DNA at the entry and exit points from the nucleosome and thus controls the angle of the linker DNA that connects adjacent nucleosomes. In that way, H1 is at a crucial position to regulate chromatin structure (Izzo and Schneider 2016). The histone tails, together with H1, regulate structural transitions in chromatin, which facilitate access of transcription factors, RNA polymerases, and other proteins to particular DNA sequences. Many histone PTMs serve to organize chromatin according to the functional needs of the cell.

By converting arginine residues to citrulline residues, PAD4 removes the positive charge from the amino termini of core histones (Fig. 2.1) and thus diminishes the attractive forces between histones and DNA. As a result, histone deimination loosens the structure of chromatin. Yanming Wang and colleagues used defined chromatin templates to show their structural relaxation upon treatment with PAD4 (Wang et al. 2009), and a similar transition may provide the force that expands the nucleus and ultimately ruptures the nuclear envelope to release NETs. In their sequence, H1 histones contain a large number of lysine residues but only a few arginine residues. Therefore, we asked whether one or more of these arginines are deiminated by PAD (Dwivedi et al. 2014). We prepared highly modified histone H1 and subjected it to tandem mass spectrometry. Peptide analysis determined that the most highly conserved arginine, R53, in the winged helix domain of H1, is a substrate of PAD4. In independent studies by the laboratory of Tony Kouzarides in Cambridge, deimination of R53 was reported to occur during transcriptional reprogramming that accompanies the development of cell lineages from pluripotent stem cells (Christophorou et al. 2014). The deimination of H1 by PAD4 thus was shown to play a crucial and conserved role in the developmental program of higher eukaryotes. Strikingly, the same H1 modification that facilitates global changes in gene expression also was co-opted toward a unique mechanism of innate immune response.

The myriad of stimuli that lead to NET release and the potential existence of different forms of NETs make it difficult to identify the signaling pathways that participate in the activation of PAD4. Signals from Gram-positive bacteria, including lipopolysaccharide (LPS) acting on the Toll-like receptor 4 (TLR4), may transmit signals via MyD88 and its associated catalytic subunits to IRAK1 (Huang et al. 2015). Through the activation of distinct IKK subunits, the pro-inflammatory axis of NFkappaB is engaged, leading via MEK1 to the further activation of ERK1 and ERK2 (Yu et al. 2015). Alternatively, it was reported that FcgammaRIIIb, acting through TAK1, leads to the activation of ERK1/ERK2 (Aleman et al. 2016). Additional feed-forward signals may involve activation of PLCgamma and the formation of its messenger IP3, followed by calcium release from endogenous ER stores (Numaga et al. 2010). Alternatively, a calcium-activated potassium channel may directly engage signals leading to NETosis (Douda et al. 2015). Calcium could act as an additional signal by activating PKC subunits, which have been shown to have a direct effect on NET release. Our own studies revealed an unexpected complexity of PKC involvement in NETosis. Experiments with an inhibitor of classical PKC, chelerythrine, as well as a structurally related compound, sanguinarine, demonstrated that classical PKC enzymes may block activation of PAD4, yet an atypical PKC, most likely PKCzeta, exerts an activating role upstream of PAD4 (Neeli and Radic 2013). The opposing effects of two PKC isoforms argue for very precise regulation of PAD4 in neutrophils. Through as yet incompletely understood mechanisms, these enzymes contribute to the disruption of granule and nuclear membranes, chromatin relaxation, and, ultimately, NET release. Some of the remaining questions regarding the regulation of NET release are summarized in Box 2.2.

The activation of PAD4 can most easily be seen by monitoring histone deimination (citrullination), for which specific antibody reagents are commercially available. However, discovery of new substrates will require the availability of additional approaches. One method to detect the activation of PAD4 in cells is by isolating neutrophil proteins and exposing them to a compound that selectively reacts with the ureido group of citrulline. Under acidic conditions, phenylglyoxal reacts with citrullines to form a covalent bond that is stable during subsequent manipulations at neutral pH (Lewallen et al. 2015). By conjugating rhodamine to the phenylglyoxal probe (Rh-PG), citrulline-containing proteins become fluorescently labeled. We have used such an approach to explore the diversity of neutrophil proteins that are substrates for PAD4 (Fig. 2.3).

The results demonstrate that, in purified human neutrophils, certain proteins react with Rh-PG following incubation in calcium-containing buffer (lanes 1). However, the intensity of Rh-PG reactivity increases following neutrophil activation with ionophore (lanes 2). In particular, histones in ionophore-activated neutrophils become quite reactive, indicating deimination by PAD4. An increased intensity of reactivity is observed following ionophore stimulation in the presence of a classical PKC inhibitor (lanes 3). Purified calf thymus histones that were deiminated in vitro by recombinant PAD4 (kind gift from Paul Thompson) indicate the migration of histones (lane labeled H). The specificity of Rh-PG labeling is tested by comparison to the reactivity of antibodies reactive against modified citrulline (lane MC). There is extensive agreement between these two approaches (compare lanes 3 to MC). An alternative method could use a mouse monoclonal antibody, F95, which was raised against a deca-citrulline peptide and which reacts with citrulline-containing antigens (Nicholas and Whitaker 2002). Each of these should be added to the list of reagents that will play a crucial role in the further identification of PAD4 functions in NETosis.

Box 2.2: Unsolved Issues of PAD4 Activation in Classical NETosis

1. By using a defined NETosis stimulus, the precise series of contingent activation steps leading to PAD4 activity need to be established.
2. The precise relation between PAD4 activation and nuclear chromatin relaxation should be demonstrated.
3. The possible activation of PAD4 during the mitochondrial NET release and/or swarming should be examined.

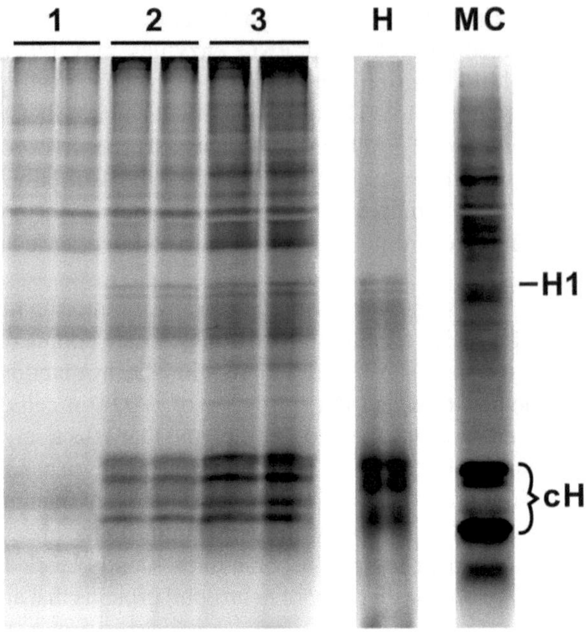

Fig. 2.3 Detection of deiminated proteins in neutrophils. Purified neutrophils were incubated with calcium buffer (lanes 1), calcium plus ionophore A23187 (lanes 2), or as in 2 with the addition of chelerythrine for 2 h (lanes 3). Cells were frozen and thawed, incubated with 20% TCA, sonicated, and reacted with Rh-PG as described by Bicker et al. Proteins were resolved on SDS-PAGE, and citrulline-containing bands were visualized on a Typhoon imager. For comparison a lane with purified deiminated histones is shown (H), and total neutrophil proteins were also reacted with a modified citrulline antibody, as recommended by supplier (Millipore). Approximate positions of H1 histone and core histones (cH) are indicated

2.6 NETs in Autoimmunity

Work in our laboratory, with essential contributions from collaborators, established the first indications that deiminated histones are important autoantigens in various autoimmune diseases that affect connective tissues. By using in vitro deiminated or unmodified histones for screening by ELISA or Western blots, Dwivedi et al. found that certain autoimmune IgG prefers to bind deiminated histones, often showing additional preference for specific core or linker histones (Dwivedi et al. 2012). This observation was confirmed with subsets of sera from SLE and Sjögren's syndrome patients (Dwivedi et al. 2014). Most notably, sera from patients with Felty's syndrome, a more severe variant of arthritis with neutropenia, splenomegaly, and anti-histone antibodies, showed a dramatic preference for deiminated histones

(Dwivedi et al. 2012). Thus, immunochemistry accounted for the intense preference of autoimmune sera for activated neutrophils and NETs over freshly isolated blood neutrophils that is observed in immunofluorescence microscopy. Histones are well-known autoantigens in SLE, and anti-histone antibodies also arise in Felty's, Sjögren's, and RA. Surprisingly, investigators had not previously evaluated relative binding to deiminated histones. Following the publication of our study, the lab of Paola Migliorini confirmed that a certain subgroup of RA patients express autoantibodies that preferentially, even exclusively, bind to the deiminated antigens (Pratesi et al. 2014).

The list of PAD substrates which react more vigorously with autoantibodies upon deimination continues to expand and currently includes over one dozen proteins (Muller and Radic 2015). Similar to peptides from deiminated filaggrin that show over 70% sensitivity and 96–98% specificity in the diagnosis of RA (Schellekens et al. 1998), autoantibodies from RA sera also bind to deiminated peptides from fibrin and from histone H4 with similar sensitivity and specificity (Pratesi et al. 2014). Other autoantigens that are recognized in citrullinated form include additional extracellular substrates, including collagen type II (Haag et al. 2014). One important MS autoantigen that forms part of the axon sheath and insulates neurons to facilitate their function over distance is myelin basic protein, which is modified by PAD2 (Bradford et al. 2014). Cytokines, in particular CXCL8 (Proost et al. 2008) and CXCL10 (Loos et al. 2008) that stimulate inflammation in vivo, are additional substrates of deiminases. A consequence of deimination is the decrease of the specific activity of the cytokines, suggesting that PAD activity also has the capacity to dampen an inflammatory response. Cytoplasmic substrates such as vimentin (Van Steendam et al. 2010) and f-actin-capping protein (Matsuo et al. 2006) are modified by PADs, and their deimination likely affects the proper function of the cytoskeleton. Notably, nuclear and cytoplasmic enzymes provide additional substrates for PAD, which include enolase (Wegner et al. 2010) and PAD4 itself (Andrade et al. 2010). These examples serve as evidence for the diversity of cellular functions that are affected by deiminases. Moreover, the initial hypothesis that citrullination provides new epitopes which break tolerance is strongly supported by the accumulated evidence.

The role of NETs in the pathogenesis of RA is an area of research that has yielded much information to date, although NETs are suspected as important contributors to disease manifestations in an increasing number of disorders, including SLE, vasculitis (Yoshida et al. 2013), colitis (Chen et al. 2008), multiple sclerosis (Bradford et al. 2014), Felty's syndrome, and gout (Schauer et al. 2014). In RA, neutrophils infiltrate into affected joints just before, or concurrent with, the first signs of pain, stiffening, and inflammation (Patel and Haynes 2001). The long-term consequences of RA are damage to the cartilage and bones of the affected joints, leading to permanent deformation and loss of function. The influx of neutrophils into the fluid-containing capsule surrounding the joints, which is called the synovium, therefore deserves close scrutiny. Early studies of joints from animal models of RA revealed that neutrophils represent the most numerous cellular infiltrate and that sections of the inflamed joints demonstrate extracellular PAD (Damgaard et al. 2014) and citrulline-containing immunoreactive material (van Beers et al. 2013). These observations motivated efforts to more closely characterize RA synovial fluids and their cellular infiltrates.

Synovial fluids can be prepared from swollen joints of RA patients, a procedure that is clinically effective because it provides a temporary relief of pain. The fluid is rich in pro-inflammatory cytokines, including TNF and IL-17, and contains high numbers of activated neutrophils that migrate to this location. A comparison between synovial neutrophils and those from peripheral blood of the same individual, or synovial neutrophils of osteoarthritis patients, revealed that RA synovial neutrophils are much more prone to release NETs (Khandpur et al. 2013). This tendency held true whether additional NET-inducing cytokines were added to the cultures or not. As expected, RA neutrophils displayed increased PAD activity, and purified NETs contained citrullinated enolase and vimentin. Importantly, NETosis of RA neutrophils was enhanced by addition of synovial fluids or autologous patient serum, and antibodies present in RA sera alone had stimulatory activity. Release of NETs from patient neutrophils was dependent on deiminase activity and reactive oxygen species, as shown by the use of inhibitors of PAD4 or NADPH oxidase.

A separate study reported increased deimination in neutrophils exposed to cytolytic agents and high extracellular calcium. Incubation of neutrophils with perforin and granzymes, proteins expressed by cytolytic T cells, induced extensive deimination, resembling synovial cell extracts prepared from RA patients with active disease (Romero et al. 2013). Proteomic analysis of synovial extracts identified deiminated cytoskeletal proteins, such as actin, tubulin, myosin, vinculin, vimentin, talin, and coronin, secreted proteins, such as serpin B6, heat-shock protein 60, and myeloperoxidase, along with histones H2A and H4. Interestingly, the analysis also identified deiminated plasminogen, fibrinogen, and collagen type 1, suggesting that the experimental treatment also led to the release of active PAD to the extracellular space, consistent with cell rupture. This finding would explain the observed deimination of extracellular substrates, many of which are recognized as autoantigens in autoimmune diseases.

A set of recent studies explored neutrophil NETs in the pathogenesis of lupus. Both studies (Caielli et al. 2016; Lood et al. 2016) examined the response of neutrophils to anti-ribonucleoprotein autoantibodies and observed that the NETs released under these conditions consisted mainly of oxidized mitochondrial DNA. Perhaps SLE-specific stimuli determine the type of neutrophil response, and, in turn, the components of NETs affect the clinical manifestations of the autoimmune response. Previously, oxidized mitochondrial DNA was shown to be particularly effective at stimulating the Toll-like receptor pathways and inducing the activation of dendritic cells (Pazmandi et al. 2014). The recent neutrophil analyses clarified features of neutrophil mitophagy and possible deviations in this process that may lead to lupus (Muller and Radic 2016). Overall, the study of PAD4 contributions to autoimmunity portends many additional insights and raises new questions, some of which are listed in Box 2.3.

Box 2.3: Questions Concerning the Role of PAD4 in Autoimmunity

1. What is the comprehensive proteome of PAD4 substrates following specific stimuli?
2. Which autoantigens contain binding determinants that are focused on and depend on citrullines?
3. What is the contribution of environmental factors that predispose to autoimmunity (e.g., diet, obesity, injuries, or infections) toward the activation of PAD4?

2.7 Model of PAD4 Activation in the Pathogenesis of Autoimmune Disease

The release of NETs represents an immune mechanism that may act as a final barrier to disseminated disease. However, NET release carries with it a major risk, which is the release of extracellular proteases and reactive oxygen species that contribute to tissue damage. In addition, the enzymatic reactions that make essential contributions to NET release also lead to the release of modified autoantigens to the exterior of the cells. Thus, sequestration of nuclear autoantigens is disrupted, and chromatin and other nuclear autoantigens become accessible to cell surface receptors of antigen-presenting cells. Moreover, the release of NETs occurs in close apposition with immunostimulatory cytokines and pathogen-derived, adjuvant-like molecules that may enhance the development of autoreactivity. As argued by others, environmental triggers associated with choices in lifestyle (Makrygiannakis et al. 2008) and diet (Mohanan et al. 2013) also may promote activation of PADs along with the release of NETs. Given the currently available evidence, it appears likely that, over time, chronic stimulation of neutrophils by infections or inflammation provides conditions that enhance the production of deiminated autoantigens both inside and outside of tissue-infiltrating neutrophils. A good example of the progression of disease is provided by patients with Felty's syndrome. Usually, Felty's develops after several years of progressive RA (Dwivedi and Radic 2014). Patients are likely to receive medication that leads to immunosuppression, leaving the individuals more prone to infections. The recurrent infections likely lead to frequent neutrophil activation and aggravated NET production. Additional stimulation of the adaptive immune system by deiminated autoantigens gives rise to ACPA. In turn, these autoantibodies promote NET release. Secondary effects will likely include splenomegaly due to the clearance of spent neutrophils and additional stimulation of the adaptive immune response. Overall, these interactions mutually enhance each other, and the individually protective adaptations mutually synergize to drive autoimmunity.

References

Aleman, O. R., Mora, N., Cortes-Vieyra, R., Uribe-Querol, E., & Rosales, C. (2016). Transforming growth factor-beta-activated kinase 1 is required for human FcgammaRIIIb-induced neutrophil extracellular trap formation. *Frontiers in Immunology, 7*, 277.

Andrade, F., Darrah, E., Gucek, M., Cole, R. N., Rosen, A., & Zhu, X. (2010). Autocitrullination of human peptidyl arginine deiminase type 4 regulates protein citrullination during cell activation. *Arthritis and Rheumatism, 62*(6), 1630–1640.

Anzilotti, C., Pratesi, F., Tommasi, C., & Migliorini, P. (2010). Peptidylarginine deiminase 4 and citrullination in health and disease. *Autoimmunity Reviews, 9*(3), 158–160.

Arita, K., Hashimoto, H., Shimizu, T., Nakashima, K., Yamada, M., & Sato, M. (2004). Structural basis for Ca(2+)-induced activation of human PAD4. *Nature Structural & Molecular Biology, 11*(8), 777–783.

Arita, K., Shimizu, T., Hashimoto, H., Hidaka, Y., Yamada, M., & Sato, M. (2006). Structural basis for histone N-terminal recognition by human peptidylarginine deiminase 4. *Proceedings of the National Academy of Sciences of the United States of America, 103*(14), 5291–5296.

Bicker, K. L., & Thompson, P. R. (2013). The protein arginine deiminases: Structure, function, inhibition, and disease. *Biopolymers, 99*(2), 155–163.

Bradford, C. M., Ramos, I., Cross, A. K., Haddock, G., McQuaid, S., Nicholas, A. P., & Woodroofe, M. N. (2014). Localisation of citrullinated proteins in normal appearing white matter and lesions in the central nervous system in multiple sclerosis. *Journal of Neuroimmunology, 273*(1–2), 85–95.

Branzk, N., Lubojemska, A., Hardison, S. E., Wang, Q., Gutierrez, M. G., Brown, G. D., & Papayannopoulos, V. (2014). Neutrophils sense microbe size and selectively release neutrophil extracellular traps in response to large pathogens. *Nature Immunology, 15*(11), 1017–1025.

Brinkmann, V., Reichard, U., Goosmann, C., Fauler, B., Uhlemann, Y., Weiss, D. S., Weinrauch, Y., & Zychlinsky, A. (2004). Neutrophil extracellular traps kill bacteria. *Science, 303*(5663), 1532–1535.

Buchanan, J. T., Simpson, A. J., Aziz, R. K., Liu, G. Y., Kristian, S. A., Kotb, M., Feramisco, J., & Nizet, V. (2006). DNase expression allows the pathogen group A Streptococcus to escape killing in neutrophil extracellular traps. *Current Biology, 16*(4), 396–400.

Byrd, A. S., O'Brien, X. M., Johnson, C. M., Lavigne, L. M., & Reichner, J. S. (2013). An extracellular matrix-based mechanism of rapid neutrophil extracellular trap formation in response to Candida albicans. *Journal of Immunology, 190*(8), 4136–4148.

Caielli, S., Athale, S., Domic, B., Murat, E., Chandra, M., Banchereau, R., Baisch, J., Phelps, K., Clayton, S., Gong, M., Wright, T., Punaro, M., Palucka, K., Guiducci, C., Banchereau, J., & Pascual, V. (2016). Oxidized mitochondrial nucleoids released by neutrophils drive type I interferon production in human lupus. *The Journal of Experimental Medicine, 213*(5), 697–713.

Chavanas, S., Mechin, M. C., Takahara, H., Kawada, A., Nachat, R., Serre, G., & Simon, M. (2004). Comparative analysis of the mouse and human peptidylarginine deiminase gene clusters reveals highly conserved non-coding segments and a new human gene, PADI6. *Gene, 330*, 19–27.

Chen, C. C., Isomoto, H., Narumi, Y., Sato, K., Oishi, Y., Kobayashi, T., Yanagihara, K., Mizuta, Y., Kohno, S., & Tsukamoto, K. (2008). Haplotypes of PADI4 susceptible to rheumatoid arthritis are also associated with ulcerative colitis in the Japanese population. *Clinical Immunology, 126*(2), 165–171.

Christophorou, M. A., Castelo-Branco, G., Halley-Stott, R. P., Oliveira, C. S., Loos, R., Radzisheuskaya, A., Mowen, K. A., Bertone, P., Silva, J. C., Zernicka-Goetz, M., Nielsen, M. L., Gurdon, J. B., & Kouzarides, T. (2014). Citrullination regulates pluripotency and histone H1 binding to chromatin. *Nature, 507*(7490), 104–108.

Cuthbert, G. L., Daujat, S., Snowden, A. W., Erdjument-Bromage, H., Hagiwara, T., Yamada, M., Schneider, R., Gregory, P. D., Tempst, P., Bannister, A. J., & Kouzarides, T. (2004). Histone deimination antagonizes arginine methylation. *Cell, 118*(5), 545–553.

Damgaard, D., Senolt, L., Nielsen, M. F., Pruijn, G. J., & Nielsen, C. H. (2014). Demonstration of extracellular peptidylarginine deiminase (PAD) activity in synovial fluid of patients with rheumatoid arthritis using a novel assay for citrullination of fibrinogen. *Arthritis Research & Therapy, 16*(6), 498.

Douda, D. N., Khan, M. A., Grasemann, H., & Palaniyar, N. (2015). SK3 channel and mitochondrial ROS mediate NADPH oxidase-independent NETosis induced by calcium influx. *Proceedings of the National Academy of Sciences of the United States of America, 112*(9), 2817–2822.

Doyle, H. A., Aswad, D. W., & Mamula, M. J. (2013). Autoimmunity to isomerized histone H2B in systemic lupus erythematosus. *Autoimmunity, 46*(1), 6–13.

Dreyton, C. J., Anderson, E. D., Subramanian, V., Boger, D. L., & Thompson, P. R. (2014). Insights into the mechanism of streptonigrin-induced protein arginine deiminase inactivation. *Bioorganic & Medicinal Chemistry, 22*(4), 1362–1369.

Dwivedi, N., & Radic, M. (2014). Citrullination of autoantigens implicates NETosis in the induction of autoimmunity. *Annals of the Rheumatic Diseases, 73*(3), 483–491.

Dwivedi, N., Upadhyay, J., Neeli, I., Khan, S., Pattanaik, D., Myers, L., Kirou, K. A., Hellmich, B., Knuckley, B., Thompson, P. R., Crow, M. K., Mikuls, T. R., Csernok, E., & Radic, M. (2012). Felty's syndrome autoantibodies bind to deiminated histones and neutrophil extracellular chromatin traps. *Arthritis and Rheumatism, 64*(4), 982–992.

Dwivedi, N., Neeli, I., Schall, N., Wan, H., Desiderio, D. M., Csernok, E., Thompson, P. R., Dali, H., Briand, J. P., Muller, S., & Radic, M. (2014). Deimination of linker histones links neutrophil extracellular trap release with autoantibodies in systemic autoimmunity. *The FASEB Journal, 28*(7), 2840–2851.

Fuchs, T. A., Abed, U., Goosmann, C., Hurwitz, R., Schulze, I., Wahn, V., Weinrauch, Y., Brinkmann, V., & Zychlinsky, A. (2007). Novel cell death program leads to neutrophil extracellular traps. *The Journal of Cell Biology, 176*(2), 231–241.

Fujisaki, M., & Sugawara, K. (1981). Properties of peptidylarginine deiminase from the epidermis of newborn rats. *Journal of Biochemistry, 89*(1), 257–263.

Gazendam, R. P., van Hamme, J. L., Tool, A. T., Hoogenboezem, M., van den Berg, J. M., Prins, J. M., Vitkov, L., van de Veerdonk, F. L., van den Berg, T. K., Roos, D., & Kuijpers, T. W. (2016). Human neutrophils use different mechanisms to kill Aspergillus fumigatus conidia and hyphae: Evidence from phagocyte defects. *Journal of Immunology, 196*(3), 1272–1283.

Gupta, A. K., Hasler, P., Holzgreve, W., Gebhardt, S., & Hahn, S. (2005). Induction of neutrophil extracellular DNA lattices by placental microparticles and IL-8 and their presence in preeclampsia. *Human Immunology, 66*(11), 1146–1154.

Haag, S., Schneider, N., Mason, D. E., Tuncel, J., Andersson, I. E., Peters, E. C., Burkhardt, H., & Holmdahl, R. (2014). Identification of new citrulline-specific autoantibodies, which bind to human arthritic cartilage, by mass spectrometric analysis of citrullinated type II collagen. *Arthritis & Rhematology, 66*(6), 1440–1449.

Heit, B., Tavener, S., Raharjo, E., & Kubes, P. (2002). An intracellular signaling hierarchy determines direction of migration in opposing chemotactic gradients. *The Journal of Cell Biology, 159*(1), 91–102.

Huang, H., Tohme, S., Al-Khafaji, A. B., Tai, S., Loughran, P., Chen, L., Wang, S., Kim, J., Billiar, T., Wang, Y., & Tsung, A. (2015). Damage-associated molecular pattern-activated neutrophil extracellular trap exacerbates sterile inflammatory liver injury. *Hepatology, 62*(2), 600–614.

Izzo, A., & Schneider, R. (2016). The role of linker histone H1 modifications in the regulation of gene expression and chromatin dynamics. *Biochimica et Biophysica Acta, 1859*(3), 486–495.

Jenne, C. N., Wong, C. H., Zemp, F. J., McDonald, B., Rahman, M. M., Forsyth, P. A., McFadden, G., & Kubes, P. (2013). Neutrophils recruited to sites of infection protect from virus challenge by releasing neutrophil extracellular traps. *Cell Host & Microbe, 13*(2), 169–180.

Johansson, L., Pratesi, F., Brink, M., Arlestig, L., D'Amato, C., Bartaloni, D., Migliorini, P., & Rantapaa-Dahlqvist, S. (2016). Antibodies directed against endogenous and exogenous citrullinated antigens pre-date the onset of rheumatoid arthritis. *Arthritis Research & Therapy, 18*(1), 127.

Kanno, T., Kawada, A., Yamanouchi, J., Yosida-Noro, C., Yoshiki, A., Shiraiwa, M., Kusakabe, M., Manabe, M., Tezuka, T., & Takahara, H. (2000). Human peptidylarginine deiminase type III: Molecular cloning and nucleotide sequence of the cDNA, properties of the recombinant enzyme, and immunohistochemical localization in human skin. *The Journal of Investigative Dermatology, 115*(5), 813–823.

Kearney, P. L., Bhatia, M., Jones, N. G., Yuan, L., Glascock, M. C., Catchings, K. L., Yamada, M., & Thompson, P. R. (2005). Kinetic characterization of protein arginine deiminase 4: A transcriptional corepressor implicated in the onset and progression of rheumatoid arthritis. *Biochemistry, 44*(31), 10570–10582.

Khandpur, R., Carmona-Rivera, C., Vivekanandan-Giri, A., Gizinski, A., Yalavarthi, S., Knight, J. S., Friday, S., Li, S., Patel, R. M., Subramanian, V., Thompson, P., Chen, P., Fox, D. A., Pennathur, S., & Kaplan, M. J. (2013). NETs are a source of citrullinated autoantigens and stimulate inflammatory responses in rheumatoid arthritis. *Science Translational Medicine, 5*(178), 178ra140.

Knuckley, B., Jones, J. E., Bachovchin, D. A., Slack, J., Causey, C. P., Brown, S. J., Rosen, H., Cravatt, B. F., & Thompson, P. R. (2010). A fluopol-ABPP HTS assay to identify PAD inhibitors. *Chemical Communications, 46*(38), 7175–7177.

Kobayashi, S. D., Voyich, J. M., Buhl, C. L., Stahl, R. M., & DeLeo, F. R. (2002). Global changes in gene expression by human polymorphonuclear leukocytes during receptor-mediated phagocytosis: Cell fate is regulated at the level of gene expression. *Proceedings of the National Academy of Sciences of the United States of America, 99*(10), 6901–6906.

Kolaczkowska, E., Jenne, C. N., Surewaard, B. G., Thanabalasuriar, A., Lee, W. Y., Sanz, M. J., Mowen, K., Opdenakker, G., & Kubes, P. (2015). Molecular mechanisms of NET formation and degradation revealed by intravital imaging in the liver vasculature. *Nature Communications, 6*, 6673.

Lammermann, T., Afonso, P. V., Angermann, B. R., Wang, J. M., Kastenmuller, W., Parent, C. A., & Germain, R. N. (2013). Neutrophil swarms require LTB4 and integrins at sites of cell death in vivo. *Nature, 498*(7454), 371–375.

Lewallen, D. M., Bicker, K. L., Madoux, F., Chase, P., Anguish, L., Coonrod, S., Hodder, P., & Thompson, P. R. (2014). A FluoPol-ABPP PAD2 high-throughput screen identifies the first calcium site inhibitor targeting the PADs. *ACS Chemical Biology, 9*(4), 913–921.

Lewallen, D. M., Bicker, K. L., Subramanian, V., Clancy, K. W., Slade, D. J., Martell, J., Dreyton, C. J., Sokolove, J., Weerapana, E., & Thompson, P. R. (2015). Chemical proteomic platform to identify citrullinated proteins. *ACS Chemical Biology, 10*(11), 2520–2528.

Lewis, H. D., Liddle, J., Coote, J. E., Atkinson, S. J., Barker, M. D., Bax, B. D., Bicker, K. L., Bingham, R. P., Campbell, M., Chen, Y. H., Chung, C. W., Craggs, P. D., Davis, R. P., Eberhard, D., Joberty, G., Lind, K. E., Locke, K., Maller, C., Martinod, K., Patten, C., Polyakova, O., Rise, C. E., Rudiger, M., Sheppard, R. J., Slade, D. J., Thomas, P., Thorpe, J., Yao, G., Drewes, G., Wagner, D. D., Thompson, P. R., Prinjha, R. K., & Wilson, D. M. (2015). Inhibition of PAD4 activity is sufficient to disrupt mouse and human NET formation. *Nature Chemical Biology, 11*(3), 189–191.

Li, P., Li, M., Lindberg, M. R., Kennett, M. J., Xiong, N., & Wang, Y. (2010). PAD4 is essential for antibacterial innate immunity mediated by neutrophil extracellular traps. *The Journal of Experimental Medicine, 207*(9), 1853–1862.

Liao, K. P., Batra, K. L., Chibnik, L., Schur, P. H., & Costenbader, K. H. (2008). Anti-cyclic citrullinated peptide revised criteria for the classification of rheumatoid arthritis. *Annals of the Rheumatic Diseases, 67*(11), 1557–1561.

Lin, R. H., Mamula, M. J., Hardin, J. A., & Janeway, C. A., Jr. (1991). Induction of autoreactive B cells allows priming of autoreactive T cells. *The Journal of Experimental Medicine, 173*(6), 1433–1439.

Lood, C., Blanco, L. P., Purmalek, M. M., Carmona-Rivera, C., De Ravin, S. S., Smith, C. K., Malech, H. L., Ledbetter, J. A., Elkon, K. B., & Kaplan, M. J. (2016). Neutrophil extracellular traps enriched in oxidized mitochondrial DNA are interferogenic and contribute to lupus-like disease. *Nature Medicine, 22*(2), 146–153.

Loos, T., Mortier, A., Gouwy, M., Ronsse, I., Put, W., Lenaerts, J. P., Van Damme, J., & Proost, P. (2008). Citrullination of CXCL10 and CXCL11 by peptidylarginine deiminase: A naturally occurring posttranslational modification of chemokines and new dimension of immunoregulation. *Blood, 112*(7), 2648–2656.

Luo, Y., Knuckley, B., Lee, Y. H., Stallcup, M. R., & Thompson, P. R. (2006). A fluoroacetamidine-based inactivator of protein arginine deiminase 4: Design, synthesis, and in vitro and in vivo evaluation. *Journal of the American Chemical Society, 128*(4), 1092–1093.

Makrygiannakis, D., Hermansson, M., Ulfgren, A. K., Nicholas, A. P., Zendman, A. J., Eklund, A., Grunewald, J., Skold, C. M., Klareskog, L., & Catrina, A. I. (2008). Smoking increases peptidylarginine deiminase 2 enzyme expression in human lungs and increases citrullination in BAL cells. *Annals of the Rheumatic Diseases, 67*(10), 1488–1492.

Martin, R., McFarland, H. F., & McFarlin, D. E. (1992). Immunological aspects of demyelinating diseases. *Annual Review of Immunology, 10*, 153–187.

Matsuo, K., Xiang, Y., Nakamura, H., Masuko, K., Yudoh, K., Noyori, K., Nishioka, K., Saito, T., & Kato, T. (2006). Identification of novel citrullinated autoantigens of synovium in rheumatoid arthritis using a proteomic approach. *Arthritis Research & Therapy, 8*(6), R175.

McColl, B. W., Rothwell, N. J., & Allan, S. M. (2008). Systemic inflammation alters the kinetics of cerebrovascular tight junction disruption after experimental stroke in mice. *The Journal of Neuroscience, 28*(38), 9451–9462.

McDonald, B., Pittman, K., Menezes, G. B., Hirota, S. A., Slaba, I., Waterhouse, C. C., Beck, P. L., Muruve, D. A., & Kubes, P. (2010). Intravascular danger signals guide neutrophils to sites of sterile inflammation. *Science, 330*(6002), 362–366.

Mechin, M. C., Coudane, F., Adoue, V., Arnaud, J., Duplan, H., Charveron, M., Schmitt, A. M., Takahara, H., Serre, G., & Simon, M. (2010). Deimination is regulated at multiple levels including auto-deimination of peptidylarginine deiminases. *Cellular and Molecular Life Sciences, 67*(9), 1491–1503.

Michels, A. W., & Nakayama, M. (2010). The anti-insulin trimolecular complex in type 1 diabetes. *Current Opinion in Endocrinology, Diabetes, and Obesity, 17*(4), 329–334.

Mocsai, A., Walzog, B., & Lowell, C. A. (2015). Intracellular signalling during neutrophil recruitment. *Cardiovascular Research, 107*(3), 373–385.

Moelants, E. A., Mortier, A., Grauwen, K., Ronsse, I., Van Damme, J., & Proost, P. (2013). Citrullination of TNF-alpha by peptidylarginine deiminases reduces its capacity to stimulate the production of inflammatory chemokines. *Cytokine, 61*(1), 161–167.

Mohanan, S., Horibata, S., McElwee, J. L., Dannenberg, A. J., & Coonrod, S. A. (2013). Identification of macrophage extracellular trap-like structures in mammary gland adipose tissue: A preliminary study. *Frontiers in Immunology, 4*, 67.

Moudrianakis, E. N., & Arents, G. (1993). Structure of the histone octamer core of the nucleosome and its potential interactions with DNA. *Cold Spring Harbor Symposia on Quantitative Biology, 58*, 273–279.

Muller, S., & Radic, M. (2015). Citrullinated autoantigens: From diagnostic markers to pathogenetic mechanisms. *Clinical Reviews in Allergy & Immunology, 49*(2), 232–239.

Muller, S., & Radic, M. (2016). Oxidation and mitochondrial origin of NET DNA in the pathogenesis of lupus. *Nature Medicine, 22*(2), 126–127.

Nakashima, K., Hagiwara, T., & Yamada, M. (2002). Nuclear localization of peptidylarginine deiminase V and histone deimination in granulocytes. *The Journal of Biological Chemistry, 277*(51), 49562–49568.

Nathan, C. (2006). Neutrophils and immunity: Challenges and opportunities. *Nature Reviews. Immunology, 6*(3), 173–182.

Neeli, I., & Radic, M. (2013). Opposition between PKC isoforms regulates histone deimination and neutrophil extracellular chromatin release. *Frontiers in Immunology, 4*, 38.

Neeli, I., Khan, S. N., & Radic, M. (2008). Histone deimination as a response to inflammatory stimuli in neutrophils. *Journal of Immunology, 180*(3), 1895–1902.

Neeli, I., Dwivedi, N., Khan, S., & Radic, M. (2009). Regulation of extracellular chromatin release from neutrophils. *Journal of Innate Immunity, 1*(3), 194–201.

Nicholas, A. P., & Whitaker, J. N. (2002). Preparation of a monoclonal antibody to citrullinated epitopes: Its characterization and some applications to immunohistochemistry in human brain. *Glia, 37*(4), 328–336.

Numaga, T., Nishida, M., Kiyonaka, S., Kato, K., Katano, M., Mori, E., Kurosaki, T., Inoue, R., Hikida, M., Putney, J. W., Jr., & Mori, Y. (2010). Ca2+ influx and protein scaffolding via TRPC3 sustain PKCbeta and ERK activation in B cells. *Journal of Cell Science, 123*(Pt 6), 927–938.

Palmer, E. (2003). Negative selection—Clearing out the bad apples from the T-cell repertoire. *Nature Reviews. Immunology, 3*(5), 383–391.

Papayannopoulos, V., Metzler, K. D., Hakkim, A., & Zychlinsky, A. (2010). Neutrophil elastase and myeloperoxidase regulate the formation of neutrophil extracellular traps. *The Journal of Cell Biology, 191*(3), 677–691.

Patel, D. D., & Haynes, B. F. (2001). Leukocyte homing to synovium. *Current Directions in Autoimmunity, 3*, 133–167.

Pazmandi, K., Agod, Z., Kumar, B. V., Szabo, A., Fekete, T., Sogor, V., Veres, A., Boldogh, I., Rajnavolgyi, E., Lanyi, A., & Bacsi, A. (2014). Oxidative modification enhances the immunostimulatory effects of extracellular mitochondrial DNA on plasmacytoid dendritic cells. *Free Radical Biology & Medicine, 77*, 281–290.

Pratesi, F., Dioni, I., Tommasi, C., Alcaro, M. C., Paolini, I., Barbetti, F., Boscaro, F., Panza, F., Puxeddu, I., Rovero, P., & Migliorini, P. (2014). Antibodies from patients with rheumatoid arthritis target citrullinated histone 4 contained in neutrophils extracellular traps. *Annals of the Rheumatic Diseases, 73*(7), 1414–1422.

Proost, P., Loos, T., Mortier, A., Schutyser, E., Gouwy, M., Noppen, S., Dillen, C., Ronsse, I., Conings, R., Struyf, S., Opdenakker, G., Maudgal, P. C., & Van Damme, J. (2008). Citrullination of CXCL8 by peptidylarginine deiminase alters receptor usage, prevents proteolysis, and dampens tissue inflammation. *The Journal of Experimental Medicine, 205*(9), 2085–2097.

Radic, M., & Muller, S. (2013). Epigenetics of autoantigens: New opportunities for therapy of autoimmune diseases. *Genetics & Epigenetics, 5*, 63–70.

Radic, M. Z., & Zouali, M. (1996). Receptor editing, immune diversification, and self-tolerance. *Immunity, 5*(6), 505–511.

Raijmakers, R., Zendman, A. J., Egberts, W. V., Vossenaar, E. R., Raats, J., Soede-Huijbregts, C., Rutjes, F. P., van Veelen, P. A., Drijfhout, J. W., & Pruijn, G. J. (2007). Methylation of arginine residues interferes with citrullination by peptidylarginine deiminases in vitro. *Journal of Molecular Biology, 367*(4), 1118–1129.

Riese, S. B., Kuehne, C., Tedder, T. F., Hallmann, R., Hohenester, E., & Buscher, K. (2014). Heterotropic modulation of selectin affinity by allosteric antibodies affects leukocyte rolling. *Journal of Immunology, 192*(4), 1862–1869.

Rogers, G. E., & Simmonds, D. H. (1958). Content of citrulline and other amino-acids in a protein of hair follicles. *Nature, 182*(4629), 186–187.

Rohrbach, A. S., Hemmers, S., Arandjelovic, S., Corr, M., & Mowen, K. A. (2012a). PAD4 is not essential for disease in the K/BxN murine autoantibody-mediated model of arthritis. *Arthritis Research & Therapy, 14*(3), R104.

Rohrbach, A. S., Slade, D. J., Thompson, P. R., & Mowen, K. A. (2012b). Activation of PAD4 in NET formation. *Frontiers in Immunology, 3*, 360.

Romero, V., Fert-Bober, J., Nigrovic, P. A., Darrah, E., Haque, U. J., Lee, D. M., van Eyk, J., Rosen, A., & Andrade, F. (2013). Immune-mediated pore-forming pathways induce cellular hypercitrullination and generate citrullinated autoantigens in rheumatoid arthritis. *Science Translational Medicine, 5*(209), 209ra150.

Rorvig, S., Ostergaard, O., Heegaard, N. H., & Borregaard, N. (2013). Proteome profiling of human neutrophil granule subsets, secretory vesicles, and cell membrane: Correlation with transcriptome profiling of neutrophil precursors. *Journal of Leukocyte Biology, 94*(4), 711–721.

Rowley, M. J., Nandakumar, K. S., & Holmdahl, R. (2008). The role of collagen antibodies in mediating arthritis. *Modern Rheumatology, 18*(5), 429–441.

Rubin, R. L., Tang, F. L., Tsay, G., & Pollard, K. M. (1990). Pseudoautoimmunity in normal mice: Anti-histone antibodies elicited by immunization versus induction during graft-versus-host reaction. *Clinical Immunology and Immunopathology, 54*(2), 320–332.

Schauer, C., Janko, C., Munoz, L. E., Zhao, Y., Kienhofer, D., Frey, B., Lell, M., Manger, B., Rech, J., Naschberger, E., Holmdahl, R., Krenn, V., Harrer, T., Jeremic, I., Bilyy, R., Schett, G., Hoffmann, M., & Herrmann, M. (2014). Aggregated neutrophil extracellular traps limit inflammation by degrading cytokines and chemokines. *Nature Medicine, 20*(5), 511–517.

Schellekens, G. A., de Jong, B. A., van den Hoogen, F. H., van de Putte, L. B., & van Venrooij, W. J. (1998). Citrulline is an essential constituent of antigenic determinants recognized by rheumatoid arthritis-specific autoantibodies. *The Journal of Clinical Investigation, 101*(1), 273–281.

Slack, J. L., Jones, L. E., Jr., Bhatia, M. M., & Thompson, P. R. (2011). Autodeimination of protein arginine deiminase 4 alters protein-protein interactions but not activity. *Biochemistry, 50*(19), 3997–4010.

Stollar, B. D. (1971). Reactions of systemic lupus erythematosus sera with histone fractions and histone-DNA complexes. *Arthritis and Rheumatism, 14*(4), 485–492.

Terakawa, H., Takahara, H., & Sugawara, K. (1991). Three types of mouse peptidylarginine deiminase: Characterization and tissue distribution. *Journal of Biochemistry, 110*(4), 661–666.

Theilgaard-Monch, K., Jacobsen, L. C., Borup, R., Rasmussen, T., Bjerregaard, M. D., Nielsen, F. C., Cowland, J. B., & Borregaard, N. (2005). The transcriptional program of terminal granulocytic differentiation. *Blood, 105*(4), 1785–1796.

Thompson, P. R., & Fast, W. (2006). Histone citrullination by protein arginine deiminase: Is arginine methylation a green light or a roadblock? *ACS Chemical Biology, 1*(7), 433–441.

van Beers, J. J., Schwarte, C. M., Stammen-Vogelzangs, J., Oosterink, E., Bozic, B., & Pruijn, G. J. (2013a). The rheumatoid arthritis synovial fluid citrullinome reveals novel citrullinated epitopes in apolipoprotein E, myeloid nuclear differentiation antigen, and beta-actin. *Arthritis and Rheumatism, 65*(1), 69–80.

van Beers, J. J., Zendman, A. J., Raijmakers, R., Stammen-Vogelzangs, J., & Pruijn, G. J. (2013b). Peptidylarginine deiminase expression and activity in PAD2 knock-out and PAD4-low mice. *Biochimie, 95*(2), 299–308.

Van Steendam, K., Tilleman, K., De Ceuleneer, M., De Keyser, F., Elewaut, D., & Deforce, D. (2010). Citrullinated vimentin as an important antigen in immune complexes from synovial fluid of rheumatoid arthritis patients with antibodies against citrullinated proteins. *Arthritis Research & Therapy, 12*(4), R132.

Vossenaar, E. R., Zendman, A. J., van Venrooij, W. J., & Pruijn, G. J. (2003). PAD, a growing family of citrullinating enzymes: Genes, features and involvement in disease. *BioEssays, 25*(11), 1106–1118.

Vossenaar, E. R., Radstake, T. R., van der Heijden, A., van Mansum, M. A., Dieteren, C., de Rooij, D. J., Barrera, P., Zendman, A. J., & van Venrooij, W. J. (2004). Expression and activity of citrullinating peptidylarginine deiminase enzymes in monocytes and macrophages. *Annals of the Rheumatic Diseases, 63*(4), 373–381.

Wang, Y., Wysocka, J., Sayegh, J., Lee, Y. H., Perlin, J. R., Leonelli, L., Sonbuchner, L. S., McDonald, C. H., Cook, R. G., Dou, Y., Roeder, R. G., Clarke, S., Stallcup, M. R., Allis, C. D., & Coonrod, S. A. (2004). Human PAD4 regulates histone arginine methylation levels via demethylimination. *Science, 306*(5694), 279–283.

Wang, Y., Li, M., Stadler, S., Correll, S., Li, P., Wang, D., Hayama, R., Leonelli, L., Han, H., Grigoryev, S. A., Allis, C. D., & Coonrod, S. A. (2009). Histone hypercitrullination mediates chromatin decondensation and neutrophil extracellular trap formation. *The Journal of Cell Biology, 184*(2), 205–213.

Wegner, N., Wait, R., Sroka, A., Eick, S., Nguyen, K. A., Lundberg, K., Kinloch, A., Culshaw, S., Potempa, J., & Venables, P. J. (2010). Peptidylarginine deiminase from Porphyromonas gingivalis citrullinates human fibrinogen and alpha-enolase: Implications for autoimmunity in rheumatoid arthritis. *Arthritis and Rheumatism, 62*(9), 2662–2672.

Weigle, W. O. (1962). Termination of acquired immunological tolerance to protein antigens following immunization with altered protein antigens. *The Journal of Experimental Medicine, 116*, 913–928.

Yipp, B. G., Petri, B., Salina, D., Jenne, C. N., Scott, B. N., Zbytnuik, L. D., Pittman, K., Asaduzzaman, M., Wu, K., Meijndert, H. C., Malawista, S. E., de Boisfleury Chevance, A., Zhang, K., Conly, J., & Kubes, P. (2012). Infection-induced NETosis is a dynamic process involving neutrophil multitasking in vivo. *Nature Medicine, 18*(9), 1386–1393.

Yoshida, M., Sasaki, M., Sugisaki, K., Yamaguchi, Y., & Yamada, M. (2013). Neutrophil extracellular trap components in fibrinoid necrosis of the kidney with myeloperoxidase-ANCA-associated vasculitis. *Clinical Kidney Journal, 6*(3), 308–312.

Yousefi, S., Mihalache, C., Kozlowski, E., Schmid, I., & Simon, H. U. (2009). Viable neutrophils release mitochondrial DNA to form neutrophil extracellular traps. *Cell Death and Differentiation, 16*(11), 1438–1444.

Yu, Y., Koehn, C. D., Yue, Y., Li, S., Thiele, G. M., Hearth-Holmes, M. P., Mikuls, T. R., O'Dell, J. R., Klassen, L. W., Zhang, Z., & Su, K. (2015). Celastrol inhibits inflammatory stimuli-induced neutrophil extracellular trap formation. *Current Molecular Medicine, 15*(4), 401–410.

Yurttas, P., Vitale, A. M., Fitzhenry, R. J., Cohen-Gould, L., Wu, W., Gossen, J. A., & Coonrod, S. A. (2008). Role for PADI6 and the cytoplasmic lattices in ribosomal storage in oocytes and translational control in the early mouse embryo. *Development, 135*(15), 2627–2636.

Zhang, X., Bolt, M., Guertin, M. J., Chen, W., Zhang, S., Cherrington, B. D., Slade, D. J., Dreyton, C. J., Subramanian, V., Bicker, K. L., Thompson, P. R., Mancini, M. A., Lis, J. T., & Coonrod, S. A. (2012). Peptidylarginine deiminase 2-catalyzed histone H3 arginine 26 citrullination facilitates estrogen receptor alpha target gene activation. *Proceedings of the National Academy of Sciences of the United States of America, 109*(33), 13331–13336.

Chapter 3
Structures and Functions of Peptidylarginine Deiminases

Masaki Unno, Kenji Kizawa, and Hidenari Takahara

3.1 Introduction

Peptidylarginine deiminase (PAD; EC 3.5.3.15) catalyzes the conversion of peptidylarginine to peptidylcitrulline in a Ca^{2+}-dependent manner (Vossenaar et al. 2003; Bicker and Thompson 2013; Jones et al. 2009). Citrullination is an irreversible process because no de-citrullinase has been found to date (Gyorgy et al. 2006). This posttranslational modification alters both inter- and intramolecular interactions of target proteins, thereby changing their conformation and molecular assembly. Citrullinated proteins perform potential roles in important biological processes such as skin keratinization, neuron insulation, central nervous system plasticity, and neutrophil extracellular trap formation (Li et al. 2010). Furthermore, citrullination of histone core protein is important for the epigenetic regulation of gene transcription (Baka et al. 2012; Nijenhuis et al. 2004; Zhang et al. 2012; Slade et al. 2014a, b).

Five PAD isozymes (PAD1–PAD4 and PAD6) have been described in humans. These isozymes are highly conserved, with 50–55% amino acid sequence identity. They show tissue-specific expression and citrullinate various substrate proteins. PAD1 is found in the skin epidermis and citrullinates keratins and filaggrins (Guerrin et al. 2003). PAD2 is found in several tissues, including the brain and muscles, and citrullinates myelin basic protein (Ishigami et al. 2002). PAD3 is found in hair

M. Unno (✉)
Graduate School of Science and Engineering, Ibaraki University, Hitachi, Japan

Frontier Research Center for Applied Atomic Sciences, Ibaraki University, Tokai-mura, Japan
e-mail: masaki.unno.19@vc.ibaraki.ac.jp

K. Kizawa
Kao Corporation, Biological Science Research Laboratory, Odawara, Japan

H. Takahara
Frontier Research Center for Applied Atomic Sciences, Ibaraki University, Tokai-mura, Japan

College of Agriculture, Ibaraki University, Ami Inashiki, Japan

A.P. Nicholas et al. (eds.), *Protein Deimination in Human Health and Disease*,
DOI 10.1007/978-3-319-58244-3_3

follicles and citrullinates trichohyalin and S100A3 protein (Kizawa et al. 2008; Kanno et al. 2000). PAD4 is found in granulocytes, monocytes, and macrophages and citrullinates histones H1, H2A, H3, and H4 and nucleophosmin/B23 (Liu et al. 2011; Nakashima et al. 1999; Vossenaar et al. 2004; Christophorou et al. 2014; Dwivedi et al. 2014). PAD6 is present in embryonic stem cells and oocytes (Chavanas et al. 2004); however, it appears to be a pseudo-enzyme with no detectable catalytic activity (Taki et al. 2011). Although several protein substrates of human PAD isozymes have been identified, citrullination sites in natural substrate proteins do not generally share a specific consensus motif. This suggests that PAD isozymes use highly complex mechanism(s) to selectively recognize arginines to be modified in natural substrate proteins.

The first high-resolution structure of PAD was reported in 2004 (Arita et al. 2004). Professor M. Sato and coworkers obtained an X-ray crystal structure of human PAD4. Analysis of the X-ray crystal structure showed that PAD4 has five Ca^{2+}-binding sites and that its active site undergoes a conformational change after binding of Ca^{2+} ions. Professor M. Sato and coworkers also determined the structures of PAD4 complexed with histone N-terminal peptides (Arita et al. 2006), which provided important structural information regarding mechanisms underlying its catalysis and substrate recognition. In 2015, Professor P. R. Thompson and coworkers reported the structure of human PAD2 (Slade et al. 2015). Because the number of Ca^{2+}-binding sites is different between PAD2 and PAD4, they suggested the presence of a Ca^{2+}-dependent molecular switch mechanism. Recently, we determined the structure of PAD1 (Saijo et al. 2016). Structural analysis of PAD1 indicated that it has an extended N-terminal tail and that it exists as a monomer. In contrast, PAD2 and PAD4 form head-to-tail dimers. The structure of PAD from periodontal pathogen *Porphyromonas gingivalis* (PPAD) has also been determined (Montgomery et al. 2016; Goulas et al. 2015). Structures of PAD isozymes provide novel insights into their functions, reaction mechanisms, and substrate selectivity and can help in designing PAD isozyme-specific inhibitors. In this review, we discuss the structures and functions of PAD isozymes.

3.2 PAD Isozymes: Similarities and Differences

3.2.1 Overall Folds

Human PAD isozymes share a common overall structure that is reminiscent of a rubber boot (Fig. 3.1a) (Arita et al. 2004). This structure is divided into two domains, i.e., N- and C-terminal domains. The N-terminal domain is further divided into two immunoglobulin (Ig)-like domains (IgG1 and IgG2 domains). The C-terminal domain consists of five $\beta\beta\alpha\beta$ modules that form an α/β propeller motif (Humm et al. 1997). This domain contains the active site of the enzyme (therefore, we have referred to this domain as catalytic domain [CD] hereafter). The amino acid sequence of the C-terminal CD is well conserved, and the known 3D structures

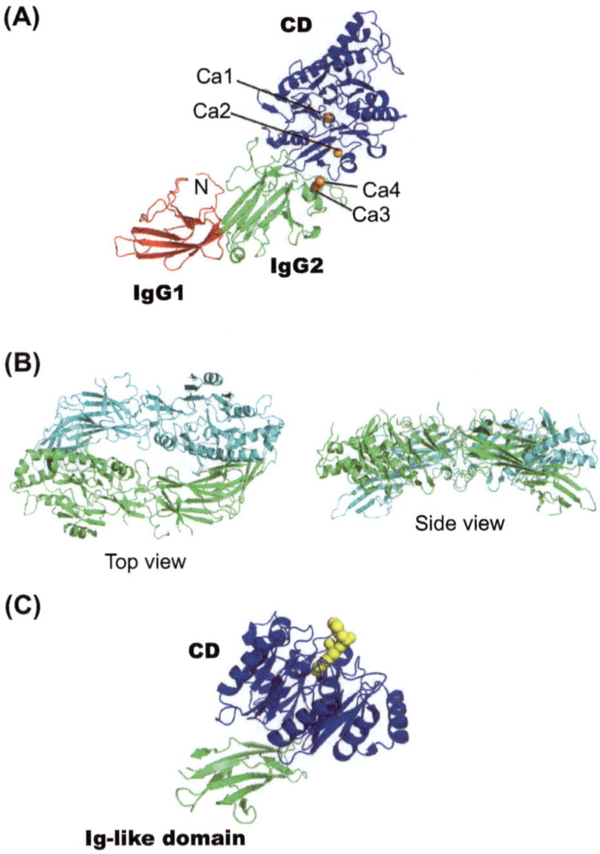

Fig. 3.1 Overall structures of human PAD isozymes. (**a**) PAD1 (PDB ID; 5HP5) as a representative of the common PAD fold and Ca^{2+}-binding sites (*orange*). (**b**) Structure of PAD4 dimer (PDB ID; 1WD8), which is similar to PAD2 dimer. (**c**) Overall structure of PPAD and the substrate (*yellow*) (PDB ID; 4YTG)

of the CDs of the three human PAD isozymes for which structures are known, i.e., PAD1, PAD2, and PAD4, are well superimposed (Saijo et al. 2016). However, the N-terminal domains of these isozymes show lower sequence similarity, and the 3D structures of the N-terminal IgG1 and IgG2 domains of these three PAD isozymes show significant structural variation (Saijo et al. 2016). The N-terminal IgG1 and IgG2 domains are proposed to be important for protein–protein interaction and substrate selection (Arita et al. 2004; Saijo et al. 2016; Fuhrmann et al. 2015).

PAD2 and PAD4 form head-to-tail homodimers (Fig. 3.1b), whereas PAD1 exists as a monomer (Arita et al. 2004; Slade et al. 2015; Saijo et al. 2016). PAD3 also forms a homodimer similar to PAD2 and PAD4 (Saijo et al. 2016). Analysis of the dimeric structure of PAD isozymes indicates that both active site cavities are located on the same face of the dimer structure and are separated by a distance of ~65 Å (Fuhrmann et al. 2015). Dimerization of PAD4 is suggested to be crucial to its catalytic activity, and disruption of the dimerization surface decreases the catalytic activity of PAD4 (Liu et al. 2011; Arita et al. 2004). PAD1 has broad substrate specificity compared with other PAD isozymes. The monomeric form of PAD1 seems to be structurally less constrained than the dimeric structures of other PAD isozymes.

The flexible PAD1 molecule might confer its tolerant substrate specificity. Arg8 of human PAD4, a key amino acid residue that is suggested to stabilize the homodimeric structure of PAD4 (Liu et al. 2011), is conserved in human PAD2 and PAD3 but is substituted by glutamine in human PAD1. This substitution might disrupt the formation of a PAD1 dimer. Relatively low stability of dimeric PAD4 (K_d = ~500 nM) and conserved activity of its dissociated monomeric mutant protein (25% for R8E/D547E) (Liu et al. 2011) suggest that some portion of dimeric PAD isozymes exist as monomers depending on the cellular microenvironmental condition (Saijo et al. 2016; Fuhrmann et al. 2015).

PPAD has four domains, namely, an N-terminal signal peptide domain, CD domain, Ig-like domain, and C-terminal domain (Montgomery et al. 2016). Deletion of the N-terminal signal peptide and the C-terminal domain is tolerated and results in highly soluble enzyme with significantly increased catalytic activity. The truncated form was successfully crystallized and its high-resolution structure was determined. Structural analysis of PPAD indicates that its CD and Ig-like domains are tightly packed (Fig. 3.1c).

3.2.2 Structures and the Functions of the Catalytic Domains

The CDs of human PADs share a common 3D structure. Moreover, these domains are flexible in the absence of Ca^{2+} binding and form active sites after binding of Ca^{2+} ions (Arita et al. 2004; Slade et al. 2015; Saijo et al. 2016). Cys645 (PAD4 numbering), present in the active site pocket, plays a critical role in nucleophilic catalysis (Fig. 3.2a). Structural analysis of PAD4 (C645A mutant) complexed with a synthetic substrate benzoyl-L-arginine amide (BAA) indicates the presence of an interesting binding mode in which hydrophobic Trp347 and Val469 form a hydrophobic wall for the substrate arginine (Arita et al. 2004). Trp347 may be an important residue because substitution of this residue (Trp348 in PAD2) by alanine in PAD2 decreases k_{cat}/K_M by 16,500-fold (Slade et al. 2015). Asp350 and Asp473 form hydrogen bonds with the side-chain nitrogen atoms of substrate arginine. Another catalytic residue, i.e., His471, is located near these residues to form the active site. His471 is important for the protonation of ammonia leaving group and subsequent activation of incoming water (Kearney et al. 2005; Knuckley et al. 2007, 2010a). Hydrogen bonding occurs between two main-chain peptide oxygen atoms (O1 and O2) of substrate and Arg374 as well as between N2 atom of substrate and main-chain carbonyl O of Arg639 (Fig. 3.2a). These residues of PAD4 are commonly involved in BAA binding as well as in the binding of peptides mimicking the N-termini of histones H3 and H4 (Arita et al. 2006).

Structural analysis of PAD4 complexed with histone peptides indicates that substrate recognition by PAD4 is mainly mediated by interactions with the substrate peptide backbone (Arita et al. 2006). In addition to BAA-binding residues, Gln346 and Trp347 form hydrogen bonds with the main-chain carbonyl O of histone peptides. Further, the main chain of Val469 binds to histone peptides. These residues,

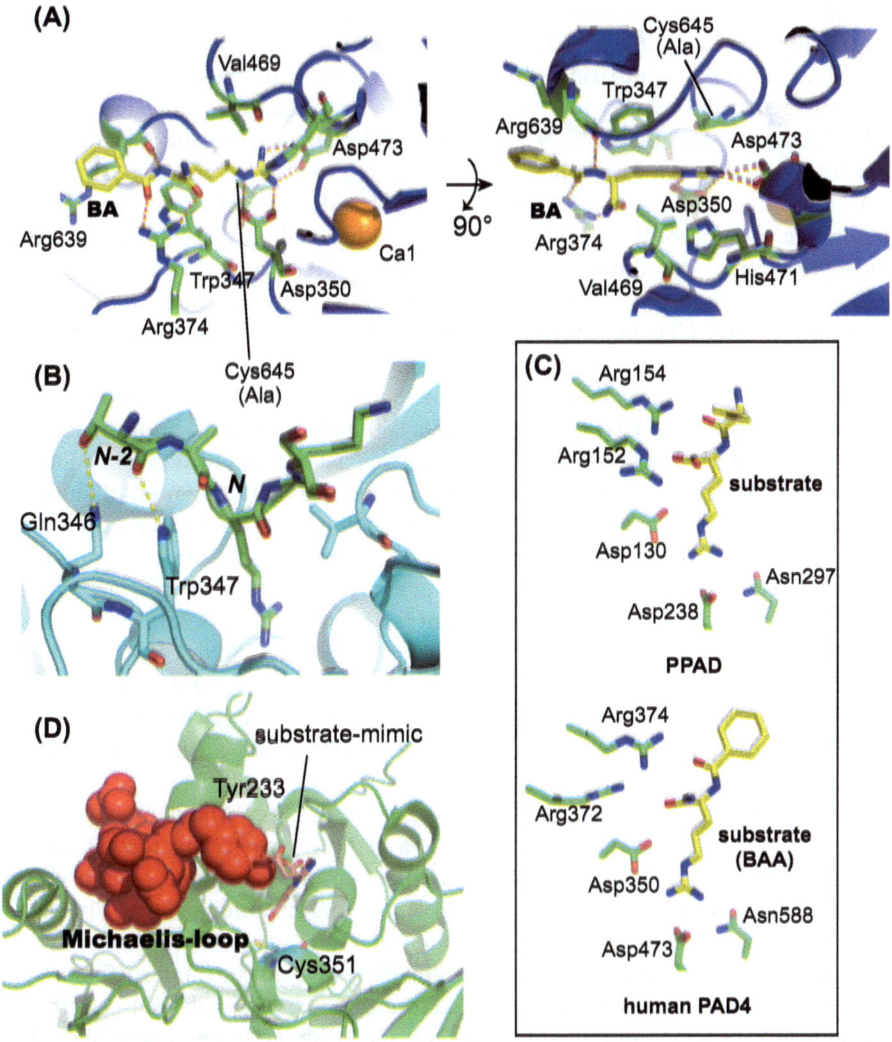

Fig. 3.2 Structure of the catalytic domain of human PAD isozymes. (**a**) Magnified view of the substrate-binding pocket in human PAD4. *Left and right panels* are viewed from two directions. Important residues are shown. In this structure (PDB ID; 1WDA), Cys645 is replaced by alanine to obtain the substrate complex. (**b**) PAD4 (*cyan*-based chain) in complex with histone H3 peptide (*green*-based chain) (PDB ID; 2DEW). (**c**) Comparison of the active sites between PPAD (*top*, PDB ID; 4YTG) and human PAD4 (*bottom*). Carbon atoms are colored with *yellow* for the substrates and *green* for proteins. Nitrogens and oxygens are colored with *blue* and *red*, respectively, and sulfur is colored with *ocher*. (**d**) Magnified view of the active site of PPAD (PDB ID; 4YTB). The *red spheres* are amino acids consisting of Michaelis loop

except Gln346, are conserved in PAD1, PAD2, and PAD3. Gln346, seen in PAD4, is substituted by arginine in PAD1, PAD2, and PAD3 at the corresponding position (as determined by comparing the primary structures of these isozymes). This difference may be related to substrate specificity. Structural analysis of human PAD4 complexed with histone N-terminal peptides showed that the substrate had a locally induced β-turn-like bent conformation (Fig. 3.2b). Detailed analysis of the substrate-binding region of PAD4 suggested that the side chain at two residues before the position of the target arginine should be small (Arita et al. 2006). Also, the highly similar structures of CDs of the three known isozymes, PAD1, PAD2, and PAD4, suggest that substrate specificity of these enzymes is determined by both the structure of their CDs and N-terminal Ig-like domain, as proposed earlier (Arita et al. 2004; Saijo et al. 2016).

The CD of PPAD has an α/β propeller structure consists of five ββαβ modules, which is similar to the CDs of human PADs and other guanidino-group-modifying enzymes (GMEs) (Goulas et al. 2015). Catalytic cysteine and histidine residues, as well as two aspartate residues interacting with the guanidinium group of substrate arginine residue, are conserved among PPAD and human PADs (Goulas et al. 2015). In addition, two arginine residues that bind to the main chain of the substrate are conserved in PPAD and human PADs (Arita et al. 2004, 2006; Goulas et al. 2015). Interestingly, Asn297 of PPAD is important for catalysis (Fig. 3.2c). Residues corresponding this residue are conserved in human PADs (Asn588 in PAD1, Asn590 in PAD2, Asn589 in PAD3, and Asn588 in PAD4) and in citrullinating GMEs. Interestingly, PAD6, which does not show any catalytic activity in vitro, lacks this crucial asparagine residue (Arg619 in human PAD6 exists at the corresponding position when the primary structures are aligned). PPAD citrullinates free L-arginine residues, but mammalian PADs do not (Gyorgy et al. 2006; Abdullah et al. 2013). The propeller shaft of PPAD is solid and has a shallow cavity compared with that of human PADs; this may be because PPAD binds to the C-terminal region of substrates. This structural feature is common among arginine deiminases. The active site pocket of PPAD, which contains a Michaelis loop, including a bulky tyrosine residue, prevents binding of the C-terminally extended substrates (Fig. 3.2d). Furthermore, Arg152 and Arg154 in PPAD forms ionic bonds with the free carboxyl group of a C-terminal arginine of substrates; therefore, an internal arginine would not be catalyzed by PPAD (Montgomery et al. 2016). PPAD also works along with an arginine-specific protease arginine gingipain to generate C-terminally citrullinated peptides (McGraw et al. 1999; Wegner et al. 2010).

3.2.3 Ca^{2+}-Binding Sites and Functions

PPAD citrullinates target proteins in a Ca^{2+}-independent manner, whereas human PADs citrullinate target proteins in a Ca^{2+}-dependent manner. To obtain insights on Ca^{2+}-dependent citrullination catalyzed by human PADs, structures of Ca^{2+}-bound

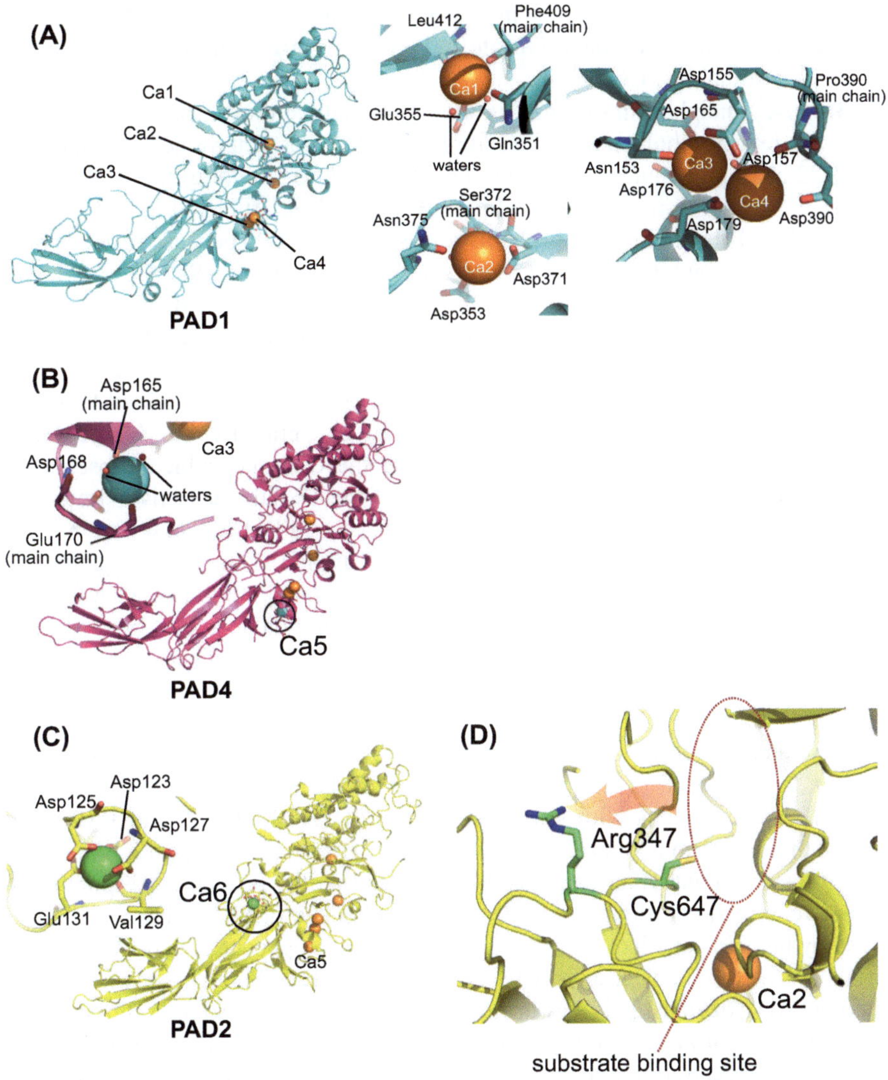

Fig. 3.3 Ca^{2+}-binding sites and their close-up view. (**a**) PAD1 contains four Ca^{2+}-binding sites (Ca1–Ca4) (PDB ID; 5HP5). (**b**) PAD4 contains five Ca^{2+}-binding sites (PDB ID; 1WDA). A *cyan ball* represents Ca^{2+} at Ca5. (**c**) PAD2 (F221/222A) contains six Ca^{2+}-binding sites (PDB ID; 4N2C). A *green ball* represents Ca^{2+} at Ca6. (**d**) Upon Ca^{2+} binding, Arg347 moves to open the space for the substrate

human PAD1, PAD2, and PAD4 were determined. Recently, we found four Ca^{2+}-binding sites in human PAD1 (Saijo et al. 2016) (Fig. 3.3a). By contrast, PAD4 has five Ca^{2+}-binding sites (Arita et al. 2004) (Fig. 3.3b), of which two sites, i.e., Ca1 and Ca2, are present in the CD. The remaining three Ca^{2+}-binding sites, namely,

Ca3, Ca4, and Ca5, are present in the N-terminal Ig-like domain. Ca^{2+}-binding induces the formation of the active site and promotes large conformational changes in the N-terminal Ig-like domain.

Because a neighboring subunit in the crystal lattice prevented Ca^{2+}-binding at Ca2, the reported holoenzyme structure of PAD2 was determined by analyzing F221/222A double mutant. PAD2 has an additional Ca^{2+}-binding site (Ca6) beside Ca1–Ca5, and this site is suggested to have the highest affinity (Fig. 3.3c) (Slade et al. 2015). This binding site probably exists only in PAD2 because residues present in this site are not conserved in other PAD isozymes (Saijo et al. 2016). Structural analysis of human PAD2 indicates the binding order by Ca^{2+} titrations. The Ca1 site of PAD2 has the second-highest affinity. Moreover, the Ca3–Ca5 sites of PAD2 do not show electron density until the concentration of Ca^{2+} is >100 μM (Slade et al. 2015). Calcium titrations by X-ray crystallography yielded Hill coefficients ranging from 3.3 to 3.8 indicating strong cooperativity. Ca^{2+} binding at the Ca3–Ca5 sites may promote Ca^{2+} binding at the Ca2 site, which subsequently causes the movement of Arg347 out of and Cys647 into the active site pocket (Fig. 3.3d). Arg347 shields the active site in Ca^{2+}-free PAD2. These findings and the results of biochemical, biophysical, and mutational studies led the authors to propose the "calcium switch" mechanism. The enzyme first binds Ca^{2+} at Ca6 and Ca1 sites then at the Ca3–Ca5 sites. The latter binding induces Ca^{2+} binding at the Ca2 site, which induces active site formation. The Ca2–Ca5 sites act as a calcium switch to regulate the overall activity of PAD2 and possibly of other PAD isozymes (Slade et al. 2015).

In human PAD1, the site corresponding to Ca5 showed no electron density even though the structure of PAD1 was obtained in the presence of 200 mM Ca^{2+}. Sequence comparisons with known structures of PAD1, PAD2, and PAD4 suggested that PAD1 may bind Ca^{2+} at the Ca5 site. However, superposition of the 3D structures of human PAD1 and PAD4 indicates that the residue corresponding to aspartate, which coordinates with Ca^{2+} at the Ca5 site of PAD4, has been substituted by histidine and its side chain was placed to the opposite side in PAD1 (Saijo et al. 2016). It is unclear as to why the number of Ca^{2+}-binding sites is different among PAD isozymes. This may be related to their physiological roles. For instance, different PAD isozymes modify substrate proteins at multiple sites. Therefore, cellular Ca^{2+} concentration may be a key factor that regulates the catalytic activity of each PAD isozyme.

3.2.4 Structures and the Functions of N-Terminal Ig-Like Domain

The C-terminal CDs of human PADs are very similar to each other, but their N-terminal Ig-like domains show structural variation. The N-terminal Ig-like domain is subdivided into IgG1 and IgG2 domains. The IgG1 domain of PAD2 and

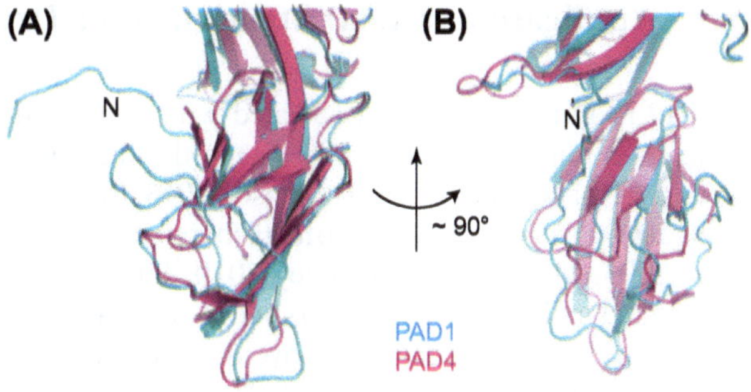

Fig. 3.4 Flexible N-terminal tail of PAD1. Superpositions of the IgG1 domains of PAD1 (*cyan*) and PAD4 (*magenta*) are shown. The direction is different between (**a**) and (**b**) by ~90°

PAD4 contains nine β-strands, whereas that of human PAD1 lacks the first β-strand but contains a flexible tail structure (Fig. 3.4) (Saijo et al. 2016). This elongated N-terminal tail of human PAD1 prevents it from forming a stable homodimer similar to that formed by human PAD2 and PAD4. Although the role of the extended N-terminal tail of human PAD1 is unclear, its functional role can be elucidated based on the role of a corresponding region of PAD6 that contains Ser10 and that undergoes phosphorylation. Although it is unclear whether PAD1 undergoes phosphorylation, the flexibility of this region may allow PAD1 to undergo modification or protein interaction.

The β5 strand of the IgG1 domain shows the highest variation among different PAD isozymes. The β-strand of PAD1 is twisted and faces a different direction compared with the β-strand of PAD2 and PAD4. The following loop has not been observed in the human PAD4 structure (Arita et al. 2004, 2006). In human PAD4, this region contains a nuclear localization signal. Functional variants of the gene encoding PAD4 in Japanese rheumatoid arthritis (RA) population are present in the region encoding the IgG1 domain which is located away from the CD, implying the importance of the IgG1 domain. The loop connecting β5 and β6 strands is present in both PAD1 and PAD2; however, the conformation of this loop is different in PAD1 and PAD2. This variation in the loop structure suggests its flexibility in PAD isozymes.

The IgG2 domain connects IgG1 to the CD. This domain contains ten β-strands and four short α-helices. IgG2 domain of PAD1 contains at least two Ca^{2+}-binding sites, that of PAD2 contains four Ca^{2+}-binding sites, and that of PAD4 contains three Ca^{2+}-binding sites. The difference in the number of Ca^{2+}-binding sites results in structural differences among the three PAD isozymes. Binding of Ca^{2+} to this domain is suggested to induce structural changes in the Ca2 site through a loop, which promotes the binding of Ca^{2+} to this site (Slade et al. 2015). Therefore, this domain may be indispensable for the functions of PAD isozymes.

The Ig-like domain of PPAD has a β-sandwich motif like human PADs. However, this domain in PPAD is C-terminal to the CD and is structurally different from human PADs. Further, intermolecular interactions in PPAD are distinct from that in human PADs. Moreover, CD and Ig-like domains of PPAD are tightly packed (Fig. 3.1c).

3.3 Application of PAD Structures for Drug Design

Human PAD isozymes are associated with some pathologies, including psoriasis (PAD1), multiple sclerosis (PAD2), neurodegeneration (PAD3), and RA and tumorigenesis (PAD4) (Bicker and Thompson 2013; Stadler et al. 2013; Ishida-Yamamoto et al. 2000; Wood et al. 1996; Bhattacharya et al. 2006; Chang et al. 2009; Suzuki et al. 2003; Mohlake and Whiteley 2010; Jang et al. 2013; Lange et al. 2011). To understand molecular mechanisms and biological roles of each PAD isozyme and to develop drugs for treating the above diseases, it is important to develop isozyme-specific inhibitors. Potent and selective peptidic inhibitors have been identified; however, these peptidic agents are easily degraded in living cells and their sizes are large, thus limiting their application in animal cells (Knuckley et al. 2008, 2010b; Slack et al. 2011; Jones et al. 2012).

Cl-amidine is the most advanced and irreversible PAD inhibitor (Luo et al. 2006; Jamali et al. 2015). The use of this inhibitor has increased our understanding of the biological roles of PAD isozymes in different diseases. However, Cl-amidine shows modest isozyme selectivity. By continuous effort to develop bioavailable inhibitors, tetrazole analogs of Cl-amidine showed highly potent inhibitory activity with selectivity toward particular isozymes. Biphenyl tetrazole *tert*-butyl Cl-amidine (BB-Cl-amidine) exhibited enhanced cell killing in PAD4 expressing osteosarcoma bone marrow cell line and was able to block the formation neutrophil extracellular traps (Subramanian et al. 2015). In addition, a non-peptidic PAD3-selective inhibitor was developed by screening a guanidine substrate library (Jamali et al. 2015). Allosteric activation has been proposed based on the structural analysis of PAD2 (Slade et al. 2015). Based on structural insights on the Ca^{2+} switch, it should be possible to develop an allosteric inhibitor as was recently described for PAD4 (Lewis et al. 2015). Knowledge of the molecular structures of PAD isozymes will undoubtedly aid the rational design of new PAD isozyme-selective drugs with improved potency and bioavailability (Teo et al. 2012).

Periodontitis is a risk factor of RA. Sequence homology between PPAD and human PADs is limited to key residues in active sites. Therefore, a 3D structure is distinctly different from that of human PAD isozymes. Analysis of the structure of PPAD is crucial to understand its reaction mechanism by comparing with human PADs and to develop PPAD-specific inhibitors as a prophylactic drug for RA.

3.4 Conclusions

Recent protein X-ray crystallographic studies combined with small-angle X-ray scattering analyses substantially clarified the tertiary and quaternary structures of PAD isozymes. PAD1 exists as a monomer in solution, while PAD3 forms a homodimer similar to PAD2 and PAD4 (Saijo et al. 2016). PAD isozymes show different substrate specificity for peptidylarginine-harboring peptides (Bhattacharya et al. 2006;

Suzuki et al. 2003). The degree of tolerance is different between partial peptides and an intact molecule of S100A3, whose structure has been determined previously (Kizawa et al. 2008, 2014; Unno et al. 2011). Our results imply that the monomeric structure of PAD1 is advantageous for its tolerant substrate specificity. These results offer substantial insights on the structures and functions of PAD isozymes. However, it is unclear whether PAD2, PAD3, PAD4, and PAD6 always form a homodimer and PAD1 may be capable of undergoing dimerization. We cannot exclude the possibility that unknown posttranslational modifications or some cellular conditions may force these isozymes to form monomers. Alternatively, other protein ligands may bind to regulatory site(s) on PAD isozymes and may act as chaperones to activate or inactivate these enzymes. Thus, regulation of catalytic activities of PAD isozymes and net regulation of naturally occurring citrullination in vivo may be more complex than postulated previously. Future studies should be performed by paying attention to intermolecular interactions that activate these enzymes and mutual recognition mechanisms between intact substrate proteins and PAD isozymes. Studies elucidating differences in molecular mechanisms underlying substrate recognition by PAD isozymes are ongoing.

References

Abdullah, S. N., Farmer, E. A., Spargo, L., Logan, R., & Gully, N. (2013). Porphyromonas gingivalis peptidylarginine deiminase substrate specificity. *Anaerobe, 23*, 102–108.

Arita, K., et al. (2004). Structural basis for Ca^{2+}-induced activation of human PAD4. *Nature Structural & Molecular Biology, 11*(8), 777–783.

Arita, K., et al. (2006). Structural basis for histone N-terminal recognition by human peptidylarginine deiminase 4. *Proceedings of the National Academy of Sciences of the United States of America, 103*(14), 5291–5296.

Baka, Z., et al. (2012). Citrullination under physiological and pathological conditions. *Joint, Bone, Spine, 79*(5), 431–436.

Bhattacharya, S. K., et al. (2006). Proteomics implicates peptidyl arginine deiminase 2 and optic nerve citrullination in glaucoma pathogenesis. *Investigative Ophthalmology & Visual Science, 47*(6), 2508–2514.

Bicker, K. L., & Thompson, P. R. (2013). The protein arginine deiminases: Structure, function, inhibition, and disease. *Biopolymers, 99*(2), 155–163.

Chang, X., et al. (2009). Increased PADI4 expression in blood and tissues of patients with malignant tumors. *BMC Cancer, 9*, 40.

Chavanas, S., et al. (2004). Comparative analysis of the mouse and human peptidylarginine deiminase gene clusters reveals highly conserved non-coding segments and a new human gene, PADI6. *Gene, 330*, 19–27.

Christophorou, M. A., et al. (2014). Citrullination regulates pluripotency and histone H1 binding to chromatin. *Nature, 507*(7490), 104–108.

Dwivedi, N., et al. (2014). Deimination of linker histones links neutrophil extracellular trap release with autoantibodies in systemic autoimmunity. *The FASEB Journal, 28*(7), 2840–2851.

Fuhrmann, J., Clancy, K. W., & Thompson, P. R. (2015). Chemical biology of protein arginine modifications in epigenetic regulation. *Chemical Reviews, 115*(11), 5413–5461.

Goulas, T., et al. (2015). Structure and mechanism of a bacterial host-protein citrullinating virulence factor, Porphyromonas gingivalis peptidylarginine deiminase. *Scientific Reports, 5*, 11969.

Guerrin, M., et al. (2003). cDNA cloning, gene organization and expression analysis of human peptidylarginine deiminase type I. *The Biochemical Journal, 370*(Pt 1), 167–174.
Gyorgy, B., Toth, E., Tarcsa, E., Falus, A., & Buzas, E. I. (2006). Citrullination: A posttranslational modification in health and disease. *The International Journal of Biochemistry & Cell Biology, 38*(10), 1662–1677.
Humm, A., Fritsche, E., Steinbacher, S., & Huber, R. (1997). Crystal structure and mechanism of human L-arginine:Glycine amidinotransferase: A mitochondrial enzyme involved in creatine biosynthesis. *The EMBO Journal, 16*(12), 3373–3385.
Ishida-Yamamoto, A., et al. (2000). Decreased deiminated keratin K1 in psoriatic hyperproliferative epidermis. *The Journal of Investigative Dermatology, 114*(4), 701–705.
Ishigami, A., et al. (2002). Human peptidylarginine deiminase type II: Molecular cloning, gene organization, and expression in human skin. *Archives of Biochemistry and Biophysics, 407*(1), 25–31.
Jamali, H., Khan, H. A., Stringer, J. R., Chowdhury, S., & Ellman, J. A. (2015). Identification of multiple structurally distinct, nonpeptidic small molecule inhibitors of protein arginine deiminase 3 using a substrate-based fragment method. *Journal of the American Chemical Society, 137*(10), 3616–3621.
Jang, B., et al. (2013). Peptidylarginine deiminase and protein citrullination in prion diseases: Strong evidence of neurodegeneration. *Prion, 7*(1), 42–46.
Jones, J. E., Causey, C. P., Knuckley, B., Slack-Noyes, J. L., & Thompson, P. R. (2009). Protein arginine deiminase 4 (PAD4): Current understanding and future therapeutic potential. *Current Opinion in Drug Discovery & Development, 12*(5), 616–627.
Jones, J. E., et al. (2012). Synthesis and screening of a haloacetamidine containing library to identify PAD4 selective inhibitors. *ACS Chemical Biology, 7*(1), 160–165.
Kanno, T., et al. (2000). Human peptidylarginine deiminase type III: Molecular cloning and nucleotide sequence of the cDNA, properties of the recombinant enzyme, and immunohistochemical localization in human skin. *The Journal of Investigative Dermatology, 115*(5), 813–823.
Kearney, P. L., et al. (2005). Kinetic characterization of protein arginine deiminase 4: A transcriptional corepressor implicated in the onset and progression of rheumatoid arthritis. *Biochemistry, 44*(31), 10570–10582.
Kizawa, K., et al. (2008). Specific citrullination causes assembly of a globular S100A3 homotetramer: A putative Ca^{2+} modulator matures human hair cuticle. *The Journal of Biological Chemistry, 283*(8), 5004–5013.
Kizawa, K., Unno, M., Heizmann, C. W., & Takahara, H. (2014). Chapter 8: Importance of citrullination of hair protein molecular assembly during trichocytic differentiation. In A. P. Nicholas & S. K. Bhattacharya (Eds.), *Protein deimination in human health and disease* (pp. 129–148). New York: Springer.
Knuckley, B., Bhatia, M., & Thompson, P. R. (2007). Protein arginine deiminase 4: Evidence for a reverse protonation mechanism. *Biochemistry, 46*(22), 6578–6587.
Knuckley, B., Luo, Y., & Thompson, P. R. (2008). Profiling protein arginine deiminase 4 (PAD4): A novel screen to identify PAD4 inhibitors. *Bioorganic & Medicinal Chemistry, 16*(2), 739–745.
Knuckley, B., Causey, C. P., Pellechia, P. J., Cook, P. F., & Thompson, P. R. (2010a). Haloacetamidine-based inactivators of protein arginine deiminase 4 (PAD4): Evidence that general acid catalysis promotes efficient inactivation. *Chembiochem, 11*(2), 161–165.
Knuckley, B., et al. (2010b). A fluopol-ABPP HTS assay to identify PAD inhibitors. *Chemical Communications, 46*(38), 7175–7177.
Lange, S., et al. (2011). Protein deiminases: New players in the developmentally regulated loss of neural regenerative ability. *Developmental Biology, 355*(2), 205–214.
Lewis, H. D., et al. (2015). Inhibition of PAD4 activity is sufficient to disrupt mouse and human NET formation. *Nature Chemical Biology, 11*(3), 189–191.
Li, P., et al. (2010). PAD4 is essential for antibacterial innate immunity mediated by neutrophil extracellular traps. *The Journal of Experimental Medicine, 207*(9), 1853–1862.

Liu, Y. L., Chiang, Y. H., Liu, G. Y., & Hung, H. C. (2011). Functional role of dimerization of human peptidylarginine deiminase 4 (PAD4). *PloS One, 6*(6), e21314.

Luo, Y., et al. (2006). Inhibitors and inactivators of protein arginine deiminase 4: Functional and structural characterization. *Biochemistry, 45*(39), 11727–11736.

McGraw, W. T., Potempa, J., Farley, D., & Travis, J. (1999). Purification, characterization, and sequence analysis of a potential virulence factor from Porphyromonas gingivalis, peptidylarginine deiminase. *Infection and Immunity, 67*(7), 3248–3256.

Mohlake, P., & Whiteley, C. G. (2010). Arginine metabolising enzymes as therapeutic tools for Alzheimer's disease: Peptidyl arginine deiminase catalyses fibrillogenesis of beta-amyloid peptides. *Molecular Neurobiology, 41*(2–3), 149–158.

Montgomery, A. B., et al. (2016). Crystal structure of Porphyromonas gingivalis peptidylarginine deiminase: Implications for autoimmunity in rheumatoid arthritis. *Annals of the Rheumatic Diseases, 75*(6), 1255–1261.

Nakashima, K., et al. (1999). Molecular characterization of peptidylarginine deiminase in HL-60 cells induced by retinoic acid and 1alpha,25-dihydroxyvitamin D(3). *The Journal of Biological Chemistry, 274*(39), 27786–27792.

Nijenhuis, S., Zendman, A. J., Vossenaar, E. R., Pruijn, G. J., & vanVenrooij, W. J. (2004). Autoantibodies to citrullinated proteins in rheumatoid arthritis: Clinical performance and biochemical aspects of an RA-specific marker. *Clinica Chimica Acta, 350*(1–2), 17–34.

Saijo, S., et al. (2016). Monomeric form of peptidylarginine deiminase type I revealed by X-ray crystallography and small-angle X-ray scattering. *Journal of Molecular Biology, 428*(15), 3058–3073.

Slack, J. L., Causey, C. P., Luo, Y., & Thompson, P. R. (2011). Development and use of clickable activity based protein profiling agents for protein arginine deiminase 4. *ACS Chemical Biology, 6*(5), 466–476.

Slade, D. J., Subramanian, V., Fuhrmann, J., & Thompson, P. R. (2014a). Chemical and biological methods to detect post-translational modifications of arginine. *Biopolymers, 101*(2), 133–143.

Slade, D. J., Subramanian, V., & Thompson, P. R. (2014b). Pluripotency: Citrullination unravels stem cells. *Nature Chemical Biology, 10*(5), 327–328.

Slade, D. J., et al. (2015). Protein arginine deiminase 2 binds calcium in an ordered fashion: Implications for inhibitor design. *ACS Chemical Biology, 10*(4), 1043–1053.

Stadler, S. C., et al. (2013). Dysregulation of PAD4-mediated citrullination of nuclear GSK3beta activates TGF-beta signaling and induces epithelial-to-mesenchymal transition in breast cancer cells. *Proceedings of the National Academy of Sciences of the United States of America, 110*(29), 11851–11856.

Subramanian, V., et al. (2015). Design, synthesis, and biological evaluation of tetrazole analogs of Cl-amidine as protein arginine deiminase inhibitors. *Journal of Medicinal Chemistry, 58*(3), 1337–1344.

Suzuki, A., et al. (2003). Functional haplotypes of PADI4, encoding citrullinating enzyme peptidylarginine deiminase 4, are associated with rheumatoid arthritis. *Nature Genetics, 34*(4), 395–402.

Taki, H., et al. (2011). Purification of enzymatically inactive peptidylarginine deiminase type 6 from mouse ovary that reveals hexameric structure different from other dimeric isoforms. *Advances in Bioscience and Biotechnology, 2*(4), 7.

Teo, C. Y., et al. (2012). Discovery of a new class of inhibitors for the protein arginine deiminase type 4 (PAD4) by structure-based virtual screening. *BMC Bioinformatics, 13*(Suppl 17), S4.

Unno, M., Kawasaki, T., Takahara, H., Heizmann, C. W., & Kizawa, K. (2011). Refined crystal structures of human Ca^{2+}/Zn^{2+}-binding S100A3 protein characterized by two disulfide bridges. *Journal of Molecular Biology, 408*(3), 477–490.

Vossenaar, E. R., Zendman, A. J., van Venrooij, W. J., & Pruijn, G. J. (2003). PAD, a growing family of citrullinating enzymes: Genes, features and involvement in disease. *BioEssays, 25*(11), 1106–1118.

Vossenaar, E. R., Zendman, A. J., & Van Venrooij, W. J. (2004). Citrullination, a possible functional link between susceptibility genes and rheumatoid arthritis. *Arthritis Research & Therapy, 6*(1), 1–5.

Wegner, N., et al. (2010). Peptidylarginine deiminase from Porphyromonas gingivalis citrullinates human fibrinogen and alpha-enolase: Implications for autoimmunity in rheumatoid arthritis. *Arthritis and Rheumatism, 62*(9), 2662–2672.

Wood, D. D., Bilbao, J. M., O'Connors, P., & Moscarello, M. A. (1996). Acute multiple sclerosis (Marburg type) is associated with developmentally immature myelin basic protein. *Annals of Neurology, 40*(1), 18–24.

Zhang, X., et al. (2012). Peptidylarginine deiminase 2-catalyzed histone H3 arginine 26 citrullination facilitates estrogen receptor alpha target gene activation. *Proceedings of the National Academy of Sciences of the United States of America, 109*(33), 13331–13336.

Chapter 4
The Use of Genetically Engineered Mice to Study PAD Biology and Pathology

Chinatsu Mukai, Brooke A. Marks, and Scott A. Coonrod

4.1 Introduction

Peptidylarginine deiminases (PADs or PADIs) are a family of calcium-dependent enzymes that post-translationally convert positively charged arginine residues to neutrally charged citrulline in a process called citrullination or deimination. There are five PAD family members (PAD1–PAD4 and PAD6). PAD genes arose by duplication and are clustered within a ~300-kb region on chromosome 1p36 in humans and within a ~230-kb region on chromosome 4 in mice. In both species PAD1, PAD3, PAD4, and PAD6 are grouped closely together and are oriented in the same direction, while PAD2 is set apart from the other PADs by at least 60 kb and is oriented in the opposite direction (Vossenaar et al. 2003). PAD isozymes are expressed in a range of tissues in mammals, with PAD2 being broadly expressed in numerous tissues, while PAD4 is highly represented in immune cells. PAD6 expression, on the other hand, is primarily limited to oocytes and early embryos, whereas PAD1 and PAD3 appear to be mainly expressed in the epidermis. While still coming to light, functional roles for PADs in mammalian physiology and pathology are diverse and include cellular differentiation, nerve growth, apoptosis, inflammation, gene regulation, and early embryonic development. Over the last several years, investigators have generated genetically engineered mice (GEM) for PAD2, PAD3, PAD4, and PAD6 to investigate the functions of these unique enzymes at the organismal level. Outcomes from these studies are highlighted in Fig. 4.1, and the goal of this chapter is to provide a broad summary of findings obtained from these animals.

C. Mukai • B.A. Marks • S.A. Coonrod (✉)
Baker Institute for Animal Health, College of Veterinary Medicine, Cornell University, Ithaca, NY, USA
e-mail: sac269@cornell.edu

A.P. Nicholas et al. (eds.), *Protein Deimination in Human Health and Disease*,
DOI 10.1007/978-3-319-58244-3_4

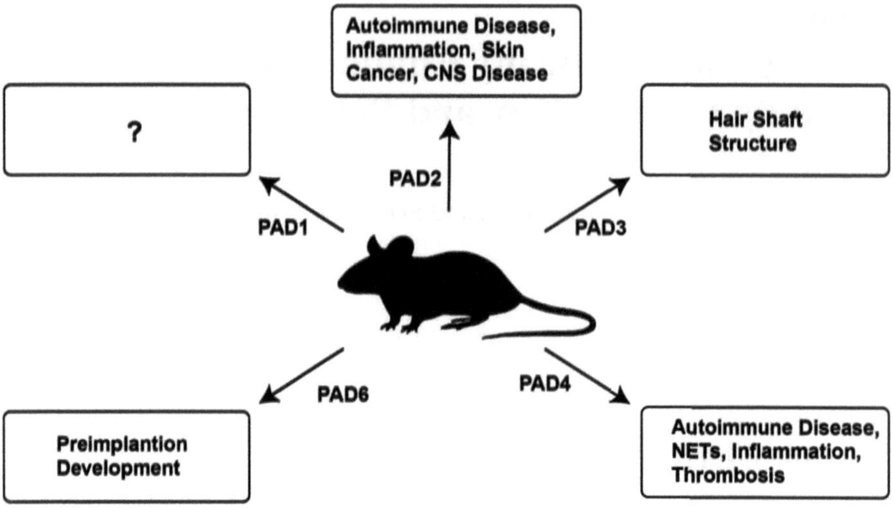

Fig. 4.1 Specific physiological and pathological functions that have been given to PAD isozymes based on outcomes from studies using genetically engineered mice

4.2 PAD2

PAD2 appears to be the most broadly expressed of the PADs, with its expression in humans being reported in muscles, nervous tissue, skin, immune cells, and secretory epithelia (including the mammary gland, uterus, and salivary glands) (Horibata et al. 2012). Most of the earlier reports on PAD2, however, focused on its role in citrullinating myelin basic protein (MBP) during the inflammatory phase of multiple sclerosis (MS) (Akiyama et al. 1999). To determine the extent to which PAD2-mediated protein citrullination contributes to this process, Raijmakers et al. tested the effects of PAD2 deletion on induced experimental autoimmune encephalomyelitis (EAE) in mice (Raijmakers et al. 2006). These mice were generated by disrupting a 164-bp region of exon 1 using a targeting vector which was derived from the Lambda KOS system (Raijmakers et al. 2006). The investigators first validated the knockout and demonstrated that PAD2 protein was lost in the brain and spinal cord of the null mice. While not shown, the investigators also noted that the expression levels of PAD1, PAD3, and PAD4 were not affected by PAD2 deletion. They then tested the effects of loss of PAD2 on EAE and found that, while PAD2 is required for citrullination in nervous tissue, PAD2 does not appear to be required for experimentally induced EAE. Importantly, PAD2 deletion also did not appear to affect brain or spinal cord development, suggesting that, while PAD2 is abundantly expressed in the nervous system, it is not required for normal function in these tissues (Raijmakers et al. 2006). These studies support the hypothesis that while PAD2 is the main PAD isozyme that is responsible for citrullination in the CNS in both healthy and EAE mice, this activity does not play a critical role in experimentally induced EAE. In a separate study using the same PAD2-null mouse line, investigators demonstrated that deimination is seen in isolated myelin of PAD2 knockout mice and predict that this citrullination activity may actually be catalyzed by PAD4 (Wood et al. 2008).

In another study aimed at investigating the role of PAD2-mediated citrullination of MBP in MS, Musse et al. (2008) developed a mouse line that contained multiple copies of rat PAD2 cDNA under the control of the MBP promoter. Following validation of transgene overexpression in the CNS, the investigators documented clinical disease development over a 6-month period following birth. They found that clinical signs of demyelinating disease occurred at a younger age in PAD2-overexpressing mice when compared to controls. The investigators then tested whether the symptoms were caused by increased PAD2 activity in the brain. They found that PAD2 protein levels were elevated in the white matter of the PAD2-overexpressing mice and that citrullination of MBP was increased at these sites as well. Additionally, the investigators found that subtle changes in axon and myelin thickness were seen in the PAD2-overexpressing mice. Interestingly, PAD4 mRNA levels were found to be increased by fourfold in PAD2 heterozygous mice and a further fivefold in PAD2 homozygous mice. The investigators also reported that histone citrullination and nuclear PAD4 were significantly elevated in the PAD2-overexpressing mice, as were tumor necrosis factor alpha (TNFα) levels (Musse et al. 2008). Taken together, these findings suggest that PAD2 may play a role in myelin destabilization during MS progression and also raise the possibility that PAD2 cooperates with PAD4 to mediate this activity.

PAD2 and PAD4 are often found to be co-expressed in specific cell types, particularly immune cells, and both of these PADs are implicated in various pathologies including inflammation, autoimmune disease, and cancer. In order to begin to tease out specific functions for these two family members, a recent study made use of PAD2 and PAD4 mutant mice to investigate the expression and activity of these isozymes in various murine tissues (van Beers et al. 2013). In this particular study, the PAD4 mutant line was generated via insertional mutagenesis, which resulted in the insertion of a 156-bp cassette in intron 1 of the PAD4 gene. Subsequent PCR and Western blot analysis found that this mutant line was actually a partial knockout, and these mice retained a low level of PAD4 protein (van Beers et al. 2013). The PAD2-null mice used in this study were the same line as that used by Raijmakers et al. and were true knockouts. The investigators went on to document citrullination levels in these mutant mice in various tissues (including the brain, lung, spleen, etc.) using an antibody-based assay for PAD activity (ABAP) and an anti-modified citrulline antibody. Results show that PAD activity was virtually absent in a range of tissues from PAD2-null mice while PAD activity in the PAD4-low mice was similar to that of controls. The investigators then measured PAD activity in peripheral blood mononuclear cells and in polymorphonuclear leukocytes and found that loss of PAD2 only moderately affected citrullination levels in immune cells. These results suggest that the contribution of PAD2 to overall PAD activity in white blood cells is not as pronounced as that seen in other tissues. Interestingly, these investigators also found that PAD6 levels were increased PAD2-null and PAD4-low mouse tissue. This surprising result raises the possibility that PAD6 (which is normally expressed in oocytes and early embryos) may compensate for the loss of other PADs in somatic tissues. This prediction does not seem likely, however, since PAD6 does not appear to be catalytically active (Taki et al. 2011).

Similar to PAD4, PAD2 has been documented to be expressed in T cells (Ferrari-Lacraz et al. 2012), neutrophils (Darrah et al. 2012), and macrophages (Vossenaar et al. 2004). While the role of PAD4 in immunological processes such as inflammation and autoimmune disease has received most of the attention, PAD2 has also been implicated in these processes (Foulquier et al. 2007; Damgaard et al. 2015). In order to more directly investigate and compare the roles of PAD2 and PAD4 in inflammation, a recent study investigated the effect of PAD2 or PAD4 deletion on inflammation using a TNF-overexpressing mouse model of inflammatory arthritis (Bawadekar et al. 2017). For the study, the investigators crossed TNF-overexpressing mice with either PAD2- or PAD4-null lines and then investigated the effects of PAD deletion on citrullination and ankle joint inflammation. The PAD2- and PAD4-null mouse lines used for this study were generated by other groups (Raijmakers et al. 2006; Li et al. 2010), and both appear to be complete knockouts. Surprisingly, results show that PAD2, but not PAD4, is required for gross protein citrullination in the inflamed arthritic joints of these mice. Further, they found that PAD2 is required for maximal severity of TNF-induced arthritis in their mouse model. PAD4 has previously been found to be required for neutrophil extracellular trap (NET) production via its role in histone citrullination (Li et al. 2010; Wang et al. 2009). Given the previously established links between NETs and inflammation, the investigators tested to see if PAD2 might also be required for NET production (note: a more detailed discussion on NETs is found in Sect. 4.4 below). Results show that, while the PAD4-null mice could not produce NETs, production of this inflammation-promoting structure was not affected by PAD2 deletion. Taken together, these studies provide the most direct evidence that PAD2 contributes to inflammatory arthritis. Additionally, these findings suggest that NETs may not be the main source of citrullinated proteins in arthritic mice and raise the possibility that PAD4's role in inflammation may be something other than direct citrullination of antigens. Lastly, given the requirement of PAD2 for generating citrullinated proteins in this model system, these results suggest that PAD2 may play an even larger role in autoimmune diseases than previously expected.

In addition to nervous tissue and immune cells, PAD2 is also expressed in the skin. Previous reports have found that PAD1, PAD2, and PAD3 are expressed in the epidermis, with PAD2 expression being primarily found within the spinous and granular layers of the mouse epidermis (Ying et al. 2009). Interestingly, a recent analysis of the skin in PAD2-null mice found that loss of PAD2 does not appear to affect citrullination levels in the skin of mice, nor did PAD2 deletion appear to affect PAD1 and PAD3 levels in the epidermis (Coudane et al. 2011).

Over the last several years, we have been investigating the role of PAD2 in breast cancer using cell lines and mouse xenografts. In order to further investigate these links, we recently generated a FVB/N mouse that overexpressed human FLAG-PAD2 under control of the mammary tumor virus (MMTV) promoter (McElwee et al. 2014). Our results demonstrated that, similar to previous studies, MMTV drove the expression of PAD2 in a range of tissues including the mammary gland, salivary gland, ovaries, and skin. Interestingly, we found that, while this mouse line did not develop overt signs of mammary cancer, ~40% of the transgenic mice developed skin lesions between 4 and 12 months, with a subset of these lesions progressing to squamous cell carcinoma (SCC). At a more mechanistic level, we found that the expression of inflammatory markers such as IL6 and Cox2 was significantly increased in the skin of PAD2-overexpressing mice and also in PAD2-overexpressing SCC cell lines (McElwee et al. 2014). As noted earlier, previous studies have shown that, in some tissues, either deletion or overexpression of one PAD can alter the expression of other PADs. In our study, we found that the expression of PAD1–PAD4 appeared to be significantly dysregulated in several tissues in the PAD2-overexpressing mice. These results provided the first genetic evidence that PAD enzymes can function as oncogenes and also suggested that PAD2 may promote oncogenesis via its role as an inflammatory mediator. Given that PAD2 overexpression altered the expression of other PAD family members, we cannot rule out the possibility that some of the effects observed in our study were due to compensation by other PADs.

4.3 PAD3

As noted above, PAD3 appears to primarily be expressed in the epidermis and hair follicles. Recent studies have found that mutation of the PAD3 and TGM3 (transglutaminase 3) genes in humans causes uncomfortable hair syndrome (UHS), also known as “spun hair glass syndrome” (Ü Basmanav et al. 2016). This rare disease is an anomaly of the hair shaft and is characterized by dry frizzy hair that is difficult to comb flat. The prediction that PAD3 mutations cause UHS in humans was supported by the investigators finding that deletion of PAD3 resulted in subtle alterations in the hair shaft in these animals. Results showed that, while these mice were viable and had grossly normal skin, SEM analysis of their skin found that hair shafts from these mice were irregular, rough, and appeared as if “hammered” (Ü Basmanav et al. 2016). The investigators then generated a mutant construct for cell culture studies and found that the putative causative mutation significantly reduced PAD3 activity in vitro, suggesting that PAD3 activity was required for the observed defect. The investigators also noted that, while PAD3 is highly expressed in the hair follicle and upper epidermis, there was no observable phenotype in the epidermis of the null mice. While not tested, the investigators speculated that this might be due to compensation by other PADs.

4.4 PAD4

Given PAD4’s close ties with rheumatoid arthritis (RA), which affects ~1% of the world’s population (Begovich et al. 2004), this PAD family member has received most of the attention in the biomedical arena. Individuals with RA frequently have autoantibodies to citrullinated peptides (van Gaalen et al. 2004) and several studies have identified PAD4 as a susceptibility locus for RA (Suzuki et al. 2003). Additionally, as noted above, PAD4 has also been found to be involved in neutrophil extracellular trap (NET) production (Wang et al. 2009). These chromatin-based structures are released by neutrophils as part of the innate immune system to bind to and kill pathogens (Brinkmann and Zychlinsky 2012). Importantly, NETs have also been found to be a source of citrullinated autoantigens, which can promote RA symptoms (Khandpur et al. 2013). Aside from RA, dysregulated NET activity has been associated with a range of disease states, including vasculitis, atherosclerosis, and thrombosis (Kaplan and Radic 2012). Recently, investigators have begun studying PAD4-null mouse lines to strengthen links between PAD4 and these disease states and also to investigate the mechanisms by which PAD4 modulates the immune system.

In one of the first reports using PAD4 knockout mice, Li et al. (2010) generated a PAD4-null line by deleting exon II from PAD4, thus causing premature termination of the PAD4 transcript. The investigators then confirmed that PAD4 protein was absent from neutrophils in these mice and found that, despite the loss of PAD4, the null mice were viable with no detectable gross physical abnormalities. The investigators then demonstrated that histone H3 citrullination was abolished in neutrophils from PAD4-null mice and that these neutrophils were unable to undergo NET production following stimulation with chemokines or bacteria. The investigators then utilized a mouse model of necrotizing fasciitis to demonstrate that PAD4-null mice are more susceptible to bacterial infection. The investigators concluded that this deficiency was likely due to a failure to produce NETs at the site of infection (Li et al. 2010). These important studies provided the first genetic evidence that PAD4 is an important immune mediator that is required for NET-mediated antibacterial innate immunity.

A separate study by Hemmers et al. generated another PAD4 knockout line by introducing loxP sites that flanked exons 9 and 10 (which encode the active site of PAD4), and the site was removed by mating the mice with CMV-Cre deleter mice. Similar to the study by Li et al., the investigators found that histone citrullination and NET production was not detectable in PAD4-null neutrophils. These investigators also found that PAD4 deficiency in neutrophils does not impact leukocyte recruitment to the lungs of influenza-challenged mice. These results suggest that, while PAD4-mediated NET production plays an important role in antibacterial immunity, PAD4-mediated NET formation appears to be dispensable during viral infection (Hemmers et al. 2011).

A report by Rohrbach et al. used the PAD4 knockout line generated by Hemmers et al. to test the requirement of PAD4 for autoimmune disease using a K/BxN autoantibody-mediated model of arthritis (Rohrbach et al. 2012). In this model, autoantibody-containing serum from K/BxN mice (which spontaneously develop a progressive inflammatory joint disease) was injected into the peritoneum of wild-type or PAD4-null mice, to stimulate autoantibody-mediated arthritis. Results show that, similar to previous studies, PAD4 activity and NET formation was absent in neutrophils from PAD4-null mice. However, both wild-type and PAD4-null mice developed K/BxN-induced inflammatory arthritis, suggesting that PAD4 is dispensable for the effector phase of this disease (Rohrbach et al. 2012).

TNFα is known to play a critical role in RA and, more generally, in inflammatory arthritis (Feldmann and Maini 2003). Additionally, TNFα can also induce the nuclear translocation of PAD4 , histone citrullination, and NET formation (Mastronardi et al. 2006). It has also been shown that citrullinated proteins and anti-citrullinated protein antibodies (ACPAs) can induce TNFα production by macrophages (Sokolove et al. 2011). Given these interconnections, investigators recently crossed TNFα-overexpressing mice with wild-type and PAD4-null mice in order to test the hypothesis that PAD4 may regulate TNFα-mediated autoantibody production and inflammatory arthritis (Shelef et al. 2014). Results showed that TNFα-overexpressing wild-type mice displayed increased levels of autoantibodies that were reactive against native and citrullinated antigens when compared to the TNFα-overexpressing PAD4-null mice. Additionally, the TNFα-overexpressing PAD4-null mice also displayed reduced inflammation and arthritis when compared to TNFα-overexpressing wild-type PAD4 mice. These results suggest that PAD4 mediates autoantibody production and inflammatory arthritis downstream of TNFα (Shelef et al. 2014). The important role that PAD4 plays in promoting inflammatory arthritis has also been supported by several recent studies showing that arthritis severity is significantly reduced in PAD4 knockout mice using glucose-6-phosphate (Seri et al. 2015) and collagen-induced (Suzuki et al. 2016) models of inflammation.

Given that PAD4-null mice do not produce NETs, these mice have also been utilized to investigate the role of NETs in placentation disorders (Erpenbeck et al. 2016). Results from these studies found that overexpression of soluble fms-like tyrosine kinase (sFlt-1, which has been associated with abnormal placentation disorders during early gestation) resulted in miscarriage and the accumulation of neutrophils and NETs in the placentas of wild-type PAD4 mice. However, neutrophil invasion, NET production, inflammation, and pregnancy losses were significantly abrogated in PAD4-null mice that overexpressed sFlt-1. The investigators then went on to show that neutrophil invasion and NET production was significantly higher in preeclamptic women compared to non-hypertensive controls (Erpenbeck et al. 2016). These findings suggest that PAD4-specific inhibitors may have therapeutic potential for the treatment of preecplampsia in women.

NETs have recently been shown to be involved in thrombosis formation via their role in generating the thrombus scaffold and promoting coagulation (Martinod and Wagner 2014). To investigate the mechanism by which PAD4-mediated citrullination mediates this process, Martinod et al. recently tested the effects of PAD4 deletion on thrombus formation using a mouse venous stenosis model of deep vein thrombosis. The study found that <10% of the PAD4-null mice produced a thrombus, compared to 90% of wild-type mice which generated a thrombus. Interestingly, the investigators found that thrombosis could be rescued by infusion of wild-type neutrophils, suggesting that neutrophil PAD4 was sufficient for the observed effect. These results strongly suggests that NETs represent a crucial component of the thrombus scaffold (Martinod et al. 2013).

Acute myocardial infarction (AMI) is a major component of cardiovascular disease and is caused by intraluminal coronary thrombosis. AMI can be modeled in mice using an ischemia and reperfusion (I/R) technique. To test whether PAD4-mediated citrullination may also play a role in this type of thrombosis, Savchenko et al. examined the effect of I/R on thrombus production in PAD4-null mice (Savchenko et al. 2014). Results showed that the mice (which do not produce NETs) were significantly protected from I/R treatment. The study also found that PAD4 deficiency reduced leukocyte recruitment to the infarcted myocardium and prevented nuclear histone citrullination at this site (Savchenko et al. 2014).

Outcomes from these studies suggested that PAD4 inhibitors could have therapeutic benefit for patients with thrombosis and ischemic/reperfusion injury. A follow-up study by Martinod et al. (Martinod et al. 2015) investigated the role of NETs in sepsis, with respect to the balance between their antimicrobial and cytotoxic actions (Martinod et al. 2015). Given that PAD4 is required for NET production and that NETs may have important antimicrobial capabilities, this study was carried out because there was concern that the therapeutic benefit of PAD4 inhibitors may be offset by the possibility that these inhibitors might promote sepsis. The investigators addressed this concern by inducing sepsis in PAD4 KO mice using cecal ligation and puncture. Results showed that survival was comparable between PAD4 KO and wild-type mice. They also found that neutrophil functions involved in bacterial killing (other than NETosis) remained intact in the PAD4 KO mice. Outcomes from these studies suggested that preventing NET formation by PAD4 inhibition in inflammatory or thrombotic diseases will not likely increase the patients risk for bacterial infections (Martinod et al. 2015).

In another study aimed at investigating antimicrobial effects of PAD4 and NETs, Kolaczkowska et al. found that, during bloodstream infection with *S. aureus*, most bacteria are sequestered within liver sinusoids by Kupffer cells and that this sequestration promotes ischemia and neutrophil infiltration into the area (Kolaczkowska et al. 2015). The investigators then found that the sequestered neutrophils release NETs (which contain high levels of neutrophil elastase) into the sinusoids and these NETs then become anchored to the endothelium by binding to von Willebrand factor (VWF). Next, the investigators showed that, while DNase is highly efficient at removing DNA from the NETs, elastase and histones (which are highly cytotoxic) remain associated with the NETs and likely promoted severe tissue damage. Importantly, however, the report shows that, inhibition of NET production, as modeled by the PAD4-null mice, prevents collateral host tissue damage, suggesting that therapeutic PAD4 inhibitors are not likely to cause host tissue damage during infection (Kolaczkowska et al. 2015).

4.5 PAD6

PAD6 expression is primarily limited to mammalian oocytes and early embryos and was first cloned and characterized because of its high expression levels in mouse eggs (Wright et al. 2003). In order to study the function of PAD6, we investigated the effects of PAD6 deletion (using a somatic knockout) on fertility and development. We found that, while PAD6-null males were fertile, PAD6-null females were sterile (Esposito et al. 2007). Further analysis found that oocytes from these females could be fertilized at normal rates; however, the resulting embryos underwent developmental arrest at the two-cell stage (Esposito et al. 2007). This result indicated that PAD6 functions as a maternal effect gene. We then went on to investigate the potential mechanisms that caused the developmental arrest and found that the PAD-null embryos appeared to have defects in their ribosomal machinery leading to defective protein synthesis and a failure to undergo embryonic genome activation (Yurttas et al. 2008). Additionally, EM analysis of the null oocytes and early embryos found that PAD6 is required for the formation of an abundant cytoskeletal structure known as the cytoplasmic lattices (Esposito et al. 2007). We also found that microtubule dynamics were defective in the PAD6-null oocytes and that these oocytes could not properly reposition mitochondria and endoplasmic reticulum during oocyte maturation (Kan et al. 2011). Taken together, results from our PAD6 knockout studies suggested that PAD6 is a component of a large supramolecular complex (i.e., the lattices) and that this microtubule-based complex plays a critical role in protein synthesis and organelle positioning in the oocyte and early embryo. The precise mechanism by which PAD6 regulates cytoplasmic lattice assembly and function remains unclear. As opposed to the other PADs, PAD6 does not appear to be catalytically active (Taki et al. 2011). Therefore, it seems likely the role of PAD6 in these processes is structural. Given the strong association between PADs and histone citrullination, we have also tested whether histone citrullination may play a role in early development. Results showed that the pan-PAD inhibitor, Cl-amidine, suppresses histone H3 and H4 tail citrullination and, similar to that seen in PAD6-null oocytes, potently suppresses early cleavage divisions (Kan et al. 2012). We then investigated histone citrullination levels in PAD6- and PAD4-null oocytes and found that deletion of these two family members did not affect histone citrullination, suggesting that another PAD was likely catalyzing this activity in eggs and early embryos. We recently found that PAD1-specific inhibitors and morpholinos both suppressed histone citrullination and early-stage cleavage divisions, suggesting that PAD1 may play a critical role in early development via its role in histone citrullination (Zhang et al. 2016).

While we have shown that PAD6 is expressed human oocytes, until recently, it has remained unclear as to whether PAD6 plays an important role in fertility in women. However, Xu et al. recently found that PAD6 mutations appear to be the cause of infertility in several women who have failed to become pregnant following multiple IVF and ICSI cycles (Xu et al. 2016). These mutations occurred within a consanguineous family affected by a homozygous premature nonsense mutation and also in two females with compound heterozygous mutations. Consistent with what we had found in our PAD6 knockout mice, their study also found that PAD6 protein was absent from affected oocytes and that, following fertilization, all of the affected embryos arrested at the 2–4 cell stage due to embryonic genome activation failure (Xu et al. 2016). This finding was recently supported by another case reporting that a PAD6 mutation was the cause of infertility in a women that had unsuccessfully undergone multiple rounds of ICSI only to have the resulting embryos arrest at the two-cell stage (Maddirevula et al. 2017). Taken together, these findings suggest that PAD6 mutations may be a significant cause of infertility in women.

4.6 Conclusions

The critical role that PAD enzymes play in mammalian development, physiology, and pathology is now coming to light. Our new understanding of PAD biology has been significantly advanced in recent years, in part, through the use of genetically engineered mice. Outcomes from PAD4 knockout mice indicate that PAD4 is required for citrullination in immune cells and that this activity plays an important physiological role in innate immunity. These studies also demonstrate that dysregulated PAD4 activity can lead to various pathologies. For example, the knockout studies have supported previous clinical findings that PAD4 activity plays a critical role in chronic inflammatory diseases such as rheumatoid arthritis. Additionally, these studies also highlight a novel and important role for PAD4 in thrombosis formation via its role in NET production within the thrombus scaffold. Given the prevalence of autoimmune diseases and thrombotic disorders in society, it is understandable that there is currently considerable interest in developing PAD4-specific compounds and then testing the efficacy of these compounds in a clinical setting.

Outcomes from PAD2-overexpressing and knockout mice have also proved to be informative. Studies investigating the role of PAD2 in CNS function found that, while PAD2 activity appears to be the main driver of citrullination in this tissue in both healthy and in autoimmune encephalomyelitis, it does not appear to play a critical role in CNS development or in the development of experimental autoimmune encephalomyelitis. These findings indicate that the precise functional role of PAD2 within the CNS remains to be determined. With respect to immune cells, the observation that citrullination activity was only moderately reduced in immune cells from PAD2 knockout mice suggests that the overall contribution of PAD2 to immune cell function may be less than that of PAD4. Regarding the role of PAD2 in the skin, our finding that PAD2 overexpression in the skin promotes inflammation and tumor growth demonstrates, for the first time, that PADs can function as oncogenes. Given the fact that PAD activity is strongly associated with inflammatory events, our findings also raise the possibility that PAD-mediated inflammation may promote the growth of a range of cancers.

A surprising outcome from the totality of PAD transgenic/knockout studies is the lack of a strong grossly observable phenotype in these lines. All of the PAD2, PAD3, and PAD4 mutant strains generated to date appear normal at the whole body level and are fertile. PAD6 mutant mice are also grossly normal; however, null females are infertile, due to an arrest at the two-cell stage of development. Given that, aside from PAD6, PADs are expressed in a broad range of tissues including the CNS, muscle tissue, exocrine glands, and immune cells, one would predict that the PAD mutant mice would likely acquire developmental defects that would render the offspring nonviable. A likely explanation for this lack of effect is that one or several of the PAD isozymes compensate for the loss of another family member. In fact, several of the studies mentioned in this review noted that deletion or overexpression of one family member resulted in the up- or downregulation of another family member. Therefore, future studies aimed at deciphering how PAD isozymes compensate for each other and also how these family members potentially regulate each other's expression will be critical for understanding how PADs function at the organism level. Additionally the use of new CRISPR technologies to generate mouse lines that lack more than one PAD family member will also likely greatly contribute to our understanding of PAD physiology and pathology.

References

Akiyama, K., Sakurai, Y., Asou, H., & Senshu, T. (1999). Localization of peptidylarginine deiminase type II in a stage-specific immature oligodendrocyte from rat cerebral hemisphere. *Neuroscience Letters, 274*(1), 53–55.

Bawadekar, M., Shim, D., Johnson, C. J., Warner, T. F., Rebernick, R., Damgaard, D., Nielsen, C. H., Pruijn, G. J. M., Nett, J. E., & Shelef, M. A. (2017). Peptidylarginine deiminase 2 is required for tumor necrosis factor alpha-induced citrullination and arthritis, but not neutrophil extracellular trap formation. *Journal of Autoimmunity, 80*, 39–47.

Begovich, A. B., Carlton, V. E. H., Honigberg, L. A., Schrodi, S. J., Chokkalingam, A. P., Alexander, H. C., Ardlie, K. G., Huang, Q., Smith, A. M., Spoerke, J. M., Conn, M. T., Chang, M., Chang, S.-Y. P., Saiki, R. K., Catanese, J. J., Leong, D. U., Garcia, V. E., McAllister, L. B., Jeffery, D. A., Lee, A. T., Batliwalla, F., Remmers, E., Criswell, L. A., Seldin, M. F., Kastner, D. L., Amos, C. I., Sninsky, J. J., & Gregersen, P. K. (2004). A missense single-nucleotide polymorphism in a gene encoding a protein tyrosine phosphatase (PTPN22) is associated with rheumatoid arthritis. *The American Journal of Human Genetics, 75*(2), 330–337.

Brinkmann, V., & Zychlinsky, A. (2012). Neutrophil extracellular traps: Is immunity the second function of chromatin? *The Journal of Cell Biology, 198*(5), 773–783.

Coudane, F., Mechin, M.-C., Huchenq, A., Henry, J., Nachat, R., Ishigami, A., Adoue, V., Sebbag, M., Serre, G., & Simon, M. (2011). Deimination and expression of peptidylarginine deiminases during cutaneous wound healing in mice. *European Journal of Dermatology, 21*(3), 376–384. [cited 2017 Feb 10]. Retrieved from PMID: 21697043. http://www.ncbi.nlm.nih.gov/pubmed/21697043.

Damgaard, D., Friberg, M., Nielsen, B., Quisgaard Gaunsbaek, M., Palarasah, Y., Svane-Knudsen, V., Nielsen, C. H., Friberg, M., Gaunsbaek, M. Q., Palarasah, Y., Svane-Knudsen, V., & Nielsen, C. H. (2015). Smoking is associated with increased levels of extra-cellular peptidylarginine deiminase 2 (PAD2) in the lungs. *Clinical and Experimental Rheumatology, 33*(3), 405–408.

Darrah, E., Rosen, A., Giles, J. T., & Andrade, F. (2012). Peptidylarginine deiminase 2, 3 and 4 have distinct specificities against cellular substrates: Novel insights into autoantigen selection in rheumatoid arthritis. *Annals of the Rheumatic Diseases, 71*(1), 92–98. BMJ Publishing Group Ltd. [cited 2017 Feb 21]. PMID: 21859690. Retrieved from http://www.ncbi.nlm.nih.gov/pubmed/21859690.

Erpenbeck, L., Chowdhury, C. S., Zsengeller, Z. K., Gallant, M., Burke, S. D., Cifuni, S., Hahn, S., Wagner, D. D., & Karumanchi, S. A. (2016). PAD4 deficiency decreases inflammation and susceptibility to pregnancy loss in a mouse model. *Biology of Reproduction, 95*(6), 132. Oxford University Press. [cited 2017 Feb 27]. Retrieved from https://academic.oup.com/biolreprod/article-lookup/doi/10.1095/biolreprod.116.140293.

Esposito, G., Vitale, A. M., Leijten, F. P. J., Strik, A. M., Koonen-Reemst, A. M. C. B., Yurttas, P., Robben, T. J. A. A., Coonrod, S., & Gossen, J. A. (2007). Peptidylarginine deiminase (PAD)

6 is essential for oocyte cytoskeletal sheet formation and female fertility. *Molecular and Cellular Endocrinology, 273*(1–2), 25–31. [cited 2017 Feb 22]. PMID: 17587491. Retrieved from http://www.ncbi.nlm.nih.gov/pubmed/17587491.

Feldmann, M., & Maini, R. N. (2003). TNF defined as a therapeutic target for rheumatoid arthritis and other autoimmune diseases. *Nature Medicine, 9*(10), 1245–1250. Nature Publishing Group. [cited 2017 Mar 1]. Retrieved from http://www.nature.com/doifinder/10.1038/nm939.

Ferrari-Lacraz, S., Sebbag, M., Chicheportiche, R., Foulquier, C., Serre, G., & Dayer, J.-M. (2012). Contact with stimulated T cells up-regulates expression of peptidylarginine deiminase 2 and 4 by human monocytes. *European Cytokine Network, 23*(2), 36–44. [cited 2017 Feb 21]. PMID: 22614825. Retrieved from http://www.ncbi.nlm.nih.gov/pubmed/22614825.

Foulquier, C., Sebbag, M., Clavel, C., Chapuy-Regaud, S., Al Badine, R., Méchin, M.-C., Vincent, C., Nachat, R., Yamada, M., Takahara, H., Simon, M., Guerrin, M., & Serre, G. (2007). Peptidyl arginine deiminase type 2 (PAD-2) and PAD-4 but not PAD-1, PAD-3, and PAD-6 are expressed in rheumatoid arthritis synovium in close association with tissue inflammation. *Arthritis and Rheumatism, 56*(11), 3541–3553. Wiley Subscription Services, Inc., A Wiley Company. [cited 2017 Feb 21]. Retrieved from http://doi.wiley.com/10.1002/art.22983.

Hemmers, S., Teijaro, J. R., Arandjelovic, S., & Mowen, K. A. (2011). PAD4-mediated neutrophil extracellular trap formation is not required for immunity against influenza infection. *PLoS One, 6*(7), e22043. Coonrod SA, editor. Public Library of Science. [cited 2017 Feb 23]. Retrieved from http://dx.plos.org/10.1371/journal.pone.0022043.

Horibata, S., Coonrod, S. A., & Cherrington, B. D. (2012). Role for peptidylarginine deiminase enzymes in disease and female reproduction. *The Journal of Reproduction and Development, 58*, 274–282.

Kan, R., Yurttas, P., Kim, B., Jin, M., Wo, L., Lee, B., Gosden, R., & Coonrod, S. A. (2011). Regulation of mouse oocyte microtubule and organelle dynamics by PADI6 and the cytoplasmic lattices. *Developmental Biology, 350*(2), 311–322.

Kan, R., Jin, M., Subramanian, V., Causey, C. P., Thompson, P. R., & Coonrod, S. A. (2012). Potential role for PADI-mediated histone citrullination in preimplantation development. *BMC Developmental Biology, 12*(1), 19. BioMed Central. [cited 2017 Feb 24]. Retrieved from http://bmcdevbiol.biomedcentral.com/articles/10.1186/1471-213X-12-19.

Kaplan, M. J., & Radic, M. (2012). Neutrophil extracellular traps: Double-edged swords of innate immunity. *Journal of Immunology, 189*(6), 2689–2695. NIH Public Access. [cited 2017 Feb 23]. PMID: 22956760. Retrieved from http://www.ncbi.nlm.nih.gov/pubmed/22956760.

Khandpur, R., Carmona-Rivera, C., Vivekanandan-Giri, A., Gizinski, A., Yalavarthi, S., Knight, J. S., Friday, S., Li, S., Patel, R. M., Subramanian, V., Thompson, P., Chen, P., Fox, D. A., Pennathur, S., & Kaplan, M. J. (2013). NETs are a source of citrullinated autoantigens and stimulate inflammatory responses in rheumatoid arthritis. *Science Translational Medicine, 5*(178), 178ra40.

Kolaczkowska, E., Jenne, C. N., Surewaard, B. G. J., Thanabalasuriar, A., Lee, W.-Y., Sanz, M.-J., Mowen, K., Opdenakker, G., & Kubes, P. (2015). Molecular mechanisms of NET formation and degradation revealed by intravital imaging in the liver vasculature. *Nature Communications, 6*, 6673. Nature Publishing Group. [cited 2017 Feb 24]. Retrieved from http://www.nature.com/doifinder/10.1038/ncomms7673.

Li, P., Li, M., Lindberg, M. R., Kennett, M. J., Xiong, N., & Wang, Y. (2010). PAD4 is essential for antibacterial innate immunity mediated by neutrophil extracellular traps. *The Journal of Experimental Medicine, 207*(9), 1853–1862. The Rockefeller University Press. [cited 2017 Feb 23]. PMID: 20733033. Retrieved from http://www.ncbi.nlm.nih.gov/pubmed/20733033.

Maddirevula, S., Coskun, S., Awartani, K., Alsaif, H., Abdulwahab, F. M., & Alkuraya, F. S. (2017). The human knockout phenotype of *PADI6* is female sterility caused by cleavage failure of their fertilized eggs. *Clinical Genetics, 91*(2), 344–345. Blackwell Publishing Ltd. [cited 2017 Feb 22]. Retrieved from http://doi.wiley.com/10.1111/cge.12866.

Martinod, K., & Wagner, D. D. (2014). Thrombosis: Tangled up in NETs. *Blood, 123*(18), 2768–2776.

Martinod, K., Demers, M., Fuchs, T. A., Wong, S. L., Brill, A., Gallant, M., Hu, J., Wang, Y., & Wagner, D. D. (2013). Neutrophil histone modification by peptidylarginine deiminase 4 is critical for deep vein thrombosis in mice. *Proceedings of the National Academy of Sciences of the United States of America, 110*(21), 8674–8679. National Academy of Sciences. [cited 2017 Feb 23]. PMID: 23650392. Retrieved from http://www.ncbi.nlm.nih.gov/pubmed/23650392.

Martinod, K., Fuchs, T. A., Zitomersky, N. L., Wong, S. L., Demers, M., Gallant, M., Wang, Y., & Wagner, D. D. (2015). PAD4-deficiency does not affect bacteremia in polymicrobial sepsis and ameliorates endotoxemic shock. *Blood, 125*(12), 1948–1956.

Mastronardi, F. G., Wood, D. D., Mei, J., Raijmakers, R., Tseveleki, V., Dosch, H.-M., Probert, L., Casaccia-Bonnefil, P., & Moscarello, M. A. (2006). Increased citrullination of histone H3 in multiple sclerosis brain and animal models of demyelination: A role for tumor necrosis factor-induced peptidylarginine deiminase 4 translocation. *The Journal of Neuroscience, 26*(44), 11387–11396.

McElwee, J. L., Mohanan, S., Horibata, S., Sams, K. L., Anguish, L. J., McLean, D., Cvita, I., Wakshlag, J. J., & Coonrod, S. A. (2014). PAD2 overexpression in transgenic mice promotes spontaneous skin neoplasia. *Cancer Research, 74*(21), 6306–6317. [cited 2017 Feb 1]. PMID: 25213324. Retrieved from http://www.ncbi.nlm.nih.gov/pubmed/25213324.

Musse, A. A., Li, Z., Ackerley, C. A., Bienzle, D., Lei, H., Poma, R., Harauz, G., Moscarello, M. A., & Mastronardi, F. G. (2008). Peptidylarginine deiminase 2 (PAD2) overexpression in transgenic mice leads to myelin loss in the central nervous system. *Disease Models & Mechanisms, 1*(4–5), 229–240. Company of Biologists. [cited 2017 Feb 10]. PMID: 19093029. Retrieved from http://www.ncbi.nlm. nih.gov/pubmed/19093029.

Raijmakers, R., Vogelzangs, J., Raats, J., Panzenbeck, M., Corby, M., Jiang, H., Thibodeau, M., Haynes, N., Van Venrooij, W. J., Pruijn, G. J. M., & Werneburg, B. (2006). Experimental autoimmune encephalomyelitis induction in peptidylarginine deiminase 2 knockout mice. *Journal of Comparative Neurology, 498*(2), 217–226. Wiley Subscription Services, Inc., A Wiley Company. [cited 2017 Feb 10]. Retrieved from http://doi.wiley.com/10.1002/cne.21055.

Rohrbach, A. S., Hemmers, S., Arandjelovic, S., Corr, M., & Mowen, K. A. (2012). PAD4 is not essential for disease in the K/BxN murine autoantibody-mediated model of arthritis. *Arthritis Research & Therapy, 14*, R104.

Savchenko, A. S., Borissoff, J. I., Martinod, K., De Meyer, S. F., Gallant, M., Erpenbeck, L., Brill, A., Wang, Y., & Wagner, D. D. (2014). VWF-mediated leukocyte recruitment with chromatin decondensation by PAD4 increases myocardial ischemia/reperfusion injury in mice. *Blood, 123*(1), 141–148.

Seri, Y., Shoda, H., Suzuki, A., Matsumoto, I., Sumida, T., Fujio, K., & Yamamoto, K. (2015). Peptidylarginine deiminase type 4 deficiency reduced arthritis severity in a glucose-6-phosphate isomerase-induced arthritis model. *Scientific Reports, 5*, 13041. Nature Publishing Group. [cited 2017 Feb 24]. Retrieved from http://www.nature.com/articles/srep13041.

Shelef, M. A., Sokolove, J., Lahey, L. J., Wagner, C. A., Sackmann, E. K., Warner, T. F., Wang, Y., Beebe, D. J., Robinson, W. H., & Huttenlocher, A. (2014). Peptidylarginine deiminase 4 contributes to tumor necrosis factor α-induced inflammatory arthritis. *Arthritis & Rhematology, 66*(6), 1482–1491. [cited 2017 Feb 23]. Retrieved from http://doi.wiley.com/10.1002/art.38393.

Sokolove, J., Zhao, X., Chandra, P. E., & Robinson, W. H. (2011). Immune complexes containing citrullinated fibrinogen costimulate macrophages via Toll-like receptor 4 and Fcγ receptor. *Arthritis and Rheumatism, 63*(1), 53–62. [cited 2017 Feb 23]. PMID: 20954191. Retrieved from http://www.ncbi.nlm.nih.gov/pubmed/20954191.

Suzuki, A., Yamada, R., Chang, X., Tokuhiro, S., Sawada, T., Suzuki, M., Nagasaki, M., Nakayama-Hamada, M., Kawaida, R., Ono, M., Ohtsuki, M., Furukawa, H., Yoshino, S., Yukioka, M., Tohma, S., Matsubara, T., Wakitani, S., Teshima, R., Nishioka, Y., Sekine, A., Iida, A., Takahashi, A., Tsunoda, T., Nakamura, Y., & Yamamoto, K. (2003). Functional haplotypes of PADI4, encoding citrullinating enzyme peptidylarginine deiminase 4, are associated with rheumatoid arthritis. *Nature Genetics, 34*(4), 395–402. Nature Publishing Group. [cited 2017 Feb 23]. Retrieved from http://www.nature.com/doifinder/10.1038/ng1206.

Suzuki, A., Kochi, Y., Shoda, H., Seri, Y., Fujio, K., Sawada, T., Yamada, R., & Yamamoto, K. (2016). Decreased severity of experimental autoimmune arthritis in peptidylarginine deiminase type 4 knockout mice. *BMC Musculoskeletal Disorders, 17*(1), 205. BioMed Central. [cited 2017 Feb 24]. Retrieved from http://bmcmusculoskeletdisord.biomedcentral.com/articles/10.1186/s12891-016-1055-2.

Taki, H., Gomi, T., Knuckley, B., Thompson, P. R., Vugrek, O., Hirata, K., Miyahara, T., Shinoda, K., Hounoki, H., Sugiyama, E., Usui, I., Urakaze, M., Tobe, K., Ishimoto, T., Inoue, R., Tanaka, A., Mano, H., Ogawa, H., & Mori, H. (2011). Purification of enzymatically inactive peptidylarginine deiminase type 6 from mouse ovary that reveals hexameric structure different from other dimeric isoforms. *Advances in Bioscience and Biotechnology, 2*(4), 304–310. Scientific Research Publishing. [cited 2017 Feb 24]. Retrieved from http://www.scirp.org/journal/PaperDownload.aspx? DOI=10.4236/abb.2011.24044.

Ü Basmanav, F. B., Cau, L., Tafazzoli, A., Méchin, M.-C., Wolf, S., Romano, M. T., Valentin, F., Wiegmann, H., Huchenq, A., Kandil, R., Garcia Bartels, N., Kilic, A., George, S., Ralser, D. J., Bergner, S., Ferguson, D. J. P., Oprisoreanu, A.-M., Wehner, M., Thiele, H., Altmüller, J., Nürnberg, P., Swan, D., Houniet, D., Büchner, A., Weibel, L., Wagner, N., Grimalt, R., Bygum, A., Serre, G., Blume-Peytavi, U., Sprecher, E., Schoch, S., Oji, V., Hamm, H., Farrant, P., Simon, M., & Betz, R. C. (2016). Mutations in three genes encoding proteins involved in hair shaft formation cause uncombable hair syndrome. *American Journal of Human Genetics, 99*(6), 1292–1304.

van Beers, J. J. B. C., Zendman, A. J. W., Raijmakers, R., Stammen-Vogelzangs, J., & Pruijn, G. J. M. (2013). Peptidylarginine deiminase expression and activity in PAD2 knock-out and PAD4-low mice. *Biochimie, 95*(2), 299–308.

van Gaalen, F. A., Linn-Rasker, S. P., van Venrooij, W. J., de Jong, B. A., Breedveld, F. C., Verweij, C. L., Toes, R. E. M., & Huizinga, T. W. J. (2004). Autoantibodies to cyclic citrullinated peptides predict progression to rheumatoid arthritis in patients with undifferentiated arthritis: A prospective cohort study. *Arthritis and Rheumatism, 50*(3), 709–715. Wiley Subscription Services, Inc., A Wiley Company. [cited 2017 Mar 1]. Retrieved from http://doi.wiley.com/10.1002/art.20044.

Vossenaar, E. R., Zendman, A. J. W., van Venrooij, W. J., & Pruijn, G. J. M. (2003). PAD, a growing family of citrullinating enzymes: Genes, features and involvement in disease. *BioEssays, 25*(11), 1106–1118. Wiley Subscription Services, Inc., A Wiley Company. Retrieved from http://doi.wiley.com/10.1002/bies.10357.

Vossenaar, E. R., Radstake, T. R. D., van der Heijden, A., van Mansum, M. A. M., Dieteren, C., de Rooij, D.-J., Barrera, P., Zendman, A. J. W., & van Venrooij, W. J. (2004). Expression and activity of citrullinating peptidylarginine deiminase enzymes in monocytes and macrophages. *Annals of the Rheumatic Diseases, 63*(4), 373–381. [cited 2017 Feb 21]. PMID: 15020330. Retrieved from http://www.ncbi.nlm.nih.gov/pubmed/15020330.

Wang, Y., Li, M., Stadler, S., Correll, S., Li, P., Wang, D., Hayama, R., Leonelli, L., Han, H., Grigoryev, S. A., Allis, C. D., & Coonrod, S. A. (2009). Histone hypercitrullination mediates chromatin decondensation and neutrophil extracellular trap formation. *The Journal of Cell Biology, 184*(2), 205–213. [cited 2017 Feb 23]. PMID: 19153223. Retrieved from http://www.ncbi.nlm.nih.gov/pubmed/19153223.

Wood, D. D., Ackerley, C. A., Van Den Brand, B., Zhang, L., Raijmakers, R., Mastronardi, F. G., & Moscarello, M. A. (2008). Myelin localization of peptidylarginine deiminases 2 and 4: Comparison of PAD2 and PAD4 activities. *Laboratory Investigation, 88*, 354–364.

Wright, P. W., Bolling, L. C., Calvert, M. E., Sarmento, O. F., Berkeley, E. V., Shea, M. C., Hao, Z., Jayes, F. C., Bush, L. A., Shetty, J., Shore, A. N., Reddi, P. P., Tung, K. S., Samy, E., Allietta, M. M., Sherman, N. E., Herr, J. C., & Coonrod, S. A. (2003). ePAD, an oocyte and early embryo-abundant peptidylarginine deiminase-like protein that localizes to egg cytoplasmic sheets. *Developmental Biology, 256*(1), 73–88. [cited 2017 Feb 22]. PMID: 12654293. Retrieved from http://www.ncbi.nlm.nih.gov/pubmed/12654293.

Xu, Y., Shi, Y., Fu, J., Yu, M., Feng, R., Sang, Q., Liang, B., Chen, B., Qu, R., Li, B., Yan, Z., Mao, X., Kuang, Y., Jin, L., He, L., Sun, X., & Wang, L. (2016). Mutations in PADI6 cause female infertility characterized by early embryonic arrest. *American Journal of Human Genetics, 99*(3), 744–752. [cited 2017 Feb 22]. Retrieved from http://linkinghub.elsevier.com/retrieve/pii/S0002929716302282.

Ying, S., Dong, S., Kawada, A., Kojima, T., Phane Chavanas, S., Mé Chin, M.-C., Ronique Adoue, V., Serre, G., Simon, M., & Takahara, H. (2009). Transcriptional regulation of peptidylarginine deiminase expression in human keratinocytes. *Journal of Dermatological Science, 53*(1), 2–9.

Yurttas, P., Vitale, A. M., Fitzhenry, R. J., Cohen-Gould, L., Wu, W., Gossen, J. A., & Coonrod, S. A. (2008). Role for PADI6 and the cytoplasmic lattices in ribosomal storage in oocytes and translational control in the early mouse embryo. *Development, 135*(15), 2627–2636.

Zhang, X., Liu, X., Zhang, M., Li, T., Muth, A., Thompson, P. R., Coonrod, S. A., & Zhang, X. (2016). Peptidylarginine deiminase 1-catalyzed histone citrullination is essential for early embryo development. *Scientific Reports, 6*, 38727. Nature Publishing Group. [cited 2017 Feb 24]. PMID: 27929094. Retrieved from http://www.ncbi.nlm.nih.gov/pubmed/27929094.

Chapter 5
PAD Activation in Arthritis

Dres Damgaard and Ger J.M. Pruijn

5.1 Introduction

The findings that citrullinated proteins are targeted by the most specific immune response in rheumatoid arthritis, and that citrullination also plays an important role in neurodegenerative diseases and certain cancers, have triggered many researchers to study various aspects of this form of posttranslational modification. This chapter is focused on the conditions that are needed for peptidylarginine deiminases to become active citrullinating enzymes, in particular in relation to joint inflammation as observed in rheumatoid arthritis. After briefly discussing the available methods to study the activity of peptidylarginine deiminases and their substrate specificity, the isoforms that are most relevant for citrullination during inflammation and the factors that are mediating their activation are addressed. Citrullination is crucial for one of the processes that are tightly associated with inflammation, NETosis, which is more extensively discussed in Chap. 8. Finally, peptidylarginine deiminase activation and the resulting citrullination in the context of the inflamed joints in rheumatoid arthritis are described.

Although protein deimination (citrullination) is widespread in biological systems, the finding that a major autoimmune response in rheumatoid arthritis (RA) is directed to citrullinated antigens boosted research aimed at understanding both physiological and pathophysiological aspects of deimination. Anti-citrullinated protein antibodies (ACPA) are found in up to 75% of RA patients and in less than 5% of patients with other (autoimmune) diseases and in healthy subjects. Many

D. Damgaard
Institute for Inflammation Research, Center for Rheumatology and Spine Diseases, Copenhagen University Hospital, Rigshospitalet, Copenhagen, Denmark

G.J.M. Pruijn (✉)
Department of Biomolecular Chemistry, Radboud Institute for Molecular Life Sciences and Institute for Molecules and Materials, Radboud University, Nijmegen, The Netherlands
e-mail: G.Pruijn@ncmls.ru.nl

A.P. Nicholas et al. (eds.), *Protein Deimination in Human Health and Disease*,
DOI 10.1007/978-3-319-58244-3_5

autoimmune diseases, including RA, are characterized by chronic inflammation. Inflammation-associated events, such as chemical or enzyme-mediated modification of proteins, may play a crucial role in the formation of neoepitopes, which can be recognized by the immune system as non-self. In genetically susceptible individuals, such neoepitopes may trigger an immune response, which in time may spread also to other epitopes of the antigens involved. The enzymes involved in citrullination, peptidylarginine deiminases (PADs), were characterized in more detail; for some of them, crystal structures were determined, substrate-based activation mechanisms were proposed, and conformational changes associated with their activation have been elucidated. New methods to study PAD activity and protein citrullination have been developed, and these have been applied in a variety of studies addressing the activation of PADs and the extent of protein citrullination in various cells and tissues. More insight into the factors required for their activation and into the substrate specificity of PAD isoforms have been obtained. The finding that one of the PAD isoforms plays an essential role in NETosis, a process important for the function of neutrophils in anti-microbial responses and inflammation, emphasized the role of PADs in both health and disease. The PAD isoforms that are most relevant for citrullination in arthritis have been determined, and many proteins that are citrullinated in inflamed joints have been identified. Several PAD inhibitors have been generated, and their applicability is being explored, not only in research but also in preclinical therapeutic approaches.

So far, five PAD isotypes have been identified in mammals, (PADs 1–4 and 6). These isozymes are highly conserved, with 50–55% overall sequence homology and close to 70% identity within the catalytic domain. PAD1 is mainly expressed in epidermis and hair follicles. PAD2 is the most ubiquitously expressed PAD enzyme and is present in a.o. macrophages and astrocytes. PAD3 is primarily expressed in the upper layers of epidermis and in hair follicles, and PAD4 is mainly expressed in white blood cells (granulocytes and monocytes). PAD6 is expressed mainly in the ovary, testis and peripheral blood leukocytes. This chapter is focused on the activation of PADs in arthritis with a focus on PAD2 and PAD4. Methods available for studying PAD activity are summarized. The evidence for the involvement of PAD isoforms in arthritis is described, as well as the factors that are important for the enzymatic activity of PADs. Finally, the importance of PAD activation for the inflammation in RA is discussed.

5.2 Methods to Determine PAD Activity

The ability to identify and quantify citrullinated proteins is the key to understanding the involvement of PADs and the role of this type of posttranslational modification in physiological and pathophysiological processes. The specific detection of peptidylcitrulline in complex biological samples is challenging due to the small chemical difference between citrulline and arginine. Moreover, many methods used to detect

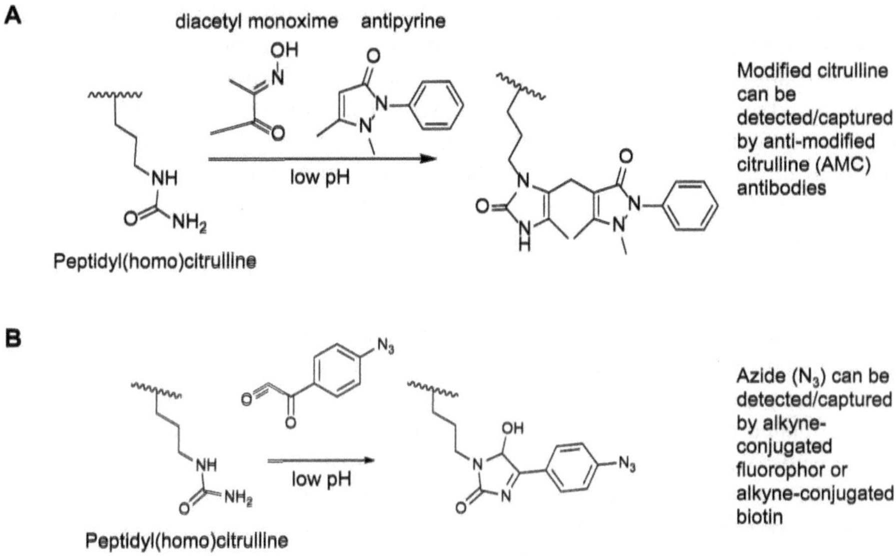

Fig. 5.1 Chemical tools for peptidylcitrulline detection. (**a**) At low pH the ureido group of (homo)citrulline reacts with diacetyl monoxime and antipyrine, resulting in modified citrulline. Anti-modified citrulline antibodies can subsequently be used for visualization or isolation of the citrullinated proteins. (**b**) At low pH (homo)citrulline is also reactive with glyoxal, e.g., in the form of azido-phenylglyoxal, which facilitates the detection of citrullinated proteins by subsequent reactions with alkyne-containing compounds. Fluorophores or biotin may also be conjugated to phenylglyoxal prior to the reaction with peptidyl(homo)citrulline

peptidylcitrulline do not discriminate between citrullinated and homocitrullinated proteins. Homocitrulline, which differs from citrulline by the presence of one additional carbon in the side chain, can occur in proteins as a result of the carbamylation of lysine residues, a process that can spontaneously occur in the presence of (iso) cyanate. The most widely applied method for peptidylcitrulline detection is the so-called anti-modified citrulline (AMC) approach, originally developed by Dr. Tatsuo Senshu in Japan (Senshu et al. 1992). In this method, the proteins are incubated with compounds that are specifically reactive with the ureido side chain, resulting in an adduct that can be detected with AMC antibodies (Fig. 5.1a). Many results obtained with this elegant approach need to be interpreted with some caution, because this method will detect carbamylated proteins as efficiently as citrullinated proteins. There is increasing evidence that carbamylation occurs under (patho)physiological conditions as well (Hensen and Pruijn 2014).

Unfortunately, the availability of anti-modified citrulline antibodies from immunized rabbits became problematic due to several failed attempts to elicit new anti-modified citrulline antibodies. The lack of these antibodies could be solved by using antibodies that target the citrullinated isoforms of specific proteins, e.g., anti-citrullinated fibrinogen and anti-citrullinated chemokine antibodies. For studies that are aimed at the global detection of citrullinated or carbamylated proteins, such antibodies are not suitable for obvious reasons, but recently monoclonal anti-modified citrulline antibodies have been generated, which can be applied in the procedure originally developed by Dr. Senshu. An alternative antibody that is widely used in citrullination studies is the monoclonal antibody F95, which was raised against a deca-citrullinated peptide linked to the activated carrier protein keyhole limpet hemocyanin (Nicholas and Whitaker 2002). Although this antibody appears to be reactive with a wide spectrum of citrullinated molecules, the results of a number of studies suggest that F95 displays a restricted recognition pattern (De et al. 2005; Makrygiannakis et al. 2012).

The antibody-based assay for PAD activity (ABAP) is based on the detection of citrullinated peptides by an antibody that reacts in a citrulline-dependent manner with these peptides. Arginine-containing peptides, e.g., a peptide corresponding to a citrullinated filaggrin epitope recognized by RA autoantibodies, are immobilized in the wells of a 96-well microtiter plate and incubated with PAD containing samples. The conversion of peptidylarginine to peptidylcitrulline can subsequently be detected with an antibody specifically reactive with the citrullinated peptide, followed by an ELISA-like staining procedure (Zendman et al. 2007).

An alternative method for the detection of (homo)citrulline containing proteins is based on the chemical reaction of the ureido group with phenylglyoxal under highly acidic conditions (Tutturen et al. 2010). Under these conditions, phenylglyoxal specifically reacts with (homo)citrulline (Fig. 5.1b), and fluorophore-conjugated or click-chemistry-based variants of phenylglyoxal have been used to detect (homo)citrullinated proteins in biological samples (Hensen et al. 2015; Bicker et al. 2012). Methods to detect citrullinated/carbamylated proteins in complex samples based on this compound, however, appeared to be hampered by relatively high background reactivities (Hensen and Pruijn 2014; Hensen et al. 2015).

Finally, the analysis of proteins by mass spectrometry provides very attractive possibilities to identify and characterize citrullinated proteins. For more information on the applicability of these methods, we refer the reader to recent review articles (Hensen and Pruijn 2014; Slade et al. 2014). Compared to other techniques, one of the main advantages of mass spectrometry is that it discriminates between citrullination (mass increase 0.98 Da) and carbamylation (mass increase 43.02 Da). In addition, the modification site can be mapped using tandem mass spectrometry. Collision-induced dissociation during tandem mass spectrometry has been shown to lead to the conversion of peptidylcitrulline to peptidyl-ornithine concomitant with the loss of 43.02 Da (Wang et al. 2016; Hao et al. 2009) which thus can be considered a citrulline signature. However, it should be noted that the loss of the carbamyl group during this procedure is also expected for homocitrulline. We have used mass spectrometric approaches to get more information on the substrate specificity of PAD enzymes and to characterize citrullinated proteins in synovial fluid samples of RA patients, which will be addressed below.

5.3 PAD Substrate Specificity

Like many other enzyme-catalyzed posttranslational modifications, citrullination is restricted to specific substrate arginines, and the amino acid context plays an important role in the selection of citrullination sites. Several studies have addressed the mechanism of action and the substrate requirements for PAD enzymes (Kearney et al. 2005; Knuckley et al. 2010; Dreyton et al. 2014), in particular for PAD2 and PAD4. The conversion of free L-arginine appeared to be nearly undetectable, but the results of experiments with benzoylated arginines indicated that an N-terminal amide is critical and sufficient for recognition (Knuckley et al. 2010; Dreyton et al. 2014). Kinetic and mechanistic studies of PAD2 suggested that, unlike the other members of the PAD family, which were proposed to use a reverse-protonation mechanism, PAD2 uses a substrate-assisted mechanism of catalysis (Dreyton et al. 2014).

By global analyses of citrullinated proteins and by investigating the efficiency of citrullination of synthetic peptides, the influence of neighboring amino acids was assessed. Although both favored and disfavored amino acids at positions flanking the deiminated arginine have been identified, the results demonstrated that it is difficult to derive a consensus sequence for citrullination sites (Stensland et al. 2009; Assohou-Luty et al. 2014). Amino acid residues most commonly found in citrullination sites for both isotypes are glycine at +1 and tyrosine at +3 relative to the target arginine. Nevertheless, based upon the results obtained with synthetic peptide substrates and mixtures of cellular proteins, a consensus sequence comprising four amino acids flanking the human PAD4 citrullination site, two on each side, was proposed: (M/K)-(D/S)-R-(G/D)-(H/W). Neighboring lysines negatively influenced the deimination by PAD4. In addition, it was demonstrated that the human PAD4 displays more pronounced substrate specificity than the human PAD2 (Assohou-Luty et al. 2014), which is confirmed by studies looking at PAD2 versus PAD4 deimination of specific proteins, such as fibrinogen (Nakayama-Hamada et al. 2005; van Beers et al. 2010).

5.4 PAD Isoforms Involved in Arthritis

5.4.1 PAD2 and PAD4 Expression in Hematopoietic Cells

PAD2 and PAD4 are encoded in various cells of hematopoietic origin (Vossenaar et al. 2003) and are the main PAD isoforms that have been demonstrated in the synovium of RA patients (Foulquier et al. 2007). Both isoforms are expressed by peripheral blood mononuclear cells (PBMCs), granulocytes, fibroblast-like cells, and osteoclasts. There are some discrepancies between the data obtained by mRNA analyses and protein analyses, which suggests that the expression of PADs, at least in part, may be regulated at the translational level. PADs have been reported to be active in the nucleus, the cytoplasm and the extracellular space, although there are differences between the various PAD isoforms. It was initially thought that PAD4 was solely the nuclear isoform (Nakashima et al. 2002), whereas PAD2 was cytoplasmic. Indeed, nuclear localization of PAD4 is essential for the formation of neutrophil extracellular traps (NETs; see below) (Lewis et al. 2015). However, PAD4 is also localized within cytoplasmic granules (Asaga et al. 2001), and PAD2 was found within the nucleus of macrophages (Mohanan et al. 2013) and regulates genes within mammary gland epithelial cells through nuclear functioning (Cherrington et al. 2012). Thus, it is not completely clear yet which isoform is responsible for citrullination of proteins with a particular intracellular localization; the same applies for extracellularly located citrullinated proteins (van Beers et al. 2013; Wang et al. 2016), as both PAD2 and PAD4 are found extracellularly in a synovial fluid (SF) (Kinloch et al. 2008).

PAD3 expression has been demonstrated in neutrophils in a single study (Darrah et al. 2012), and PAD3 mRNA was recently demonstrated in synovial tissue obtained from RA patients (Olivares-Martinez et al. 2016). It is unclear if PAD3 is indeed expressed as protein in the synovium and thus contributes to the production of citrullinated autoantigens in the joints.

PAD2 is expressed in mature macrophages after differentiation from monocytes (Vossenaar et al. 2004a); although using a different protocol, PAD2 has also been demonstrated in monocytes (Foulquier et al. 2007). Also the observations made for PAD4 expression differed between these two studies. The conflicting results on PAD4 and PAD2 expression at either the mRNA or protein level in monocytes and macrophages indicate that more work has to be done to clarify this situation.

Notably, in RA synovium sections, macrophage-like cells account for a substantial part of the synovial tissue (Salisbury et al. 1987) making them a considerable source for PAD2 and possibly also PAD4. PAD2 and PAD4 are also expressed by fibroblast-like cells, which are also a dominant cell type in the inflamed synovial membrane (Chang et al. 2005; Badillo-Soto et al. 2016). With respect to granulocytes, both PAD2 and PAD4 are expressed by neutrophils (Darrah et al. 2012; Spengler et al. 2015), which are the most abundant immune cell in SF aspirated from the joint of RA patients (Bjelle et al. 1982). Neutrophils appear to be the major source of PAD4 in the inflamed joints but also importantly contribute to PAD2 levels. PAD4 is also expressed in eosinophils (Asaga et al. 2001), and PAD2 expression has been demonstrated in mast cells (Arandjelovic et al. 2012). Osteoclasts are another cell type of hematopoietic origin expressing PADs (Harre et al. 2012). PAD2 and PAD4 are expressed in osteoclast precursor cells, but PAD4 expression decreases with further differentiation into osteoclasts, while PAD2 expression increases during this process (Harre et al. 2012). Citrullinated proteins, such as vimentin, are targeted on the surface of osteoclasts by ACPAs, which are frequently found in the sera of RA patients, and this interaction induces osteoclastogenesis. This process has recently been shown to be IL-8 dependent, and PAD4 was also found to be expressed in mature osteoclasts in this latter study and thus may indeed contribute this process (Krishnamurthy et al. 2016). For a more detailed description of deamination and bone loss, we refer the reader to Chap. 6 of this volume.

5.4.2 PAD2 and PAD4 Expression in Joints

In accordance with the expression of PAD2 and PAD4 in various inflammatory cells, both isoforms have been demonstrated in the synovium of RA patients (Nakayama-Hamada et al. 2005; Foulquier et al. 2007; Kinloch et al. 2008). The presence of PAD2 and PAD4 in synovial tissue and SF is not exclusive for RA, as they are also detected in other inflammatory (joint) diseases, such as ankylosing spondylitis and psoriatic arthritis (Foulquier et al. 2007; Kinloch et al. 2008). Neither is the presence of citrullinated proteins, which are found in other forms of synovitis and also other inflammatory diseases, as exemplified by patients with inflammatory bowel disease and polymyositis (Vossenaar et al. 2004b; Raijmakers et al. 2012; Chapuy-Regaud et al. 2005; Kinloch et al. 2008; Makrygiannakis et al. 2006). Both PAD2 and PAD4 have been demonstrated in the synovium of osteoarthritis (OA) patients but at lower levels compared to RA patients (Damgaard et al. 2016b; Kinloch et al. 2008; Foulquier et al. 2007; Chang et al. 2013). Accordingly, PAD activity assays reveal significantly higher catalytic activity in SF from RA patients than from OA patients (Spengler et al. 2015; Damgaard et al. 2016b), and citrullinated proteins are almost absent in most OA samples (Chang et al. 2005; Kinloch et al. 2008). A recent study identified 200 citrullination sites, by mass spectrometry, in SF obtained from ACPA-positive RA patients and only four sites in SF from OA patients (Wang et al. 2016). These findings probably reflect the high degree of inflammatory cell infiltration into RA joints compared to a generally non-inflammatory process in OA, in which is substantiated by the 50-fold higher synovial fluid cell count in RA (Damgaard et al. 2016b). The various inflammatory stimuli in RA are then likely to promote citrullination (Blachere et al. 2015; Arandjelovic et al. 2012). It is still unknown if one PAD isoform is predominantly responsible for citrullination of the proteins that are found in the inflamed joints of RA patients (van Beers et al. 2013; Wang et al. 2016). Certain variations in the PADI4 gene have been identified as a risk factor for the development of RA (Suzuki et al. 2003; Plenge et al. 2005). In European populations, most studies do not support this association (Barton et al. 2004; Martinez et al. 2005; Burr et al. 2010) while a few studies did observe such an association (Okada et al. 2014; Hoppe et al. 2006). A single study reported PADI2 genetic variations to be associated with RA (Chang et al. 2013). No functional studies on the reported gene variations have identified a mechanism for these increased risk ratios (Cantaert et al. 2005). The above-listed studies support the notion that PAD2, PAD4, and citrullinated proteins are not specifically associated with RA but instead with inflammation in general, although the intensity of citrullination may be much higher in RA than in other inflammatory conditions.

5.5 PAD Activation

Members of the PAD family require the presence of Ca^{2+} and reducing conditions for their catalytic activity; these requirements were first demonstrated in vitro using rabbit or murine PADs with various concentrations of $CaCl_2$ and the reducing agent dithiothreitol (DTT) (Takahara et al. 1986; Terakawa et al. 1991). PAD4, and likely also other PAD isoforms, is more active as a dimer, although it is capable of citrullinating target substrates as a monomer (Liu et al. 2011). The normal intracellular Ca^{2+} concentration appears insufficient for PAD activity (Fig. 5.2), and this is also the case for the levels that can be achieved during apoptosis (Davies and Hallett 1998; Neeli et al. 2008). In contrast, the extracellular levels of Ca^{2+} are sufficient, but paradoxically in this environment, the levels of reducing agents, such as reduced glutathione (GSH), are orders of magnitude lower (Griffith 1999; Jones 2002) than what is required for PAD activity (Damgaard et al. 2016a). Thus, citrullination seems to require a combination of intracellular (high levels of reducing capacity) and extracellular (high levels of Ca^{2+}) conditions (Fig. 5.2). This raises the question of how PADs are activated in inflamed tissues and if unidentified activating factors inside cells or in complex body fluids may promote citrullination. Possibilities might be polyamines (Brooks 2013) and activating autoantibodies (Darrah et al. 2013).

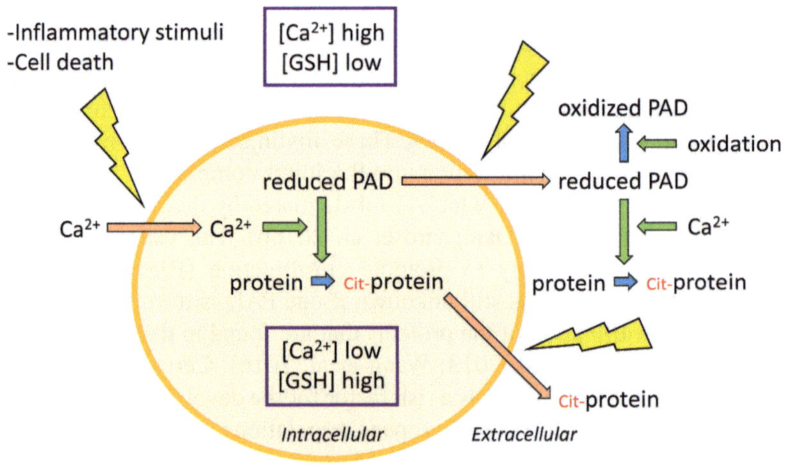

Fig. 5.2 Regulation of PAD activity in the intra- and extracellular space. Extracellular PADs are likely to exist in an inactivated form due to the oxidative extracellular environment; extracellular Ca^{2+} levels are sufficient for activity. Intracellular PADs exist in a reduced state but are inactive due to the low intracellular Ca^{2+}-levels. Upon activation of cells, e.g., by inflammatory stimuli or uncontrolled cell death, Ca^{2+}-influx can activate intracellular PADs followed by citrullination of intracellular proteins. Depending on the state of the stimulated/dying cells these citrullinated proteins and PADs can be released to the extracellular space. The high levels of extracellular Ca^{2+} concentration may lead to further activation of PADs, followed by citrullination of extracellular proteins. Extracellular PADs will in turn be inactivated by the extracellular oxidative environment

5.5.1 Calcium

Citrullination can be induced by treating cells (neutrophils, monocytes, or macrophages) with calcium ionophores such as ionomycin (Nakashima et al. 2002; Vossenaar et al. 2004a). Ca^{2+} is an essential cofactor of PADs, whereas other physiologically relevant divalent cations fail to induce catalytic activity in vitro and can inhibit the activation by Ca^{2+} (Kearney et al. 2005). Therefore, zinc has been proposed as a physiological regulator of PAD activation (Stenberg and Roth 2015). Structural analysis of the PAD4 enzyme revealed five calcium-binding sites which are fairly conserved among all PADs except PAD6 (Arita et al. 2004). Mutation of the functional residues at the catalytic site has been shown to abolish enzymatic activity. One of these is Cys645 (Cys647 in PAD2), which contains an essential thiol group (Knuckley et al. 2007). Binding of Ca^{2+} leads to a conformational change, which positions this thiol into the active site (Arita et al. 2004). A more recent study demonstrated how a "Ca^{2+} switch", initiated by the binding of Ca^{2+} in three of the five positions, is responsible for the proper positioning of the active site cysteine (Slade et al. 2015). This switch was found to occur at Ca^{2+} concentrations around 0.25 mM. Depending on the method of detection, half-maximal PAD activities have been reported at Ca^{2+} concentrations ranging from 40 μM (Zendman et al. 2007) to as high as 3.3 mM (Darrah et al. 2013). Most studies find the Ca^{2+} concentration required for half-maximal activity to be around 0.2–0.5 mM and demonstrate full enzymatic activity at Ca^{2+} concentrations above 2–2.5 mM—comparable to what is found in synovial fluid (Kearney et al. 2005; Damgaard et al. 2014). Inflammatory stimuli, but not apoptosis, have been shown to induce citrullination of histones and fibrinogen (Blachere et al. 2015; Neeli et al. 2008). ATP is another inflammatory signal that has been shown to induce PAD activation and robust protein citrullination (Arandjelovic et al. 2012), and this process required extracellular Ca^{2+}.

5.5.2 Reducing Conditions

Most studies of protein citrullination and PAD activity are based on in vitro experiments using the reducing agent DTT in the reaction buffer. DTT is a non-physiological synthetic molecule which provides the required reducing conditions for PAD activity in vitro (Terakawa et al. 1991). It is apparent that reduction of the active site thiol precedes the attack on the guanidinium carbon of arginine. In a recent study we showed that reduced glutathione (GSH), the most abundant intracellular small molecule thiol, is capable of reducing PADs to promote their catalytic activity (Damgaard et al. 2016a) at concentrations comparable to those found within cells (Dixon et al. 2008). The extracellular GSH levels are, on the other hand, two to three orders of magnitude lower than the requirements for PAD activity (Griffith 1999). In the same study, we showed that PADs in synovial fluids are catalytically inactive, unless a reducing agent is added, and that PAD is active when 10 mM exogenous GSH is used. Interestingly, PAD released from phorbol 12-myristate 13-acetate (PMA)-stimulated leukocytes was capable of citrullinating fibrinogen without applying a reducing agent. In agreement with the findings in SFs, PAD released to leukocyte culture supernatants required a reducing agent to restore catalytic activity. The evidence for GSH being important in reducing PADs in vivo came from the observation that the glutathione reductase inhibitor 2-acetylamino-3-[4-(2-acetylamino-2-carboxyethylsulfanylthiocarbonylamino)phenylthiocarbamoylsulfanyl]propionic acid (2-AAPA), which decreases intracellular free GSH levels (Seefeldt et al. 2009), abolished the activity of the released PAD enzymes from leukocytes. Further studies will be necessary to fully dissect the role of GSH in PAD activation and to confirm its relevance. Other reducing agents such as thioredoxin may also be able to reduce and activate PADs (Yoshida et al. 1999). The balance between reducing conditions and reactive oxygen species seems important for activation/inactivation of PADs as also purposed by Dreyton and colleagues in their report on the PAD2 mechanism of action (Dreyton et al. 2014).

5.6 PAD and NETosis

During inflammation, e.g., caused by infections with bacteria, fungi, or viruses, neutrophils initiate a form of programmed cell death termed NETosis. This process results in the release of decondensed chromatin from the cells, termed neutrophil extracellular traps (NETs). Other inflammatory cells have been demonstrated to produce extracellular traps upon activation by pro-inflammatory stimuli. Most interestingly, PAD4 activity has been found to be required for NETosis, and citrullinated histones are incorporated into NETs (Muller and Radic 2015). It is likely that activated PAD4 is also released from the neutrophils during NETosis, and this may lead to the citrullination of additional proteins in the extracellular space. In addition to the association with citrullinated histones, extracellular traps are decorated with anti-microbial compounds originating from neutrophil granules (Parker and Winterbourn 2012). Thus, extracellular traps represent macromolecular assemblies, which contain citrullinated proteins, and therefore it is tempting to speculate that they play a role in the initiation of the citrulline-specific immune response in RA or in the progression of this response. Several lines of evidence support this relationship. Prominent NET autoreactivity has been detected in the sera from patients with RA, systemic lupus erythematosus (SLE), and Felty's syndrome (FS). Autoantibodies to citrullinated histones appeared to be produced in the majority of FS patients and in a subset of RA and SLE patients (Dwivedi et al. 2012; Dwivedi and Radic 2014; Muller and Radic 2015). Additional evidence for a functional relationship between NETosis and autoimmunity came from experiments showing that circulating neutrophils from RA or SLE patients and neutrophils in SF from RA patients display enhanced NETosis compared to neutrophils from healthy controls and from OA patients, respectively (Garcia-Romo et al. 2011; Khandpur et al. 2013). Moreover,

RA sera and immunoglobulin fractions from RA patients, with high levels of ACPA and/or rheumatoid factor, significantly enhance NETosis (Khandpur et al. 2013). Finally, SLE sera were reported to have a reduced capacity to degrade NETs (Hakkim et al. 2010). Taken together, these data indicate that autoimmunity, at least in RA and SLE, is associated with changes in NET formation and/or degradation, but the involvement of NET-associated citrullinated and carbamylated proteins is yet unknown. For a more detailed description of the relationship between protein deamination and NETosis, we refer the reader to Chap. 8 of this volume.

5.7 PAD Activation and Citrullination in Relation to Inflammation in RA

The levels of synovial citrullinated proteins and PAD enzymes appeared to correlate with the degree of joint inflammation (Foulquier et al. 2007; Makrygiannakis et al. 2012). In immunohistochemistry of synovial tissue sections, intracellular citrullinated proteins co-localized with PAD2 in 59% of RA samples versus 17% in controls, i.e., patients with spondyloarthropathy, OA, juvenile chronic arthritis, and villonodular synovitis (De et al. 2005). A later study detected intracellular citrullinated proteins in 85.7% of RA tissue samples and the complete absence in healthy synovium (Makrygiannakis et al. 2012). The same percentage was positive for extracellular citrullinated proteins, which were only found in one out of eight healthy donors. All RA samples stained positive for PAD2 and PAD4, which was also the case for a vast majority of the samples from healthy controls, although their levels were significantly lower (Makrygiannakis et al. 2012). It is not surprising that PAD2 and PAD4 are also expressed in tissues from healthy individuals due to the presence of various cells of hematopoietic origin, but PAD activation appears to be specific for inflammatory conditions. Protein citrullination, as well as PAD2 expression, correlated with cell infiltration and vascularity in synovial tissue (Makrygiannakis et al. 2012), and local administration of glucocorticoids decreased cell infiltration, levels of citrullinated proteins, and PAD4 levels. A high degree of inflammation and inflammatory cell infiltration are likely to promote protein citrullination, because high levels of PAD2 and PAD4 and the activation and release of these enzymes due to cell death/activation are associated with inflammation. Recently, we found significantly higher extracellular PAD2 levels in SF from ACPA-positive than from ACPA-negative RA patients (Damgaard et al. 2016b). In agreement with other studies showing a link between citrullination and inflammation, PAD2 in SF correlated with the cell count, as well as with IL-6, IL-8, and IL-10 levels in SF, and with circulating levels of CRP and ACPA. Notably, the patients' SF PAD2 levels also correlated with the disease activity, as assessed by DAS28. The cell counts were generally higher in ACPA-positive than in ACPA-negative RA patients, although the differences did not reach statistical significance (Damgaard et al. 2016b). High levels of citrullinated proteins promote persistent inflammation in patients with ACPA-positive RA (van Venrooij and Pruijn 2008) carrying HLA-DRB1 alleles containing the shared epitope (Hill et al. 2003; Scally et al. 2013). Thus, citrullinated fibrin induces a higher inflammatory response in cultured RA synovial fibroblasts than non-citrullinated fibrin, as reflected by elevated IL-6 and IL-8 production (Sanchez-Pernaute et al. 2013).

Immune complexes containing citrullinated fibrinogen stimulate macrophages to produce more TNFα than immune complexes containing native fibrinogen (Sokolove et al. 2011). Thus, ACPAs and citrullinated proteins in immune complexes may account for the enhanced stimulation of macrophages observed among ACPA-positive patients. Strong evidence for the pathogenic role of citrullinated proteins in ACPA-positive RA came from the finding that around 30% of DR4-transgenic mice develop arthritis following immunization with citrullinated fibrinogen but not after immunization with non-citrullinated fibrinogen (Hill et al. 2008). Another genetic risk factor for development of RA is the C1858T polymorphism in the gene encoding protein tyrosine phosphatase, non-receptor type 22 (PTPN22), which results in an amino acid change from arginine to tryptophan at position 620 (Begovich et al. 2004). Conflicting data exist regarding an effect of this polymorphism on the activation of lymphocytes, which may contribute to its impact on the development of RA, but another important aspect is that the major allelic variant R620 of PTPN22 physically interacts with PAD4 and inhibits citrullination (and thereby NET formation), whereas the W620 variant of PTPN22 fails to do so (Chang et al. 2015). Citrullination of E2F-1 assists its association with chromatin, specifically with genes encoding cytokine production in granulocytes, which is mediated by the enhanced binding citrullinated E2F-1 to bromodomain-containing protein 4 (BRD4) (Ghari et al. 2016). Accordingly, the combined inhibition of PAD4 and BRD4 disrupts the chromatin-bound complex and suppresses cytokine gene expression. Disrupted chromatin association and suppression of cytokine gene expression was also observed after inhibition of PAD in dendritic cells (Jang et al. 2015).

In addition to the effects on cytokines at the transcriptional level, citrullination may also affect their function by direct modification of these signaling molecules. A number of studies have shown that cytokines and chemokines are likely to alter functionality upon citrullination, as, for example, CXCL8 and TNFα (Proost et al. 2008; Moelants et al. 2013). Citrullination of the chemokine CXCL5 appeared to be significantly higher in RA sera and SF than in normal sera and in SF from patients with other rheumatic diseases (Yoshida et al. 2014). Moreover, citrullinated CXCL5 induced more severe inflammation and recruited more monocytes than its non-citrullinated counterpart in a mouse model of inflammatory arthritis. On the other hand, citrullinated TNFα was shown less potent to stimulate cultured human fibroblasts to produce chemokines (Moelants et al. 2013). Taken together, these results indicate that citrullination, in addition to antigen production, may affect the disease process in RA at several other levels, varying from the induction of NETosis to the expression and functional activity of molecules modulating inflammation (Fig. 5.3). Each of these phenomena may be targeted therapeutically. One of the approaches, the use of PAD inhibitors, will be addressed in Chapter 25 of this volume.

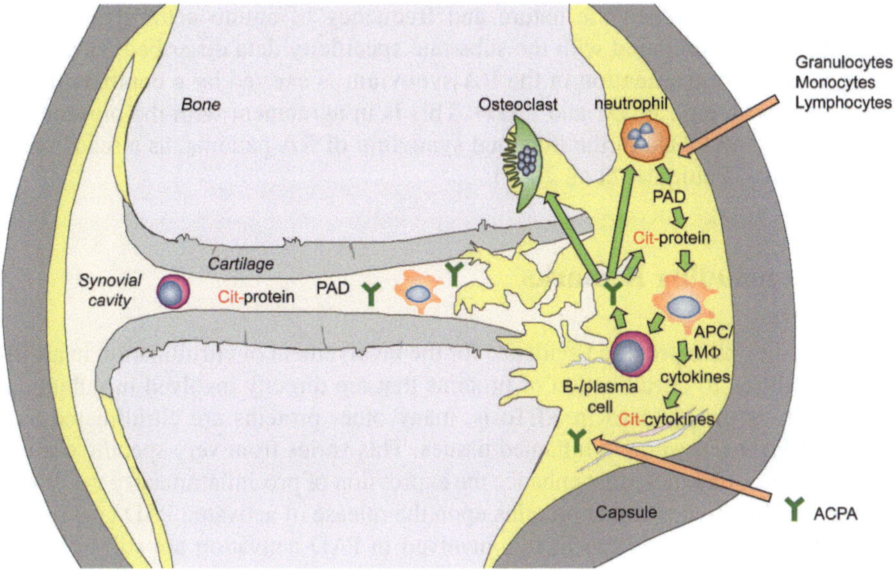

Fig. 5.3 PAD activation and protein citrullination in the inflamed joint. Upon infiltration of inflammatory cells, including neutrophils, in the synovial membrane, NETosis can lead to the release of PAD (and intracellular citrullinated proteins) in the extracellular space, where the relatively high Ca^{2+} concentrations may lead to further PAD activation and citrullination of extracellular proteins. Antigen-presenting cells (APC), e.g., macrophages (MΦ), process, and present citrullinated epitopes to the T cells (not shown), which subsequently will activate B cells, resulting in their differentiation into plasma cells and ACPA production. ACPA may also originate from activated B cells in other tissues. ACPA may form immune complexes with the citrullinated proteins, which can enhance the inflammatory response via various mechanisms, can enhance NETosis, and can activate osteoclasts (not discussed in this chapter). Immune cells and proteins can also migrate to the synovial fluid in the synovial cavity

5.8 Citrullinated Proteins in Synovial Fluid

In view of the putative pathophysiologic role of citrullinated protein-specific immune complexes in RA, it is important to obtain a comprehensive view of the citrullinated proteins present in the inflamed joints of patients with RA. A systematic analysis of citrullinated proteins present in the synovial fluid of RA patients by mass spectrometry led to the identification of 53 polypeptides containing one or more citrulline residues, which comprises 28% of all proteins identified (van Beers et al. 2013). The only partially overlapping data obtained with material from different patients suggests that this is only a minority of the citrullinated proteins occurring in the inflamed joints of RA patients, implying that many proteins will be citrullinated in these tissues. This was recently confirmed by the results of independent studies (Tutturen et al. 2014; Wang et al. 2016). Among the proteins that were found to be citrullinated were fibrinogen, vimentin, fibronectin, and histones, proteins that were reported to be modified in inflamed tissues in previous studies. When the nature and frequency of amino acids flanking the citrulline were compared with the substrate specificity data described above, this suggested that citrullination in the RA synovium is exerted by a combination of PAD enzymes, e.g., PAD2 and PAD4. This is in agreement with the presence of both PAD2 and PAD4 in the inflamed synovium of RA patients, as already mentioned above (Foulquier et al. 2007).

5.9 Concluding Remarks

Many studies have provided evidence for the involvement of citrullination in arthritis. In addition to citrullination of proteins that are directly involved in inflammation, such as the histones in NETosis, many other proteins are citrullinated as a result of PAD activation in inflamed tissues. This varies from very specific citrullination events, like those that enhance the expression of pro-inflammatory cytokines, to the citrullination of many proteins upon the release of activated PAD enzymes in the extracellular space. The factors involved in PAD activation are still not completely understood. The requirement for calcium has been well established, but it is not completely clear yet how intracellular calcium levels can be changed (locally) to levels that are needed for PAD activation, e.g., within nucleus. Up to now, the requirement for reducing conditions has received much less attention, and this is particularly relevant for extracellular PAD activity. Glutathione released from dying cells, possibly concomitant with PADs, may at least temporarily provide a (local) reducing environment allowing PAD to citrullinate extracellular proteins. More research will be required to fully understand how the necessary conditions for PAD activity are generated, both in arthritic tissues and in other cells/tissues. The specific immune response to citrullinated proteins in RA, in combination with the elevated levels of citrullination and the citrullination-dependent immune complexes formed, has opened several avenues for therapeutic strategies aimed at interference with protein citrullination and PAD activation. New methods to determine PAD activity and more insight in the substrate specificity of the most relevant PAD isoforms will be helpful to assess whether such strategies will be utilizable.

Acknowledgments The authors would like to thank Claus H. Nielsen (Copenhagen University Hospital, Rigshospitalet, Copenhagen, Denmark) for the critical reading of and helpful comments on this chapter.

References

Arandjelovic, S., McKenney, K. R., Leming, S. S., & Mowen, K. A. (2012). ATP induces protein arginine deiminase 2-dependent citrullination in mast cells through the P2X7 purinergic receptor. *Journal of Immunology, 189*(8), 4112–4122.

Arita, K., Hashimoto, H., Shimizu, T., Nakashima, K., Yamada, M., & Sato, M. (2004). Structural basis for Ca(2+)-induced activation of human PAD4. *Nature Structural & Molecular Biology, 11*(8), 777–783.

Asaga, H., Nakashima, K., Senshu, T., Ishigami, A., & Yamada, M. (2001). Immunocytochemical localization of peptidylarginine deiminase in human eosinophils and neutrophils. *Journal of Leukocyte Biology, 70*(1), 46–51.

Assohou-Luty, C., Raijmakers, R., Benckhuijsen, W. E., Stammen-Vogelzangs, J., de, R. A., van Veelen, P. A., Franken, K. L., Drijfhout, J. W., & Pruijn, G. J. (2014). The human peptidylarginine deiminases type 2 and type 4 have distinct substrate specificities. *Biochimica et Biophysica Acta, 1844*(4), 829–836.

Badillo-Soto, M. A., Rodríguez-Rodríguez, M., Pérez-Pérez, M. E., Daza-Benitez, L., Bollain-y-Goytia, J. J., Carrillo-Jiménez, M. A., Avalos-Díaz, E., & Herrera-Esparza, R. (2016). Potential protein targets of the peptidylarginine deiminase 2 and peptidylarginine deiminase 4 enzymes in rheumatoid synovial tissue and its possible meaning. *European Journal of Rheumatology, 3*(2), 44–49. doi:10.5152/eurjrheum.2015.0055.

Barton, A., Bowes, J., Eyre, S., Spreckley, K., Hinks, A., John, S., & Worthington, J. (2004). A functional haplotype of the PADI4 gene associated with rheumatoid arthritis in a Japanese population is not associated in a United Kingdom population. *Arthritis and Rheumatism, 50*(4), 1117–1121.

Begovich, A. B., Carlton, V. E., Honigberg, L. A., Schrodi, S. J., Chokkalingam, A. P., Alexander, H. C., Ardlie, K. G., Huang, Q., Smith, A. M., Spoerke, J. M., Conn, M. T., Chang, M., Chang, S. Y., Saiki, R. K., Catanese, J. J., Leong, D. U., Garcia, V. E., McAllister, L. B., Jeffery, D. A., Lee, A. T., Batliwalla, F., Remmers, E., Criswell, L. A., Seldin, M. F., Kastner, D. L., Amos, C. I., Sninsky, J. J., & Gregersen, P. K. (2004). A missense single-nucleotide polymorphism in a gene encoding a protein tyrosine phosphatase (PTPN22) is associated with rheumatoid arthritis. *American Journal of Human Genetics, 75*(2), 330–337.

Bicker, K. L., Subramanian, V., Chumanevich, A. A., Hofseth, L. J., & Thompson, P. R. (2012). Seeing citrulline: Development of a phenylglyoxal-based probe to visualize protein citrullination. *Journal of the American Chemical Society, 134*(41), 17015–17018.

Bjelle, A., Norberg, B., & Sjogren, G. (1982). The cytology of joint exudates in rheumatoid arthritis. Morphology and preparation techniques. *Scandinavian Journal of Rheumatology, 11*(2), 124–128.

Blachere, N. E., Parveen, S., Fak, J., Frank, M. O., & Orange, D. E. (2015). Inflammatory but not apoptotic death of granulocytes citrullinates fibrinogen. *Arthritis Research & Therapy, 17*(1), 369.

Brooks, W. H. (2013). Increased polyamines alter chromatin and stabilize autoantigens in autoimmune diseases. *Frontiers in Immunology, 4*, 91.

Burr, M. L., Naseem, H., Hinks, A., Eyre, S., Gibbons, L. J., Bowes, J., Wilson, A. G., Maxwell, J., Morgan, A. W., Emery, P., Steer, S., Hocking, L., Reid, D. M., Wordsworth, P., Harrison, P., Thomson, W., Worthington, J., & Barton, A. (2010). PADI4 genotype is not associated with rheumatoid arthritis in a large UK Caucasian population. *Annals of the Rheumatic Diseases, 69*(4), 666–670.

Cantaert, T., Coucke, P., De, R. L., Veys, E. M., De, K. F., & Baeten, D. (2005). Functional haplotypes of PADI4: Relevance for rheumatoid arthritis specific synovial intracellular citrullinated proteins and anticitrullinated protein antibodies. *Annals of the Rheumatic Diseases, 64*(9), 1316–1320.

Chang, X., Yamada, R., Suzuki, A., Sawada, T., Yoshino, S., Tokuhiro, S., & Yamamoto, K. (2005). Localization of peptidylarginine deiminase 4 (PADI4) and citrullinated protein in synovial tissue of rheumatoid arthritis. *Rheumatology (Oxford), 44*(1), 40–50.

Chang, X., Xia, Y., Pan, J., Meng, Q., Zhao, Y., & Yan, X. (2013). PADI2 is significantly associated with rheumatoid arthritis. *PloS One, 8*(12), e81259.

Chang, H. H., Dwivedi, N., Nicholas, A. P., & Ho, I. C. (2015). The W620 polymorphism in PTPN22 disrupts its interaction with peptidylarginine deiminase type 4 and enhances citrullination and NETosis. *Arthritis & Rhematology, 67*(9), 2323–2334.

Chapuy-Regaud, S., Sebbag, M., Baeten, D., Clavel, C., Foulquier, C., De, K. F., & Serre, G. (2005). Fibrin deimination in synovial tissue is not specific for rheumatoid arthritis but commonly occurs during synovitides. *Journal of Immunology, 174*(8), 5057–5064.

Cherrington, B. D., Zhang, X., McElwee, J. L., Morency, E., Anguish, L. J., & Coonrod, S. A. (2012). Potential role for PAD2 in gene regulation in breast cancer cells. *PloS One, 7*(7), e41242.

Damgaard, D., Senolt, L., Nielsen, M., Pruijn, G., & Nielsen, C. H. (2014). Demonstration of extracellular peptidylarginine deiminase (PAD) activity in synovial fluid of patients with rheumatoid arthritis using a novel assay for citrullination of fibrinogen. *Arthritis Research & Therapy, 16*(6), 498.

Damgaard, D., Bjorn, M. E., Steffensen, M. A., Pruijn, G. J., & Nielsen, C. H. (2016a). Reduced glutathione as a physiological co-activator in the activation of peptidylarginine deiminase. *Arthritis Research & Therapy, 18*(1), 102.

Damgaard, D., Senolt, L., & Nielsen, C. H. (2016b). Increased levels of peptidylarginine deiminase 2 in synovial fluid from anti-CCP-positive rheumatoid arthritis patients: Association with disease activity and inflammatory markers. *Rheumatology (Oxford), 55*(5), 918–927.

Darrah, E., Rosen, A., Giles, J. T., & Andrade, F. (2012). Peptidylarginine deiminase 2, 3 and 4 have distinct specificities against cellular substrates: Novel insights into autoantigen selection in rheumatoid arthritis. *Annals of the Rheumatic Diseases, 71*(1), 92–98.

Darrah, E., Giles, J. T., Ols, M. L., Bull, H. G., Andrade, F., & Rosen, A. (2013). Erosive rheumatoid arthritis is associated with antibodies that activate PAD4 by increasing calcium sensitivity. *Science Translational Medicine, 5*(186), 186ra65.

Davies, E. V., & Hallett, M. B. (1998). High micromolar Ca2+ beneath the plasma membrane in stimulated neutrophils. *Biochemical and Biophysical Research Communications, 248*(3), 679–683.

De, R. L., Nicholas, A. P., Cantaert, T., Kruithof, E., Echols, J. D., Vandekerckhove, B., Veys, E. M., De, K. F., & Baeten, D. (2005). Synovial intracellular citrullinated proteins colocalizing with peptidyl arginine deiminase as pathophysiologically relevant antigenic determinants of rheumatoid arthritis-specific humoral autoimmunity. *Arthritis and Rheumatism, 52*(8), 2323–2330.

Dixon, B. M., Heath, S. H., Kim, R., Suh, J. H., & Hagen, T. M. (2008). Assessment of endoplasmic reticulum glutathione redox status is confounded by extensive ex vivo oxidation. *Antioxidants & Redox Signaling, 10*(5), 963–972.

Dreyton, C. J., Knuckley, B., Jones, J. E., Lewallen, D. M., & Thompson, P. R. (2014). Mechanistic studies of protein arginine deiminase 2: Evidence for a substrate-assisted mechanism. *Biochemistry, 53*(27), 4426–4433.

Dwivedi, N., & Radic, M. (2014). Citrullination of autoantigens implicates NETosis in the induction of autoimmunity. *Annals of the Rheumatic Diseases, 73*(3), 483–491.

Dwivedi, N., Upadhyay, J., Neeli, I., Khan, S., Pattanaik, D., Myers, L., Kirou, K. A., Hellmich, B., Knuckley, B., Thompson, P. R., Crow, M. K., Mikuls, T. R., Csernok, E., & Radic, M. (2012). Felty's syndrome autoantibodies bind to deiminated histones and neutrophil extracellular chromatin traps. *Arthritis and Rheumatism, 64*(4), 982–992.

Foulquier, C., Sebbag, M., Clavel, C., Chapuy-Regaud, S., Al, B. R., Mechin, M. C., Vincent, C., Nachat, R., Yamada, M., Takahara, H., Simon, M., Guerrin, M., & Serre, G. (2007). Peptidyl arginine deiminase type 2 (PAD-2) and PAD-4 but not PAD-1, PAD-3, and PAD-6 are expressed in rheumatoid arthritis synovium in close association with tissue inflammation. *Arthritis and Rheumatism, 56*(11), 3541–3553.

Garcia-Romo, G. S., Caielli, S., Vega, B., Connolly, J., Allantaz, F., Xu, Z., Punaro, M., Baisch, J., Guiducci, C., Coffman, R. L., Barrat, F. J., Banchereau, J., & Pascual, V. (2011). Netting neutrophils are major inducers of type I IFN production in pediatric systemic lupus erythematosus. *Science Translational Medicine, 3*(73), 73ra20.

Ghari, F., Quirke, A. M., Munro, S., Kawalkowska, J., Picaud, S., McGouran, J., Subramanian, V., Muth, A., Williams, R., Kessler, B., Thompson, P. R., Fillipakopoulos, P., Knapp, S., Venables, P. J., & La Thangue, N. B. (2016). Citrullination-acetylation interplay guides E2F-1 activity during the inflammatory response. *Science Advances, 2*(2), e1501257.

Griffith, O. W. (1999). Biologic and pharmacologic regulation of mammalian glutathione synthesis. *Free Radical Biology & Medicine, 27*(9–10), 922–935.

Hakkim, A., Furnrohr, B. G., Amann, K., Laube, B., Abed, U. A., Brinkmann, V., Herrmann, M., Voll, R. E., & Zychlinsky, A. (2010). Impairment of neutrophil extracellular trap degradation is associated with lupus nephritis. *Proceedings of the National Academy of Sciences of the United States of America, 107*(21), 9813–9818.

Hao, G., Wang, D., Gu, J., Shen, Q., Gross, S. S., & Wang, Y. (2009). Neutral loss of isocyanic acid in peptide CID spectra: A novel diagnostic marker for mass spectrometric identification of protein citrullination. *Journal of the American Society for Mass Spectrometry, 20*(4), 723–727.

Harre, U., Georgess, D., Bang, H., Bozec, A., Axmann, R., Ossipova, E., Jakobsson, P. J., Baum, W., Nimmerjahn, F., Szarka, E., Sarmay, G., Krumbholz, G., Neumann, E., Toes, R., Scherer, H. U., Catrina, A. I., Klareskog, L., Jurdic, P., & Schett, G. (2012). Induction of osteoclastogenesis and bone loss by human autoantibodies against citrullinated vimentin. *The Journal of Clinical Investigation, 122*(5), 1791–1802.

Hensen, S. M., & Pruijn, G. J. (2014). Methods for the detection of peptidylarginine deiminase (PAD) activity and protein citrullination. *Molecular & Cellular Proteomics, 13*(2), 388–396.

Hensen, S. M., Boelens, W. C., Bonger, K. M., van Cruchten, R. T., van Delft, F. L., & Pruijn, G. J. (2015). Phenylglyoxal-based visualization of citrullinated proteins on Western blots. *Molecules, 20*(4), 6592–6600.

Hill, J. A., Southwood, S., Sette, A., Jevnikar, A. M., Bell, D. A., & Cairns, E. (2003). Cutting edge: The conversion of arginine to citrulline allows for a high-affinity peptide interaction with the rheumatoid arthritis-associated HLA- DRB1*0401 MHC class II molecule. *Journal of Immunology, 171*(2), 538–541.

Hill, J. A., Bell, D. A., Brintnell, W., Yue, D., Wehrli, B., Jevnikar, A. M., Lee, D. M., Hueber, W., Robinson, W. H., & Cairns, E. (2008). Arthritis induced by posttranslationally modified (citrullinated) fibrinogen in DR4-IE transgenic mice. *The Journal of Experimental Medicine, 205*(4), 967–979.

Hoppe, B., Haupl, T., Gruber, R., Kiesewetter, H., Burmester, G. R., Salama, A., & Dorner, T. (2006). Detailed analysis of the variability of peptidylarginine deiminase type 4 in German patients with rheumatoid arthritis: A case-control study. *Arthritis Research & Therapy, 8*(2), R34.

Jang, B., Kim, H. W., Kim, J. S., Kim, W. S., Lee, B. R., Kim, S., Kim, H., Han, S. J., Ha, S. J., & Shin, S. J. (2015). Peptidylarginine deiminase inhibition impairs Toll-like receptor agonist-induced functional maturation of dendritic cells, resulting in the loss of T cell-proliferative capacity: A partial mechanism with therapeutic potential in inflammatory settings. *Journal of Leukocyte Biology, 97*(2), 351–362.

Jones, D. P. (2002). Redox potential of GSH/GSSG couple: Assay and biological significance. *Methods in Enzymology, 348*, 93–112.

Kearney, P. L., Bhatia, M., Jones, N. G., Yuan, L., Glascock, M. C., Catchings, K. L., Yamada, M., & Thompson, P. R. (2005). Kinetic characterization of protein arginine deiminase 4: A transcriptional corepressor implicated in the onset and progression of rheumatoid arthritis. *Biochemistry, 44*(31), 10570–10582.

Khandpur, R., Carmona-Rivera, C., Vivekanandan-Giri, A., Gizinski, A., Yalavarthi, S., Knight, J. S., Friday, S., Li, S., Patel, R. M., Subramanian, V., Thompson, P., Chen, P., Fox, D. A., Pennathur, S., & Kaplan, M. J. (2013). NETs are a source of citrullinated autoantigens and stimulate inflammatory responses in rheumatoid arthritis. *Science Translational Medicine, 5*(178), 178ra40.

Kinloch, A., Lundberg, K., Wait, R., Wegner, N., Lim, N. H., Zendman, A. J., Saxne, T., Malmstrom, V., & Venables, P. J. (2008). Synovial fluid is a site of citrullination of autoantigens in inflammatory arthritis. *Arthritis and Rheumatism, 58*(8), 2287–2295.

Knuckley, B., Bhatia, M., & Thompson, P. R. (2007). Protein arginine deiminase 4: Evidence for a reverse protonation mechanism. *Biochemistry, 46*(22), 6578–6587.

Knuckley, B., Causey, C. P., Jones, J. E., Bhatia, M., Dreyton, C. J., Osborne, T. C., Takahara, H., & Thompson, P. R. (2010). Substrate specificity and kinetic studies of PADs 1, 3, and 4 identify potent and selective inhibitors of protein arginine deiminase 3. *Biochemistry, 49*(23), 4852–4863.

Krishnamurthy, A., Joshua, V., Haj, H. A., Jin, T., Sun, M., Vivar, N., Ytterberg, A. J., Engstrom, M., Fernandes-Cerqueira, C., Amara, K., Magnusson, M., Wigerblad, G., Kato, J., Jimenez-Andrade, J. M., Tyson, K., Rapecki, S., Lundberg, K., Catrina, S. B., Jakobsson, P. J., Svensson, C., Malmstrom, V., Klareskog, L., Wahamaa, H., & Catrina, A. I. (2016). Identification of a novel chemokine-dependent molecular mechanism underlying rheumatoid arthritis-associated autoantibody-mediated bone loss. *Annals of the Rheumatic Diseases, 75*(4), 721–729.

Lewis, H. D., Liddle, J., Coote, J. E., Atkinson, S. J., Barker, M. D., Bax, B. D., Bicker, K. L., Bingham, R. P., Campbell, M., Chen, Y. H., Chung, C. W., Craggs, P. D., Davis, R. P., Eberhard, D., Joberty, G., Lind, K. E., Locke, K., Maller, C., Martinod, K., Patten, C., Polyakova, O., Rise, C. E., Rudiger, M., Sheppard, R. J., Slade, D. J., Thomas, P., Thorpe, J., Yao, G., Drewes, G., Wagner, D. D., Thompson, P. R., Prinjha, R. K., & Wilson, D. M. (2015). Inhibition of PAD4 activity is sufficient to disrupt mouse and human NET formation. *Nature Chemical Biology, 11*(3), 189–191.

Liu, Y. L., Chiang, Y. H., Liu, G. Y., & Hung, H. C. (2011). Functional role of dimerization of human peptidylarginine deiminase 4 (PAD4). *PloS One, 6*(6), e21314.

Makrygiannakis, D., af, K. E., Lundberg, I. E., Lofberg, R., Ulfgren, A. K., Klareskog, L., & Catrina, A. I. (2006). Citrullination is an inflammation-dependent process. *Annals of the Rheumatic Diseases, 65*(9), 1219–1222.

Makrygiannakis, D., Revu, S., Engstrom, M., af, K. E., Nicholas, A. P., Pruijn, G. J., & Catrina, A. I. (2012). Local administration of glucocorticoids decreases synovial citrullination in rheumatoid arthritis. *Arthritis Research & Therapy, 14*(1), R20.

Martinez, A., Valdivia, A., Pascual-Salcedo, D., Lamas, J. R., Fernandez-Arquero, M., Balsa, A., Fernandez-Gutierrez, B., de la Concha, E. G., & Urcelay, E. (2005). PADI4 polymorphisms are not associated with rheumatoid arthritis in the Spanish population. *Rheumatology (Oxford), 44*(10), 1263–1266.

Moelants, E. A., Mortier, A., Grauwen, K., Ronsse, I., Van, D. J., & Proost, P. (2013). Citrullination of TNF-alpha by peptidylarginine deiminases reduces its capacity to stimulate the production of inflammatory chemokines. *Cytokine, 61*(1), 161–167.

Mohanan, S., Horibata, S., McElwee, J. L., Dannenberg, A. J., & Coonrod, S. A. (2013). Identification of macrophage extracellular trap-like structures in mammary gland adipose tissue: A preliminary study. *Frontiers in Immunology, 4*, 67.

Muller, S., & Radic, M. (2015). Citrullinated autoantigens: From diagnostic markers to pathogenetic mechanisms. *Clinical Reviews in Allergy & Immunology, 49*(2), 232–239.

Nakashima, K., Hagiwara, T., & Yamada, M. (2002). Nuclear localization of peptidylarginine deiminase V and histone deimination in granulocytes. *The Journal of Biological Chemistry, 277*(51), 49562–49568.

Nakayama-Hamada, M., Suzuki, A., Kubota, K., Takazawa, T., Ohsaka, M., Kawaida, R., Ono, M., Kasuya, A., Furukawa, H., Yamada, R., & Yamamoto, K. (2005). Comparison of enzymatic properties between hPADI2 and hPADI4. *Biochemical and Biophysical Research Communications, 327*(1), 192–200.

Neeli, I., Khan, S. N., & Radic, M. (2008). Histone deimination as a response to inflammatory stimuli in neutrophils. *Journal of Immunology, 180*(3), 1895–1902.

Nicholas, A. P., & Whitaker, J. N. (2002). Preparation of a monoclonal antibody to citrullinated epitopes: Its characterization and some applications to immunohistochemistry in human brain. *Glia, 37*(4), 328–336.

Okada, Y., Wu, D., Trynka, G., Raj, T., Terao, C., Ikari, K., Kochi, Y., Ohmura, K., Suzuki, A., Yoshida, S., Graham, R. R., Manoharan, A., Ortmann, W., Bhangale, T., Denny, J. C., Carroll, R. J., Eyler, A. E., Greenberg, J. D., Kremer, J. M., Pappas, D. A., Jiang, L., Yin, J., Ye, L., Su, D. F., Yang, J., Xie, G., Keystone, E., Westra, H. J., Esko, T., Metspalu, A., Zhou, X., Gupta, N., Mirel, D., Stahl, E. A., Diogo, D., Cui, J., Liao, K., Guo, M. H., Myouzen, K., Kawaguchi, T., Coenen, M. J., van Riel, P. L., van de Laar, M. A., Guchelaar, H. J., Huizinga, T. W., Dieude,

P., Mariette, X., Bridges, S. L., Jr., Zhernakova, A., Toes, R. E., Tak, P. P., Miceli-Richard, C., Bang, S. Y., Lee, H. S., Martin, J., Gonzalez-Gay, M. A., Rodriguez-Rodriguez, L., Rantapaa-Dahlqvist, S., Arlestig, L., Choi, H. K., Kamatani, Y., Galan, P., Lathrop, M., Eyre, S., Bowes, J., Barton, A., de, V. N., Moreland, L. W., Criswell, L. A., Karlson, E. W., Taniguchi, A., Yamada, R., Kubo, M., Liu, J. S., Bae, S. C., Worthington, J., Padyukov, L., Klareskog, L., Gregersen, P. K., Raychaudhuri, S., Stranger, B. E., De Jager, P. L., Franke, L., Visscher, P. M., Brown, M. A., Yamanaka, H., Mimori, T., Takahashi, A., Xu, H., Behrens, T. W., Siminovitch, K. A., Momohara, S., Matsuda, F., Yamamoto, K., & Plenge, R. M. (2014). Genetics of rheumatoid arthritis contributes to biology and drug discovery. *Nature, 506*(7488), 376–381.

Olivares-Martinez, E., Hernandez-Ramirez, D. F., Nunez-Alvarez, C. A., Cabral, A. R., & Llorente, L. (2016). The amount of citrullinated proteins in synovial tissue is related to serum anti-cyclic citrullinated peptide (anti-CCP) antibody levels. *Clinical Rheumatology, 35*(1), 55–61.

Parker, H., & Winterbourn, C. C. (2012). Reactive oxidants and myeloperoxidase and their involvement in neutrophil extracellular traps. *Frontiers in Immunology, 3*, 424.

Plenge, R. M., Padyukov, L., Remmers, E. F., Purcell, S., Lee, A. T., Karlson, E. W., Wolfe, F., Kastner, D. L., Alfredsson, L., Altshuler, D., Gregersen, P. K., Klareskog, L., & Rioux, J. D. (2005). Replication of putative candidate-gene associations with rheumatoid arthritis in >4,000 samples from North America and Sweden: association of susceptibility with PTPN22, CTLA4, and PADI4. *American Journal of Human Genetics, 77*(6), 1044–1060.

Proost, P., Loos, T., Mortier, A., Schutyser, E., Gouwy, M., Noppen, S., Dillen, C., Ronsse, I., Conings, R., Struyf, S., Opdenakker, G., Maudgal, P. C., & Van, D. J. (2008). Citrullination of CXCL8 by peptidylarginine deiminase alters receptor usage, prevents proteolysis, and dampens tissue inflammation. *The Journal of Experimental Medicine, 205*(9), 2085–2097.

Raijmakers, R., van Beers, J. J., El-Azzouny, M., Visser, N. F., Bozic, B., Pruijn, G. J., & Heck, A. J. (2012). Elevated levels of fibrinogen-derived endogenous citrullinated peptides in synovial fluid of rheumatoid arthritis patients. *Arthritis Research & Therapy, 14*(3), R114.

Salisbury, A. K., Duke, O., & Poulter, L. W. (1987). Macrophage-like cells of the pannus area in rheumatoid arthritic joints. *Scandinavian Journal of Rheumatology, 16*(4), 263–272.

Sanchez-Pernaute, O., Filkova, M., Gabucio, A., Klein, M., Maciejewska-Rodrigues, H., Ospelt, C., Brentano, F., Michel, B. A., Gay, R. E., Herrero-Beaumont, G., Gay, S., Neidhart, M., & Juengel, A. (2013). Citrullination enhances the pro-inflammatory response to fibrin in rheumatoid arthritis synovial fibroblasts. *Annals of the Rheumatic Diseases, 72*(8), 1400–1406.

Scally, S. W., Petersen, J., Law, S. C., Dudek, N. L., Nel, H. J., Loh, K. L., Wijeyewickrema, L. C., Eckle, S. B., van, H. J., Pike, R. N., McCluskey, J., Toes, R. E., La Gruta, N. L., Purcell, A. W., Reid, H. H., Thomas, R., & Rossjohn, J. (2013). A molecular basis for the association of the HLA-DRB1 locus, citrullination, and rheumatoid arthritis. *The Journal of Experimental Medicine, 210*(12), 2569–2582.

Seefeldt, T., Zhao, Y., Chen, W., Raza, A. S., Carlson, L., Herman, J., Stoebner, A., Hanson, S., Foll, R., & Guan, X. (2009). Characterization of a novel dithiocarbamate glutathione reductase inhibitor and its use as a tool to modulate intracellular glutathione. *The Journal of Biological Chemistry, 284*(5), 2729–2737.

Senshu, T., Sato, T., Inoue, T., Akiyama, K., & Asaga, H. (1992). Detection of citrulline residues in deiminated proteins on polyvinylidene difluoride membrane. *Analytical Biochemistry, 203*(1), 94–100.

Slade, D. J., Subramanian, V., Fuhrmann, J., & Thompson, P. R. (2014). Chemical and biological methods to detect post-translational modifications of arginine. *Biopolymers, 101*(2), 133–143.

Slade, D. J., Fang, P., Dreyton, C. J., Zhang, Y., Fuhrmann, J., Rempel, D., Bax, B. D., Coonrod, S. A., Lewis, H. D., Guo, M., Gross, M. L., & Thompson, P. R. (2015). Protein arginine deiminase 2 binds calcium in an ordered fashion: Implications for inhibitor design. *ACS Chemical Biology, 10*(4), 1043–1053.

Sokolove, J., Zhao, X., Chandra, P. E., & Robinson, W. H. (2011). Immune complexes containing citrullinated fibrinogen costimulate macrophages via Toll-like receptor 4 and Fcgamma receptor. *Arthritis and Rheumatism, 63*(1), 53–62.

Spengler, J., Lugonja, B., Jimmy, Y. A., Zubarev, R. A., Creese, A. J., Pearson, M. J., Grant, M. M., Milward, M., Lundberg, K., Buckley, C. D., Filer, A., Raza, K., Cooper, P. R., Chapple, I. L., & Scheel-Toellner, D. (2015). Release of active peptidyl arginine deiminases by neutrophils can explain production of extracellular citrullinated autoantigens in rheumatoid arthritis synovial fluid. *Arthritis & Rhematology, 67*(12), 3135–3145.

Stenberg, P., & Roth, B. (2015). Zinc is the modulator of the calcium-dependent activation of post-translationally acting thiol-enzymes in autoimmune diseases. *Medical Hypotheses, 84*(4), 331–335.

Stensland, M. E., Pollmann, S., Molberg, O., Sollid, L. M., & Fleckenstein, B. (2009). Primary sequence, together with other factors, influence peptide deimination by peptidylarginine deiminase-4. *Biological Chemistry, 390*(2), 99–107.

Suzuki, A., Yamada, R., Chang, X., Tokuhiro, S., Sawada, T., Suzuki, M., Nagasaki, M., Nakayama-Hamada, M., Kawaida, R., Ono, M., Ohtsuki, M., Furukawa, H., Yoshino, S., Yukioka, M., Tohma, S., Matsubara, T., Wakitani, S., Teshima, R., Nishioka, Y., Sekine, A., Iida, A., Takahashi, A., Tsunoda, T., Nakamura, Y., & Yamamoto, K. (2003). Functional haplotypes of PADI4, encoding citrullinating enzyme peptidylarginine deiminase 4, are associated with rheumatoid arthritis. *Nature Genetics, 34*(4), 395–402.

Takahara, H., Okamoto, H., & Sugawara, K. (1986). Calcium-dependent properties of peptidylarginine deiminase from rabbit skeletal muscle. *Agricultural and Biological Chemistry, 50*, 2899–2904.

Terakawa, H., Takahara, H., & Sugawara, K. (1991). Three types of mouse peptidylarginine deiminase: Characterization and tissue distribution. *Journal of Biochemistry, 110*(4), 661–666.

Tutturen, A. E., Fleckenstein, B., & de Souza, G. A. (2014). Assessing the citrullinome in rheumatoid arthritis synovial fluid with and without enrichment of citrullinated peptides. *Journal of Proteome Research, 13*(6), 2867–2873.

Tutturen, A. E., Holm, A,, Jorgensen, M,, Stadtmuller, P., Rise, F., & Fleckenstein, B. (2010). A technique for the specific enrichment of citrulline-containing peptides. *Analytical Biochemistry, 403*, 43–51.

van Beers, J. J., Raijmakers, R., Alexander, L. E., Stammen-Vogelzangs, J., Lokate, A. M., Heck, A. J., Schasfoort, R. B., & Pruijn, G. J. (2010). Mapping of citrullinated fibrinogen B-cell epitopes in rheumatoid arthritis by imaging surface plasmon resonance. *Arthritis Research & Therapy, 12*(6), R219.

van Beers, J. J., Schwarte, C. M., Stammen-Vogelzangs, J., Oosterink, E., Bozic, B., & Pruijn, G. J. (2013). The rheumatoid arthritis synovial fluid citrullinome reveals novel citrullinated epitopes in apolipoprotein E, myeloid nuclear differentiation antigen, and beta-actin. *Arthritis and Rheumatism, 65*(1), 69–80.

van Venrooij, W. J., & Pruijn, G. J. (2008). An important step towards completing the rheumatoid arthritis cycle. *Arthritis Research & Therapy, 10*(5), 117.

Vossenaar, E. R., Zendman, A. J., van Venrooij, W. J., & Pruijn, G. J. (2003). PAD, a growing family of citrullinating enzymes: Genes, features and involvement in disease. *BioEssays, 25*(11), 1106–1118.

Vossenaar, E. R., Radstake, T. R., van der Heijden, A., van Mansum, M. A., Dieteren, C., de Rooij, D. J., Barrera, P., Zendman, A. J., & van Venrooij, W. J. (2004a). Expression and activity of citrullinating peptidylarginine deiminase enzymes in monocytes and macrophages. *Annals of the Rheumatic Diseases, 63*(4), 373–381.

Vossenaar, E. R., Smeets, T. J., Kraan, M. C., Raats, J. M., van Venrooij, W. J., & Tak, P. P. (2004b). The presence of citrullinated proteins is not specific for rheumatoid synovial tissue. *Arthritis and Rheumatism, 50*(11), 3485–3494.

Wang, F., Chen, F. F., Gao, W. B., Wang, H. Y., Zhao, N. W., Xu, M., Gao, D. Y., Yu, W., Yan, X. L., Zhao, J. N., & Li, X. J. (2016). Identification of citrullinated peptides in the synovial fluid of patients with rheumatoid arthritis using LC-MALDI-TOF/TOF. *Clinical Rheumatology, 35*(9), 2185–2194.

Yoshida, S., Katoh, T., Tetsuka, T., Uno, K., Matsui, N., & Okamoto, T. (1999). Involvement of thioredoxin in rheumatoid arthritis: Its costimulatory roles in the TNF-alpha-induced production of IL-6 and IL-8 from cultured synovial fibroblasts. *Journal of Immunology, 163*(1), 351–358.

Yoshida, K., Korchynskyi, O., Tak, P. P., Isozaki, T., Ruth, J. H., Campbell, P. L., Baeten, D. L., Gerlag, D. M., Amin, M. A., & Koch, A. E. (2014). Citrullination of epithelial neutrophil-activating peptide 78/CXCL5 results in conversion from a non-monocyte-recruiting chemokine to a monocyte-recruiting chemokine. *Arthritis & Rhematology, 66*(10), 2716–2727.

Zendman, A. J., Raijmakers, R., Nijenhuis, S., Vossenaar, E. R., Tillaart, M., Chirivi, R. G., Raats, J. M., van Venrooij, W. J., Drijfhout, J. W., & Pruijn, G. J. (2007). ABAP: Antibody-based assay for peptidylarginine deiminase activity. *Analytical Biochemistry, 369*(2), 232–240.

Chapter 6
Rheumatoid Arthritis: Transition from Systemic Autoimmunity to Joint Inflammation and Bone Loss

Bence Rethi, Akilan Krishnamurthy, and Anca I. Catrina

6.1 Introduction

Rheumatoid arthritis (RA) is a chronic inflammatory joint disease resulting from the complex interaction between genetic and environmental factors, with at least two distinct clinical phenotypes that are differentiated by the presence or absence of anti-citrullinated protein antibodies (ACPAs), i.e., ACPA-positive RA and ACPA-negative RA (Klareskog et al. 2009). These two phenotypes differ both in terms of risk factors and disease mechanisms. Specific environment-gene interactions (such as smoking and presence of the HLA-DRB1 risk allele variants) (Huizinga et al. 2005a; Klareskog et al. 2006a; Pedersen et al. 2007; Ding et al. 2009; Lundstrom et al. 2009) only confer risk for developing of ACPA-positive RA but not ACPA-negative RA. ACPA-positive RA has a more severe disease course, and more frequently associates with bone destruction (Ronnelid et al. 2005; van der Helm-van Mil et al. 2005; Syversen et al. 2010a) as compared to ACPA-negative RA. More recent findings suggest that ACPA might play an important role in the transition from systemic autoimmunity (that develops before joint inflammation) to joint disease through specific targeting of the bone and activation of osteoclasts. The current chapter will review existing evidence showing that ACPA might be generated at extra-articular mucosal sites and contribute to the initiation of chronic joint inflammation.

B. Rethi • A. Krishnamurthy • A.I. Catrina (✉)
Rheumatology Unit, D2:01, Department of Medicine, Karolinska University Hospital and Karolinska Institute, Stockholm, Sweden
e-mail: anca.catrina@ki.se

A.P. Nicholas et al. (eds.), *Protein Deimination in Human Health and Disease*,
DOI 10.1007/978-3-319-58244-3_6

6.2 The Immune System Might Be Primed Against Citrullinated Autoantigens at Mucosal Extra-Articular Sites

ACPAs are detected in the peripheral blood, years before the onset of symptomatic RA (Aho et al. 1991; Kurki et al. 1992; Rantapaa-Dahlqvist et al. 2003; Nielen et al. 2004; Chibnik et al. 2009; Majka et al. 2008; Shi et al. 2014), suggesting that the antibody response against citrullinated antigens might originate from extra-articular sites. Smoking, a major environmental risk factor for developing RA, is associated with the presence of ACPAs, and a robust gene-environment interaction has been demonstrated between the expression of the RA susceptibility HLA-DR alleles carrying the characteristic amino acid motif HLA-shared epitope (SE) and smoking history in ACPA-positive RA (Padyukov et al. 2004; Klareskog et al. 2006b; Huizinga et al. 2005b; Pedersen et al. 2006; Karlson et al. 2010; Too et al. 2012). The mechanistic link between HLA SE gene expression and ACPA-positive RA has been proposed by structural studies on MHC-peptide interactions, which revealed a preferential binding of peptides with an uncharged citrulline residue to HLA-DRβ chain shared susceptibility epitope that contains a conserved positively charged lysine at position 71 (Scally et al. 2013). Interestingly, the RA resistance allele HLA-DRB1*04:02 contains a glutamate in the same position that enabled contacts with both Arg- and Cit-containing peptides. In addition to differential MHC binding, citrullination has also been shown to affect peptide processing providing protection for otherwise degraded citrullinated regions that can potentially be presented to autoreactive T cells (Scally et al. 2013). Importantly, the use of smokeless oral tobacco (moist snuff) does not increase the risk of RA, indicating that it is smoke itself, instead of nicotine, which increases RA susceptibility (Jiang et al. 2014). In line with this finding, exposure to other environmental airway irritants including silica and textile dust has been also associated with ACPA-positive RA (Too et al. 2016; Stolt et al. 2005).

Taken together these epidemiological findings suggest that environmental exposures (such as smoking and potentially other airway pollutants) to the lungs might contribute to the priming of the immune system against citrullinated targets (Fig. 6.1). The exact mechanisms by which lung injuries induced by these exposures eventually trigger anti-citrulline responses are incompletely understood. It has been shown that smoking induces inflammation and increases the expression of citrullinated proteins in the lungs (Makrygiannakis et al. 2008; Reynisdottir et al. 2014; Lugli et al. 2015). Other indirect contributions might be alterations in antigen-presenting dendritic cell functions (Arnson et al. 2010), activation of pattern recognition molecules (Arnson et al. 2010), and changes in the composition of the lung microbiome (Larsen et al. 2012), all of them events that can also influence immune responses. Further proof for a potential role of the lungs in triggering of the immune response against citrullinated proteins came from detection of lung abnormalities in arthritis-free individuals at risk for developing RA (Demoruelle et al. 2012) and the association between ACPA and bronchiectasis in the absence of clinically evident RA (Perry et al. 2014; Janssen et al. 2015). ACPAs have been detected, not only in the circulation, but also in the

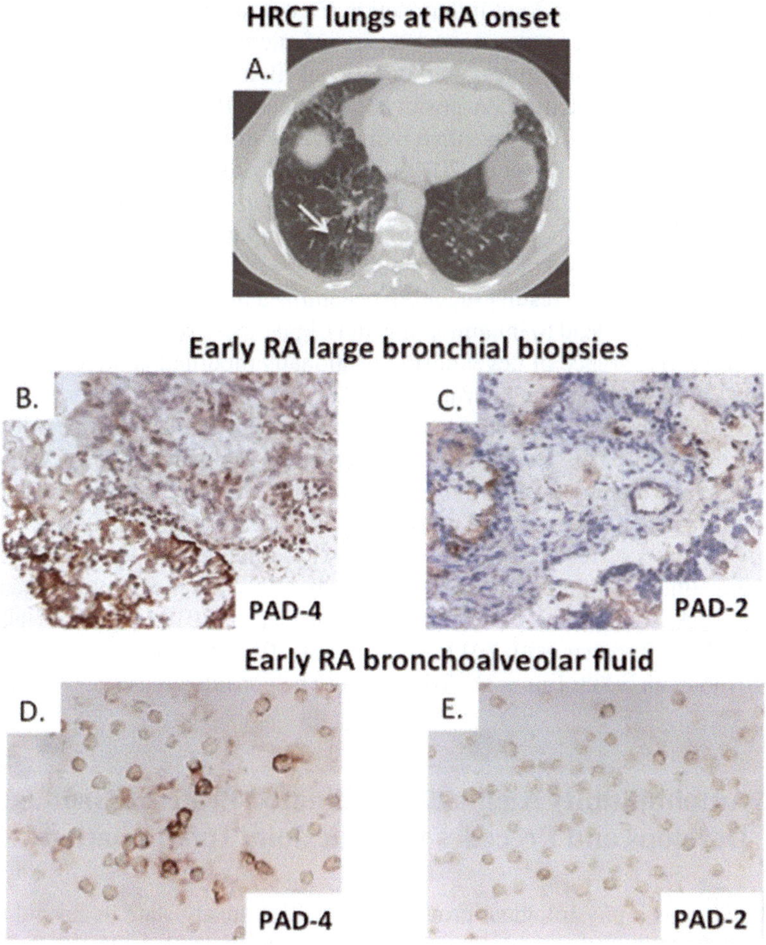

Fig. 6.1 Lung subclinical inflammatory changes are present in rheumatoid arthritis already at disease onset. Images showing fibrosis on high-resolution computer tomography (**a**) in a patient newly diagnosed with RA and red immunohistochemical staining of PAD-2 and PAD-4 enzymes in the large bronchial tissue (**b**, **c**) and bronchoalveolar fluid (**d**, **e**)

sputum of some individuals at risk for developing RA, due to previously recorded ACPA seropositivity, indicating the production of these antibodies in the airways already before RA onset (Willis et al. 2013). Identical citrullinated epitopes from the ACPA targeted proteins such as vimentin, actin, or annexin II have been detected in both synovial and bronchial biopsies obtained from RA patients, indicating that not only the antibodies but their citrullinated targets can also be present in the lung compartment (Ytterberg et al. 2015). High ACPA titers associated with airway abnormalities were occasionally linked to RA development during relatively short follow-up periods, further supporting a scenario where an extra-articular anti-citrulline response could precede and initiate joint inflammation (Demoruelle et al. 2012; Fischer et al. 2012). Further evidence comes from detection of microscopic signs of inflammation and ectopic lymphoid structures, with occasional iBALT formation in the bronchial biopsies of early untreated ACPA-positive RA patients (Reynisdottir et al. 2016) and in the lung biopsies of ACPA-positive individuals with lung disease but no signs of joint inflammation (Fischer et al. 2012). Interestingly these changes were observed not only in smokers, but also in a large majority of non-smoking ACPA-positive RA patients, suggesting that cigarette smoke while important is not the only factor responsible for generation of ACPA immunity.

Taken together, these results suggest a generalized model where various forms of airway insults could lead to inflammation and trigger increased citrullination, which results in the appearance of novel posttranslationally modified peptide epitopes that are under- or non-represented during thymic T cell education and gives rise to autoreactive T cell responses (Klareskog et al. 2006b; Catrina et al. 2014; Chatzidionisyou and Catrina 2016). This leads to tolerance breaking and generation of ACPA with consecutive development of pathogenic immunity and ACPA-positive RA in the subgroup of genetic susceptible individuals. While most of the current evidence strongly suggests a role for the lungs as an extra-articular site for the initiation of pathogenic immunity, in part because this link has been the most investigated, other mucosal sites, such as the oral mucosa (Wegner et al. 2010; Mikuls et al. 2012; Nesse et al. 2012; Scher et al. 2012; Harvey et al. 2013) and potentially also the gut (Vaahtovuo et al. 2008; Liu et al. 2013), should be considered as sites of initiation.

6.3 Autoimmunity Against Citrullinated Proteins Gradually Develops and Precedes Chronic Joint Inflammation

Citrullination is a physiological process occurring at steady state in several tissues, including the skin or the central nervous system (Senshu et al. 1996, 1999; Harding and Scott 1983; Pearton et al. 2002; Scott et al. 1982; Moscarello et al. 1994; Beniac et al. 2000) or at various sites of inflammation (Makrygiannakis et al. 2006; Vossenaar et al. 2004a; Cantaert et al. 2006). Protein citrullination regulates differentiation and functions of diverse cell types, such as stem cells where histone citrullination regulates pluripotency (Nakashima et al. 2013; Christophorou et al. 2014; Slade et al. 2014; Wang et al. 2009), or neutrophil granulocytes, which require histone citrullination for the formation of neutrophil extracellular traps, a network of extracellular fibers that immobilize pathogens (Wang et al. 2009). Immune response against citrullinated self-proteins, on the other hand, is highly specific for RA, and ACPA-positive RA is unique in its association with genetic and environmental factors (Huizinga et al. 2005a; Klareskog et al. 2006a; Pedersen et al. 2007; Ding et al. 2009; Lundstrom et al. 2009) and with more joint destruction (Innala et al. 2008; Mustila et al. 2011; Syversen et al. 2010b; van Steenbergen et al. 2015; Hecht et al. 2015) as compared to ACPA-negative RA. The sequence of events leading to ACPA-positive RA is most probably initiated with an immune response against a very limited number of citrullinated epitopes, with variable specificities observed in different individuals (Brink et al. 2013). Apart from the already identified genetic and environmental risk factors, multiple further signals can contribute to the initiation anti-citrulline responses, such as acute infections or tissue damage, which lower the threshold for priming antigen-specific responses and/or contribute to increased citrullination. Antibody responses against citrullinated epitopes increase slowly both in intensity and in the number of recognized targets, without inducing joint inflammation. Interestingly, antibody responses against certain citrullinated targets expand more robustly, whereas others remain permanently at low levels suggesting that distinct ACPA specificities might have more important contribution to RA progression than others (Brink et al. 2013). In line with these results, association with genetic risk factors or smoking was more a characteristic for patients with specific pattern of ACPA reactivities, which included antibodies against citrullinated α-enolase and vimentin, instead of anti-citrulline responses in general (Lundberg et al. 2013). A rapid increase in epitope spreading and in the intensity of ACPA responses typically marks the onset of RA. It is interesting to note that the progressive diversification of ACPA reactivities, with the underlying mutations affecting hypervariability regions and, atypically, the framework sites of the antibodies as well, together with the cross-reactivity of individual ACPA clones between diverse citrullinated antigens resemble broadly neutralizing antibodies developed against the persistently evolving virus HIV-1 in a small number of chronically infected individuals (Klein et al. 2013). The similarity between antibody responses against citrullinated autoantigens and a highly variable virus that is characterized by efficient immune evasion strategies indicates the complicated nature of distinguishing citrullinated antigens from the naturally occurring protein sequences and, potentially, a progressive change in citrullinated targets.

6.4 ACPAs Target Maturing Osteoclasts Leading to Bone Loss and Pain in the Absence of Inflammation

It is yet to be fully understood how ACPA responses translocate to the joints and contribute to the pathogenesis of RA; however, recent findings have highlighted a unique role of osteoclasts in the initiation of joint inflammation in response to antibody binding to citrullinated cell surface molecules (Harre et al. 2012; Krishnamurthy et al. 2016).

Uncontrolled disease progression leads to bone damage in RA, suggesting an important role for inflammatory mediators in driving osteoclast activation and bone resorption. Nevertheless, the presence of magnetic resonance changes showing signs of bone marrow inflammation, which precedes development of bone erosion, has been detected by several studies in patients at very early stages of RA, suggesting that bones might be targeted already before the onset of a detectable synovial inflammation (Hetland et al. 2009; Haavardsholm et al. 2008; Boyesen et al. 2011; McQueen et al. 2003). Further, healthy individuals characterized by the presence of ACPA in the circulation presented detectable bone loss and altered bone architecture, which appeared mostly in the cortical bone that became thinner and more porous due to fenestrations and small lesions underneath (Kleyer et al. 2014). The key mechanisms that link circulating ACPAs to bone erosion likely include a direct antibody-mediated induction of cytokines in osteoclasts or their precursors, which leads to an autocrine stimulation of osteoclast differentiation and to other pathological events associated with RA. OC myeloid precursor cells (such as monocytes and macrophages) express the PAD2 and PAD4 enzymes, suggesting a physiological role for protein citrullination in the development, homeostasis, or functions of these cell types (Vossenaar et al. 2004b), and we have recently characterized the presence of both citrullinated proteins and PAD2 and PAD4 enzymes in maturing OCs (Fig. 6.2) showing that PAD inhibition strongly interfered with OC differentiation in vitro (Krishnamurthy et al. 2016). Protein citrullination occurring as part of the physiological differentiation process in osteoclasts (but not others cells) could explain the initial specific osteoclast targeting by ACPAs. In line with this finding, developing osteoclasts express the citrullinated form of the cytoskeletal protein vimentin on the cell surface, and anti-citrullinated vimentin antibodies induce bone erosion in vivo, when injected into mice together with minute amounts of LPS and altered myeloid cell phenotype in the spleen (Harre et al. 2012). We have recently identified a crucial role for the chemokine IL-8 in the monoclonal and polyclonal ACPA-induced autocrine stimulation of osteoclasts in vitro and bone loss in vivo in mice (Krishnamurthy et al. 2016). Importantly, ACPAs injected into mice efficiently induced, not only bone erosion, but also joint pain, as detected by an increased mechanical and thermal sensitivity in the paws of ACPA-treated mice (Wigerblad et al. 2016). IL-8 blockade interfered with both ACPA-induced bone loss and joint pain in mice (Krishnamurthy et al. 2016; Wigerblad et al. 2016) and ACPA-dependent osteoclastogenesis in vitro (Krishnamurthy et al. 2016), indicating key roles for IL-8 in the ACPA-mediated pathological events.

Despite these recent advances, the exact molecular mechanisms responsible for the effect of ACPAs on myeloid cells still remain to be elucidated. As discussed earlier, OCs and their precursors express citrullinated proteins on the cell surface, which are recognized by ACPAs (Harre et al. 2012; Krishnamurthy et al. 2016; Wigerblad et al. 2016), suggesting the possibility that APCAs can directly stimulate signaling processes through triggering their target molecules on the cell surface. Such a scenario is also supported by the increased OC differentiation induced in vitro in response to Fab fragments generated from polyclonal ACPAs (Krishnamurthy et al. 2016). The types and amounts of different citrullinated proteins exposed on the cell surface could greatly influence how ACPAs influence cellular functions. Upon inflammatory stimuli or PMA-mediated activation, the glycolytic enzyme α-enolase can be redistributed to the cell surface on various immune cells, where it acts as a plasminogen-binding receptor. Circulating monocytes have been shown to express α-enolase on the cell surface in RA patients, and the binding of anti-α-enolase antibodies, or the natural ligand plasminogen, to these molecules induced the production of various inflammatory cytokines (Bae et al. 2012). Citrullinated α-enolase is targeted by several ACPAs, indicating that the citrullination and cell surface exposure of this molecule might provide important

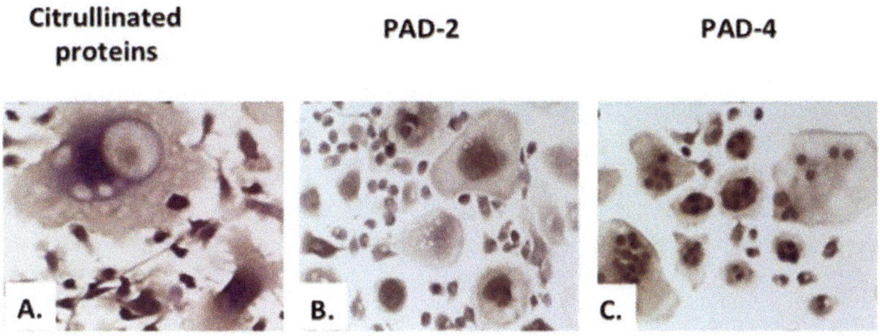

Fig. 6.2 Expression of citrullinated proteins (**a**), PAD-2 (**b**), and PAD-4 (**c**) in mature osteoclasts

pro-inflammatory targets for autoantibodies. The intermediary filament protein vimentin can also be expressed on the surface of macrophages and developing OCs (Harre et al. 2012; Mor-Vaknin et al. 2003), and vimentin secretion was increased in macrophages by tumor necrosis factor (TNF), suggesting increased target availability for vimentin-binding antibodies in inflamed tissues (Mor-Vaknin et al. 2003). As discussed earlier, ACPA binding to citrullinated vimentin increases the differentiation of OCs through induced cytokine production; however, the signaling mechanisms triggered by vimentin binding have not yet been clarified. In addition to direct effects on cell surface molecules, ACPAs might also present in immune complexes and act through Fc receptor-mediated mechanisms. Immune complexes containing citrullinated fibrinogen and ACPAs have been detected in a large number of RA patients (Zhao et al. 2008), and citrullinated fibrinogen-ACPA complexes were shown to increase TNF production in macrophage cultures through a mechanism mediated by FcγRs and TLR4 (Sokolove et al. 2011). The combination of ACPAs and IgM rheumatoid factor antibodies was associated with higher serum concentration of several inflammatory cytokines in RA patients, as compared to single-positive individuals, and rheumatoid factors could further increase macrophage activation induced by citrullinated fibrinogen-ACPA complexes in vitro (Sokolove et al. 2014). The several target molecules and signaling mechanisms potentially triggered by ACPAs suggest a complex relationship between anti-citrulline responses and RA development (Fig. 6.3). ACPAs might trigger diverse pro-inflammatory and bone erosive pathways in the myeloid cells, which might be variable between patients or disease stages due to differences in citrullination levels and in the targeted proteins. Nevertheless, the existing data points to similar functional effects of the ACPAs on the cells of the macrophage-OC lineage, namely, an increase in inflammatory cytokine production that subsequently leads to increased OC differentiation, bone erosion, and pain.

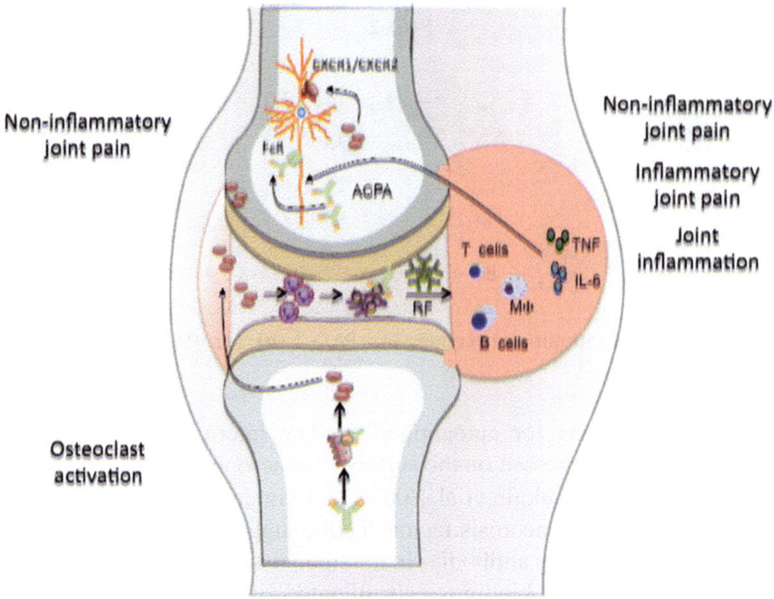

Fig. 6.3 The figure presents a hypothetical model on the pathogenic role of anti-citrullinated antibodies in rheumatoid arthritis. *ACPA* anti-citrullinated proteins antibodies, *OC* osteoclasts, *TNF* tumor necrosis factor, *IL-8* interleukin 8, *MF* macrophages

6.5 Summary

The specific interaction between genes and environment is crucial for the development of ACPA and ACPA-positive RA. In genetically susceptible hosts, environmental challenges at mucosal sites (such as smoking) might lead to posttranslational changes and the formation of neoantigens with tolerance breaking and the generation of autoantibodies. These antibodies emerge several years before disease onset to later become pathogenic once localized in the joint compartment. Recent findings suggest that osteoclasts might be responsible for this localization due to their dependency on citrullination for their physiological maturation and development. Osteoclast targeting by ACPA initiates an IL-8 autocrine loop leading first to bone loss and pain and only later on synovial inflammation. Future research needs to address the molecular mechanisms and intracellular signals responsible for the ACPA effects and to further dissect the transition from ACPA-dependent bone loss and pain to overt joint inflammation.

References

Aho, K., Heliovaara, M., Maatela, J., Tuomi, T., & Palosuo, T. (1991). Rheumatoid factors antedating clinical rheumatoid arthritis. *The Journal of Rheumatology, 18*(9), 1282–1284.

Arnson, Y., Shoenfeld, Y., & Amital, H. (2010). Effects of tobacco smoke on immunity, inflammation and autoimmunity. *Journal of Autoimmunity, 34*(3), J258–J265.

Bae, S., Kim, H., Lee, N., Won, C., Kim, H. R., Hwang, Y. I., et al. (2012). Alpha-enolase expressed on the surfaces of monocytes and macrophages induces robust synovial inflammation in rheumatoid arthritis. *Journal of Immunology, 189*(1), 365–372.

Beniac, D. R., Wood, D. D., Palaniyar, N., Ottensmeyer, F. P., Moscarello, M. A., & Harauz, G. (2000). Cryoelectron microscopy of protein-lipid complexes of human myelin basic protein charge isomers differing in degree of citrullination. *Journal of Structural Biology, 129*(1), 80–95.

Boyesen, P., Haavardsholm, E. A., van der Heijde, D., Ostergaard, M., Hammer, H. B., Sesseng, S., et al. (2011). Prediction of MRI erosive progression: A comparison of modern imaging modalities in early rheumatoid arthritis patients. *Annals of the Rheumatic Diseases, 70*(1), 176–179.

Brink, M., Hansson, M., Mathsson, L., Jakobsson, P. J., Holmdahl, R., Hallmans, G., et al. (2013). Multiplex analyses of antibodies against citrullinated peptides in individuals prior to development of rheumatoid arthritis. *Arthritis and Rheumatism, 65*(4), 899–910.

Cantaert, T., De Rycke, L., Bongartz, T., Matteson, E. L., Tak, P. P., Nicholas, A. P., et al. (2006). Citrullinated proteins in rheumatoid arthritis: Crucial...But not sufficient! *Arthritis and Rheumatism, 54*(11), 3381–3389.

Catrina, A. I., Ytterberg, A. J., Reynisdottir, G., Malmstrom, V., & Klareskog, L. (2014). Lungs, joints and immunity against citrullinated proteins in rheumatoid arthritis. *Nature Reviews Rheumatology, 10*(11), 645–653.

Chatzidionisyou, A., & Catrina, A. I. (2016). The lung in rheumatoid arthritis, cause or consequence? *Current Opinion in Rheumatology, 28*(1), 76–82.

Chibnik, L. B., Mandl, L. A., Costenbader, K. H., Schur, P. H., & Karlson, E. W. (2009). Comparison of threshold cutpoints and continuous measures of anti-cyclic citrullinated peptide antibodies in predicting future rheumatoid arthritis. *The Journal of Rheumatology, 36*(4), 706–711.

Christophorou, M. A., Castelo-Branco, G., Halley-Stott, R. P., Oliveira, C. S., Loos, R., Radzisheuskaya, A., et al. (2014). Citrullination regulates pluripotency and histone H1 binding to chromatin. *Nature, 507*(7490), 104–108.

Demoruelle, M. K., Weisman, M. H., Simonian, P. L., Lynch, D. A., Sachs, P. B., Pedraza, I. F., et al. (2012). Brief report: Airways abnormalities and rheumatoid arthritis-related autoantibodies in subjects without arthritis: Early injury or initiating site of autoimmunity? *Arthritis and Rheumatism, 64*(6), 1756–1761.

Ding, B., Padyukov, L., Lundstrom, E., Seielstad, M., Plenge, R. M., Oksenberg, J. R., et al. (2009). Different patterns of associations with anti-citrullinated protein antibody-positive and anti-citrullinated protein antibody-negative rheumatoid arthritis in the extended major histocompatibility complex region. *Arthritis and Rheumatism, 60*(1), 30–38.

Fischer, A., Solomon, J. J., du Bois, R. M., Deane, K. D., Olson, A. L., Fernandez-Perez, E. R., et al. (2012). Lung disease with anti-CCP antibodies but not rheumatoid arthritis or connective tissue disease. *Respiratory Medicine, 106*(7), 1040–1047.

Haavardsholm, E. A., Boyesen, P., Ostergaard, M., Schildvold, A., & Kvien, T. K. (2008). Magnetic resonance imaging findings in 84 patients with early rheumatoid arthritis: Bone marrow oedema predicts erosive progression. *Annals of the Rheumatic Diseases, 67*(6), 794–800.

Harding, C. R., & Scott, I. R. (1983). Histidine-rich proteins (filaggrins): Structural and functional heterogeneity during epidermal differentiation. *Journal of Molecular Biology, 170*(3), 651–673.

Harre, U., Georgess, D., Bang, H., Bozec, A., Axmann, R., Ossipova, E., et al. (2012). Induction of osteoclastogenesis and bone loss by human autoantibodies against citrullinated vimentin. *The Journal of Clinical Investigation, 122*(5), 1791–1802.

Harvey, G. P., Fitzsimmons, T. R., Dhamarpatni, A. A., Marchant, C., Haynes, D. R., & Bartold, P. M. (2013). Expression of peptidylarginine deiminase-2 and -4, citrullinated proteins and anti-citrullinated protein antibodies in human gingiva. *Journal of Periodontal Research, 48*(2), 252–261.

Hecht, C., Englbrecht, M., Rech, J., Schmidt, S., Araujo, E., Engelke, K., et al. (2015). Additive effect of anti-citrullinated protein antibodies and rheumatoid factor on bone erosions in patients with RA. *Annals of the Rheumatic Diseases, 74*(12), 2151–2156.

Hetland, M. L., Ejbjerg, B., Horslev-Petersen, K., Jacobsen, S., Vestergaard, A., Jurik, A. G., et al. (2009). MRI bone oedema is the strongest predictor of subsequent radiographic progression in early rheumatoid arthritis. Results from a 2-year randomised controlled trial (CIMESTRA). *Annals of the Rheumatic Diseases, 68*(3), 384–390.

Huizinga, T. W., Amos, C. I., van der Helm-van Mil, A. H., Chen, W., van Gaalen, F. A., Jawaheer, D., et al. (2005a). Refining the complex rheumatoid arthritis phenotype based on specificity of the HLA-DRB1 shared epitope for antibodies to citrullinated proteins. *Arthritis and Rheumatism, 52*(11), 3433–3438.

Innala, L., Kokkonen, H., Eriksson, C., Jidell, E., Berglin, E., & Dahlqvst, S. R. (2008). Antibodies against mutated citrullinated vimentin are a better predictor of disease activity at 24 months in early rheumatoid arthritis than antibodies against cyclic citrullinated peptides. *The Journal of Rheumatology, 35*(6), 1002–1008.

Janssen, K. M., de Smit, M. J., Brouwer, E., de Kok, F. A., Kraan, J., Altenburg, J., et al. (2015). Rheumatoid arthritis-associated autoantibodies in non-rheumatoid arthritis patients with mucosal inflammation: A case-control study. *Arthritis Research & Therapy, 17*, 174.

Jiang, X., Alfredsson, L., Klareskog, L., & Bengtsson, C. (2014). Smokeless tobacco (moist snuff) use and the risk of developing rheumatoid arthritis: Results from a case-control study. *Arthritis Care & Research, 66*(10), 1582–1586.

Karlson, E. W., Chang, S. C., Cui, J., Chibnik, L. B., Fraser, P. A., De Vivo, I., et al. (2010). Gene-environment interaction between HLA-DRB1 shared epitope and heavy cigarette smoking in predicting incident rheumatoid arthritis. *Annals of the Rheumatic Diseases, 69*(1), 54–60.

Klareskog, L., Stolt, P., Lundberg, K., Kallberg, H., Bengtsson, C., Grunewald, J., et al. (2006a). A new model for an etiology of rheumatoid arthritis: Smoking may trigger HLA-DR (shared epitope)-restricted immune reactions to autoantigens modified by citrullination. *Arthritis and Rheumatism, 54*(1), 38–46.

Klareskog, L., Catrina, A. I., & Paget, S. (2009). Rheumatoid arthritis. *Lancet, 373*(9664), 659–672.

Klein, F., Diskin, R., Scheid, J. F., Gaebler, C., Mouquet, H., Georgiev, I. S., et al. (2013). Somatic mutations of the immunoglobulin framework are generally required for broad and potent HIV-1 neutralization. *Cell, 153*(1), 126–138.

Kleyer, A., Finzel, S., Rech, J., Manger, B., Krieter, M., Faustini, F., et al. (2014). Bone loss before the clinical onset of rheumatoid arthritis in subjects with anticitrullinated protein antibodies. *Annals of the Rheumatic Diseases, 73*(5), 854–860.

Krishnamurthy, A., Joshua, V., Haj Hensvold, A., Jin, T., Sun, M., Vivar, N., et al. (2016). Identification of a novel chemokine-dependent molecular mechanism underlying rheumatoid arthritis-associated autoantibody-mediated bone loss. *Annals of the Rheumatic Diseases, 75*(4), 721–729.

Kurki, P., Aho, K., Palosuo, T., & Heliovaara, M. (1992). Immunopathology of rheumatoid arthritis. Antikeratin antibodies precede the clinical disease. *Arthritis and Rheumatism, 35*(8), 914–917.

Larsen, J. M., Steen-Jensen, D. B., Laursen, J. M., Sondergaard, J. N., Musavian, H. S., Butt, T. M., et al. (2012). Divergent pro-inflammatory profile of human dendritic cells in response to commensal and pathogenic bacteria associated with the airway microbiota. *PloS One, 7*(2), e31976.

Liu, X., Zou, Q., Zeng, B., Fang, Y., & Wei, H. (2013). Analysis of fecal lactobacillus community structure in patients with early rheumatoid arthritis. *Current Microbiology, 67*(2), 170–176.

Lugli, E. B., Correia, R., Fischer, R., Lundberg, K., Bracke, K. R., Montgomery, A. B., et al. (2015). Expression of citrulline and homocitrulline residues in the lungs of non-smokers and smokers: Implications for autoimmunity in rheumatoid arthritis. *Arthritis Research & Therapy, 17*(1), 9.

Lundberg, K., Bengtsson, C., Kharlamova, N., Reed, E., Jiang, X., Kallberg, H., et al. (2013). Genetic and environmental determinants for disease risk in subsets of rheumatoid arthritis defined by the anticitrullinated protein/peptide antibody fine specificity profile. *Annals of the Rheumatic Diseases, 72*(5), 652–658.

Lundstrom, E., Kallberg, H., Alfredsson, L., Klareskog, L., & Padyukov, L. (2009). Gene-environment interaction between the DRB1 shared epitope and smoking in the risk of anti-citrullinated protein antibody-positive rheumatoid arthritis. *Arthritis and Rheumatism, 60*(6), 1597–1603.

Majka, D. S., Deane, K. D., Parrish, L. A., Lazar, A. A., Baron, A. E., Walker, C. W., et al. (2008). Duration of preclinical rheumatoid arthritis-related autoantibody positivity increases in subjects with older age at time of disease diagnosis. *Annals of the Rheumatic Diseases, 67*(6), 801–807.

Makrygiannakis, D., af Klint, E., Lundberg, I. E., Lofberg, R., Ulfgren, A. K., Klareskog, L., et al. (2006). Citrullination is an inflammation-dependent process. *Annals of the Rheumatic Diseases, 65*(9), 1219–1222.

Makrygiannakis, D., Hermansson, M., Ulfgren, A. K., Nicholas, A. P., Zendman, A. J., Eklund, A., et al. (2008). Smoking increases peptidylarginine deiminase 2 enzyme expression in human lungs and increases citrullination in BAL cells. *Annals of the Rheumatic Diseases, 67*(10), 1488–1492.

McQueen, F. M., Benton, N., Perry, D., Crabbe, J., Robinson, E., Yeoman, S., et al. (2003). Bone edema scored on magnetic resonance imaging scans of the dominant carpus at presentation predicts radiographic joint damage of the hands and feet six years later in patients with rheumatoid arthritis. *Arthritis and Rheumatism, 48*(7), 1814–1827.

Mikuls, T. R., Thiele, G. M., Deane, K. D., Payne, J. B., O'Dell, J. R., Yu, F., et al. (2012). Porphyromonas gingivalis and disease-related autoantibodies in individuals at increased risk of rheumatoid arthritis. *Arthritis and Rheumatism, 64*(11), 3522–3530.

Mor-Vaknin, N., Punturieri, A., Sitwala, K., & Markovitz, D. M. (2003). Vimentin is secreted by activated macrophages. *Nature Cell Biology, 5*(1), 59–63.

Moscarello, M. A., Wood, D. D., Ackerley, C., & Boulias, C. (1994). Myelin in multiple sclerosis is developmentally immature. *The Journal of Clinical Investigation, 94*(1), 146–154.

Mustila, A., Korpela, M., Haapala, A. M., Kautiainen, H., Laasonen, L., Mottonen, T., et al. (2011). Anti-citrullinated peptide antibodies and the progression of radiographic joint erosions in patients with early rheumatoid arthritis treated with FIN-RACo combination and single disease-modifying antirheumatic drug strategies. *Clinical and Experimental Rheumatology, 29*(3), 500–505.

Nakashima, K., Arai, S., Suzuki, A., Nariai, Y., Urano, T., Nakayama, M., et al. (2013). PAD4 regulates proliferation of multipotent haematopoietic cells by controlling c-myc expression. *Nature Communications, 4*, 1836.

Nesse, W., Westra, J., van der Wal, J. E., Abbas, F., Nicholas, A. P., Vissink, A., et al. (2012). The periodontium of periodontitis patients contains citrullinated proteins which may play a role in ACPA (anti-citrullinated protein antibody) formation. *Journal of Clinical Periodontology, 39*(7), 599–607.

Nielen, M. M., van Schaardenburg, D., Reesink, H. W., van de Stadt, R. J., van der Horst-Bruinsma, I. E., de Koning, M. H., et al. (2004). Specific autoantibodies precede the symptoms of rheumatoid arthritis: A study of serial measurements in blood donors. *Arthritis and Rheumatism, 50*(2), 380–386.

Padyukov, L., Silva, C., Stolt, P., Alfredsson, L., & Klareskog, L. (2004). A gene-environment interaction between smoking and shared epitope genes in HLA-DR provides a high risk of seropositive rheumatoid arthritis. *Arthritis and Rheumatism, 50*(10), 3085–3092.

Pearton, D. J., Dale, B. A., & Presland, R. B. (2002). Functional analysis of the profilaggrin N-terminal peptide: Identification of domains that regulate nuclear and cytoplasmic distribution. *The Journal of Investigative Dermatology, 119*(3), 661–669.

Pedersen, M., Jacobsen, S., Klarlund, M., Pedersen, B. V., Wiik, A., Wohlfahrt, J., et al. (2006). Environmental risk factors differ between rheumatoid arthritis with and without auto-antibodies against cyclic citrullinated peptides. *Arthritis Research & Therapy, 8*(4), R133.

Pedersen, M., Jacobsen, S., Garred, P., Madsen, H. O., Klarlund, M., Svejgaard, A., et al. (2007). Strong combined gene-environment effects in anti-cyclic citrullinated peptide-positive rheumatoid arthritis: A nationwide case-control study in Denmark. *Arthritis and Rheumatism, 56*(5), 1446–1453.

Perry, E., Stenton, C., Kelly, C., Eggleton, P., Hutchinson, D., & De Soyza, A. (2014). RA autoantibodies as predictors of rheumatoid arthritis in non-cystic fibrosis bronchiectasis patients. *The European Respiratory Journal, 44*(4), 1082–1085.

Rantapaa-Dahlqvist, S., de Jong, B. A., Berglin, E., Hallmans, G., Wadell, G., Stenlund, H., et al. (2003). Antibodies against cyclic citrullinated peptide and IgA rheumatoid factor predict the development of rheumatoid arthritis. *Arthritis and Rheumatism, 48*(10), 2741–2749.

Reynisdottir, G., Karimi, R., Joshua, V., Olsen, H., Hensvold, A. H., Harju, A., et al. (2014). Structural changes and antibody enrichment in the lungs are early features of anti-citrullinated protein antibody-positive rheumatoid arthritis. *Arthritis & Rheumatology, 66*(1), 31–39.

Reynisdottir, G., Olsen, H., Joshua, V., Engström, M., Forsslund, H., Karimi, R., Sköld, C. M., Nyren, S., Eklund, A., Grunewald, J., Catrina, A. I., et al. (2016). Signs of immune activation and local inflammation are present in the bronchial tissue of patients with untreated early rheumatoid arthritis. *Annals of the Rheumatic Diseases, 75*(9), 1722–1727.

Ronnelid, J., Wick, M. C., Lampa, J., Lindblad, S., Nordmark, B., Klareskog, L., et al. (2005). Longitudinal analysis of citrullinated protein/peptide antibodies (anti-CP) during 5 year follow up in early rheumatoid arthritis: Anti-CP status predicts worse disease activity and greater radiological progression. *Annals of the Rheumatic Diseases, 64*(12), 1744–1749.

Scally, S. W., Petersen, J., Law, S. C., Dudek, N. L., Nel, H. J., Loh, K. L., et al. (2013). A molecular basis for the association of the HLA-DRB1 locus, citrullination, and rheumatoid arthritis. *The Journal of Experimental Medicine, 210*(12), 2569–2582.

Scher, J. U., Ubeda, C., Equinda, M., Khanin, R., Buischi, Y., Viale, A., et al. (2012). Periodontal disease and the oral microbiota in new-onset rheumatoid arthritis. *Arthritis and Rheumatism, 64*(10), 3083–3094.

Scott, I. R., Harding, C. R., & Barrett, J. G. (1982). Histidine-rich protein of the keratohyalin granules. Source of the free amino acids, urocanic acid and pyrrolidone carboxylic acid in the stratum corneum. *Biochimica et Biophysica Acta, 719*(1), 110–117.

Senshu, T., Kan, S., Ogawa, H., Manabe, M., & Asaga, H. (1996). Preferential deimination of keratin K1 and filaggrin during the terminal differentiation of human epidermis. *Biochemical and Biophysical Research Communications, 225*(3), 712–719.

Senshu, T., Akiyama, K., & Nomura, K. (1999). Identification of citrulline residues in the V subdomains of keratin K1 derived from the cornified layer of newborn mouse epidermis. *Experimental Dermatology, 8*(5), 392–401.

Shi, J., van de Stadt, L. A., Levarht, E. W., Huizinga, T. W., Hamann, D., van Schaardenburg, D., et al. (2014). Anti-carbamylated protein (anti-CarP) antibodies precede the onset of rheumatoid arthritis. *Annals of the Rheumatic Diseases, 73*(4), 780–783.

Slade, D. J., Subramanian, V., & Thompson, P. R. (2014). Pluripotency: Citrullination unravels stem cells. *Nature Chemical Biology, 10*(5), 327–328.

Sokolove, J., Zhao, X., Chandra, P. E., & Robinson, W. H. (2011). Immune complexes containing citrullinated fibrinogen costimulate macrophages via toll-like receptor 4 and Fcgamma receptor. *Arthritis and Rheumatism, 63*(1), 53–62.

Sokolove, J., Johnson, D. S., Lahey, L. J., Wagner, C. A., Cheng, D., Thiele, G. M., et al. (2014). Rheumatoid factor as a potentiator of anti-citrullinated protein antibody-mediated inflammation in rheumatoid arthritis. *Arthritis & Rheumatology, 66*(4), 813–821.

Stolt, P., Källberg, H., Lundberg, I., Sjögren, B., Klareskog, L., Alfredsson, L., et al. (2005). EIRA study group. *Annals of the Rheumatic Diseases, 64*(4), 582–586.

Syversen, S. W., Goll, G. L., van der Heijde, D., Landewe, R., Lie, B. A., Odegard, S., et al. (2010a). Prediction of radiographic progression in rheumatoid arthritis and the role of antibodies against mutated citrullinated vimentin: Results from a 10-year prospective study. *Annals of the Rheumatic Diseases, 69*(2), 345–351.

Too, C. L., Yahya, A., Murad, S., Dhaliwal, J. S., Larsson, P. T., Muhamad, N. A., et al. (2012). Smoking interacts with HLA-DRB1 shared epitope in the development of anti-citrullinated protein antibody-positive rheumatoid arthritis: Results from the Malaysian epidemiological Investigation of rheumatoid arthritis (MyEIRA). *Arthritis Research & Therapy, 14*(2), R89.

Too, C.L., Muhamad, N.A., Ilar, A., Padyukov, L,, Alfredsson, L., Klareskog, L., Murad, S., Bengtsson, C., et al. (2016). MyEIRA study group. *Annals of the Rheumatic Diseases, 75*(6), 997–1002.

Vaahtovuo, J., Munukka, E., Korkeamaki, M., Luukkainen, R., & Toivanen, P. (2008). Fecal microbiota in early rheumatoid arthritis. *The Journal of Rheumatology, 35*(8), 1500–1505.

van der Helm-van Mil, A. H., Verpoort, K. N., Breedveld, F. C., Toes, R. E., & Huizinga, T. W. (2005). Antibodies to citrullinated proteins and differences in clinical progression of rheumatoid arthritis. *Arthritis Research & Therapy, 7*(5), R949–R958.

van Steenbergen, H. W., Ajeganova, S., Forslind, K., Svensson, B., & van der Helm-van Mil, A. H. (2015). The effects of rheumatoid factor and anticitrullinated peptide antibodies on bone erosions in rheumatoid arthritis. *Annals of the Rheumatic Diseases, 74*(1), e3.

Vossenaar, E. R., Smeets, T. J. M., Kraan, M. C., Raats, J. M., van Venrooij, W. J., & Tak, P. P. (2004a). The presence of citrullinated proteins is not specific for rheumatoid synovial tissue. *Arthritis and Rheumatism, 50*(11), 3485–3494.

Vossenaar, E. R., Radstake, T. R., van der Heijden, A., van Mansum, M. A., Dieteren, C., de Rooij, D.-J., et al. (2004b). Expression and activity of citrullinating peptidylarginine deiminase enzymes in monocytes and macrophages. *Annals of the Rheumatic Diseases, 63*(4), 373–381.

Wang, Y., Li, M., Stadler, S., Correll, S., Li, P., Wang, D., et al. (2009). Histone hypercitrullination mediates chromatin decondensation and neutrophil extracellular trap formation. *The Journal of Cell Biology, 184*(2), 205–213.

Wegner, N., Wait, R., Sroka, A., Eick, S., Nguyen, K. A., Lundberg, K., et al. (2010). Peptidylarginine deiminase from Porphyromonas gingivalis citrullinates human fibrinogen and alpha-enolase: Implications for autoimmunity in rheumatoid arthritis. *Arthritis and Rheumatism, 62*(9), 2662–2672.

Wigerblad, G., Bas, D. B., Fernades-Cerqueira, C., Krishnamurthy, A., Nandakumar, K. S., Rogoz, K., et al. (2016). Autoantibodies to citrullinated proteins induce joint pain independent of inflammation via a chemokine-dependent mechanism. *Annals of the Rheumatic Diseases, 75*(4), 730–738.

Willis, V. C., Demoruelle, M. K., Derber, L. A., Chartier-Logan, C. J., Parish, M. C., Pedraza, I. F., et al. (2013). Sputum autoantibodies in patients with established rheumatoid arthritis and subjects at risk of future clinically apparent disease. *Arthritis and Rheumatism, 65*(10), 2545–2554.

Ytterberg, A. J., Joshua, V., Reynisdottir, G., Tarasova, N. K., Rutishauser, D., Ossipova, E., et al. (2015). Shared immunological targets in the lungs and joints of patients with rheumatoid arthritis: Identification and validation. *Annals of the Rheumatic Diseases, 74*(9), 1772–1777.

Zhao, X., Okeke, N. L., Sharpe, O., Batliwalla, F. M., Lee, A. T., Ho, P. P., et al. (2008). Circulating immune complexes contain citrullinated fibrinogen in rheumatoid arthritis. *Arthritis Research & Therapy, 10*(4), R94.

Chapter 7
Porphyromonas gingivalis Peptidyl Arginine Deiminase: A Unique Bacterial PAD with Implications for Periodontal Disease and Rheumatoid Arthritis

Katarzyna Gawron, Anna Montgomery, Katarzyna Łazarz-Bartyzel, Grzegorz Bereta, Maria Chomyszyn-Gajewska, Patrick Venables, and Jan Potempa

Abbreviations

ACPA	Anti-citrullinated protein antibodies
ADI	Arginine deiminase
AgDI	Agmatine deiminase
APD	Aggressive periodontitis
AST	Arginine succinyltransferase
AT	Antithrombin
CD	Catalytic domain
CEP-1	Citrullinated enolase peptide-1
CK	Cytokeratin
CNS	Central nervous system
COX-2	Cyclooxygenase-2

K. Gawron (✉) • G. Bereta • J. Potempa
Department of Microbiology, Faculty of Biochemistry, Biophysics and Biotechnology, Jagiellonian University, Gronostajowa 7, 30-387 Krakow, Poland
e-mail: katarzyna.gawron@uj.edu.pl; jan.potempa@uj.edu.pl

A. Montgomery • P. Venables
Kennedy Institute of Rheumatology, Nuffield Department of Orthopaedics, Rheumatology and Musculoskeletal Sciences, University of Oxford, Oxford, United Kingdom
e-mail: anna.montgomery@stx.ox.ac.uk; patrick.venables@kennedy.ox.ac.uk

K. Łazarz-Bartyzel • M. Chomyszyn-Gajewska
Department of Periodontology and Oral Medicine, Jagiellonian University, Medical College, Krakow, Poland

A.P. Nicholas et al. (eds.), *Protein Deimination in Human Health and Disease*, DOI 10.1007/978-3-319-58244-3_7

CTD	C-terminal domain
CXCL	Chemokine
DDAH	N^{ω},N^{ω}-dimethylarginine dimethylaminohydrolase
EGF	Epidermal growth factor
ER	Estrogen receptor
GATM	L-arginine-glycine amidinotransferase
GCF	Gingival crevicular fluid
GFAP	Glial fibrillary acidic protein
GMEs	Guanidino-group modifying superfamily of enzymes
HEL	Hen egg lysozyme
HLA	Human leukocyte antigen
IgLF	Immunoglobulin-like fold domain
IL	Interleukin
KB	Oral epithelial cell line
LPS	Lipopolysaccharide
MBP	Myelin basic protein
MHC	Major histocompatibility complex
mPGES-1	Microsomal prostaglandin E synthase-1
MS	Multiple sclerosis
NET	Neutrophil extracellular trap
NHANES	National Health and Nutrition Examination Survey
NSAIDs	Nonsteroidal anti-inflammatory drugs
NtSP	N-terminal signal peptide
OMVs	Outer membrane vesicles
OR	Odds ratio
P. gingivalis	*Porphyromonas gingivalis*
PAD/PADI	Mammalian peptidyl arginine deiminase
PARs	Protease-activated receptors
PD	Periodontitis
PGE2	Prostaglandin E2
PHGF	Primary human gingival fibroblasts
PorSS	Por secretion system of *P. gingivalis*
PorU	Sortase
PPAD	Peptidyl arginine deiminase from *P. gingivalis*
PTPN22	Protein-tyrosine phosphatase, non-receptor type 22
RA	Rheumatoid arthritis
REP-1	Arginine control peptide
Rgps	Cysteine proteases (gingipains) from *P. gingivalis*
SLPI	Secretory leukocyte protease inhibitor
TERT-2	Immortalized human oral keratinocytes
TNFα	Tumor necrosis factor α
TR	Thyroid receptor

7.1 Introduction

Recently, peptidyl arginine deiminase from *Porphyromonas gingivalis* (PPAD) has become a subject of an intense research (Lundberg et al. 2010; Wegner et al. 2010; Pyrc et al. 2013; Gawron et al. 2014, Quirke et al. 2014; Goulas et al. 2015; Montgomery et al. 2016). Despite catalyzing the reaction of citrullination (deimination) and being a member of the guanidino-group modifying superfamily of enzymes (GMEs), PPAD and the mammalian peptidyl arginine deiminases (PAD/PADI) are phylogenetically unrelated proteins (McGraw et al. 1999). The citrullination of host and bacterial proteins by PPAD has been implicated in the pathogenesis of periodontitis (PD) and postulated as the link between PD and rheumatoid arthritis (RA). In this chapter, we discuss the role of protein citrullination by PPAD in the pathogenesis of PD and RA.

7.2 Epidemiology, Etiology, and Clinical Features of PD

PD is an inflammatory disease of tooth-supporting tissues, which affects 5–20% adults worldwide (Petersen et al. 2005; Toh et al. 2011), and is the most common chronic inflammatory disease caused by bacterial infection (Oliver and Brown 1993; Eke et al. 2012). PD is manifested by gingival bleeding, periodontal ligament degradation, attachment loss, and alveolar bone resorption, which results in formation of periodontal pocket(s) and tooth mobility. If left untreated, the disease may lead to tooth exfoliation (Toh et al. 2011; Oliver and Brown 1993; Armitage 1995; Yucel-Lindberg and Båge 2013). The clinical and histological presentations of chronic PD are shown in Fig. 7.1.

Considering the heterogeneity of the clinical course and fluctuation of disease progression, PD is characterized as a complex and nonlinear entity. The current concept on PD etiology assumes that the disease involves multiple causal components, which interact on each other simultaneously (Heaton and Dietrich 2012). Among these are (a) subgingival bacterial biofilm on the tooth root surface and on the pocket epithelial lining (Sokransky and Haffajee 2005); (b) genetic risk factors (Heaton and Dietrich 2012; Laine et al. 2012) and epigenetic regulation (Lindroth and Park 2013); (c) age (Shaik-Dasthagirisaheb et al. 2010; Van der Velden 1984; Russell 1967); (d) lifestyle-related factors, i.e., stress, inappropriate diet, and smoking (Akcali et al. 2013; Bergström 2003; Van der Velden et al. 2011); (e) systemic diseases (Pihlstrom 2001; Chee et al. 2013); and (f) iatrogenic factors yet unknown (Gher 1998; Van der Velden et al. 2006) (Fig. 7.2). It is important to note that the relative contribution of each of the causal factors varies from patient to patient. In general, environmental and lifestyle factors, such as many years of biofilm accumulation, smoking, poor diet, and none or irregular visits to dental professionals, have a major contribution to PD in older patients. Conversely, PD in the younger population is to a greater extent dependent on genetic factors, which play an important role

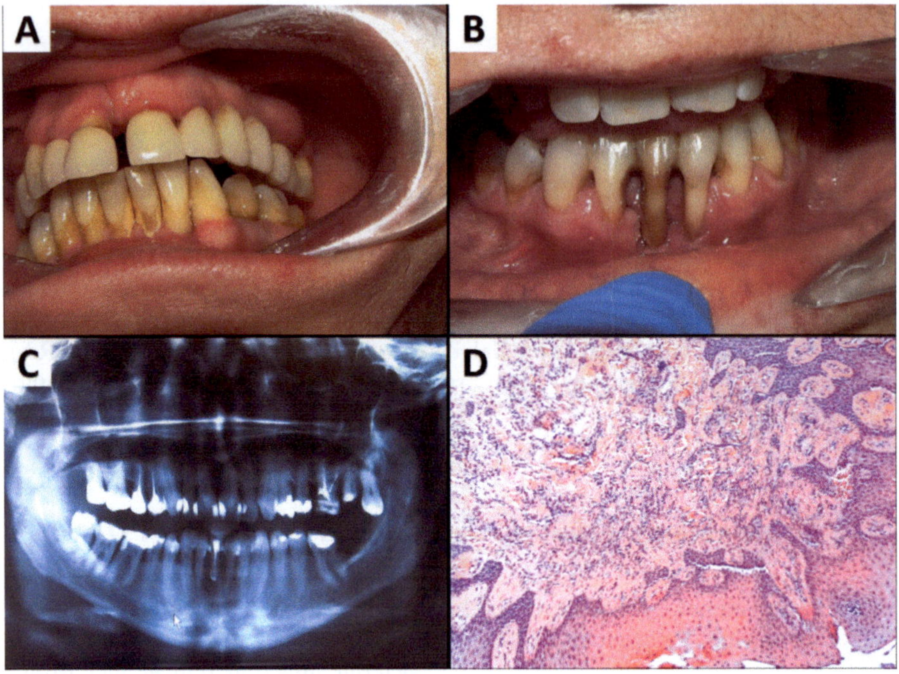

Fig. 7.1 Chronic periodontitis in two patients treated at the Department of Periodontology and Oral Medicine, Jagiellonian University, Medical College, Krakow, Poland: (**a**) clinical view of periodontium from patient 1 and (**b**) from patient 2, (**c**) X-ray showing resorption of alveolar bone (patient 2), (**d**) inflammatory infiltrates in connective tissue of periodontium (patient 2) stained by H/E, original magnific. ×100

in the pathogenesis of aggressive periodontitis (APD) (Laine et al. 2012; Mucci et al. 2005; Stabholz et al. 2010).

7.3 Oral Cavity Colonization and PD

It is estimated that the oral cavity of healthy human contains approximately six billion of microbes representing 300–500 species which constitute almost half the species of commensal bacteria in the human body. The composition of bacterial biofilm in oral cavity is site dependent and results from specific interactions between bacterial cell adhesins and complementary receptors on oral surfaces. This is further influenced by the ultrastructure and function of the oral surface (Gibbons 1989; Gibbons et al. 1976; Mager et al. 2003). In certain conditions, some opportunistic pathogens within oral cavity induce local and/or systemic infections.

Primary colonizers of the tooth surface include streptococci, such as *Streptococcus oralis*, *Streptococcus mutans*, and *Streptococcus sanguis,* as well as the *Neisseria*

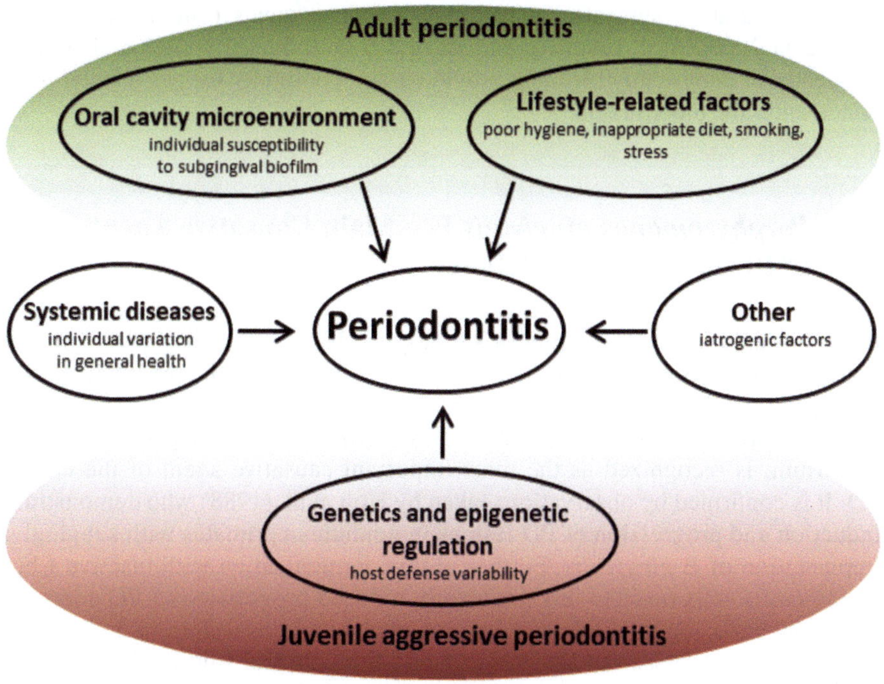

Fig. 7.2 A complex nature of periodontitis. In the onset and progression of periodontitis simultaneously interact multiple casual factors, such as oral cavity microenvironment, lifestyle, genetic and epigenetic regulation, systemic diseases, iatrogenic factors

sp. Primary colonizers are mostly Gram-positive aerobes or facultative anaerobes which provide substrates and greatly decrease the local availability of oxygen to form a suitable background for the attachment and growth of secondary colonizers that are predominantly anaerobic. Among secondary colonizers are actinomycetes (*Actinomyces odontolyticus* and *Actinomyces naeslundii*), *Veillonella* sp., and the species of the so-called orange complex, i.e., *Fusobacterium nucleatum*, *Prevotella intermedia*, *Peptostreptococcus micros*, and *Campylobacter rectus* (Socransky et al. 1998; Signat et al. 2011; Rickard et al. 2003). Further to this, the bacterial composition is also influenced by pH, diet, oral hygiene, and interactions between microbes. Overall, the oral microbial ecosystem is highly dynamic and faces a constant challenge of opportunistic infections if the symbiosis is disturbed (Jin et al. 2003). An example is the development of gingivitis, which may lead to PD.

In gingivitis, the subgingival plaque microflora is enriched with Gram-negative obligate anaerobes (Moore et al. 1987). The deepening of the gingival sulcus caused by dental plaque growth, local inflammation, and bacterially mediated destruction of the gingival fiber attachment leads to the formation of a pathological space between the tooth root and the gingiva, termed the periodontal pocket. The periodontal pocket releases gingival crevicular fluid (GCF), an inflammatory exudate containing cell debris, bacterial degradation products, inflammatory mediators, connective tissue fragments, enzymes, and other proteins (Subrahmanyam and Sangeetha 2003; Yucel-Lindberg and Båge 2013). Inflamed periodontal tissue is frequently or persistently bleeding providing an opportunity for periodontal infection to enter the bloodstream.

7.4 *Porphyromonas gingivalis* Is a Main Causative Agent of PD in Adult Population

PD is initiated by the spreading of subgingival biofilm and enhanced colonization of periodontal gingival pockets by bacterial species of the red complex, i.e., *P. gingivalis*, *Tannerella forsythia*, and *Treponema denticola* (Socransky et al. 1998; Signat et al. 2011). Among them, *P. gingivalis*, a Gram-negative, anaerobic bacterium, is recognized as the most important causative agent of the chronic PD. It is confirmed by observations taken by Holt et al. (1988) who demonstrated induction and progression of PD lesions in nonhuman primates with subgingival implantation of *P. gingivalis*. Further to this, immunization with inactive whole *P. gingivalis* cells resulted in significantly reduced progression of PD compared with indigenous or *P. gingivalis*-superinfected subgingival flora in a nonhuman primate model (Persson et al. 1994). In addition, clinical studies showed a proportional increase in the level of *P. gingivalis* in subgingival plaque which correlated with disease severity, as assessed by attachment loss, periodontal pocket depth, and bleeding on probing (Mayrand and Holt 1988; Socransky et al. 1991). Accordingly, eradication of this microorganism from the cultivable subgingival microbial population has shown to completely resolve the disease (Loesche et al. 1981; Van Dyke et al. 1988). Taken together, these and other data emanating from animal models and human clinical studies suggest that enhanced colonization of periodontal gingival pockets by *P. gingivalis*, acting possibly in concert with the absence of beneficial species and certain immunological deficiencies in the host, appears to be essential for the development, progression, and severity of the chronic PD in a middle-age population (Slots 1977; White and Mayrand 1981; Zambon et al. 1981; Holt et al. 1988; Socransky and Haffajee 1992; Socransky et al. 1998; Cekici et al. 2014; Hasturk and Kantarci 2015; Meyle and Chapple 2015).

P. gingivalis expresses several factors, i.e., lipopolysaccharide (LPS) (Wang et al. 2002), cysteine proteases (Kadowaki et al. 1994, 2004; Pathirana et al. 2007; Potempa et al. 2000; Smalley et al. 1989), outer membrane vesicles (OMVs) (Grenier and Mayrand 1987; Imamura et al. 1995), fimbriae (Weinberg et al. 1997; Davey et al. 2008), and hemolysins, which via colonization of periodontal pockets and invasion of gingival tissue cause destruction of periodontal supporting tissues (Tatakis and Kumar 2005; Hajishengallis and Lamont 2012; Lamont and Jenkinson 1998; Bostanci and Belibasakis 2012). Extracellular hydrolytic (cysteine) proteases produced by *P. gingivalis*, also called gingipains (Rgps), are the most broadly characterized virulence factor (Cutler et al. 1995; Chen et al. 1992; Imamura et al. 1994, 1995; Pike et al. 1994; Potempa et al. 1995). Rgps consist of two arginine residue-specific enzymes, termed RgpA and RgpB, and another lysine residue-specific enzyme, termed Kgp (O'Brien-Simpson et al. 2003). These enzymes play a critical role in the onset of periodontal inflammation through their proteolytic activities, which cause, among others, enhancement of vascular permeability by activation of the kallikrein/kinin pathway; dysregulation of plasma clot formation; evasion of host defense mechanisms through the degradation of immunoglobulins; activation of complement components; inactivation of secretory leukocyte protease inhibitor (SLPI), an endogenous inhibitor for neutrophil-derived proteases; and degradation of epithelial cell-cell junctional complexes (Grenier 1992; Imamura et al. 1994, 1995, 1997; Wingrove et al. 1992; Lourbakos et al. 2001b; Into et al. 2006; Katz et al. 2002). Degradation of periodontal tissue (Grenier and Mayrand 1987; Travis et al. 1997) and plasma constituents, particularly iron-binding proteins (Henskens et al. 1993; Carlsson et al. 1984) by bacterial proteinases, provides nutrients to sustain the asaccharolytic growth of *P. gingivalis*. Moreover, these enzymes have been shown to affect eukaryotic cells via cleavage of the protease-activated receptors (PARs), which are the members of the seven-transmembrane superfamily of cell surface G protein-coupled receptors (Coughlin 2000; Ossovskaya and Bunnett 2004). Hence, Rgps cleave and activate PARs on neutrophils (Lourbakos et al. 1998), platelets (Lourbakos et al. 2001b), and oral epithelial cell line (KB) (Lourbakos et al. 2001a), while selective cleavage of PAR-1 on immortalized human oral keratinocytes (TERT-2) upregulated expression of interleukins (IL), i.e., IL-1α, IL-1β, and IL-6, and tumor necrosis factor α (TNFα) (Giacaman et al. 2009). Moreover, Rgps activation of PAR-2 has been linked to inflammation and induction of alveolar bone loss in PD (Holzhausen et al. 2006). Collectively, these data have shown that the proteolytic activity from *P. gingivalis* represents one of the major virulence traits contributing to the severity and progression of PD. Recently, however, a non-proteolytic enzyme from *P. gingivalis*, named peptidyl arginine deiminase (PPAD), has attracted an interest as a novel factor of virulence of this periodontal pathogen.

7.5 Citrullination Is a Posttranslational Modification Catalyzed by Peptidyl Arginine Deiminases

7.5.1 Citrullination Catalyzed by PAD/PADI Is a Common Physiological Protein Modification

Citrullination, also called arginine deimination, is the posttranslational conversion of arginine to citrulline by PAD/PADI enzymes. In mammals, there are five known PAD isozymes, i.e., PAD-1, PAD-2, PAD-3, PAD-4, and PAD-6, which are highly conserved (~50%) (Jones et al. 2009; Vossenaar et al. 2003; Arita et al. 2004).

PAD activity is dependent on calcium, which induces a conformational change upon ligation of numerous calcium-binding sites to enable enzymatic capability of the enzyme. With the exception of PAD-6, these calcium-binding sites are conserved between PAD-1, PAD-2, PAD-3, and PAD-4 (Jones et al. 2009; Vossenaar et al. 2003; Bicker and Thompson 2013).

Given that PAD isozymes possess similar but not identical substrate specificities, it is perhaps not surprising that the major distinguishing feature of individual isoforms is their tissue distribution patterns (Knuckley et al. 2010; Vossenaar et al. 2003; Raijmakers et al. 2007). Namely, PAD-2 is found in muscle tissue and nervous system, whereas PAD-1, PAD-3, PAD-4, and PAD-6 are most predominantly expressed in the epidermis, hair follicles, immune cells, and oocytes, respectively (Vossenaar et al. 2003; Jones et al. 2009). While all of the PADs are found in the cytoplasm of a cell, PAD-4 possesses a nuclear localization domain and so is expressed in the cell nucleus, which enables the enzyme to participate in gene regulation (Vossenaar et al. 2003). It is established that PAD-4 modifies histones H3 and H4, and this modification is associated with the decreased transcription of genes under the control of a number of transcription factors, including the estrogen receptor (ER), thyroid receptor (TR), and p53 (Cuthbert et al. 2004; Wang et al. 2004; Li et al. 2008). Moreover, PAD-4-catalyzed histone hypercitrullination is essential in antibacterial neutrophil extracellular trap (NET) formation (Li et al. 2010). Evidence is emerging to suggest that PAD-4 is also present in granules, while PAD-2 in mitochondria (Jang et al. 2011; Asaga et al. 2001). Considering such a broad distribution of PADs within organs and tissues, it is not surprising in that these enzymes participate in a multitude of physiological processes, i.e., gene regulation, embryonic development, nerve myelination, cornification of the skin, fertility, and immune system regulation (Vossenaar et al. 2003; Ishida-Yamamoto et al. 2000; Steinert et al. 2003; Lamensa and Moscarello 1993; Pritzker et al. 2000; Wright et al. 2003).

In the central nervous system (CNS), mostly PAD-2 is expressed, mainly by oligodendrocytes, astrocytes, and microglial cells, and the enzyme citrullinates myelin basic protein (MBP; see Chap. 19) and glial fibrillary acidic protein (GFAP; see Chap. 20), among others (György et al. 2006). It is hypothesized that citrullination of MBP is essential in the function of the myelin sheath, as well as in the plasticity of the CNS in young age as the ratio of citrullinated MBP and total MBP changes rapidly after postnatal life (György et al. 2006).

During terminal differentiation of keratinocytes, several proteins are altered by cleavage, covalent cross-linking and citrullination that help to create a protective matrix in the skin. Cytokeratin (CK) is the main intermediate filament of keratinocytes building up keratin filaments in the skin, hair, and nail. Due to citrullination, the structure of CK is altered, which enables proteins to bind to it (Baka et al. 2012).

PADs also have been shown to regulate the innate immune response, for instance, PAD-2 and PAD-4 deiminate chemokines (CXCL), i.e., CXCL10, CXCL11, and CXCL12, which reduces their ability to trigger chemotaxis and cell signaling in vitro (Loos et al. 2008; Proost et al. 2008; Struyf et al. 2009). Citrullination of CXCL8 only moderately alters in vitro activities of this chemokine but results in a considerable reduction of glycosaminoglycan-binding properties and prevents proteolysis by plasmin or thrombin into CXCL8 (6–77). Furthermore, this modification abrogates the capacity of CXCL8 to recruit neutrophils into the peritoneal cavity, whereas it does not affect angiogenic properties (Proost et al. 2008).

7.5.2 Citrullination Catalyzed by PAD/PADI Is Implicated in the Pathogenesis of Several Diseases

In the psoriatic hyperproliferative epidermis, decreased CK1 deimination has been reported (György et al. 2006). However, a small percent of psoriatic patients with arthritis have anti-cyclic citrullinated peptide (anti-CCP) antibodies, and their affected joints often show a polyarthritic pattern similar to RA (Alenius et al. 2006).

Moscarello et al. (1994) have proposed that myelin damage in multiple sclerosis (MS) results from a failure to maintain the myelin sheath due to abnormally enhanced citrullination of MBP. The cause of hypercitrullination may be increased expression of PAD-2 and PAD-4 (Mastronardi et al. 2006) and/or hypomethylation of the PAD-2 promoter (Mastronardi et al. 2007). In PAD-2 knockout mice, CNS citrullination is diminished, and demyelination is not seen (Raijmakers et al. 2006), while in a transgenic mouse line containing multiple copies of the PAD-2 cDNA, increased severity of clinical symptoms of MS is observed in line with increased expression and activity of PAD-2, as well as MBP citrullination (Musse et al. 2008).

There is a growing body of evidence that PAD-4 and citrullination play a role in tumorigenesis. For example, Chang and Han (2006) have found high tissue expression of PAD-4 and citrullination in various malignant tumors, but not in benign tumors or non-tumorous tissues. They have also reported elevated serum PAD-4 and citrullinated antithrombin (AT) levels in patients with variable malignancies (Chang et al. 2009). The serum levels were associated with tumor markers and considerably dropped after tumor excision therapy. Citrullination of AT abolishes its activity to inhibit thrombin (Ordóñez et al. 2009); therefore, increased thrombin activity may promote angiogenesis and tumor cell invasion. Additionally, Omary et al. (1998) have found that the posttranslational modifications of CK alter the physical and chemical properties of CK. Indeed, citrullinated CK found in cancerous tissues may also interfere with cell homeostasis (Chang and Han 2006; Baka et al. 2011). Therefore, citrullination may promote tumorigenesis through PAD-4's interference with the p53 pathway, attenuating the activity of AT, and altering CK functions.

In RA, the main pathogenic autoantibodies are reactive with citrullinated proteins, termed anti-citrullinated protein antibodies (ACPA). Although the trigger for ACPA production remains unknown, it is thought that interactions between environmental and genetic factors are involved, i.e., citrullination triggered by smoking or infection may lead to ACPA formation in individuals carrying specific genetic susceptibility alleles such as the human leukocyte antigen molecules (HLA-DRB1) termed the "shared epitope" and/or general autoimmunity marker protein-tyrosine phosphatase, non-receptor type 22 (PTPN22) allele (Klareskog et al. 2006; Lee et al. 2009).

7.5.3 Citrullination Catalyzed by PPAD Is a Unique Feature of P. gingivalis

Recently, an increasing interest has focused on PPAD and its mechanisms of citrullination (Lundberg et al. 2010; Wegner et al. 2010; Pyrc et al. 2013; Maresz et al. 2013, Gawron et al. 2014; Bielecka et al. 2014; Quirke et al. 2014; Goulas et al. 2015; Montgomery et al. 2016). PPAD was first identified and purified by McGraw et al. (1999) and subsequently cloned and expressed by Rodriguez et al. (2009). To date, *P. gingivalis* expresses the only known prokaryotic PAD enzyme, which is genetically unrelated to the mammalian PADs despite functional similarity and belonging to the GMEs. PPAD catalyzes the same citrullination reaction as mammalian PADs, but in contrast to mammalian PADs, it preferentially catalyzes carboxy-terminal arginine and can also deiminate free L-arginine (McGraw et al. 1999). Moreover, PPAD does not require calcium or any specific cofactors for activity and is reportedly capable of autocitrullination (McGraw et al. 1999; Rodriguez et al. 2009; Quirke et al. 2014). Interestingly, the latter observation suggests that citrullination of internal arginine residues in a polypeptide chain cannot be excluded in spite of the preference for arginine positioned at the carboxy-terminus.

7.5.4 Biochemical Properties of PPAD

Recently the 3D crystal structure of PPAD was solved by Goulas et al. (2015) and Montgomery et al. (2016), which provided a structural perspective to understand the biochemical properties of the enzyme. As with gingipains, PPAD is processed and secreted from *P. gingivalis* via the type IX secretion system (T9SS) also referred to as "PerioGate" or the Por secretion system (PorSS). PPAD comprises the same domain structure as RgpB, which has been well characterized structurally (Zhou et al. 2013), namely, a profragment/N-terminal signal peptide (NtSP), a catalytic domain (CD), an immunoglobulin-like fold (IgLF) domain, and a C-terminal domain (CTD), which is well conserved among PorSS proteins (de Diego et al. 2016) (Fig. 7.3a). Mutagenesis and enzymatic studies have indicated key active site residues, Asp130, His236, Asp238, Asn297, and Cys351 (Rodríguez et al. 2010), which are highly conserved among GMEs (Shirai et al. 2001) (Fig. 7.3b). Other GME members include mammalian PADs, agmatine deiminase (AgDI), arginine deiminase (ADI), L-arginine-glycine amidinotransferase (GATM), N^{ω},N^{ω}-dimethylarginine dimethylaminohydrolase (DDAH), and arginine succinyltransferase (AST). Both Goulas et al. (2015) and Montgomery et al. (2016)

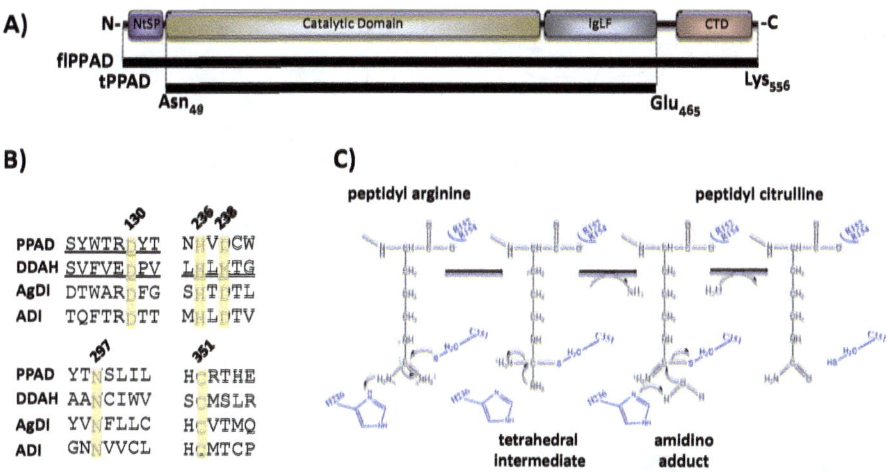

Fig. 7.3 Structural determination of PPAD: (**a**) the subdomain architecture of full length (flPPAD) and truncated (tPPAD) used for structural determination. PPAD comprises a prodomain/N-terminal signal peptide (NtSP), catalytic domain (CD), immunoglobulin-like fold (IgLF), and C-terminal domain (CTD), (**b**) the active site residues of PPAD are conserved among members of the guanidino-group modifying superfamily of enzymes (GMEs), N^{ω},N^{ω}-dimethylarginine dimethylaminohydrolase (DDAH), agmatine deiminase (AgDI), and arginine deiminase (ADI), (**c**) citrullination is catalysis of peptidyl arginine to peptidyl citrulline residues with production of an amidino adduct

determined the 3D structure of a soluble, processed form of PPAD, lacking the NtSP and CTD, which is thus comparable to mature PPAD which is secreted from the bacterium via T9SS. Posttranslational processing occurring during the secretion via T9SS is well characterized for RgpB (Zhou et al. 2013), and due to similarities in subdomain structure, PPAD is thought to undergo the same process (Bielecka et al. 2014). Briefly, the NtSP targets PPAD for translocation across the inner membrane via the Sec system. Next, the pro-peptide is cleaved off to activate the latent enzyme (Veillard et al. 2015). The CTD then targets the active protein to the T9SS, and the enzyme is translocated through the outer membrane (Sato et al. 2010; Shoji et al. 2011), during which the CTD is removed by a specialized proteolytic enzyme, sortase (PorU, PG0026) (Glew et al. 2012). This generates the 46 kDa mature enzyme (McGraw et al. 1999). The molecule of PPAD is comprised of the CD (aa 44–359) and IgLF (aa 360–465) domains (Fig. 7.4a). This two-domain structure is 55 Å × 57 Å × 50 Å ($h \times w \times d$) and has an overall shape reminiscent of Rgp and Kgp (Eichinger et al. 1999; de Diego et al. 2014). The IgLF is a 4- and 5-antiparallel strand sandwich and may play a nonenzymatic role in protein stability or anchoring to the membrane surface. The CD comprises 8 α-helices and 20 β-sheets and forms a disfigured "propeller" structure of 5 α-β folds or "blades" (Fig. 7.4b). The "blades" are connected by a series of loops, forming a teroid around central axes. This manufactures a narrow binding grove for the arginine substrate on one face and interactions with the IgLF on the other. The entrance

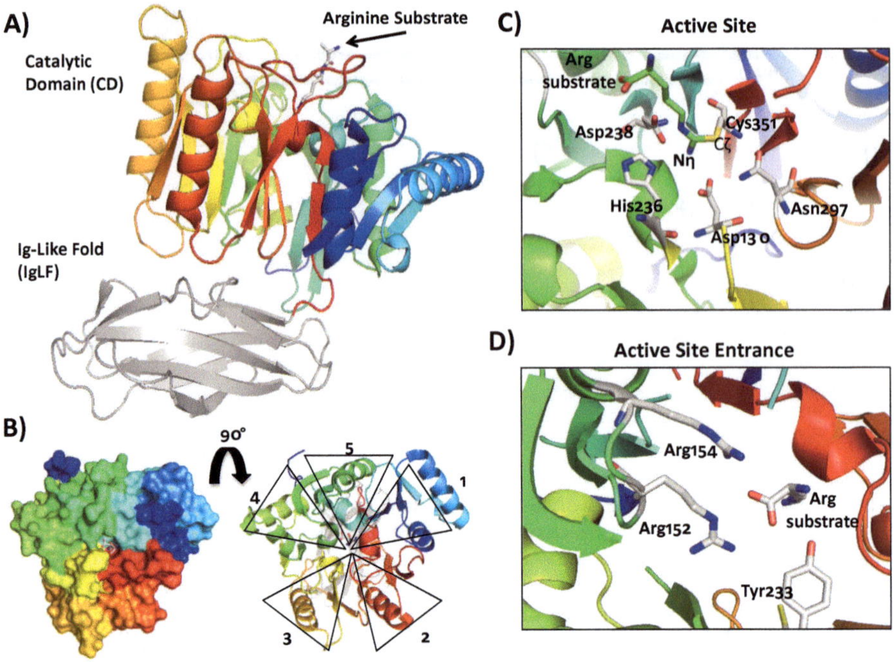

Fig. 7.4 Three-dimensional structure of PPAD: (**a**) the crystallized structure comprises catalytic domain (CD) aa 49–360 (*colored ribbons*) and immunoglobulin-like fold (IgLF) aa 361–461 (*gray ribbons*), (**b**) the catalytic domain viewed from the top face forms a disfigured five-blade propeller, with substrate binding grove at the center, (**c**) the active site with an arginine-bound ligand, (**d**) the active site entrance viewed from the top face of the catalytic domain

to the active site is gated by a surface loop (V226–V237) which appears to change enzyme confirmation to "open" or "closed" with substrate binding, demonstrated using substrate bound and unbound confirmations (Goulas et al. 2015).

Goulas et al. (2015) and Montgomery et al. (2016) reported the structures representing different states of the PPAD enzyme catalysis. The structures of a catalytically inactive mutant PPAD bound with arginine substrate (Montgomery et al. 2016) or with substrate mimic (Goulas et al. 2015) reveal ligand binding in the active site and provide information of stabilization of substrate peptide before turnover. From the structural information one can surmise, to bind substrate, the surface loop adopts the open conformation, and the arginine residue of the peptide substrate enters the active site grove, using its hydrocarbon moiety to pack against the hydrophobic cavity wall (residues Trp127 and Ile234). The guanidino group of the arginine is stabilized within the active site, comprising Asp130, His236, Asp238, Asn297, and Cys351, by ionic interactions between the two terminal guanidino nitrogens (Nη1 and Nη2) and Asp238 and Asp130 (Fig. 7.4c). Further ionic interactions between Arg152 and Arg154 with the substrate carboxyl group stabilize the arginine at the entrance to the active site (Fig. 7.4d), and the surface loop forms a "closed" conformation with further interactions between Y233 and the carboxyl group. This conformation positions the guanidino group between His236 and Cys351, where the Cζ atom of the guanidino group undergoes nucleophilic attack by the Cys351 sulfuryl group to form a tetrahedral reaction intermediate. His236 acts as a general acid/base for proton transfer and first abstracts a proton from guanidino Nη1, which in turn takes a proton from the catalytic thiol group leaving His236 deprotonated. The tetrahedral intermediate collapses to a covalent thiol intermediate with production of ammonia, which receives a proton from His236 and leaves the active site. A free water molecule enters the space left by ammonia. His236 abstracts a proton from the water molecule, which in turn performs nucleophilic attack on the thiol intermediate and then collapses to form a citrullinated product. Although not known to have a direct role in catalysis, the position of residue Asn297 within the active site suggests a role in substrate stabilization for turnover.

The key roles for active site residues identified in the 3D structure were subsequently confirmed by mutagenesis in vitro, whereby mutation of any of the five residues composing the active site to Ala, i.e., C351A, D238A, H236A, N297A, and D130A, completely abolishes the enzyme activity (Rodríguez et al. 2010; Goulas et al. 2015). PPAD displays a strong preference for C-terminal arginine, which is well documented (McGraw et al. 1999; Goulas et al. 2015; Montgomery et al. 2016). The structural basis for this specificity is due to ionic interactions between residues Arg152 and Arg154 with the arginine carboxyl group. Following identification of interactions between Arg152 and Arg154 with the arginine substrate in the structure, involvement of these residues in catalysis was confirmed using protein mutations in vitro. When Arg152 and Arg154 were mutated to Ala, both mutants have reduced activity on C-terminal arginine-containing peptides compared to wild-type enzyme. R152A displayed <20% of wild-type PPAD, and activity of R154A was reduced by 50% (Goulas et al. 2015; Montgomery et al. 2016). Thus, interactions between Arg152 and Arg154 and the arginine substrate carboxyl group are essential for full enzymatic capability, and should the arginine be located internally or at the N-terminus of the substrate peptide, these key interactions would not occur. In addition, it is proposed that Tyr233 would sterically clash with any C-terminal extension to the peptide.

When comparing the properties of PPAD to other GMEs, moderate structural conservation between the α-β propeller domains is observed, with RMSD of 2–2.4 Å. However, in the structure of both PAD-2 and PAD-4, the CD is preceded by two immunoglobulin-like domains (unrelated to the IgLF of PPAD) (Slade et al. 2015; Arita et al. 2004). PAD-2 and PAD-4 also use arginine residues to stabilize the main body of the peptide, but instead of binding with free carboxyl group as in PPAD, they interact with the upstream peptide carbonyl. The PPAD entrance loop is also lacking in PAD-2/PAD-4, with key residue Tyr233 replaced by Thr468 (PAD-2) or Ser468 (PAD-4). Taken together, these augmentations enable the binding of internal peptidyl arginine by PPAD, which is in contrast to the in vitro observation that PAD-2 and PAD-4 prefer internal arginine substrates (Nomura 1992). Conversely, the fully closed formation of other GMEs, i.e., AgDI and ADI, enables the catalysis of free L-arginine only.

7.5.5 The Facts and Hypotheses on PPAD Implication in the Pathogenesis of PD

P. gingivalis utilizes a multitude of virulence factors to evade host defenses (reviewed in Sect. 7.4), and PPAD is yet another attractive addition to this list. Citrullination of free arginine provides energy during anaerobic growth, while ammonia generated by this system enables *P. gingivalis* survival during acid cleansing cycles in the oral cavity (Casiano-Colón and Marquis 1988; Kanapka and Kleinberg 1983). In line with this finding, it has been reported that the production of ammonia and neutralization of acidity enhance the survival of *P. gingivalis* within the periodontal pocket (McGraw et al. 1999; Mangat et al. 2010) and are associated with a pathogenic effect on host cell function (Niederman et al. 1990). Further, the complementary mechanisms of Rgp (via degradation) and PPAD have been proposed as a mechanism responsible for inactivation of several plasma constituents, increased GCF flow, and *P. gingivalis* persistence in gingival pockets. In brief, Imamura et al. (1994) demonstrated that Rgps activate prekallikrein leading to the release of bradykinin. The presence of bradykinin in host tissue results in increased vascular permeability (Imamura et al. 1994), as evidenced by the development of edema and, in the case of PD, an increase in GCF flow (Travis et al. 1997). The latter effect correlates with the presence of *P. gingivalis* (Grenier and Mayrand 1987; Marsh et al. 1989) and the increased levels of proteinases from this organism in crevicular fluid (Eley and Cox 1996; Gazi et al. 1996; Potempa and Travis 1996). Therefore, an intriguing question has arisen regarding "the unknown element" of the pathologic mechanism by which *P. gingivalis* regulates the rate of the GCF flow so that it is not flushed out from the pocket due to excessive exudation of plasma. An answer was found during the experiments of digestion of peptides with Rgps preparations. Briefly, Chen et al. (1992) discovered "an activity" which altered the retention times, but not the composition of arginine carboxy-terminal products based on the Rgp digestion reactions. Moreover, the same preparations of Rgp were used for incubation with complement component C5. The recovery of C5a activity was only about 25% for C5 utilized in the experiment, suggesting some alteration to the functionally crucial carboxy-terminal arginine of C5a (Wingrove et al. 1992). Next, Hayashi et al. (1993) observed in their experiments with hemagglutinin preparation a trypsin-like proteolytic activity as well as "an activity" that produces a citrulline residue from the carboxy-terminal arginine. The peptide was cleaved by the trypsin-like activity at the internal arginine of the hemagglutinin, and the resultant peptide with the arginine at its carboxy-terminal retained an inhibitory potential. While longer incubations were conducted, the anti-hemagglutination activity of this fragment was completely inactivated due to citrullination of the arginine at the carboxy-terminus by a non-proteolytic activity within the hemagglutinin preparation.

The observations described above indicate that the deimination of arginine at the carboxy-terminus of peptides generated via the cleavage by Rgps could affect the functionality of biologically active peptides. In keeping with this hypothesis,

McGraw et al. (1999) proposed that the cause of the diminished vascular permeability enhancement response to vesicles, compared to the Rgp response (Imamura et al. 1994, 1995), resulted from the action of PPAD. This was confirmed by several recent studies. For instance, Pyrc et al. (2013) reported that PPAD efficiently deiminates the carboxy-terminal arginine of epidermal growth factor (EGF), and this modification subsequently impairs biological activity of the cytokine. Decreased activity of EGF in gingival pockets may at least partially contribute to the tissue damage and delayed healing in the periodontium in PD.

Bielecka et al. (2014) studied the effect of bacterial citrullination on pro-inflammatory functions of anaphylatoxin C5a. Although, previous studies (Wingrove et al. 1992) showed a decrease of C5a activity after incubation with Rgps, no functional explanation of this phenomenon was offered. Anaphylatoxin C5a is released from C5 by C5 convertase during complement activation. Widespread expression of two C5a receptors, i.e., C5aR and C5L2, in the human body ensures a wide spectrum of biological responses, including chemotaxis of inflammatory cells, phagocytosis, respiratory burst, vascular permeability, and releases of pro-inflammatory cytokines and chemokines. The carboxy-terminal arginine residue is crucial for the C5a function, and in vivo, the molecule is rapidly converted by carboxypeptidases to the far less potent C5a-desArg with significantly lower affinity for C5aR (Monk et al. 2007). It has been reported that preincubation of C5a with PPAD strongly reduces its chemotactic activity for neutrophils, in a concentration-dependent manner, whereas it has no effect on the activity of C5a-desArg. This result suggests that citrullinated form of C5a, similar to C5a-desArg, has significantly lower affinity for C5aR on neutrophils. In agreement with this finding, treatment of C5a with PPAD also impairs its ability to induce calcium influx in a myeloid-derived cell line transfected with C5aR. Notably, at higher concentrations and/or prolonged incubation with bacterial enzyme, the capacity of C5a to activate neutrophils and U937-C5aR cells has been completely abrogated. In contrast, C5a-desArg treated with PPAD did not show altered potential to stimulate neutrophil chemotaxis and calcium release in U937-C5aR cells. As expected, native C5a-desArg had much lower activity than C5a in these assays. Overall, it has been hypothesized that this reaction can be facilitated by close proximity of Rgps and PPAD on both the *P. gingivalis* cell surface and in OMVs released by this bacterium, in the manner similar to that found during citrullination of carboxy-terminal arginine residues of fibrinogen-derived peptides generated during incubation of *P. gingivalis* with fibrinogen (Wegner et al. 2010).

In the oral cavity, the gingival epithelium acts as a barrier that prevents the intrusion of oral bacteria into subepithelial tissues (Amano 2007). Several studies showed that *P. gingivalis* is able to invade various cell lines, including epithelial cells (Sandros et al. 1993, 1994; Lamont et al. 1995; Belton et al. 1999; Rudney et al. 2001), endothelial cells (Deshpande et al. 1998; Dorn et al. 2000), and gingival fibroblasts (Amornchat et al. 2003). Recently, Gawron et al. (2014) evaluated the interaction of *P. gingivalis* with primary human gingival fibroblasts (PHGF). Authors hypothesized that the absence of citrullinated proteins on the *P. gingivalis* surface in a PPAD-null mutant and/or inability to citrullinate host cell surface proteins may affect its interaction with cells. To verify this hypothesis, first the efficiency of infection of PHGF was tested using wild-type *P. gingivalis* and an isogenic PPAD knockout strain. In all tested conditions, wild-type *P. gingivalis* adhered to and invaded PHGF more efficiently than PPAD mutant (Fig. 7.5, Table 7.1). Moreover, the addition of purified, active PPAD to the culture medium in an amount equivalent to that carried by wild-type *P. gingivalis* reestablished the adhesive/invasive phenotype of the mutant (Fig. 7.6), eliminating the possibility that the attenuated adhesion and invasion are due to a polar effect of the deletion in PPAD gene. These data indicate that PPAD is involved in the interaction of *P. gingivalis* with gingival fibroblasts and contributes to the effective adhesion to and invasion of these cells by this pathogen.

Adhesion of *P. gingivalis* to host cells is multimodal (Lamont and Jenkinson 1998) and involves a variety of cell surface and extracellular components (Cutler et al. 1995). Several groups have provided clear evidence to support the key role of *P. gingivalis* major fimbriae in adhesion to and invasion of host cells, including epithelial cells (Nakagawa et al. 2006; Isogai et al. 1988; Njoroge et al. 1997; Sojar et al. 1999, 2002). The most recent data from our collaborating laboratories showed that PPAD knockout strain attenuation of adhesion and invasion of PHGF by *P. gingivalis* is apparently dependent on citrullination of a protein(s) on the bacterial surface (unpublished data).

Gawron et al. (2014) also showed that the infection of fibroblasts with wild-type *P. gingivalis* upregulated the expression of two key enzymes that participate in prostaglandin E2 (PGE2) synthesis, cyclooxygenase-2 (COX-2) (Fig. 7.7a), and microsomal prostaglandin E synthase-1 (mPGES-1) (Fig. 7.7b). Remarkably, upregulation of both enzymes was strongly reduced in cells infected with PPAD knockout strain and a mutated strain expressing the catalytically inactive PPAD. In agreement with this, the level of synthesized PGE2 was significantly lower in the conditioned media of cells infected with the mutant strains compared to the media of cells infected with wild-type *P. gingivalis* (Fig. 7.7c). Significantly, the addition of purified PPAD into the culture media during incubation of the mutant strain with fibroblasts restored the expression of COX-2 and mPGES-1 as well as the synthesis of PGE2 to the levels comparable to those observed during infection with wild-type *P. gingivalis*. Increased levels of PGE2 are present in periodontal tissue and the GCF of patients suffering from PD (Offenbacher et al. 1986, 1993; Preshaw and Heasman 2002) and strongly correlate with disease severity as measured by attachment loss. Furthermore, the involvement of PGE2 in PD is supported by findings that treatment with nonsteroidal anti-inflammatory drugs (NSAIDs), as well as selective COX-2 inhibitors known to inhibit PGE2 synthesis, decreased PD severity as measured by alveolar bone resorption (Offenbacher et al. 1986, 1993; Noguchi and Ishikawa 2007). Therefore, citrullination-dependent activation of gingival fibroblasts, the predominant cell type in periodontal connective tissue, produces excessive amounts of prostaglandins, in particular PGE2, which may affect homeostasis in the periodontium. Taking under consideration the proven role of PGE2 in bone resorption, data published by Gawron et al. (2014) represent yet another example which implicates PPAD as an important virulence factor and valid target for drug development.

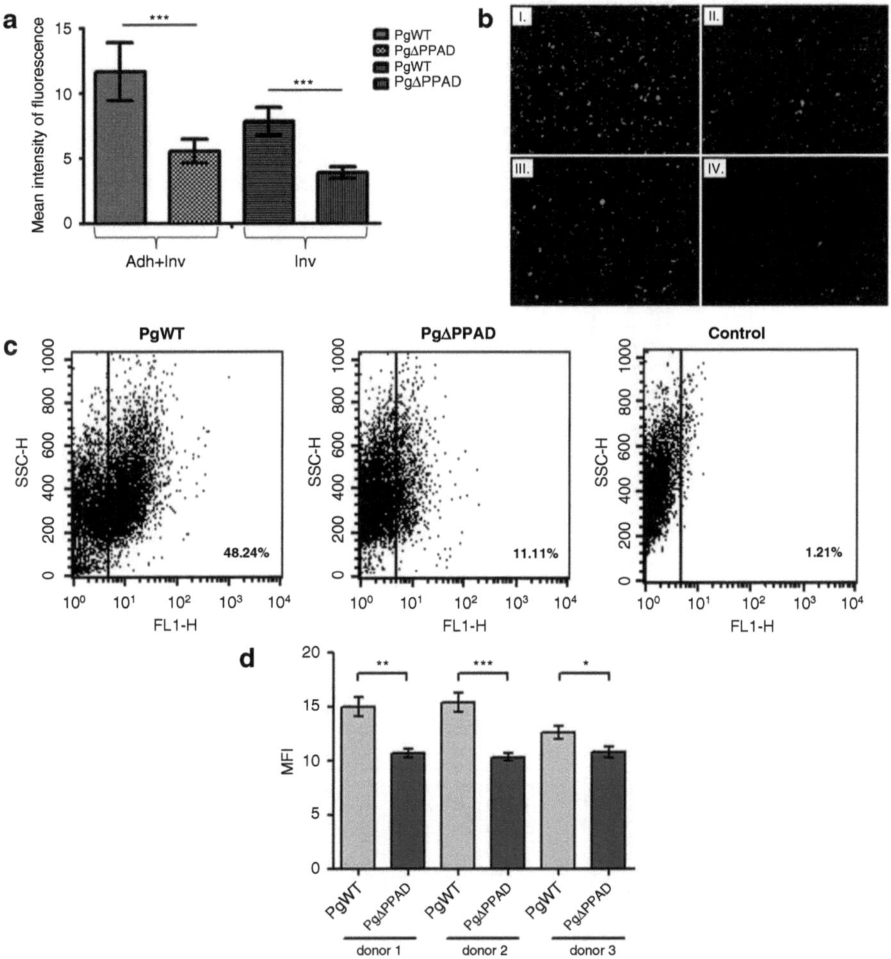

Fig. 7.5 *P. gingivalis* adheres to and invades primary human gingival fibroblasts (PHGF) more efficiently than PPAD-null mutant. Human fibroblasts were infected with CFSE-labeled wild-type *P. gingivalis* (PgWT) or the isogenic PPAD knockout strain (PgΔPPAD): (**a**) mean intensity of fluorescence for adhesion and invasion (Adh + Inv) or invasion only (Inv) within a defined area of interest (AOI) was determined and expressed as arbitrary units [*I/pix*], (**b**) representative images from microscopic analysis of Adh + Inv (I and II) and Inv (III and IV) by PgWT (I and III) and PgΔPPAD (II and IV) are shown, original magnific. ×200, (**c**) dot plots of the percentage of infected cells, and (**d**) the mean fluorescence intensity (MFI) of infection (Adh + Inv) determined by flow cytometry analysis of PHGF infected with CFSE-labeled PgWT and PgΔPPAD strains. In all experiments, PHGF were infected for 3 h at an MOI (multiplicity of infection) of 100, and all experiments were performed three times. Results are expressed as means ± SD (***, $p < 0.001$; **, $p < 0.01$; and *, $p < 0.05$) (*Reproduced from Molecular Oral Microbiology, 29(6), Gawron K, Bereta G, Nowakowska Z, Lazarz-Bartyzel K, Lazarz M, Szmigielski B, Mizgalska D, Buda A, Koziel J, Oruba Z, Chomyszyn-Gajewska M, Potempa J. Peptidylarginine deiminase from Porphyromonas gingivalis contributes to infection of gingival fibroblasts and induction of prostaglandin E2-signaling pathway. pp. 321–32, 2014, with permission from John Wiley & Sons Inc.*)

Table 7.1 Comparison of adhesion/invasion or invasion alone of primary human gingival fibroblasts (PHGF) by *P. gingivalis* (ATCC 33277): cells were infected with 10^8 CFU (MOI 100) of wild-type *P. gingivalis* (PgWT) and an isogenic PPAD knockout strain (PgΔPPAD) for 1.5, 3, and 6 h

Strain	Inoculum [CFU]	Infection time	Adhering and invading bacteria [CFU]	Adhering and invading bacteria: Student *t* test	Invading bacteria [CFU]	Invading bacteria: Student *t* test
PgWT	$1 \times 10^8 \pm 3.3 \times 10^6$	1.5 h	$1.8 \times 10^7 \pm 2.2 \times 10^6$	$p = 0.005$	$1.4 \times 10^7 \pm 2.1 \times 10^6$	$p = 0.0001$
PgΔPPAD	$1 \times 10^8 \pm 4.3 \times 10^6$	1.5 h	$1.4 \times 10^7 \pm 2.1 \times 10^6$		$0.9 \times 10^7 \pm 1.4 \times 10^6$	
PgWT	$1 \times 10^8 \pm 5 \times 10^6$	3 h	$2.6 \times 10^7 \pm 2.2 \times 10^6$	$p = 0.0004$	$2.1 \times 10^7 \pm 1.5 \times 10^6$	$p = 6.3\text{E}{-}07$
PgΔPPAD	$1 \times 10^8 \pm 6.3 \times 10^6$	3 h	$1.8 \times 10^7 \pm 2.2 \times 10^6$		$1.2 \times 10^7 \pm 0.7 \times 10^6$	
PgWT	$1 \times 10^8 \pm 6.6 \times 10^6$	6 h	$2.1 \times 10^7 \pm 1.8 \times 10^6$	$p = 0.0005$	$1.7 \times 10^7 \pm 2 \times 10^6$	$p = 0.003$
PgΔPPAD	$1 \times 10^8 \pm 6.6 \times 10^6$	6 h	$1.4 \times 10^7 \pm 2.1 \times 10^6$		$1.1 \times 10^7 \pm 1.2 \times 10^6$	

Infected cell monolayers were washed and cells were lysed. Lysates were serially diluted and plated on blood agar, and CFU were counted after 10 days of cultivation. Invasion was assessed using an antibiotic protection assay. Results are expressed as means ± SD and represent three independent experiments (*Reproduced from Molecular Oral Microbiology, 29(6), Gawron K, Bereta G, Nowakowska Z, Lazarz-Bartyzel K, Lazarz M, Szmigielski B, Mizgalska D, Buda A, Koziel J, Oruba Z, Chomyszyn-Gajewska M, Potempa J. Peptidylarginine deiminase from Porphyromonas gingivalis contributes to infection of gingival fibroblasts and induction of prostaglandin E2-signaling pathway. pp. 321–32, 2014, with permission from John Wiley & Sons Inc.)*

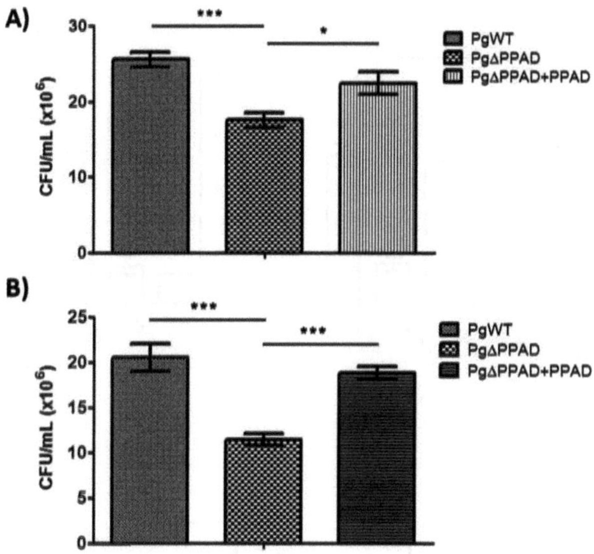

Fig. 7.6 PPAD supplementation restores the ability of *P. gingivalis* ΔPPAD to adhere to and invade primary human gingival fibroblasts (PHGF). Purified PPAD (0.066 mU/μL, total 66 mU) was added to a culture medium of PHGF cells during infection with an isogenic PPAD knockout strain (PgΔPPAD), and (**a**) adherence and invasion, and (**b**) invasion alone were determined. In all experiments, PHGF were infected for 3 h at an MOI of 100. Data represent three independent experiments and are expressed as means ± SD (***, $p < 0.001$; **, $p < 0.01$; and *, $p < 0.05$) (*Reproduced from Molecular Oral Microbiology, 29(6), Gawron K, Bereta G, Nowakowska Z, Lazarz-Bartyzel K, Lazarz M, Szmigielski B, Mizgalska D, Buda A, Koziel J, Oruba Z, Chomyszyn-Gajewska M, Potempa J. Peptidylarginine deiminase from Porphyromonas gingivalis contributes to infection of gingival fibroblasts and induction of prostaglandin E2-signaling pathway. pp. 321–32, 2014, with permission from John Wiley & Sons Inc.*)

7.6 The Associations of PD with Systemic Diseases: A General Concept

In PD, continuing destruction of the periodontal attachment and deep periodontal pockets may develop with significant loss of tooth-supporting tissues and alveolar bone. Under these conditions, the thin, highly permeable, and frequently ulcerated periodontal pocket epithelium is the only barrier between the bacterial biofilms and the underlying connective tissue. Apparently, in these conditions, subgingival bacterial plaque constitutes a substantial infectious burden for the entire body capable of releasing bacteria, bacterial toxins, and other inflammatory mediators into the bloodstream thus affecting the other parts of the body. This notion represents a paradigm shift in thinking about the directionality of oral and systemic associations (Scannapieco 1998; Li et al. 2000; Hayashi et al. 2010). Accordingly, a mounting evidence has accumulated supporting a role for periodontal infection as a risk factor for several systemic diseases, including RA (Wegner et al. 2010; Mangat et al. 2010;

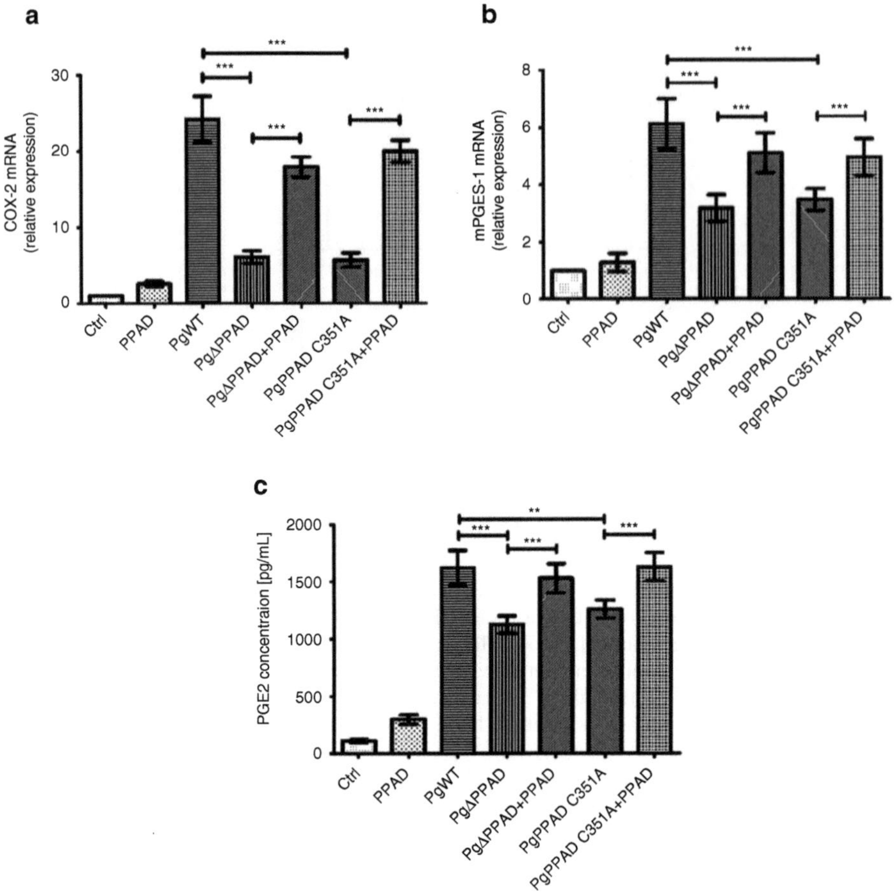

Fig. 7.7 PPAD induces prostaglandin E2 (PGE2) signaling in primary human gingival fibroblasts (PHGF). Relative expression levels of (**a**) cyclooxygenase-2 (COX-2), (**b**) microsomal prostaglandin E synthase-1 (mPGES-1), and (**c**) the concentration of PGE2 in PHGF infected with wild-type *P. gingivalis* (PgWT), an isogenic PPAD knockout strain (PgΔPPAD),and a mutated strain expressing the catalytically inactive PPAD (PgPPAD C351A) for 3 h at an MOI of 100. Infection with mutant strains was also performed in the presence of purified PPAD (0.066 mU/μL, total 66 mU). Quantitative real-time PCR was performed using β-actin as a reference gene. PGE2 concentrations were evaluated by EIA test. Data represent three independent experiments and are expressed as means ± SD (***, $p < 0.001$; **, $p < 0.01$; and *, $p < 0.05$) (*Reproduced from Molecular Oral Microbiology, 29(6), Gawron K, Bereta G, Nowakowska Z, Lazarz-Bartyzel K, Lazarz M, Szmigielski B, Mizgalska D, Buda A, Koziel J, Oruba Z, Chomyszyn-Gajewska M, Potempa J. Peptidylarginine deiminase from Porphyromonas gingivalis contributes to infection of gingival fibroblasts and induction of prostaglandin E2-signaling pathway. pp. 321–32, 2014, with permission from John Wiley & Sons Inc.*)

Lundberg et al. 2010; Maresz et al. 2013; Quirke et al. 2014), atherosclerotic cardiovascular disease (Genco et al. 2002; Gibson et al. 2004, 2008; Padilla et al. 2006), respiratory disease (Scannapieco et al. 2003; Azarpazhooh and Leake 2006; Benedyk et al. 2016), diabetes mellitus (Grossi and Genco 1998; Mealey and Ocampo 2007), adverse pregnancy outcomes (Vergnes and Sixou 2007), and other systemic conditions, which continue to appear in the literature.

7.6.1 The Links Between PD and RA

The association between PD and RA dates back over 2400 years (reviewed in Mayo 1922) with an apocryphal report from the days of Hippocrates, which claimed to cure arthritis by removing "a bad tooth." Similarly, in 1818, the American physician Benjamin Rush cured rheumatism by dental extraction. The concept that dental sepsis caused systemic inflammation, including arthritis, led to the "septic focus" theory, developed by William Hunter and Edward C. Rosenow in the early 1900s. The "septic focus" concept remained popular through the early twentieth century, and total dental clearance was widely practiced for the treatment of RA and other systemic diseases. The widespread use of dental X-rays and the introduction of "conservative dentistry" by Dr. C. Edmund Kells (Kracher 2000) resulted in negation of benefits of the "septic focus" theory to the point that removing teeth in an attempt to treat systemic disease was tantamount to "criminal behavior." Nevertheless, total dental clearance to treat RA continued to be practiced until the 1970s and only ceased with the advent of the first effective disease-modifying drugs such as gold and penicillamine.

At the beginning of the twenty first century, epidemiological links between RA and dental disease started to be investigated systematically, with PD being the specific area of interest (de Pablo et al. 2009). Further momentum for such investigation was provided by the discovery of a PPAD, leading to the hypothesis that autoimmunity to citrullinated proteins, which characterize RA, can be primed by the activity of this unique bacterial enzyme.

Therefore, to draw together these suggestive links between PD and RA, there are three factors to consider: (a) Is there an epidemiologic association between the two diseases? (b) Is the link due *P. gingivalis* infection? (c) If so, what is the mechanism?

7.6.2 Epidemiologic Associations Between PD and RA

The earliest studies examined the prevalence of RA in patients with PD. Mercado et al. (2000) studied 1412 individuals referred to a dental clinic divided into two groups, i.e., those that had PD ($n = 809$) and those that attended the clinic for other treatment ($n = 603$). The prevalence of RA in the PD group was 32 out of 809 subjects (3.95%), compared to 4 out of 603 (0.66%) in the general group ($p < 0.05$).

Within the PD group, patients with RA were more likely to have moderate-to-severe PD (62.5% of all RA patients) than patients without RA (43.8%) ($p < 0.05$). Obvious limitations to this study were the reliance on self-reported RA; the definition of PD using a nonvalidated parameter, i.e., referral; and the failure to document and adjust for smoking.

More recent studies have focused on the prevalence of PD in patients with established RA. In one small but reasonably robust investigation of 57 Caucasian subjects with RA and 52 non-RA controls (Pischon et al. 2008), an increased risk of PD was found in subjects with RA compared to controls; the odds ratio (OR) was 8.05 (95% CI 2.93–22.09), independent of smoking, gender, education, alcohol consumption, and BMI, and adjusted for age. When further adjusted for oral hygiene measures (plaque, gingival bleeding), the OR for the association of RA with PD was 6.09 (95% CI 1.72–21.58). Another study on data from 4461 subjects from the American National Health and Nutrition Examination Survey III (NHANES III) investigated the associations between RA and periodontal measures (de Pablo et al. 2008); 2.3% of all subjects were defined as having RA, which is considerably higher than the average (0.5–1%). Compared with non-RA subjects, participants with RA had a higher prevalence of PD (16% vs. 10%), edentulism (56% vs. 34%), and number of missing teeth (20 vs. 16, $p = 0.0001$). Since then, a number of additional studies of PD in patients with RA have been published (Potikuri et al. 2012; Scher et al. 2012; Dissick et al. 2010; de Smit et al. 2012), and all have found an approximately 2-fold higher prevalence of PD in patients with RA. All of these studies can be criticized for relatively small numbers of subjects and poor matching of RA patients and controls. However, a well-controlled study of 287 patients with RA and 330 patients with osteoarthritis by Mikuls et al. (2014) confirmed a significant increase in the frequency of PD, though in this case, the relative risk was around 1.5-fold. In a separate report in the same group of patients, the severity of the PD, as measured by alveolar bone loss, correlated with several measures of disease activity and severity of RA. Importantly the strongest correlations were seen with the titer of anti-CCP antibodies, suggesting a direct link between PD and the autoimmune response in RA (Coburn et al. 2015).

7.6.3 **P. gingivalis** *Infection Implicates to the Pathogenesis of RA*

Because *P. gingivalis* is an obligate anaerobe, culture from clinical samples is difficult and not sufficiently robust for epidemiological analysis. Therefore, investigations have had to rely on detection of the bacterial genome by PCR or by measuring the serological response. In two studies, using PCR for a multicopy gene from *P. gingivalis* encoding 5s RNA, the frequency of detection in RA was the same as controls (Scher et al. 2012; Mikuls et al. 2014). Interestingly, in one of these studies, there was an increased frequency of antibodies to *P. gingivalis* in ACPA-positive RA using an ELISA with a semipurifed membrane fraction of the bacterium (Mikuls et al. 2014). This assay must be interpreted with caution, since the lysates of *P. gingivalis* contain multiple citrullinated polypeptides (Wegner et al. 2010) and it is possible that the assay was also detecting ACPA reacting with citrullinated epitopes on bacterial proteins. Our collaborating laboratories attempted to overcome this problem by developing an improved assay using the purified *P. gingivalis*-specific enzyme, RgpB, as antigen. Using ELISA, there was no difference in the levels of anti-RgpB antibodies between 82 patients with RA and 80 controls (Quirke et al. 2014). Using the same assay in serum samples taken before the onset of RA in a Southern European cohort, it showed that levels of anti-RgpB antibodies in 103 people destined to get RA did not differ for levels found in over 300 age- and sex-matched controls (Fisher et al. 2016). In contrast, a much larger epidemiologic analysis of nearly 2000 Swedish patients with early RA found that raised antibodies to RgpB conferred an OR of 2.96 for RA, higher than that of the well-documented environmental factor of smoking, which in this population had an OR of 1.37 (Kharlamova et al. 2016). Adding CCP positivity, the major histocompatibility complex (MHC) "shared epitope" and smoking to anti-RgpB generated astonishing ORs of over 15, suggesting that the immune response to *P. gingivalis* interacts with other environmental and genetic risk factors in promoting the ACPA response in RA. Considering the Swedish study, in the question as to whether the link between RA and PD is due to *P. gingivalis*, the answer would appear to be a resounding "yes," but the discrepancy with other studies of different populations and with different assays suggests there is much more work to be done to resolve this issue.

7.6.4 PPAD as a Main Mechanistic Link Between PD and RA

If it is accepted that the newly established link between PD and RA is largely due to *P. gingivalis*, PPAD remains the prime candidate underlying the mechanisms for tolerance breakdown in RA. PPAD differs from mammalian PADs in its ability to citrullinate free L-arginine and its preference for carboxy-terminal peptidyl arginines. Given that the same subset of carboxy-terminally citrullinated peptides are less likely to be generated by the host PAD enzymes such as PAD-2 and PAD-4, these formed by PPAD may act as neo-epitopes and thus potentially react with T cells to drive an antibody response against both, bacterially derived and host proteins. One problem with this hypothesis is the lack of evidence that carboxy-terminally citrullinated peptides bind to MHC molecules or to responding T cell receptors. However, one group has shown that a highly immunogenic peptide from hen egg lysozyme (HEL) was citrullinated by endogenous PADs in a penultimate C-terminal position (flanked only by a downstream carboxy-terminal glycine) and bound to the P9 pocket of the MHC. This interaction resulted in autoantibodies to uncitrullinated HEL and autophagy in HEL transgenic mice (Ireland et al. 2006; Ireland and Unanue 2011). Thus, it remains possible for carboxy-terminally citrullinated peptides from enolase or fibrinogen to engage with the P9 or P10 pocket of the "shared epitope" and break tolerance by a similar mechanism. In support of this idea is the observation that patients with PD, as in HEL transgenic mice, react mainly with uncitrullinated variants of the peptides (Lappin et al. 2013; de Pablo et al. 2014).

A second mechanism for PPAD breaking tolerance to citrullinated proteins is via autocitrullination, which occurs during the production and purification of recombinant PPAD (Quirke et al. 2014). Citrullinated peptides from PPAD are targeted by antibodies in RA (Quirke et al. 2014; Harvey et al. 2013), but autocitrullination of PPAD has yet to be demonstrated in vivo, so it is possible that the antibodies to citrullinated PPAD peptides are simply a part of the cross-reactive ACPA response. However, if autocitrullination of PPAD does prove to be a mechanism for inducing autoimmunity in RA, it has profound therapeutic implications in that PPAD inhibitors could be a novel and targeted approach to treating patients whose RA has been induced by PD. Although PPAD has a similar overall fold and active site architecture to mammalian PADs, it should be possible to develop a PPAD-specific inhibitor because these two enzymes share low sequence homology with numerous divergent residues lining or adjacent to the active site.

A third mechanism for citrullination breaking tolerance may not involve PPAD at all but the host PADs, such as PAD-2 and PAD-4. In PD, citrullinated proteins and enzymes PAD-2 and PAD-4 have been detected in gingival tissues and have been shown in higher levels in inflamed tissues, correlating with clinical parameters of PD such as depth of periodontal pockets (Nesse et al. 2012; Harvey et al. 2013). Both PAD-2 and PAD-4 are important in the formation of NETs. It has been proposed that NET formation may induce autoimmunity to citrullinated proteins in RA through internalization of bacteria "coated with" citrullinated proteins, inducing an antibody response to both bacterial and native proteins (Dwivedi and Radic 2014).

A fourth possible mechanism for inducing ACPA may not involve citrullination by PPAD, at least not in the initial stages. Of relevance in this respect is the immunodominant epitope of citrullinated alpha-enolase known as citrullinated enolase peptide-1 (CEP-1). The sequence KIHA-cit-EIFDS-cit-GNPTVE is 97% identical to the corresponding sequence in citrullinated *P. gingivalis* enolase (Lundberg et al. 2008). Thus, *P. gingivalis* infection could initiate tolerance breakdown through cross-reactivity of antibacterial enolase immunity with mammalian enolase. Such a possibility was demonstrated in an animal model by Kinloch et al. (2011), where immunization of DR4 transgenic mice with both, citrullinated and uncitrullinated *P. gingivalis* enolase induced autoantibodies to mammalian enolase. Importantly, the antibody response was not citrulline specific as judged by reactivity of serum from the immunized mice with the arginine control peptide (REP-1) (Kinloch et al. 2011). This indicates that citrullination is not required for inducing an immune response, and when the response occurs, the antibodies are not citrulline specific. This is in keeping with the findings of largely not citrulline specific autoantibody response to CEP-1 and REP-1 in PD (Lappin et al. 2013; de Pablo et al. 2014) and a very similar pattern of noncitrulline specific ACPA responses in another chronic bacterial infection, bronchiectasis (Quirke et al. 2015).

7.7 Concluding Remarks and Future Perspectives

In summary, infection with *P. gingivalis* is strongly associated with adult PD, one of the most common chronic, bacterial infections. This is apparent from a number of clinical studies where a proportional increase of the level of this oral pathogen in subgingival plaque correlated with disease severity evaluated by attachment loss, periodontal pocket depth, and bleeding on probing. Furthermore, the successful eradication of *P. gingivalis* from periodontal pockets was often accompanied by the resolution of the disease. *P. gingivalis* and other dysbiotic bacteria in subgingival plaque drive chronic inflammatory reaction resulting in irreversible destruction of tooth-supporting tissues, and if untreated, the disease may lead to tooth loss. Recently, a non-proteolytic enzyme from *P. gingivalis*, named PPAD, attracted a considerable interest as a novel factor of virulence. The cooperative effects of Rgps and PPAD have been proposed as a mechanism responsible for inactivation of several plasma constituents, increased GCF flow, and *P. gingivalis* persistence in gingival pockets. In addition, citrullination of bacterial cell surface proteins seems to contribute to adhesion and invasion of gingival cells and activation of the PGE2 pathway. Moreover, the pathogenicity of ACPA in RA, and the expression of PPAD by *P. gingivalis* in PD, has implicated citrullination by PPAD as a mechanistic link in the development of RA. Taken together, inhibition of PPAD could represent a completely novel approach to treat PD and prevent development of RA in susceptible individuals.

Acknowledgments Many thanks to the National Science Centre of Poland (2012/07/B/NZ6/03524 to K.G.) for providing funding to many of the experiments reviewed in this chapter.

References

Akcali, A., Huck, O., Tenenbaum, H., Davideau, J. L., & Buduneli, N. (2013). Periodontal diseases and stress: A brief review. *Journal of Oral Rehabilitation, 40*(1), 60–68.

Alenius, G. M., Berglin, E., & Rantapää Dahlqvist, S. (2006). Antibodies against cyclic citrullinated peptide (CCP) in psoriatic patients with or without joint inflammation. *Annals of the Rheumatic Diseases, 65*(3), 398–400.

Amano, A. (2007). Disruption of epithelial barrier and impairment of cellular function by Porphyromonas gingivalis. *Frontiers in Bioscience, 12*, 3965–3974.

Amornchat, C., Rassameemasmaung, S., Sripairojthikoon, W., & Swasdison, S. (2003). Invasion of Porphyromonas gingivalis into human gingival fibroblasts in vitro. *Journal of the International Academy of Periodontology, 5*(4), 98–105.

Arita, K., Hashimoto, H., Shimizu, T., Nakashima, K., Yamada, M., & Sato, M. (2004). Structural basis for ca(2+)-induced activation of human PAD4. *Nature Structural & Molecular Biology, 11*(8), 777–783.

Armitage, G. C. (1995). Clinical evaluation of periodontal diseases. *Periodontology 2000, 7*, 39–53.

Asaga, H., Nakashima, K., Senshu, T., Ishigami, A., & Yamada, M. (2001). Immunocytochemical localization of peptidylarginine deiminase in human eosinophils and neutrophils. *Journal of Leukocyte Biology, 70*(1), 46–51.

Azarpazhooh, A., & Leake, J. L. (2006). Systematic review of the association between respiratory diseases and oral health. *Journal of Periodontology, 77*(9), 1465–1482.

Baka, Z., Barta, P., Losonczy, G., Krenács, T., Pápay, J., Szarka, E., Sármay, G., Babos, F., Magyar, A., Géher, P., Buzás, E. I., & Nagy, G. (2011). Specific expression of PAD4 and citrullinated proteins in lung cancer is not associated with anti-CCP antibody production. *International Immunology, 23*(6), 405–414.

Baka, Z., György, B., Géher, P., Buzás, E. I., Falus, A., & Nagy, G. (2012). Citrullination under physiological and pathological conditions. *Joint, Bone, Spine, 79*(5), 431–436.

Belton, C. M., Izutsu, K. T., Goodwin, P. C., Park, Y., & Lamont, R. J. (1999). Fluorescence image analysis of the association between Porphyromonas gingivalis and gingival epithelial cells. *Cellular Microbiology, 1*(3), 215–223.

Benedyk, M., Mydel, P. M., Delaleu, N., Płaza, K., Gawron, K., Milewska, A., Maresz, K., Koziel, J., Pyrc, K., & Potempa, J. (2016). Gingipains: Critical factors in the development of aspiration pneumonia caused by Porphyromonas gingivalis. *Journal of Innate Immunity, 8*(2), 185–198.

Bergström, J. (2003). Tobacco smoking and risk for periodontal disease. *Journal of Clinical Periodontology, 30*(2), 107–113.

Bicker, K. L., & Thompson, P. R. (2013). The protein arginine deiminases (PADs): Structure, function, inhibition, and disease. *Biopolymers, 99*(2), 155–163.

Bielecka, E., Scavenius, C., Kantyka, T., Jusko, M., Mizgalska, D., Szmigielski, B., Potempa, B., Enghild, J. J., Prossnitz, E. R., Blom, A. M., & Potempa, J. (2014). Peptidyl arginine deiminase from Porphyromonas gingivalis abolishes anaphylatoxin C5a activity. *The Journal of Biological Chemistry, 289*(47), 32481–32487.

Bostanci, N., & Belibasakis, G. N. (2012). Porphyromonas gingivalis: An invasive and evasive opportunistic oral pathogen. *FEMS Microbiology Letters, 333*(1), 1–9.

Carlsson, J., Höfling, J. F., & Sundqvist, G. K. (1984). Degradation of albumin, haemopexin, haptoglobin and transferrin, by black-pigmented Bacteroides species. *Journal of Medical Microbiology, 18*(1), 39–46.

Casiano-Colón, A., & Marquis, R. E. (1988). Role of the arginine deiminase system in protecting oral bacteria and an enzymatic basis for acid tolerance. *Applied and Environmental Microbiology, 54*(6), 1318–1324.

Cekici, A., Kantarci, A., Hasturk, H., & Van Dyke, T. E. (2014). Inflammatory and immune pathways in the pathogenesis of periodontal disease. *Periodontology 2000, 64*(1), 57–80.

Chang, X., & Han, J. (2006). Expression of peptidylarginine deiminase type 4 (PAD4) in various tumors. *Molecular Carcinogenesis, 45*(3), 183–196.

Chang, X., Han, J., Pang, L., Zhao, Y., Yang, Y., & Shen, Z. (2009). Increased PADI4 expression in blood and tissues of patients with malignant tumors. *BMC Cancer, 9*, 40.

Chee, B., Park, B., & Bartold, P. M. (2013). Periodontitis and type II diabetes: A two-way relationship. *International Journal of Evidence-Based Healthcare, 11*(4), 317–329.

Chen, Z., Potempa, J., Polanowski, A., Wikstrom, M., & Travis, J. (1992). Purification and characterization of a 50-kDa cysteine proteinase (gingipain) from Porphyromonas gingivalis. *The Journal of Biological Chemistry, 267*(26), 18896–18901.

Coburn, B. W., Sayles, H. R., Payne, J. B., Redman, R. S., Markt, J. C., Beatty, M. W., Griffiths, G. R., McGowan, D. J., & Mikuls, T. R. (2015). Performance of self-reported measures for periodontitis in rheumatoid arthritis and osteoarthritis. *Journal of Periodontology, 86*(1), 16–26.

Coughlin, S. R. (2000). Thrombin signalling and protease-activated receptors. *Nature, 407*(6801), 258–264.

Cuthbert, G. L., Daujat, S., Snowden, A. W., Erdjument-Bromage, H., Hagiwara, T., Yamada, M., Schneider, R., Gregory, P. D., Tempst, P., Bannister, A. J., & Kouzarides, T. (2004). Histone deimination antagonizes arginine methylation. *Cell, 118*(5), 545–553.

Cutler, C. W., Kalmar, J. R., & Genco, C. A. (1995). Pathogenic strategies of the oral anaerobe, Porphyromonas gingivalis. *Trends in Microbiology, 3*(2), 45–51.

Davey, M., Liu, X., Ukai, T., Jain, V., Gudino, C., Gibson, F. C., Golenbock, D., Visintin, A., & Genco, C. A. (2008). Bacterial fimbriae stimulate proinflammatory activation in the endothelium through distinct TLRs. *Journal of Immunology, 180*(4), 2187–2195.

de Diego, I., Veillard, F., Sztukowska, M. N., Guevara, T., Potempa, B., Pomowski, A., Huntington, J. A., Potempa, J., & Gomis-Rüth, F. X. (2014). Structure and mechanism of cysteine peptidase gingipain K (Kgp), a major virulence factor of Porphyromonas gingivalis in periodontitis. *The Journal of Biological Chemistry, 289*(46), 32291–32302.

de Diego, I., Ksiazek, M., Mizgalska, D., Koneru, L., Golik, P., Szmigielski, B., Nowak, M., Nowakowska, Z., Potempa, B., Houston, J. A., Enghild, J. J., Thøgersen, I. B., Gao, J., Kwan, A. H., Trewhella, J., Dubin, G., Gomis-Rüth, F. X., Nguyen, K. A., & Potempa, J. (2016). The outer-membrane export signal of Porphyromonas gingivalis type IX secretion system (T9SS) is a conserved C-terminal β-sandwich domain. *Scientific Reports, 6*, 23123.

de Pablo, P., Dietrich, T., & McAlindon, T. E. (2008). Association of periodontal disease and tooth loss with rheumatoid arthritis in the US population. *The Journal of Rheumatology, 35*(1), 70–76.

de Pablo, P., Chapple, I. L., Buckley, C. D., & Dietrich, T. (2009). Periodontitis in systemic rheumatic diseases. *Nature Reviews. Rheumatology, 5*(4), 218–224.

de Pablo, P., Dietrich, T., Chapple, I. L., Milward, M., Chowdhury, M., Charles, P. J., Buckley, C. D., & Venables, P. J. (2014). The autoantibody repertoire in periodontitis: A role in the induction of autoimmunity to citrullinated proteins in rheumatoid arthritis? *Annals of the Rheumatic Diseases, 73*(3), 580–586.

de Smit, M., Westra, J., Vissink, A., Doornbos-van der Meer, B., Brouwer, E., & van Winkelhoff, A. J. (2012). Periodontitis in established rheumatoid arthritis patients: A cross-sectional clinical, microbiological and serological study. *Arthritis Research & Therapy, 14*(5), R222.

Deshpande, R. G., Khan, M. B., & Genco, C. A. (1998). Invasion of aortic and heart endothelial cells by Porphyromonas gingivalis. *Infection and Immunity, 66*(11), 5337–5343.

Dissick, A., Redman, R. S., Jones, M., Rangan, B. V., Reimold, A., Griffiths, G. R., Mikuls, T. R., Amdur, R. L., Richards, J. S., & Kerr, G. S. (2010). Association of periodontitis with rheumatoid arthritis: A pilot study. *Journal of Periodontology, 81*(2), 223–230.

Dorn, B. R., Burks, J. N., Seifert, K. N., & Progulske-Fox, A. (2000). Invasion of endothelial and epithelial cells by strains of Porphyromonas gingivalis. *FEMS Microbiology Letters, 187*(2), 139–144.

Dwivedi, N., & Radic, M. (2014). Citrullination of autoantigens implicates NETosis in the induction of autoimmunity. *Annals of the Rheumatic Diseases, 73*(3), 483–491.

Eichinger, A., Beisel, H. G., Jacob, U., Huber, R., Medrano, F. J., Banbula, A., Potempa, J., Travis, J., & Bode, W. (1999). Crystal structure of gingipain R: An Arg-specific bacterial cysteine proteinase with a caspase-like fold. *The EMBO Journal, 18*(20), 5453–5462.

Eke, P. I., Dye, B. A., Wei, L., Thornton-Evans, G. O., Genco, R. J., Beck, J., Douglass, G., Page, R., Slade, G., Taylor, G. W., & Borgnakke, W. (2012). Prevalence of periodontitis in adults in the United States: 2009 and 2010. *Journal of Dental Research, 91*, 914–920.

Eley, B. M., & Cox, S. W. (1996). Correlation between gingivain/gingipain and bacterial dipeptidyl peptidase activity in gingival crevicular fluid and periodontal attachment loss in chronic periodontitis patients. A 2-year longitudinal study. *Journal of Periodontology, 67*(7), 703–716.

Fisher, B. A., Cartwright, A. J., Quirke, A. M., de Pablo, P., Romaguera, D., Panico, S., Mattiello, A., Gavrila, D., Navarro, C., Sacerdote, C., Vineis, P., Tumino, R., Lappin, D. F., Apatzidou, D., Culshaw, S., Potempa, J., Michaud, D. S., Riboli, E., & Venables, P. J. (2016). Erratum to: Smoking, Porphyromonas gingivalis and the immune response to citrullinated autoantigens before the clinical onset of rheumatoid arthritis in a southern European nested case-control study. *BMC Musculoskeletal Disorders, 17*(1), 62.

Gawron, K., Bereta, G., Nowakowska, Z., Lazarz-Bartyzel, K., Lazarz, M., Szmigielski, B., Mizgalska, D., Buda, A., Koziel, J., Oruba, Z., Chomyszyn-Gajewska, M., & Potempa, J. (2014). Peptidylarginine deiminase from *Porphyromonas gingivalis* contributes to infection of gingival fibroblasts and induction of prostaglandin E2-signaling pathway. *Molecular Oral Microbiology, 29*(6), 321–332.

Gazi, M. I., Cox, S. W., Clark, D. T., & Eley, B. M. (1996). A comparison of cysteine and serine proteinases in human gingival crevicular fluid with tissue, saliva and bacterial enzymes by analytical isoelectric focusing. *Archives of Oral Biology, 41*(5), 393–400.

Genco, R., Offenbacher, S., & Beck, J. (2002). Periodontal disease and cardiovascular disease: Epidemiology and possible mechanisms. *Journal of the American Dental Association (1939), 133*(Suppl), 14S–22S.

Gher, M. E. (1998). Changing concepts. The effects of occlusion on periodontitis. *Dental Clinics of North America, 42*(2), 285–299.

Giacaman, R. A., Asrani, A. C., Ross, K. F., & Herzberg, M. C. (2009). Cleavage of protease-activated receptors on an immortalized oral epithelial cell line by Porphyromonas gingivalis gingipains. *Microbiology, 155*(Pt 10), 3238–3246.

Gibbons, R. J. (1989). Bacterial adhesion to oral tissues: A model for infectious diseases. *Journal of Dental Research, 68*(5), 750–760.

Gibbons, R. J., Spinell, D. M., & Skobe, Z. (1976). Selective adherence as a determinant of the host tropisms of certain indigenous and pathogenic bacteria. *Infection and Immunity, 13*(1), 238–246.

Gibson, F. C., 3rd, Hong, C., Chou, H. H., Yumoto, H., Chen, J., Lien, E., Wong, J., & Genco, C. A. (2004). Innate immune recognition of invasive bacteria accelerates atherosclerosis in apolipoprotein E-deficient mice. *Circulation, 109*(22), 2801–2806.

Gibson, F. C., 3rd, Ukai, T., & Genco, C. A. (2008). Engagement of specific innate immune signaling pathways during Porphyromonas gingivalis induced chronic inflammation and atherosclerosis. *Frontiers in Bioscience, 13*, 2041–2059.

Glew, M. D., Veith, P. D., Peng, B., Chen, Y. Y., Gorasia, D. G., Yang, Q., Slakeski, N., Chen, D., Moore, C., Crawford, S., & Reynolds, E. C. (2012). PG0026 is the C-terminal signal peptidase of a novel secretion system of Porphyromonas gingivalis. *The Journal of Biological Chemistry, 287*(29), 24605–24617.

Goulas, T., Mizgalska, D., Garcia-Ferrer, I., Kantyka, T., Guevara, T., Szmigielski, B., Sroka, A., Millán, C., Usón, I., Veillard, F., Potempa, B., Mydel, P., Solà, M., Potempa, J., & Gomis-Rüth, F. X. (2015). Structure and mechanism of a bacterial host-protein citrullinating virulence factor, Porphyromonas gingivalis peptidylarginine deiminase. *Scientific Reports, 5*, 11969.

Grenier, D. (1992). Inactivation of human serum bactericidal activity by a trypsinlike protease isolated from Porphyromonas gingivalis. *Infection and Immunity, 60*(5), 1854–1857.

Grenier, D., & Mayrand, D. (1987). Functional characterization of extracellular vesicles produced by Bacteroides Gingivalis. *Infection and Immunity, 55*(1), 111–117.

Grossi, S. G., & Genco, R. J. (1998). Periodontal disease and diabetes mellitus: A two-way relationship. *Annals of Periodontology, 3*(1), 51–61.

György, B., Tóth, E., Tarcsa, E., Falus, A., & Buzás, E. I. (2006). Citrullination: A posttranslational modification in health and disease. *The International Journal of Biochemistry & Cell Biology, 38*(10), 1662–1677.

Hajishengallis, G., & Lamont, R. J. (2012). Beyond the red complex and into more complexity: The polymicrobial synergy and dysbiosis (PSD) model of periodontal disease etiology. *Molecular Oral Microbiology, 27*(6), 409–419.

Harvey, G. P., Fitzsimmons, T. R., Dhamarpatni, A. A., Marchant, C., Haynes, D. R., & Bartold, P. M. (2013). Expression of peptidylarginine deiminase-2 and -4, citrullinated proteins and anti-citrullinated protein antibodies in human gingiva. *Journal of Periodontal Research, 48*(2), 252–261.

Hasturk, H., & Kantarci, A. (2015). Activation and resolution of periodontal inflammation and its systemic impact. *Periodontology 2000, 69*(1), 255–273.

Hayashi, H., Morioka, M., Ichimiya, S., Yamato, K., Hinode, D., Nagata, A., & Nakamura, R. (1993). Participation of an arginyl residue of insulin chain B in the inhibition of hemagglutination by Porphyromonas gingivalis. *Oral Microbiology and Immunology, 8*(6), 386–389.

Hayashi, C., Gudino, C. V., Gibson, F. C., 3rd, & Genco, C. A. (2010). Pathogen-induced inflammation at sites distant from oral infection: Bacterial persistence and induction of cell-specific innate immune inflammatory pathways. *Molecular Oral Microbiology, 25*(5), 305–316.

Heaton, B., & Dietrich, T. (2012). Causal theory and the etiology of periodontal diseases. *Periodontology 2000, 58*(1), 26–36.

Henskens, Y. M., van der Velden, U., Veerman, E. C., & Nieuw Amerongen, A. V. (1993). Protein, albumin and cystatin concentrations in saliva of healthy subjects and of patients with gingivitis or periodontitis. *Journal of Periodontal Research, 28*(1), 43–48.

Holt, S. C., Ebersole, J., Felton, J., Brunsvold, M., & Kornman, K. S. (1988). Implantation of Bacteroides Gingivalis in nonhuman primates initiates progression of periodontitis. *Science, 239*(4835), 55–57.

Holzhausen, M., Spolidorio, L. C., Ellen, R. P., Jobin, M. C., Steinhoff, M., Andrade-Gordon, P., & Vergnolle, N. (2006). Protease-activated receptor-2 activation: A major role in the pathogenesis of Porphyromonas gingivalis infection. *American Journal of Pathology, 168*(4), 1189–1199.

Imamura, T., Pike, R. N., Potempa, J., & Travis, J. (1994). Pathogenesis of periodontitis: A major arginine-specific cysteine proteinase from Porphyromonas gingivalis induces vascular permeability enhancement through activation of the kallikrein/kinin pathway. *The Journal of Clinical Investigation, 94*(1), 361–367.

Imamura, T., Potempa, J., Pike, R. N., & Travis, J. (1995). Dependence of vascular permeability enhancement on cysteine proteinases in vesicles of Porphyromonas gingivalis. *Infection and Immunity, 63*(5), 1999–2003.

Imamura, T., Potempa, J., Tanase, S., & Travis, J. (1997). Activation of blood coagulation factor X by arginine-specific cysteine proteinases (gingipain-Rs) from *Porphyromonas gingivalis*. *The Journal of Biological Chemistry, 272*(25), 16062–16067.

Into, T., Inomata, M., Kanno, Y., Matsuyama, T., Machigashira, M., Izumi, Y., Imamura, T., Nakashima, M., Noguchi, T., & Matsushita, K. (2006). Arginine-specific gingipains from Porphyromonas gingivalis deprive protective functions of secretory leucocyte protease inhibitor in periodontal tissue. *Clinical and Experimental Immunology, 145*(3), 545–554.

Ireland, J. M., & Unanue, E. R. (2011). Autophagy in antigen-presenting cells results in presentation of citrullinated peptides to CD4 T cells. *The Journal of Experimental Medicine, 208*(13), 2625–2632.

Ireland, J., Herzog, J., & Unanue, E. R. (2006). Cutting edge: Unique T cells that recognize citrullinated peptides are a feature of protein immunization. *Journal of Immunology, 177*(3), 1421–1425.

Ishida-Yamamoto, A., Senshu, T., Takahashi, H., Akiyama, K., Nomura, K., & Iizuka, H. (2000). Decreased deiminated keratin K1 in psoriatic hyperproliferative epidermis. *The Journal of Investigative Dermatology, 114*(4), 701–705.

Isogai, H., Isogai, E., Yoshimura, F., Suzuki, T., Kagota, W., & Takano, K. (1988). Specific inhibition of adherence of an oral strain of Bacteroides Gingivalis 381 to epithelial cells by monoclonal antibodies against the bacterial fimbriae. *Archives of Oral Biology, 33*(7), 479–485.

Jang, B., Shin, H. Y., Choi, J. K., Nguyen, d. P. T., Jeong, B. H., Ishigami, A., Maruyama, N., Carp, R. I., Kim, Y. S., & Choi, E. K. (2011). Subcellular localization of peptidylarginine deiminase 2 and citrullinated proteins in brains of scrapie-infected mice: Nuclear localization of PAD2 and membrane fraction-enriched citrullinated proteins. *Journal of Neuropathology and Experimental Neurology, 70*(2), 116–124.

Jin, L. J., Chiu, G. K., & Corbet, E. F. (2003). Are periodontal diseases risk factors for certain systemic disorders—What matters to medical practitioners? *Hong Kong Medical Journal, 9*(1), 31–37.

Jones, J. E., Causey, C. P., Knuckley, B., Slack-Noyes, J. L., & Thompson, P. R. (2009). Protein arginine deiminase 4 (PAD4): Current understanding and future therapeutic potential. *Current Opinion in Drug Discovery & Development, 12*(5), 616–627.

Kadowaki, T., Yoneda, M., Okamoto, K., Maeda, K., & Yamamoto, K. (1994). Purification and characterization of a novel arginine-specific cysteine proteinase (argingipain) involved in the pathogenesis of periodontal disease from the culture supernatant of Porphyromonas gingivalis. *The Journal of Biological Chemistry, 269*(33), 21371–21378.

Kadowaki, T., Baba, A., Abe, N., Takii, R., Hashimoto, M., Tsukuba, T., Okazaki, S., Suda, Y., Asao, T., & Yamamoto, K. (2004). Suppression of pathogenicity of Porphyromonas gingivalis by newly developed gingipain inhibitors. *Molecular Pharmacology, 66*(6), 1599–1606.

Kanapka, J. A., & Kleinberg, I. (1983). Catabolism of arginine by the mixed bacteria in human salivary sediment under conditions of low and high glucose concentration. *Archives of Oral Biology, 28*(11), 1007–1015.

Katz, J., Yang, Q. B., Zhang, P., Potempa, J., Travis, J., Michalek, S. M., & Balkovetz, D. F. (2002). Hydrolysis of epithelial junctional proteins by Porphyromonas gingivalis gingipains. *Infection and Immunity, 70*(5), 2512–2518.

Kharlamova, N., Jiang, X., Sherina, N., Potempa, B., Israelsson, L., Quirke, A. M., Eriksson, K., Yucel-Lindberg, T., Venables, P. J., Potempa, J., Alfredsson, L., & Lundberg, K. (2016). Antibodies to Porphyromonas gingivalis indicate interaction between oral infection, smoking, and risk genes in rheumatoid arthritis etiology. *Arthritis & Rhematology, 68*(3), 604–613.

Kinloch, A. J., Alzabin, S., Brintnell, W., Wilson, E., Barra, L., Wegner, N., Bell, D. A., Cairns, E., & Venables, P. J. (2011). Immunization with Porphyromonas gingivalis enolase induces autoimmunity to mammalian α enolase and arthritis in DR4-IE-transgenic mice. *Arthritis & Rheumatology, 63*(12), 3818–3823.

Klareskog, L., Stolt, P., Lundberg, K., Källberg, H., Bengtsson, C., Grunewald, J., Rönnelid, J., Harris, H. E., Ulfgren, A. K., Rantapää-Dahlqvist, S., Eklund, A., Padyukov, L., & Alfredsson, L. (2006). A new model for an etiology of rheumatoid arthritis: Smoking may trigger HLA-DR (shared epitope)-restricted immune reactions to autoantigens modified by citrullination. *Arthritis and Rheumatism, 54*(1), 38–46.

Knuckley, B., Causey, C. P., Jones, J. E., Bhatia, M., Dreyton, C. J., Osborne, T. C., Takahara, H., & Thompson, P. R. (2010). Substrate specificity and kinetic studies of PADs 1, 3, and 4 identify potent and selective inhibitors of protein arginine deiminase 3. *Biochemistry, 49*(23), 4852–4863.

Kracher, C. M. (2000). C. Edmund Kells (1856-1928). *Journal of the History of Dentistry, 48*(2), 65–69.

Laine, M. L., Crielaard, W., & Loos, B. G. (2012). Genetic susceptibility to periodontitis. *Periodontology 2000, 58*(1), 37–68.

Lamensa, J. W., & Moscarello, M. A. (1993). Deimination of human myelin basic protein by a peptidylarginine deiminase from bovine brain. *Journal of Neurochemistry, 61*(3), 987–996.

Lamont, R. J., & Jenkinson, H. F. (1998). Life below the gum line: Pathogenic mechanisms of Porphyromonas gingivalis. *Microbiology and Molecular Biology Reviews, 62*(4), 1244–1263.

Lamont, R. J., Chan, A., Belton, C. M., Izutsu, K. T., Vasel, D., & Weinberg, A. (1995). Porphyromonas gingivalis invasion of gingival epithelial cells. *Infection and Immunity, 63*(10), 3878–3885.

Lappin, D. F., Apatzidou, D., Quirke, A. M., Oliver-Bell, J., Butcher, J. P., Kinane, D. F., Riggio, M. P., Venables, P., McInnes, I. B., & Culshaw, S. (2013). Influence of periodontal disease, Porphyromonas gingivalis and cigarette smoking on systemic anti-citrullinated peptide antibody titres. *Journal of Clinical Periodontology, 40*(10), 907–915.

Lee, D. M., Phillips, R., Hagan, E. M., Chibnik, L. B., Costenbader, K. H., & Schur, P. H. (2009). Quantifying anti-cyclic citrullinated peptide titres: Clinical utility and association with tobacco exposure in patients with rheumatoid arthritis. *Annals of the Rheumatic Diseases, 68*(2), 201–208.

Li, X., Kolltveit, K. M., Tronstad, L., & Olsen, I. (2000). Systemic diseases caused by oral infection. *Clinical Microbiology Reviews, 13*(4), 547–558.

Li, P., Yao, H., Zhang, Z., Li, M., Luo, Y., Thompson, P. R., Gilmour, D. S., & Wang, Y. (2008). Regulation of p53 target gene expression by peptidylarginine deiminase 4. *Molecular and Cellular Biology, 28*(15), 4745–4758.

Li, P., Li, M., Lindberg, M. R., Kennett, M. J., Xiong, N., & Wang, Y. (2010). PAD4 is essential for antibacterial innate immunity mediated by neutrophil extracellular traps. *The Journal of Experimental Medicine, 207*(9), 1853–1862.

Lindroth, A. M., & Park, Y. J. (2013). Epigenetic biomarkers: A step forward for understanding periodontitis. *Journal of Periodontal & Implant Science, 43*(3), 111–120.

Loesche, W. J., Syed, S. A., Morrison, E. C., Laughon, B., & Grossman, N. S. (1981). Treatment of periodontal infections due to anaerobic bacteria with short-term treatment with metronidazole. *Journal of Clinical Periodontology, 8*(1), 29–44.

Loos, T., Mortier, A., Gouwy, M., Ronsse, I., Put, W., Lenaerts, J. P., Van Damme, J., & Proost, P. (2008). Citrullination of CXCL10 and CXCL11 by peptidylarginine deiminase: A naturally occurring posttranslational modification of chemokines and new dimension of immunoregulation. *Blood, 112*(7), 2648–2656.

Lourbakos, A., Chinni, C., Thompson, P., Potempa, J., Travis, J., Mackie, E. J., & Pike, R. N. (1998). Cleavage and activation of proteinase-activated receptor-2 on human neutrophils by gingipain-R from Porphyromonas gingivalis. *FEBS Letters, 435*(1), 45–48.

Lourbakos, A., Potempa, J., Travis, J., D'Andrea, M. R., Andrade-Gordon, P., Santulli, R., Mackie, E. J., & Pike, R. N. (2001a). Arginine-specific protease from Porphyromonas gingivalis activates protease-activated receptors on human oral epithelial cells and induces interleukin-6 secretion. *Infection and Immunity, 69*(8), 5121–5130.

Lourbakos, A., Yuan, Y. P., Jenkins, A. L., Travis, J., Andrade-Gordon, P., Santulli, R., Potempa, J., & Pike, R. N. (2001b). Activation of protease-activated receptors by gingipains from Porphyromonas gingivalis leads to platelet aggregation: A new trait in microbial pathogenicity. *Blood, 97*(12), 3790–3797.

Lundberg, K., Kinloch, A., Fisher, B. A., Wegner, N., Wait, R., Charles, P., Mikuls, T. R., & Venables, P. J. (2008). Antibodies to citrullinated alpha-enolase peptide 1 are specific for rheumatoid arthritis and cross-react with bacterial enolase. *Arthritis and Rheumatism, 58*(10), 3009–3019.

Lundberg, K., Wegner, N., Yucel-Lindberg, T., & Venables, P. J. (2010). Periodontitis in RA-the citrullinated enolase connection. *Nature Reviews. Rheumatology, 6*(12), 727–730.

Mager, D. L., Ximenez-Fyvie, L. A., Haffajee, A. D., & Socransky, S. S. (2003). Distribution of selected bacterial species on intraoral surfaces. *Journal of Clinical Periodontology, 30*(7), 644–654.

Mangat, P., Wegner, N., Venables, P. J., & Potempa, J. (2010). Bacterial and human peptidylarginine deiminases: Targets for inhibiting the autoimmune response in rheumatoid arthritis? *Arthritis Research & Therapy, 12*(3), 209.

Maresz, K. J., Hellvard, A., Sroka, A., Adamowicz, K., Bielecka, E., Koziel, J., Gawron, K., Mizgalska, D., Marcinska, K. A., Benedyk, M., Pyrc, K., Quirke, A. M., Jonsson, R., Alzabin, S., Venables, P. J., Nguyen, K. A., Mydel, P., & Potempa, J. (2013). Porphyromonas gingivalis facilitates the development and progression of destructive arthritis through its unique bacterial peptidylarginine deiminase (PAD). *PLoS Pathogens, 9*(9), e1003627.

Marsh, P. D., McKee, A. S., McDermid, A. S., & Dowsett, A. B. (1989). Ultrastructure and enzyme activities of a virulent and an avirulent variant of Bacteroides Gingivalis W50. *FEMS Microbiology Letters, 50*(1–2), 181–185.

Mastronardi, F. G., Wood, D. D., Mei, J., Raijmakers, R., Tseveleki, V., Dosch, H. M., Probert, L., Casaccia-Bonnefil, P., & Moscarello, M. A. (2006). Increased citrullination of histone H3 in multiple sclerosis brain and animal models of demyelination: A role for tumor necrosis factor-induced peptidylarginine deiminase 4 translocation. *The Journal of Neuroscience, 26*(44), 11387–11396.

Mastronardi, F. G., Noor, A., Wood, D. D., Paton, T., & Moscarello, M. A. (2007). Peptidyl arginine deiminase 2 CpG island in multiple sclerosis white matter is hypomethylated. *Journal of Neuroscience Research, 85*(9), 2006–2016.

Mayo, C. H. (1922). Focal infection of dental origin. *Dental Cosmos, 64*, 1206–1208.

Mayrand, D., & Holt, S. C. (1988). Biology of asaccharolytic black-pigmented Bacteroides species. *Microbiological Reviews, 52*(1), 134–152.

McGraw, W. T., Potempa, J., Farley, D., & Travis, J. (1999). Purification, characterization, and sequence analysis of a potential virulence factor from Porphyromonas gingivalis, peptidylarginine deiminase. *Infection and Immunity, 67*, 3248–3256.

Mealey, B. L., & Ocampo, G. L. (2007). Diabetes mellitus and periodontal disease. *Periodontology 2000, 44*, 127–153.

Mercado, F., Marshall, R. I., Klestov, A. C., & Bartold, P. M. (2000). Is there a relationship between rheumatoid arthritis and periodontal disease? *Journal of Clinical Periodontology, 27*(4), 267–272.

Meyle, J., & Chapple, I. (2015). Molecular aspects of the pathogenesis of periodontitis. *Periodontology 2000, 69*(1), 7–17.

Mikuls, T. R., Payne, J. B., Yu, F., Thiele, G. M., Reynolds, R. J., Cannon, G. W., Markt, J., McGowan, D., Kerr, G. S., Redman, R. S., Reimold, A., Griffiths, G., Beatty, M., Gonzalez, S. M., Bergman, D. A., Hamilton, B. C., 3rd, Erickson, A. R., Sokolove, J., Robinson, W. H., Walker, C., Chandad, F., & O'Dell, J. R. (2014). Periodontitis and Porphyromonas gingivalis in patients with rheumatoid arthritis. *Arthritis & Rhematology, 66*(5), 1090–1100.

Monk, P. N., Scola, A. M., Madala, P., & Fairlie, D. P. (2007). Function, structure and therapeutic potential of complement C5a receptors. *British Journal of Pharmacology, 152*(4), 429–448.

Montgomery, A. B., Kopec, J., Shrestha, L., Thezenas, M. L., Burgess-Brown, N. A., Fischer, R., Yue, W. W., & Venables, P. J. (2016). Crystal structure of Porphyromonas gingivalis peptidylarginine deiminase: Implications for autoimmunity in rheumatoid arthritis. *Annals of the Rheumatic Diseases, 75*(6), 1255–1261.

Moore, L. V., Moore, W. E., Cato, E. P., Smibert, R. M., Burmeister, J. A., Best, A. M., & Ranney, R. R. (1987). Bacteriology of human gingivitis. *Journal of Dental Research, 66*(5), 989–995.

Moscarello, M. A., Wood, D. D., Ackerley, C., & Boulias, C. (1994). Myelin in multiple sclerosis is developmentally immature. *The Journal of Clinical Investigation, 94*(1), 146–154.

Mucci, L. A., Björkman, L., Douglass, C. W., & Pedersen, N. L. (2005). Environmental and heritable factors in the etiology of oral diseases-a population-based study of Swedish twins. *Journal of Dental Research, 84*(9), 800–805.

Musse, A. A., Li, Z., Ackerley, C. A., Bienzle, D., Lei, H., Poma, R., Harauz, G., Moscarello, M. A., & Mastronardi, F. G. (2008). Peptidylarginine deiminase 2 (PAD2) overexpression in transgenic mice leads to myelin loss in the central nervous system. *Disease Models & Mechanisms, 1*(4–5), 229–240.

Nakagawa, I., Inaba, H., Yamamura, T., Kato, T., Kawai, S., Ooshima, T., & Amano, A. (2006). Invasion of epithelial cells and proteolysis of cellular focal adhesion components by distinct types of Porphyromonas gingivalis fimbriae. *Infection and Immunity, 74*(7), 3773–3782.

Nesse, W., Westra, J., van der Wal, J. E., Abbas, F., Nicholas, A. P., Vissink, A., & Brouwer, E. (2012). The periodontium of periodontitis patients contains citrullinated proteins which may play a role in ACPA (anti-citrullinated protein antibody) formation. *Journal of Clinical Periodontology, 39*(7), 599–607.

Niederman, R., Brunkhorst, B., Smith, S., Weinreb, R. N., & Ryder, M. I. (1990). Ammonia as a potential mediator of adult human periodontal infection: Inhibition of neutrophil function. *Archives of Oral Biology, 35*(Suppl), 205S–209S.

Njoroge, T., Genco, R. J., Sojar, H. T., Hamada, N., & Genco, C. A. (1997). A role for fimbriae in Porphyromonas gingivalis invasion of oral epithelial cells. *Infection and Immunity, 65*(5), 1980–1984.

Noguchi, K., & Ishikawa, I. (2007). The roles of cyclooxygenase-2 and prostaglandin E2 in periodontal disease. *Periodontology 2000, 43*, 85–101.

Nomura, K. (1992). Specificity and mode of action of the muscle-type protein-arginine deiminase. *Archives of Biochemistry and Biophysics, 293*(2), 362–369.

O'Brien-Simpson, N. M., Veith, P. D., Dashper, S. G., & Reynolds, E. C. (2003). Porphyromonas gingivalis gingipains: The molecular teeth of a microbial vampire. *Current Protein & Peptide Science, 4*(6), 409–426.

Offenbacher, S., Odle, B. M., & Van Dyke, T. E. (1986). The use of crevicular fluid prostaglandin E2 levels as a predictor of periodontal attachment loss. *Journal of Periodontal Research, 21*(2), 101–112.

Offenbacher, S., Heasman, P. A., & Collins, J. G. (1993). Modulation of host PGE2 secretion as a determinant of periodontal disease expression. *Journal of Periodontology, 64*(5 Suppl), 432–444.

Oliver, R. C., & Brown, L. J. (1993). Periodontal diseases and tooth loss. *Periodontology 2000, 2*, 117–127.

Omary, M. B., Ku, N. O., Liao, J., & Price, D. (1998). Keratin modifications and solubility properties in epithelial cells and in vitro. *Sub-Cellular Biochemistry, 31*, 105–140.

Ordóñez, A., Martínez-Martínez, I., Corrales, F. J., Miqueo, C., Miñano, A., Vicente, V., & Corral, J. (2009). Effect of citrullination on the function and conformation of antithrombin. *The FEBS Journal, 276*(22), 6763–6772.

Ossovskaya, V. S., & Bunnett, N. W. (2004). Protease-activated receptors: Contribution to physiology and disease. *Physiological Reviews, 84*(2), 579–621.

Padilla, C., Lobos, O., Hubert, E., González, C., Matus, S., Pereira, M., Hasbun, S., & Descouvieres, C. (2006). Periodontal pathogens in atheromatous plaques isolated from patients with chronic periodontitis. *Journal of Periodontal Research, 41*(4), 350–353.

Pathirana, R. D., O'Brien-Simpson, N. M., Brammar, G. C., Slakeski, N., & Reynolds, E. C. (2007). Kgp and RgpB, but not RgpA, are important for Porphyromonas gingivalis virulence in the murine periodontitis model. *Infection and Immunity, 75*(3), 1436–1442.

Persson, G. R., Engel, D., Whitney, G., Darveau, R., Weinberg, A., Brunsvold, M., & Page, R. C. (1994). Immunization against Porphyromonas gingivalis inhibits progression of experimental periodontitis in nonhuman primates. *Infection and Immunity, 62*(3), 1026–1031.

Petersen, P. E., Bourgeois, D., Ogawa, H., Estupinan-Day, S., & Ndiaye, C. (2005). The global burden of oral diseases and risks to oral health. *Bulletin of the World Health Organization, 83*(9), 661–669.

Pihlstrom, B. L. (2001). Periodontal risk assessment, diagnosis and treatment planning. *Periodontology 2000, 25*, 37–58.

Pike, R., McGraw, W., Potempa, J., & Travis, J. (1994). Lysine- and arginine specific proteinases from Porphyromonas gingivalis. Isolation, characterization, and evidence for the existence of complexes with hemagglutinins. *The Journal of Biological Chemistry, 269*(1), 406–411.

Pischon, N., Pischon, T., Kröger, J., Gülmez, E., Kleber, B. M., Bernimoulin, J. P., Landau, H., Brinkmann, P. G., Schlattmann, P., Zernicke, J., Buttgereit, F., & Detert, J. (2008). Association among rheumatoid arthritis, oral hygiene, and periodontitis. *Journal of Periodontology, 79*(6), 979–986.

Potempa, J., & Travis, J. (1996). Porphyromonas gingivalis proteinases in periodontitis, a review. *Acta Biochimica Polonica, 43*(3), 455–465.

Potempa, J., Pike, R., & Travis, J. (1995). The multiple forms of trypsin-like activity present in various strains of Porphyromonas gingivalis are due to the presence of either Arg-gingipain or Lys-gingipain. *Infection and Immunity, 63*(4), 1176–1182.

Potempa, J., Banbula, A., & Travis, J. (2000). Role of bacterial proteinases in matrix destruction and modulation of host responses. *Periodontology 2000, 24*, 153–192.

Potikuri, D., Dannana, K. C., Kanchinadam, S., Agrawal, S., Kancharla, A., Rajasekhar, L., Pothuraju, S., & Gumdal, N. (2012). Periodontal disease is significantly higher in non-smoking

treatment-naive rheumatoid arthritis patients: Results from a case-control study. *Annals of the Rheumatic Diseases, 71*(9), 1541–1544.
Preshaw, P. M., & Heasman, P. A. (2002). Prostaglandin E2 concentrations in gingival crevicular fluid: Observations in untreated chronic periodontitis. *Journal of Clinical Periodontology, 29*(1), 15–20.
Pritzker, L. B., Joshi, S., Gowan, J. J., Harauz, G., & Moscarello, M. A. (2000). Deimination of myelin basic protein.1. Effect of deimination of arginyl residues of myelin basic protein on its structure and susceptibility to digestion by cathepsin D. *Biochemistry, 39*(18), 5374–5381.
Proost, P., Loos, T., Mortier, A., Schutyser, E., Gouwy, M., Noppen, S., Dillen, C., Ronsse, I., Conings, R., Struyf, S., Opdenakker, G., Maudgal, P. C., & Van Damme, J. (2008). Citrullination of CXCL8 by peptidylarginine deiminase alters receptor usage, prevents proteolysis, and dampens tissue inflammation. *The Journal of Experimental Medicine, 205*(9), 2085–2097.
Pyrc, K., Milewska, A., Kantyka, T., Sroka, A., Maresz, K., Koziel, J., Nguyen, K. A., Enghild, J. J., Knudsen, A. D., & Potempa, J. (2013). Inactivation of epidermal growth factor by Porphyromonas gingivalis as a potential mechanism for periodontal tissue damage. *Infection and Immunity, 81*(1), 55–64.
Quirke, A. M., Lugli, E. B., Wegner, N., Hamilton, B. C., Charles, P., Chowdhury, M., Ytterberg, A. J., Zubarev, R. A., Potempa, J., Culshaw, S., Guo, Y., Fisher, B. A., Thiele, G., Mikuls, T. R., & Venables, P. J. (2014). Heightened immune response to autocitrullinated Porphyromonas gingivalis peptidylarginine deiminase: A potential mechanism for breaching immunologic tolerance in rheumatoid arthritis. *Annals of the Rheumatic Diseases, 73*(1), 263–269.
Quirke, A. M., Perry, E., Cartwright, A., Kelly, C., De Soyza, A., Eggleton, P., Hutchinson, D., & Venables, P. J. (2015). Bronchiectasis is a model for chronic bacterial infection inducing autoimmunity in rheumatoid arthritis. *Arthritis & Rhematology, 67*(9), 2335–2342.
Raijmakers, R., Vogelzangs, J., Raats, J., Panzenbeck, M., Corby, M., Jiang, H., Thibodeau, M., Haynes, N., van Venrooij, W. J., Pruijn, G. J., & Werneburg, B. (2006). Experimental autoimmune encephalomyelitis induction in peptidylarginine deiminase 2 knockout mice. *The Journal of Comparative Neurology, 498*(2), 217–226.
Raijmakers, R., Zendman, A. J., Egberts, W. V., Vossenaar, E. R., Raats, J., Soede-Huijbregts, C., Rutjes, F. P., van Veelen, P. A., Drijfhout, J. W., & Pruijn, G. J. (2007). Methylation of arginine residues interferes with citrullination by peptidylarginine deiminases in vitro. *Journal of Molecular Biology, 367*(4), 1118–1129.
Rickard, A. H., Gilbert, P., High, N. J., Kolenbrander, P. E., & Handley, P. S. (2003). Bacterial coaggregation: An integral process in the development of multi-species biofilms. *Trends in Microbiology, 11*(2), 94–100.
Rodriguez, S. B., Stitt, B. L., & Ash, D. E. (2009). Expression of peptidylarginine deiminase from Porphyromonas gingivalis in Escherichia Coli: Enzyme purification and characterization. *Archives of Biochemistry and Biophysics, 488*(1), 14–22.
Rodríguez, S. B., Stitt, B. L., & Ash, D. E. (2010). Cysteine 351 is an essential nucleophile in catalysis by Porphyromonas gingivalis peptidylarginine deiminase. *Archives of Biochemistry and Biophysics, 504*(2), 190–196.
Rudney, J. D., Chen, R., & Sedgewick, G. J. (2001). Intracellular Actinobacillus actinomycetemcomitans and Porphyromonas gingivalis in buccal epithelial cells collected from human subjects. *Infection and Immunity, 69*(4), 2700–2707.
Russell, A. L. (1967). Epidemiology of periodontal disease. *International Dental Journal, 17*(2), 282–296.
Sandros, J., Papapanou, P., & Dahlén, G. (1993). Porphyromonas gingivalis invades oral epithelial cells in vitro. *Journal of Periodontal Research, 28*(3), 219–226.
Sandros, J., Papapanou, P. N., Nannmark, U., & Dahlén, G. (1994). Porphyromonas gingivalis invades human pocket epithelium in vitro. *Journal of Periodontal Research, 29*(1), 62–69.
Sato, K., Naito, M., Yukitake, H., Hirakawa, H., Shoji, M., McBride, M. J., Rhodes, R. G., & Nakayama, K. (2010). A protein secretion system linked to bacteroidete gliding motility

and pathogenesis. *Proceedings of the National Academy of Sciences of the United States of America, 107*(1), 276–281.

Scannapieco, F. A. (1998). Position paper of the American Academy of Periodontology: Periodontal disease as a potential risk factor for systemic diseases. *Journal of Periodontology, 69*(7), 841–850.

Scannapieco, F. A., Bush, R. B., & Paju, S. (2003). Association between periodontal disease and risk for nosocomial bacterial pneumonia and chronic obstructive pulmonary disease: A systematic review. *Annals of Periodontology, 8*(1), 54–69.

Scher, J. U., Ubeda, C., Equinda, M., Khanin, R., Buischi, Y., Viale, A., Lipuma, L., Attur, M., Pillinger, M. H., Weissmann, G., Littman, D. R., Pamer, E. G., Bretz, W. A., & Abramson, S. B. (2012). Periodontal disease and the oral microbiota in new-onset rheumatoid arthritis. *Arthritis and Rheumatism, 64*(10), 3083–3094.

Shaik-Dasthagirisaheb, Y. B., Kantarci, A., & Gibson, F. C., 3rd. (2010). Immune response of macrophages from young and aged mice to the oral pathogenic bacterium Porphyromonas gingivalis. *Immunity & Ageing, 7*, 15.

Shirai, H., Blundell, T. L., & Mizuguchi, K. (2001). A novel superfamily of enzymes that catalyze the modification of guanidino groups. *Trends in Biochemical Sciences, 26*(8), 465–468.

Shoji, M., Sato, K., Yukitake, H., Kondo, Y., Narita, Y., Kadowaki, T., Naito, M., & Nakayama, K. (2011). Por secretion system-dependent secretion and glycosylation of Porphyromonas gingivalis hemin-binding protein 35. *PloS One, 6*(6), e21372.

Signat, B., Roques, C., Poulet, P., & Duffaut, D. (2011). Fusobacterium nucleatum in periodontal health and disease. *Current Issues in Molecular Biology, 13*(2), 25–36.

Slade, D. J., Fang, P., Dreyton, C. J., Zhang, Y., Fuhrmann, J., Rempel, D., Bax, B. D., Coonrod, S. A., Lewis, H. D., Guo, M., Gross, M. L., & Thompson, P. R. (2015). Protein arginine deiminase 2 binds calcium in an ordered fashion: Implications for inhibitor design. *ACS Chemical Biology, 10*(4), 1043–1053.

Slots, J. (1977). The predominant cultivable microflora of advanced periodontitis. *Scandinavian Journal of Dental Research, 85*(2), 114–121.

Smalley, J. W., Birss, A. J., Kay, H. M., McKee, A. S., & Marsh, P. D. (1989). The distribution of trypsin-like enzyme activity in cultures of a virulent and an avirulent strain of Bacteroides Gingivalis W50. *Oral Microbiology and Immunology, 4*(3), 178–181.

Socransky, S. S., & Haffajee, A. D. (1992). The bacterial etiology of destructive periodontal disease: Current concepts. *Journal of Periodontology, 63*(4 Suppl), 322–331.

Socransky, S. S., Haffajee, A. D., Smith, C., & Dibart, S. (1991). Relation of counts of microbial species to clinical status at the sampled site. *Journal of Clinical Periodontology, 18*(10), 766–775.

Socransky, S. S., Haffajee, A. D., Cugini, M. A., Smith, C., & Kent, R. L., Jr. (1998). Microbial complexes in subgingival plaque. *Journal of Clinical Periodontology, 25*(2), 134–144.

Sojar, H. T., Han, Y., Hamada, N., Sharma, A., & Genco, R. J. (1999). Role of the amino-terminal region of Porphyromonas gingivalis fimbriae in adherence to epithelial cells. *Infection and Immunity, 67*(11), 6173–6176.

Sojar, H. T., Sharma, A., & Genco, R. J. (2002). Porphyromonas gingivalis fimbriae bind to cytokeratin of epithelial cells. *Infection and Immunity, 70*(1), 96–101.

Sokransky, S. S., & Haffajee, A. D. (2005). Periodontal microbial ecology. *Periodontology 2000, 38*, 135–187.

Stabholz, A., Soskolne, W. A., & Shapira, L. (2010). Genetic and environmental risk factors for chronic periodontitis and aggressive periodontitis. *Periodontology 2000, 53*, 138–153.

Steinert, P. M., Parry, D. A., & Marekov, L. N. (2003). Trichohyalin mechanically strengthens the hair follicle: Multiple cross-bridging roles in the inner root sheath. *The Journal of Biological Chemistry, 278*(42), 41409–41419.

Struyf, S., Noppen, S., Loos, T., Mortier, A., Gouwy, M., Verbeke, H., Huskens, D., Luangsay, S., Parmentier, M., Geboes, K., Schols, D., Van Damme, J., & Proost, P. (2009). Citrullination of

CXCL12 differentially reduces CXCR4 and CXCR7 binding with loss of inflammatory and anti-HIV-1 activity via CXCR4. *Journal of Immunology, 182*(1), 666–674.

Subrahmanyam, M. V., & Sangeetha, M. (2003). Gingival crevicular fluid a marker of the periodontal disease activity. *Indian Journal of Clinical Biochemistry, 18*(1), 5–7.

Tatakis, D. N., & Kumar, P. S. (2005). Etiology and pathogenesis of periodontal diseases. *Dental Clinics of North America, 49*(3), 491–516, v.

Toh, E. C., Dashper, S. G., Huq, N. L., Attard, T. J., O'Brien-Simpson, N. M., Chen, Y. Y., Cross, K. J., Stanton, D. P., Paolini, R. A., & Reynolds, E. C. (2011). Porphyromonas gingivalis cysteine proteinase inhibition by κ-casein peptides. *Antimicrobial Agents and Chemotherapy, 55*, 1155–1161.

Travis, J., Pike, R., Imamura, T., & Potempa, J. (1997). Porphyromonas gingivalis proteinases as virulence factors in the development of periodontitis. *Journal of Periodontal Research, 32*(1 Pt 2), 120–125.

Van der Velden, U. (1984). Effect of age on the periodontium. *Journal of Clinical Periodontology, 11*(5), 281–294.

Van der Velden, U., Abbas, F., Armand, S., Loos, B. G., Timmerman, M. F., Van der Weijden, G. A., Van Winkelhoff, A. J., & Winkel, E. G. (2006). Java project on periodontal diseases. The natural development of periodontitis: Risk factors, risk predictors and risk determinants. *Journal of Clinical Periodontology, 33*(8), 540–548.

Van der Velden, U., Kuzmanova, D., & Chapple, I. L. (2011). Micronutritional approaches to periodontal therapy. *Journal of Clinical Periodontology, 38*(Suppl 11), 142–158.

Van Dyke, T. E., Offenbacher, S., Place, D., Dowell, V. R., & Jones, J. (1988). Refractory periodontitis: Mixed infection with Bacteroides Gingivalis and other unusual Bacteroides species. *Journal of Periodontology, 59*(3), 184–189.

Veillard, F., Potempa, B., Guo, Y., Ksiazek, M., Sztukowska, M. N., Houston, J. A., Koneru, L., Nguyen, K. A., & Potempa, J. (2015). Purification and characterisation of recombinant his-tagged RgpB gingipain from Porphyromonas gingivalis. *Biological Chemistry, 396*(4), 377–384.

Vergnes, J. N., & Sixou, M. (2007). Preterm low birth weight and maternal periodontal status: A meta-analysis. *American Journal of Obstetrics and Gynecology, 196*(2), 135.e1–135.e7.

Vossenaar, E. R., Zendman, A. J., van Venrooij, W. J., & Pruijn, G. J. (2003). PAD, a growing family of citrullinating enzymes: Genes, features and involvement in disease. *BioEssays, 25*(11), 1106–1118.

Wang, P. L., Oido-Mori, M., Fujii, T., Kowashi, Y., Kikuchi, M., Suetsugu, Y., Tanaka, J., Azuma, Y., Shinohara, M., & Ohura, K. (2002). Effect of anti-CD14 antibody on experimental periodontitis induced by Porphyromonas gingivalis lipopolysaccharide. *Japanese Journal of Pharmacology, 89*(2), 176–183.

Wang, Y., Wysocka, J., Sayegh, J., Lee, Y. H., Perlin, J. R., Leonelli, L., Sonbuchner, L. S., McDonald, C. H., Cook, R. G., Dou, Y., Roeder, R. G., Clarke, S., Stallcup, M. R., Allis, C. D., & Coonrod, S. A. (2004). Human PAD4 regulates histone arginine methylation levels via demethylimination. *Science, 306*(5694), 279–283.

Wegner, N., Wait, R., Sroka, A., Eick, S., Nguyen, K. A., Lundberg, K., Kinloch, A., Culshaw, S., Potempa, J., & Venables, P. J. (2010). Peptidylarginine deiminase from Porphyromonas gingivalis citrullinates human fibrinogen and α-enolase: Implications for autoimmunity in rheumatoid arthritis. *Arthritis and Rheumatism, 62*(9), 2662–2672.

Weinberg, A., Belton, C. M., Park, Y., & Lamont, R. J. (1997). Role of fimbriae in Porphyromonas gingivalis invasion of gingival epithelial cells. *Infection and Immunity, 65*(1), 313–316.

White, D., & Mayrand, D. (1981). Association of oral Bacteroides with gingivitis and adult periodontitis. *Journal of Periodontal Research, 16*(3), 259–265.

Wingrove, J. A., DiScipio, R. G., Chen, Z., Potempa, J., Travis, J., & Hugli, T. E. (1992). Activation of complement components C3 and C5 by a cysteine proteinase (gingipain-1) from Porphyromonas (Bacteroides) gingivalis. *The Journal of Biological Chemistry, 267*(26), 18902–18907.

Wright, P. W., Bolling, L. C., Calvert, M. E., Sarmento OF, Berkeley, E. V., Shea, M. C., Hao, Z., Jayes, F. C., Bush, L. A., Shetty, J., Shore, A. N., Reddi, P. P., Tung, K. S., Samy, E., Allietta, M. M., Sherman, N. E., Herr, J. C., & Coonrod, S. A. (2003). ePAD, an oocyte and early embryo-abundant peptidylarginine deiminase-like protein that localizes to egg cytoplasmic sheets. *Developmental Biology, 256*(1), 73–88.

Yucel-Lindberg, T., & Båge, T. (2013). Inflammatory mediators in the pathogenesis of periodontitis. *Expert Reviews in Molecular Medicine, 15*, e7

Zambon, J. J., Reynolds, H. S., & Slots, J. (1981). Black-pigmented Bacteroides spp. in the human oral cavity. *Infection and Immunity, 32*(1), 198–203.

Zhou, X. Y., Gao, J. L., Hunter, N., Potempa, J., & Nguyen, K. A. (2013). Sequence-independent processing site of the C-terminal domain (CTD) influences maturation of the RgpB protease from Porphyromonas gingivalis. *Molecular Microbiology, 89*(5), 903–917.

Chapter 8
Citrullination and Neutrophil Extracellular Traps

Nishant Dwivedi, Hui-Hsin Chang, and I-Cheng Ho

8.1 Introduction

Neutrophil extracellular traps (NETs) were first reported in 2004 as a meshwork of chromatin entangling Gram-positive (*Staphylococcus aureus*) and Gram-negative (*Salmonella typhimurium*) bacteria (Brinkmann et al. 2004). The threads of chromatin wrapped around the bacteria are about 17–30 nm in diameter, suggesting active uncoiling of condensed chromatin, a feature that establishes the formation of NETs (or NETosis) as an active and regulated form of neutrophil death. In recent years, NETosis has attracted significant attention-studies have attested to both the uniqueness and importance of NETosis in defending the host from infectious onslaught. NETs are laced with many granule-derived antimicrobial proteins such as neutrophil elastase (NE), myeloperoxidase (MPO), cathepsin G, and LL-37, which help in killing pathogens after trapping them in the chromatin meshwork. Notably, cathepsin G and NE are also involved in the formation of NETs after sensing invading pathogens by dissolving both nuclear and cellular membranes.

N. Dwivedi • H.-H. Chang • I.-C. Ho (✉)
Division of Rheumatology, Immunology, and Allergy, Department of Medicine, Brigham and Women's Hospital, Boston, MA 02115, USA

Harvard Medical School, Boston, MA 02115, USA
e-mail: IHO@partners.org

A.P. Nicholas et al. (eds.), *Protein Deimination in Human Health and Disease*,
DOI 10.1007/978-3-319-58244-3_8

8.2 Formation of NETs

8.2.1 Essential Role of Histone Citrullination in NETosis

One important step prior to the release of DNA during NETosis is decondensation of chromatin, which is usually kept within the nuclear envelop in a higher-ordered configuration by interaction between positively charged histones and negatively charged DNA together with various nuclear stabilizing proteins (Chen and Dent 2014). Decondensation of chromatin requires the citrullination of histones. Citrullination, also called deimination, is a unique posttranslational modification in which positively charged arginines are converted to neutral citrullines. This modification is catalyzed by the peptidylarginine deiminases (PADs), including PAD1–4 and PAD6, in a calcium-dependent manner. Histone citrullination is detected in decondensed chromatin in neutrophils (Wang et al. 2009). Overexpression of PAD4 results in histone hypercitrullination, chromatin decondensation, and the formation of NET-like structures in both granulocyte-like cells and non-granulocytes (Wang et al. 2009; Leshner et al. 2012). This effect is dependent on the enzymatic activity of PAD4. PAD4-mediated citrullination of arginine residues in the core (H2A, H3, and H4) and linker histones (H1) is essential for the unraveling of chromatin during NET formation (Wang et al. 2009; Dwivedi et al. 2014). Accordingly, pan-PAD inhibitors reduce histone citrullination and suppress the formation of NETs (Wang et al. 2009), and neutrophils deficient in PAD4 are also deficient in forming NETs (Leshner et al. 2012; Li et al. 2010; Hemmers et al. 2011). Neutrophils express the highest level of PAD4 among all hematopoietic cells, and its transcript is almost 100-fold higher than that of other PADs. Such a high level of PAD4 probably explains why only neutrophils are endowed with the unique ability to form NETs even though invading pathogens can be recognized by almost all cells through pathogen-associated molecular pattern (PAMP) recognition receptors.

How PAD4 promotes chromatin decondensation is still not fully understood. PAD4 does not appear to affect the in vitro formation of nucleosome core particle assembled with core histones H3, H2B, H2A, and H4. Instead, PAD4 activity is required for the disassembly of higher-order chromatin structure orchestrated by linker histone H1 (Wang et al. 2009). It is possible that citrullination leads to a global neutralization of the positive charge of chromatin, thereby disrupting the electrostatic interaction between DNA and histones. Chromatin condensation and the formation of heterochromatin require the binding of heterochromatin protein 1 (HP1) to methylated K9 of histone H3. This binding is disrupted by the citrullination of the neighboring R8 residues (Leshner et al. 2012). Thus, citrullination of histones very likely also inhibits the binding of chromatin condensation factors.

8.2.2 Other Critical Elements for the Formation of NETs

The dynamics of chromatin extrusion from neutrophils during NETosis is uniquely fascinating. From recognition of external stimulus to expulsion of fine threadlike chromatin, a complex set of events are orchestrated in neutrophils in a synchronized as well as hierarchical manner. Though invading pathogens can be recognized by almost all cells through PAMP receptors, only neutrophils are endowed with the special set of characteristics that facilitate NET formation. The chromatin of neutrophils at basal state is relatively decondensed facilitating a quick release of chromatin threads. Much higher expression of enzymes such as NADPH, PAD4, PR3, and MPO in neutrophils also empowers neutrophils for NET formation.

NET Stimuli

A wide variety of inflammatory stimuli have been identified for triggering NETs. Other than microbial agents such as bacteria and fungi, NETs have been induced by pharmacological agents (phorbol 12-myristate 13-acetate [PMA], calcium ionophore, and tamoxifen) (Parker et al. 2012; Corriden et al. 2015), inflammatory cytokines (tumor necrosis factor-α, IL-1β, IL-8) (Keshari et al. 2012; Gupta et al. 2005; Mitroulis et al. 2011), oxidized low-density lipoprotein (Awasthi et al. 2016), activated endothelial cells and platelets (Clark et al. 2007; Gupta et al. 2010), immune complexes (Behnen et al. 2014), and even monosodium urate crystals (Mitroulis et al. 2011). The sensors triggering NETosis by some of these stimuli have been identified. For example, oxidized LDL acts through TLR 2 and 6 (Awasthi et al. 2016) and platelets through P-selectin binding to P-selectin glycoprotein ligand-1 (Etulain et al. 2015). Such a wide range of stimuli culminating in NET formation suggests a common downstream mechanism that is not yet fully deciphered. In contrast, prostaglandin E2 (PGE2) can inhibit NETosis through EP2 and EP4 receptors in vitro and in vivo (Shishikura et al. 2016; Domingo-Gonzalez et al. 2016), providing an additional layer of regulation of NETosis.

NET Signaling Pathways

Most commonly, NET formation has been studied by using PMA as stimulus where generation of reactive oxygen species (ROS) by NADPH oxidase plays a key role in NET production (Parker et al. 2012; Kirchner et al. 2012). Patients with chronic granulomatous disease, who have decreased NADPH oxidase activity, show impaired NET formation (Fuchs et al. 2007). PMA is a strong driver of NETosis for mouse and human neutrophils and induces histone citrullination and NETosis in a PAD4-dependent manner in mouse neutrophils (Li et al. 2010). We have also confirmed the induction of histone citrullination by PMA with immunocytochemistry in human neutrophils (data not shown). Surprisingly, at least two groups have independently demonstrated that PMA does not induce citrullination in human neutrophils (Zhou et al. 2015; Neeli and Radic 2013). The cause of this discrepancy is still unclear.

Signaling from PMA and other inducers of NETs leads to the activation of NADPH oxidase through protein kinase C (Neeli and Radic 2013; Gray et al. 2013). Various isoforms of PKC play contrasting roles in the activation of PAD4 and NET formation. While most isoforms of PKC enhance citrullination and NETosis, PKC-alpha was uniquely identified to inhibit PMA-induced PAD activation and NET formation (Neeli and Radic 2013). In addition, PGE2 binding to EP2 and EP4, two G protein-coupled receptors, inhibits PMA-induced NETosis through the c-AMP-PKA pathway (Shishikura et al. 2016; Domingo-Gonzalez et al. 2016).

NETosis shares some features with necroptosis, but whether the pathways involved in necroptosis have a role in NETosis is not clear as conflicting reports of the role of receptor-interacting protein kinase-1 and the mixed lineage kinase domain-like (RIPK-1/MLKL) pathway were published recently (Amini et al. 2016; Desai et al. 2016). The discrepancies can be due to differences in the way stimulation was given to neutrophils or the way NETosis was quantitated. Future studies with more robust methodologies will be needed to resolve these discrepancies. Signaling through the RAF/MEK/ ERK pathway is also found important for citrullination and NETosis, but the full mechanism is not yet understood (Hakkim et al. 2011).

NADPH oxidase-independent PAD4 activation and NET formation is also reported where ionomycin-induced opening of SK3 potassium channels mediates NET formation (Douda et al. 2015). In this report, mitochondrial ROS production was found important for NETosis (Douda et al. 2015). Intracellular scavenging of ROS produced by NADPH oxidase by a chemical inhibitor is able to inhibit NET formation in response to PMA but not when whole bacteria are used to stimulate human neutrophils (Parker et al. 2012), indicating that though NADPH oxidase-generated ROS is important for PAD4-mediated NETosis, the role of other factors in making of NETs is equally important.

Enzymes

Various neutrophil-associated serine proteases such as NE, proteinase 3 (PR3), and MPO are found entangled in NETs (Urban et al. 2009) and are thought to play an important role in NETosis. Mice deficient in elastase are less potent in making NETs as compared to the wild-type controls (Papayannopoulos et al. 2010; Martinod et al. 2016) as are neutrophils from MPO-deficient individuals when subjected either to PMA or bacteria (Parker et al. 2012). Though the exact role of these enzymes in NETosis is not understood, it is believed that neutrophil activation directs the vesicular-coated serine proteases toward the nucleus where they digest nuclear membrane (Parker et al. 2012; Papayannopoulos et al. 2010; Munoz-Caro et al. 2015).

Cytoskeleton

The release of chromatin threads during NETosis also requires cytoskeleton elements. Chemical inhibitors of actin microfilaments assembly (cytochalasin D) as well as microtubular polymerization (nocodazole) reduce NET formation as well as histone citrullination (Neeli et al. 2009). It is possible that these cytoskeleton elements play a role in trafficking of PAD4 as well as serine proteases to the nucleus and also provide a meshwork to disentangle chromatin threads for spreading.

8.2.3 Regulation of PAD Activity

How the activity of PADs is regulated is still not fully understood. PADs require a high concentration of calcium to be enzymatically active. Ionophore, pore-forming complex, and even transfection are known to activate PADs by increasing intracellular calcium concentration (Romero et al. 2013). In addition, PAD4 is known to form homodimers, and the dimerization is critical for its optimal calcium binding and full enzymatic activity (Liu et al. 2011). PAD4 can undergo auto-citrullination *in vivo* and *in vitro* (Andrade et al. 2010; Slack et al. 2011). However, it is still controversial as to whether auto-citrullination has any impact on the enzymatic activity of PAD4. PAD4 is tyrosyl phosphorylated in primary mouse macrophages, but the phosphorylation sites and the role of tyrosine phosphorylation in regulating the activity of PADs are unknown (Chang et al. 2015). A recent study indicates that a subset of anti-PAD4 autoantibodies found in patients with RA can enhance the catalytic activity of PAD4 (Darrah et al. 2013), suggesting that its activity can be modulated through interacting with other proteins.

8.3 Protective NETs: A Mechanism to Counter Infection

Immune system has evolved various strategies to protect humans from thousands of microbial organisms they interact with. Being the most numerous circulating immune cells in the body as well as being the first responders to infections, neutrophils are equipped with diverse mechanisms to fight infections. Before NETosis was discovered, phagocytosis followed by reactive oxygen species generation was considered to be the dominant neutrophil mechanism to counter infections. NETosis probably plays an equally important role in limiting bacteria and fungi infections. NETs are known to be effective in trapping Gram-positive (*S. aureus* (Brinkmann et al. 2004; Pilsczek et al. 2010), *Listeria monocytogenes* (Munafo et al. 2009)) and Gram-negative (*Shigella flexneri* (Brinkmann et al. 2004), *Klebsiella pneumoniae* (Papayannopoulos et al. 2010) bacteria and fungi (*Candida albicans* (Urban et al. 2006), *Aspergillus fumigatus* (McCormick et al. 2010)).

It was suggested that the decision of whether to phagocytose or to make NETs is determined by pathogen size as NETosis is commonly observed with larger fungal hyphae of *C. albicans* or aggregates of *Mycobacterium bovis* (Branzk et al. 2014). To the contrary, reports have suggested release of human and mouse NETs in response to intracellular parasites, such as *Toxoplasma gondii* (Abi Abdallah et al. 2012) and *Plasmodium falciparum* (Baker et al. 2008). NETs were also formed in response to human immunodeficiency virus (Saitoh et al. 2012). The ability of neutrophils to cast their NETs in response to such a diverse array of infectious agents makes NETs an important marker for inflammation, and as such NET-associated components (DNA, MPO, and PR3) in serum have come up as markers for NETosis. While NETs are enriched with bactericidal molecules, it remains controversial if

NETs can indeed kill entrapped pathogens. *S. aureus and C. albicans* recovered from NETs are fully viable in vitro (Menegazzi et al. 2012), suggesting that NETs have limited intrinsic killing activity.

Neutrophils from patients with chronic granulomatous disease, which lack functional NADPH oxidase, are less efficient in clearing *S. aureus* (Fuchs et al. 2007). However, it is still unclear how much of the impaired bacteria clearing can be attributed to impaired NETosis. In vivo animal studies provide us with a glimpse into the role of NETs in defense against bacterial infection. *S. pyogenes* is capable of secreting DNase to dissolve DNA and escape NETs, thereby causing necrotizing fasciitis upon subcutaneous inoculation in mice. A mutant strain of *S. pyogenes* that is deficient in such DNase is unable to escape NETs and therefore less virulent in wild-type mice (Li et al. 2010). However, this mutant strain is almost as virulent as wild-type strain in PAD4-deficient mice, which are unable to form NETs. In contrast, PAD4-deficient mice have a comparable survival as wild-type mice after cecal ligation and puncture (Martinod et al. 2015), a severe polymicrobial sepsis model mediated mainly by Gram-negative bacteria. It remains to be determined if the discrepancy is due to the difference in the extent of infection (local versus systemic) or in strains of bacteria (single Gram-positive bacteria versus multiple Gram-negative bacteria). The in vivo roles of NETs in immune defense against other pathogens, including virus and fungi, remain to be determined. PAD4-deficient mice display no apparent difference in viral load, weight loss, and survival after influenza infection when compared to wild-type mice (Hemmers et al. 2011).

8.4 Deleterious NETs: Adverse Effects of Uncontrolled NETosis

Neutrophils are very reactive immune cells laced with a plethora of pro-inflammatory substances. When stimulated, their ability to make inflammatory cytokines and various antibacterial proteases increases manifold. Thus, sudden disruption of cellular and nuclear membranes while undergoing NETosis results in massive inflammation in the local environment. This might be a welcome mechanism while fighting local invading pathogens, but unregulated NETosis can cause significant harm to hosts. NETosis has been implicated in the pathogenesis of many human diseases, such as diabetes, asthma, and cystic fibrosis, which are discussed in the other chapters. This chapter will focus on the pathogenic role of citrullination-dependent NETosis in thrombosis and autoimmune diseases, including rheumatoid arthritis and systemic erythematosus.

8.4.1 NETs and Thrombosis

NETs provide a structural scaffold for platelets and RBC entrapment and aggregation. The negatively charged DNA backbone further promotes thrombosis by activating plasma proteins such as fibrinogen, fibronectin, and von Willebrand factor (VWF)

(Fuchs et al. 2010). Expectedly, disruption of NETs by DNase 1 reduces deep vein thrombosis in experimental mouse models (Brill et al. 2012). PAD enzyme-induced citrullination is critical for NET formation, and the use of PAD inhibitors in mouse models of atherosclerosis results in reduced plaque formation and thrombus (Knight et al. 2014a). In a stenosis model of deep vein thrombosis, only 10% of PAD4-deficient mice showed thrombus formation after inferior vena cava stenosis as compared to above 90% of control mice (Martinod et al. 2013). Apart from the DNA, histones in NETs can also activate platelets and result in thrombin generation (Semeraro et al. 2011). Other components of NETs such as granule-derived serine proteases (elastase, cathepsin G) can degrade tissue factor pathway inhibitor, which serves as an important antagonist of coagulation (Massberg et al. 2010).

Antiphospholipid antibody syndrome (APLAS) is characterized by thromboembolic events induced by antiphospholipid antibodies, such as anticardiolipin and anti-β2 glycoprotein I. Serum from patients with APLAS has a decreased ability to degrade NETs and contains more cell-free DNA and NETs (Leffler et al. 2014; Yalavarthi et al. 2015). In addition, their neutrophils are prone to undergo NETosis spontaneously. Furthermore, purified antiphospholipid antibodies or serum IgG fraction of patients with APLAS triggers NETosis of control neutrophils (Yalavarthi et al. 2015). Thus dysregulated NETosis is a potential mechanism of thromboembolism seen in APLAS.

8.4.2 NETs and Rheumatoid Arthritis

The role of citrullination in driving autoimmune inflammation is more clearly defined in rheumatoid arthritis (RA) where anti-citrullinated protein antibodies (ACPA) have been included into the revised 2010 American College of Rheumatology criteria for diagnosing RA (Aletaha et al. 2010). ACPA are a unique feature of RA and are detected in about 70% of RA patients and show marked sensitivity and specificity in diagnosing RA (Payet et al. 2014). Cigarette smoking and periodontitis, two major risk factors of RA, have been shown to increase the level of citrullinated proteins (Makrygiannakis et al. 2008; Damgaard et al. 2015; McGraw et al. 1999; Rosenstein et al. 2004); a haplotype of *padi4*, which encodes PAD4, stabilizes PAD4 transcript and is associated with a higher risk of RA (Suzuki et al. 2003; Kang et al. 2006; Hou et al. 2012; Hoppe et al. 2009). These observations suggest that dysregulated citrullination is at the root of RA pathogenesis.

Hypercitrullination not only expands the pool of citrullinated autoantigens, thereby adding fuel to the fire of RA autoimmunity, but also propagates RA pathogenesis through NETosis. Neutrophils are abundant in synovial fluid of inflamed RA joints. Spontaneous NETosis upon cultivation in vitro is seen in neutrophils from RA patients as compared to control subjects (Sur Chowdhury et al. 2014). NETosis contributes to the pathogenesis of RA in at least two ways. NETs are a rich source of citrullinated RA antigens. Both citrullinated vimentin and citrullinated histones have been shown to decorate NETs (Khandpur et al. 2013). A recent study demonstrates that as high as 40% of monoclonal antibodies derived from synovial ectopic lymphoid B cells recognize citrullinated histone H2A/H2B and bind to NETs (Corsiero et al. 2015). Interestingly, antibodies to citrullinated histones are more abundant in Felty's syndrome, a severe form of arthritis, implying the deleterious consequences of neutrophil activation in joint destruction (Dwivedi et al. 2012). Additionally, NETs can induce the expression of inflammatory cytokines, such as IL-6 and IL-18, as well as CCL20 and ICAM-1 from fibroblast-like synoviocytes (Khandpur et al. 2013). This effect of NETs can be partly abrogated by DNase treatment, suggesting that the architecture of NETs is critical for its pro-inflammatory property.

While hypercitrullination caused by various environmental and genetic risk factors surely contributes to enhanced NETosis, RA neutrophils readily contain more NET-inducing molecules, such as ROS, MPO, and NE, and have more nuclear PAD4 despite a normal level of total PAD4. One possible explanation for the "readiness" of NETosis is that inflammatory cytokines and ACPA are present in the serum and synovial fluid of RA patients (Khandpur et al. 2013). Neutrophils of healthy donors can be induced to form NETs by incubating with IL-17 and TNFα, two key inflammatory cytokines of RA, or serum from RA patients. The NET-inducing effect of RA serum is abrogated by depleting its IgG fraction, suggesting that ACPA is one of the serum factors responsible for conditioning neutrophils for NETosis. In addition, neutrophils from inflamed RA joints display hypercitrullination of proteins across a broad range of molecular weight (Romero et al. 2013). This pattern of hypercitrullination is caused by two immune-mediated pore-forming pathways: perforin and the membrane-attacking complement complex.

In agreement with the strong in vitro data, a pan-PAD inhibitor is therapeutic in the murine model of collagen-induced arthritis (Willis et al. 2011). Similarly, deficiency of PAD4 reduces the severity of arthritis in GPI-immunized mice or TNFα-transgenic mice (Seri et al. 2015; Shelef et al. 2014). The development of arthritis in these three models depends on concerted action of both adaptive and innate immune cells. However, in arthritis models that do not require adaptive immune cells, the PAD inhibitor or PAD4 deficiency has little impact on the severity of arthritis (Willis et al. 2011; Rohrbach et al. 2012). Thus, it remains to be demonstrated if NETs can indeed contribute to the inflammatory phase of arthritis *in vivo*.

8.4.3 NETs and Systemic Lupus Erythematosus

Systemic lupus erythematosus (SLE) is a systemic autoimmune disease characterized by antibodies against many nuclear autoantigens, including ribonucleoprotein (RNP), dsDNA and histones, and deposition of immune complex in tissues. Many of the lupus nuclear autoantigens are also abundantly exposed during NETosis, and antibodies against citrullinated linker histones are detected in a subset of SLE patients (Dwivedi et al. 2014). Furthermore, netting neutrophils are readily detected in affected skin and organs of lupus patients (Villanueva et al. 2011); pan-PAD inhibitors attenuate disease activity in lupus-prone mice (Knight et al. 2013, 2014b). These observations further strengthen the link between citrullination-dependent NETosis and SLE.

Extracellular DNA, released through NETosis or from apoptotic cells, very likely plays a pathogenic role in SLE. Deficiency in DNase 1 results in spontaneous development of lupus-like features in mice (Napirei et al. 2000), and genetic variations at the DNase 1 locus are associated with a higher risk of SLE (Bodano et al. 2006; AlFadhli et al. 2010). In addition, serum from as high as 30% of SLE patients shows defective DNase activity with impaired ability to clear NETs (Hakkim et al. 2010; Leffler et al. 2013, 2015).

Type 1 interferon (IFN), including IFNα and IFNβ, has been implicated in the pathogenesis of lupus (Wahren-Herlenius and Dorner 2013). It has been demonstrated that NETs released by neutrophils of lupus patients, but not healthy donors, induce the production of type 1 IFN by plasmacytoid DC via TLR9 (Villanueva et al. 2011; Lande et al. 2011; Garcia-Romo et al. 2011). This unique property of lupus NETs can be recapitulated by inducing NETosis of control neutrophils with immune complex containing RNP (RNP-IC) (Lood et al. 2016), which can be detected in lupus patients. Lupus NETs or RNP-IC-induced NETs from healthy donors are enriched with oxidized mitochondrial DNA in addition to nuclear DNA (Lood et al. 2016; Wang et al. 2015). Mitochondrial DNA obtained from lupus or RNP-IC-induced NETs can promote the expression of type 1 IFN and several other inflammatory cytokines, such as IL-1β and IL-6, in human PBMC or plasmacytoid DC in vitro or after injection into mice (Lood et al. 2016). The induction of type 1 IFN by lupus NETs is dependent on the DNA sensor STING but not Myd88. This observation is consistent with a report showing that oxidized DNA induces the expression of IFNα through STING but not TLR9 (Gehrke et al. 2013). The release of oxidized mitochondrial DNA into NETs requires ROS bursts from mitochondria. Accordingly, administration of a mitochondrial ROS scavenger attenuates lupus-like disease in MRL/*lpr* mice, a commonly used murine lupus model (Lood et al. 2016). In addition, treatment with metformin, which inhibits mitochondrial ROS production, reduces the content of mitochondrial DNA in NETs and results in decreases in clinical flares and prednisone exposure in lupus patients in a small clinical trial (Wang et al. 2015).

The Role of Low-Density Granulocytes

A subset of neutrophils that segregated with mononuclear cells on density gradient centrifugation of whole blood is known as low-density granulocytes (LDGs) (Denny et al. 2010). These LDGs carry an inflammatory phenotype and are capable of secreting high levels of type 1 interferon, TNFα, and IFN-γ (Denny et al. 2010). Elevated numbers of LDGs are found in juvenile-onset SLE (Midgley and Beresford 2016) as compared to controls. LDGs show increased propensity to make NETs and very likely are partly responsible for the unique property of lupus neutrophils and NETs. Elevated numbers of LDGs are also detected in patients with anti-neutrophil cytoplasmic antigen (ANCA)-associated vasculitis (Grayson et al. 2015), a disease characterized by antibodies against two NET-associated proteins PR3 and MPO, suggesting overlapping pathogenic mechanisms between ANCA-associated vasculitis and SLE.

8.5 PTPN22: A Novel Inhibitor of Citrullination and NETosis

8.5.1 Phosphatase-Dependent and Phosphatase-Independent Function of PTPN22

A major advance in our understanding of the pathogenic role of citrullination and NETosis in autoimmune diseases comes from the discovery that PTPN22 is a natural inhibitor of PAD activity. PTPN22 is a non-receptor protein tyrosine phosphatase preferentially expressed in hematopoietic cells (Cohen et al. 1999). It has been shown to attenuate the activation signals in lymphocytes by interacting with several signaling molecules, such as Csk, Zap70, and Grb2 (Gjorloff-Wingren et al. 1999; Hill et al. 2002; Ghose et al. 2001; Wu et al. 2006). Indeed, PTPN22-deficient mice develop splenomegaly at ages older than 6 months due to hyperactivation of T cells (Hasegawa et al. 2004). It also negatively regulates the homeostasis of Treg cells partly by inhibiting the expression and signaling of glucocorticoid-induced TNFR family-related protein (GITFR) (Brownlie et al. 2012; Maine et al. 2012; Nowakowska and Kissler 2016). In addition, PTPN22 is detected in the cytoplasm and nucleus of macrophages (Chang et al. 2013). Cytoplasmic PTPN22 suppresses the IFNγ/LPS-mediated polarization of macrophage to the M1 phenotype, which is characterized by the expression of inflammatory cytokines. In contrast, nuclear PTPN22 promotes IL-4-/IL-13-mediated polarization of macrophages to the M2 phenotype, featuring the production of anti-inflammatory cytokines. Furthermore, PTPN22 can attenuate IFNα-induced signals in myeloid cells (Holmes et al. 2015). A recent study further demonstrates that PTPN22 dephosphorylates NLRP3 and activates this key component of inflammasome, thereby enhancing the production of processed IL-1β (Spalinger et al. 2016). All these functions depend on the phosphatase activity of PTPN22.

In addition to the conserved PTP domain in its N-terminal half, the C-terminal half of PTPN22 contains no known structural domain with the exception of four proline-rich regions. Thus, PTPN22 is capable of carrying out phosphatase-independent function. Indeed, it was recently discovered that PTPN22 promotes TLR-induced expression of type 1 IFN by myeloid cells (Wang et al. 2013). This function is independent of its phosphatase activity. However this observation is somewhat counterintuitive given the report that PTPN22 actually inhibits the signaling of type 1 IFN (Holmes et al. 2015). In addition, we found that PTPN22-deficient macrophages produced more IFNβ in response to LPS/IFNγ stimulation (Fig. 8.1). The cause of this discrepancy is still unknown.

8.5.2 PTPN22 Is an Autoimmune Gene

The huge interest in PTPN22 is due to the observation that a C-to-T single nucleotide polymorphism (SNP), located at position 1858 of human PTPN22 cDNA, is associated with a high risk of developing RA (Kunz and Ibrahim 2011; Begovich et al.

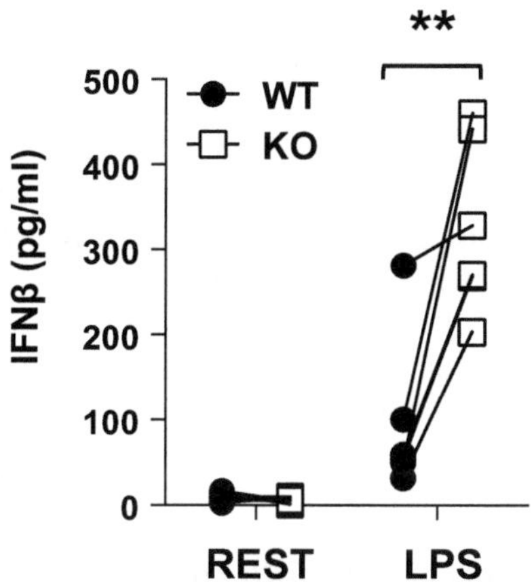

Fig. 8.1 PTPN22 inhibits LPS-induced expression of IFNβ by macrophages. Splenic macrophages (one million cells) of wild type (WT) and PTPN22-deficient (KO) mice were left unstimulated (rest) or stimulated with LPS (1 μg/mL) for 24 h. The concentration of IFNβ in supernatant was measured with ELISA. The data shown is cumulated results of six experiments (one pair of mice per experiment). The statistical analysis was performed with Student's *t* test

2004; Orozco et al. 2005; Lee et al. 2005). This C1858T SNP carries an odds ratio of nearly 2, which is the highest among all non-HLA genetic variations. In addition, it has synergistic effects with several RA-prone HLA alleles in raising the risk of RA (Kallberg et al. 2007). Interestingly, this synergy is observed only in RA patients who are positive for ACPA. This SNP also increases the risk of other autoimmune diseases, including lupus and type 1 diabetes, which are not characterized by ACPA. It converts a conserved arginine (R620) to a tryptophan (W620) and is very likely a causative SNP of RA. Knock-in mice carrying this R-to-W mutation spontaneously develop several autoimmune features (Carlton et al. 2005; Zhang et al. 2011; Dai et al. 2013). This phenotype is in sharp contrast to that of PTPN22-deficient mice, which do not develop autoimmune features (Hasegawa et al. 2004). In addition, both PTPN22 knockdown and transgenic NOD mice are resistant to diabetes (Zheng and Kissler 2013; Yeh et al. 2013), an animal model of type 1 diabetes. These animal data strongly indicate that the W620-PTPN22 is neither a gain nor a loss of function variant, but instead very likely functions as a dominant-negative mutant.

However, how the C1858T SNP has such a unique impact on ACPA+ RA and also increases the risk of other autoimmune diseases is still not fully understood. Its effect on the activation of lymphocytes is subtle and controversial. Both hypo-activation and hyperactivation have been observed in lymphocytes from donors carrying this SNP (Zhang et al. 2011; Vang et al. 2005; Aarnisalo et al. 2008; Thompson et al. 2014). Although abnormal homeostasis of B cells, including a decrease in percentage of memory B cells and an increase in transitional B cells, has been observed in healthy donors who are heterozygous for this SNP (Rieck et al. 2007; Habib et al. 2012), this finding cannot be confirmed in homozygous donors (Thompson et al. 2014). In addition, this SNP does not affect thymic selection, which is highly sensitive to even a subtle change in the strength in activation signals, and has little impact on the polarization of macrophages (Chang et al. 2013; Wu et al. 2014). A recent study indicates that this SNP results in a gain-of-function variant of PTPN22 that is more potent in dephosphorylating NLRP3 and in enhancing the production of processed IL-1β (Spalinger et al. 2016), which can be pathogenic in human RA. However, this model does not offer an explanation for the synergy between the C1858T SNP and RA-prone HLA in ACPA+ RA. In addition, uncontrolled activation of the inflammasome and excessive production of IL-1β often lead to auto-inflammatory diseases; however, the C1858T SNP is not associated with auto-inflammatory diseases.

8.5.3 PTPN22 Interacts with and Inhibits the Activity of PAD4

It was recently discovered that PTPN22 interacts with and suppresses the activity of PAD4 (Chang et al. 2015). PTPN22-deficient macrophages or lymph node cells display hypercitrullination; knocking down PTPN22 in human Jurkat T cells also results in hypercitrullination. Therefore, both human and mouse PTPN22 can suppress protein citrullination. Human PTPN22 is expressed as several alternatively spliced forms, each of which lacks a part of the protein (Chang et al. 2012, 2013). With the exception of PTPN22.6, which lacks the nearly entire PTP domain, all other isoforms retain full phosphatase activity. Surprisingly, only the full-length PTPN22 is capable of suppressing citrullination (Chang et al. 2015). In addition, a phosphatase-dead mutant PTPN22 is as efficient as wild-type PTPN22 in suppressing citrullination. These observations demonstrate that PTPN22 suppresses protein citrullination independently of its phosphatase activity. Despite the strong effect of PTPN22 deficiency on citrullination in lymphocytes and macrophages, PTPN22-deficient murine neutrophils have a normal level of citrullinated proteins and do not show increased tendency to form NETs, suggesting the presence of a compensatory mechanism that is sufficient to suppress citrullination and NETosis in the absence of PTPN22 in murine neutrophils. The transcript level of PAD4 in neutrophils is much higher than that in other hematopoietic cells according to the Immunological Genome Project (https://www.immgen.org). Thus, it would not be surprising if multiple mechanisms are needed to keep PAD4 in check in neutrophils.

8.5.4 The W620-PTPN22 Is a Dominant-Negative Variant of PTPN22

More importantly, the R-to-W conversion renders PTPN22 unable to interact with PAD4 and suppress citrullination (Chang et al. 2015). Accordingly healthy donors carrying the C1858T SNP display hypercitrullination in their PBMC. In addition, overexpression of W620-PTPN22, but not R620-PTPN22, in wild-type T cells, which already express a physiological level of R620-PTPN22, is sufficient to cause hypercitrullination, indicating that W620-PTPN22 is a dominant-negative variant. While deficiency of PTPN22 has little impact on mouse neutrophils, neutrophils from C1858T donors, which express both R620- and W620-PTPN22, contain a higher level of citrullinated proteins and have a high propensity to form NETs spontaneously ex vivo (Fig. 8.2), further demonstrating the dominant-negative nature of W620-PTPN22. Bayley et al. also discovered that neutrophils from donors carrying the C1858T SNP have a higher transmigration capacity in response to TNFα, stronger calcium influx

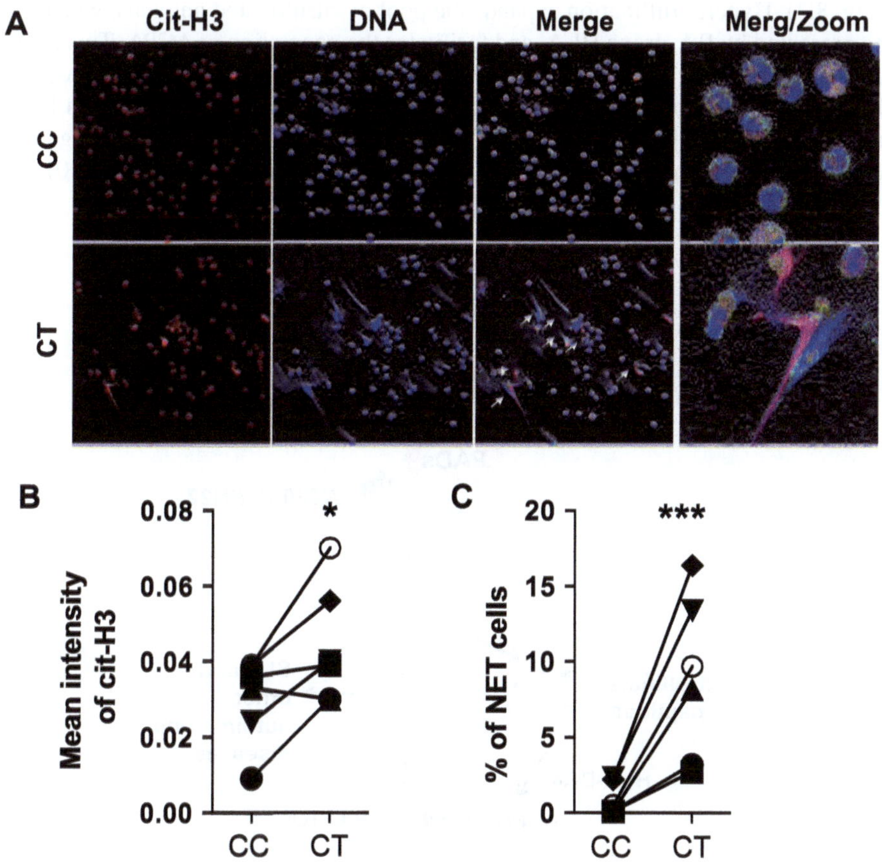

Fig. 8.2 The C1858T SNP is associated with enhanced NETosis. Neutrophils from healthy donors carrying (CT) or not carrying (CC) the C1858T SNP were subjected to immunocytochemistry. Representative images of immunocytochemistry are shown in (**A**). Cells undergoing NETosis are marked with *white arrows*. The mean intensity of cit-H3 is shown in (**B**). The percentage of NET-forming cells is shown in (**C**). In each experiment, neutrophils from one CC and one CT donors were analyzed simultaneously. Each pair is connected with a line and denoted with the same symbol in (**B, C**). (Adapted from Chang et al. (2015))

after fMLP stimulation, and more ROS production after stimulation sequentially with TNFα and fMLP (Bayley et al. 2014). However, it is unclear if these features are caused by altered phosphatase activity of PTPN22 or by hypercitrullination.

8.5.5 A Working Model for the Increased Risk of Autoimmune Diseases Associated with the C1858T SNP of PTPN22

In essence, the C1858T SNP results in the production of a dominant-negative variant of PTPN22, leading to hypercitrullination and heightened propensity to form NETs (Fig. 8.3). Hypercitrullination expands the pool of citrullinated antigens, which can be presented by RA-prone HLA, and facilitates the generation of ACPA. This mechanism satisfactorily explains the synergy between the C1858T SNP and RA-prone HLA in ACPA-positive RA. In addition, the heightened propensity to form NETs not only propagate joint inflammation in RA but also contribute to the pathogenesis of other autoimmune diseases, such as SLE and type 1 diabetes (see Chap. 10).

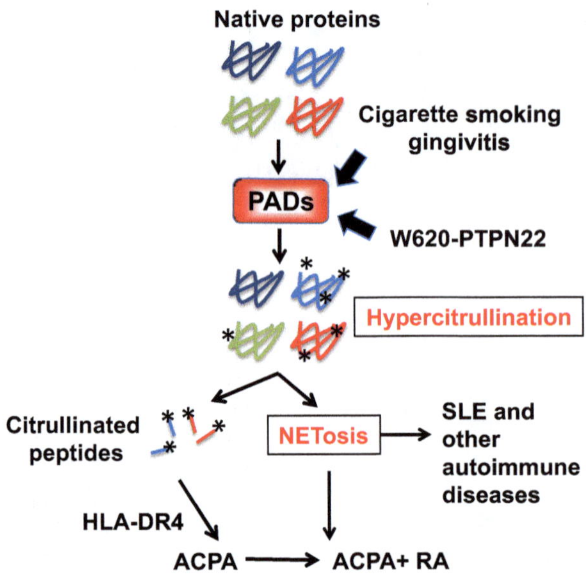

Fig. 8.3 A working model, by which the C1858T SNP of PTPN22 increases the risk of RA and other autoimmune diseases. Native proteins are citrullated by PADs, which are normally inhibited by PTPN22. The C1858T SNP creates a dominant negative variant of PTPN22, W620-PTPN22, which along with other environmental risk factors of RA, including smoking and gingivitis, leads to hypercitrullination. Hypercitrullination expands the pool of citrullinated antigens, which can be presented by RA-prone HLA, such as HLA-DR4, promoting the development of ACPA. Hypercitrullination also enhances the formation of NETs, aggravating joint inflammation and contributing to the pathogenesis of SLE and other autoimmune diseases

8.6 Conclusions

Many major advances have been made over the past few years regarding the molecular mechanism regulating the activity of PADs and the role of citrullination in controlling NETosis. Emerging data have also demonstrated that NETosis is potentially a key component of innate immunity but can also contribute to the development of autoimmune diseases in many ways. The discovery that the C1858T SNP renders PTPN22 unable to interact with PAD4 and inhibit citrullination further strengthens the link between PAD4-dependent NETosis and autoimmune diseases. However, many questions remain unanswered. For example, what is the in vivo role of NETs in immune defense against Gram-negative bacteria, viruses, and fungi? How does dysregulated citrullination and NETosis results in such diverse disease manifestations? Are the components of NETs useful biomarkers of autoimmune diseases? How does PTPN22 inhibit the activity of PADs? Answers to these questions will pave the road for future clinical trials using PAD inhibitors in treating RA, SLE, or other human diseases.

References

Aarnisalo, J., Treszl, A., Svec, P., Marttila, J., Oling, V., Simell, O., Knip, M., Korner, A., Madacsy, L., Vasarhelyi, B., Ilonen, J., & Hermann, R. (2008). Reduced CD4+T cell activation in children with type 1 diabetes carrying the PTPN22/Lyp 620Trp variant. *Journal of Autoimmunity, 31*, 13–21.

Abi Abdallah, D. S., Lin, C., Ball, C. J., King, M. R., Duhamel, G. E., & Denkers, E. Y. (2012). Toxoplasma gondii triggers release of human and mouse neutrophil extracellular traps. *Infection and Immunity, 80*, 768–777.

Aletaha, D., Neogi, T., Silman, A. J., Funovits, J., Felson, D. T., Bingham, C. O., 3rd, Birnbaum, N. S., Burmester, G. R., Bykerk, V. P., Cohen, M. D., Combe, B., Costenbader, K. H., Dougados, M., Emery, P., Ferraccioli, G., Hazes, J. M., Hobbs, K., Huizinga, T. W., Kavanaugh, A., Kay, J., Kvien, T. K., Laing, T., Mease, P., Menard, H. A., Moreland, L. W., Naden, R. L., Pincus, T., Smolen, J. S., Stanislawska-Biernat, E., Symmons, D., Tak, P. P., Upchurch, K. S., Vencovsky, J., Wolfe, F., & Hawker, G. (2010). 2010 rheumatoid arthritis classification criteria: An American College of Rheumatology/European league against Rheumatism collaborative initiative. *Arthritis and Rheumatism, 62*, 2569–2581.

AlFadhli, S., AlTamimy, B., Kharrat, N., AlSaeid, K., Haider, M. Z., & Rebai, A. (2010). Molecular analysis of HumDN1 VNTR polymorphism of the human deoxyribonuclease I in systemic lupus erythematosus. *International Journal of Immunogenetics, 37*, 5–8.

Amini, P., Stojkov, D., Wang, X., Wicki, S., Kaufmann, T., Wong, W. W., Simon, H. U., & Yousefi, S. (2016). NET formation can occur independently of RIPK3 and MLKL signaling. *European Journal of Immunology, 46*, 178–184.

Andrade, F., Darrah, E., Gucek, M., Cole, R. N., Rosen, A., & Zhu, X. (2010). Autocitrullination of human peptidyl arginine deiminase type 4 regulates protein citrullination during cell activation. *Arthritis and Rheumatism, 62*, 1630–1640.

Awasthi, D., Nagarkoti, S., Kumar, A., Dubey, M., Singh, A. K., Pathak, P., Chandra, T., Barthwal, M. K., & Dikshit, M. (2016). Oxidized LDL induced extracellular trap formation in human neutrophils via TLR-PKC-IRAK-MAPK and NADPH-oxidase activation. *Free Radical Biology & Medicine, 93*, 190–203.

Baker, V. S., Imade, G. E., Molta, N. B., Tawde, P., Pam, S. D., Obadofin, M. O., Sagay, S. A., Egah, D. Z., Iya, D., Afolabi, B. B., Baker, M., Ford, K., Ford, R., Roux, K. H., & Keller, T. C., 3rd. (2008). Cytokine-associated neutrophil extracellular traps and antinuclear antibodies in plasmodium falciparum infected children under six years of age. *Malaria Journal, 7*, 41.

Bayley, R., Kite, K. A., McGettrick, H. M., Smith, J. P., Kitas, G. D., Buckley, C. D., & Young, S. P. (2014). The autoimmune-associated genetic variant PTPN22 R620W enhances neutrophil activation and function in patients with rheumatoid arthritis and healthy individuals. *Annals of the Rheumatic Diseases, 74*(8), 1588–1595.

Begovich, A. B., Carlton, V. E., Honigberg, L. A., Schrodi, S. J., Chokkalingam, A. P., Alexander, H. C., Ardlie, K. G., Huang, Q., Smith, A. M., Spoerke, J. M., Conn, M. T., Chang, M., Chang, S. Y., Saiki, R. K., Catanese, J. J., Leong, D. U., Garcia, V. E., McAllister, L. B., Jeffery, D. A., Lee, A. T., Batliwalla, F., Remmers, E., Criswell, L. A., Seldin, M. F., Kastner, D. L., Amos, C. I., Sninsky, J. J., & Gregersen, P. K. (2004). A missense single-nucleotide polymorphism in a gene encoding a protein tyrosine phosphatase (PTPN22) is associated with rheumatoid arthritis. *American Journal of Human Genetics, 75*, 330–337.

Behnen, M., Leschczyk, C., Moller, S., Batel, T., Klinger, M., Solbach, W., & Laskay, T. (2014). Immobilized immune complexes induce neutrophil extracellular trap release by human neutrophil granulocytes via FcgammaRIIIB and mac-1. *Journal of Immunology, 193*, 1954–1965.

Bodano, A., Gonzalez, A., Ferreiros-Vidal, I., Balada, E., Ordi, J., Carreira, P., Gomez-Reino, J. J., & Conde, C. (2006). Association of a non-synonymous single-nucleotide polymorphism of DNASEI with SLE susceptibility. *Rheumatology (Oxford, England), 45*, 819–823.

Branzk, N., Lubojemska, A., Hardison, S. E., Wang, Q., Gutierrez, M. G., Brown, G. D., & Papayannopoulos, V. (2014). Neutrophils sense microbe size and selectively release neutrophil extracellular traps in response to large pathogens. *Nature Immunology, 15*, 1017–1025.

Brill, A., Fuchs, T. A., Savchenko, A. S., Thomas, G. M., Martinod, K., De Meyer, S. F., Bhandari, A. A., & Wagner, D. D. (2012). Neutrophil extracellular traps promote deep vein thrombosis in mice. *Journal of Thrombosis and Haemostasis, 10*, 136–144.

Brinkmann, V., Reichard, U., Goosmann, C., Fauler, B., Uhlemann, Y., Weiss, D. S., Weinrauch, Y., & Zychlinsky, A. (2004). Neutrophil extracellular traps kill bacteria. *Science, 303*, 1532–1535.

Brownlie, R. J., Miosge, L. A., Vassilakos, D., Svensson, L. M., Cope, A., & Zamoyska, R. (2012). Lack of the phosphatase PTPN22 increases adhesion of murine regulatory T cells to improve their immunosuppressive function. *Science Signaling, 5*, ra87.

Carlton, V. E., Hu, X., Chokkalingam, A. P., Schrodi, S. J., Brandon, R., Alexander, H. C., Chang, M., Catanese, J. J., Leong, D. U., Ardlie, K. G., Kastner, D. L., Seldin, M. F., Criswell, L. A., Gregersen, P. K., Beasley, E., Thomson, G., Amos, C. I., & Begovich, A. B. (2005). PTPN22 genetic variation: Evidence for multiple variants associated with rheumatoid arthritis. *American Journal of Human Genetics, 77*, 567–581.

Chang, H. H., Tai, T. S., Lu, B., Iannaccone, C., Cernadas, M., Weinblatt, M., Shadick, N., Miaw, S. C., & Ho, I. C. (2012). PTPN22.6, a dominant negative isoform of PTPN22 and potential biomarker of rheumatoid arthritis. *PloS One, 7*, e33067.

Chang, H. H., Miaw, S. C., Tseng, W., Sun, Y. W., Liu, C. C., Tsao, H. W., & Ho, I. C. (2013). PTPN22 modulates macrophage polarization and susceptibility to dextran sulfate sodium-induced colitis. *Journal of Immunology, 191*, 2134–2143.

Chang, H. H., Dwivedi, N., Nicholas, A. P., & Ho, I. C. (2015). The W620 polymorphism in PTPN22 disrupts its interaction with peptidylarginine deiminase type 4 and enhances citrullination and NETosis. *Arthritis & Rhematology, 67*, 2323–2334.

Chen, T., & Dent, S. Y. (2014). Chromatin modifiers and remodellers: Regulators of cellular differentiation. *Nature Reviews Genetics, 15*, 93–106.

Clark, S. R., Ma, A. C., Tavener, S. A., McDonald, B., Goodarzi, Z., Kelly, M. M., Patel, K. D., Chakrabarti, S., McAvoy, E., Sinclair, G. D., Keys, E. M., Allen-Vercoe, E., Devinney, R., Doig, C. J., Green, F. H., & Kubes, P. (2007). Platelet TLR4 activates neutrophil extracellular traps to ensnare bacteria in septic blood. *Nature Medicine, 13*, 463–469.

Cohen, S., Dadi, H., Shaoul, E., Sharfe, N., & Roifman, C. M. (1999). Cloning and characterization of a lymphoid-specific, inducible human protein tyrosine phosphatase, Lyp. *Blood, 93*, 2013–2024.

Corriden, R., Hollands, A., Olson, J., Derieux, J., Lopez, J., Chang, J. T., Gonzalez, D. J., & Nizet, V. (2015). Tamoxifen augments the innate immune function of neutrophils through modulation of intracellular ceramide. *Nature Communications, 6*, 8369.

Corsiero, E., Bombardieri, M., Carlotti, E., Pratesi, F., Robinson, W., Migliorini, P., & Pitzalis, C. (2015). Single cell cloning and recombinant monoclonal antibodies generation from RA synovial B cells reveal frequent targeting of citrullinated histones of NETs. *Annals of the Rheumatic Diseases, 75*(10), 1866–1875.

Dai, X., James, R. G., Habib, T., Singh, S., Jackson, S., Khim, S., Moon, R. T., Liggitt, D., Wolf-Yadlin, A., Buckner, J. H., & Rawlings, D. J. (2013). A disease-associated PTPN22 variant promotes systemic autoimmunity in murine models. *The Journal of Clinical Investigation, 123*, 2024–2036.

Damgaard, D., Friberg Bruun Nielsen, M., Quisgaard Gaunsbaek, M., Palarasah, Y., Svane-Knudsen, V., & Nielsen, C. H. (2015). Smoking is associated with increased levels of extracellular peptidylarginine deiminase 2 (PAD2) in the lungs. *Clinical and Experimental Rheumatology, 33*, 405–408.

Darrah, E., Giles, J. T., Ols, M. L., Bull, H. G., Andrade, F., & Rosen, A. (2013). Erosive rheumatoid arthritis is associated with antibodies that activate PAD4 by increasing calcium sensitivity. *Science Translational Medicine, 5*, 186ra165.

Denny, M. F., Yalavarthi, S., Zhao, W., Thacker, S. G., Anderson, M., Sandy, A. R., McCune, W. J., & Kaplan, M. J. (2010). A distinct subset of proinflammatory neutrophils isolated from patients with systemic lupus erythematosus induces vascular damage and synthesizes type I IFNs. *Journal of Immunology, 184*, 3284–3297.

Desai, J., Kumar, S. V., Mulay, S. R., Konrad, L., Romoli, S., Schauer, C., Herrmann, M., Bilyy, R., Muller, S., Popper, B., Nakazawa, D., Weidenbusch, M., Thomasova, D., Krautwald, S., Linkermann, A., & Anders, H. J. (2016). PMA and crystal-induced neutrophil extracellular trap formation involves RIPK1-RIPK3-MLKL signaling. *European Journal of Immunology, 46*, 223–229.

Domingo-Gonzalez, R., Martinez-Colon, G. J., Smith, A. J., Smith, C. K., Ballinger, M. N., Xia, M., Murray, S., Kaplan, M. J., Yanik, G. A., & Moore, B. B. (2016). Inhibition of neutrophil extracellular trap formation after stem cell transplant by prostaglandin E2. *American Journal of Respiratory and Critical Care Medicine, 193*, 186–197.

Douda, D. N., Khan, M. A., Grasemann, H., & Palaniyar, N. (2015). SK3 channel and mitochondrial ROS mediate NADPH oxidase-independent NETosis induced by calcium influx. *Proceedings of the National Academy of Sciences of the United States of America, 112*, 2817–2822.

Dwivedi, N., Upadhyay, J., Neeli, I., Khan, S., Pattanaik, D., Myers, L., Kirou, K. A., Hellmich, B., Knuckley, B., Thompson, P. R., Crow, M. K., Mikuls, T. R., Csernok, E., & Radic, M. (2012). Felty's syndrome autoantibodies bind to deiminated histones and neutrophil extracellular chromatin traps. *Arthritis and Rheumatism, 64*, 982–992.

Dwivedi, N., Neeli, I., Schall, N., Wan, H., Desiderio, D. M., Csernok, E., Thompson, P. R., Dali, H., Briand, J. P., Muller, S., & Radic, M. (2014). Deimination of linker histones links neutrophil extracellular trap release with autoantibodies in systemic autoimmunity. *FASEB Journal, 28*, 2840–2851.

Etulain, J., Martinod, K., Wong, S. L., Cifuni, S. M., Schattner, M., & Wagner, D. D. (2015). P-selectin promotes neutrophil extracellular trap formation in mice. *Blood, 126*, 242–246.

Fuchs, T. A., Abed, U., Goosmann, C., Hurwitz, R., Schulze, I., Wahn, V., Weinrauch, Y., Brinkmann, V., & Zychlinsky, A. (2007). Novel cell death program leads to neutrophil extracellular traps. *The Journal of Cell Biology, 176*, 231–241.

Fuchs, T. A., Brill, A., Duerschmied, D., Schatzberg, D., Monestier, M., Myers, D. D., Jr., Wrobleski, S. K., Wakefield, T. W., Hartwig, J. H., & Wagner, D. D. (2010). Extracellular DNA traps promote thrombosis. *Proceedings of the National Academy of Sciences of the United States of America, 107*, 15880–15885.

Garcia-Romo, G. S., Caielli, S., Vega, B., Connolly, J., Allantaz, F., Xu, Z., Punaro, M., Baisch, J., Guiducci, C., Coffman, R. L., Barrat, F. J., Banchereau, J., & Pascual, V. (2011). Netting neutrophils are major inducers of type I IFN production in pediatric systemic lupus erythematosus. *Science Translational Medicine, 3*, 73ra20.

Gehrke, N., Mertens, C., Zillinger, T., Wenzel, J., Bald, T., Zahn, S., Tuting, T., Hartmann, G., & Barchet, W. (2013). Oxidative damage of DNA confers resistance to cytosolic nuclease TREX1 degradation and potentiates STING-dependent immune sensing. *Immunity, 39*, 482–495.

Ghose, R., Shekhtman, A., Goger, M. J., Ji, H., & Cowburn, D. (2001). A novel, specific interaction involving the Csk SH3 domain and its natural ligand. *Nature Structural Biology, 8*, 998–1004.

Gjorloff-Wingren, A., Saxena, M., Williams, S., Hammi, D., & Mustelin, T. (1999). Characterization of TCR-induced receptor-proximal signaling events negatively regulated by the protein tyrosine phosphatase PEP. *European Journal of Immunology, 29*, 3845–3854.

Gray, R. D., Lucas, C. D., Mackellar, A., Li, F., Hiersemenzel, K., Haslett, C., Davidson, D. J., & Rossi, A. G. (2013). Activation of conventional protein kinase C (PKC) is critical in the generation of human neutrophil extracellular traps. *Journal of Inflammation, 10*, 12.

Grayson, P. C., Carmona-Rivera, C., Xu, L., Lim, N., Gao, Z., Asare, A. L., Specks, U., Stone, J. H., Seo, P., Spiera, R. F., Langford, C. A., Hoffman, G. S., Kallenberg, C. G., Clair, E. W. S., Tchao, N. K., Ytterberg, S. R., Phippard, D. J., Merkel, P. A., Kaplan, M. J., Monach, P. A., & Rituximab in A.-A. V.-I. T. N. R. G. (2015). Neutrophil-related gene expression and low-density granulocytes associated with disease activity and response to treatment in antineutrophil cytoplasmic antibody-associated vasculitis. *Arthritis & Rheumatology, 67*, 1922–1932.

Gupta, A. K., Hasler, P., Holzgreve, W., Gebhardt, S., & Hahn, S. (2005). Induction of neutrophil extracellular DNA lattices by placental microparticles and IL-8 and their presence in preeclampsia. *Human Immunology, 66*, 1146–1154.

Gupta, A. K., Joshi, M. B., Philippova, M., Erne, P., Hasler, P., Hahn, S., & Resink, T. J. (2010). Activated endothelial cells induce neutrophil extracellular traps and are susceptible to NETosis-mediated cell death. *FEBS Letters, 584*, 3193–3197.

Habib, T., Funk, A., Rieck, M., Brahmandam, A., Dai, X., Panigrahi, A. K., Luning Prak, E. T., Meyer-Bahlburg, A., Sanda, S., Greenbaum, C., Rawlings, D. J., & Buckner, J. H. (2012). Altered B cell homeostasis is associated with type I diabetes and carriers of the PTPN22 allelic variant. *Journal of Immunology, 188*, 487–496.

Hakkim, A., Furnrohr, B. G., Amann, K., Laube, B., Abed, U. A., Brinkmann, V., Herrmann, M., Voll, R. E., & Zychlinsky, A. (2010). Impairment of neutrophil extracellular trap degradation is associated with lupus nephritis. *Proceedings of the National Academy of Sciences of the United States of America, 107*, 9813–9818.

Hakkim, A., Fuchs, T. A., Martinez, N. E., Hess, S., Prinz, H., Zychlinsky, A., & Waldmann, H. (2011). Activation of the Raf-MEK-ERK pathway is required for neutrophil extracellular trap formation. *Nature Chemical Biology, 7*, 75–77.

Hasegawa, K., Martin, F., Huang, G., Tumas, D., Diehl, L., & Chan, A. C. (2004). PEST domain-enriched tyrosine phosphatase (PEP) regulation of effector/memory T cells. *Science, 303*, 685–689.

Hemmers, S., Teijaro, J. R., Arandjelovic, S., & Mowen, K. A. (2011). PAD4-mediated neutrophil extracellular trap formation is not required for immunity against influenza infection. *PloS One, 6*, e22043.

Hill, R. J., Zozulya, S., Lu, Y. L., Ward, K., Gishizky, M., & Jallal, B. (2002). The lymphoid protein tyrosine phosphatase Lyp interacts with the adaptor molecule Grb2 and functions as a negative regulator of T-cell activation. *Experimental Hematology, 30*, 237–244.

Holmes, D. A., Suto, E., Lee, W. P., Ou, Q., Gong, Q., Smith, H. R., Caplazi, P., & Chan, A. C. (2015). Autoimmunity-associated protein tyrosine phosphatase PEP negatively regulates IFN-alpha receptor signaling. *The Journal of Experimental Medicine, 212*, 1081–1093.

Hoppe, B., Haupl, T., Egerer, K., Gruber, R., Kiesewetter, H., Salama, A., Burmester, G. R., & Dorner, T. (2009). Influence of peptidylarginine deiminase type 4 genotype and shared epitope on clinical characteristics and autoantibody profile of rheumatoid arthritis. *Annals of the Rheumatic Diseases, 68*, 898–903.

Hou, S., Gao, G. P., Zhang, X. J., Sun, L., Peng, W. J., Wang, H. F., Ge, X. J., Huang, W., & Sun, Y. H. (2012). PADI4 polymorphisms and susceptibility to rheumatoid arthritis: A meta-analysis. *Modern Rheumatology, 23*(1), 50–60.

Kallberg, H., Padyukov, L., Plenge, R. M., Ronnelid, J., Gregersen, P. K., van der Helm-van Mil, A. H., Toes, R. E., Huizinga, T. W., Klareskog, L., Alfredsson, L., & Epidemiological Investigation of Rheumatoid Arthritis study group. (2007). Gene-gene and gene-environment interactions involving HLA-DRB1, PTPN22, and smoking in two subsets of rheumatoid arthritis. *American Journal of Human Genetics, 80*, 867–875.

Kang, C. P., Lee, H. S., Ju, H., Cho, H., Kang, C., & Bae, S. C. (2006). A functional haplotype of the PADI4 gene associated with increased rheumatoid arthritis susceptibility in Koreans. *Arthritis and Rheumatism, 54*, 90–96.

Keshari, R. S., Jyoti, A., Dubey, M., Kothari, N., Kohli, M., Bogra, J., Barthwal, M. K., & Dikshit, M. (2012). Cytokines induced neutrophil extracellular traps formation: Implication for the inflammatory disease condition. *PloS One, 7*, e48111.

Khandpur, R., Carmona-Rivera, C., Vivekanandan-Giri, A., Gizinski, A., Yalavarthi, S., Knight, J. S., Friday, S., Li, S., Patel, R. M., Subramanian, V., Thompson, P., Chen, P., Fox, D. A., Pennathur, S., & Kaplan, M. J. (2013). NETs are a source of citrullinated autoantigens and stimulate inflammatory responses in rheumatoid arthritis. *Science Translational Medicine, 5*, 178ra140.

Kirchner, T., Moller, S., Klinger, M., Solbach, W., Laskay, T., & Behnen, M. (2012). The impact of various reactive oxygen species on the formation of neutrophil extracellular traps. *Mediators of Inflammation, 2012*, 849136.

Knight, J. S., Zhao, W., Luo, W., Subramanian, V., O'Dell, A. A., Yalavarthi, S., Hodgin, J. B., Eitzman, D. T., Thompson, P. R., & Kaplan, M. J. (2013). Peptidylarginine deiminase inhibition is immunomodulatory and vasculoprotective in murine lupus. *The Journal of Clinical Investigation, 123*, 2981–2993.

Knight, J. S., Luo, W., O'Dell, A. A., Yalavarthi, S., Zhao, W., Subramanian, V., Guo, C., Grenn, R. C., Thompson, P. R., Eitzman, D. T., & Kaplan, M. J. (2014a). Peptidylarginine deiminase inhibition reduces vascular damage and modulates innate immune responses in murine models of atherosclerosis. *Circulation Research, 114*, 947–956.

Knight, J. S., Subramanian, V., O'Dell, A. A., Yalavarthi, S., Zhao, W., Smith, C. K., Hodgin, J. B., Thompson, P. R., & Kaplan, M. J. (2014b). Peptidylarginine deiminase inhibition disrupts NET formation and protects against kidney, skin and vascular disease in lupus-prone MRL/lpr mice. *Annals of the Rheumatic Diseases, 74*(12), 2199–2206.

Kunz, M., & Ibrahim, S. M. (2011). Non-major histocompatibility complex rheumatoid arthritis susceptibility genes. *Critical Reviews in Immunology, 31*, 99–114.

Lande, R., Ganguly, D., Facchinetti, V., Frasca, L., Conrad, C., Gregorio, J., Meller, S., Chamilos, G., Sebasigari, R., Riccieri, V., Bassett, R., Amuro, H., Fukuhara, S., Ito, T., Liu, Y. J., & Gilliet, M. (2011). Neutrophils activate plasmacytoid dendritic cells by releasing self-DNA-peptide complexes in systemic lupus erythematosus. *Science Translational Medicine, 3*, 73ra19.

Lee, A. T., Li, W., Liew, A., Bombardier, C., Weisman, M., Massarotti, E. M., Kent, J., Wolfe, F., Begovich, A. B., & Gregersen, P. K. (2005). The PTPN22 R620W polymorphism associates with RF positive rheumatoid arthritis in a dose-dependent manner but not with HLA-SE status. *Genes and Immunity, 6*, 129–133.

Leffler, J., Gullstrand, B., Jonsen, A., Nilsson, J. A., Martin, M., Blom, A. M., & Bengtsson, A. A. (2013). Degradation of neutrophil extracellular traps co-varies with disease activity in patients with systemic lupus erythematosus. *Arthritis Research & Therapy, 15*, R84.

Leffler, J., Stojanovich, L., Shoenfeld, Y., Bogdanovic, G., Hesselstrand, R., & Blom, A. M. (2014). Degradation of neutrophil extracellular traps is decreased in patients with antiphospholipid syndrome. *Clinical and Experimental Rheumatology, 32*, 66–70.

Leffler, J., Ciacma, K., Gullstrand, B., Bengtsson, A. A., Martin, M., & Blom, A. M. (2015). A subset of patients with systemic lupus erythematosus fails to degrade DNA from multiple clinically relevant sources. *Arthritis Research & Therapy, 17*, 205.

Leshner, M., Wang, S., Lewis, C., Zheng, H., Chen, X. A., Santy, L., & Wang, Y. (2012). PAD4 mediated histone hypercitrullination induces heterochromatin decondensation and chromatin unfolding to form neutrophil extracellular trap-like structures. *Frontiers in Immunology, 3*, 307.

Li, P., Li, M., Lindberg, M. R., Kennett, M. J., Xiong, N., & Wang, Y. (2010). PAD4 is essential for antibacterial innate immunity mediated by neutrophil extracellular traps. *The Journal of Experimental Medicine, 207*, 1853–1862.

Liu, Y. L., Chiang, Y. H., Liu, G. Y., & Hung, H. C. (2011). Functional role of dimerization of human peptidylarginine deiminase 4 (PAD4). *PloS One, 6*, e21314.

Lood, C., Blanco, L. P., Purmalek, M. M., Carmona-Rivera, C., De Ravin, S. S., Smith, C. K., Malech, H. L., Ledbetter, J. A., Elkon, K. B., & Kaplan, M. J. (2016). Neutrophil extracellular traps enriched in oxidized mitochondrial DNA are interferogenic and contribute to lupus-like disease. *Nature Medicine, 22*, 146–153.

Maine, C. J., Hamilton-Williams, E. E., Cheung, J., Stanford, S. M., Bottini, N., Wicker, L. S., & Sherman, L. A. (2012). PTPN22 alters the development of regulatory T cells in the thymus. *Journal of Immunology, 188*, 5267–5275.

Makrygiannakis, D., Hermansson, M., Ulfgren, A. K., Nicholas, A. P., Zendman, A. J., Eklund, A., Grunewald, J., Skold, C. M., Klareskog, L., & Catrina, A. I. (2008). Smoking increases peptidylarginine deiminase 2 enzyme expression in human lungs and increases citrullination in BAL cells. *Annals of the Rheumatic Diseases, 67*, 1488–1492.

Martinod, K., Demers, M., Fuchs, T. A., Wong, S. L., Brill, A., Gallant, M., Hu, J., Wang, Y., & Wagner, D. D. (2013). Neutrophil histone modification by peptidylarginine deiminase 4 is critical for deep vein thrombosis in mice. *Proceedings of the National Academy of Sciences of the United States of America, 110*, 8674–8679.

Martinod, K., Fuchs, T. A., Zitomersky, N. L., Wong, S. L., Demers, M., Gallant, M., Wang, Y., & Wagner, D. D. (2015). PAD4-deficiency does not affect bacteremia in polymicrobial sepsis and ameliorates endotoxemic shock. *Blood, 125*, 1948–1956.

Martinod, K., Witsch, T., Farley, K., Gallant, M., Remold-O'Donnell, E., & Wagner, D. D. (2016). Neutrophil elastase-deficient mice form neutrophil extracellular traps in an experimental model of deep vein thrombosis. *Journal of Thrombosis and Haemostasis, 14*, 551–558.

Massberg, S., Grahl, L., von Bruehl, M. L., Manukyan, D., Pfeiler, S., Goosmann, C., Brinkmann, V., Lorenz, M., Bidzhekov, K., Khandagale, A. B., Konrad, I., Kennerknecht, E., Reges, K., Holdenrieder, S., Braun, S., Reinhardt, C., Spannagl, M., Preissner, K. T., & Engelmann, B. (2010). Reciprocal coupling of coagulation and innate immunity via neutrophil serine proteases. *Nature Medicine, 16*, 887–896.

McCormick, A., Heesemann, L., Wagener, J., Marcos, V., Hartl, D., Loeffler, J., Heesemann, J., & Ebel, F. (2010). NETs formed by human neutrophils inhibit growth of the pathogenic mold aspergillus fumigatus. *Microbes and Infection, 12*, 928–936.

McGraw, W. T., Potempa, J., Farley, D., & Travis, J. (1999). Purification, characterization, and sequence analysis of a potential virulence factor from Porphyromonas gingivalis, peptidylarginine deiminase. *Infection and Immunity, 67*, 3248–3256.

Menegazzi, R., Decleva, E., & Dri, P. (2012). Killing by neutrophil extracellular traps: Fact or folklore? *Blood, 119*, 1214–1216.

Midgley, A., & Beresford, M. W. (2016). Increased expression of low density granulocytes in juvenile-onset systemic lupus erythematosus patients correlates with disease activity. *Lupus, 25*, 407–411.

Mitroulis, I., Kambas, K., Chrysanthopoulou, A., Skendros, P., Apostolidou, E., Kourtzelis, I., Drosos, G. I., Boumpas, D. T., & Ritis, K. (2011). Neutrophil extracellular trap formation is associated with IL-1beta and autophagy-related signaling in gout. *PloS One, 6*, e29318.

Munafo, D. B., Johnson, J. L., Brzezinska, A. A., Ellis, B. A., Wood, M. R., & Catz, S. D. (2009). DNase I inhibits a late phase of reactive oxygen species production in neutrophils. *Journal of Innate Immunity, 1*, 527–542.

Munoz-Caro, T., Lendner, M., Daugschies, A., Hermosilla, C., & Taubert, A. (2015). NADPH oxidase, MPO, NE, ERK1/2, p38 MAPK and Ca2+ influx are essential for Cryptosporidium Parvum-induced NET formation. *Developmental and Comparative Immunology, 52*, 245–254.

Napirei, M., Karsunky, H., Zevnik, B., Stephan, H., Mannherz, H. G., & Moroy, T. (2000). Features of systemic lupus erythematosus in Dnase1-deficient mice. *Nature Genetics, 25*, 177–181.

Neeli, I., & Radic, M. (2013). Opposition between PKC isoforms regulates histone deimination and neutrophil extracellular chromatin release. *Frontiers in Immunology, 4*, 38.

Neeli, I., Dwivedi, N., Khan, S., & Radic, M. (2009). Regulation of extracellular chromatin release from neutrophils. *Journal of Innate Immunity, 1*, 194–201.

Nowakowska, D. J., & Kissler, S. (2016). Ptpn22 modifies regulatory T cell homeostasis via GITR upregulation. *Journal of Immunology, 196*, 2145–2152.

Orozco, G., Sanchez, E., Gonzalez-Gay, M. A., Lopez-Nevot, M. A., Torres, B., Caliz, R., Ortego-Centeno, N., Jimenez-Alonso, J., Pascual-Salcedo, D., Balsa, A., de Pablo, R., Nunez-Roldan, A., Gonzalez-Escribano, M. F., & Martin, J. (2005). Association of a functional single-nucleotide polymorphism of PTPN22, encoding lymphoid protein phosphatase, with rheumatoid arthritis and systemic lupus erythematosus. *Arthritis and Rheumatism, 52*, 219–224.

Papayannopoulos, V., Metzler, K. D., Hakkim, A., & Zychlinsky, A. (2010). Neutrophil elastase and myeloperoxidase regulate the formation of neutrophil extracellular traps. *The Journal of Cell Biology, 191*, 677–691.

Parker, H., Dragunow, M., Hampton, M. B., Kettle, A. J., & Winterbourn, C. C. (2012). Requirements for NADPH oxidase and myeloperoxidase in neutrophil extracellular trap formation differ depending on the stimulus. *Journal of Leukocyte Biology, 92*, 841–849.

Payet, J., Goulvestre, C., Biale, L., Avouac, J., Wipff, J., Job-Deslandre, C., Batteux, F., Dougados, M., Kahan, A., & Allanore, Y. (2014). Anticyclic citrullinated peptide antibodies in rheumatoid and nonrheumatoid rheumatic disorders: Experience with 1162 patients. *The Journal of Rheumatology, 41*, 2395–2402.

Pilsczek, F. H., Salina, D., Poon, K. K., Fahey, C., Yipp, B. G., Sibley, C. D., Robbins, S. M., Green, F. H., Surette, M. G., Sugai, M., Bowden, M. G., Hussain, M., Zhang, K., & Kubes, P. (2010). A novel mechanism of rapid nuclear neutrophil extracellular trap formation in response to Staphylococcus aureus. *Journal of Immunology, 185*, 7413–7425.

Rieck, M., Arechiga, A., Onengut-Gumuscu, S., Greenbaum, C., Concannon, P., & Buckner, J. H. (2007). Genetic variation in PTPN22 corresponds to altered function of T and B lymphocytes. *Journal of Immunology, 179*, 4704–4710.

Rohrbach, A. S., Hemmers, S., Arandjelovic, S., Corr, M., & Mowen, K. A. (2012). PAD4 is not essential for disease in the K/BxN murine autoantibody-mediated model of arthritis. *Arthritis Research & Therapy, 14*, R104.

Romero, V., Fert-Bober, J., Nigrovic, P. A., Darrah, E., Haque, U. J., Lee, D. M., van Eyk, J., Rosen, A., & Andrade, F. (2013). Immune-mediated pore-forming pathways induce cellular hypercitrullination and generate citrullinated autoantigens in rheumatoid arthritis. *Science Translational Medicine, 5*, 209ra150.

Rosenstein, E. D., Greenwald, R. A., Kushner, L. J., & Weissmann, G. (2004). Hypothesis: The humoral immune response to oral bacteria provides a stimulus for the development of rheumatoid arthritis. *Inflammation, 28*, 311–318.

Saitoh, T., Komano, J., Saitoh, Y., Misawa, T., Takahama, M., Kozaki, T., Uehata, T., Iwasaki, H., Omori, H., Yamaoka, S., Yamamoto, N., & Akira, S. (2012). Neutrophil extracellular traps mediate a host defense response to human immunodeficiency virus-1. *Cell Host & Microbe, 12*, 109–116.

Semeraro, F., Ammollo, C. T., Morrissey, J. H., Dale, G. L., Friese, P., Esmon, N. L., & Esmon, C. T. (2011). Extracellular histones promote thrombin generation through platelet-dependent mechanisms: Involvement of platelet TLR2 and TLR4. *Blood, 118*, 1952–1961.

Seri, Y., Shoda, H., Suzuki, A., Matsumoto, I., Sumida, T., Fujio, K., & Yamamoto, K. (2015). Peptidylarginine deiminase type 4 deficiency reduced arthritis severity in a glucose-6-phosphate isomerase-induced arthritis model. *Scientific Reports, 5*, 13041.

Shelef, M. A., Sokolove, J., Lahey, L. J., Wagner, C. A., Sackmann, E. K., Warner, T. F., Wang, Y., Beebe, D. J., Robinson, W. H., & Huttenlocher, A. (2014). Peptidylarginine deiminase 4 contributes to tumor necrosis factor alpha-induced inflammatory arthritis. *Arthritis & Rhematology, 66*, 1482–1491.

Shishikura, K., Horiuchi, T., Sakata, N., Trinh, D. A., Shirakawa, R., Kimura, T., Asada, Y., & Horiuchi, H. (2016). Prostaglandin E2 inhibits neutrophil extracellular trap formation through production of cyclic AMP. *British Journal of Pharmacology, 173*, 319–331.

Slack, J. L., Jones, L. E., Jr., Bhatia, M. M., & Thompson, P. R. (2011). Autodeimination of protein arginine deiminase 4 alters protein-protein interactions but not activity. *Biochemistry, 50*, 3997–4010.

Spalinger, M. R., Kasper, S., Gottier, C., Lang, S., Atrott, K., Vavricka, S. R., Scharl, S., Gutte, P. M., Grutter, M. G., Beer, H. D., Contassot, E., Chan, A. C., Dai, X., Rawlings, D. J., Mair, F., Becher, B., Falk, W., Fried, M., Rogler, G., & Scharl, M. (2016). NLRP3 tyrosine phosphorylation is controlled by protein tyrosine phosphatase PTPN22. *The Journal of Clinical Investigation, 126*(11), 4388.

Sur Chowdhury, C., Giaglis, S., Walker, U. A., Buser, A., Hahn, S., & Hasler, P. (2014). Enhanced neutrophil extracellular trap generation in rheumatoid arthritis: Analysis of underlying signal transduction pathways and potential diagnostic utility. *Arthritis Research & Therapy, 16*, R122.

Suzuki, A., Yamada, R., Chang, X., Tokuhiro, S., Sawada, T., Suzuki, M., Nagasaki, M., Nakayama-Hamada, M., Kawaida, R., Ono, M., Ohtsuki, M., Furukawa, H., Yoshino, S., Yukioka, M., Tohma, S., Matsubara, T., Wakitani, S., Teshima, R., Nishioka, Y., Sekine, A., Iida, A., Takahashi, A., Tsunoda, T., Nakamura, Y., & Yamamoto, K. (2003). Functional haplotypes of PADI4, encoding citrullinating enzyme peptidylarginine deiminase 4, are associated with rheumatoid arthritis. *Nature Genetics, 34*, 395–402.

Thompson, W. S., Pekalski, M. L., Simons, H. Z., Smyth, D. J., Castro-Dopico, X., Guo, H., Guy, C., Dunger, D. B., Arif, S., Peakman, M., Wallace, C., Wicker, L. S., Todd, J. A., & Ferreira, R. C. (2014). Multi-parametric flow cytometric and genetic investigation of the peripheral B cell compartment in human type 1 diabetes. *Clinical and Experimental Immunology, 177*, 571–585.

Urban, C. F., Reichard, U., Brinkmann, V., & Zychlinsky, A. (2006). Neutrophil extracellular traps capture and kill Candida albicans yeast and hyphal forms. *Cellular Microbiology, 8*, 668–676.

Urban, C. F., Ermert, D., Schmid, M., Abu-Abed, U., Goosmann, C., Nacken, W., Brinkmann, V., Jungblut, P. R., & Zychlinsky, A. (2009). Neutrophil extracellular traps contain calprotectin, a cytosolic protein complex involved in host defense against Candida albicans. *PLoS Pathogens, 5*, e1000639.

Vang, T., Congia, M., Macis, M. D., Musumeci, L., Orru, V., Zavattari, P., Nika, K., Tautz, L., Tasken, K., Cucca, F., Mustelin, T., & Bottini, N. (2005). Autoimmune-associated lymphoid tyrosine phosphatase is a gain-of-function variant. *Nature Genetics, 37*, 1317–1319.

Villanueva, E., Yalavarthi, S., Berthier, C. C., Hodgin, J. B., Khandpur, R., Lin, A. M., Rubin, C. J., Zhao, W., Olsen, S. H., Klinker, M., Shealy, D., Denny, M. F., Plumas, J., Chaperot, L., Kretzler, M., Bruce, A. T., & Kaplan, M. J. (2011). Netting neutrophils induce endothelial damage, infiltrate tissues, and expose immunostimulatory molecules in systemic lupus erythematosus. *Journal of Immunology, 187*, 538–552.

Wahren-Herlenius, M., & Dorner, T. (2013). Immunopathogenic mechanisms of systemic autoimmune disease. *Lancet, 382*, 819–831.

Wang, Y., Li, M., Stadler, S., Correll, S., Li, P., Wang, D., Hayama, R., Leonelli, L., Han, H., Grigoryev, S. A., Allis, C. D., & Coonrod, S. A. (2009). Histone hypercitrullination mediates chromatin decondensation and neutrophil extracellular trap formation. *The Journal of Cell Biology, 184*, 205–213.

Wang, Y., Shaked, I., Stanford, S. M., Zhou, W., Curtsinger, J. M., Mikulski, Z., Shaheen, Z. R., Cheng, G., Sawatzke, K., Campbell, A. M., Auger, J. L., Bilgic, H., Shoyama, F. M., Schmeling, D. O., Balfour, H. H., Jr., Hasegawa, K., Chan, A. C., Corbett, J. A., Binstadt, B. A., Mescher, M. F., Ley, K., Bottini, N., & Peterson, E. J. (2013). The autoimmunity-associated gene PTPN22 potentiates toll-like receptor-driven, type 1 interferon-dependent immunity. *Immunity, 39*, 111–122.

Wang, H., Li, T., Chen, S., Gu, Y., & Ye, S. (2015). Neutrophil extracellular trap mitochondrial DNA and its autoantibody in systemic lupus erythematosus and a proof-of-concept trial of metformin. *Arthritis & Rhematology, 67*, 3190–3200.

Willis, V. C., Gizinski, A. M., Banda, N. K., Causey, C. P., Knuckley, B., Cordova, K. N., Luo, Y., Levitt, B., Glogowska, M., Chandra, P., Kulik, L., Robinson, W. H., Arend, W. P., Thompson, P. R., & Holers, V. M. (2011). N-alpha-benzoyl-N5-(2-chloro-1-iminoethyl)-L-ornithine amide, a protein arginine deiminase inhibitor, reduces the severity of murine collagen-induced arthritis. *Journal of Immunology, 186*, 4396–4404.

Wu, J., Katrekar, A., Honigberg, L. A., Smith, A. M., Conn, M. T., Tang, J., Jeffery, D., Mortara, K., Sampang, J., Williams, S. R., Buggy, J., & Clark, J. M. (2006). Identification of substrates of human protein-tyrosine phosphatase PTPN22. *The Journal of Biological Chemistry, 281*, 11002–11010.

Wu, D. J., Zhou, W., Enouz, S., Orru, V., Stanford, S. M., Maine, C. J., Rapini, N., Sawatzke, K., Engel, I., Fiorillo, E., Sherman, L. A., Kronenberg, M., Zehn, D., Peterson, E., & Bottini, N. (2014). Autoimmunity-associated LYP-W620 does not impair thymic negative selection of autoreactive T cells. *PloS One, 9*, e86677.

Yalavarthi, S., Gould, T. J., Rao, A. N., Mazza, L. F., Morris, A. E., Nunez-Alvarez, C., Hernandez-Ramirez, D., Bockenstedt, P. L., Liaw, P. C., Cabral, A. R., & Knight, J. S. (2015). Release of neutrophil extracellular traps by neutrophils stimulated with antiphospholipid antibodies: A newly identified mechanism of thrombosis in the antiphospholipid syndrome. *Arthritis & Rhematology, 67*, 2990–3003.

Yeh, L. T., Miaw, S. C., Lin, M. H., Chou, F. C., Shieh, S. J., Chuang, Y. P., Lin, S. H., Chang, D. M., & Sytwu, H. K. (2013). Different modulation of Ptpn22 in effector and regulatory T cells leads to attenuation of autoimmune diabetes in transgenic nonobese diabetic mice. *Journal of Immunology, 191*, 594–607.

Zhang, J., Zahir, N., Jiang, Q., Miliotis, H., Heyraud, S., Meng, X., Dong, B., Xie, G., Qiu, F., Hao, Z., McCulloch, C. A., Keystone, E. C., Peterson, A. C., & Siminovitch, K. A. (2011). The autoimmune disease-associated PTPN22 variant promotes calpain-mediated Lyp/pep degradation associated with lymphocyte and dendritic cell hyperresponsiveness. *Nature Genetics, 43*, 902–907.

Zheng, P., & Kissler, S. (2013). PTPN22 silencing in the NOD model indicates the type 1 diabetes-associated allele is not a loss-of-function variant. *Diabetes, 62*, 896–904.

Zhou, Y., Di Pucchio, T., Sims, G. P., Mittereder, N., & Mustelin, T. (2015). Characterization of the hypercitrullination reaction in human neutrophils and other leukocytes. *Mediators of Inflammation, 2015*, 236451.

Chapter 9
Citrullination and Autophagy

Guido Valesini, Tania Colasanti, Cristiano Alessandri, Michele Bombardieri, Fabrizio Conti, and Maurizio Sorice

9.1 Introduction

During evolution, the eukaryotic cells develop different processes in order to adapt themselves to environmental changes, as well as to die or to survive. These processes include apoptosis, NETosis, and autophagy, and citrullination is implicated in all of these physiological mechanisms (Mohammed et al. 2013), as summarized in Fig. 9.1.

This chapter will focus on the role of autophagy and citrullination in the pathogenesis of autoimmune diseases.

9.2 Autophagy: General Aspects

The term "autophagy" derives from the ancient Greek words "αὐτός φαγεῖν," which means "self-eating." It was first coined by Christian de Duve over 40 years ago and was largely based on the observed degradation of mitochondria and other intracellular structures within lysosomes of rat liver perfused with the pancreatic hormone glucagon (Glick et al. 2010). In recent years, the scientific world has reevaluated autophagy, in order to better understand the molecular mechanisms of its regulation. In this sense, many molecular studies in delineating how autophagy is regulated and

G. Valesini (✉) • T. Colasanti • C. Alessandri • F. Conti
Rheumatology, Department of Internal Medicine and Medical Specialties, Sapienza University of Rome, Rome, Italy
e-mail: guido.valesini@uniroma1.it

M. Bombardieri
Centre for Experimental Medicine and Rheumatology, William Harvey Research Institute, Queen Mary University of London, London, UK

M. Sorice
Department of Experimental Medicine, Sapienza University of Rome, Rome, Italy

A.P. Nicholas et al. (eds.), *Protein Deimination in Human Health and Disease*,
DOI 10.1007/978-3-319-58244-3_9

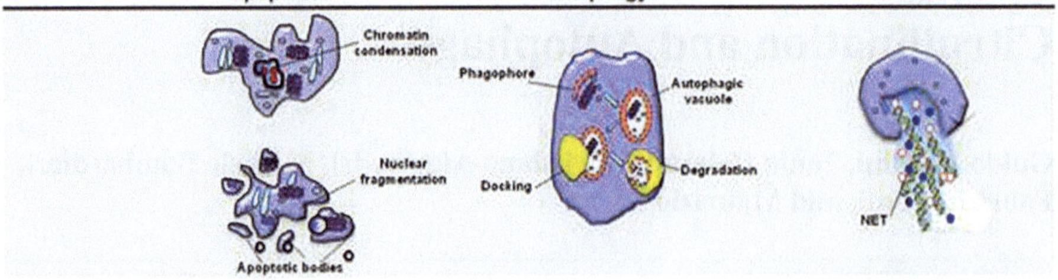

	Apoptosis	*Autophagy*	*Netosis*
Morphology	Chromatin condensation; nuclear fragmentation; apoptotic bodies	Autophagic vacuoles	Extrusion of NET: DNA, histones, granular proteins (myeloperoxidase, elastase, lactoferrin, etc.), cytoplasmic proteins (calprotectin, catalase)
Triggers	Extrinsic pathway → death receptors (FasL). Intrinsic pathway → viral infection, DNA damage, mitochondrial release of cytochrome c	Amino acid starvation; grown factor withdrawal; energy withdrawal; environmental stress (intracellula reactive oxygen species)	PMA, LPS, IL-8, TNF-α, LL-37, H_2O_2, LPS-activated platelets, ANCAs, immune complexes, ANAs, activated endothelium, *Streptococcus* species, *Pseudomonas aeruginosa*, *Staphylococcus* species, *Candida albicans*, *Escherichia coli*, *Mycobacterium tuberculosis*, HIV
Mediators	Extrinsic pathway → caspase 8,10 Intrinsic pathway → caspase 9	Autophagy–related proteins 1, 5, 6, 8, 12, 17, 13	MPO, elastase, NOX2, NADPH, PAD4, H_2O_2
Inhibitors	Caspase inhibitors	Bcl-2?; anabolic metabolism (class I PI3K)	Catalase, DNase, DPI
Citrullination	Citrullination is needed to prepare the intracellular proteins for degradation, with a complete loss of polymerization competence of the intermediate filament protein and of filament forming ability	The presentation of citrullinated peptides by the APC was associated with autophagy and citrullination could be considered as a biochemical marker of autophagy	Citrullination of histones is responsible of chromatin decondensation that leads to the release of histones in the NET and prevents histone methylation and gene transcription

Fig. 9.1 The role of citrullination in cell death modalities. *ANAs* anti-neutrophil antibodies, *ANCAs* anti-neutrophil cytoplasmic antibodies, *APC* antigen-presenting cell, *Bcl-2* B cell lymphoma 2, *DPI* diphenyleneiodonium, *HIV* human immunodeficiency virus, *IL* interleukin, *LL-37* cathelicidin LL-37, *LPS* lipopolysaccharide, *NET* neutrophil extracellular trap, *MPO* myeloperoxidase, *NAPDH* nicotinamide adenine dinucleotide phosphate, *NOX2* NADPH oxidase 2, *PAD4* peptidylarginine deiminase 4, *PI3K* phosphatidylinositol 3 kinase, *PMA* phorbol 12-myristate 13-acetate, *TNF-α* tumor necrosis factor-α (Modified from Valesini et al., Autoimmun Rev., 2015)

executed have been made in yeast (*S. cerevisiae*), but the process is highly conserved, so that the importance of autophagy is well recognized in mammalian systems. At present, about 32 autophagy-related genes (*atg*) in yeast have been characterized, and their mammalian equivalents are now known (Glick et al. 2010).

Autophagy can be considered a cellular surveillance process, which in physiological conditions works to remove misfolded or aggregated proteins, damaged organelles (such as mitochondria or endoplasmic reticulum (ER), and intracellular pathogens. Moreover, autophagy is involved in cellular senescence and antigen presentation, playing a role in many diseases such as cancer (Durrant et al. 2016), neurodegeneration, autoimmune diseases, and infections. When autophagy is deregulated or altered, such as in pathological conditions, the process may be linked to non-apoptotic cell death (Glick et al. 2010).

9.2.1 Primary Autophagy Mechanisms

Generally, three main types of autophagy have been described: macroautophagy, microautophagy, and chaperone-mediated autophagy. They share the principal outlines, with some differences.

During macroautophagy (hereafter referred to as autophagy), parts of the cytoplasm and intracellular organelles are sequestered within characteristic double- or multi-membraned autophagic vacuoles (named autophagosomes) and are finally delivered to lysosomes to form what is known as the autophagolysosome for bulk degradation. Autophagy is a highly regulated process that can either be involved in the turnover of long-lived proteins and whole organelles in a generalized fashion or can specifically target distinct organelles, thereby eliminating supernumerary or damaged organelles. Thus, apoptosis and autophagy constitute the two processes through which superfluous, damaged, or aged cells or organelles are eliminated. Beyond this homeostatic function, autophagy is also a process by which cells adapt their metabolism to starvation, imposed by decreased extracellular nutrients or by decreased intracellular metabolite concentrations that result from the loss of growth factor signaling, and which often governs the uptake of nutrients. By the catabolism of macromolecules, autophagy generates metabolic substrates that meet the bioenergetic needs of cells and thereby allows for adaptive protein synthesis (Maiuri et al. 2007).

In microautophagy, by contrast, cytosolic components are directly taken up by the lysosome itself through invagination of the lysosomal membrane. Both macro- and microautophagy are able to engulf large structures through both selective and nonselective mechanisms (Glick et al. 2010).

In chaperone-mediated autophagy, target proteins are translocated across the lysosomal membrane in a complex with chaperone proteins, such as heat shock protein-70 (Glick et al. 2010). These complexes are then recognized by the lysosomal membrane receptor lysosome-associated membrane protein-2A (LAMP-2A), resulting in their unfolding and degradation (Glick et al. 2010). On the other hand, as reported in a recent review on autophagy monitoring (Klionsky et al. 2016), several other selective autophagy mechanisms have been described, which involve a huge heterogeneous amount of organelles.

The cytoplasm-to-vacuole targeting (Cvt) pathway is a biosynthetic route that utilizes the autophagy-related protein machinery, whereas other types of selective autophagy are degradative. The latter include pexophagy, mitophagy, reticulophagy, ribophagy, and xenophagy, and each of these processes has its own marker proteins. To note, the Cvt pathway has been demonstrated to occur only in yeast (Klionsky et al. 2016).

9.2.2 Selective Autophagy Mechanisms

As previously mentioned, several types of selective autophagy have also been described. For example, aggrephagy is the selective removal of aggregates by macroautophagy. This process can be followed by monitoring the levels of an aggregate-prone protein such as mutant α-synuclein, whose role in autophagy regulation (i.e., in lymphocytes) was the object of some intensive studies (Colasanti et al. 2014; Klionsky et al. 2016).

In *C. elegans*, mitochondria and hence mitochondrial DNA from sperm, are eliminated by an autophagic process. This method of allogeneic (nonself) organelle autophagy is termed allophagy. During allophagy in *C. elegans*, both paternal mitochondria and membranous organelles (a sperm-specific membrane compartment) are eliminated by the 16-cell stage (100–120 min postfertilization). The degradation process can be monitored in living embryos using ubiquitin labeled with green fluorescent protein (Klionsky et al. 2016).

Regarding selective degradation of mitochondria (mitophagy), as with any organelle-specific form of autophagy, it is necessary to demonstrate (a) increased levels of autophagosomes containing mitochondria; (b) maturation of these autophagosomes that culminates with mitochondrial degradation, which can be blocked by specific inhibitors of autophagy or of lysosomal degradation; and (c) whether the changes are due to selective mitophagy or increased mitochondrial degradation during nonselective autophagy. Recently, studies of mitophagy in yeast identified a mitochondrial membrane protein, Atg32, which plays an essential role in this organelle's turnover (Ohsumi 2014).

Although the process of pexophagy is prominent, relatively little work has been done in the area of selective mammalian peroxisome degradation by autophagy. Typically, peroxisomes are induced by treatment with hypolipidemic drugs, such as clofibrate or dioctyl phthalate, which bind to a subfamily of nuclear receptors, referred to as peroxisome proliferator-activated receptors. Indeed, the selective degradation of peroxisomes in methylotrophic yeasts, such as *P. pastoris* and *H. polymorpha*, had been well studied. The peroxisomes of these yeasts proliferate tremendously when the microorganisms are cultured in medium containing methanol as the sole carbon source. When these cells are shifted to ethanol or glucose, proliferated peroxisomes are degraded by autophagy. Tuttle (Tuttle and Dunn 1995) showed that *P. pastoris* utilizes different modes of autophagy, macroautophagy, and microautophagy, depending upon the carbon source.

Besides functioning as the primary energy source for plants, chloroplasts represent a major reservoir of fixed carbon and nitrogen to be remobilized from senescing leaves to storage organs. The turnover of these organelles has long been considered to occur via an autophagy mechanism, called chlorophagy. So, while the detection of chloroplasts within autophagic body-like vesicles or within vacuole-like compartments has been observed for decades, only recent studies described a direct link between chloroplast turnover and autophagy, through the analysis of *atg* mutants (Ohsumi 2014; Klionsky et al. 2016).

Starvation in yeast induces a type of selective macroautophagy of the ER, which depends on the autophagy receptors Atg39 and Atg40. ER stress also triggers an autophagic response, which includes the formation of multi-lamellar ER whorls and their degradation by a microautophagic mechanism. ER-selective autophagy has been termed ER-phagy or reticulophagy. Selective autophagy of the ER has also been observed in mammalian cells. Since reticulophagy is selective, it should be able to act in ER quality control, sequester parts of the ER that are damaged, and eliminate protein aggregates that cannot be removed in other ways. It may also serve to limit stress-induced ER expansion, for example, by reducing the ER to a normal level after a particular stress condition has ended (Klionsky et al. 2016).

9.2.3 *Macroautophagy and Immune Mechanisms*

The macroautophagy pathway has emerged as an important cellular factor in both innate and adaptive immunity. Many in vitro and in vivo studies have demonstrated that genes encoding macroautophagy components are required for host defense against infection by bacteria, parasites, and viruses.

Xenophagy is often used as a term to describe autophagy of microbial pathogens, mediating their capture and delivery to lysosomes for degradation. Since xenophagy presents an immune defense, it is not surprising that microbial pathogens have evolved strategies to overcome it. The interactions of such pathogens with the autophagy system of host cells are complex and have been the subject of several excellent reviews (Klionsky et al. 2016). Here, we will make note of a few key considerations when studying interactions of microbial pathogens with the autophagy system. Importantly, autophagy should no longer be considered as strictly antibacterial, and several studies have described the fact that autophagy may serve to either restrict or promote bacterial replication both in vivo and in vitro (Klionsky et al. 2016). Minor components of the autophagic process also include zimophagy, lipophagy, ferritinophagy, and some others (He and Klionsky 2009).

9.3 Autophagy and Autoimmune Diseases

Growing evidence supports the importance of autophagy in physiology and pathophysiology, such as aging, infectious diseases, cancer, and neurodegenerative diseases (Choi et al. 2013; Durrant et al. 2016). According to an emerging hypothesis, perturbations in autophagy have also been implicated in autoimmune diseases.

Autophagy participates in several aspects of immunity, affecting both innate and adaptive immunity processes. Indeed, autophagy is known to have a role in thymic selection of T cells, survival of B cells, immune tolerance, and antigen presentation (Pierdominici et al. 2012). A crosstalk between autophagy and inflammatory mechanisms has been recently suggested (Netea-Maier et al. 2016), in part since autophagy displays also a pivotal role in the processing of antigen. Moreover, autophagy can also mediate processing of both extracellular and nonconventional intracellular antigens (Dengjel et al. 2005). Like apoptosis, autophagy is a genetically programmed process that requires the activity of Atg proteins. Genome-wide association studies have linked polymorphisms in genes codifying Atg proteins with autoimmune diseases, such as systemic lupus erythematosus (SLE), inflammatory bowel disease (see Chap. 23), and multiple sclerosis (see Chap. 18).

The involvement of the lysosomal compartment in autoimmunity was suggested for the first time in 1964, by a pioneering study that associated lysosomal functions with SLE (Weissmann 1964). In a recent study, we described a significant disparity in the autophagic propensity between T lymphocytes from healthy donors and patients with SLE, the latter being resistant to autophagy induction (Alessandri et al. 2012). Indeed, whereas no significant differences were seen in spontaneous autophagy of T lymphocytes from patients with SLE compared to healthy donors, autophagic resistance in SLE T cells may result in increased apoptosis and could be associated with the defective removal of apoptotic bodies favoring the persistence of autoimmune phenomena (Alessandri et al. 2012). Of note, a deregulation of autophagy has also been described in T cells from lupus-prone mice. More recently, Clarke and colleagues (Clarke et al. 2015) demonstrate enhanced autophagy in murine and human lupus B cells. Requirement for autophagy in B cell survival and differentiation during early B cell development is therefore an immunological checkpoint for the formation of plasmablasts (immature antibody-secreting plasma cells). Although the precise mechanisms leading to autophagic dysregulation in SLE are still not understood, this pathway has been implicated in promoting survival of autoimmune T and B cells.

It is well known that the majority of patients with SLE develop autoantibodies to lymphocyte surface antigens able to inhibit T cell activation and proliferation. We added further insights in this scenario, since we discovered that serum IgG from patients with SLE were able to induce autophagy in T lymphocytes from healthy donors, suggesting a role for anti-lymphocyte antibodies as autophagy inducers. We also identified the small GTPase family inhibitor D4GDI as a possible key antigenic determinant of anti-lymphocyte antibodies implicated in the pathogenesis of SLE disease (Barbati et al. 2015). These autoantibodies, once bound to D4GDI at the cell surface, can "unlock" Rho small GTPases and activate actin network remodeling. Furthermore, anti-D4GDI autoantibodies could contribute to the selection of SLE T cell clones that are resistant to autophagy. Interestingly, we found a significant association between the presence of anti-D4GDI antibodies and hematologic manifestations (i.e., leukopenia and thrombocytopenia) occurring in SLE patients (Barbati et al. 2015).

Actually, genetic studies have linked some mutations of autophagic regulators with SLE disease (International Consortium for Systemic Lupus Erythematosus Genetics SLEGEN et al. 2008). T lymphocytes from patients with SLE showed overexpression of genes negatively regulating autophagy, such as α-synuclein and single-nucleotide polymorphisms (SNPs) of *atg5* (Colasanti et al. 2014). In addition, activation of the mammalian target of rapamycin (mTOR), a key player in autophagy regulation, has been demonstrated in SLE (Pierdominici et al. 2012), and blockade of mTOR with rapamycin improved the clinical conditions of SLE patients (Fernandez et al. 2006).

Finally, modulation of autophagy may represent a promising therapeutic approach for a wide range of autoimmune diseases. Several drugs that have been demonstrated to act as autophagy modulators are already used or are under preclinical development in SLE, as well as in other similar autoimmune disorders. Thus, new perspectives in the development of therapeutic strategies aimed at modulation of autophagic pathways would be mandatory in autoimmune disease research.

For example, in the last few years, autophagy has been shown to play a role in the pathogenesis of rheumatoid arthritis (RA), one of the most common autoimmune disorders in humans (Ireland and Unanue 2012; Valesini et al. 2015; Dai and Hu 2016). This includes several observations supporting the hypothesis that continued removal of unfolded and misfolded proteins by the proteasome pathway and by autophagy is more active in RA synovial fibroblasts, as compared to normal cells (Clausen et al. 2010). In addition, RA synovium exhibits a highly increased ER stress, and TNF-α has been shown to increase the expression of ER stress markers in RA synovial fibroblasts (Connor et al. 2012). Indeed, autophagy induction by either proteasome inhibition or ER stress is higher in RA synovial fibroblasts than in those from control patients with osteoarthritis. Thus, a dual role of autophagy in the regulation of stress-induced cell death in RA synovial fibroblasts has been reported (Kato et al. 2014).

Autophagy activation exhibited a protective role in MG132-induced apoptosis and contributed to the apoptosis-resistant phenotype (Kato et al. 2014). In contrast, fibroblasts were hypersensitive to autophagy under conditions of severe ER stress induced by thapsigargin, which was associated with imbalance in p62/sequestosome (also known as the ubiquitin-binding protein p62, an autophagosome cargo protein that targets other proteins that bind to it for selective autophagy) and autophagy-linked FYVE protein (ALFY) expression, leading to the formation of polyubiquitinated protein aggregates and non-apoptotic cell death (Clausen et al. 2010; Kato et al. 2014).

9.4 Autophagy and Citrullination

A central role for autophagy in citrullination of peptides by antigen-presenting cells (APCs) has been hypothesized. Ireland and Unanue (2011) were the first to demonstrate that autophagy was involved in the generation of citrullinated peptides by APCs, with increased peptidyl arginine deiminase (PAD) activity detected in purified autophagosomes.

In fact, citrullination has been shown to alter the structure of immunogenic peptides that can lead to increased accessibility to proteolysis and expand the repertoire of presented peptides in a process known as epitope spreading (Hanyecz et al. 2014). For example, Ireland and Unanue (2011) demonstrated that APC presentation of citrullinated peptides, but not of similar unmodified peptides, was associated with autophagy.

As a result, it can be theorized that self-peptides posttranslationally modified by a process such as citrullination would possibly form neo-antigens that are recognized by APCs of the immune system and thus represent a target for autoimmunity. The presentation of citrullinated peptides then would be included as a biochemical marker of an autophagy response in APCs.

Supporting this idea, our recent results (Sorice et al. 2016) revealed a role for autophagy in the citrullination process in vitro. Human synoviocytes treated with a potent ER stress inducer such as tunicamycin (Sakaki et al. 2008; Matarrese et al. 2014), as well as with the mTOR activator rapamycin (Fleming et al. 2011), revealed an activation of PAD4, with consequent generation of citrullinated proteins (Fig. 9.2). PADs are a family of enzymes that mediate posttranslational modifications of protein arginine residues by deimination or demethylimination to produce peptidyl-citrulline.

Although the exact processes for regulation of PAD activity in vivo remain largely elusive, there is growing evidence that the deregulation of PADs is involved in the etiology of multiple human diseases, including RA (Suzuki et al. 2003; Yamada et al. 2005). To further support the role of autophagy in the citrullination process, we investigated whether PAD4 may be present in autophagy vesicles. We observed that, during the autophagic process, the autophagy marker microtubule-associated protein light chain 3 (LC3)-II (one of the mammalian homologues of Atg8 that undergo lipid conjugation, leading to the conversion of the soluble form of LC3, named LC3-I, to the autophagic vesicle-associated form LC3-II) is recruited into autophagosomes, where it strictly interacts with PAD4. These findings are in agreement with previous data reporting that PAD activity can be detected in isolated autophagosomes with LC3-II enrichment (Ireland and Unanue 2011), supporting the view that citrullination may occur in these compartments. This concurs with our observations that generation of citrullinated proteins was a consequence of PAD4 activation following autophagic stimuli (Fig. 9.2). Interestingly, the protein citrullination was significantly increased in fibroblast-like synoviocytes from RA patients, as compared to control cells from osteoarthritis, a noninflammatory arthritis without anti-citrullinated protein antibodies. In particular, we demonstrated that the main citrullinated RA candidate antigens (Snir et al. 2010), including vimentin, α-enolase, filaggrin, and fibrinogen β, were processed, following autophagic stimulus. In vivo, a significant association between levels of autophagy and anti-citrullinated protein antibodies (ACPA) was observed in "naïve" RA patients with early active disease. These findings support the view that processing of proteins in autophagy generates citrullinated peptides recognized by the immune system in RA (Fig. 9.3), prompting a hypothesis that this similarly also occurs in vivo. As a result, it can be suggested that in RA patients who display a significant increase of both protein citrullination and LC3-II expression in synovial fibroblasts, autophagy may be able to trigger biochemical pathway(s) leading to PAD activation, with consequent processing of defined RA-associated citrullinated antigens, which in turn may be responsible for the presence of ACPA.

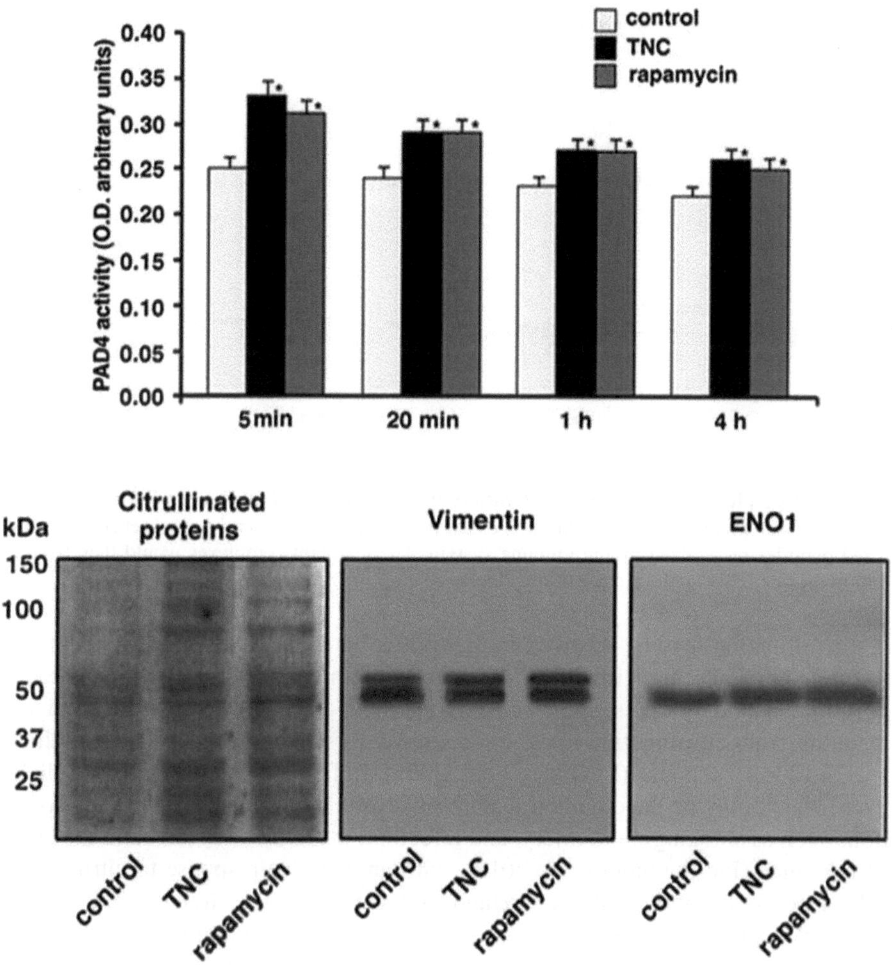

Fig. 9.2 Role of autophagy in activation of PAD. *Upper panel*: fibroblast-like synoviocytes from RA patients were treated with either tunicamycin or rapamycin. The results showed that the enzyme was active in a time-dependent manner, even after just 5 min. Statistical analysis: $*P < 0.01$ *vs*. control, as determined using a *t*-test. Results are expressed as mean ± SD of three independent experiments. *Lower panel*: the analysis of fibroblast-like synoviocytes from anti-citrullinated protein antibodies (ACPA)-positive RA patients by Western blot showed the appearance of numerous bands (corresponding to citrullinated proteins) following autophagic stimulus. To better characterize these bands, each polyvinylidene difluoride membrane was stripped and re-probed with specific anti-vimentin or anti-α-enolase antibodies. As expected, the main citrullinated bands were also stained by these antibodies, indicating that autophagic stimuli were able to induce citrullination of vimentin and α-enolase [Modified from Sorice et al., Rheumatology (Oxford), 2016]

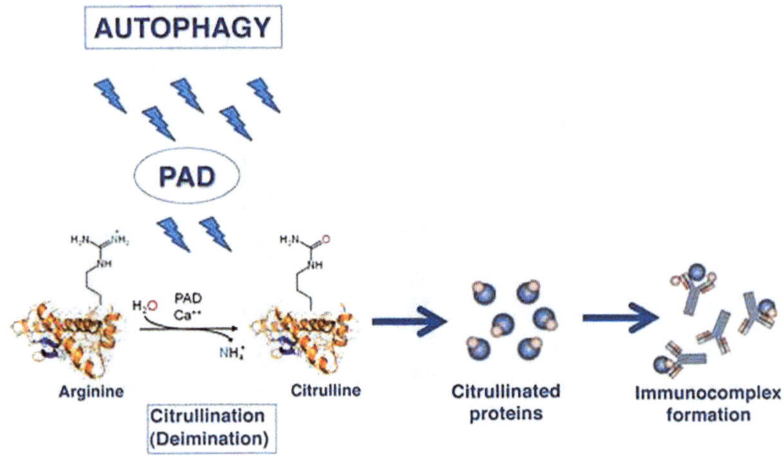

Fig. 9.3 Autophagy-induced PAD activation and protein citrullination. In RA patients, processing of autophagic proteins induces PAD activation and consequently may generate citrullinated peptides recognized by the immune system. The resultant complexes would then have an important role in the pathogenesis of the disease

Citrullination may be involved in the formation of neo-antigens as a result of a posttranslational protein modification where protein-bound arginine is converted to citrulline in antigen-presenting cells (Romero et al. 2013). These neo-antigens resulting from citrullination may be recognized by the immune system, eliciting a specific T cell response (Lundberg et al. 2005). In fact, autophagy is a key cellular event which may be the common feature of several factors, including stress conditions such as smoking, joint injury, and infection, that may be involved in the break of tolerance (Pierdominici et al. 2012), with an adaptive response to citrullinated self-proteins, triggering ACPA (Ireland and Unanue 2011; Ireland and Unanue 2012). Whether citrullination really occurs in APCs, is not completely understood; however, autophagy may play a role in the generation and presentation of citrullinated peptides by APCs. Indeed, autophagy was associated with the presentation of citrullinated peptides, but not unmodified peptides, and PAD activity was also detected in autophagosomes. Dendritic cells and macrophages show constitutive levels of autophagy; thus, these cells are able to present citrullinated peptides via autophagy. Presentation of these peptides can also be inhibited by autophagy-blocking compounds, such as 3-methyladenine (3-MA). As a result, B cells may present citrullinated peptides following autophagy induction through serum starvation or B cell antigen receptor engagement, and this presentation was shown to be blocked by 3-MA or by Atg5 inhibition (Romero et al. 2013).

In conclusion, these collective observations support the view that processing of proteins in autophagy may generate citrullinated peptides recognized by the immune system in RA patients and thus provide evidence that autophagy may also play a role in the pathogenesis of this and other autoimmune diseases.

References

Alessandri, C., Barbati, C., Vacirca, D., et al. (2012). T lymphocytes from patients with systemic lupus erythematosus are resistant to induction of autophagy. *The FASEB Journal, 26*(11), 4722–4732.

Barbati, C., Alessandri, C., Vomero, M., et al. (2015). Autoantibodies specific to D4GDI modulate Rho GTPase mediated cytoskeleton remodeling and induce autophagy in T lymphocytes. *Journal of Autoimmunity, 58*, 78–89.

Choi, A. M., Ryter, S. W., & Levine, B. (2013). Autophagy in human health and disease. *The New England Journal of Medicine, 368*(19), 1845–1846.

Clarke, A. J., Ellinghaus, U., Cortini, A., et al. (2015). Autophagy is activated in systemic lupus erythematosus and required for plasmablast development. *Annals of the Rheumatic Diseases, 74*(5), 912–920.

Clausen, T. H., Lamark, T., Isakson, P., et al. (2010). p62/SQSTM1 and ALFY interact to facilitate the formation of p62 bodies/ALIS and their degradation by autophagy. *Autophagy, 6*, 330–344.

Colasanti, T., Vomero, M., Alessandri, C., et al. (2014). Role of alpha-synuclein in autophagy modulation of primary human T lymphocytes. *Cell Death & Disease, 5*, e1265.

Connor, A. M., Mahomed, N., Gandhi, R., et al. (2012). TNFα modulates protein degradation pathways in rheumatoid arthritis synovial fibroblasts. *Arthritis Research & Therapy, 14*, R62.

Dai, Y., & Hu, S. (2016). Recent insights into the role of autophagy in the pathogenesis of rheumatoid arthritis. *Rheumatology (Oxford), 55*(3), 403–410.

Dengjel, J., Schoor, O., Fischer, R., et al. (2005). Autophagy promotes MHC class II presentation of peptides from intracellular source proteins. *Proceedings of the National Academy of Sciences of the United States of America, 102*(22), 7922–7927.

Durrant, L. G., Metheringham, R. L., & Brentville, V. A. (2016). Autophagy, citrullination and cancer. *Autophagy, 12*(6), 1055–1056.

Fernandez, D., Bonilla, E., Mirza, N., et al. (2006). Rapamycin reduces disease activity and normalizes T cell activation-induced calcium fluxing in patients with systemic lupus erythematosus. *Arthritis and Rheumatism, 54*(9), 2983–2988.

Fleming, A., Noda, T., Yoshimori, T., et al. (2011). Chemical modulators of autophagy as biological probes and potential therapeutics. *Nature Chemical Biology, 7*, 9–17.

Glick, D., Barth, S., & Macleod, K. F. (2010). Autophagy: Cellular and molecular mechanisms. *The Journal of Pathology, 221*(1), 3–12.

Hanyecz, A., Olasz, K., Tarjanyi, O., et al. (2014). Proteoglycan aggrecan conducting T cell activation and apoptosis in a murine model of rheumatoid arthritis. *BioMed Research International, 2014*, 942148.

He, C., & Klionsky, D. J. (2009). Regulation mechanisms and signaling pathways of autophagy. *Annual Review of Genetics, 43*, 67–93.

International Consortium for Systemic Lupus Erythematosus Genetics (SLEGEN), Harley, J. B., Alarcón-Riquelme, M. E., et al. (2008). Genome-wide association scan in women with systemic lupus erythematosus identifies susceptibility variants in ITGAM, PXK, KIAA1542 and other loci. *Nature Genetics, 40*(2), 204–210.

Ireland, J. M., & Unanue, E. R. (2011). Autophagy in antigen-presenting cells results in presentation of citrullinated peptides to CD4 T cells. *The Journal of Experimental Medicine, 208*, 2625–2632.

Ireland, J. M., & Unanue, E. R. (2012). Processing of proteins in autophagy vesicles of antigen-presenting cells generates citrullinated peptides recognized by the immune system. *Autophagy, 8*, 429–430.

Kato, M., Ospelt, C., Gay, R. E., et al. (2014). Dual role of autophagy in stress-induced cell death in rheumatoid arthritis synovial fibroblasts. *Arthritis & Rhematology, 66*, 40–48.

Klionsky, D. J., Abdelmohsen, K., Abe, A., et al. (2016). Guidelines for the use and interpretation of assays for monitoring autophagy (3rd edition). *Autophagy, 12*(1), 1–222.

Lundberg, K., Nijenhuis, S., Vossenaar, E. R., et al. (2005). Citrullinated proteins have increased immunogenicity and arthritogenicity and their presence in arthritic joints correlates with disease severity. *Arthritis Research & Therapy, 7*, 458–467.

Maiuri, M. C., Zalckvar, E., Kimchi, A., et al. (2007). Self-eating and self-killing: Crosstalk between autophagy and apoptosis. *Nature Reviews. Molecular Cell Biology, 8*(9), 741–752.

Matarrese, P., Garofalo, T., Manganelli, V., et al. (2014). Evidence for the involvement of GD3 ganglioside in autophagosome formation and maturation. *Autophagy, 10*, 750–765.

Mohammed, B. M., Fisher, B. J., Kraskauskas, D., et al. (2013). Vitamin C: A novel regulator of neutrophil extracellular trap formation. *Nutrients, 5*(8), 3131–3151.

Netea-Maier, R. T., Plantinga, T. S., van de Veerdonk, F. L., et al. (2016). Modulation of inflammation by autophagy: Consequences for human disease. *Autophagy, 12*(2), 245–260.

Ohsumi, Y. (2014). Historical landmarks of autophagy research. *Cell Research, 24*(1), 9–23.

Pierdominici, M., Vomero, M., Barbati, C., et al. (2012). Role of autophagy in immunity and autoimmunity, with a special focus on systemic lupus erythematosus. *The FASEB Journal, 26*(4), 1400–1412.

Romero, V., Fert-Bober, J., Nigrovic, P. A., et al. (2013). Immune-mediated pore-forming pathways induce cellular hypercitrullination and generate citrullinated autoantigens in rheumatoid arthritis. *Science Translational Medicine, 5*, 209ra150.

Sakaki, K., Wu, J., & Kaufman, R. J. (2008). Protein kinase Ctheta is required for autophagy in response to stress in the endoplasmic reticulum. *The Journal of Biological Chemistry, 283*, 15370–15380.

Snir, O., Widhe, M., Hermansson, M., et al. (2010). Antibodies to several citrullinated antigens are enriched in the joints of rheumatoid arthritis patients. *Arthritis and Rheumatism, 62*, 44–52.

Sorice, M., Iannuccelli, C., Manganelli, V., et al. (2016). Autophagy generates citrullinated peptides in human synoviocytes: A possible trigger for anti-citrullinated peptide antibodies. *Rheumatology (Oxford), 55*(8), 1374–1385.

Suzuki, A., Yamada, R., Chang, X., et al. (2003). Functional haplotypes of PADI4, encoding citrullinating enzyme peptidylarginine deiminase 4, are associated with rheumatoid arthritis. *Nature Genetics, 34*(4), 395–402.

Tuttle, D. L., & Dunn, W. A., Jr. (1995). Divergent modes of autophagy in the methylotrophic yeast *Pichia pastoris*. *Journal of Cell Science, 108*(Pt 1), 25–35.

Valesini, G., Gerardi, M. C., Iannuccelli, C., et al. (2015). Citrullination and autoimmunity. *Autoimmunity Reviews, 14*(6), 490–497.

Weissmann, G. (1964). Lysosomes, autoimmune phenomena, and diseases of connective tissue. *Lancet, 2*(7374), 1373–1375.

Yamada, R., Suzuki, A., Chang, X., et al. (2005). Citrullinated proteins in rheumatoid arthritis. *Frontiers in Bioscience, 10*, 54–64.

Chapter 10
Antigen Deimination in Human Type 1 Diabetes and Nonobese Diabetic Mice

Hai Nguyen and Eddie A. James

10.1 Introduction

Type 1 diabetes (T1D) is an immune-mediated disease in which the insulin-producing beta cells are selectively targeted for destruction, leading to lifelong insulin deficiency (Chiang et al. 2014). Successive genetic studies indicate that T1D arises due to a diverse combination of genetic risk factors that combine with incompletely defined environmental triggers to precipitate disease (Barrett et al. 2009; Couper 2001; Peng and Hagopian 2006). T1D is typically diagnosed by the occurrence of hyperglycemia in conjunction with autoantibodies ("Diagnosis and classification of diabetes mellitus" 2012). The disease poses an acute risk of diabetic ketoacidosis and engenders a host of long-term complications that are only averted by intensive dietary management and treatment with exogenous insulin (Chiang et al. 2014). In total, T1D has estimated age-related incidences ranging from 5 to 37/100,000 per year in US and European populations, with the highest rates occurring in Finland and Sardinia (Maahs et al. 2010). Development of T1D is most common in children and young adults, but the age of onset covers a broad spectrum, sometime ranging into the later decades of life (Hawa et al. 2013). Imminent risk of the disease is well predicted by an accumulation of islet cell antibodies (ICA), but the destruction of insulin producing beta cells is thought to be orchestrated by autoreactive CD4+ and CD8+ T cells (Roep and Tree 2014). The best evidence of T cell-mediated beta cell destruction is the detection of T cells in pancreatic biopsies from newly diagnosed patients and their relative abundance in islet infiltrates (Coppieters et al. 2012; Willcox et al. 2009). Although characterization of the antigen specificity of islet-infiltrating T cells remains incomplete, these include cells specific for beta cell-associated proteins such as insulin (Pathiraja et al. 2015). Following onset, individuals with the disease exhibit some residual capacity to produce insulin, but at a broad spectrum of levels which

H. Nguyen • E.A. James (✉)
Benaroya Research Institute, Seattle, WA, USA
e-mail: ejames@benaroyaresearch.org

A.P. Nicholas et al. (eds.), *Protein Deimination in Human Health and Disease*,
DOI 10.1007/978-3-319-58244-3_10

diminish with time. Such residual insulin secretion is clinically significant, as patients with detectable C-peptide levels require lower insulin doses and exhibit fewer chronic complications (Almeida et al. 2013). Therefore, T1D can be considered to be a heterogeneous disease with significant diversity in its age and rapidity of onset and in its severity and risk of complications.

Human T1D is recapitulated in many respects by the nonobese diabetic (NOD) mouse model. NOD mice develop spontaneous diabetes that shares important immunological features with human T1D, including development of islet antigen-specific autoantibodies and expansion of autoreactive CD4+ and CD8+ T cells (Anderson and Bluestone 2005). Further, the NOD model shares genetic features with T1D, including susceptibility at the MHC and other loci (Driver et al. 2012). These similarities make this model an important experimental tool for investigating the mechanisms that underlie disease especially as these relate to attributes that are not amenable to direct study in human subjects. The model's key limitation is the relative ease with which disease can be prevented (in comparison to human disease), as evidenced by numerous potential therapies that showed promise in the NOD model but provided no benefit in humans (Kachapati et al. 2012). Although important differences exist between the NOD model and human T1D (Reed and Herold 2015), use of this animal model has led to important mechanistic insights and the elucidation of important environmental factors that appear to contribute to disease susceptibility (Pearson et al. 2016).

10.2 Islet Antibodies and Determinant Spreading in Type 1 Diabetes

Individuals who develop T1D experience a progressive loss of tolerance to beta cell antigens, a process that is most clearly evidenced by the development of ICA. Multiple beta cell-associated antigens are recognized by ICA, and the accumulation of additional specificities can be used as a diagnostic indicator of risk. Subjects who are positive for more than one autoantibody specificity have a significantly higher risk of developing diabetes, exhibit evidence of more aggressive beta cell destruction following diagnosis, and require more exogenous insulin to maintain glycemic control (Sabbah et al. 1999). ICA typically arise sequentially, with antibodies that recognize insulin appearing most often at early time points (in some cases within the first year of life), followed most commonly (and sometimes almost immediately) by GAD65 and then other specificities (including IA-2 and ZNT8), which tend to appear at later times (Yu et al. 2013). The specific pattern of autoantibody formation can vary for individuals, but in general, subjects who are positive for a single ICA continue to develop additional ICA as they approach the onset of overt diabetes, a phenomenon known as determinant spreading. Given the immunologic link between antibody formation and T cell help, it is thought that determinant spreading encompasses both T cell and B cell responses and arises as a consequence of multiple bursts of autoimmune activity that are separated by periods of relative quiescence. However, the initial loss of tolerance in T1D and subsequent determinant spreading and waves of autoimmune activity that lead to beta cell destruction and loss of glycemic control are silent processes. Therefore, many unanswered questions remain about the events that underlie the disease. Indeed, several models have been proposed to explain the process of beta cell loss in T1D (van Belle et al. 2011). Our favored model is summarized in Fig. 10.1. Although diverse in some of their details, each model agrees that genetic, environmental, and immunologic factors play crucial roles in the disease and that an increasing autoimmune burden, evidenced by expanding specificities of self-reactive antibodies and T cells, leads to beta cell loss that eventually results in a critical decrease in insulin secretion and loss of glycemic control.

Although ICA are a useful prognostic indicator of diabetes risk, these antibodies are generally not considered to be pathogenic because blocking B cell function has no effect on disease development in experimental models (Bendelac et al. 1988) and because maternal transmission of islet autoantibodies apparently decreases the risk of autoimmune diabetes (Koczwara et al. 2004). However, because of the conceptual and demonstrated overlap between antigens targeted by B cells (antibodies) and CD4+ T cells (Sette et al. 2008), ICA are thought to give an important clue about the specificity of autoreactive CD4+ T cells in T1D. Indeed, many studies have successfully defined CD4+ T cell epitopes within the beta cell proteins that are targeted by antibodies. Early studies implicated certain antigens as "primary" beta cell antigens (Schloot et al. 1998; Lohmann et al. 1994). In particular, elegant studies in the NOD mouse model demonstrated that insulin is a crucial antigen for disease development (Nakayama et al. 2005). Other human studies have demonstrated epitope-specific T cell responses against each of the diagnostic antibody targets in T1D: insulin, GAD65, IA-2, IGRP, and ZNT8 (Yang et al. 2008, 2014; Durinovic-Bello et al. 2004; Reijonen et al. 2004; Herzog et al. 2004; Scotto et al. 2012; Dang et al. 2011). However, studies of both the NOD model and human diabetes also support the importance of additional antigens, including IGRP, chromogranin A, and IAPP in various stages of the disease (Lieberman et al. 2003; Yang et al. 2006; Stadinski et al. 2010; Li et al. 2015; Delong et al. 2011). These studies also indicate that CD4+ T cells recognize diverse antigens in T1D and that the most prevalent specificities can vary for different individuals.

10.3 T1D Susceptible HLA and Enhanced Presentation of Citrullinated Peptides

The significant association between T1D risk and a limited number of HLA haplotypes implies that disease-associated HLA-DQ and HLA-DR proteins promote the selection of a potentially autoreactive T cell repertoire, such that even unaffected subjects have circulating T cells with autoreactive specificities (Danke et al. 2004).

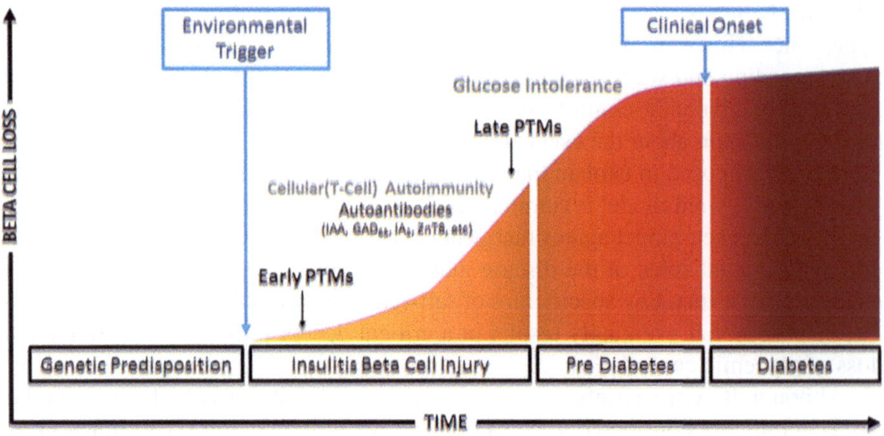

Fig. 10.1 Proposed timeline for the development of type 1 diabetes, adapted from the model of linear beta cell decay proposed by Eisenbarth (1986). The first stage of this model includes a suspected environmental trigger that initiates autoimmunity and gradual beta cell loss in genetically susceptible individuals. Additional events generate inflammation, leading to determinant spreading and more aggressive beta cell loss. Finally, suboptimally restrained immune damage of beta cells crosses a threshold, leading to impaired glucose tolerance (prediabetes) and eventually to the onset of diabetes. During this process, posttranslational modifications (such as citrullination) can be envisioned to happen during an early time window (*first black arrow*) as part of the initiation of autoimmunity or during a later time window (*second black arrow*) as a key component of determinant spreading

Several complimentary theories have been proposed to explain incomplete self-tolerance in T cell-mediated autoimmune disease in spite of thymic selection (Gough and Simmonds 2007). Among these, the favored presentation of posttranslationally modified self-peptides is a concept that has garnered increasing support and attention (Petersen et al. 2009). As described in greater detail elsewhere within this volume, the recognition of citrullinated self-antigens has a clear role in the etiology of rheumatoid arthritis (RA) (Snir et al. 2009; Scally et al. 2013). For HLA-DRB1*04:01 in particular, it is well documented that conversion of specific arginine residues into citrulline by PAD enzymes leads to the improved binding and presentation of self-peptides (Hill et al. 2003; James et al. 2014). This preferential presentation is dictated by the size and charge characteristics of binding pockets (most notably pocket 4) that are shared by homologous HLA class II proteins which preferentially accommodate citrulline because of its increased flexibility and neutral charge (James et al. 2010). This creates a scenario in which peptides that are incapable of being bound and presented by HLA-DRB1*04:01 in their native form can become high-affinity neo-epitopes in their citrullinated form (Fig. 10.2a). In a complimentary fashion, these same differences in flexibility and charge can alter interactions between the peptide and variable residues within T cell receptors (TCR) leading to increased responsiveness to the citrullinated peptide. Indeed, published work has documented that both of these scenarios occur in patients with RA (James et al. 2014).

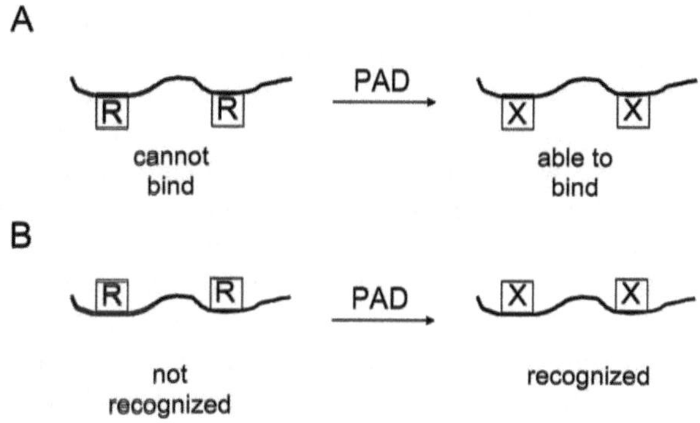

Fig. 10.2 Enhanced recognition of citrullinated peptides. (**a**) Autoimmune susceptible HLA proteins such as DRB1*04:01 are able to accommodate citrulline but not arginine within key binding pockets because of its flexibility and neutral charge. This can cause up to 500-fold improvement in the binding affinity of citrullinated peptides if arginine residues in the native sequence coincide with binding anchor positions. (**b**) T cell receptors (TCR) can also have increased recognition of citrulline through these same changes in flexibility and neutral charge if arginine residues in the native sequence coincide with T cell receptor contact residues

Thinking about translating this information into the context of autoimmune diabetes, it is notable then that HLA-DRB1*04:01 is part of the highest-risk haplotype for T1D. Therefore, a major proportion of subjects who develop T1D also have this propensity to present and recognize citrullinated peptides.

10.4 Citrullination of Antigens in Model Systems

As outlined in subsequent sections of this chapter, recent studies have begun to validate the importance of citrullinated antigens in human T1D. However, many important early observations have originated from the hen egg lysozyme (HEL) experimental mouse system and the NOD mouse model of spontaneous diabetes. These systems have provided evidence indicating that protein citrullination generates unique epitopes in autoimmune-prone animals and have provided important insights about the mechanisms through which citrullinated peptides are generated in vivo and presented to T cells.

10.4.1 Protein Citrullination in Hen Egg Lysozyme System

Some of the most detailed work investigating the importance of citrullination in eliciting immune responses utilized the immunization of mice with hen egg lysozyme (HEL) as a protein antigen (Ireland et al. 2006). These experiments utilized B10.BR-Tg mice expressing membrane-associated HEL under the class II Eα promoter as an artificial self-antigen. T cell hybridomas generated from these mice selectively recognized citrullinated HEL epitopes presented by I-Ak with no cross-reactivity to unmodified peptide. Furthermore, different lineages of antigen presenting cells (macrophages and bone marrow derived dendritic cells) could activate these citrulline selective hybridomas when pulsed with unmodified HEL, indicating that these professional APC can process and present citrullinated peptides. This is consistent with observations that PAD2 and PAD4 mRNAs are present within antigen presenting cells (Vossenaar et al. 2004). Notably, presentation of citrullinated peptides could be blocked by 3-methyladenine or by shRNA targeting Atg5 (a protein essential for autophagy) whereas the presentation of unmodified peptides by APC that received either treatment remained intact. Therefore, presentation of citrullinated epitopes to T cells appears to be an autophagy-dependent process, at least in this experimental system. However, these observations contrast somewhat with recent findings suggesting that hypercitrullination in RA patients occurs as a result of immune-mediated membranolytic pathways (Romero et al. 2013).

10.4.2 Protein Citrullination in the Nonobese Diabetic Mouse Model

As outlined above, NOD mice develop spontaneous autoimmune diabetes that shares many important features with human T1D. Therefore, researchers have utilized the NOD mouse model to investigate mechanistic factors that contribute to the progression of T1D. Seminal studies with this mouse model identified many important beta cell antigens that are also targeted by antibodies and T cells in human subjects, including proinsulin (Wegmann et al. 1994), glutamic acid decarboxylase (GAD65) (Baekkeskov et al. 1990), islet-specific glucose-6-phosphatase catalytic subunit-related protein (Lieberman et al. 2003), chromogranin A (Stadinski et al. 2010), and zinc transporter 8 (Dang et al. 2011). Utilizing this model, a recent study demonstrated that citrullinated glucose-regulated protein 78 (GRP78) is recognized in mice that develop spontaneous diabetes (Ronda et al. 2015). This study observed GRP78-specific autoantibodies, providing the first evidence of antibody recognition of a citrullinated protein in autoimmune diabetes. Splenocytes isolated from these mice produced IFN-γ in response to citrullinated but not native GRP78, indicating a citrulline-specific T cell response. Furthermore, this study documented an upregulation of PAD2 in pancreatic islets, suggesting a mechanism for increased protein citrullination in inflamed beta cells. Notably, GRP78, also known as binding immunoglobulin protein or BiP, has been implicated as a citrullinated antigen in RA (Shoda et al. 2015), suggesting a possible mechanistic overlap between these diseases. GRP78 is a major ER chaperone with anti-apoptotic properties (Lee 2005) and functions as a regulator of the unfolded protein response (UPR), which is a key biological process for secretory cells such as pancreatic beta cells (Wu and Kaufman 2006). Therefore, the presence, citrullination, and subsequent loss of tolerance to GRP78 make sense from a mechanistic point of view.

10.5 Recognition of Citrullinated Beta Cell Antigens in T1D

In spite of the significant overlap in the genetic risk between RA and T1D, including overlapping high-risk HLA haplotypes (Burton et al. 2007), surprisingly, few studies have investigated the relevance and role of citrullinated antigens in T1D. However, it would be reasonable to assert that these diseases are likely to have commonalities in their etiology. Like RA, type 1 diabetes (T1D) is a T cell-mediated autoimmune disease for which the appearance of autoantibodies is a relevant indicator for the risk of onset (Sabbah et al. 1999). Furthermore, at a population level, there is a specific association between T1D and CCP+ but not CCP-RA (Liao et al. 2009). Beyond this, a recent study by Wong et al. (2015) established an interesting link between hyperglycemia and protein citrullination. In this study, neutrophils from subjects with diabetes (either type 1 or type 2) were shown to have elevated PAD4 expression and were primed to produce NETs. Correspondingly, diabetic mice exhibited higher levels of citrullinated H3 and exhibited delayed wound healing, whereas would healing was accelerated in PAD4 knockout mice and NET formation was not observed in their wounds. Cumulatively, these findings imply that hyperglycemia elicits increased PAD4 expression and activity, leading to increased protein citrullination. Therefore, it is not surprising that recent data is beginning to affirm the recognition of citrullinated antigens in subjects with T1D.

10.5.1 Recognition of Citrullinated Beta Cell Peptides by CD4+ T Cells

It has been known for decades that an arginine residue of the insulin B chain can be citrullinated (Hayashi et al. 1993) and a diversity of published evidence implies a possible role for the recognition of modified epitopes in T1D (Mannering et al. 2005; Delong et al. 2012; van Lummel et al. 2014; Doyle and Mamula 2012). However, it was only recently demonstrated that citrullinated beta cell antigens are targeted by autoreactive T cells in subjects with autoimmune diabetes. Our group observed that a citrullinated peptide derived from GAD65 elicited functional T cell responses in T1D patients (McGinty et al. 2014). More importantly, this work went on to show that T cell clones isolated from T1D patients preferentially responded to citrullinated peptide and to GAD65 protein that has been incubated in vitro with PAD enzyme. Direct ex vivo analysis with the corresponding HLA-DR0401 tetramer revealed that CD4+ T cells that recognize this epitope were present at elevated frequencies in subjects with T1D and exhibited an antigen-experienced cell surface phenotype. In follow-up experiments, we have gone on to show that additional citrullinated peptides derived from beta cell antigens are recognized by T cells. As summarized in Table 10.1, some but not all of these peptides are preferentially recognized in their citrullinated form.

Table 10.1 Sequences and binding affinity for citrullinated beta cell peptides

Peptide	Modification	Amino acid sequence[a,b]	Modified IC_{50} (μmol/L)[c]	Wild-type IC_{50} (μmol/L)[d]
GAD_{89-108}[e]	105 Cit	YAFLHATDLLPACDGE**X**PTL	2.9	3.2
$GAD_{265-284}$	272 Cit	KGMAALP**X**LIAFTSEHSHFS	0.8	0.3
$GAD_{473-492}$	488 Cit	KGMAALP**X**LIAFTSEHSHFS	0.9	10
$GAD_{553-572}$[e]	558 Cit	KVNFF**X**MVISNPAATHQDID	0.04	0.01
$IAPP_{65-84}$	73Cit, 81Cit	VGSNTYGK**X**NAVEVLK**X**EPL	2.0	n.b.

[a]Each modified residue is indicated in boldface
[b]X indicates citrulline
[c]IC50 represents the peptide concentration that displaces half of the reference peptide
[d]n.b. indicates nonbinding
[e]Indicates peptides that are not preferentially recognized when citrullinated

The two peptides which are not preferentially recognized are bound and presented by DRB1*04:01 in their unmodified form, indicating that the arginine/citrulline residue does not coincide with an HLA anchor position (see Fig. 10.2). Binding predictions suggest that the arginine/citrulline within GAD_{89-108} occurs at a flanking residue that also does not interact with TCR. The arginine/citrulline within $GAD_{553-572}$ is predicted to occur at a TCR-interacting residue, but citrullination does not lead to increased recognition in this instance.

The preferential recognition of citrullinated peptides by T cells parallels observations of enhanced recognition of citrullinated GRP78 in NOD mice. It remains to be verified whether citrullinated beta cell proteins are recognized by autoantibodies, as they are in in mice that develop spontaneous diabetes.

10.5.2 *Beta Cell Stress and Upregulation of Modifying Enzymes*

As already mentioned in the context of GRP78 and its role regulating the UPR and related cellular stress pathways, pancreatic beta cells naturally undergo high levels of ER stress as a result of their normal secretory physiology. In particular, beta cells undergo significant ER stress during normal glucose-stimulated insulin synthesis and secretion (Lipson et al. 2006). In addition to the stress that is inherent to normal physiologic function, environmental triggers, including some which are thought to be associated with development of T1D (see Fig. 10.1), can further elevate ER stress in beta cells. For example, viral infection can disrupt the ER membrane leading to calcium release (van Kuppeveld et al. 1997, 2005), exposure of beta cells to reactive oxygen species causes oxidation and protein misfolding (Bhandary et al. 2013), and cytokine exposure can elicit ER calcium release through the c-Jun N-terminal kinase pathway (Wang et al. 2009). Furthermore, the hyperglycemia and increased glucose sensing that occurs during the progression of autoimmune diabetes significantly increase insulin demand, thereby heightening physiologic beta cell stress (Defronzo et al. 1979). Therefore, heightened ER stress has been linked to several events that occur during the development of T1D.

There is the obvious link between cellular stress and calcium release. Beyond this, calcium levels are known to have a crucial influence on the activity of calcium-dependent enzymes. Notably, PAD enzymes are calcium dependent, becoming activated when cytosolic calcium concentrations increase (Vossenaar et al. 2003). As such it can be reasonably hypothesized that beta cell stress (induced by environmental and/or physiological conditions) increases the activity of PAD enzymes, which can be recruited to subcellular compartments to modify self-proteins. In subjects with susceptible HLA haplotypes, abnormally modified proteins can be preferentially presented to and recognized by autoreactive T cells, eliciting high-affinity autoimmune responses against these neo-epitopes. Analogously, a recent study demonstrated that chemical induction of ER stress (via thapsigargin treatment) led to increased tissue transglutaminase 2 activity, culminating in elevated T cell responses, as compared with untreated murine islets and insulinoma cells (Marre et al. 2016).

Placing this back in the context of the overall disease process (Fig. 10.1), there is sufficient rationale to support the view that an early environmental triggering event, such as viral infection, could generate beta cell stress, leading to increased activity of calcium-dependent PAD enzymes (presumably PAD2 or PAD4) in pancreatic beta cells. The resulting citrullination of beta cell proteins could then generate neo-epitopes, eliciting autoreactive T cell responses that play a key role in initiating disease. However, it is equally plausible that immune responses that occur close to the initiating events of the disease could be confined mainly to the unmodified epitopes and antigens such as insulin that have been implicated through studies in the NOD model. In this scenario, the inflammation and hyperglycemia caused by successive waves of autoimmune destruction could then lead to increased PAD activity and citrullination of beta cell proteins, eliciting autoreactive T cell responses that play a key role in disease progression. Published data on epitope-specific responses in subjects at risk of developing T1D is extremely limited, but one study included limited data indicating that the frequency of T cells that recognize citrullinated GAD65 is not significantly elevated in at risk subjects who are single autoantibody positive (McGinty et al. 2014). However, this study was not adequately powered to detect small differences between at risk subjects and controls.

10.6 Prospects for Diagnostic Detection of Anti-citrulline Responses in T1D

As noted earlier in this chapter, ICA are routinely used in prevention studies as a prognostic indicator of diabetes risk. As a result, robust platforms are available for autoantibody testing, including a recently developed electrochemiluminescence-based assay that may be suitable for wider screening of the at risk population

(Zhao et al. 2016). Antibody responses that target citrullinated beta cell antigens have not yet been observed in human subjects. However, based on recent studies documenting antibodies that recognize citrullinated GRP78 in NOD mice and T cell responses to citrullinated GAD65 in patients with T1D, it seems reasonable to assert that such antibody responses are likely to exist. Indeed, posttranslationally modified oxidized IAA antibodies have recently been detected in T1D patients (Strollo et al. 2015). If the relevance of such autoantibodies can be confirmed, it should be technically feasible to design diagnostic assays for their detection based either on established methodologies that are used for ICA assays or on the highly effective CCP assays that are utilized as laboratory diagnostics in the setting of RA (Szekanecz et al. 2008). Autoantibodies against PTM, including citrulline-specific antibodies, could then provide a more comprehensive picture of the immune status of at risk subjects and of patients who are largely negative for unmodified antigens that are currently monitored in prevention studies and clinical trials.

Detection of T cell responses directed against citrullinated antigens poses a greater technical challenge. Several complimentary technologies exist for measuring antigen-specific T cell responses, including cytokine release assays such as ELISPOT, cell proliferation assays such as 5,6-carboxyfluorescein diacetate succinimidyl ester (CFSE) dilution, and HLA multimer-based assays (Mannering et al. 2010). However, none of these options has progressed to the point of being accepted as an approved diagnostic. Furthermore, it has been shown that CD4+ T cells recognize diverse antigens and epitopes in T1D (Yang et al. 2013) and occur at relatively low frequencies (McGinty et al. 2014; Chow et al. 2014). The same appears to be true for citrullinated antigens in subjects with RA (James et al. 2014). Therefore, any successful assay must have the capacity to efficiently characterize multiple specificities. One encouraging new prospect is a recently developed combinatorial class II tetramer methodology that would enable parallel analysis of up to six specificities (or six groups of specificities) within a single staining tube (Uchtenhagen et al. 2016). A combined approach utilizing a multiplex T cell assay in conjunction with ICA and citrullinated antibody data could allow the early identification of at risk subjects who are progressing toward the onset of autoimmune diseases. One possible alternative or complement to such sophisticated T cell assays would be proteomic detection of modified beta cell proteins utilizing established methods for detecting citrullinated proteins (Hensen and Pruijn 2014; Clancy et al. 2016), which could be taken to imply an increased likelihood for the loss of tolerance to these antigens. However, acquisition of appropriate samples for such analysis may not be feasible in the setting of human disease.

10.7 Summary and Significance of Citrullination in T1D

Given all of the information that we have presented here in this chapter, citrullination of beta cell proteins can unmask modified self-epitopes, promoting the loss of self-tolerance either through increased presentation of these peptides via disease susceptible HLA class II proteins such as DRB1*04:01 (through citrullination of arginine residues at anchor positions) or through increased recognition of these peptides by autoreactive TCR (through citrullination of arginine residues at TCR contact positions). Accumulating evidence suggests that protein citrullination in the beta cell occurs as a result of biochemical stresses and inflammation, initiating an immunological process that plays a role in loss of tolerance in T1D. Our favored model is that inflammation and hyperglycemia that occur during successive waves of autoimmune destruction lead to increased PAD activity and citrullination of beta cell proteins, eliciting autoreactive T cell responses that play a key role in disease progression and accompanying antibody responses that could have utility as an additional diagnostic indicator of disease risk. However, it could be posited that recognition of citrullinated epitopes happens early in disease in response to the environmental insults that are thought to serve as an early trigger for autoimmunity.

More generally, the enhanced recognition of citrullinated self-proteins appears to be a shared mechanism that is relevant to several autoimmune diseases, as outlined in the corresponding chapters of this volume on the topics of RA, multiple sclerosis, and Alzheimer's. In each of these settings, the deimination of arginine residues alters peptide presentation and recognition, thereby expanding epitope diversity and, perhaps, activating a set of immune cells that is suboptimally tolerized because of reduced (or absent) PAD activity in non-inflamed tissues. Given that immune recognition of citrullinated epitopes reflects a shared disease pathway among diverse autoimmune diseases, the pursuit of treatment options such as specific PAD inhibitors, as described in the final chapter of this volume, may provide an important avenue for therapeutic intervention.

References

Almeida, M. H., Dantas, J. R., Barone, B., Serfaty, F. M., Kupfer, R., Albernaz, M., Bencke, M. R., Zajdenverg, L., Rodacki, M., & Oliveira, J. E. (2013). Residual C-peptide in patients with type 1 diabetes and multiethnic backgrounds. *Clinics (São Paulo, Brazil), 68*, 123–126.

Anderson, M. S., & Bluestone, J. A. (2005). The NOD mouse: A model of immune dysregulation. *Annual Review of Immunology, 23*, 447–485.

Baekkeskov, S., Aanstoot, H. J., Christgau, S., Reetz, A., Solimena, M., Cascalho, M., Folli, F., Richterolesen, H., & Camilli, P. D. (1990). Identification of the 64k autoantigen in insulin-dependent diabetes as the GABA-synthesizing enzyme glutamic-acid decarboxylase. *Nature, 347*, 151–156.

Barrett, J. C., Clayton, D. G., Concannon, P., Akolkar, B., Cooper, J. D., Erlich, H. A., Julier, C., Morahan, G., Nerup, J., Nierras, C., Plagnol, V., Pociot, F., Schuilenburg, H., Smyth, D. J., Stevens, H., Todd, J. A., Walker, N. M., Rich, S. S., & Consortium Type 1 Diabetes Genetics. (2009). Genome-wide association study and meta-analysis find that over 40 loci affect risk of type 1 diabetes. *Nature Genetics, 41*, 703–707.

Bendelac, A., Boitard, C., Bedossa, P., Bazin, H., Bach, J. F., & Carnaud, C. (1988). Adoptive T cell transfer of autoimmune nonobese diabetic mouse diabetes does not require recruitment of host B lymphocytes. *Journal of Immunology, 141*(8), 2625.

Bhandary, B., Marahatta, A., Kim, H. R., & Chae, H. J. (2013). An involvement of oxidative stress in endoplasmic reticulum stress and its associated diseases. *International Journal of Molecular Sciences, 14*, 434–456.

Burton, P. R., Clayton, D. G., Cardon, L. R., Craddock, N., Deloukas, P., Duncanson, A., Kwiatkowski, D. P., McCarthy, M. I., Ouwehand, W. H., Samani, N. J., Todd, J. A., Donnelly, P., Barrett, J. C., Burton, P. R., Davison, D., Donnelly, P., Easton, D., Evans, D., Leung, H. T., Marchini, J. L., Morris, A. P., Spencer, C. C., Tobin, M. D., Cardon, L. R., Clayton, D. G., Attwood, A. P., Boorman, J. P., Cant, B., Everson, U., Hussey, J. M., Jolley, J. D., Knight, A. S., Koch, K., Meech, E., Nutland, S., Prowse, C. V., Stevens, H. E., Taylor, N. C., Walters, G. R., Walker, N. M., Watkins, N. A., Winzer, T., Todd, J. A., Ouwehand, W. H., Jones, R. W., McArdle, W. L., Ring, S. M., Strachan, D. P., Pembrey, M., Breen, G., St Clair, D., Caesar, S., Gordon-Smith, K., Jones, L., Fraser, C., Green, E. K., Grozeva, D., Hamshere, M. L., Holmans, P. A., Jones, I. R., Kirov, G., Moskvina, V., Nikolov, I., O'Donovan, M. C., Owen, M. J., Craddock, N., Collier, D. A., Elkin, A., Farmer, A., Williamson, R., McGuffin, P., Young, A. H., Ferrier, I. N., Ball, S. G., Balmforth, A. J., Barrett, J. H., Bishop, D. T., Iles, M. M., Maqbool, A., Yuldasheva, N., Hall, A. S., Braund, P. S., Burton, P. R., Dixon, R. J., Mangino, M., Suzanne, S., Tobin, M. D., Thompson, J. R., Samani, N. J., Bredin, F., Tremelling, M., Parkes, M., Drummond, H., Lees, C. W., Nimmo, E. R., Satsangi, J., Fisher, S. A., Forbes, A., Lewis, C. M., Onnie, C. M., Prescott, N. J., Sanderson, J., Mathew, C. G., Barbour, J., Mohiuddin, M. K., Todhunter, C. E., Mansfield, J. C., Ahmad, T., Cummings, F. R., Jewell, D. P., Webster, J., Brown, M. J., Clayton, D. G., Lathrop, G. M., Connell, J., Dominczak, A., Samani, N. J., Marcano, C. A., Burke, B., Dobson, R., Gungadoo, J., Lee, K. L., Munroe, P. B., Newhouse, S. J., Onipinla, A., Wallace, C., Xue, M., Caulfield, M., Farrall, M., Barton, A., Bruce, I. N., Donovan, H., Eyre, S., Gilbert, P. D., Hider, S. L., Hinks, A. M., John, S. L., Potter, C., Silman, A. J., Symmmons, D. P., Thomson, W., Worthington, J., Clayton, D. G., Dunger, D. B., Nutland, S., Stevens, H. E., Walker, N. M., Widmer, B., Todd, J. A., Frayling, T. A., Freathy, R. M., Lango, H., Perry, J. R., Shields, B. M., Weedon, M. N., Hattersley, A. T., Hitman, G. A., Walker, M., Elliott, K. S., Groves, C. J., Lindgren, C. M., Rayner, N. W., Timpson, N. J., Zeggini, E., McCarthy, M. I., Newport, M., Sirugo, G., Lyons, E., Vannberg, F., Hill, A. V., Bradbury, L. A., Farrar, C., Pointon, J. J., Wordsworth, P., Brown, M. A., Franklyn, J. A., Heward, J. M., Simmonds, M. J., Gough, S. C., Seal, S., Stratton, M. R., Rahman, N., Ban, M., Goris, A., Sawcer, S. J., Compston, A., Conway, D., Jallow, M., Newport, M., Sirugo, G., Rockett, K. A., Kwiatowski, D. P., Bumpstead, S. J., Chaney, A., Downes, K., Ghori, M. J., Gwilliam, R., Hunt, S. E., Inouye, M., Keniry, A., King, E., McGinnis, R., Potter, S., Ravindrarajah, R., Whittaker, P., Widden, C., Withers, D., Deloukas, P., Leung, H. T., Nutland, S., Stevens, H. E., Walker, N. M., Todd, J. A., Easton, D., Clayton, D. G., Burton, P. R., Tobin, M. D., Barrett, J. C., Evans, D., Morris, A. P., Cardon, L. R., Cardin, N. J., Davison, D., Ferreira, T., Pereira-Gale, J., Hallgrimsdottir, I. B., Howie, B. N., Marchini, J. L., Spencer, C. C., Su, Z., Teo, Y. Y., Vukcevic, D., Donnelly, P., Bentley, D., Brown, M. A., Gordon, L. R., Caulfield, M., Clayton, D. G., Compston, A., Craddock, N., Deloukas, P., Donnelly, P., Farrall, M., Gough, S. C., Hall, A. S., Hattersley, A. T., Hill, A. V., Kwiatkowski, D. P., Mathew, C., McCarthy, M. I., Ouwehand, W. H., Parkes, M., Pembrey, M., Rahman, N., Samani, N. J., Stratton, M. R., Todd, J. A., & Worthington, J. (2007). Genome-wide association study of 14,000 cases of seven common diseases and 3,000 shared controls. *Nature, 447*, 661–678.

Chiang, J. L., Kirkman, M. S., Laffel, L. M. B., & Peters, A. L. (2014). Type 1 diabetes through the life span: A position statement of the American Diabetes Association. *Diabetes Care, 37*, 2034–2054.

Chow, I. T., Yang, J., Gates, T. J., James, E. A., Mai, D. T., Greenbaum, C., & Kwok, W. W. (2014). Assessment of CD4+ T cell responses to glutamic acid decarboxylase 65 using DQ8 tetramers reveals a pathogenic role of GAD65 121-140 and GAD65 250-266 in T1D development. *PloS One, 9*, e112882.

Clancy, K. W., Weerapana, E., & Thompson, P. R. (2016). Detection and identification of protein citrullination in complex biological systems. *Current Opinion in Chemical Biology, 30*, 1–6.

Coppieters, K. T., Dotta, F., Amirian, N., Campbell, P. D., Kay, T. W., Atkinson, M. A., Roep, B. O., & von Herrath, M. G. (2012). Demonstration of islet-autoreactive CD8 T cells in insulitic lesions from recent onset and long-term type 1 diabetes patients. *The Journal of Experimental Medicine, 209*, 51–60.

Couper, J. J. (2001). Environmental triggers of type 1 diabetes. *Journal of Paediatrics and Child Health, 37*, 218–220.

Dang, M. L., Rockell, J., Wagner, R., Wenzlau, J. M., Yu, L., Hutton, J. C., Gottlieb, P. A., & Davidson, H. W. (2011). Human type 1 diabetes is associated with t cell autoimmunity to zinc transporter 8. *Journal of Immunology, 186*, 6056–6063.

Danke, N. A., Koelle, D. M., Yee, C., Beheray, S., & Kwok, W. W. (2004). Autoreactive T cells in healthy individuals. *Journal of Immunology, 172*, 5967–5972.

Defronzo, R. A., Tobin, J. D., & Andres, R. (1979). Glucose clamp technique - method for quantifying insulin-secretion and resistance. *American Journal of Physiology, 237*, E214–EE23.

Delong, T., Baker, R. L., Reisdorph, N., Reisdorph, R., Powell, R. L., Armstrong, M., Barbour, G., Bradley, B., & Haskins, K. (2011). Islet amyloid polypeptide is a target antigen for diabetogenic CD4+ T cells. *Diabetes, 60*, 2325–2330.

Delong, T., Baker, R. L., He, J., Barbour, G., Bradley, B., & Haskins, K. (2012). Diabetogenic T-cell clones recognize an altered peptide of chromogranin A. *Diabetes, 61*, 3239–3246.

Diagnosis and classification of diabetes mellitus. (2012). *Diabetes Care, 35*(Suppl 1), S64–S71.

Doyle, H. A., & Mamula, M. J. (2012). Autoantigenesis: The evolution of protein modifications in autoimmune disease. *Current Opinion in Immunology, 24*, 112–118.

Driver, J. P., Chen, Y.-G., & Mathews, C. E. (2012). Comparative genetics: Synergizing human and NOD mouse studies for identifying genetic causation of type 1 diabetes. *The Review of Diabetic Studies: RDS, 9*, 169–187.

Durinovic-Bello, I., Schlosser, M., Riedl, M., Maisel, N., Rosinger, S., Kalbacher, H., Deeg, M., Ziegler, M., Elliott, J., Roep, B. O., Karges, W., & Boehm, B. O. (2004). Pro- and anti-inflammatory cytokine production by autoimmune T cells against preproinsulin in HLA-DRB1*04, DQ8 type 1 diabetes. *Diabetologia, 47*, 439–450.

Eisenbarth, G. S. (1986). Type I diabetes mellitus. A chronic autoimmune disease. *The New England Journal of Medicine, 314*, 1360–1368.

Gough, S. C., & Simmonds, M. J. (2007). The HLA region and autoimmune disease: Associations and mechanisms of action. *Current Genomics, 8*, 453–465.

Hawa, M. I., Kolb, H., Schloot, N., Beyan, H., Paschou, S. A., Buzzetti, R., Mauricio, D., De Leiva, A., Yderstraede, K., Beck-Neilsen, H., Tuomilehto, J., Sarti, C., Thivolet, C., Hadden, D., Hunter, S., Schernthaner, G., Scherbaum, W. A., Williams, R., Brophy, S., Pozzilli, P., & Leslie, R. D. (2013). Adult-onset autoimmune diabetes in Europe is prevalent with a broad clinical phenotype: Action LADA 7. *Diabetes Care, 36*, 908–913.

Hayashi, H., Morioka, M., Ichimiya, S., Yamato, K., Hinode, D., Nagata, A., & Nakamura, R. (1993). Participation of an arginyl residue of insulin chain-B in the inhibition of hemagglutination by porphyromonas-gingivalis. *Oral Microbiology and Immunology, 8*, 386–389.

Hensen, S. M. M., & Pruijn, G. J. M. (2014). Methods for the detection of peptidylarginine deiminase (PAD) activity and protein citrullination. *Molecular & Cellular Proteomics, 13*, 388–396.

Herzog, B. A., Ott, P. A., Dittrich, M. T., Quast, S., Karulin, A. Y., Kalbacher, H., Karges, W., Tary-Lehmann, M., Lehmann, P. V., Boehm, B. O., & Durinovic-Bello, I. (2004). Increased in vivo frequency of IA-2 peptide-reactive IFN gamma(+)/IL-4(−) T cells in type 1 diabetic subjects. *Journal of Autoimmunity, 23*, 45–54.

Hill, J. A., Southwood, S., Sette, A., Jevnikar, A. M., Bell, D. A., & Cairns, E. (2003). Cutting edge: The conversion of arginine to citrulline allows for a high-affinity peptide interaction with the rheumatoid arthritis-associated HLA-DRB1*0401 MHC class II molecule. *Journal of Immunology, 171*, 538–541.

Ireland, J., Herzog, J., & Unanue, E. R. (2006). Cutting edge: Unique T cells that recognize citrullinated peptides are a feature of protein immunization. *Journal of Immunology, 177*, 1421–1425.

James, E. A., Moustakas, A. K., Bui, J., Papadopoulos, G. K., Bondinas, G., Buckner, J. H., & Kwok, W. W. (2010). HLA-DR1001 presents "altered-self" peptides derived from joint-associated proteins by accepting citrulline in three of its binding pockets. *Arthritis and Rheumatism, 62*, 2909–2918.

James, E. A., Rieck, M., Pieper, J., Gebe, J. A., Yue, B. B., Tatum, M., Peda, M., Sandin, C., Klareskog, L., Malmstrom, V., & Buckner, J. H. (2014). Citrulline-specific Th1 cells are increased in rheumatoid arthritis and their frequency is influenced by disease duration and therapy. *Arthritis & Rhematology, 66*, 1712–1722.

Kachapati, K., Adams, D., Bednar, K., & Ridgway, W. M. (2012). The non-obese diabetic (NOD) mouse as a model of human type 1 diabetes. *Methods in Molecular Biology, 933*, 3–16.

Koczwara, K., Bonifacio, E., & Ziegler, A. G. (2004). Transmission of maternal islet antibodies and risk of autoimmune diabetes in offspring of mothers with type 1 diabetes. *Diabetes, 53*, 1–4.

Lee, A. S. (2005). The ER chaperone and signaling regulator GRP78/BiP as a monitor of endoplasmic reticulum stress. *Methods, 35*, 373–381.

Li, Y., Zhou, L., Li, Y., Zhang, J., Guo, B., Meng, G., Chen, X., Zheng, Q., Zhang, L., Zhang, M., & Wang, L. (2015). Identification of autoreactive CD8(+) T cell responses targeting chromogranin a in humanized NOD mice and type 1 diabetes patients. *Clinical Immunology, 159*, 63–71.

Liao, K. P., Gunnarsson, M., Kallberg, H., Ding, B., Plenge, R. M., Padyukov, L., Karlson, E. W., Klareskog, L., Askling, J., & Alfredsson, L. (2009). Specific association of type 1 diabetes mellitus with anti-cyclic citrullinated peptide-positive rheumatoid arthritis. *Arthritis and Rheumatism, 60*, 653–660.

Lieberman, S. M., Evans, A. M., Han, B. Y., Takaki, T., Vinnitskaya, Y., Caldwell, J. A., Serreze, D. V., Shabanowitz, J., Hunt, D. F., Nathenson, S. G., Santamaria, P., & DiLorenzo, T. P. (2003). Identification of the beta cell antigen targeted by a prevalent population of pathogenic CD8(+) T cells in autoimmune diabetes. *Proceedings of the National Academy of Sciences of the United States of America, 100*, 8384–8388.

Lipson, K. L., Fonseca, S. G., & Urano, F. (2006). Endoplasmic reticulum stress-induced apoptosis and auto-immunity in diabetes. *Current Molecular Medicine, 6*, 71–77.

Lohmann, T., Leslie, R. D. G., Hawa, M., Geysen, M., Rodda, S., & Londei, M. (1994). Immunodominant epitopes of glutamic-acid decarboxylase-65 and decarboxylase-67 in insulin-dependent diabetes-mellitus. *Lancet, 343*, 1607–1608.

Maahs, D. M., West, N. A., Lawrence, J. M., & Mayer-Davis, E. J. (2010). Epidemiology of type 1 diabetes. *Endocrinology and Metabolism Clinics of North America, 39*, 481–497.

Mannering, S. I., Harrison, L. C., Williamson, N. A., Morris, J. S., Thearle, D. J., Jensen, K. P., Kay, T. W., Rossjohn, J., Falk, B. A., Nepom, G. T., & Purcell, A. W. (2005). The insulin A-chain epitope recognized by human T cells is posttranslationally modified. *The Journal of Experimental Medicine, 202*, 1191–1197.

Mannering, S. I., Wong, F. S., Durinovic-Bello, I., Brooks-Worrell, B., Tree, T. I., Cilio, C. M., Schloot, N. C., & Mallone, R. (2010). Current approaches to measuring human islet-antigen specific T cell function in type 1 diabetes. *Clinical and Experimental Immunology, 162*, 197–209.

Marre, M. L., Profozich, J. L., Coneybeer, J. T., Geng, X., Bertera, S., Ford, M. J., Trucco, M., & Piganelli, J. D. (2016). Inherent ER stress in pancreatic islet beta cells causes self-recognition by autoreactive T cells in type 1 diabetes. *Journal of Autoimmunity, 72*, 33–46.

McGinty, J. W., Chow, I. T., Greenbaum, C., Odegard, J., Kwok, W. W., & James, E. A. (2014). Recognition of posttranslationally modified GAD65 epitopes in subjects with type 1 diabetes. *Diabetes, 63*, 3033–3040.

Nakayama, M., Abiru, N., Moriyama, H., Babaya, N., Liu, E., Miao, D. M., Yu, L. P., Wegmann, D. R., Hutton, J. C., Elliott, J. F., & Eisenbarth, G. S. (2005). Prime role for an insulin epitope in the development of type 1 diabetes in NOD mice. *Nature, 435*, 220–223.

Pathiraja, V., Kuehlich, J. P., Campbell, P. D., Krishnamurthy, B., Loudovaris, T., Coates, P. T., Brodnicki, T. C., O'Connell, P. J., Kedzierska, K., Rodda, C., Bergman, P., Hill, E., Purcell, A. W., Dudek, N. L., Thomas, H. E., Kay, T. W., & Mannering, S. I. (2015). Proinsulin-specific, HLA-DQ8, and HLA-DQ8-transdimer-restricted CD4+ T cells infiltrate islets in type 1 diabetes. *Diabetes, 64*, 172–182.

Pearson, J. A., Wong, F. S., & Wen, L. (2016). The importance of the non obese diabetic (NOD) mouse model in autoimmune diabetes. *Journal of Autoimmunity, 66*, 76–88.

Peng, H., & Hagopian, W. (2006). Environmental factors in the development of type 1 diabetes. *Reviews in Endocrine & Metabolic Disorders, 7*, 149–162.

Petersen, J., Purcell, A. W., & Rossjohn, J. (2009). Post-translationally modified T cell epitopes: Immune recognition and immunotherapy. *Journal of Molecular Medicine (Berlin), 87*, 1045–1051.

Reed, J. C., & Herold, K. C. (2015). Thinking bedside at the bench: The NOD mouse model of T1DM. *Nature Reviews Endocrinology, 11*, 308–314.

Reijonen, H., Mallone, R., Heninger, A. K., Laughlin, E. M., Kochik, S. A., Falk, B., Kwok, W. W., Greenbaum, C., & Nepom, G. T. (2004). GAD65-specific CD4(+) T-cells with high antigen avidity are prevalent in peripheral blood of patients with type 1 diabetes. *Diabetes, 53*, 1987–1994.

Roep, B. O., & Tree, T. I. M. (2014). Immune modulation in humans: Implications for type 1 diabetes mellitus. *Nature Reviews Endocrinology, 10*, 229–242.

Romero, V., Fert-Bober, J., Nigrovic, P. A., Darrah, E., Haque, U. J., Lee, D. M., van Eyk, J., Rosen, A., & Andrade, F. (2013). Immune-mediated pore-forming pathways induce cellular hypercitrullination and generate citrullinated autoantigens in rheumatoid arthritis. *Science Translational Medicine, 5*, 209ra150.

Ronda, D., Crèvecoeur, I., D'Hertog, W., Ferreira, G. B., Staes, A., Garg, A. D., Eizirik, D. L., Agostinis, P., Gevaert, K., Overbergh, L., & Mathieu, C. (2015). Citrullinated glucose-regulated protein 78 is an autoantigen in type 1 diabetes. *Diabetes, 64*, 573–586.

Sabbah, E., Savola, K., Kulmala, P., Veijola, R., Vahasalo, P., Karjalainen, J., Akerblom, H. K., Knip, M., & Grp Childhood Diabetes Finland Study. (1999). Diabetes-associated autoantibodies in relation to clinical characteristics and natural course in children with newly diagnosed type 1 diabetes. The childhood diabetes in Finland study group. *Journal of Clinical Endocrinology & Metabolism, 84*, 1534–1539.

Scally, S. W., Petersen, J., Law, S. C., Dudek, N. L., Nel, H. J., Loh, K. L., Wijeyewickrema, L. C., Eckle, S. B. G., van Heemst, J., Pike, R. N., McCluskey, J., Toes, R. E., La Gruta, N. L., Purcell, A. W., Reid, H. H., Thomas, R., & Rossjohn, J. (2013). A molecular basis for the association of the HLA-DRB1 locus, citrullination, and rheumatoid arthritis. *Journal of Experimental Medicine, 210*, 2569–2582.

Schloot, N. C., Willemen, S., Duinkerken, G., de Vries, R. R. P., & Roep, B. O. (1998). Cloned T cells from a recent onset IDDM patient reactive with insulin B-chain. *Journal of Autoimmunity, 11*, 169–175.

Scotto, M., Afonso, G., Larger, E., Raverdy, C., Lemonnier, F. A., Carel, J. C., Dubois-Laforgue, D., Baz, B., Levy, D., Gautier, J. F., Launay, O., Bruno, G., Boitard, C., Sechi, L. A., Hutton, J. C., Davidson, H. W., & Mallone, R. (2012). Zinc transporter (ZnT)8(186-194) is an immunodominant CD8(+) T cell epitope in HLA-A2(+) type 1 diabetic patients. *Diabetologia, 55*, 2026–2031.

Sette, A., Moutaftsi, M., Moyron-Quiroz, J., McCausland, M. M., Davies, D. H., Johnston, R. J., Peters, L., Rafii-El-Idrissi Benhnia, M., Hoffmann, J., Su, H.-P., Singh, K., Garboczi, D. N., Head, S., Grey, H., Felgner, P. L., & Crotty, S. (2008). Selective CD4(+) T cell help for antibody responses to a large viral pathogen: Deterministic linkage of specificities. *Immunity, 28*, 847–858.

Shoda, H., Fujio, K., Sakurai, K., Ishigaki, K., Nagafuchi, Y., Shibuya, M., Sumitomo, S., Okamura, T., & Yamamoto, K. (2015). Autoantigen BiP-derived HLA-DR4 epitopes differentially recognized by effector and regulatory T cells in rheumatoid arthritis. *Arthritis & Rhematology, 67*, 1171–1181.

Snir, O., Widhe, M., von Spee, C., Lindberg, J., Padyukov, L., Lundberg, K., Engstrom, A., Venables, P. J., Lundeberg, J., Holmdahl, R., Klareskog, L., & Malmstrom, V. (2009). Multiple antibody reactivities to citrullinated antigens in sera from patients with rheumatoid arthritis: Association with HLA-DRB1 alleles. *Annals of the Rheumatic Diseases, 68*, 736–743.

Stadinski, B. D., Delong, T., Reisdorph, N., Reisdorph, R., Powell, R. L., Armstrong, M., Piganelli, J. D., Barbour, G., Bradley, B., Crawford, F., Marrack, P., Mahata, S. K., Kappler, J. W., & Haskins, K. (2010). Chromogranin a is an autoantigen in type 1 diabetes. *Nature Immunology, 11*, 225–2U5.

Strollo, R., Vinci, C., Arshad, M. H., Perrett, D., Tiberti, C., Chiarelli, F., Napoli, N., Pozzilli, P., & Nissim, A. (2015). Antibodies to post-translationally modified insulin in type 1 diabetes. *Diabetologia, 58*, 2851–2860.

Szekanecz, Z., Soos, L., Szabo, Z., Fekete, A., Kapitany, A., Vegvari, A., Sipka, S., Szucs, G., Szanto, S., & Lakos, G. (2008). Anti-citrullinated protein antibodies in rheumatoid arthritis: As good as it gets? *Clinical Reviews in Allergy & Immunology, 34*, 26–31.

Uchtenhagen, H., Rims, C., Blahnik, G., Chow, I. T., Kwok, W. W., Buckner, J. H., & James, E. A. (2016). Efficient ex vivo analysis of CD4+ T-cell responses using combinatorial HLA class II tetramer staining. *Nature Communications, 7*, 12614.
van Belle, T. L., Coppieters, K. T., & von Herrath, M. G. (2011). Type 1 diabetes: Etiology, immunology, and therapeutic strategies. *Physiological Reviews, 91*, 79–118.
van Kuppeveld, F. J. M., Hoenderop, J. G. J., Smeets, R. L. L., Willems, P., Dijkman, H., Galama, J. M. D., & Melchers, W. J. G. (1997). Coxsackievirus protein 2B modifies endoplasmic reticulum membrane and plasma membrane permeability and facilitates virus release. *EMBO Journal, 16*, 3519–3532.
van Kuppeveld, F. J., de Jong, A. S., Melchers, W. J., & Willems, P. H. (2005). Enterovirus protein 2B po(u)res out the calcium: A viral strategy to survive? *Trends in Microbiology, 13*, 41–44.
van Lummel, M., Duinkerken, G., van Veelen, P. A., de Ru, A., Cordfunke, R., Zaldumbide, A., Gomez-Touriño, I., Arif, S., Peakman, M., Drijfhout, J. W., & Roep, B. O. (2014). Posttranslational modification of HLA-DQ binding islet autoantigens in type 1 diabetes. *Diabetes, 63*, 237–247.
Vossenaar, E. R., Zendman, A. J. W., van Venrooij, W. J., & Pruijn, G. J. M. (2003). PAD, a growing family of citrullinating enzymes: Genes, features and involvement in disease. *BioEssays, 25*, 1106–1118.
Vossenaar, E. R., Radstake, T. R., van der Heijden, A., van Mansum, M. A., Dieteren, C., de Rooij, D. J., Barrera, P., Zendman, A. J., & van Venrooij, W. J. (2004). Expression and activity of citrullinating peptidylarginine deiminase enzymes in monocytes and macrophages. *Annals of the Rheumatic Diseases, 63*, 373–381.
Wang, Q., Zhang, H., Zhao, B., & Fei, H. (2009). IL-1beta caused pancreatic beta-cells apoptosis is mediated in part by endoplasmic reticulum stress via the induction of endoplasmic reticulum Ca2+ release through the c-Jun N-terminal kinase pathway. *Molecular and Cellular Biochemistry, 324*, 183–190.
Wegmann, D. R., Norburyglaser, M., & Daniel, D. (1994). Insulin-specific T-cells are a predominant component of islet infiltrates in prediabetic NOD mice. *European Journal of Immunology, 24*, 1853–1857.
Willcox, A., Richardson, S. J., Bone, A. J., Foulis, A. K., & Morgan, N. G. (2009). Analysis of islet inflammation in human type 1 diabetes. *Clinical and Experimental Immunology, 155*, 173–181.
Wong, S. L., Demers, M., Martinod, K., Gallant, M., Wang, Y., Goldfine, A. B., Kahn, C. R., & Wagner, D. D. (2015). Diabetes primes neutrophils to undergo NETosis, which impairs wound healing. *Nature Medicine, 21*, 815–819.
Wu, J., & Kaufman, R. J. (2006). From acute ER stress to physiological roles of the unfolded protein response. *Cell Death and Differentiation, 13*, 374–384.
Yang, J. B., Danke, N. A., Berger, D., Reichstetter, S., Reijonen, H., Greenbaum, C., Pihoker, C., James, E. A., & Kwok, W. W. (2006). Islet-specific glucose-6-phosphatase catalytic subunit-related protein-reactive CD4(+) T cells in human subjects. *Journal of Immunology, 176*, 2781–2789.
Yang, J., Danke, N., Roti, M., Huston, L., Greenbaum, C., Pihoker, C., James, E. A., & Kwok, W. W. (2008). CD4+ T cells from type 1 diabetic and healthy subjects exhibit different thresholds of activation to a naturally processed proinsulin epitope. *Journal of Autoimmunity, 31*, 30–41.
Yang, J., James, E. A., Sanda, S., Greenbaum, C., & Kwok, W. W. (2013). CD4+ T cells recognize diverse epitopes within GAD65: Implications for repertoire development and diabetes monitoring. *Immunology, 138*, 269–279.
Yang, J., Chow, I. T., Sosinowski, T., Torres-Chinn, N., Greenbaum, C. J., James, E. A., Kappler, J. W., Davidson, H. W., & Kwok, W. W. (2014). Autoreactive T cells specific for insulin B:11-23 recognize a low-affinity peptide register in human subjects with autoimmune diabetes. *Proceedings of the National Academy of Sciences of the United States of America, 111*, 14840–14845.

Yu, L. P., Dong, F., Miao, D. M., Fouts, A. R., Wenzlau, J. M., & Steck, A. K. (2013). Proinsulin/insulin autoantibodies measured with electrochemiluminescent assay are the earliest indicator of prediabetic islet autoimmunity. *Diabetes Care, 36*, 2266–2270.

Zhao, Z., Miao, D., Michels, A., Steck, A., Dong, F., Rewers, M., & Yu, L. (2016). A multiplex assay combining insulin, GAD, IA-2 and transglutaminase autoantibodies to facilitate screening for pre-type 1 diabetes and celiac disease. *Journal of Immunological Methods, 430*, 28–32.

Chapter 11
Citrullinated Autoantigen Targets as Markers of Extra-Articular Disease in Rheumatoid Arthritis

Vinitha Ganesan and Dana P. Ascherman

11.1 Introduction

Citrullination represents an increasingly recognized posttranslational modification stemming from underlying physiological stressors that dysregulate intracellular calcium flux. Although this enzymatic process mediated by various isoforms of peptidylarginine deiminase (PAD) is fairly ubiquitous (occurring in normal as well as pathological states), the immune response to citrullinated proteins is heavily influenced by underlying HLA status and therefore highly associated with rheumatoid arthritis (RA) (Szodoray et al. 2010). Given that the humoral immune responses to citrullinated proteins may serve as an immunological "fingerprint" in RA, the question is whether delineating targets of anti-citrullinated protein antibodies (ACPAs) can provide insight regarding the site where immune tolerance is bypassed/broken or clarify the underlying pathophysiology of articular and extra-articular manifestations in this systemic autoimmune disease—even if the relative contribution of protein deimination versus citrulline-targeted immunity remains unresolved. In fact, extensive investigation over the last 10–15 years has yielded an expanded repertoire of ACPA specificities potentially linked with defined extra-articular manifestations, such as premature atherosclerosis, myocardial dysfunction, and interstitial lung disease (ILD), which negatively impact clinical outcome. Fueling these discoveries, novel approaches for identifying citrullinated autoantigen/autoantibody combinations have supported the search for additional biomarkers of extra-articular involvement that should further elucidate the immunobiology of relevant systemic disease pathways in RA.

V. Ganesan
Carnegie Mellon University, Pittsburgh, PA, USA

D.P. Ascherman, M.D. (✉)
University of Miami, Miami, FL, USA
e-mail: DAscherman@med.miami.edu

A.P. Nicholas et al. (eds.), *Protein Deimination in Human Health and Disease*,
DOI 10.1007/978-3-319-58244-3_11

11.2 Relationship Between Anti-CCP Antibodies and Global Extra-Articular Manifestations

Previous studies have assessed the relationship between anti-cyclic citrullinated peptide (anti-CCP) antibodies and different extra-articular manifestations of RA, with varying results depending on cohort size, population characteristics, and designated criteria for grading the presence/severity of extra-articular disease. Coupled with the relative lack of specificity of anti-CCP antibodies for defined extra-articular organ complications, these considerations have limited the use of commercial anti-CCP assays as diagnostic biomarkers of extra-articular disease. Nevertheless, correlations between extra-articular disease manifestations and anti-CCP titer suggest that this serologic marker may have significant prognostic value. In a Korean study, for example, investigators demonstrated that anti-CCP antibodies (along with other variables such as age and smoking) were predictive of a wide constellation of extra-articular manifestations, a conclusion supported by quantitative correlations with anti-CCP titers (Kim et al. 2008). Similarly, Turesson et al. found that rheumatoid factor (RF) and, to a lesser extent, anti-CCP antibodies were associated with "severe" extra-articular disease manifestations including Felty's syndrome and cutaneous vasculitis (Turesson et al. 2007). While a number of other analyses have also suggested a relationship between anti-CCP antibody levels and extra-articular complications such as rheumatoid nodules, peripheral neuropathy (predominantly entrapment neuropathy), hearing loss, and secondary Sjogren's syndrome (Goeldner et al. 2011; Lobo et al. 2016; Sim et al. 2014; Vignesh and Srinivasan 2015), at least one prominent study failed to substantiate these associations (Korkmaz et al. 2006).

11.3 Anti-CCP Antibodies as Biomarkers of Cardiovascular Risk in RA

Despite these limitations, more focused examination has suggested a relationship between anti-CCP titers and different indicators of cardiovascular disease, which is a major contributor to the heightened morbidity and mortality of RA. These observations may reflect the increase in myocardial citrullination that has been demonstrated in RA (further discussed in Chap. 12) and linked to lower myocardial mass (Giles et al. 2012), although the identity of protein(s) specifically citrullinated in cardiac tissue has not been fully established. Even in the absence of defined antibodies targeting citrullinated versions of cardiac antigens, investigators have shown that anti-CCP antibody positivity is independently associated with markers of ischemic heart disease (such as carotid intima-media thickness (CIMT)) and all-cause mortality after controlling for traditional risk factors and pro-inflammatory cytokine levels (Lopez-Longo et al. 2009; Vazquez-Del Mercado et al. 2015). Extending these findings that mesh with immunohistochemical demonstration of citrullination in coronary artery plaques (Sokolove et al. 2013), multivariate regression modeling has indicated that anti-CCP titers correlate with diastolic dysfunction (Marasovic-Krstulovic et al. 2011). While other studies have also shown that anti-CCP antibodies are linked to different parameters of myocardial dysfunction (Arnab et al. 2013), results have not uniformly supported the link with cardiovascular disease (Mackey et al. 2015), in part because of confounding associations with inflammation and disease duration (Arnab et al. 2013; Mackey et al. 2015). As an illustration of the discordance stemming from these factors, studies involving anti-citrullinated apoE versus anti-citrullinated fibrinogen/CCP2 have shown conflicting associations with cardiovascular risk and accepted measures of subclinical atherosclerosis that include vascular calcification, carotid plaque, and CIMT (Montes et al. 2015; Solow et al. 2015). Such discrepancies highlight the need for tissue-specific serologic markers of early cardiac involvement, particularly given the potential mechanistic relationship between citrullination and/or anti-citrullinated protein immune responses and dysregulation of vascular tone that has been suggested by statistical associations linking anti-CCP positivity with diminished levels of eNOS (Hjeltnes et al. 2011).

11.4 Citrullination and Pulmonary Disease

Beyond cardiovascular disease, pulmonary manifestations represent a significant source of morbidity in rheumatoid arthritis (discussed further in Chap. 13), contributing to a standardized mortality ratio of 2.5–5.0 (Bongartz et al. 2010; Brown 2007). While multiple tissues within the lung can be involved in the inflammatory processes associated with rheumatoid arthritis (including the pleura, airways, and/or lung parenchyma), citrullination has been most strongly linked with airway abnormalities (such as bronchiolitis) and interstitial lung disease (ILD) (Bongartz et al. 2007; Reynisdottir et al. 2014; Ytterberg et al. 2015). Supportive evidence for enhanced deimination in the lungs stems from immunohistochemical staining of lung biopsy specimens as well as analysis of bronchoalveolar lavage cells/tissue which collectively demonstrate increased extracellular PAD activity (particularly in smokers) and associated protein citrullination (Bongartz et al. 2007; Makrygiannakis et al. 2008; Damgaard et al. 2015). Intriguingly, citrullination of lung tissue is not unique to rheumatoid arthritis, occurring in the setting of chronic obstructive pulmonary disease as well as idiopathic pulmonary fibrosis (which shares usual interstitial pneumonia (UIP) histopathology with subsets of RA-associated ILD) (Bongartz et al. 2007; Lugli et al. 2015). What generally distinguishes RA-ILD from these other disease processes, however, is the degree of cellular infiltration and the unique humoral immune response to citrullinated antigens that is highly characteristic of RA.

Because citrulline-targeted immune responses can precede clinically evident articular disease, the lung may, in fact, serve as the site where immunologic tolerance is initially breached in RA. Several pieces of evidence circumstantially support this paradigm, including the preferential expression of ACPAs in bronchoalveolar lavage fluid (relative to serum) of early RA patients (Reynisdottir et al. 2014) as well as the presence of anti-CCP antibodies in sputum samples of individuals prior to *systemic* seroconversion and/or development of synovitis/inflammatory arthritis (Willis et al. 2013). Moreover, the emergence of anti-CCP versus anti-citrullinated vimentin antibodies in subsets of patients with COPD, bronchiectasis, or cystic fibrosis (Gerardi et al. 2013; Janssen et al. 2015)—even in the absence of overt articular involvement—indicates that chronic/persistent lung inflammation can serve as a potent stimulus for PAD-mediated protein citrullination which, in the appropriate genetic context, leads to ACPA formation. Consistent with this hypothesis, immunohistochemical staining has demonstrated a significant increase in citrullination of bronchial tissue derived from ACPA+ relative to ACPA- RA patients (Reynisdottir et al. 2014). At the same time, histopathologic studies in patients with RA-ILD have revealed inducible bronchus-associated lymphoid tissue (iBALT) containing plasma cells that specifically recognize citrullinated proteins (Rangel-Moreno et al. 2006).

11.5 ACPAs as Predictors of RA-Associated Interstitial Lung Disease

Based on these considerations, a number of investigators have explored the relationship between anti-CCP/CCP2 antibody levels and the development/presence of RA-ILD. While some of the earlier studies yielded equivocal results (Inui et al. 2008), more recent reports have generally supported an association between ACPA repertoire and this extra-articular disease manifestation (Alexiou et al. 2008; Kelly et al. 2014; Perry et al. 2014; Restrepo et al. 2015; Reynisdottir et al. 2014). In fact, multivariate analysis of a large Swedish cohort of early RA patients indicated that ACPA positivity is strongly linked to radiographically evident interstitial lung disease, with a predictive capacity exceeding that of smoking, age, and articular disease activity (as measured by DAS28) (Reynisdottir et al. 2014). Other studies demonstrating that higher titers of anti-CCP antibodies portend a greater risk of RA-ILD provide additional quantitative support for the relationship between citrulline-targeted immune responses and the development of ILD (Kelly et al. 2014; Restrepo et al. 2015). Perhaps even more compelling, detailed autoantibody profiling based on recognition of citrullinated peptide target arrays has shown that the breadth of ACPA repertoire is strongly associated with the extent/severity of ILD (with a discriminatory capacity exceeding that of high titer anti-CCP2 alone), particularly in subsets of patients with RA-UIP (Giles et al. 2014).

11.6 Identification of Novel ACPAs Through Reverse Immunophenotyping: Anti-citHSP90

Although these studies collectively support a pathogenic link between citrullination of lung tissue, formation of ACPAs, and the development of RA-ILD, their diagnostic utility is limited by the lack of lung-specific target antigen(s) and the general

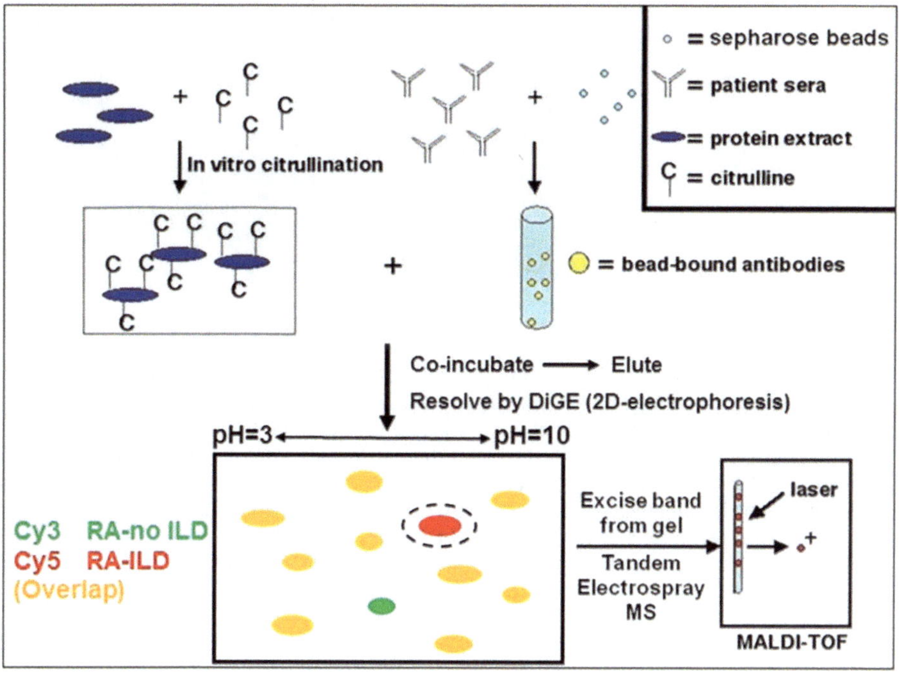

Fig. 11.1 Reverse immunophenotyping approach. This schematic summarizes the steps of in vitro protein citrullination (cell vs. tissue extracts), immunoprecipitation with patient sera, 2D difference in gel electrophoresis (DiGE), band excision, and protein identification through tandem electrospray mass spectrometry. *Band colors* correspond to proteins immunoprecipitated by RA-ILD sera (*red*), RA-no ILD sera (*green*), or both (*orange*)

overlap of anti-CCP profiles in RA patients with/without ILD. To circumvent this problem (that is compounded by the relative inaccessibility of lung tissue for direct analysis of citrullinated protein repertoire), we have devised a "reverse immunophenotyping" approach (Fig. 11.1) which is based on the concept that comparison of anti-citrullinated protein antibody profiles in RA patients with or without ILD can provide a fingerprint of lung-specific protein citrullination (Harlow et al. 2013). Application of this methodology has identified citrullinated HSP90 (citHSP90) as a candidate autoantigen associated with RA-ILD (Harlow et al. 2013). In turn, high throughput ELISAs have (partially) validated this finding by showing that antibodies targeting citrullinated isoforms of HSP90 can identify RA-ILD patients with modest sensitivity (25–30%) and distinguish them from RA patients lacking ILD (as well as from individuals with alternative disease states such as IPF) with high specificity (>90%) (Harlow et al. 2013). More detailed epitope mapping studies have facilitated comparative antibody profiling of serum and bronchoalveolar lavage fluid from selected patients with varying stages of RA-ILD, demonstrating both qualitative and quantitative differences that further implicate the lung as an

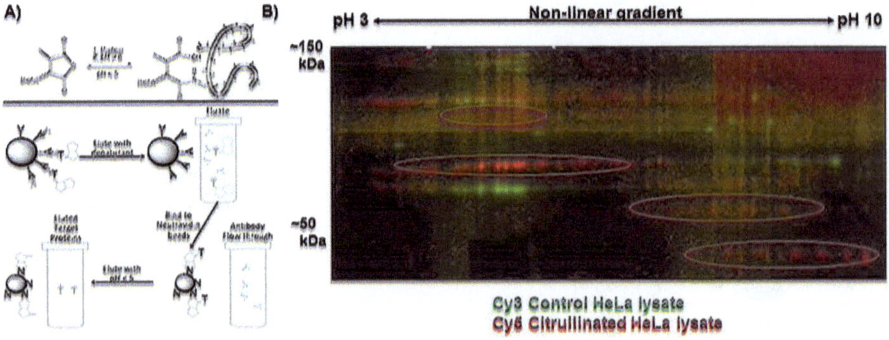

Fig. 11.2 Modified difference in gel electrophoresis (DiGE). (**a**) The schematic outlines the modified DiGE strategy, highlighting the key steps of Biotin-CDM coupling to chromophore-labeled protein extracts, post-elution binding of immunoprecipitated proteins to a streptavidin-coated platform, and subsequent release of purified proteins through low pH buffering prior to 2D gel electrophoresis. (**b**) DiGE separation of citrullinated (*orange*) versus uncitrullinated (*green*) HeLa cell-derived proteins immunoprecipitated by RA-ILD serum and purified using the methodology outlined in panel (**a**). Note that encircled protein "trains" reflect different levels of citrullination that impact isoelectric focusing of individual immunoprecipitated proteins

immunologically active site of repertoire development and antigen-specific autoantibody production (Harlow et al. 2014).

As shown by these analyses, the use of differential immunoprecipitation represents a viable approach for the identification of autoantigen targets associated with the development of RA-ILD. Although this work has focused primarily on citHSP90 and derivative peptides, additional studies involving two-dimensional electrophoretic separation techniques indicate that multiple citrullinated proteins are differentially immunoprecipitated by sera obtained from RA patients with versus without ILD (Ganesan and Ascherman, unpublished data). However, because immunoglobulin components represent a significant fraction of the proteins eluted from protein A sepharose columns, we have modified our strategic approach (as outlined in Fig. 11.2; Ganesan, unpublished data) to eliminate labeling of donor-derived proteins and more effectively separate unlabeled immunoglobulins from immunoprecipitation eluates prior to electrophoretic resolution. In short, labeling cell extracts with specific chromophores as well as a pH-dependent biotinylated conjugate (Biotin-CDM) before the immunoprecipitation step facilitates separation of putative autoantigen targets from immunoglobulin components which can obscure subsequent mass spectrometric sequence identification (Ganesan et al. 2015, 2016). Incorporation of these novel methodologic alterations should therefore enhance the capacity to identify additional autoantibody/autoantigen interactions that can serve as biomarkers of RA-ILD (versus alternative sub-phenotypes of RA) and improve our understanding of disease pathways contributing to this extra-articular complication.

11.7 Modeling Protein Citrullination in the Lung Through In Vitro Stimulation of Human Bronchial Epithelial Cells

Beyond these indirect, immunoprecipitation-based approaches, development of in vitro lung epithelial cell culture systems has provided the opportunity to explore the mechanistic contribution of smoking/oxidative stress to deimination and other posttranslational modifications potentially linked to the emergence of RA-ILD. For example, human bronchial epithelial (HBE) cells isolated from the major bronchi of lung tissue donors can be expanded and differentiated at an air-liquid interface (ALI), effectively serving to model the in vivo air-epithelial/air-alveolar interface (Bernacki et al. 1999; Nlend et al. 2002). Exposure of cultured HBEs (or alveolar epithelial cells) to smoke and other oxidative stressors (heat shock, oxygen sparging, and/or chemical agents such as menadione, hydrogen peroxide (H_2O_2), antimycin A, or mitomycin C)) therefore has the potential to elucidate the in vivo impact of these injurious stimuli on protein citrullination, protein–protein interactions, oxidative pathway induction (including peroxynitrite activation/3-nitrotryrosine formation and protein carbonylation), and pro-inflammatory cytokine production.

Demonstrating the value of this approach in deriving alternative biomarkers and providing relevant pathogenic insight, preliminary studies have shown that exposure of HBEs to cigarette smoke enhances protein carbonylation (a measure of oxidative stress) and augments the expression of citrullinated proteins (Ascherman, unpublished data). Whether smoke/oxidative stress directly activates different PAD isoforms remains unclear, but these findings indicate that in vitro stimulation of HBE/alveolar epithelial cell cultures represents a viable test system for rigorously assessing the mechanistic link between smoking, tissue damage, and posttranslational protein modifications such as citrullination (a paradigm supported by data showing increased PAD2 activity and protein citrullination in the bronchoalveolar fluid/tissue of smokers) (Makrygiannakis et al., 2008). As an extension of this system, co-immunoprecipitation studies involving extracts of HBE cells exposed to varying doses of cigarette smoke indicate that citrullination alters protein–protein interactions (Fig. 11.3; Ganesan and Ascherman, unpublished data), with the potential to significantly impact the RA-ILD disease process by enhancing deleterious pro-inflammatory pathways or disrupting protective signaling networks. Based on these preliminary observations, identifying additional modifications in protein–protein interaction through similar co-immunoprecipitation approaches holds great promise for clarifying the functional impact of citrulline-induced structural modifications on intracellular signaling pathways contributing to the interstitial inflammation/fibrosis characteristic of RA-ILD.

From a biomarker standpoint, use of the in vitro HBE culture system will also be critical in defining novel autoantigens that are targets of citrullination in disease pathways leading to RA-ILD. In fact, immunoblotting experiments employing HBE cell extracts and anti-modified citrulline antibody have shown that smoke exposure is associated with citrullination of numerous proteins of varying molecular weight

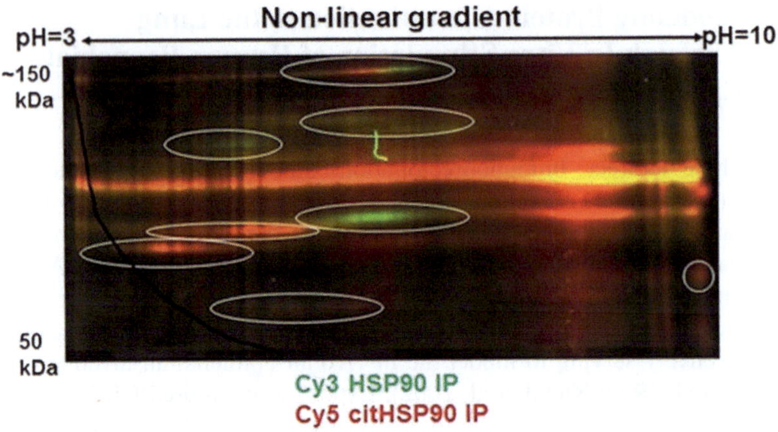

Fig. 11.3 Differential HSP90/citHSP90 co-immunoprecipitation. Co-incubation of HSP90 vs. citHSP90 with cigarette smoke-exposed HBE cell extracts demonstrates unique interactions (*circled*) between different proteins and HSP90 (Cy3; *green*) vs. citHSP90 (Cy5; *red*). Samples were immunoprecipitated with anti-HSP90 antibodies (Santa Cruz Biotech) prior to sepharose bead binding, washing, elution, and resolution by DiGE

(Ascherman, unpublished data). Mass spectrometric identification of candidate autoantigens has permitted preliminary ELISA-based screening of sera derived from cohorts of RA patients with different stages of ILD; while these studies have shown some preferential recognition by RA-ILD sera, the diagnostic sensitivity of individual autoantibody/autoantigen combinations has thus far been low. Therefore, these results collectively suggest that panels of citrullinated substrate antigens (in the form of a lung-specific array) will ultimately be required to enhance the diagnostic utility of this method. Despite such limitations, this approach is likely to yield meaningful functional data given the capacity of heat shock proteins (e.g., HSP70) and other citrullinated ligands identified by these profiling methods to directly/indirectly activate Toll-like receptors—key components of the innate immune response linked to downstream NF-κB activation and pro-inflammatory cytokine production that are highly relevant to the progression of ILD/pulmonary fibrosis (Go et al. 2014; He et al. 2009; Kovach and Standiford 2011; Margaritopoulos et al. 2010; Papanikolaou et al. 2015). Viewed more broadly, then, development of these "functional" biomarker discovery systems will serve the dual role of defining clinically relevant biomarkers and clarifying disease pathogenesis in extra-articular complications of RA such as ILD.

11.8 Conclusions

Overall, the weight of existing clinical/epidemiological evidence indicates that citrulline-targeted immune responses in RA are associated with a more aggressive disease course marked by structurally damaging arthritis, significant extra-articular disease manifestations, and increased mortality (Jilani and Mackworth-Young 2015;

Kuller et al. 2014; Lopez-Longo et al. 2009; Farragher et al. 2008). Yet, these data are largely associative, potentially confounded by immunogenetic background, host demographic characteristics (age, gender), environmental factors such as smoking, and duration/extent of the underlying inflammatory process. These considerations highlight the need for tissue-specific markers of citrulline-induced pathology to solidify the connection between this posttranslational modification and various subphenotypes of RA. Although the relative inaccessibility of cardiac, pulmonary, and other non-synovial tissues complicates this type of analysis, the unique humoral immune response to citrullinated antigens in RA has provided the opportunity to discover novel biomarkers of internal organ involvement that, ultimately, must be demonstrated in situ through immunohistochemical, PCR, and mass spectrometric peptide profiling-based approaches. Given the structural/functional consequences of citrullination that range from altered protein–protein interactions to initiation of potentially deleterious innate immune signaling cascades, the importance of these investigations clearly extends beyond the discovery of clinically meaningful biomarkers to include elucidation of disease-relevant pathways contributing to extra-articular complications of RA. As an illustration of this "translational" potential, a growing body of evidence suggests that citrullinated antigens are capable of engaging TLRs, either directly or via immune complexes that co-activate Fcγ receptors (Sokolove et al. 2011; Sanchez-Pernaute et al. 2013). Notwithstanding such progress, however, advancing our understanding of deimination in the context of disease pathogenesis and defining novel therapeutic targets related to aberrant protein citrullination will hinge on further refinement of the described biomarker discovery methods combining serologic/proteomic characterization and in vitro functional modeling.

References

Alexiou, I., Germenis, A., Koutroumpas, A., Kontogianni, A., Theodoridou, K., & Sakkas, L. I. (2008). Anti-cyclic citrullinated peptide-2 (CCP2) autoantibodies and extra-articular manifestations in Greek patients with rheumatoid arthritis. *Clinical Rheumatology, 27*(4), 511–513. doi:10.1007/s10067-007-0800-1.

Arnab, B., Biswadip, G., Arindam, P., Shyamash, M., Anirban, G., & Rajan, P. (2013). Anti-CCP antibody in patients with established rheumatoid arthritis: Does it predict adverse cardiovascular profile? *Journal of Cardiovascular Disease Research, 4*(2), 102–106. doi:10.1016/j.jcdr.2012.09.003.

Bernacki, S. H., Nelson, A. L., Abdullah, L., Sheehan, J. K., Harris, A., Davis, C. W., & Randell, S. H. (1999). Mucin gene expression during differentiation of human airway epithelia in vitro. Muc4 and muc5b are strongly induced. *American Journal of Respiratory Cell and Molecular Biology, 20*(4), 595–604. doi:10.1165/ajrcmb.20.4.3442.

Bongartz, T., Cantaert, T., Atkins, S. R., Harle, P., Myers, J. L., Turesson, C., Ryu, J. H., Baeten, D., & Matteson, E. L. (2007). Citrullination in extra-articular manifestations of rheumatoid arthritis. *Rheumatology (Oxford), 46*(1), 70–75. doi:10.1093/rheumatology/kel202. kel202 [pii].

Bongartz, T., Nannini, C., Medina-Velasquez, Y. F., Achenbach, S. J., Crowson, C. S., Ryu, J. H., Vassallo, R., Gabriel, S. E., & Matteson, E. L. (2010). Incidence and mortality of interstitial lung disease in rheumatoid arthritis: A population-based study. *Arthritis & Rheumatology, 62*(6), 1583–1591. doi:10.1002/art.27405.

Brown, K. K. (2007). Rheumatoid lung disease. *Proceedings of the American Thoracic Society, 4*(5), 443–448. doi:10.1513/pats.200703-045MS. 4/5/443 [pii].

Damgaard, D., Friberg Bruun Nielsen, M., Quisgaard Gaunsbaek, M., Palarasah, Y., Svane-Knudsen, V., & Nielsen, C. H. (2015). Smoking is associated with increased levels of extracellular peptidylarginine deiminase 2 (PAD2) in the lungs. *Clinical and Experimental Rheumatology, 33*(3), 405–408.

Farragher, T. M., Goodson, N. J., Naseem, H., Silman, A. J., Thomson, W., Symmons, D., & Barton, A. (2008). Association of the HLA-DRB1 gene with premature death, particularly from cardiovascular disease, in patients with rheumatoid arthritis and inflammatory polyarthritis. *Arthritis and Rheumatism, 58*(2), 359–369. doi:10.1002/art.23149.

Ganesan, V., Schmidt, B., Avula, R., Cooke, D., Maggiacomo, T., Tellin, L., Ascherman, D. P., Bruchez, M. P., & Minden, J. (2015). Immuno-proteomics: Development of a novel reagent for separating antibodies from their target proteins. *Biochimica et Biophysica Acta, 1854*(6), 592–600. doi:10.1016/j.bbapap.2014.10.011.

Ganesan, V., Ascherman, D. P., & Minden, J. S. (2016). Immunoproteomics technologies in the discovery of autoantigens in autoimmune diseases. *Biomolecular Concepts, 7*(2), 133–143. doi:10.1515/bmc-2016-0007.

Gerardi, M. C., De Luca, N., Alessandri, C., Iannuccelli, C., Valesini, G., & Di Franco, M. (2013). Frequency of antibodies to mutated citrullinated vimentin in chronic obstructive pulmonary disease: Comment on the article by Demoruelle et al. *Arthritis and Rheumatism, 65*(6), 1672–1673. doi:10.1002/art.37905.

Giles, J. T., Fert-Bober, J., Park, J. K., Bingham, C. O., 3rd, Andrade, F., Fox-Talbot, K., Pappas, D., Rosen, A., van Eyk, J., Bathon, J. M., & Halushka, M. K. (2012). Myocardial citrullination in rheumatoid arthritis: A correlative histopathologic study. *Arthritis Research & Therapy, 14*(1), R39. doi:10.1186/ar3752.

Giles, J. T., Danoff, S. K., Sokolove, J., Wagner, C. A., Winchester, R., Pappas, D. A., Siegelman, S., Connors, G., Robinson, W. H., & Bathon, J. M. (2014). Association of fine specificity and repertoire expansion of anticitrullinated peptide antibodies with rheumatoid arthritis associated interstitial lung disease. *Annals of the Rheumatic Diseases, 73*(8), 1487–1494. doi:10.1136/annrheumdis-2012-203160.

Go, H., Koh, J., Kim, H. S., Jeon, Y. K., & Chung, D. H. (2014). Expression of toll-like receptor 2 and 4 is increased in the respiratory epithelial cells of chronic idiopathic interstitial pneumonia patients. *Respiratory Medicine, 108*(5), 783–792. doi:10.1016/j.rmed.2013.12.007.

Goeldner, I., Skare, T. L., de Messias Reason, I. T., Nisihara, R. M., Silva, M. B., & da Rosa Utiyama, S. R. (2011). Association of anticyclic citrullinated peptide antibodies with extra-articular manifestations, gender, and tabagism in rheumatoid arthritis patients from southern Brazil. *Clinical Rheumatology, 30*(7), 975–980. doi:10.1007/s10067-011-1711-8.

Harlow, L., Rosas, I. O., Gochuico, B. R., Mikuls, T. R., Dellaripa, P. F., Oddis, C. V., & Ascherman, D. P. (2013). Identification of citrullinated hsp90 isoforms as novel autoantigens in rheumatoid arthritis-associated interstitial lung disease. *Arthritis and Rheumatism, 65*(4), 869–879. doi:10.1002/art.37881.

Harlow, L., Gochuico, B. R., Rosas, I. O., Doyle, T. J., Osorio, J. C., Travers, T. S., Camacho, C. C., Oddis, C. V., & Ascherman, D. P. (2014). Anti-citrullinated heat shock protein 90 antibodies identified in bronchoalveolar lavage fluid are a marker of lung-specific immune responses. *Clinical Immunology, 155*(1), 60–70. doi:10.1016/j.clim.2014.08.004.

He, Z., Zhu, Y., & Jiang, H. (2009). Toll-like receptor 4 mediates lipopolysaccharide-induced collagen secretion by phosphoinositide3-kinase-Akt pathway in fibroblasts during acute lung injury. *Journal of Receptor and Signal Transduction Research, 29*(2), 119–125. doi:10.1080/10799890902845690.

Hjeltnes, G., Hollan, I., Forre, O., Wiik, A., Mikkelsen, K., & Agewall, S. (2011). Anti-CCP and RF IgM: Predictors of impaired endothelial function in rheumatoid arthritis patients. *Scandinavian Journal of Rheumatology, 40*(6), 422–427. doi:10.3109/03009742.2011.585350.

Inui, N., Enomoto, N., Suda, T., Kageyama, Y., Watanabe, H., & Chida, K. (2008). Anti-cyclic citrullinated peptide antibodies in lung diseases associated with rheumatoid arthritis. *Clinical Biochemistry, 41*(13), 1074–1077. doi:10.1016/j.clinbiochem.2008.06.014. S0009-9120(08)00246-4 [pii].

Janssen, K. M., de Smit, M. J., Brouwer, E., de Kok, F. A., Kraan, J., Altenburg, J., Verheul, M. K., Trouw, L. A., van Winkelhoff, A. J., Vissink, A., & Westra, J. (2015). Rheumatoid arthritis-associated autoantibodies in non-rheumatoid arthritis patients with mucosal inflammation: A case-control study. *Arthritis Research & Therapy, 17*, 174. doi:10.1186/s13075-015-0690-6.

Jilani, A. A., & Mackworth-Young, C. G. (2015). The role of citrullinated protein antibodies in predicting erosive disease in rheumatoid arthritis: A systematic literature review and meta-analysis. *International journal of rheumatology, 2015*, 728610. doi:10.1155/2015/728610.

Kelly, C. A., Saravanan, V., Nisar, M., Arthanari, S., Woodhead, F. A., Price-Forbes, A. N., Dawson, J., Sathi, N., Ahmad, Y., Koduri, G., Young, A., & British Rheumatoid Interstitial Lung N. (2014). Rheumatoid arthritis-related interstitial lung disease: Associations, prognostic factors and physiological and radiological characteristics--a large multicentre UK study. *Rheumatology (Oxford), 53*(9), 1676–1682. doi:10.1093/rheumatology/keu165.

Kim, S. K., Park, S. H., Shin, I. H., & Choe, J. Y. (2008). Anti-cyclic citrullinated peptide antibody, smoking, alcohol consumption, and disease duration as risk factors for extraarticular manifestations in Korean patients with rheumatoid arthritis. *The Journal of Rheumatology, 35*(6), 995–1001.

Korkmaz, C., Us, T., Kasifoglu, T., & Akgun, Y. (2006). Anti-cyclic citrullinated peptide (CCP) antibodies in patients with long-standing rheumatoid arthritis and their relationship with extra-articular manifestations. *Clinical Biochemistry, 39*(10), 961–965. doi:10.1016/j.clinbiochem.2006.06.004.

Kovach, M. A., & Standiford, T. J. (2011). Toll like receptors in diseases of the lung. *International Immunopharmacology, 11*(10), 1399–1406. doi:10.1016/j.intimp.2011.05.013.

Kuller, L. H., Mackey, R. H., Walitt, B. T., Deane, K. D., Holers, V. M., Robinson, W. H., Sokolove, J., Chang, Y., Liu, S., Parks, C. G., Wright, N. C., & Moreland, L. W. (2014). Determinants of mortality among postmenopausal women in the women's health initiative who report rheumatoid arthritis. *Arthritis & Rheumatology, 66*(3), 497–507. doi:10.1002/art.38268.

Lobo, F. S., Dossi, M. O., Batista, L., & Shinzato, M. M. (2016). Hearing impairment in patients with rheumatoid arthritis: Association with anti-citrullinated protein antibodies. *Clinical Rheumatology, 35*(9), 2327–2332. doi:10.1007/s10067-016-3278-x.

Lopez-Longo, F. J., Oliver-Minarro, D., de la Torre, I., Gonzalez-Diaz de Rabago, E., Sanchez-Ramon, S., Rodriguez-Mahou, M., Paravisini, A., Monteagudo, I., Gonzalez, C. M., Garcia-Castro, M., Casas, M. D., & Carreno, L. (2009). Association between anti-cyclic citrullinated peptide antibodies and ischemic heart disease in patients with rheumatoid arthritis. *Arthritis and Rheumatism, 61*(4), 419–424. doi:10.1002/art.24390.

Lugli, E. B., Correia, R. E., Fischer, R., Lundberg, K., Bracke, K. R., Montgomery, A. B., Kessler, B. M., Brusselle, G. G., & Venables, P. J. (2015). Expression of citrulline and homocitrulline residues in the lungs of non-smokers and smokers: Implications for autoimmunity in rheumatoid arthritis. *Arthritis Research & Therapy, 17*, 9. doi:10.1186/s13075-015-0520-x.

Mackey, R. H., Kuller, L. H., Deane, K. D., Walitt, B. T., Chang, Y. F., Holers, V. M., Robinson, W. H., Tracy, R. P., Hlatky, M. A., Eaton, C. B., Liu, S., Freiberg, M. S., Talabi, M. B., Schelbert, E. B., & Moreland, L. W. (2015). Rheumatoid arthritis, anti-cyclic citrullinated peptide positivity, and cardiovascular disease risk in the women's health initiative. *Arthritis & Rheumatology, 67*(9), 2311–2322. doi:10.1002/art.39198.

Makrygiannakis, D., Hermansson, M., Ulfgren, A. K., Nicholas, A. P., Zendman, A. J., Eklund, A., Grunewald, J., Skold, C. M., Klareskog, L., & Catrina, A. I. (2008). Smoking increases peptidylarginine deiminase 2 enzyme expression in human lungs and increases citrullination in BAL cells. *Annals of the Rheumatic Diseases, 67*(10), 1488–1492. doi:10.1136/ard.2007.075192.

Marasovic-Krstulovic, D., Martinovic-Kaliterna, D., Fabijanic, D., & Morovic-Vergles, J. (2011). Are the anti-cyclic citrullinated peptide antibodies independent predictors of myocardial involvement in patients with active rheumatoid arthritis? *Rheumatology (Oxford), 50*(8), 1505–1512. doi:10.1093/rheumatology/ker121. ker121 [pii].

Margaritopoulos, G. A., Antoniou, K. M., Karagiannis, K., Samara, K. D., Lasithiotaki, I., Vassalou, E., Lymbouridou, R., Koutala, H., & Siafakas, N. M. (2010). Investigation of Toll-like receptors in the pathogenesis of fibrotic and granulomatous disorders: A bronchoalveolar lavage study. *Fibrogenesis & Tissue Repair, 3*, 20. doi:10.1186/1755-1536-3-20.

Montes, A., Corrales, A., Calaza, M., Lopez-Mejias, R., Parra, J. A., Gonzalez-Gay, M. A., & Gonzalez, A. (2015). Brief report: Lack of replication of an association between anti-citrullinated fibrinogen and subclinical atherosclerosis in patients with rheumatoid arthritis. *Arthritis & Rheumatology, 67*(11), 2861–2865. doi:10.1002/art.39302.

Nlend, M. C., Bookman, R. J., Conner, G. E., & Salathe, M. (2002). Regulator of G-protein signaling protein 2 modulates purinergic calcium and ciliary beat frequency responses in airway epithelia. *American Journal of Respiratory Cell and Molecular Biology, 27*(4), 436–445. doi:10.1165/rcmb.2002-0012OC.

Papanikolaou, I. C., Boki, K. A., Giamarellos-Bourboulis, E. J., Kotsaki, A., Kagouridis, K., Karagiannidis, N., & Polychronopoulos, V. S. (2015). Innate immunity alterations in idiopathic interstitial pneumonias and rheumatoid arthritis-associated interstitial lung diseases. *Immunology Letters, 163*(2), 179–186. doi:10.1016/j.imlet.2014.12.004.

Perry, E., Kelly, C., Eggleton, P., De Soyza, A., & Hutchinson, D. (2014). The lung in ACPA-positive rheumatoid arthritis: An initiating site of injury? *Rheumatology (Oxford), 53*(11), 1940–1950. doi:10.1093/rheumatology/keu195.

Rangel-Moreno, J., Hartson, L., Navarro, C., Gaxiola, M., Selman, M., & Randall, T. D. (2006). Inducible bronchus-associated lymphoid tissue (iBALT) in patients with pulmonary complications of rheumatoid arthritis. *The Journal of Clinical Investigation, 116*(12), 3183–3194. doi:10.1172/JCI28756.

Restrepo, J. F., del Rincon, I., Battafarano, D. F., Haas, R. W., Doria, M., & Escalante, A. (2015). Clinical and laboratory factors associated with interstitial lung disease in rheumatoid arthritis. *Clinical Rheumatology, 34*(9), 1529–1536. doi:10.1007/s10067-015-3025-8.

Reynisdottir, G., Karimi, R., Joshua, V., Olsen, H., Hensvold, A. H., Harju, A., Engstrom, M., Grunewald, J., Nyren, S., Eklund, A., Klareskog, L., Skold, C. M., & Catrina, A. I. (2014). Structural changes and antibody enrichment in the lungs are early features of anti-citrullinated protein antibody-positive rheumatoid arthritis. *Arthritis & rheumatology, 66*(1), 31–39. doi:10.1002/art.38201.

Sanchez-Pernaute, O., Filkova, M., Gabucio, A., Klein, M., Maciejewska-Rodrigues, H., Ospelt, C., Brentano, F., Michel, B. A., Gay, R. E., Herrero-Beaumont, G., Gay, S., Neidhart, M., & Juengel, A. (2013). Citrullination enhances the pro-inflammatory response to fibrin in rheumatoid arthritis synovial fibroblasts. *Annals of the Rheumatic Diseases, 72*(8), 1400–1406. doi:10.1136/annrheumdis-2012-201906.

Sim, M. K., Kim, D. Y., Yoon, J., Park, D. H., & Kim, Y. G. (2014). Assessment of peripheral neuropathy in patients with rheumatoid arthritis who complain of neurologic symptoms. *Annals of rehabilitation medicine, 38*(2), 249–255. doi:10.5535/arm.2014.38.2.249.

Sokolove, J., Zhao, X., Chandra, P. E., & Robinson, W. H. (2011). Immune complexes containing citrullinated fibrinogen costimulate macrophages via Toll-like receptor 4 and Fcgamma receptor. *Arthritis and Rheumatism, 63*(1), 53–62. doi:10.1002/art.30081.

Sokolove, J., Brennan, M. J., Sharpe, O., Lahey, L. J., Kao, A. H., Krishnan, E., Edmundowicz, D., Lepus, C. M., Wasko, M. C., & Robinson, W. H. (2013). Brief report: Citrullination within the atherosclerotic plaque: A potential target for the anti-citrullinated protein antibody response in rheumatoid arthritis. *Arthritis and Rheumatism, 65*(7), 1719–1724. doi:10.1002/art.37961.

Solow, E. B., Yu, F., Thiele, G. M., Sokolove, J., Robinson, W. H., Pruhs, Z. M., Michaud, K. D., Erickson, A. R., Sayles, H., Kerr, G. S., Gaffo, A. L., Caplan, L., Davis, L. A., Cannon, G. W., Reimold, A. M., Baker, J., Schwab, P., Anderson, D. R., & Mikuls, T. R. (2015). Vascular calcifications on hand radiographs in rheumatoid arthritis and associations with autoantibodies, cardiovascular risk factors and mortality. *Rheumatology (Oxford), 54*(9), 1587–1595. doi:10.1093/rheumatology/kev027.

Szodoray, P., Szabo, Z., Kapitany, A., Gyetvai, A., Lakos, G., Szanto, S., Szucs, G., & Szekanecz, Z. (2010). Anti-citrullinated protein/peptide autoantibodies in association with genetic and environmental factors as indicators of disease outcome in rheumatoid arthritis. *Autoimmunity Reviews, 9*(3), 140–143. doi:10.1016/j.autrev.2009.04.006.

Turesson, C., Jacobsson, L. T., Sturfelt, G., Matteson, E. L., Mathsson, L., & Ronnelid, J. (2007). Rheumatoid factor and antibodies to cyclic citrullinated peptides are associated with severe extra-articular manifestations in rheumatoid arthritis. *Annals of the Rheumatic Diseases, 66*(1), 59–64. doi:10.1136/ard.2006.054445.

Vazquez-Del Mercado, M., Nunez-Atahualpa, L., Figueroa-Sanchez, M., Gomez-Banuelos, E., Rocha-Munoz, A. D., Martin-Marquez, B. T., Corona-Sanchez, E. G., Martinez-Garcia, E. A., Macias-Reyes, H., Gonzalez-Lopez, L., Gamez-Nava, J. I., Navarro-Hernandez, R. E., Nunez-Atahualpa, M. A., & Andrade-Garduno, J. (2015). Serum levels of anticyclic citrullinated peptide antibodies, interleukin-6, tumor necrosis factor-alpha, and C-reactive protein are associated with increased carotid intima-media thickness: A cross-sectional analysis of a cohort of rheumatoid arthritis patients without cardiovascular risk factors. *BioMed Research International, 2015*, 342649. doi:10.1155/2015/342649.

Vignesh, A. P., & Srinivasan, R. (2015). Ocular manifestations of rheumatoid arthritis and their correlation with anti-cyclic citrullinated peptide antibodies. *Clinical Ophthalmology, 9*, 393–397. doi:10.2147/OPTH.S77210.

Willis, V. C., Demoruelle, M. K., Derber, L. A., Chartier-Logan, C. J., Parish, M. C., Pedraza, I. F., Weisman, M. H., Norris, J. M., Holers, V. M., & Deane, K. D. (2013). Sputum autoantibodies in patients with established rheumatoid arthritis and subjects at risk of future clinically apparent disease. *Arthritis and Rheumatism, 65*(10), 2545–2554. doi:10.1002/art.38066.

Ytterberg, A. J., Joshua, V., Reynisdottir, G., Tarasova, N. K., Rutishauser, D., Ossipova, E., Haj Hensvold, A., Eklund, A., Skold, C. M., Grunewald, J., Malmstrom, V., Jakobsson, P. J., Ronnelid, J., Padyukov, L., Zubarev, R. A., Klareskog, L., & Catrina, A. I. (2015). Shared immunological targets in the lungs and joints of patients with rheumatoid arthritis: Identification and validation. *Annals of the Rheumatic Diseases, 74*(9), 1772–1777. doi:10.1136/annrheumdis-2013-204912.

Chapter 12
The Significance of Myofilament Protein Citrullination in Heart Failure: Citrullination in Cardiovascular Diseases

J. Fert-Bober, E.L. Crowgey, J. Sokolove, J.T. Giles, J.M. Bathon, and J.E. Van Eyk

12.1 Introduction

Since cardiac conditions resulting in heart failure (HF) are among the leading causes of morbidity and mortality worldwide, it is not surprising that HF is the leading cause of hospitalization in people older than 65 years of age (Heidenreich et al. 2013; Mozaffarian et al. 2015, 2016). Mortality is >50% at 5 years after initial diagnosis, with less than 25% after 10 years (Mozaffarian et al. 2015, 2016). The prevalence of HF is predicted to increase by 46% from 2012 to 2030, according to the American Heart Association (Heidenreich et al. 2013). HF is a complex clinical syndrome and is due to many different etiologies.

HF with reduced ejection fraction (HFrEF) occurs when the efficiency of the heart as a pump is impaired due to structural and/or functional abnormalities. Recently, there has been broader recognition that HF can occur even with preserved ejection fraction (HFpEF) and that this form of HF comprises approximately 50% of individuals with this disease (Lindenfeld et al. 2010). Unfortunately, HFpEF has been poorly investigated and currently there are no known therapies (Bhuiyan and Maurer 2011). Multiple studies have identified HFrEF-associated alterations in a number of critical cellular processes including Ca^{2+} handling, energy metabolism, and sarcomeric

J. Fert-Bober (✉) • J.E. Van Eyk
Department of Medicine, Advanced Clinical Biosystems Research Institute A9229, Cedars-Sinai Medical Center, Los Angeles, CA, USA
e-mail: Justyna.FertBober@cshs.org

E.L. Crowgey
Nemours/Alfred I. duPont Hospital for Children, Wilmington, DE, USA

J. Sokolove
Immunology, AbbVie, Redwood City, CA, USA

J.T. Giles • J.M. Bathon
Division of Rheumatology, Department of Medicine, Columbia University, New York, NY, USA

A.P. Nicholas et al. (eds.), *Protein Deimination in Human Health and Disease*,
DOI 10.1007/978-3-319-58244-3_12

contractile function (Mudd and Kass 2008; Oka and Komuro 2008; Agnetti et al. 2011). Yet, how these processes, and the molecular pathways that drive and regulate them, function together to promote HFrEF pathophysiology still remains unclear (Heidenreich et al. 2013; Mudd and Kass 2008). What is clear is at the beginning, the heart compensates for reduced blood flow by enlarging the heart chamber. This causes an increase in muscle mass that occurs via hypertrophy of the contracting cells of the heart to maintain the force of cardiac contractile function. Although individuals with HFrEF have numerous clinical symptoms in common, including fatigue, shortness of breath, and edema, the clinical presentation of heart dysfunction can be heterogeneous. HF has been divided pathophysiologically into "systolic HFrEF," where contractile failure leads to ventricular dilation, and "diastolic HFrEF," where the ventricle is non-dilated with normal contraction, but the myocardium is often hypertrophied (Mudd and Kass 2008).

There are a number of different factors that can drive HF development, such as atherosclerosis, myocardial infarction, hypertension, and viral myocarditis (Mozaffarian et al. 2015). However, HF is also associated with other chronic diseases, including diabetes, HIV, hyper- and hypothyroidism, and rheumatoid arthritis (RA) (Kaplan 2010). In fact, RA has been considered to be an independent risk factor for coronary artery disease in recent years (Pujades-Rodriguez et al. 2016; Soeiro Ade et al. 2012). It appears that variables that increase CV mortality in RA are present very early during the history of the disease, as the rheumatoid factor (RF)-positive inflammatory arthritis exhibits evidence of abnormal endothelial function, which is considered as a predictor of future development of atherosclerosis (Bergholm et al. 2002; Tomasson et al. 2010). Furthermore, preliminary evidence indicates that patients with RA who are positive for anti-cyclic citrullinated peptide (anti-CCP) antibodies have higher subclinical atherosclerosis than those who are not (Gerli et al. 2008). Recent study indicates that citrullination (deimination), an irreversible posttranslational modification (PTM), may be a common link between inflammation and structural changes that contribute to HF and potentially other underlying disease that initiate HF. Citrullination, an enzymatic conversion of the amino acid arginine in a protein into the amino acid citrulline, was first described as a physiological process, including terminal differentiation of the epidermis and apoptosis (Gyorgy et al. 2006; Baka et al. 2012). Today, citrullination has also been mapped as a central event in the context of inflammation (Makrygiannakis et al. 2006) and linked to the pathogenesis of RA (Bongartz et al. 2007; Cantaert et al. 2006; Vossenaar and van Venrooij 2004), as well as several other inflammatory diseases (Baka et al. 2012; Vossenaar et al. 2003). The main characteristics of the citrullination process are outlined below, and more detailed descriptions of each of the steps can be found in specific chapters:

- Peptidylcitrulline is generated through the conversion of arginine to citrulline within a peptide chain.
- The deimination is catalyzed by a family of calcium-binding enzymes, the peptidylarginine deiminases (PADs). To date, five isoenzymes have been identified but only four have enzymatic activity (Vossenaar et al. 2003).

- The activity of PAD enzymes is known to be dependent on high concentrations of calcium (Ca^{2+}) (Takahara et al. 1986). Ca^{2+} concentrations required for PAD activity are 100-fold higher than those present in intact or unstimulated cells. Therefore, citrullination occurs in conditions where the level of free intracellular calcium is high, such as in apoptosis, necrosis, and cellular differentiation (Vossenaar and van Venrooij 2004).
- Citrullination occurs as a part of normal intracellular homeostasis (Bicker and Thompson 2013), suggesting that microdomains of high local intracellular Ca^{2+} concentration maybe involved.
- Extracellular citrullination of proteins in tissues is a characteristic of many, if not all, inflammatory conditions (Makrygiannakis et al. 2006).
- Antibodies directed against citrullinated proteins are a key feature of, and are relatively specific for, RA, first appearing in the preclinical phase of the disease (Schellekens et al. 2000; Sokolove et al. 2012).

In this chapter, we will review the known cellular processes related to citrullination in the heart. Figure 12.1 outlines our current proposal on how RA and HFrEF are mechanistically related. As a result, we will provide evidence to support this hypothesis and then discuss the critical knowledge gaps that need to be addressed to provide mechanistic insight into the potential roles of protein citrullination in HFrEF development.

12.2 Potential Role of Citrullination and Anti-cyclic Citrullinated Peptide (Anti-CCP) Antibodies in Heart Failure in RA

Anti-citrullinated protein antibodies (ACPAs) are members of the family of autoantibodies that are reactive against a number of known citrullinated proteins. The first members of this autoantibody family are anti-perinuclear factor (APF), anti-filaggrin antibodies, and anti-keratin antibodies (AKA) that recognize citrullinated epitopes of filaggrin (Vincent et al. 1999). Until recently, the presence of several citrullinated proteins has been demonstrated in the RA synovium (Szekanecz et al. 2008). The identification of citrullinated epitopes as targets for anti-filaggrin antibodies led to the development of the first- and later second-generation anti-CCP antibody assays (Girbal-Neuhauser et al. 1999). The anti-CCP ELISA is based on highly purified synthetic peptides from dedicated libraries containing modified arginine residues (citrulline) serving as antigens, has a specificity comparable with AKA, and is more specific than APF and RF testing (Vallbracht et al. 2004). These widely used anti-CCP-2 clinical assays have high diagnostic sensitivity and specificity than previous methods and also show important predictive and prognostic value in RA (Andrade et al. 2010; van Venrooij and Pruijn 2014).

Generally, life expectancy of patients with RA is 5–10 years less than in the general population, due to their greater risk for cardiovascular disease, which is 2–5

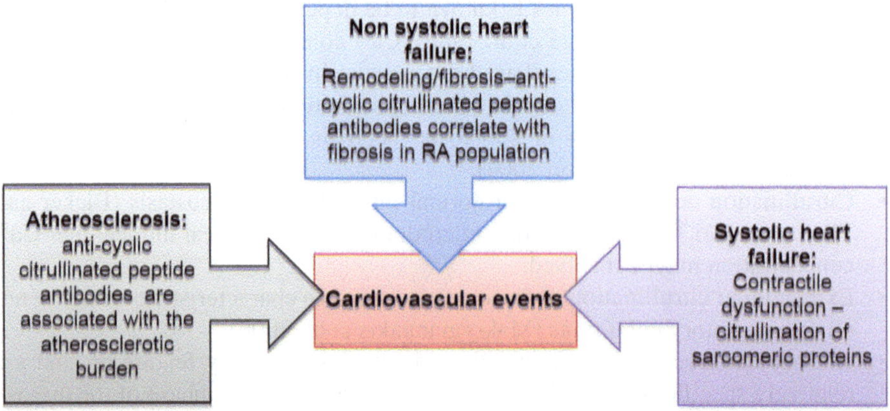

Fig. 12.1 Schematic of our working hypothesis, on the role of citrullination in cardiovascular pathophysiology, specifically atherosclerosis and two forms of HFrEF

times higher than in the general population (Soeiro Ade et al. 2012; Turesson et al. 2004). A study by Soeiro et al. (2012) showed that younger patients with RA have higher risk of cardiovascular events as compared to those without RA who are 5–10 years older (Gabriel 2008, 2010; Innala et al. 2011). The link between developing cardiovascular events and RA is continuing to be better described (Crowson et al. 2005; Davis et al. 2008; Rudominer et al. 2009; Wright et al. 2014; Solomon et al. 2003; Nicola et al. 2006), as HF correlates with the severity and chronicity of the RA disease process (Meune et al. 2010). Furthermore, certain medications used to manage RA can also contribute to heart disease, such as NSAIDs and corticosteroids (Chan et al. 2006; Trelle et al. 2011). However, so far, the exact molecular mechanism that might trigger or worsen heart disease in the context of RA is currently unknown. Interestingly, there is accumulating evidence that the phenotype of HF differs significantly between RA patients and matched non-RA controls. Importantly, several groups observed a higher prevalence (32.88%) of diastolic dysfunction with preserved ejection fraction (HFpEF) in HF patients with RA, compared with the general population (Davis et al. 2008; Di Franco et al. 2000). Interestingly, when HFrEF does occur in patients with RA, it is seen more frequently in men than women (HR 3.7, 95% CI 1.8–7.7). In contrast, no gender difference was found for those RA patients with HFpEF (0.9, 95% CI 0.5, 1.9 in males versus females) (Myasoedova et al. 2011). This is different than HFpEF in the absence of RA, where females are 62% more prevalent (Fonarow et al. 2007; Yancy et al. 2006; LeWinter and Meyer 2013). Recent literature focusing on HFpEF pathophysiology underlying this disease suggests multiple mechanisms are involved in the generation of the phenotype, such as abnormal relaxation and ventricular-vascular coupling, chronotropic incompetence, volume overload, and redistribution and/or dysfunction of endothelial cells. In view of the multiple abnormalities of diastolic function that have been identified, it may be particularly important to individualize and target treatments specific for the causative pathologic process responsible (Oktay et al. 2013). Interestingly, Giles et al. reported an association of higher serum autoantibody concentrations against citrullinated proteins (APCA) with lower myocardial mass and smaller left ventricular (LV) chamber volumes in RA patients without known cardiovascular disease (Giles et al. 2010, 2012). In addition, the mean left ventricular ejection fraction, cardiac output, and stroke volume were modestly lower in the RA group compared with controls. The results are interesting, as there were no other RA characteristics, including joint erosions, disability, systemic inflammation (i.e., elevated C-reactive protein (CRP) and interleukin-6 (IL-6) levels), or RA therapies (i.e., non-biologic disease-modifying antirheumatic drugs) that were significantly associated with mean measures of LV structure. These findings also differed by gender, such that the relative reduction in mean LV mass was greater for men than women when compared to their respective controls (20% and 14% lower, respectively; p for heterogeneity = 0.028) (Fig. 12.2) (Giles et al. 2009). These results suggest that the progression to HF in RA may occur through reduced, rather than increased, myocardial mass.

These observations raise several questions:

1. Do RA-specific autoantibodies to citrullinated proteins mediate changes in myocardial morphology? Alternatively, do changes within the myocardium in individuals with RA induce production of autoantibodies to citrullinated proteins that are released from the heart into the extracellular space/blood?
2. Do RA-specific autoantibodies against citrullinated proteins originating in the heart activate or induce interstitial inflammation and/or cardiac fibrosis, which drives HF pathophysiology?

A follow-up necropsy study of RA myocardial samples did not show myocardial inflammation or myocyte necrosis in the RA samples. However, the extent of citrullination (based on IHC using anti-citrullinated antibody staining) was higher in RA myocardium that also had interstitial fibrosis (Giles et al. 2012). Furthermore, interstitial citrullination levels based on IHC were higher in the interstitial extracellular matrix (ECM) myocardium obtained from individuals with RA than controls without heart disease, myocarditis, or scleroderma. Different isoforms of PADs (based on IHC using anti-PADI antibodies) were also observed in these hearts with different isoforms localizing based on myocardial cell type (i.e., cardiomyocytes, endothelium, etc.)

The ECM is a complex architectural network consisting of structural and nonstructural proteins, creating strength and plasticity. In healthy tissue, the ECM regenerates itself by normal remodeling, in which old or damaged proteins are broken down in a specific sequence of proteolytic events and are replaced by new proteins. During pathological conditions, naive proteins of the ECM are replaced by alternate matrix constituents or can be decorated by PTM that consequently alter the composition and quality of the matrix. Interestingly, many ECM proteins, including collagen (Sipila et al. 2014), fibronectin (Shelef et al. 2012), and fibrinogen (Hill et al. 2008; Robinson and Sokolove 2012), can be citrullinated. This raises the question as to whether myocardial citrullination and subsequent APCA generation may be a key pathological feature to HF susceptibility in RA patients. If so, detection of

Fig. 12.2 Association of RA status with cardiac MRI measures between the genders

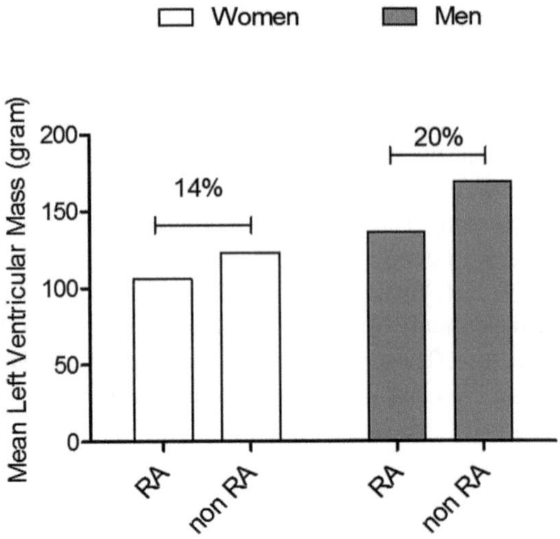

citrullination and autoantibodies could be potential early markers of myocardial dysfunction in RA. In addition, LV remodeling, particularly concentric remodeling, is a well-recognized feature of RA patients without clinical features of HF (Rudominer et al. 2009; Corrao et al. 1996; Myasoedova et al. 2013). While RA duration and other disease-related factors might promote changes in LV geometry (Rudominer et al. 2009), the biological mechanisms underlying LV remodeling in RA populations required further investigation. In line with this hypothesis, Liang et al. demonstrated that RF, antinuclear antibody, and anti-CCP antibody were associated with an increased risk of CV events and overall mortality in the RA population (Liang et al. 2009). This study suggested that anti-CCP positivity was associated with HF, myocardial infarction, and peripheral vascular disease; however, the results were not statistically significant (HR 3.11; 95% CI 0.8, 12.3), perhaps because of the small sample size.

It is important to keep in mind that our current understanding of the reactivity of anti-CCP is limited and reactivity to different proteins may vary between individuals. Anti-CCP polyclonal antibodies target specific citrullinated proteins within a host of known proteins, including fibrinogen, vimentin, enolase, type II collagen, and others (Masson-Bessiere et al. 2001; Burkhardt et al. 2005; Kinloch et al. 2005; Wegner et al. 2010; Verpoort et al. 2006). Furthermore, heterogeneity to anti-CCP antibodies across the RA population has been also demonstrated and may vary between different myocardial phenotypes in RA patients (Hueber et al. 2005; Kastbom et al. 2016). One mechanistic hypothesis involves citrullinated myocardial proteins as targets of pathogenic autoantibodies, leading to immune complex formation and/or tissue. For example, the targeting of citrullinated vimentin by anti-modified citrullinated vimentin (anti-MCV) antibodies was described as highly specific for RA (Sghiri et al. 2010). Studies from different groups showed that significant protein citrullination of vimentin occurs in patients with chronic hepatitis and that the serum concentration of anti-MCV could differentiate patients with no liver fibrosis from those with moderate to severe fibrosis (Vassiliadis et al. 2012; Abdeen et al. 2011). Although it is compelling to extrapolate this mechanism to myocardial fibrosis in RA, there is no current evidence that anti-MCV antibody status correlates with RA-related cardiomyopathy, suggesting that a clinical investigation of this possibility is warranted.

12.3 Role of Citrullinated Proteins in Atherosclerotic Plaques

Greater than 25 million people in the United States have at least one clinical manifestation of atherosclerosis, and in many more, atherosclerosis remains an occult but important harbinger of significant cardiovascular events (Barquera et al. 2015; Goff et al. 2014). The causes of atherosclerosis are complicated and still not completely understood. Atherosclerosis is thought to start when the inner lining of the artery becomes damaged. The blood vessel wall reacts to this injury by depositing fatty substances, cholesterol, calcium, and other substances on the inner lining of the artery. This plaque formation gradually narrows the blood vessels, making it harder for blood to flow. Even though atherosclerosis per se could decrease blood flow through stenosis, and thus induce cardiovascular disease, the major mechanism appears to be atherothrombosis.

Atherothrombosis is the disruption of an atherosclerotic plaque or lesion through the effects of pro-inflammatory cytokines and chemokines on the fibrous cap. In part, susceptibility of atherosclerotic plaque disruption depends on the underlying pathophysiology and status of the integrity of its fibrous cap, which prevents contact between the highly thrombogenic lipid core and the circulating blood. Sites of plaque rupture display signs of active inflammation. Although several histomorphological features indicate which plaques are at risk of rupture, the trigger factors responsible for plaque rupture are still not completely defined. In particular, there are details around the molecular mechanism through which the plaque is disrupted or about the relationship between plaque disruption and both the trigger and onset of acute disease. In recent years, evidence also suggests an important role for an epigenetics process, such as DNA methylation (Dunn et al. 2015; Cao et al. 2014), histone post-translational modification (Hastings et al. 2007; Dje N'Guessan et al. 2009), and noncoding RNA intervention (Aryal et al. 2014), in the pathogenesis of atherosclerosis and its complications. Interestingly, Sokolove et al. showed citrullinated proteins within atherosclerotic plaques and certain anti-CCP antibodies associated with atherosclerotic burden (Fig. 12.3). The subset of specific anti-Cit-fibrinogen and anti-Cit-vimentin antibodies showed greater association with subclinical aortic atherosclerosis as compared with nonspecific cyclic citrullinated peptide (Sokolove et al. 2013). This observation suggests that humoral targeting of citrullinated epitopes, specifically anti-Cit-fibrinogen and anti-Cit-vimentin within the atherosclerotic plaque, could provide a mechanism for accelerated atherosclerosis.

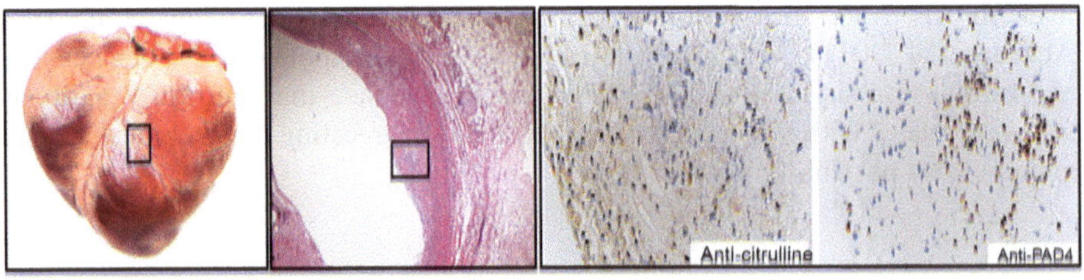

Fig. 12.3 Co-localization of citrullinated proteins (rabbit anticitrulline antibodies (Millipore)) and PAD4 (rabbit antihuman peptidylarginine deiminase type 4 antibodies (Daco)) in the atherosclerotic plaque within the right coronary artery

Similar studies by Barbarroja et al. showed correlative associations among anti-CCP, atherosclerosis, and inflammatory/oxidative stress markers, such as tumor necrosis factor-alpha (TNFα) and vascular cell adhesion molecule-1, in RA patients (Barbarroja et al. 2014; Husain et al. 2015). Despite the increased awareness of anti-CCP present within atherosclerotic plaques, only a few studies have evaluated the association between anti-CCP antibodies and subclinical atherosclerosis (Sokka et al. 2009; Gkaliagkousi et al. 2012). Vázquez-Del Mercado et al. (2015) showed elevated plasma levels of anti-CCP antibodies correlated to the size of the carotid atherosclerotic plaques in RA patients. Remarkably, anti-CCP-positive RA patients had significantly higher carotid plaque thickness than the anti-CCP-negative RA group. Furthermore, there was an independent association between serum levels of anti-CCP antibodies and thickness of carotid segments after adjustment for age, gender, and disease activity, showing an increment of 0.001 mm in the plaque formation for every unit of anti-CCP antibodies in the serum (Fig. 12.4) (Arnab et al. 2013). These findings suggest that detection of anti-CCP antibody could potentially be used as a predictor of cardiovascular risk in this group of patients. The biological mechanisms behind this process require further investigation, although it has been shown that anti-CCP-positive patients had a more atherogenic lipid profile characterized by lower HDL-c and a high atherogenic index of plasma, which correlate with patient age and RA duration (Vázquez-Del Mercado et al. 2015; Gonzalez-Juanatey et al. 2006).

Interestingly, several studies have demonstrated a genetic association of PAD4 (Harris et al. 2008) with RA. As well, PAD4 was identified as a frequent autoantigenic target in RA (Darrah et al. 2013; Halvorsen et al. 2009). PAD4 autoantibodies have been shown to be detectable prior to disease onset (Pollmann et al. 2012; Kolfenbach et al. 2010; Ferucci et al. 2013) and have been associated with more erosive RA that persists despite treatment with TNFα inhibitors (Halvorsen et al. 2009; Kolfenbach et al. 2010). Darrah et al. showed a subset of anti-PAD4 autoantibodies that cross-react with PAD3 and markedly increase the catalytic efficiency of PAD4 by decreasing the enzyme's requirement for calcium into the physiologic range. RA patients with these PAD3/PAD4 cross-reactive autoantibodies had higher baseline radiographic damage scores and a higher likelihood of radiographic progression, compared to individuals negative for these antibodies. Recent studies also

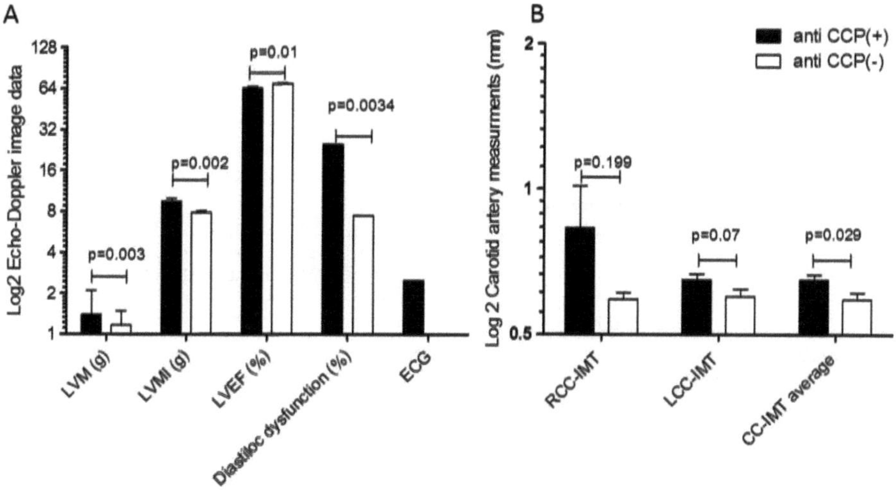

Fig. 12.4 (**a**) Echo-Doppler measurement in anti-CCP-positive and anti-CCP-negative RA patients. *LVEF* left ventricular ejection fraction, *LVM* left ventricular mass, *LVMI* left ventricular mass indexed to body surface area, *ECG* electrocardiograph (resting). (**b**) Carotid artery study. *RCC-IMT* right common carotid intima-medial thickness, *LCC-IMT* left common carotid intima-medial thickness, *CC-IMT* common carotid intima-medial thickness. *P* value statistically significant based on (Arnab et al. 2013)

have provided evidence for a unique role of citrullination and PAD4 in neutrophil extracellular trap (NET) formation, as a novel approach to targeting arterial disease (Khandpur et al. 2013; Knight et al. 2014). Knight et al. (2014) showed that inhibition of histone citrullination by PAD4 prevented chromatin decondensation, the background for NET formation, and thereby decreased atherosclerotic lesion size. Of note, Li et al. also showed that PAD4 was crucial for NET formation and bacterial killing. In a study of PAD4 knockout mice, PAD4 enzyme was required to generate NETs (Li et al. 2010). The most recent results presented by Sohn et al. indicated that citrullinated histone H2B was specifically arthritogenic and was a strong target of the anti-CCP immune response (Sohn et al. 2015) (Fig. 12.5).

12.4 Discovery of Citrullination of Cardiac Sarcomeric Proteins

It is well recognized that phosphorylation and other PTMs can act as regulators of cardiac muscle contractility in health and disease (Solaro and de Tombe 2008; Farley and Link 2009). We have recently shown that citrullination can also alter sarcomere contractility (Fert-Bober et al. 2015). Our work demonstrated that RA patients with CVD showed strong citrullination staining based on IHC in the myocardial interstitium and discovered the cellular localization of PAD isoforms in

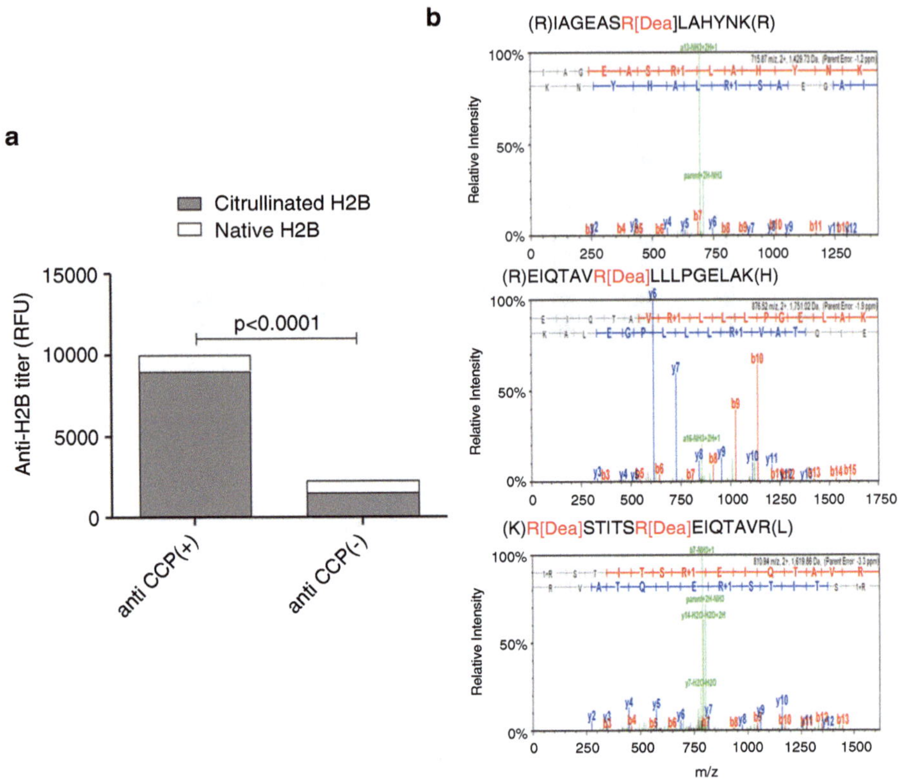

Fig. 12.5 Prominent targeting of citrullinated histone H2B by rheumatoid arthritis (RA)-associated autoantibodies. (**a**) Bead-based immunoassay analysis of anti-citrullinated H2B and anti-native H2B autoantibodies in plasma from patients with anti-cyclic citrullinated peptide 2 (anti-CCP-2)-positive RA (n = 81), anti-CCP-2-negative RA (n = 85). (**b**) Mass spectrometry-based detection of citrullinated residues in lysine C digest H2B isolated from human neutrophils

different cell types, such as cardiomyocytes (primarily PADs 1 and 3), in resident inflammatory cells (primarily PADs 2 and 4), and, to a smaller extent, in endothelial cells and vascular smooth muscle cells (Giles et al. 2012). Next, we further showed that citrullination was increased in HF patients without RA, compared with the general population using mass spectrometry techniques (Fert-Bober et al. 2015). SWATH-MS plus post search pipeline allowed for the reproducible quantification of citrullinated proteins across a large number of biological samples. In our publication (Fert-Bober et al. 2015), we showed that there is a broad cellular distribution of citrullinated proteins in healthy and HF myocardium, with enrichment in the mitochondria and sarcomeric subproteome in the diseased state. As shown, citrullinated residues were localized at the functionally significant regions of the myosin heavy chain, myosin light chains, cardiac myosin-binding protein C (cMyBP-C), actin, tropomyosin (Tm-1), cardiac troponin I (cTnI), and cardiac troponin T (cTnT). Citrullination was also found in Z-disc and M-band-related proteins, as well as in the cytoskeletal proteins such as desmin, filamen, myomesin, myozenin, LIM protein, and vimentin (Fig. 12.6). Furthermore, based on mRNA expression level, PAD2 and PAD4 were mainly expressed in cardiomyocytes, while PAD1, PAD2, and PAD4 were found in cardiac fibroblasts (Fert-Bober et al. 2015). In order to assign potential functional importance to each citrullinated site, it must first be determined if the citrulline is near or at a single nucleotide polymorphism (SNP) or mutation site that is known in the literature to be disease causing. It is estimated that ~30% of all cases of hypertrophic cardiomyopathy (HCM) are due specifically to myosin heavy chain 7 (MYH7) mutations identified in the globular head and neck domains of MYH7, encoded by exons 3 through 23 (Roberts and Sigwart 2001; Frazier et al. 2008; Yuceyar et al. 2015). The myosin head contains the actin and adenosine triphosphate-binding regions that are responsible for the force generation properties of myosin. The neck bends as the globular head is displaced throughout the power stroke. Recent data have suggested that mutations altering the charge of amino acids within critical functional domains are associated with a more severe disease phenotype than non-charge changing mutations or those altering residues outside of critical domains (Woo et al. 2003). For example, the myosin heavy chain mutation at residues Arg403Gln, Arg719Trp, Arg453Cys, Arg723Gly, or Gly716Arg in the MYH7 is associated with a high risk of sudden cardiac death (SCD) (Landstrom and Ackerman 2010; Watkins et al. 1995; Anan et al. 1994). We have identified, in MYH7, seven citrullinatable amino acid residues which are located throughout the protein (Fig. 12.7). Arg369Cit and Arg723Cit overlap with already detected mutations. These two citrullinated sites cluster around a so-called hot stop on myosin, which is a loop that forms part of the binding cleft for actin, imparting a more dramatic functional impact on the protein than mutations seen elsewhere. Furthermore, R369 and R723 citrullinated residues could be regarded as potentially pathogenic, since both result in an evolutionarily conserved sequence (Fig. 12.8). We would like to hypothesize that those variants in protein citrullination sites contribute to phenotype variation of the disease. To follow this up, we showed that citrullination of myosin decreased its intrinsic ATPase activity and inhibited the actomyosin ATPase activity up to 30% (Fert-Bober et al. 2015).

In addition, we showed that actin-myosin interaction was also affected by the citrullination of tropomyosin, independent of their citrullination state. Tropomyosin binds cooperatively to actin and forms long chains along both sides of the filament due to end-to-end overlap between adjacent molecules. Binding of tropomyosin to actin involves weak but specific electrostatic interactions called the "closed state" (Barua et al. 2013; Li et al. 2011). In this closed state, tropomyosin inhibits activation of actomyosin ATPase at low myosin concentrations; however, when the myosin head is strongly bound to actin, tropomyosin moves to its open state and is shifted away from the position occupied in the closed state (Tobacman and Butters 2000). In our study, citrullinated tropomyosin displayed enhanced binding to F-actin compared with unmodified tropomyosin. This was also observed in the presence of myosin, which suggests that citrullination of tropomyosin altered the ability of myosin to bind to the actin filament and therefore inhibits tropomyosin-actomyosin ATPase activity. We detected only one tropomyosin peptide that was citrullinated

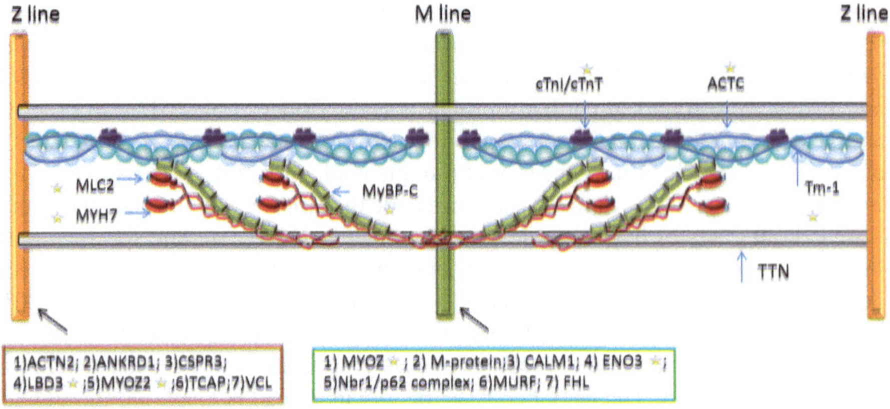

Fig. 12.6 A schematic view of the sarcomeric cytoskeleton with marked citrullinated proteins. The figure shows myosin (MYH7, *red*), actin (ACTC, *light blue*), tropomyosin (Tm-1, *dark blue*), troponin complex (cTnI/cTnT, *purple*), myosin-binding protein-C (MyBP-C, *light green*), Z-band proteins (*orange*), M-band proteins (*green*), and titin (TNN, *gray*). A *star* next to the protein indicates the protein as being citrullinated. Principal sarcomere regions are marked by Z, I, A, and M (Roberts and Sigwart 2001)

(residues 236–251), which is near the C-terminal region. This region is likely to be engaged in overlap with a contiguous tropomyosin. Unfortunately, there is no high-resolution structure of end-to-end tropomyosin molecules. Study by Sliwinska et al. showed that in the C-terminal region of tropomyosin, there are differences between tropomyosin isoforms, which determine orientations and dynamics of tropomyosin on actin. This is considered to be an important aspect of the cooperative activation of the thin filament (Sliwinska and Moraczewska 2013). Overall, these citrullinated sarcomeric peptides potentially affect thin and thick filament structure and function. The most revealing study proving the integrated effects of citrullination was carried out with single detergent-extracted ventricular myocytes incubated with vehicle or with active PAD2 (Fert-Bober et al. 2015). This study showed a decrease in Ca^{2+} sensitivity in the skinned cardiac myocyte, which reduced its ability to generate force in response to intracellular calcium.

These results support the finding that citrullination of tropomyosin induces increased binding of tropomyosin to actin filaments. This mechanistic and structural insight has revealed a wealth of important information on how citrullination can affect assembly in the sarcomeric cytoskeleton. In addition to integrating the contractile filaments in ordered sarcomeres, citrullinated proteins were also enriched in the mitochondria subproteome, thus providing a direct link between mechanical activity and cellular signaling in muscle. In order to obtain further information on the potential effect of citrullination on myofilament proteins in HF for this review, we have furthered our bioinformatics analysis and used a method of combing the network of protein-protein interactions (PPIs) with current gene ontology enrichment strategies. PPIs are commonly understood as physical contacts with molecular docking between proteins that occur in a cell or in a living organism in vivo (De Las

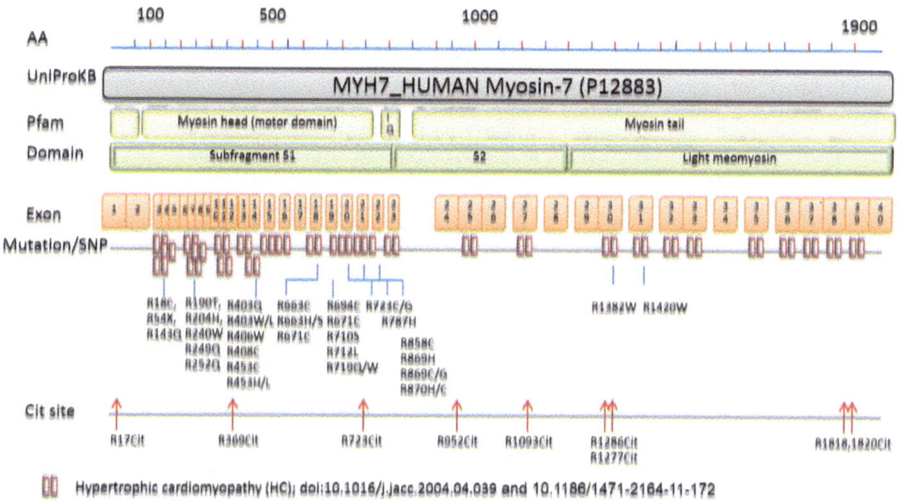

Fig. 12.7 Schematic representation of the molecular architecture of myosin-7 (MHY7) with the positions of the citrullinated residues found by Fert-Bober (data not published). The presentation emphasizes the modular construction of the proteins and distribution of mutations in the MYH7 gene based on doi:10.1016/j.jacc.2004.04.039 and 10.1186/1471–2164–11-172. Note that only the mutations on the R residues in the figure are indicated

Rivas and Fontanillo 2012). Each of these interactions is specifically adapted to carry out certain biological functions. To further understand the function of the top 54 citrullinated proteins reported by our group (Harris et al. 2008), a PPI network was built by Reactome Functional Interaction (FI), a Cytoscape plug-in (Wu et al. 2010). This plug-in accessed the Reactome FI network and allowed the construction of an FI subnetwork based on a set of genes by using linker genes. When the top 54 proteins were analyzed, without linker genes, two main modules were clustered in Reactome FI (Fig. 12.9). The first module (green) contains three proteins, and the pathway enrichment analysis revealed a Wnt signaling pathway and a nicotinic acetylcholine receptor signaling pathway. The second module (purple) also contains three proteins, and the pathway enrichment analysis annotated this module with four pathways, cardiac muscle contraction, hypertrophic cardiomyopathy, and dilated cardiomyopathy. The first module gene ontology enrichment, using cellular compartments, revealed sarcomere and stress fibers, whereas the second module was only in the sarcomere. The gene ontology enrichment, using biological processes, revealed that ventricular cardiac muscle tissue morphogenesis, sarcomere organization, cardiac muscle contraction, and muscle filament sliding were all enriched in the second module. Collectively, these data support that proteins with citrullination sites are involved in diverse pathways and that these pathways are connected through protein interactions.

A

	350	360	370	380	390	
Sp_MYH-7_Human	KLTGAIMHFG	NMKFKLKQRE	EQAEPDGTEE	ADKSAYLMGL	NSADLLKGLC	Beta strand
Sp-MYH7_Mouse	KLTGAIMHFG	NMKFKLKQRE	EQAEPDGTEE	ADKSAYLMGL	NSADLLKGLC	
Sp_MYH7_Rat	KLTGAIMHFG	NMKFKLKQRE	EQAEPDGTEE	ADKSAYLMGL	NSADLLKGLC	
Sp_MYH7_Pig	KLTGAIMHFG	NMKFKLKQRE	EQAEPDGTEE	ADKSAYLMGL	NSADLLKGLC	
Sp_MYH7_Canfl	KLTGAIMHFG	NMKFKLKQRE	EQAEPDGTEE	ADKSAYLMGL	NSADLLKGLC	
tr_A1L219_DANRE	KVVSAVLQLG	NMSFKKERNS	DQASMPDDTA	AQKVSHLLGM		
Sp_MYSN_DROME	RIVSAVLLFGS	MKFRQERNND	QATLPDNTVA	QKIAHLLGL		
	690	700	710	720	730	
Sp_MYH7_HUMAN	MHQLRCNGVL	EGIRICRKGF	PNRILYGDFR	QRYRILNPAA	IPEGQFIDSR	Helix
Sp_MYH7_MOUSE	MHQLRCNGVL	EGIRICRKGF	PNRILYGDFR	QRYRILNPAA	IPEGQFIDSR	
Sp_MYH7_RAT	MHQLRCNGVL	EGIRICRKGF	PNRILYGDFR	QRYRILNPAA	IPEGQFIDSR	
Sp_MYH7_PIG	MHQLRCNGVL	EGIRICRKGF	PNRILYGDFR	QRYRILNPAA	IPEGQFIDSR	
Sp_MYH7_CANLF	MHQLRCNGVL	EGIRICRKGF	PNRILYGDFR	QRYRILNPAA	IPEGQFIDSR	
Tr_A1L219_DANRE	LDQLRCNGVL	EGIRICRQGF	PNRIVFQEFR	QRYEILTPNA	IPKG-FMDGK	
Sp_MYSN_DROME	LDQLRCNGVL	EGIRICRQGF	PNRIPFQEFR	QRYELLTPNV	IPKG-FMDGK	

B

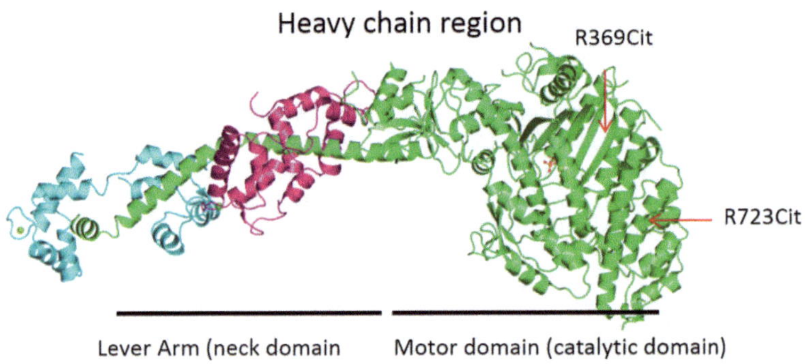

Fig. 12.8 Myosin amino acid conservation. (**a**) Homologies between myosin amino acids R369 and R723 across species. (**b**) Structure of the myosin head with two conservative citrullinated residues. Myosin heavy chain region showed in the rigor state with no nucleotide bound as solved by X-ray crystallography [http://www.ebi.ac.uk/pdbe/entry/pdb/1b7t]. The motor domain contains two detected citrullinated residues (*red arrows*)

12.5 Conclusions

There is growing evidence of a role for citrullination in HF and atherosclerosis. Our current goals are to unambiguously identify the specific proteins that are citrullinated and determine which directly effect heart structure/function and/or may cause immune responses involved in heart-specific inflammation, in order to identify early events that can be targeted with preventive or therapeutic measures. Herein, we have summarized the published data from numerous research groups, including our own, that have demonstrated the link between PAD and protein citrullination in the heart, as well as the increased immune responses to citrullinated proteins within the RA population with HF. Whether the citrullination of these proteins has physiological consequences remains unknown; however, several potential implications of the current studies can be considered. This is certainly achievable, given the huge advances

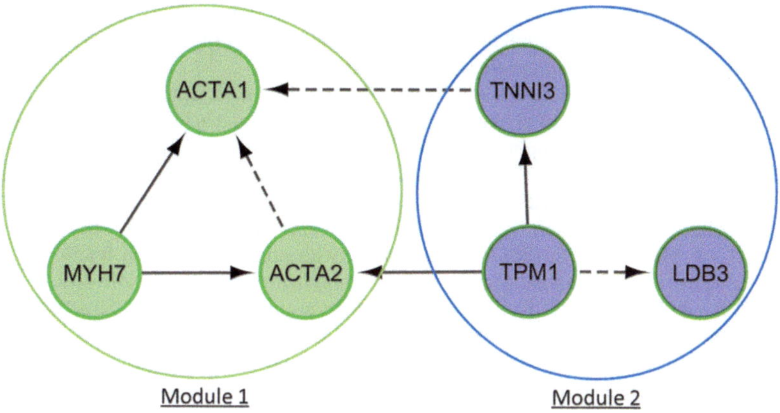

Fig. 12.9 Pathway enrichment analysis of top heart candidates with citrullination. Abbreviation: *MYH7* myosin-7; *ACTA A1/A2* actin, alpha skeletal muscle; *TNNI3* troponin I, cardiac muscle; *TPM1* tropomyosin alpha-1 chain; *LDB3* LIM domain-binding protein 3

currently taking place in both the constant development of new technologies to identify citrullinated proteins and understanding the crosstalk between genetic, environmental factors and disease pathogenesis.

References

Abdeen, S., Olusi, S. O., & George, S. (2011). Serum anti-modified citrullinated vimentin antibody concentration is associated with liver fibrosis in patients with chronic hepatitis. *Hepatic Medicine, 3*, 13–18.

Agnetti, G., Husberg, C., & Van Eyk, J. E. (2011). Divide and conquer: The application of organelle proteomics to heart failure. *Circulation Research, 108*, 512–526.

Anan, R., Greve, G., Thierfelder, L., Watkins, H., McKenna, W. J., Solomon, S., et al. (1994). Prognostic implications of novel beta cardiac myosin heavy chain gene mutations that cause familial hypertrophic cardiomyopathy. *The Journal of Clinical Investigation, 93*, 280–285.

Andrade, F., Darrah, E., Gucek, M., Cole, R. N., Rosen, A., & Zhu, X. (2010). Autocitrullination of human peptidyl arginine deiminase type 4 regulates protein citrullination during cell activation. *Arthritis and Rheumatism, 62*, 1630–1640.

Arnab, B., Biswadip, G., Arindam, P., Shyamash, M., Anirban, G., & Rajan, P. (2013). Anti-CCP antibody in patients with established rheumatoid arthritis: Does it predict adverse cardiovascular profile? *Journal of Cardiovascular Disease Research, 4*, 102–106.

Aryal, B., Rotllan, N., & Fernández-Hernando, C. (2014). Non-coding RNAs and atherosclerosis. *Current Atherosclerosis Reports, 16*, 407–407.

Baka, Z., Gyorgy, B., Geher, P., Buzas, E. I., Falus, A., & Nagy, G. (2012). Citrullination under physiological and pathological conditions. *Joint, Bone, Spine, 79*, 431–436.

Barbarroja, N., Perez-Sanchez, C., Ruiz-Limon, P., Castro-Villegas, C., Aguirre, M. A., Carretero, R., et al. (2014). Anticyclic citrullinated protein antibodies are implicated in the development of cardiovascular disease in rheumatoid arthritis. *Arteriosclerosis, Thrombosis, and Vascular Biology, 34*, 2706–2716.

Barquera, S., Pedroza-Tobias, A., Medina, C., Hernandez-Barrera, L., Bibbins-Domingo, K., Lozano, R., et al. (2015). Global overview of the epidemiology of atherosclerotic cardiovascular disease. *Archives of Medical Research, 46*, 328–338.

Barua, B., Fagnant, P. M., Winkelmann, D. A., Trybus, K. M., & Hitchcock-DeGregori, S. E. (2013). A periodic pattern of evolutionarily conserved basic and acidic residues constitutes the binding interface of actin-tropomyosin. *The Journal of Biological Chemistry, 288*, 9602–9609.

Bergholm, R., Leirisalo-Repo, M., Vehkavaara, S., Makimattila, S., Taskinen, M. R., & Yki-Jarvinen, H. (2002). Impaired responsiveness to NO in newly diagnosed patients with rheumatoid arthritis. *Arteriosclerosis, Thrombosis, and Vascular Biology, 22*, 1637–1641.

Bhuiyan, T., & Maurer, M. S. (2011). Heart failure with preserved ejection fraction: Persistent diagnosis, therapeutic enigma. *Current Cardiovascular Risk Reports, 5*, 440–449.

Bicker, K. L., & Thompson, P. R. (2013). The protein arginine deiminases: Structure, function, inhibition, and disease. *Biopolymers, 99*, 155–163.

Bongartz, T., Cantaert, T., Atkins, S. R., Harle, P., Myers, J. L., Turesson, C., et al. (2007). Citrullination in extra-articular manifestations of rheumatoid arthritis. *Rheumatology (Oxford), 46*, 70–75.

Burkhardt, H., Sehnert, B., Bockermann, R., Engstrom, A., Kalden, J. R., & Holmdahl, R. (2005). Humoral immune response to citrullinated collagen type II determinants in early rheumatoid arthritis. *European Journal of Immunology, 35*, 1643–1652.

Cantaert, T., De Rycke, L., Bongartz, T., Matteson, E. L., Tak, P. P., Nicholas, A. P., et al. (2006). Citrullinated proteins in rheumatoid arthritis: Crucial...But not sufficient! *Arthritis and Rheumatism, 54*, 3381–3389.

Cao, Q., Wang, X., Jia, L., Mondal, A. K., Diallo, A., Hawkins, G. A., et al. (2014). Inhibiting DNA methylation by 5-Aza-2′-deoxycytidine ameliorates atherosclerosis through suppressing macrophage inflammation. *Endocrinology, 155*, 4925–4938.

Chan, A. T., Manson, J. E., Albert, C. M., Chae, C. U., Rexrode, K. M., Curhan, G. C., et al. (2006). Nonsteroidal antiinflammatory drugs, acetaminophen, and the risk of cardiovascular events. *Circulation, 113*, 1578–1587.

Corrao, S., Salli, L., Arnone, S., Scaglione, R., Pinto, A., & Licata, G. (1996). Echo-Doppler left ventricular filling abnormalities in patients with rheumatoid arthritis without clinically evident cardiovascular disease. *European Journal of Clinical Investigation, 26*, 293–297.

Crowson, C. S., Nicola, P. J., Kremers, H. M., O'Fallon, W. M., Therneau, T. M., Jacobsen, S. J., et al. (2005). How much of the increased incidence of heart failure in rheumatoid arthritis is attributable to traditional cardiovascular risk factors and ischemic heart disease? *Arthritis and Rheumatism, 52*, 3039–3044.

Darrah, E., Giles, J. T., Ols, M. L., Bull, H. G., Andrade, F., & Rosen, A. (2013). Erosive rheumatoid arthritis is associated with antibodies that activate PAD4 by increasing calcium sensitivity. *Science Translational Medicine, 5*, 186ra165.

Davis, J. M., III, Roger, V. L., Crowson, C. S., Kremers, H. M., Therneau, T. M., & Gabriel, S. E. (2008). The presentation and outcome of heart failure in patients with rheumatoid arthritis differs from that in the general population. *Arthritis and Rheumatism, 58*, 2603–2611.

De Las Rivas, J., & Fontanillo, C. (2012). Protein-protein interaction networks: Unraveling the wiring of molecular machines within the cell. *Briefings in Functional Genomics, 11*, 489–496.

Di Franco, M., Paradiso, M., Mammarella, A., Paoletti, V., Labbadia, G., Coppotelli, L., et al. (2000). Diastolic function abnormalities in rheumatoid arthritis. Evaluation by echo Doppler transmitral flow and pulmonary venous flow: Relation with duration of disease. *Annals of the Rheumatic Diseases, 59*, 227–229.

Dje N'Guessan, P., Riediger, F., Vardarova, K., Scharf, S., Eitel, J., Opitz, B., et al. (2009). Statins control oxidized LDL-mediated histone modifications and gene expression in cultured human endothelial cells. *Arteriosclerosis, Thrombosis, and Vascular Biology, 29*, 380–386.

Dunn, J., Thabet, S., Jo, H., & Flow-Dependent Epigenetic, D. N. A. (2015). Methylation in endothelial gene expression and atherosclerosis. *Arteriosclerosis, Thrombosis, and Vascular Biology, 35*, 1562–1569.

Farley, A. R., & Link, A. J. (2009). Identification and quantification of protein posttranslational modifications. *Methods in Enzymology, 463*, 725–763.

Fert-Bober, J., Giles, J. T., Holewinski, R. J., Kirk, J. A., Uhrigshardt, H., Crowgey, E. L., et al. (2015). Citrullination of myofilament proteins in heart failure. *Cardiovascular Research, 108*, 232–242.

Ferucci, E. D., Darrah, E., Smolik, I., Choromanski, T. L., Robinson, D. B., Newkirk, M. M., et al. (2013). Prevalence of anti-peptidylarginine deiminase type 4 antibodies in rheumatoid arthritis and unaffected first-degree relatives in indigenous north American populations. *The Journal of Rheumatology, 40*, 1523–1528.

Fonarow, G. C., Stough, W. G., Abraham, W. T., Albert, N. M., Gheorghiade, M., Greenberg, B. H., et al. (2007). Characteristics, treatments, and outcomes of patients with preserved systolic function hospitalized for heart failure: A report from the OPTIMIZE-HF registry. *Journal of the American College of Cardiology, 50*, 768–777.

Frazier, A., Judge, D. P., Schulman, S. P., Johnson, N., Holmes, K. W., & Murphy, A. M. (2008). Familial hypertrophic cardiomyopathy associated with cardiac beta-myosin heavy chain and troponin I mutations. *Pediatric Cardiology, 29*, 846–850.

Gabriel, S. E. (2008). Cardiovascular morbidity and mortality in rheumatoid arthritis. *The American Journal of Medicine, 121*, S9–14.

Gabriel, S. E. (2010). Heart disease and rheumatoid arthritis: Understanding the risks. *Annals of the Rheumatic Diseases, 69*(Suppl 1), i61–i64.

Gerli, R., Bartoloni Bocci, E., Sherer, Y., Vaudo, G., Moscatelli, S., & Shoenfeld, Y. (2008). Association of anti-cyclic citrullinated peptide antibodies with subclinical atherosclerosis in patients with rheumatoid arthritis. *Annals of the Rheumatic Diseases, 67*, 724–725.

Giles, J. T., Szklo, M., Post, W., Petri, M., Blumenthal, R. S., Lam, G., et al. (2009). Coronary arterial calcification in rheumatoid arthritis: Comparison with the multi-ethnic study of atherosclerosis. *Arthritis Research & Therapy, 11*, R36.

Giles, J. T., Malayeri, A. A., Fernandes, V., Post, W., Blumenthal, R. S., Bluemke, D., et al. (2010). Left ventricular structure and function in patients with rheumatoid arthritis, as assessed by cardiac magnetic resonance imaging. *Arthritis and Rheumatism, 62*, 940–951.

Giles, J. T., Fert-Bober, J., Park, J. K., Bingham, C. O., III, Andrade, F., Fox-Talbot, K., et al. (2012). Myocardial citrullination in rheumatoid arthritis: A correlative histopathologic study. *Arthritis Research & Therapy, 14*, R39.

Girbal-Neuhauser, E., Durieux, J. J., Arnaud, M., Dalbon, P., Sebbag, M., Vincent, C., et al. (1999). The epitopes targeted by the rheumatoid arthritis-associated antifilaggrin autoantibodies are posttranslationally generated on various sites of (pro)filaggrin by deimination of arginine residues. *Journal of Immunology, 162*, 585–594.

Gkaliagkousi, E., Gavriilaki, E., Doumas, M., Petidis, K., Aslanidis, S., & Stella, D. (2012). Cardiovascular risk in rheumatoid arthritis: Pathogenesis, diagnosis, and management. *Journal of Clinical Rheumatology, 18*, 422–430.

Goff, D. C., Jr., Lloyd-Jones, D. M., Bennett, G., Coady, S., D'Agostino, R. B., Gibbons, R., et al. (2014). 2013 ACC/AHA guideline on the assessment of cardiovascular risk: A report of the American College of Cardiology/American Heart Association task force on practice guidelines. *Circulation, 129*, S49–S73.

Gonzalez-Juanatey, C., Llorca, J., Sanchez-Andrade, A., Garcia-Porrua, C., Martin, J., & Gonzalez-Gay, M. A. (2006). Short-term adalimumab therapy improves endothelial function in patients with rheumatoid arthritis refractory to infliximab. *Clinical and Experimental Rheumatology, 24*, 309–312.

Gyorgy, B., Toth, E., Tarcsa, E., Falus, A., & Buzas, E. I. (2006). Citrullination: A posttranslational modification in health and disease. *The International Journal of Biochemistry & Cell Biology, 38*, 1662–1677.

Halvorsen, E. H., Haavardsholm, E. A., Pollmann, S., Boonen, A., van der Heijde, D., Kvien, T. K., et al. (2009). Serum IgG antibodies to peptidylarginine deiminase 4 predict radiographic progression in patients with rheumatoid arthritis treated with tumour necrosis factor-alpha blocking agents. *Annals of the Rheumatic Diseases, 68*, 249–252.

Harris, M. L., Darrah, E., Lam, G. K., Bartlett, S. J., Giles, J. T., Grant, A. V., et al. (2008). Association of autoimmunity to peptidyl arginine deiminase type 4 with genotype and disease severity in rheumatoid arthritis. *Arthritis and Rheumatism, 58*, 1958–1967.

Hastings, N. E., Simmers, M. B., McDonald, O. G., Wamhoff, B. R., & Blackman, B. R. (2007). Atherosclerosis-prone hemodynamics differentially regulates endothelial and smooth muscle cell phenotypes and promotes pro-inflammatory priming. *American Journal of Physiology. Cell Physiology, 293*, C1824–C1833.

Heidenreich, P. A., Albert, N. M., Allen, L. A., Bluemke, D. A., Butler, J., Fonarow, G. C., et al. (2013). Forecasting the impact of heart failure in the United States: A policy statement from the American Heart Association. *Circulation. Heart Failure, 6*, 606–619.

Hill, J. A., Bell, D. A., Brintnell, W., Yue, D., Wehrli, B., Jevnikar, A. M., et al. (2008). Arthritis induced by posttranslationally modified (citrullinated) fibrinogen in DR4-IE transgenic mice. *The Journal of Experimental Medicine, 205*, 967–979.

Hueber, W., Kidd, B. A., Tomooka, B. H., Lee, B. J., Bruce, B., Fries, J. F., et al. (2005). Antigen microarray profiling of autoantibodies in rheumatoid arthritis. *Arthritis and Rheumatism, 52*, 2645–2655.

Husain, K., Hernandez, W., Ansari, R. A., & Ferder, L. (2015). Inflammation, oxidative stress and renin angiotensin system in atherosclerosis. *World Journal of Biological Chemistry, 6*, 209–217.

Innala, L., Möller, B., Ljung, L., Magnusson, S., Smedby, T., Södergren, A., et al. (2011). Cardiovascular events in early RA are a result of inflammatory burden and traditional risk factors: A five year prospective study. *Arthritis Research & Therapy, 13*, R131.

Kaplan, M. J. (2010). Cardiovascular complications of rheumatoid arthritis - assessment, prevention and treatment. *Rheumatic Diseases Clinics of North America, 36*, 405–426.

Kastbom, A., Forslind, K., Ernestam, S., Geborek, P., Karlsson, J. A., Petersson, I. F., et al. (2016). Changes in the anticitrullinated peptide antibody response in relation to therapeutic outcome in early rheumatoid arthritis: Results from the SWEFOT trial. *Annals of the Rheumatic Diseases, 75*, 356–361.

Khandpur, R., Carmona-Rivera, C., Vivekanandan-Giri, A., Gizinski, A., Yalavarthi, S., Knight, J. S., et al. (2013). NETs are a source of citrullinated autoantigens and stimulate inflammatory responses in rheumatoid arthritis. *Science Translational Medicine, 5*, 178ra140.

Kinloch, A., Tatzer, V., Wait, R., Peston, D., Lundberg, K., Donatien, P., et al. (2005). Identification of citrullinated alpha-enolase as a candidate autoantigen in rheumatoid arthritis. *Arthritis Research & Therapy, 7*, R1421–R1429.

Knight, J. S., Luo, W., O'Dell, A. A., Yalavarthi, S., Zhao, W., Subramanian, V., et al. (2014). Peptidylarginine deiminase inhibition reduces vascular damage and modulates innate immune responses in murine models of atherosclerosis. *Circulation Research, 114*, 947–956.

Kolfenbach, J. R., Deane, K. D., Derber, L. A., O'Donnell, C. I., Gilliland, W. R., Edison, J. D., et al. (2010). Autoimmunity to peptidyl arginine deiminase type 4 precedes clinical onset of rheumatoid arthritis. *Arthritis and Rheumatism, 62*, 2633–2639.

Landstrom, A. P., & Ackerman, M. J. (2010). Mutation type is not clinically useful in predicting prognosis in hypertrophic cardiomyopathy. *Circulation, 122*, 2441–2449. discussion 2450.

LeWinter, M. M., & Meyer, M. (2013). Mechanisms of diastolic dysfunction in HFpEF: If it's not one thing it's another. *Circulation. Heart Failure, 6*, 1112–1115.

Li, P., Li, M., Lindberg, M. R., Kennett, M. J., Xiong, N., & Wang, Y. (2010). PAD4 is essential for antibacterial innate immunity mediated by neutrophil extracellular traps. *The Journal of Experimental Medicine, 207*, 1853–1862.

Li, X., Tobacman, L. S., Mun, J. Y., Craig, R., Fischer, S., & Lehman, W. (2011). Tropomyosin position on F-actin revealed by EM reconstruction and computational chemistry. *Biophysical Journal, 100*, 1005–1013.

Liang, K. P., Kremers, H. M., Crowson, C. S., Snyder, M. R., Therneau, T. M., Roger, V. L., et al. (2009). Autoantibodies and the risk of cardiovascular events. *The Journal of Rheumatology, 36*, 2462–2469.

Lindenfeld, J., Albert, N. M., Boehmer, J. P., Collins, S. P., Ezekowitz, J. A., Givertz, M. M., et al. (2010). HFSA 2010 comprehensive heart failure practice guideline. *Journal of Cardiac Failure, 16*, e1–194.

Makrygiannakis, D., af Klint, E., Lundberg, I. E., Lofberg, R., Ulfgren, A. K., Klareskog, L., et al. (2006). Citrullination is an inflammation-dependent process. *Annals of the Rheumatic Diseases, 65*, 1219–1222.

Masson-Bessiere, C., Sebbag, M., Girbal-Neuhauser, E., Nogueira, L., Vincent, C., Senshu, T., et al. (2001). The major synovial targets of the rheumatoid arthritis-specific antifilaggrin autoantibodies are deiminated forms of the alpha- and beta-chains of fibrin. *Journal of Immunology, 166*, 4177–4184.

Meune, C., Touzé, E., Trinquart, L., & Allanore, Y. (2010). High risk of clinical cardiovascular events in rheumatoid arthritis: Levels of associations of myocardial infarction and stroke through a systematic review and meta-analysis. *Archives of Cardiovascular Diseases, 103*, 253–261.

Mozaffarian, D., Benjamin, E. J., Go, A. S., Arnett, D. K., Blaha, M. J., Cushman, M., et al. (2015). Heart disease and stroke statistics--2015 update: A report from the American Heart Association. *Circulation, 131*, e29–322.

Mozaffarian, D., Benjamin, E. J., Go, A. S., Arnett, D. K., Blaha, M. J., Cushman, M., et al. (2016). Heart disease and stroke statistics-2016 update: A report from the American Heart Association. *Circulation, 133*, e38–e60.

Mudd, J. O., & Kass, D. A. (2008). Tackling heart failure in the twenty-first century. *Nature, 451*, 919–928.

Myasoedova, E., Crowson, C. S., Nicola, P. J., Maradit-Kremers, H., Davis, J. M., III, Roger, V. L., et al. (2011). The influence of rheumatoid arthritis disease characteristics on heart failure. *The Journal of Rheumatology, 38*, 1601–1606.

Myasoedova, E., Davis, J. M., Crowson, C. S., Roger, V. L., Karon, B. L., Borgeson, D. D., et al. (2013). Rheumatoid arthritis is associated with left ventricular concentric remodeling: Results of a population-based cross-sectional study. *Arthritis and Rheumatism, 65*, 1713–1718.

Nicola, P. J., Crowson, C. S., Maradit-Kremers, H., Ballman, K. V., Roger, V. L., Jacobsen, S. J., et al. (2006). Contribution of congestive heart failure and ischemic heart disease to excess mortality in rheumatoid arthritis. *Arthritis and Rheumatism, 54*, 60–67.

Oka, T., & Komuro, I. (2008). Molecular mechanisms underlying the transition of cardiac hypertrophy to heart failure. *Circulation Journal, 72 Suppl A*, A13-16.

Oktay, A. A., Rich, J. D., & Shah, S. J. (2013). The emerging epidemic of heart failure with preserved ejection fraction. *Current Heart Failure Reports, 10*(4), 401–410. doi:10.1007/s11897-013-0155-7.

Pollmann, S., Stensland, M., Halvorsen, E. H., Sollid, L. M., Kvien, T. K., Fleckenstein, B., et al. (2012). Anti-PAD4 autoantibodies in rheumatoid arthritis: Levels in serum over time and impact on PAD4 activity as measured with a small synthetic substrate. *Rheumatology International, 32*, 1271–1276.

Pujades-Rodriguez, M., Duyx, B., Thomas, S. L., Stogiannis, D., Rahman, A., Smeeth, L., et al. (2016). Rheumatoid arthritis and incidence of twelve initial presentations of cardiovascular disease: A population record-linkage cohort study in England. *PloS One, 11*, e0151245.

Roberts, R., & Sigwart, U. (2001). New concepts in hypertrophic cardiomyopathies, part I. *Circulation, 104*, 2113–2116.

Robinson, W. H., & Sokolove, J. (2012). Citrullination of fibrinogen: Generation of neoepitopes and enhancement of immunostimulatory properties. *Arthritis Research & Therapy, 14*, O30.

Rudominer, R. L., Roman, M. J., Devereux, R. B., Paget, S. A., Schwartz, J. E., Lockshin, M. D., et al. (2009). Rheumatoid arthritis is independently associated with increased left ventricular mass but not reduced ejection fraction. *Arthritis and Rheumatism, 60*, 22–29.

Schellekens, G. A., Visser, H., de Jong, B. A., van den Hoogen, F. H., Hazes, J. M., Breedveld, F. C., et al. (2000). The diagnostic properties of rheumatoid arthritis antibodies recognizing a cyclic citrullinated peptide. *Arthritis and Rheumatism, 43*, 155–163.

Sghiri, R., Khalifa, O., Bouajina, E., Ben Haj Slama, F., Baccouche, K., Ben Fredj, H., et al. (2010). Presence and diagnostic performances of IgA class antibodies to mutated citrullinated vimentin in rheumatoid arthritis. *Joint, Bone, Spine, 77*, 279–280.

Shelef, M. A., Bennin, D. A., Mosher, D. F., & Huttenlocher, A. (2012). Citrullination of fibronectin modulates synovial fibroblast behavior. *Arthritis Research & Therapy, 14*, R240.

Sipila, K., Haag, S., Denessiouk, K., Kapyla, J., Peters, E. C., Denesyuk, A., et al. (2014). Citrullination of collagen II affects integrin-mediated cell adhesion in a receptor-specific manner. *The FASEB Journal, 28*, 3758–3768.

Sliwinska, M., & Moraczewska, J. (2013). Structural differences between C-terminal regions of tropomyosin isoforms. *PeerJ, 1*, e181.

Soeiro Ade, M., Haddad, M., de Almeida, M. C., Ruppert, A. D., & Serrano, C. V., Jr. (2012). Rheumatoid arthritis and cardiovascular disease: What is known about this relationship and what can currently be done for affected patients? *Revista Portuguesa de Cardiologia, 31*, 225–232.

Sohn, D. H., Rhodes, C., Onuma, K., Zhao, X., Sharpe, O., Gazitt, T., et al. (2015). Local joint inflammation and histone citrullination in a murine model of the transition from preclinical autoimmunity to inflammatory arthritis. *Arthritis & Rhematology, 67*, 2877–2887.

Sokka, T., Toloza, S., Cutolo, M., Kautiainen, H., Makinen, H., Gogus, F., et al. (2009). Women, men, and rheumatoid arthritis: Analyses of disease activity, disease characteristics, and treatments in the QUEST-RA study. *Arthritis Research & Therapy, 11*, R7.

Sokolove, J., Bromberg, R., Deane, K. D., Lahey, L. J., Derber, L. A., Chandra, P. E., et al. (2012). Autoantibody epitope spreading in the pre-clinical phase predicts progression to rheumatoid arthritis. *PloS One, 7*, e35296.

Sokolove, J., Brennan, M. J., Sharpe, O., Lahey, L. J., Kao, A. H., Krishnan, E., et al. (2013). Brief report: Citrullination within the atherosclerotic plaque: A potential target for the anti-citrullinated protein antibody response in rheumatoid arthritis. *Arthritis and Rheumatism, 65*, 1719–1724.

Solaro, R. J., & de Tombe, P. P. (2008). Review focus series: Sarcomeric proteins as key elements in integrated control of cardiac function. *Cardiovascular Research, 77*, 616–618.

Solomon, D. H., Karlson, E. W., Rimm, E. B., Cannuscio, C. C., Mandl, L. A., Manson, J. E., et al. (2003). Cardiovascular morbidity and mortality in women diagnosed with rheumatoid arthritis. *Circulation, 107*, 1303–1307.

Szekanecz, Z., Soos, L., Szabo, Z., Fekete, A., Kapitany, A., Vegvari, A., et al. (2008). Anti-citrullinated protein antibodies in rheumatoid arthritis: As good as it gets? *Clinical Reviews in Allergy & Immunology, 34*, 26–31.

Takahara, H., Okamoto, H., & Sugawara, K. (1986). Calcium-dependent properties of peptidylarginine deiminase from rabbit skeletal-muscle. *Agricultural and Biological Chemistry, 50*, 2899–2904.

Tobacman, L. S., & Butters, C. A. (2000). A new model of cooperative myosin-thin filament binding. *The Journal of Biological Chemistry, 275*, 27587–27593.

Tomasson, G., Aspelund, T., Jonsson, T., Valdimarsson, H., Felson, D. T., & Gudnason, V. (2010). Effect of rheumatoid factor on mortality and coronary heart disease. *Annals of the Rheumatic Diseases, 69*, 1649–1654.

Trelle, S., Reichenbach, S., Wandel, S., Hildebrand, P., Tschannen, B., Villiger, P. M., et al. (2011). Cardiovascular safety of non-steroidal anti-inflammatory drugs: Network meta-analysis. *BMJ, 342*, c7086.

Turesson, C., Jarenros, A., & Jacobsson, L. (2004). Increased incidence of cardiovascular disease in patients with rheumatoid arthritis: Results from a community based study. *Annals of the Rheumatic Diseases, 63*, 952–955.

Vallbracht, I., Rieber, J., Oppermann, M., Forger, F., Siebert, U., & Helmke, K. (2004). Diagnostic and clinical value of anti-cyclic citrullinated peptide antibodies compared with rheumatoid factor isotypes in rheumatoid arthritis. *Annals of the Rheumatic Diseases, 63*, 1079–1084.

Vassiliadis, E., Oliveira, C. P., Alvares-da-Silva, M. R., Zhang, C., Carrilho, F. J., Stefano, J. T., et al. (2012). Circulating levels of citrullinated and MMP-degraded vimentin (VICM) in liver fibrosis related pathology. *American Journal of Translational Research, 4*, 403–414.

Vázquez-Del Mercado, M., Nuñez-Atahualpa, L., et al. (2015). Serum levels of anticyclic citrullinated peptide antibodies, interleukin-6, tumor necrosis factor-α, and C-reactive protein are associated with increased carotid intima-media thickness: A cross-sectional analysis of a cohort of rheumatoid arthritis patients without cardiovascular risk factors. *BioMed Research International, 2015*, 342649.

van Venrooij, W. J., & Pruijn, G. J. (2014). How citrullination invaded rheumatoid arthritis research. *Arthritis Research & Therapy, 16*, 103.

Verpoort, K. N., Jol-van der Zijde, C. M., Papendrecht-van der Voort, E. A., Ioan-Facsinay, A., Drijfhout, J. W., van Tol, M. J., et al. (2006). Isotype distribution of anti-cyclic citrullinated peptide antibodies in undifferentiated arthritis and rheumatoid arthritis reflects an ongoing immune response. *Arthritis and Rheumatism, 54*, 3799–3808.

Vincent, C., de Keyser, F., Masson-Bessiere, C., Sebbag, M., Veys, E., & Serre, G. (1999). Anti-perinuclear factor compared with the so called "antikeratin" antibodies and antibodies to human epidermis filaggrin, in the diagnosis of arthritides. *Annals of the Rheumatic Diseases, 58*, 42–48.

Vossenaar, E. R., & van Venrooij, W. J. (2004). Citrullinated proteins: Sparks that may ignite the fire in rheumatoid arthritis. *Arthritis Research & Therapy, 6*, 107–111.

Vossenaar, E. R., Zendman, A. J., van Venrooij, W. J., & Pruijn, G. J. (2003). PAD, a growing family of citrullinating enzymes: Genes, features and involvement in disease. *BioEssays, 25*, 1106–1118.

Watkins, H., Seidman, J. G., & Seidman, C. E. (1995). Familial hypertrophic cardiomyopathy: A genetic model of cardiac hypertrophy. *Human Molecular Genetics, 4 Spec No*, 1721–1727.

Wegner, N., Lundberg, K., Kinloch, A., Fisher, B., Malmstrom, V., Feldmann, M., et al. (2010). Autoimmunity to specific citrullinated proteins gives the first clues to the etiology of rheumatoid arthritis. *Immunological Reviews, 233*, 34–54.

Woo, A., Rakowski, H., Liew, J. C., Zhao, M. S., Liew, C. C., Parker, T. G., et al. (2003). Mutations of the beta myosin heavy chain gene in hypertrophic cardiomyopathy: Critical functional sites determine prognosis. *Heart, 89*, 1179–1185.

Wright, K., Crowson, C. S., & Gabriel, S. E. (2014). Cardiovascular comorbidity in rheumatic diseases a focus on heart failure. *Heart Failure Clinics, 10*, 339–352.

Wu, G., Feng, X., & Stein, L. (2010). A human functional protein interaction network and its application to cancer data analysis. *Genome Biology, 11*, R53.

Yancy, C. W., Lopatin, M., Stevenson, L. W., De Marco, T., & Fonarow, G. C. (2006). Clinical presentation, management, and in-hospital outcomes of patients admitted with acute decompensated heart failure with preserved systolic function: A report from the acute decompensated heart failure National Registry (ADHERE) database. *Journal of the American College of Cardiology, 47*, 76–84.

Yuceyar, N., Ayhan, O., Karasoy, H., & Tolun, A. (2015). Homozygous MYH7 R1820W mutation results in recessive myosin storage myopathy: Scapuloperoneal and respiratory weakness with dilated cardiomyopathy. *Neuromuscular Disorders, 25*, 340–344.

Chapter 13
Protein Deimination in Protein Misfolding Disorders: Modeled in Human Induced Pluripotent Stem Cells (iPSCs)

Sigrun Lange, Selina Wray, Mike Devine, Mar Matarin, and John Hardy

13.1 Introduction

Neurodegenerative diseases are an increasing burden on society due to an aging population and limited treatment options. This highlights the need for biomarker discovery and novel drug-directed strategies to reduce disease progression and increase the life quality of sufferers (Hampel et al. 2010). Progressive neurodegenerative diseases are often referred to as protein misfolding disorders, as accumulation of protein aggregates, impaired calcium buffering, mitochondrial dysfunction, and epigenetic and posttranslational protein modifications are implicated in disease progression and lead to fatal neuronal loss (Beal 2005; Krüger et al. 2010; Mastroeni et al. 2011; Halliday and Mallucci 2014; Poorkaj et al., 1998).

Peptidylarginine deiminases are calcium-catalyzed enzymes that cause irreversible posttranslational changes of protein-bound arginines into citrullines. Resulting changes in charge and structure of target proteins lead to protein misfolding and functional loss (Vossenaar et al. 2003; Gyorgy et al. 2006). Some main targets of deimination identified are nuclear histones and intermediate filaments (Wang et al. 2004; Gyorgy et al. 2006; Lange et al. 2011, 2014). Besides being associated with various autoimmune diseases and cancer, protein deimination is increasingly being linked to neurodegenerative diseases including multiple sclerosis (MS), AD, prion disorders such as Creutzfeldt-Jakob disease (CJD), and PD (Moscarello et al. 1994; Musse et al. 2008; Wood et al. 2008; Bradford et al. 2014). Calcium dysregulation is one of the hallmark downstream factors caused by protein aggregation due to

S. Lange (✉)
University of Westminster, Department of Biomedical Sciences, London W1W 6UW, UK

UCL School of Pharmacy, London WC1N 1AX, UK
e-mail: S.Lange@westminster.ac.uk

S. Wray • M. Devine • M. Matarin • J. Hardy
UCL Institute of Neurology, Department of Molecular Neuroscience, London WC1N 3BG, UK

A.P. Nicholas et al. (eds.), *Protein Deimination in Human Health and Disease*,
DOI 10.1007/978-3-319-58244-3_13

various neurodegenerative gene mutations and also causes PAD-mediated protein deimination that contributes significantly to further protein misfolding and progressive neurodegenerative pathologies (see proposed mechanism in Fig. 13.1). The use of PAD-inhibiting drugs thus offers a novel interceptive route to reduce disease progression and improve patients' life quality.

13.2 Pharmacological PAD Inhibition Is Neuroprotective in CNS Damage In Vivo

Previously we have shown that the pan-PAD inhibitor Cl-Am (Luo et al. 2006) significantly reduced neuronal damage in animal models of acute neuronal insult models, namely, spinal cord damage (Lange et al. 2011) and neonatal hypoxic ischemic encephalopathy (HIE) (Lange et al. 2014). Administration of one dose of Cl-Am (80 mg/kg) straight after injury and up to two hours post-injury in the chick spinal cord damage model significantly reduced neuronal cell death, tissue loss, and deiminated histones, compared to non-treated control injuries (Lange et al. 2011). In two different mouse models of neonatal HIE, firstly, where permanent left carotid occlusion was followed with severe hypoxia (8% oxygen for 60 min) or, secondly, with 30 min hypoxia in combination with infection (as mimicked by LPS stimulation), one dose of Cl-Am (30 mg/kg) straight after hypoxia, or 10 min after LPS stimulation and again straight after hypoxia, significant reduction was observed in microglial activation, histone deimination, and cell death. In all brain regions, and specifically in the cerebral cortex and hippocampus, neuronal loss (as estimated by volume measurement) was significantly reduced (Lange et al. 2014). The fact that neuroprotective effects of PAD inhibition are translatable between both CNS injury models in different species is indeed promising for effective application in other cases of neuronal damage.

13.3 Deiminated CNS Target Proteins Linked to Neurodegenerative Diseases

In acute CNS damage, several deiminated proteins were identified by immunoprecipitation with the F95 antibody from CNS damaged tissue, supporting putative roles in neurodegeneration (Lange et al. 2011, 2014; Ferretti et al. 2014). This included histones and cytoskeletal proteins such as actin, tubulins, and microtubule-associated proteins. Additional deiminated proteins identified included prohibitin-2, which controls cell proliferation, functional integrity of mitochondria, and apoptosis (Xu et al. 2014); noelins that maintain undifferentiated neural precursors (Moreno and Bronner-Fraser 2005); syntaxin, which is involved in synaptic exocytosis as it binds to amyloid-β oligomers and impairs SNARE-mediated exocytosis

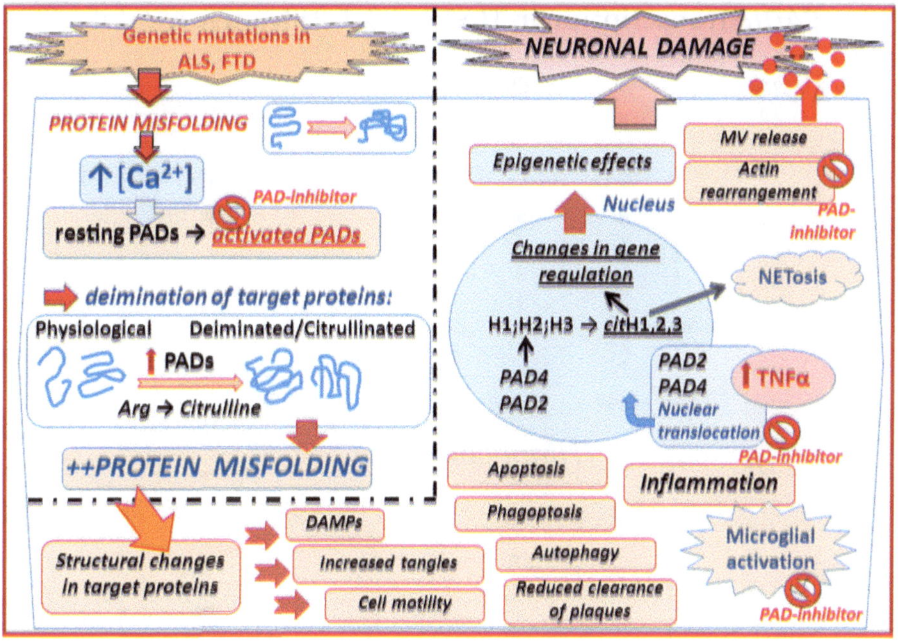

Fig. 13.1 Schematic view of peptidylarginine deiminases (PADs) in disease progression of protein misfolding disorders. Various genetic mutations, underlying neurodegenerative disorders such as Alzheimer's and Parkinson's disease, cause protein misfolding and protein aggregation, leading to impaired calcium buffering and elevated calcium levels. Calcium-activated peptidylarginine deiminases (PADs) catalyze the conversion of protein-bound arginines into citrullines, resulting in structural and functional changes in target proteins, random protein deimination, and an increased frequency of protein misfolding. In addition to PAD activation, there is also a significant increase in PAD transcription with disease progression in some mutations (see Fig. 13.2), which further contributes to protein deimination and accumulative protein misfolding in the course of disease progression. Target proteins of deimination with putative roles in neurodegenerative diseases include tubulins, microtubule-associated proteins, and proteins involved in the maintenance of neural precursors, cell proliferation, apoptosis, functional integrity of mitochondria, and perivascular drainage of amyloid-β (Lange et al. 2011; Moreno and Bronner-Fraser 2005; Durdiaková et al. 2014; Hawkes et al. 2013; Darbro et al. 2013). Inflammatory responses are caused by deiminated proteins and cellular damage. This process incudes elevated levels of tumor necrosis factor-α (TNF-α), which leads to nuclear translocation of PADs (Mastronardi et al. 2006), where they cause deimination of histones, affecting gene regulation (Wang et al. 2004). Leakage of deiminated proteins from dying cells also leads to further autoimmune responses, due to presentation of modified self-proteins, which contribute to the progressive, chronic pathology. Protein deimination is an irreversible mechanism, but targeting PAD enzyme activation using PAD inhibitors may reduce the accumulative chronic effect that is established during disease progression. PAD-inhibiting drugs pose a promising strategy to reduce pathology during neurodegenerative disease progression

(Yang et al. 2015) and has recently been linked to Asperger's syndrome (Durdiaková et al. 2014); nidogen, which is a vascular membrane protein that is associated with the loss of perivascular drainage of amyloid-β from the brain (Hawkes et al. 2013) and has also been associated with neurodegenerative Dandy-Walker syndrome

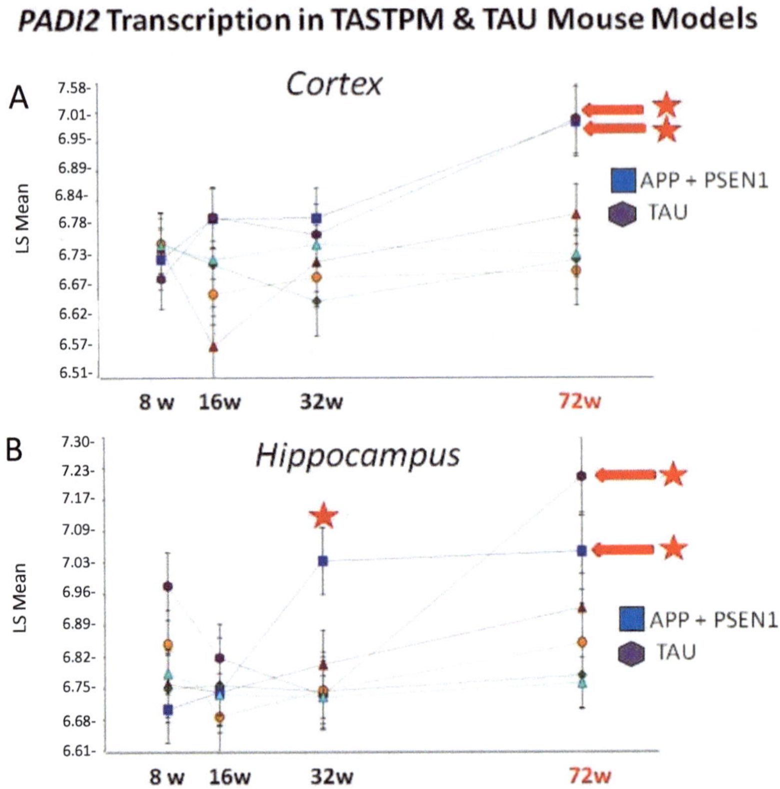

Fig. 13.2 *PADI2* transcription is significantly increased in AD and PD transgenic neurodegenerative mouse models during disease progression. Whole-genome microarrays of mouse models carrying double mutations for amyloid precursor protein and presenilin (TASTPM; APP + PSEN), as well as a mouse model for mutated TAU, showed significantly increased *PADI2* isozyme transcription in the cerebral cortex and hippocampus. (**a**) *PADI2* transcription is significantly increased in the cerebral cortex at 72 weeks in both mutants (1.4- and 1.2-fold for TAU and APP + PSEN1, respectively; $p < 0.005$; $n = 3$). (**b**) In hippocampus, the double APP + PSEN1 mutants showed a significant increase in *PADI2* transcription at 32 weeks (1.23-fold, $p < 0.001$), and both mutants showed a significant increase in PAD2 transcription at 72 weeks (1.4- and 1.2-fold for TAU and APP + PSEN1, respectively; $p < 0.005$; $n = 3$) (Matarin, Edwards and Hardy—unpublished data). This indicates that, in addition to the activation of PAD enzymes, *PADI* transcription is also increased during disease progression, which can facilitate further protein deimination (*star* indicates $p < 0.005$)

(Darbro et al. 2013); and glutamine synthetase, which is involved in glutamate homeostasis in astrocytes (Bellaver et al. 2016).

13.4 Evidence for Protein Deimination in Neurodegenerative Diseases

In cases where protein deimination has been associated with neurodegenerative diseases, including multiple sclerosis (MS) (Moscarello et al. 1994; Musse et al. 2008; Wood et al. 2008), AD, and PD, studies have mainly focused on histological analysis of postmortem human samples. AD postmortem human brain samples display increased protein deimination (Nicholas et al. 2003, 2004, 2005; Ishigami et al. 2005; Nicholas 2013; Ishigami et al. 2014), and deiminated proteins are present in amyloid-containing areas in amyloid precursor protein/presenilin1 (APP + PSEN1) transgenic AD mouse models (Borchelt et al. 1996, Nicholas et al. 2014). In AD patients, β-amyloid has been shown to be deiminated (Nicholas 2011; Nicholas et al. 2014). In hippocampal lysates from AD patients, glial fibrillary acidic protein (GFAP), an astrocyte-specific marker protein, and vimentin were identified as deiminated proteins, and the deimination of GFAP was shown to be PAD2 specific (Ishigami et al. 2015). In vitro studies demonstrated that amyloid peptides bind to PAD2, resulting in catalytic fibrillogenesis and formation of insoluble fibril aggregates (Mohlake and Whiteley 2010). In PD brain samples, increased levels of total protein deimination and deimination-positive extracellular plaques were observed (Nicholas 2011). Mutated α-synuclein protein, resulting in misfolding, has been related to increased protein deimination (Nicholas et al. 2014). Amyotrophic lateral sclerosis (ALS) spinal cords show increase in deiminated proteins (Nicholas et al. 2014), while CJD brain samples indicate roles for deiminated enolase (Jang et al. 2012). In AD brains, PTCD2 protein, an antigen target of an AD diagnostic autoantibody, is present in a deiminated form (Acharya et al. 2012). There are thus indications that disease-associated autoantibodies are generated due to the production and release of deiminated proteins and deiminated protein fragments, which may be released from damaged cells in regions of pathology (Nagele et al. 2011; Acharya et al. 2012). In AD cortex and hippocampus, both PAD2 and PAD4 were shown to be expressed in the cerebral cortex and hippocampus, the brain regions most vulnerable to AD pathology, with PAD2 localized in activated astrocytes and PAD4 selectively expressed in neurons. PAD4 expression was shown to co-localize with amyloid-β-42 in pyramidal neurons in the cerebral cortex and in hippocampal large hilar neurons of the hippocampus, which were also surrounded by activated astrocytes and microglia (Acharya et al. 2012). These neurons contained cytoplasmic accumulations of deiminated proteins. Also, the release of deiminated proteins from necrotic neurons was thought to cause an increased exposure of deiminated neuronal proteins to the immune system. In addition, the continual return of cerebrospinal fluid to the circulation via the arachnoid villi, containing modified citrullinated proteins and protein fragments, was suggested to be a key step in the ongoing pathology due to generation of autoantibodies (Acharya et al. 2012).

PADs are thus expressed in neurons residing in brain regions that are engaged in neurodegenerative pathological changes and inflammatory changes such as reactive astrogliosis and microglial migration and invasion. Although some target proteins have been described, most deiminated proteins remain to be identified. Overall, a role for PADs in neurodegenerative and inflammatory changes is supported by these studies.

13.5 Increased *PADI2* Expression with Neurodegenerative Disease Progression in Mouse Models

We analyzed whole-genome microarrays from mouse models carrying TAU and APP + PSEN1 mutations for changes in PAD expression with disease progression. In whole-genome microarrays of mouse models carrying double mutations for APP and PSEN1, and mutated TAU, *PADI2* transcription was significantly increased in the cortex at 72 weeks in both mutants (1.4- and 1.2-fold for TAU and APP + PSEN1, respectively; $p < 0.005$; $n = 3$; Fig. 13.2a). In the hippocampus, the double APP + PSEN1 mutants showed a significant increase in *PADI2* transcription at 32 weeks (1.23-fold, $p < 0.001$), and both mutants showed a significant increase in *PADI2* transcription at 72 weeks (1.4- and 1.2-fold for TAU and APP + PSEN1, respectively; $p < 0.005$; $n = 3$; Fig. 13.2b) (Matarin, Edwards and Hardy—unpublished data; Fig. 13.2).

13.6 Deiminated Proteins in Human iPSC Neuronal Models from Patient Fibroblasts

We have carried out pilot studies to test whether in vitro human iPSC neuronal models derived from patient fibroblasts are a feasible tool to model protein deimination in neurodegenerative diseases and if they may be useful for future screening of effective PAD inhibitors. Using iPSC-derived neural models alongside postmortem human brain tissue will be useful to validate molecular mechanisms that have previously been identified in postmortem human samples and rodent disease models. Such a combined approach will strengthen the relevance of previous findings to humans and may also allow drug target mechanisms to be validated in human-derived in vitro models, with hopes to increase the success rate in clinical research.

Human neuronal cell models were derived from iPSCs, generated from fibroblasts of patients carrying neurodegenerative disease mutations (Fig. 13.3) (Devine et al. 2011; Shi et al. 2012; Wray et al. 2012). Human iPSC-derived neuronal cells generated from fibroblasts of patients carrying mutations for FTD, ALS, PD, and control samples were differentiated into cortical neurons (deep- and upper-layer excitatory neurons) that in vitro were electrically active, formed functional cortical circuits, and displayed a similar pathology as in the intact brain (Shi et al. 2012). The neuronal cell lysates were analyzed by immunoblotting for PAD isozyme expression (PAD2, PAD3, PAD4, Chemicon) and for pan-protein deimination using

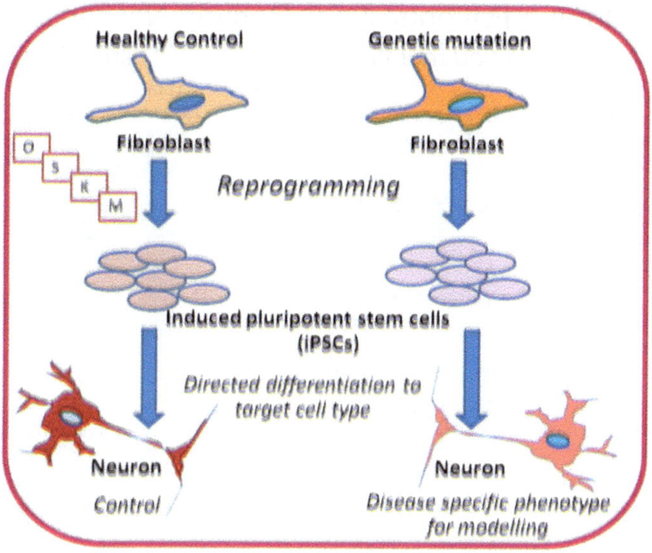

Fig. 13.3 Induced pluripotent stem cell (iPSC) models from human patients for modeling disease in vitro. Fibroblasts are obtained from the skin of patients carrying mutations for neurodegenerative diseases and healthy control individuals. These cells can be reprogrammed with specific transcription factors (OSKM: OCT4, SOX2, KLF4, and MYC) to obtain induced pluripotent stem cells that can then be directed to differentiate into the desired target cell type under study. Using this technique, the disease-specific phenotype can be generated for research into the pathological mechanisms in human models in vitro. These cultures can also be used for screening of new drug compounds (modified from Cherry and Daley 2012)

the pan-citrulline F95 antibody (Nicholas and Whitaker 2002), according to protocol by Lange et al. (2011). In addition, changes in histone 3 deimination were tested for a putative impact on gene regulation, using the citH3 antibody (Abcam ab5103). Analysis of these cell lysates from human iPSC-derived neurons (Fig. 13.4) with mutations linked to FTD and ALS (Wray et al. 2012), as well as PD (Devine et al. 2011), showed increased pan-protein deimination (Fig. 13.4a, d; F95 immunodetection) and histone H3 deimination (Fig. 13.4c, g), compared to control cell lysates. The PAD4 isozyme was found to be predominantly expressed, while PAD2 was also detected, albeit at lower levels (Fig. 13.4b, e, f).

13.7 Neurodegenerative iPSC Models for Drug-Directed PAD Inhibition

Human iPSC cell models can be used to dissect the disease mechanisms in the desired target cell types, in this case neurons, and new drug compounds can then be screened. The advantage of iPSC modeling is that pathologies that take years to accumulate in the brain can be reproduced in a relatively short time (Cherry and

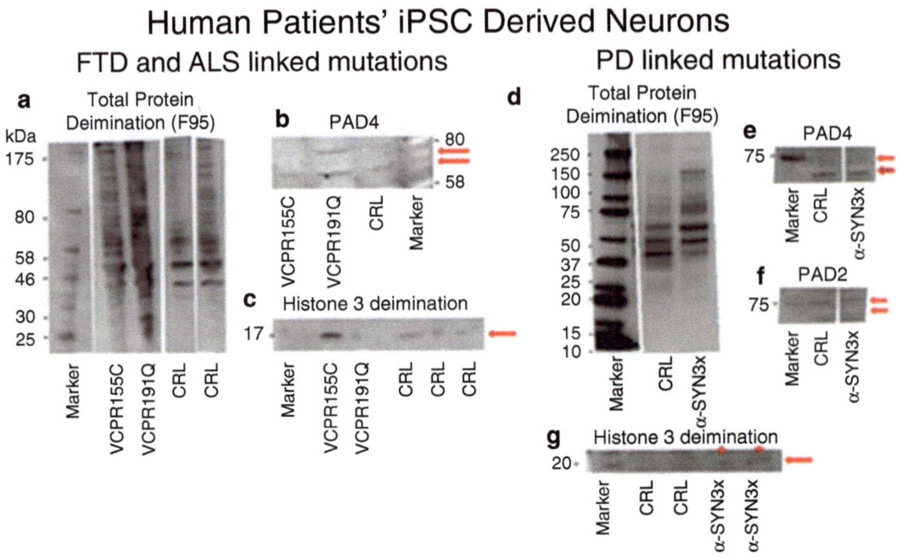

Fig. 13.4 Protein deimination is increased in iPSC-derived neurons from patients carrying mutations for frontotemporal dementia (FTD), amyotrophic lateral sclerosis (ALS), and Parkinson's disease (PD). We have shown that human iPSC-derived neuronal in vitro models do indeed mirror increased protein deimination as observed in postmortem human samples of several neurodegenerative diseases (AD, ALS, PD). Using iPSCs can provide a valuable human cell modeling tool for effects of protein deimination and PAD-inhibiting drugs in these neurodegenerative diseases. (**a**) The iPSC-derived neuronal cell lysates from patients carrying mutations VCPR155C and VCPR191Q for frontotemporal dementia (FTD) and amyotrophic lateral sclerosis (ALS) showed an increase in total protein deimination profiles as identified by immunoblotting using the pan-citrulline F95 antibody (Nicholas et al. 2014), compared to control (CRL) neuronal lysates. (**b**) PAD4 was the main isozyme expressed in the FTD and ALS iPSC-derived neurons, both in mutated and control cell lines. A double band was observed in the expected size range of 74 kDa (*arrows*). (**c**) Increased histone 3 deimination was observed in neurons derived from iPSC cell lines carrying mutation VCPR155C, while histone 3 deimination levels in VCPR191Q-derived neurons were similar to those seen in control (CRL) neuronal lysates. (**d**) The iPSC-derived neuronal cell lysates from patients carrying a triplication of α-synuclein (α-SYN3x; Devine et al. 2011) showed increased and changed pan-protein deimination profiles, compared to control (CRL) cell lysates (NAS, normal alpha-synuclein; Devine et al., 2011). (**e**) PAD4 was the main isozyme (arrows) expressed in the α-SYN3x iPSC-derived neurons, both in mutated and control (CRL) cell lines, while PAD2 isozyme was expressed (*arrows*) at lower levels (**f**). (**g**) Some histone 3 (*arrow*) deimination was observed in neurons derived from iPSC cell lines carrying α-SYN3x (*stars*), while histone 3 deimination was not detectable in the control (CRL) neuronal lysates

Daley 2012). Also, new drugs can be tested in human rather than animal models and may thus more accurately predict the efficiency of new therapeutics. Our findings are very promising, as we show that iPSC-derived neuronal models mirror findings of increased protein deimination observed in postmortem studies where protein deimination has been associated with various protein misfolding disorders including AD

(Ishigami et al. 2005; Ishigami et al. 2014; Nicholas 2011; Nicholas et al. 2014), PD (Nicholas 2011; Nicholas et al. 2014), and ALS (Nicholas et al. 2014). In addition, as protein deimination is also linked to CJD, Alexander disease (see Chap. 20), and prion diseases, it will be useful to compare PAD isozyme expression and deiminated proteins in iPSC-derived neurons from these different diseases. Interestingly, deimination of cytoskeletal proteins was both linked to acute CNS damage and was also shown to be critical to the release mechanism of cellular microvesicles (MVs) (Kholia et al. 2015). MVs are emerging as a very relevant topic in neurodegenerative conditions and are increasingly being associated with neurodegenerative disease progression (Tomlinson et al. 2015; Agosta et al. 2014; Burgos et al. 2014; Joshi et al. 2014). A novel role for PADs in the mechanism of cellular MV biogenesis is further discussed in Chap. 22.

iPSC systems are relevant human in vitro models for neurodegenerative disease progression studies, pharmaceutical screening, and understanding of how disease-related processes can be modulated to ameliorate the progression neurodegenerative mechanisms. As a functional tool, iPSC models will facilitate the examination of processes not easily observed in postmortem tissues, including changes in the processing of proteins, observation of phenotypes regarding neural connectivity, synapse formation, and the possibility of proof-of-principle experiments (Cherry and Daley 2012; Sandoe and Eggan 2013; Payne et al. 2015). Clinical trials to treat neurodegenerative diseases have a high failure rate, partly because potential drugs are derived from research in nonhuman models, while human iPSC models from patient fibroblasts may more accurately predict the efficiency of new therapeutics in those individuals.

For next step experiments of PAD inhibition in iPSC neurodegenerative disease models, the most widely studied pan-PAD inhibitor Cl-Am, which previously was shown to be effective in acute CNS damage (Lange et al. 2011, 2014) and to hinder effective microvesicular biogenesis (Kholia et al. 2015), would be the first obvious candidate to be administered to the iPSC-derived neuronal cell culture method to validate a beneficial effect of pharmacological PAD inhibition on neuronal pathology in selected mutation carrying cell lines of interest. While Cl-Am remains the most used experimental inhibitor to date, the therapeutic potential and generation of selective and isozyme-specific PAD inhibitors is receiving ever-increasing attention (Slack et al. 2011; Bozdag et al. 2013; Ferretti et al. 2013; Wei et al. 2013; Subramanian et al. 2015; Trabocchi et al. 2015). As a result, it will be of great interest for screening in iPSC-derived models of the different neurodegenerative diseases where PADs are implicated as novel key players.

PAD inhibitors offer unique interceptive drug-directed strategies to reduce neuronal pathology caused by accumulative protein misfolding downstream of calcium dysregulation, which is the common factor in many neurodegenerative disorders. PAD inhibition may also form part of a combinatory treatment strategy to maximize the benefit of treatment.

13.8 Conclusions

Our findings validate that human iPSC cultures are a novel feasible tool for modeling protein deimination in neurodegenerative protein misfolding disorders. Clinical trials to treat neurodegenerative diseases have a high failure rate partly because potential drugs are derived from research in nonhuman models, while iPSC models from patient skin cells may more accurately predict the efficiency of new therapeutics. Human patient-derived iPSC models will be useful for testing drug-directed inhibition of protein deimination for their ability to ameliorate neurodegenerative disease progression. Novel PAD-inhibiting drugs may offer a new interceptive route to reduce disease progression at the stage downstream of calcium elevation, which is a common feature of protein misfolding disorders and, via the PAD pathway, leads to further downstream protein misfolding and neurotoxic events. PAD inhibitors may also function as part of a combinatory treatment option to slow down neurodegenerative disease progression, for which there are currently few alternative treatments.

References

Acharya, N. K., Nagele, E. P., Han, M., et al. (2012). Neuronal PAD4 expression and protein citrullination: Possible role in production of autoantibodies associated with neurodegenerative disease. *Journal of Autoimmunity, 38*(4), 369–380.

Agosta, F., Dalla Libera, D., et al. (2014). Myeloid microvesicles in cerebrospinal fluid are associated with myelin damage and neuronal loss in mild cognitive impairment and Alzheimer disease. *Annals of Neurology, 76*(6), 813–825.

Beal, M. F. (2005). Mitochondria take center stage in aging and neurodegeneration. *Annals of Neurology, 58*(4), 495–505.

Bellaver, B., Souza, D. G., Souza, D. O., & Quincozes-Santos, A. (2016). Hippocampal astrocyte cultures from adult and aged rats reproduce changes in glial functionality observed in the aging brain. *Molecular Neurobiology*. doi:10.1007/s12035-016-9880-8.

Borchelt, D. R., Thinakaran, G., Eckman, C. B., et al. (1996). Familial Alzheimer's disease-linked presenilin 1 variants elevate Aβ1-42/1-40 ratio in vitro and in vivo. *Neuron, 17*, 1005–1013.

Bozdag, M., Dreker, T., Henry, C., et al. (2013). Novel small molecule protein arginine deiminase 4 (PAD4) inhibitors. *Bioorganic & Medicinal Chemistry Letters, 23*(3), 715–719.

Bradford, C. M., Ramos, I., Cross, A. K., et al. (2014). Localisation of citrullinated proteins in normal appearing white matter and lesions in the central nervous system in multiple sclerosis. *Journal of Neuroimmunology, 273*(1–2), 85–95.

Burgos, K., Malenica, I., Metpally, R., et al. (2014). Profiles of extracellular miRNA in cerebrospinal fluid and serum from patients with Alzheimer's and Parkinson's diseases correlate with disease status and features of pathology. *PloS One, 9*(5), e94839.

Cherry, A. B., & Daley, G. Q. (2012). Reprogramming cellular identity for regenerative medicine. *Cell, 148*(6), 1110–1122.

Darbro, B. W., Mahajan, V. B., Gakhar, L., et al. (2013). Mutations in extracellular matrix genes NID1 and LAMC1 cause autosomal dominant Dandy-Walker malformation and occipital cephaloceles. *Human Mutation, 34*(8), 1075–1079.

Devine, M. J., Ryten, M., Vodicka, P., et al. (2011). Parkinson's disease induced pluripotent stem cells with triplication of the α-synuclein locus. *Nature Communications, 2*, 440.

Durdiaková, J., Warrier, V., Banerjee-Basu, S., et al. (2014). STX1A and Asperger syndrome: A replication study. *Molecular Autism, 15*(1), 14.

Ferretti, P., U, K. P., Vagaska, B., et al. (2013). Discovery of a structurally novel, drug-like and potent inhibitor of peptidylarginine deiminase. *Medicinal Chemistry Communications, 4*, 1109–1113.

Ferretti, P., Lange, S., U, K. P., & Raivich, G. (2014). Chapter 15, Deimination in the developing CNS - role in its response to traumatic and hypoxic injury. In S. Bhattacharya & A. Nicholas (Eds.), *Protein deimination in human health and disease*. New York: Springer. ISBN 978-1-4614-8316-8.

Gyorgy, B., Toth, E., Tarcsa, E., et al. (2006). Citrullination: A posttranslational modification in health and disease. *The International Journal of Biochemistry & Cell Biology, 38*, 1662–1677.

Halliday, M., & Mallucci, G. R. (2014). Targeting the unfolded protein response in neurodegeneration: A new approach to therapy. *Neuropharmacology, 76*(Pt A), 169–174.

Hampel, H., Frank, R., Broich, K., et al. (2010). Biomarkers for Alzheimer's disease: Academic, industry and regulatory perspectives. *Nature Reviews. Drug Discovery, 9*(7), 560–574.

Hawkes, C. A., Gatherer, M., Sharp, M. M., et al. (2013). Regional differences in the morphological and functional effects of aging on cerebral basement membranes and perivascular drainage of amyloid-β from the mouse brain. *Aging Cell, 12*(2), 224–236.

Ishigami, A., Ohsawa, T., Hiratsuka, M., et al. (2005). Abnormal accumulation of citrullinated proteins catalyzed by peptidylarginine deiminase in hippocampal extracts from patients with Alzheimer's disease. *Journal of Neuroscience Research, 80*, 120–128.

Ishigami, A., Choi, E.-K., Kim, S., et al. (2014). Chapter 13, Deimination in Alzheimer's disease. In S. Bhattacharya & A. Nicholas (Eds.), *Protein deimination in human health and disease*. New York: Springer. ISBN 978-1-4614-8316-8.

Ishigami, A., Masutomi, H., Handa, S., et al. (2015). Mass spectrometric identification of citrullination sites and immunohistochemical detection of citrullinated glial fibrillary acidic protein in Alzheimer's disease brains. *Journal of Neuroscience Research, 93*(11), 1664–1674.

Jang, B., Jeon, Y. C., Choi, J. K., et al. (2012). Peptidylarginine deiminase modulates the physiological roles of enolase via citrullination: Links between altered multifunction of enolase and neurodegenerative diseases. *Biochemical Journal, 445*(2), 183–192.

Joshi, P., Turola, E., Ruiz, A., et al. (2014). Microglia convert aggregated amyloid-β into neurotoxic forms through the shedding of microvesicles. *Cell Death and Differentiation, 21*(4), 582–593.

Kholia, S., Jorfi, S., Thompson, P. R., et al. (2015). A novel role for peptidylarginine deiminases in microvesicle release reveals therapeutic potential of PAD inhibition in sensitizing prostate cancer cells to chemotherapy. *Journal of Extracellular Vesicles, 4*, 26192.

Krüger, J., Hinttala, R., Majamaa, K., & Remes, A. M. (2010). Mitochondrial DNA haplogroups in early-onset Alzheimer's disease and frontotemporal lobar degeneration. *Molecular Neurodegeneration, 5*, 8. doi:10.1186/1750-1326-5-8.

Lange, S., Gögel, S., Leung, K. Y., et al. (2011). Protein deiminases: New players in the developmentally regulated loss of neural regenerative ability. *Developmental Biology, 355*, 205–214.

Lange, S., Rocha-Ferreira, E., Thei, L., et al. (2014). Peptidylarginine deiminases: Novel drug targets for prevention of neuronal damage following hypoxic ischemic insult (HI) in neonates. *Journal of Neurochemistry, 130*, 555–562.

Luo, Y., Arita, K., Bhatia, M., et al. (2006). Inhibitors and inactivators of protein arginine deiminase 4: Functional and structural characterization. *Biochemistry, 45*(39), 11727–11736.

Mastroeni, D., Grover, A., Delvaux, E., et al. (2011). Epigenetic mechanisms in Alzheimer's disease. *Neurobiology of Aging, 32*(7), 1161–1180.

Mastronardi, F. G., Wood, D. D., Mei, J., et al. (2006). Increased citrullination of histone H3 in multiple sclerosis brain and animal models of demyelination: A role for tumor necrosis factor-induced peptidylarginine deiminase 4 translocation. *The Journal of Neuroscience, 26*, 11387–11396.

Mohlake, P., & Whiteley, C. G. (2010). Arginine metabolising enzymes as therapeutic tools for Alzheimer's disease: Peptidyl arginine deiminase catalyses fibrillogenesis of beta-amyloid peptides. *Molecular Neurobiology, 41*(2–3), 149–158.

Moreno, T. A., & Bronner-Fraser, M. (2005). Noelins modulate the timing of neuronal differentiation during development. *Developmental Biology, 288*(2), 434–447.

Moscarello, M. A., Wood, D. D., Ackerley, C., & Boulias, C. (1994). Myelin in multiple sclerosis is developmentally immature. *The Journal of Clinical Investigation, 94*(1), 146–154.

Musse, A. A., Li, Z., Ackerley, C. A., et al. (2008). Peptidylarginine deiminase 2 (PAD2) overexpression in transgenic mice leads to myelin loss in the central nervous system. *Disease Models & Mechanisms, 1*(4–5), 229–240.

Nagele, R. G., Clifford, P. M., Siu, G., et al. (2011). Brain-reactive autoantibodies prevalent in human sera increase intraneuronal amyloid-beta1-42 deposition. *Journal of Alzheimer's Disease, 25*, 605e22.

Nicholas, A. P. (2011). Dual immunofluorescence study of citrullinated proteins in Parkinson diseased substantia nigra. *Neuroscience Letters, 495*, 26–29.

Nicholas, A. P. (2013). Dual immunofluorescence study of citrullinated proteins in Alzheimer diseased frontal cortex. *Neuroscience Letters, 545*, 107–111.

Nicholas, A. P., & Whitaker, J. N. (2002). Preparation of a monoclonal antibody to citrullinated epitopes: Its characterization and some applications to immunohistochemistry in human brain. *Glia, 37*, 328–336.

Nicholas, A. P., King, J. L., Sambandam, T., et al. (2003). Immunohistochemical localization of citrullinated proteins in adult rat brain. *Journal of Comparative Neurology, 459*, 251–266.

Nicholas, A. P., Sambandam, T., Echols, J. D., et al. (2004). Increased citrullinated glial fibrillary acidic protein in secondary progressive multiple sclerosis. *Journal of Comparative Neurology, 473*, 128–136.

Nicholas, A. P., Sambandam, T., Echols, J. D., et al. (2005). *Journal of Comparative Neurology, 486*, 254–266.

Nicholas, A. P., Lu, L., Heaven, M., et al. (2014). Chapter 14, ongoing studies of deimination in neurodegenerative diseases using the F95 antibody. In S. Bhattacharya & A. Nicholas (Eds.), *Protein deimination in human health and disease*. New York: Springer. ISBN 978-1-4614-8316-8.

Payne, N. L., Sylvain, A., O'Brien, C., et al. (2015). Application of human induced pluripotent stem cells for modeling and treating neurodegenerative diseases. *New Biotechnology, 32*(1), 212–228.

Poorkaj, P., Bird, T. D., Wijsman, E., et al. (1998). Tau is a candidate gene for chromosome 17 frontotemporal dementia. *Annals of Neurology, 43*, 815–825.

Sandoe, J., & Eggan, K. (2013). Opportunities and challenges of pluripotent stem cell neurodegenerative disease models. *Nature Neuroscience, 16*(7), 780–789.

Shi, Y., Kirwan, P., & Livesey, F. (2012). Directed differentiation of human pluripotent stem cells to cerebral cortex neurons and neural networks. *Nature Protocols, 7*(10), 1836–1846.

Slack, J. L., Causey, C. P., & Thompson, P. R. (2011). Protein arginine deiminase 4: A target for an epigenetic cancer therapy. *Cellular and Molecular Life Sciences, 68*(4), 709–702.

Subramanian, V., Knight, J. S., Parelkar, S., et al. (2015). Design, synthesis, and biological evaluation of tetrazole analogs of cl-amidine as protein arginine deiminase inhibitors. *Journal of Medicinal Chemistry, 58*(3), 1337–1344.

Tomlinson, P. R., Zheng, Y., Fischer, R., et al. (2015). Identification of distinct circulating exosomes in Parkinson's disease. *Annals of Clinical Translational Neurology, 2*(4), 353–361.

Trabocchi, A., Pala, N., Krimmelbein, I., et al. (2015). Peptidomimetics as protein arginine deiminase 4 (PAD4) inhibitors. *Journal of Enzyme Inhibition and Medicinal Chemistry, 30*(3), 466–471.

Vossenaar, E. R., Nijenhuis, S., Helsen, M. M., et al. (2003). Citrullination of synovial proteins in murine models of rheumatoid arthritis. *Arthritis and Rheumatism, 48*, 2489–2500.

Wang, Y., Wysocka, J., Sayegh, J., et al. (2004). Human PAD4 regulates histone arginine methylation levels via demethylimination. *Science, 306*, 279–283.
Wei, L., Wasilewski, E., Chakka, S. K., et al. (2013). Novel inhibitors of protein arginine deiminase with potential activity in multiple sclerosis animal model. *Journal of Medicinal Chemistry, 56*(4), 1715–1722.
Wood, D. D., Ackerley, C. A., Brand, B., et al. (2008). Myelin localization of peptidylarginine deiminases 2 and 4: Comparison of PAD2 and PAD4 activities. *Laboratory Investigation, 88*(4), 354–364.
Wray, S., Self, M., NINDS Parkinson's Disease iPSC Consortium, et al. (2012). Creation of an open-access, mutation-defined fibroblast resource for neurological disease research. *PloS One, 7*(8), e43099.
Xu, T., Fan, X., Tan, Y., et al. (2014). Expression of PHB2 in rat brain cortex following traumatic brain injury. *International Journal of Molecular Sciences, 15*(2), 3299–3318.
Yang, Y., Kim, J., Kim, H. Y., et al. (2015). Amyloid-β oligomers may impair SNARE-mediated exocytosis by direct binding to Syntaxin 1a. *Cell Reports, 12*(8), 1244–1251.

Chapter 14
Protein Deimination in Aging and Age-Related Diseases with Ocular Manifestations

Di Ding, Mabel Enriquez-Algeciras, Sanjoy K. Bhattacharya, and Vera L. Bonilha

14.1 Introduction

Deimination refers to the conversion of protein-bound arginines into citrulline. It has been established as a posttranslational modification due to the lack of any known tRNA carrier for citrulline, as well as the presence of deiminases that are capable of catalyzing this modification in vitro. There is no known enzyme that can revert protein-bound citrulline into arginine, rendering it a relatively long-term modification. Elevated deimination has been found in neuronal tissues in a number of neurodegenerative diseases including multiple sclerosis and glaucoma. Observations in the retina, a tissue where the retinal ganglion cell layer lacks a substantial presence of astroglial cells, demonstrated that elevated and reduced deimination occurs simultaneously in astroglial cells and neurons, respectively. Such opposite effects are expected to complicate therapeutic strategies, necessitating cell-specific delivery systems for perturbation of deiminases that catalyze deimination in neuronal tissues. In this review, we will briefly discuss the occurrence of deimination with normal aging, the importance of deimination in diseases, and the effect of deimination on mRNA transport in neuronal tissue. Elevated deimination induces proteolysis via modification of protein structures, while reduced deimination affects protein synthesis and the outgrowth of dendrites in neurons.

D. Ding • M. Enriquez-Algeciras • S.K. Bhattacharya
Bascom Palmer Eye Institute, University of Miami, Miami, FL 33136, USA

V.L. Bonilha (✉)
Department of Ophthalmic Research, Cleveland Clinic, Cleveland, OH 44195, USA
e-mail: bonilhav@ccf.org

A.P. Nicholas et al. (eds.), *Protein Deimination in Human Health and Disease*,
DOI 10.1007/978-3-319-58244-3_14

14.2 Deimination

Posttranslational modifications (PTMs) are important regulators of cellular processes whose diversity varies according to the complexity of the life form being considered. Deimination and citrullination are interchangeably used terms to designate the conversion of protein-bound arginines to citrulline. It is considered to be a PTM, since no tRNA carrier for citrulline has been identified (Fig. 14.1). This PTM has been found in most mammals, frogs, and chickens (but not in lower species) and demonstrates system-specific expression levels that are catalyzed by peptidylarginine deiminase (PAD) enzymes. Five tissue-specific PADs (*PADs*1–4 and *PAD*6) have been found in mammalian systems (Enriquez-Algeciras et al. 2013; Vossenaar et al. 2003). PAD1 and PAD3 catalyze deimination in epidermal tissues; PAD2 is the major PAD in neuronal systems such as the eye and brain, and PAD4 is commonly associated with hematopoietic lineages (Bhattacharya et al. 2006a; Senshu et al. 1992; Vossenaar et al. 2003). PAD-catalyzed deimination occurring only at protein-bound arginines is calcium-dependent and generates ammonia as a by-product (Vossenaar et al. 2003). Conversely, deimination of free arginines is carried out by nitric oxide synthase (NOS) to generate nitric oxide (NO) and free citrulline. Compared to other PTMs, deimination is rare and highly regulated. There are only limited and specific proteins that undergo deimination in different tissues. Early work identified the following proteins that undergo deimination in neuronal tissues: keratin, myelin basic protein (MBP), glial fibrillary acidic protein (GFAP), vimentin, trichohyalin, histones (H2A, H3, and H4), filaggrin, and fibrinogen. Quantitative mass spectrometry has subsequently identified additional proteins (Grant et al. 2007). We will discuss the function of several proteins, RNA export binding protein (REF), 2′,3′-cyclic-nucleotide 3′-phosphodiesterase, MBP, and myelin-associated glycoprotein (MAG), that have been shown to undergo deimination in human retinal tissue (Bhattacharya et al. 2006b) and brain tissue (Wood et al. 2008) in the following section.

Protein-bound arginines can be methylated by the arginine *N*-methyltransferase (PRMT) family of enzymes. Depending on the PRMT subclass, methylation events can result in monomethylated (MMA), symmetrically dimethylated (sDMA), or asymmetrically dimethylated (aDMA) guanidino groups. Deimination or methylation has been reported on arginine residues of several histones including H2A, H3, and H4 (Nakashima et al. 2002). Because arginine residues that have been converted to citrulline can no longer be methylated, PADs can be considered inhibitors against methylation (Cuthbert et al. 2004; Wang et al. 2004). Conversely, PADs cannot catalyze the deimination of methylated arginine (both MMA and DMA) to citrulline (Raijmakers et al. 2007). Therefore, methylation and deimination are considered to be competitive posttranslational mechanisms on certain protein-bound arginines.

Arginine and citrulline have different charges, allowing for differentiation of a multiple-site deiminated protein and its non-deiminated counterpart through electrophoresis. However, the detection of endogenous deimination is difficult because of the similar structure and molecular weight of citrulline and arginine as well as the limited amount of deimination present (Fig. 14.1). Early detection attempts relied upon citrulline mod-

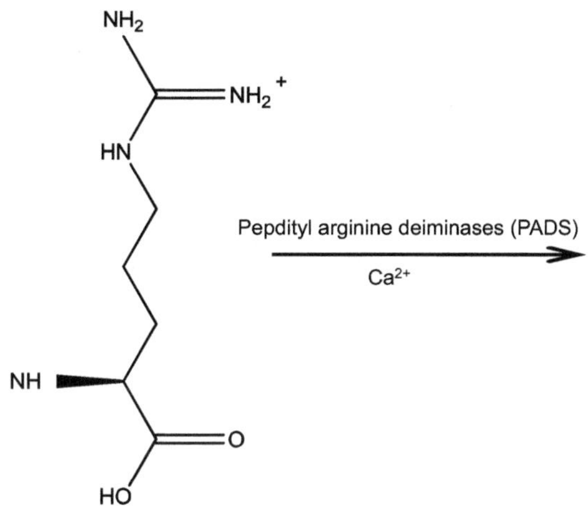

Fig. 14.1 Reaction of deimination

ification to a functionality that could be recognized by specific antibodies (Senshu et al. 1992). In this two-step reaction process, protein-bound citrulline was reacted with 2,3-butanedione monoxime and antipyrine in a strongly acidic environment to generate an antigenic adduct (Senshu et al. 1992). Antibodies such as F95 (Nicholas and Whitaker 2002) have been developed to detect protein-bound, unmodified citrulline. Presently, more antibodies are commercially available; however, the selectivity and sensitivity of these antibodies still need further characterization.

Mass spectrometry (mass spec) is the most reliable method to detect deimination thus far, and several groups have successfully applied this technique in their studies. Intact deiminated protein can be differentiated from non-deiminated protein by MALDI-TOF. However, due to the similarity of the mass-to-charge ratio between arginine and citrulline, identifying the location of deiminated residues requires high accuracy from the mass spec instrument followed by additional methods for confirmation. LTQ Orbitrap is one type of mass spec that has been previously used to detect citrulline residues. Most of the time, detection of deimination requires the combination of multiple methods for more accurate results.

14.3 Deimination in Aging and Development

Several studies have associated protein deimination with development and aging in the central nervous systems (CNS). Moscarello's group showed in 1994 that there were heavily deiminated MBPs in the infant CNS compared to the adult CNS samples (Moscarello et al. 1994). Other studies showed upregulated PAD2 expression in mice P1 compared to adults (unpublished data). An independent mouse model aging study using the F1 hybrid between Fischer 344 and Brown Norway rats

(F344BN) showed a decrease in deimination in aged rats (~24 months) when compared to young rats (~3 months) in ocular tissue (Fig. 14.2). Likewise, immunohistochemistry performed on retinal cryosections showed that the ganglion cell layer undergoes significant, age-related loss of deimination (Bhattacharya et al. 2008). Moreover, the total level of deiminated proteins in the blood serum decreased, detected by the enzyme-linked immunosorbent assay (ELISA). Both the expression level and activity of the major deiminase PAD2 decreased with aging (Bhattacharya et al. 2008), which indicated deimination plays a role in maintaining the nervous system activities during development and aging. More recently, a study reported higher amounts of deiminated proteins both in aged mice and human brains (Nicholas et al. 2014). This observation may suggest that increased citrullination may be part of the natural process of brain senescence. Alternatively, there may be tissue-specific regulation of citrullination. Future experiments studying deimination in additional aging tissue and neurological diseases are needed.

14.4 Elevated Deimination in Age-Related Diseases with Inflammation

Abnormal deimination has been detected in several age-related diseases. Multiple sclerosis (MS) is a neurodegenerative disease characterized by recurrent inflammation and the formation of localized demyelinated plaques associated with axonal degeneration in the central nervous system (Ziemssen 2005). MS has heterogeneous manifestations and is often disabling via deficits of sensory, motor, autonomic, and neurocognitive function which correlate to vision loss, extraocular muscle movement disorders, paresthesias, loss of sensation, weakness, dysarthria, spasticity, ataxia, and bladder dysfunction (Noseworthy et al. 2000). Alterations in PTMs of certain proteins may cause improper immune activation, which plays an important role in the early pathogenesis of MS (Arnon and Aharoni 2007; Ziemssen 2005). Brain tissue samples from MS patients showed hyper-deiminated regions and elevated levels of PAD2 compared to normal controls, indicating that deimination could play an important role in the development of MS (Enriquez-Algeciras et al. 2013). One protein that is heavily deiminated in MS patients is MBP (Moscarello et al. 1994), which shows enhanced susceptibility to both autocatalytic and cathepsin D-mediated proteolysis (Pritzker et al. 2000). The proteolytic products have the ability to sensitize T cells (Belogurov et al. 2008; D'Souza et al. 2005) and may also activate astrocytes that have been implicated in both innate and acquired immune responses in the CNS (Nair et al. 2008).

Age-related macular degeneration (AMD) is one of the leading causes of blindness, which is characterized by dysfunction of the retinal pigment epithelium (RPE) cells followed by the loss of photoreceptors in the macular region. Many studies suggested inflammation is implicated in the pathogenesis of AMD based on the analysis of the small yellowish deposits between the RPE layer and an adjacent basement membrane complex (Bruch's membrane) (Anderson et al. 1981; Hageman et al. 1999, 2001;

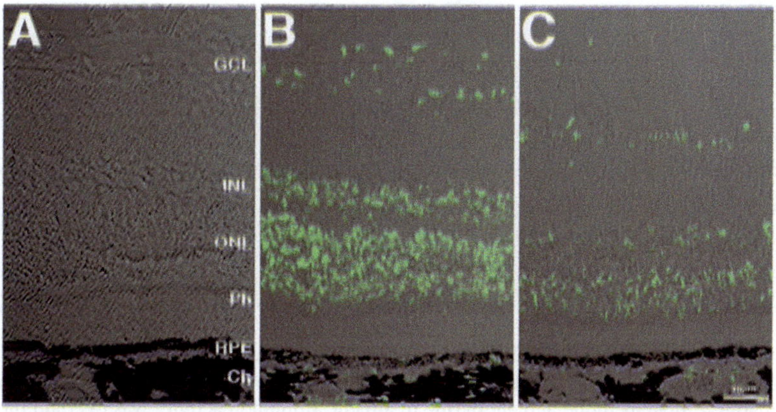

Fig. 14.2 Decreased deimination in the aging rat retina. Representative confocal images of retinal deimination on cryosections probed without (**a**) or with (**b** and **c**) modified anti-citrulline antibody (*green*) after 2,3 butanedione and antipyrine treatment. Immunoreactivity was overlaid on differential contrast images of the retina. *GCL* ganglion cell layer, *INL* inner nuclear layer, *ONL* outer nuclear layer, *Ph* photoreceptor, *RPE* retinal pigment epithelium, *Ch* choroid. Panels corresponding to 3 months' (**a** and **b**) and 24 (**c**) months' retina, respectively, suggest that deimination in the retina decreases with age. Bar = 40 μm (modified from Bhattacharya et al. (2008))

Heriot et al. 1984; Penfold et al. 1984). Deimination was detected in different layers of the retina using immunohistochemistry (IHC) in AMD, and detection of PAD2 in the retinas of human donors with AMD also revealed great cadaveric variability (Bonilha et al. 2013). Total protein deimination from retinal and RPE lysates were elevated compared to the controls, but PAD2 levels were not elevated, indicating the localized regulation of deimination levels (Figs. 14.3 and 14.4).

Elevated levels of PAD2 and protein deimination have been found in other adult-onset diseases involving inflammation at different levels, such as rheumatoid arthritis (Scofield 2004), and in several human neurological diseases including autoimmune encephalomyelitis (Nicholas et al. 2005), Alzheimer's disease (Maruyama et al. 2005; Louw et al. 2007), amyotrophic lateral sclerosis (Chou et al. 1996), and glaucoma (Bhattacharya et al. 2006a, b). Compared to decreased PAD2 expression in aging studies with F344BN rats (Bhattacharya et al. 2008), late-onset elevation in the levels of PAD2 and deiminated proteins are likely due to the pathological process of disease rather than age-associated changes.

14.5 Decreased Deimination in Age-Related Neurodegenerative Diseases

Deimination has been reported in nervous tissue cells such as astrocytes (Bhattacharya et al. 2006a), microglia, and oligodendrocytes (Senshu et al. 1992), Schwann cells (Keilhoff et al. 2008), and neurons (Enriquez-Algeciras et al. 2013).

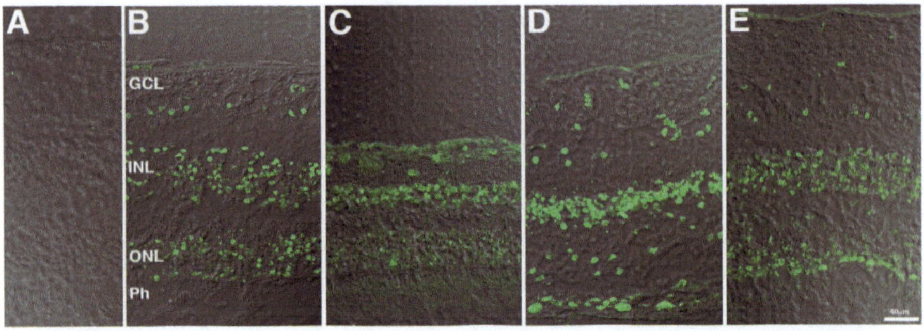

Fig. 14.3 Representative deimination in human retina. Representative confocal images of retinal deimination on cryosections probed without (**a**) or with (**b**–**e**) modified anti-citrulline antibody (*green*) after 2,3 butanedione and antipyrine treatment. Immunoreactivity was overlaid on differential contrast images of the retina. *GCL* ganglion cell layer, *INL* inner nuclear layer, *ONL* outer nuclear layer, *Ph* photoreceptor. Panels corresponding to non-age-related macular degeneration (AMD; **b**) and AMD donor retinas (**c**–**e**) demonstrated similar levels of deiminated proteins observed in both non-AMD and AMD retinas. Non-AMD retinas displayed protein deimination in the GCL, INL, and ONL. Deimination in both retinas was typically localized to the nuclei of cells. Finally, a disorganized distribution of deiminated proteins was visible in the degenerated areas of the retinas of AMD donors (*arrowheads*). Bar = 40 μm (modified from Bhattacharya et al. (2008))

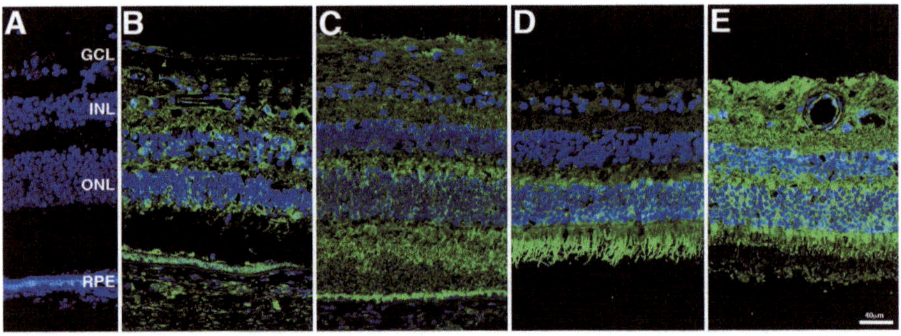

Fig. 14.4 Representative peptidylarginine deiminase type 2 (PAD2) in human retina. Representative confocal images of retinas probed without (**a**) or with (**b**–**e**) modified anti-PAD2 antibodies (*green*) and nuclei labeled by DAPI. *GCL* ganglion cell layer, *INL* inner nuclear layer, *ONL* outer nuclear layer, *RPE* retinal pigment epithelium. Panels corresponding to non-AMD (**b**) and AMD donor retinas (**c**–**e**) demonstrated similar levels of PAD2 observed in both cytoplasm and nuclei of non-AMD and AMD retinas. Bar = 40 μm (modified from Bhattacharya et al. (2008))

In comparison to the deimination levels between normal and MS donor retina sections, hypo-deimination is more discernible in the retinal ganglion cell (RGC) layer due to the lack of astroglial cells. The similar loss of deimination was observed in the RGC layer from a mouse model (ND4) of demyelinating disease (Enriquez-Algeciras et al. 2013). This is in contrast to previously known hyper-deimination in the brains of ND4 mice and deceased human MS patients (Mastronardi et al. 2007; Moscarello et al. 1994). Distinct from other tissues in the nervous system, the structure of the retina has cells arranged in different layers according to their functions. The retinal ganglion cell layer lacks a substantial presence of astroglial cells, allowing for better cell-specific measurements of aberrant deimination without the averaging effect of heterogeneous tissue. In addition to hypo-deimination in RGC layer, hyper-deimination of the inner nuclear layer and the outer nuclear layer compared to controls was also observed at the same time. Thus, elevated PAD2 and concurrently increased deimination in a large number of astrocytes render detection of neural hypo-deimination difficult in the brain.

14.6 Cell Type-Specific Deimination

Hyper- and hypo-deimination occur simultaneously in degenerating retinal tissues. Isolation of different cell types from ND4 mice confirms that deimination is decreased in RGCs and increased in astrocytes. In addition, studies of brain sections also showed hypo-deimination in neurons and hyper-deimination in astrocytes, indicating that modulation of deimination is cell type-specific (Enriquez-Algeciras et al. 2013). However, the physiological conditions that trigger PAD2 expression are still unknown. It has been shown that RGCs and astrocytes show opposite PAD2 expression patterns in response to pressure; PAD2 is elevated in response to pressure in astrocytes but is downregulated in RGCs (Bhattacharya 2009).

14.7 Consequences of Aberrant Deimination

Deimination, as a less abundant PTM, is strictly regulated. There are few identified proteins that undergo deimination, and they are highly cell type specific. Aberrant expression of deiminases would lead to complicated consequences. Previous work with chick embryos demonstrated elevated deimination in general may contribute to the loss of nerve regeneration during early development (Lange et al. 2011). Conversely, evidence from infant studies indicated that elevated deimination is important for early neuron development (Moscarello et al. 1994). Excess citrullination has been found in neurodegenerative diseases. Studies have demonstrated that MBP and MOG are hyper-deiminated in multiple sclerosis patients (Mastronardi et al. 1993; Moscarello et al. 1994). Hyper-deiminated MBP has been shown to have a loose structure which is more susceptible to proteolysis (Musse et al. 2006). Degraded MBP is antigenic and can lead to autoimmune pathogenesis. More recent studies are more closely investigating deimination based on cell types. In neuronal cells, a group of proteins including RNA-binding factor (REF) have been identified as targets of the PAD2 enzyme. Deiminated REF has a stronger affinity for RNA, which in turn facilitates protein translation in neuronal dendrites (Ding et al. 2012; Enriquez-Algeciras et al. 2013). Isolated neurons from ND4 mice with PAD2 over-expression showed elevated neurite outgrowth compared to controls without PAD2 overexpression. These ND4 mice also demonstrated visual function recovery compared to their littermates (Enriquez-Algeciras et al. 2013). Elevated deimination of histone protein in granulocytes contributes to neutrophil extracellular trap (NET) formation (Rohrbach et al. 2012). Other studies in astroglial cells also indicated that increased deimination correlates to astrocyte activation (Algeciras et al. 2008). Deimination was also investigated in retinas of human donors with AMD. Immunohistology detection of citrullination revealed great cadaveric variability in the immunoreaction of the retina (Fig. 14.3). However, we have previously detected elevated levels of deimination in retinal and RPE lysates from AMD donors (Bonilha 2008). A similar observation was detected in lysates from donors with glaucoma or MS (Enriquez-Algeciras et al. 2013).

Reduced deimination in different cell types results in divergent phenotypes. Studies have shown that reduced PAD2 expression is associated with neuronal degeneration (Ding et al. 2012; Enriquez-Algeciras et al. 2013). Several proteins such as REF have a significant loss of deimination in degenerating neurons compared to the controls and, as a consequence, lead to deficient mRNA transport. A subset of these mRNAs are related to synaptic vesicle formation and mitochondria function, and a decrease in deimination would directly interfere with dendritic protein synthesis. However, the consequences of reduced deimination in glial cells are still under investigation. In summary, deimination is highly regulated, yet global up- and downregulation of deimination may show conflicting results.

Neurons, as a specialized cell type, have multiple extensions from the soma. A neuron needs to constantly receive environmental signals such as insulin from the endocrine system to maintain its function. Efficient transport of cargo such as glucose receptor mRNA from the soma to dendritic sites is critical for responding to these signals. Posttranslational modification plays an important regulatory role for transport factors, as aberrant or absent modifications will exert a critical influence.

14.8 Concluding Remarks

In summary, PAD enzymes are responsible for carrying out deimination. These enzymes are tissue specific, and their expression levels are associated with development, aging, and neurodegenerative diseases. The consequences of deimination vary based on the specific target of the deimination and the specific cell type being considered. Interestingly, no enzymes that convert citrulline to arginine have been found to date, and deiminated proteins accumulate in cells until they are degraded. As a result, the long turnover rate of deiminated proteins infers that deimination can be considered a long-term PTM.

How is deimination regulated? There are many possibilities to account for that may contribute to deimination regulation. Local calcium concentration is a direct controlling factor for PAD activity. PAD enzyme expression may be induced by different stimuli from the environment, such as hormones (Takahara et al. 1992), pressures (Algeciras et al. 2008), or bacterial infection (Rohrbach et al. 2012). At the same time, deiminated proteins are regulated by degradation systems. Inefficient protease activity results in the accumulation of deimination. All these regulations must be kept in equilibrium with one another. Any interruption at any stage can lead to aberrant deimination; however the exact mechanism remains unclear. Understanding the homeostasis of deimination under normal conditions is of key importance in uncovering the exact mechanism of action.

References

Algeciras, M. E., Takahara, H., & Bhattacharya, S. K. (2008). Mechanical stretching elevates peptidyl arginine deiminase 2 expression in astrocytes. *Current Eye Research, 33*, 994–1001.

Anderson, S., Bankier, A. T., Barrell, B. G., de Bruijn, M. H., Coulson, A. R., Drouin, J., Eperon, I. C., Nierlich, D. P., Roe, B. A., Sanger, F., Schreier, P. H., Smith, A. J., Staden, R., & Young, I. G. (1981). Sequence and organization of the human mitochondrial genome. *Nature, 290*, 457–465.

Arnon, R., & Aharoni, R. (2007). Neurogenesis and neuroprotection in the CNS--fundamental elements in the effect of Glatiramer acetate on treatment of autoimmune neurological disorders. *Molecular Neurobiology, 36*, 245–253.

Belogurov, A. A., Jr., Kurkova, I. N., Friboulet, A., Thomas, D., Misikov, V. K., Zakharova, M. Y., Suchkov, S. V., Kotov, S. V., Alehin, A. I., Avalle, B., Souslova, E. A., Morse, H. C., III, Gabibov, A. G., & Ponomarenko, N. A. (2008). Recognition and degradation of myelin basic protein peptides by serum autoantibodies: Novel biomarker for multiple sclerosis. *Journal of Immunology, 180*, 1258–1267.

Bhattacharya, S. K. (2009). Retinal deimination in aging and disease. *IUBMB Life, 61*, 504–509.

Bhattacharya, S. K., Crabb, J. S., Bonilha, V. L., Gu, X., Takahara, H., & Crabb, J. W. (2006a). Proteomics implicates peptidyl arginine deiminase 2 and optic nerve citrullination in glaucoma pathogenesis. *Investigative Ophthalmology & Visual Science, 47*, 2508–2514.

Bhattacharya, S. K., Crabb, J. S., Bonilha, V. L., Gu, X., Takahara, H., & Crabb, J. W. (2006b). Proteomics implicates peptidyl arginine deiminase 2 and optic nerve citrullination in glaucoma pathogenesis. *Investigative Ophthalmology & Visual Science, 47*, 2508–2514.

Bhattacharya, S. K., Sinicrope, B., Rayborn, M. E., Hollyfield, J. G., & Bonilha, V. L. (2008). Age-related reduction in retinal deimination levels in the F344BN rat. *Aging Cell, 7*, 441–444.

Bonilha, V. L. (2008). Age and disease-related structural changes in the retinal pigment epithelium. *Clinical Ophthalmology, 2*, 413–424.

Bonilha, V. L., Shadrach, K. G., Rayborn, M. E., Li, Y., Pauer, G. J., Hagstrom, S. A., Bhattacharya, S. K., & Hollyfield, J. G. (2013). Retinal deimination and PAD2 levels in retinas from donors with age-related macular degeneration (AMD). *Experimental Eye Research, 111*, 71–78.

Chou, S. M., Wang, H. S., & Komai, K. (1996). Colocalization of NOS and SOD1 in neurofilament accumulation within motor neurons of amyotrophic lateral sclerosis: An immunohistochemical study. *Journal of Chemical Neuroanatomy, 10*, 249–258.

Cuthbert, G. L., Daujat, S., Snowden, A. W., Erdjument-Bromage, H., Hagiwara, T., Yamada, M., Schneider, R., Gregory, P. D., Tempst, P., Bannister, A. J., & Kouzarides, T. (2004). Histone deimination antagonizes arginine methylation. *Cell, 118*, 545–553.

Ding, D., Enriquez-Algeciras, M., Dave, K. R., Perez-Pinzon, M., & Bhattacharya, S. K. (2012). The role of deimination in ATP5b mRNA transport in a transgenic mouse model of multiple sclerosis. *EMBO Reports, 13*, 230–236.

D'Souza, C. A., Wood, D. D., She, Y. M., & Moscarello, M. A. (2005). Autocatalytic cleavage of myelin basic protein: An alternative to molecular mimicry. *Biochemistry, 44*, 12905–12913.

Enriquez-Algeciras, M., Ding, D., Mastronardi, F. G., Marc, R. E., Porciatti, V., & Bhattacharya, S. K. (2013). Deimination restores inner retinal visual function in murine demyelinating disease. *The Journal of Clinical Investigation, 123*, 646–656.
Grant, J. E., Hu, J., Liu, T., Jain, M. R., Elkabes, S., & Li, H. (2007). Post-translational modifications in the rat lumbar spinal cord in experimental autoimmune encephalomyelitis. *Journal of Proteome Research, 6*, 2786–2791.
Hageman, G. S., Mullins, R. F., Russell, S. R., Johnson, L. V., & Anderson, D. H. (1999). Vitronectin is a constituent of ocular drusen and the vitronectin gene is expressed in human retinal pigmented epithelial cells. *FASEB Journal, 13*, 477–484.
Hageman, G. S., Luthert, P. J., Victor Chong, N. H., Johnson, L. V., Anderson, D. H., & Mullins, R. F. (2001). An integrated hypothesis that considers drusen as biomarkers of immune-mediated processes at the RPE-Bruch's membrane interface in aging and age-related macular degeneration. *Progress in Retinal and Eye Research, 20*, 705–732.
Heriot, W. J., Henkind, P., Bellhorn, R. W., & Burns, M. S. (1984). Choroidal neovascularization can digest Bruch's membrane. A prior break is not essential. *Ophthalmology, 91*, 1603–1608.
Keilhoff, G., Prell, T., Langnaese, K., Mawrin, C., Simon, M., Fansa, H., & Nicholas, A. P. (2008). Expression pattern of peptidylarginine deiminase in rat and human Schwann cells. *Developmental Neurobiology, 68*, 101–114.
Lange, S., Gogel, S., Leung, K. Y., Vernay, B., Nicholas, A. P., Causey, C. P., Thompson, P. R., Greene, N. D., & Ferretti, P. (2011). Protein deiminases: New players in the developmentally regulated loss of neural regenerative ability. *Developmental Biology, 355*, 205–214.
Louw, C., Gordon, A., Johnston, N., Mollatt, C., Bradley, G., & Whiteley, C. G. (2007). Arginine deiminases: Therapeutic tools in the etiology and pathogenesis of Alzheimer's disease. *Journal of Enzyme Inhibition and Medicinal Chemistry, 22*, 121–126.
Maruyama, M., Higuchi, M., Takaki, Y., Matsuba, Y., Tanji, H., Nemoto, M., Tomita, N., Matsui, T., Iwata, N., Mizukami, H., Muramatsu, S., Ozawa, K., Saido, T. C., Arai, H., & Sasaki, H. (2005). Cerebrospinal fluid neprilysin is reduced in prodromal Alzheimer's disease. *Annals of Neurology, 57*, 832–842.
Mastronardi, F. G., Ackerley, C. A., Arsenault, L., Roots, B. I., & Moscarello, M. A. (1993). Demyelination in a transgenic mouse: A model for multiple sclerosis. *Journal of Neuroscience Research, 36*, 315–324.
Mastronardi, F. G., Noor, A., Wood, D. D., Paton, T., & Moscarello, M. A. (2007). Peptidyl argininedeiminase 2 CpG island in multiple sclerosis white matter is hypomethylated. *Journal of Neuroscience Research, 85*, 2006–2016.
Moscarello, M. A., Wood, D. D., Ackerley, C., & Boulias, C. (1994). Myelin in multiple sclerosis is developmentally immature. *The Journal of Clinical Investigation, 94*, 146–154.
Musse, A. A., Boggs, J. M., & Harauz, G. (2006). Deimination of membrane-bound myelin basic protein in multiple sclerosis exposes an immunodominant epitope. *Proceedings of the National Academy of Sciences of the United States of America, 103*, 4422–4427.
Nair, A., Frederick, T. J., & Miller, S. D. (2008). Astrocytes in multiple sclerosis: A product of their environment. *Cellular and Molecular Life Sciences, 65*, 2702–2720.
Nakashima, K., Hagiwara, T., & Yamada, M. (2002). Nuclear localization of peptidylarginine deiminase V and histone deimination in granulocytes. *The Journal of Biological Chemistry, 277*, 49562–49568.
Nicholas, A. P., & Whitaker, J. N. (2002). Preparation of a monoclonal antibody to citrullinated epitopes: Its characterization and some applications to immunohistochemistry in human brain. *Glia, 37*, 328–336.
Nicholas, A. P., Sambandam, T., Echols, J. D., & Barnum, S. R. (2005). Expression of citrullinated proteins in murine experimental autoimmune encephalomyelitis. *Journal of Comparative Neurology, 486*, 254–266.
Nicholas, A. P., Lu, L., Heaven, M., Kadish, I., van Groen, T., Accaviti-Loper, M. A., Wewering, S., Kofskey, D., Gambetti, P., & Brenner, M. (2014). Ongoing studies of demination in neurodegenerative diseases using the F95 antibody. In A. P. Nicholas & S. K. Bhattacharya (Eds.), *Protein deimination in human health and disease* (pp. 257–280). New York: Springer.

Noseworthy, J. H., Lucchinetti, C., Rodriguez, M., & Weinshenker, B. G. (2000). Multiple sclerosis. *The New England Journal of Medicine, 343*, 938–952.

Penfold, P., Killingsworth, M., & Sarks, S. (1984). An ultrastructural study of the role of leucocytes and fibroblasts in the breakdown of Bruch's membrane. *Australian Journal of Ophthalmology, 12*, 23–31.

Pritzker, L. B., Joshi, S., Gowan, J. J., Harauz, G., & Moscarello, M. A. (2000). Deimination of myelin basic protein. 1. Effect of deimination of arginyl residues of myelin basic protein on its structure and susceptibility to digestion by cathepsin D. *Biochemistry, 39*, 5374–5381.

Raijmakers, R., Zendman, A. J., Egberts, W. V., Vossenaar, E. R., Raats, J., Soede-Huijbregts, C., Rutjes, F. P., van Veelen, P. A., Drijfhout, J. W., & Pruijn, G. J. (2007). Methylation of arginine residues interferes with citrullination by peptidylarginine deiminases in vitro. *Journal of Molecular Biology, 367*, 1118–1129.

Rohrbach, A. S., Slade, D. J., Thompson, P. R., & Mowen, K. A. (2012). Activation of PAD4 in NET formation. *Frontiers in Immunology, 3*, 360.

Scofield, R. H. (2004). Autoantibodies as predictors of disease. *Lancet, 363*, 1544–1546.

Senshu, T., Sato, T., Inoue, T., Akiyama, K., & Asaga, H. (1992). Detection of citrulline residues in deiminated proteins on polyvinylidene difluoride membrane. *Analytical Biochemistry, 203*, 94–100.

Takahara, H., Kusubata, M., Tsuchida, M., Kohsaka, T., Tagami, S., & Sugawara, K. (1992). Expression of peptidylarginine deiminase in the uterine epithelial cells of mouse is dependent on estrogen. *The Journal of Biological Chemistry, 267*, 520–525.

Vossenaar, E. R., Zendman, A. J., van Venrooij, W. J., & Pruijn, G. J. (2003). PAD, a growing family of citrullinating enzymes: Genes, features and involvement in disease. *BioEssays, 25*, 1106–1118.

Wang, Y., Wysocka, J., Sayegh, J., Lee, Y. H., Perlin, J. R., Leonelli, L., Sonbuchner, L. S., McDonald, C. H., Cook, R. G., Dou, Y., Roeder, R. G., Clarke, S., Stallcup, M. R., Allis, C. D., & Coonrod, S. A. (2004). Human PAD4 regulates histone arginine methylation levels via demethylimination. *Science, 306*, 279–283.

Wood, D. D., Ackerley, C. A., Brand, B., Zhang, L., Raijmakers, R., Mastronardi, F. G., & Moscarello, M. A. (2008). Myelin localization of peptidylarginine deiminases 2 and 4: Comparison of PAD2 and PAD4 activities. *Laboratory Investigation, 88*, 354–364.

Ziemssen, T. (2005). Modulating processes within the central nervous system is central to therapeutic control of multiple sclerosis. *Journal of Neurology, 252*(Suppl 5), 38–45.

Chapter 15
Chemical Modification and Mass Spectrometric Approaches for Detection of Brain Protein Deimination

Mabel Enriquez-Algeciras, Di Ding, Dana P. Ascherman, and Sanjoy K. Bhattacharya

15.1 Introduction

The posttranslational modification of deimination refers to conversion of protein-bound arginine into citrulline. It is frequently detected by monoxime treatment (acid phase 2,3-butanedione and antipyrine reaction) resulting in an adduct formation. We present evidence that deiminated proteins are prone to aggregate formation upon prolonged monoxime exposure. Moreover, the efficiency of adduct formation is nonlinear and dramatically reduced with increasing polypeptide complexity [complete with smaller but progressively decreasing with higher (poly)peptides]. This nonlinearity results in vastly noncomparable detection among different methods such as immunohistochemistry, Western blot, and mass spectrometry. Mass spectrometric detection, based on mass addition of +238 on citrulline moiety with monoxime and +1 change due to deimination without monoxime treatment, corroborates serious limitations in monoxime adduct formation on proteins. Methodological limitations may also interfere with the identification of deiminated proteins as well as their modification sites and, as a consequence, the understanding of the biological role of deimination. We present here methods that alter the sequence of digestion and the combination of chromatographic techniques that collectively reduce complexity and help capture the deiminated peptides. Thin-layer chromatographic methods, together with different enzymatic digestions, can also be potentially routinely used for detection of changes in deimination sites within a given protein in different states of a cell or tissue.

M. Enriquez-Algeciras • D. Ding • S.K. Bhattacharya (✉)
Bascom Palmer Eye Institute (McKnight Bldg.), University of Miami,
1638 NW 10th Avenue, Suite 706A, Miami, FL 33136, USA
e-mail: SBhattacharya@med.miami.edu

D.P. Ascherman
Department of Medicine, University of Miami, Miami, FL 33136, USA

A.P. Nicholas et al. (eds.), *Protein Deimination in Human Health and Disease*,
DOI 10.1007/978-3-319-58244-3_15

Deimination refers to the process of modification of protein-bound arginines into citrulline (Fig. 15.1). The conversion of free arginine to citrulline is commonly referred to as citrullination (Marletta et al. 1998). In mammalian cells, peptidyl arginine deiminases (PADs) catalyze the conversion of protein-bound arginine into citrulline in the presence of calcium (Marletta et al. 1998; Stuehr 2004; Vossenaar et al. 2003). There exists no known carrier tRNAs for citrulline, and therefore occurrence of protein-bound citrulline is attributed entirely to PAD-catalyzed posttranslational modification (PTM). Out of the five known deiminases in mammals, PAD1-3 and PAD6 are cytosolic, while PAD4 is ubiquitous and localizes to the nucleus (Vossenaar et al. 2003). Deimination is a relative recently discovered and less explored PTM. In general, enzymes for catalysis of PTM occur in pairs, for example, kinase and phosphatase for phosphorylation and acetylase and deacetylase for acetylation (Bradbury 1992; Ryan and Bauer 2008; Zhang and Wang 2008); however, an enzyme for the reversal of deimination remains yet to be identified. Partly due to lack of a known enzymatic activity to reverse this modification, deimination can be regarded as a long-term PTM.

The role that deimination plays has not been subjected to such an intense investigation as many other modifications such as phosphorylation or N-linked glycosylation (Arnold et al. 2008; Macek et al. 2008; Nita-Lazar et al. 2008; Sims and Reinberg 2008). Only a handful of proteins have been identified to undergo deimination so far, and more protein substrates that undergo this modification remain to be identified (Algeciras and Bhattacharya 2007). As a result, the physiological role of deimination is not fully understood. Because conversion to citrulline alters the charge state of proteins, this PTM has functional consequences associated with alterations in protein interactions. For example, deimination has been implicated to play a role in transcriptional activation, apoptosis, and, somewhat controversially, in reversal of protein methylation. Elevated and aberrant deimination has been observed in rheumatoid arthritis, cutaneous diseases, and in a number of disorders of the nervous system that affect the brain and the eye, such as multiple sclerosis, Alzheimer's disease, and glaucoma (Chavanas et al. 2006; Ding et al. 2012; Enriquez-Algeciras et al. 2011, 2013; Gyorgy et al. 2006; Harauz and Musse 2007; Mechin et al. 2007). Also, the level of deimination and PAD2 appears to undergo a significant decrease with the process of normal aging (Bhattacharya et al. 2008).

Since the small 1 Da mass increase that occurs with deimination can be challenging to decipher from other hydrolytic PTMs, the detection of deiminated proteins by mass spectrometry (MS) and MS/MS is a difficult endeavor. Therefore, we tested a method for the detection of citrulline which was initially developed by organic chemists (Archibald 1944). First attempts were made for stable monoxime modification of urea, *N*-butyl urea, and esters of arginine (that bear molecular similarities to arginine) which enable spectrophotometric detection (Archibald 1944; Boyde and Rahmatullah 1980; Watanabe et al. 1988; Zarabian et al. 1987). These methods were extended to treatment of proteins with 2,3-butanedione and antipyrine in a strong acidic environment (Senshu et al. 1992), resulting in formation of an adduct that is recognized by an antibody which has facilitated detection by various methods including Western blot and immunohistochemical analysis. Treatment with

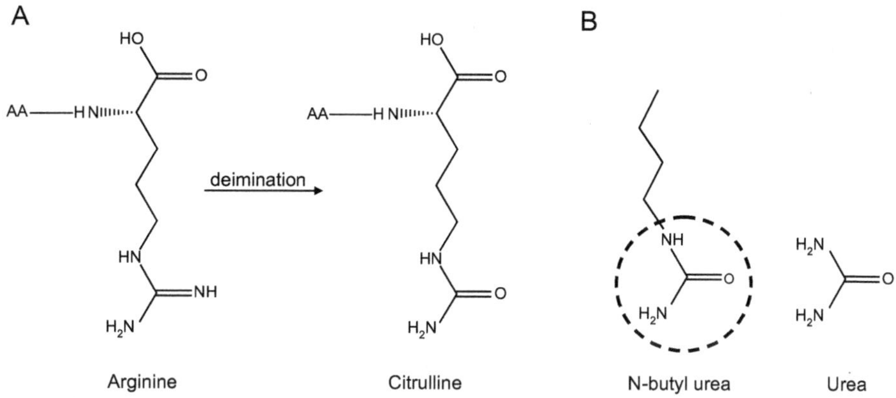

Fig. 15.1 Schematic representation of deimination and the reactive group. (**a**) Conversion of arginine into citrulline residues in a polypeptide chain is termed as deimination. AA represents polypeptide chain prior to terminal arginine. (**b**) The *N*-butyl urea possesses a ureido (a common reactive group with citrulline) group shown by a *dashed circle*; the name of the group has been derived from urea

2,3-butanedione and antipyrine in two steps was used for generation of the adduct, adding a monoisotopic mass of +238, which can be easily detected with an electrospray ionization-quadrupole ion trap (ESI-QIT) mass spectrometer (Holm et al. 2006). However, this method did not allow successful identification of deiminated peptides from myelin basic protein (MBP). Discrepancies between Western blot detection of proteins treated with 2,3-butanedione/antipyrine and identification of modified proteins through immunoprecipitation with anti-citrulline antibodies or sequencing of peptides via mass spectrometry showed that, as the complexity increases from simple peptides to proteins, the chemical monoxime modification method becomes less efficient. Our previous work has also shown that monoxime treatment results in aggregation of deiminated proteins such as MBP.. To address these problems, we have developed methodological modifications enabling more complete monoxime modification of purified deiminated proteins, enhanced separation and capture of deiminated proteins, as well as more efficient identification of deiminated proteins from a protein mixture (such as total proteins from a cell or tissues) following monoxime modification of digested peptides.

15.2 Experimental Procedures

Peptide generation. Synthetic peptides, SWGAEGQRPGFGYGG (Peptide 1) and SWGAEGQ(CIT)PGFGYGG (Peptide 2), were generated using custom peptide synthesis services (GenScript Corporation, Piscataway, NJ). The desalted peptides were subjected to 2,3-butanedione and antipyrine treatment in an acidic environment at various time points up to 24 h as indicated in the individual experiments.

Unless stated otherwise, the peptides were acetone precipitated using 4 volumes of acetone to 1 volume of aqueous phase, incubated for 15 min at room temperature, and centrifuged at 10,000 × *g* for 10 min. The final precipitate was suspended in molecular grade water and subjected to TLC or mass spectrometric analysis as stated for individual experiments.

15.2.1 Protein Purification

MBP from human brain white matter (WM) was prepared using previously published methods (Wood and Moscarello 1989). Briefly, the brain tissues after extraction of lipids using chloroform-methanol were extracted with 0.2 N sulfuric acid (H_2SO_4) containing 1 mM phenylmethylsulfonyl fluoride (PMSF; Sigma Chemical Co., St. Louis, MO) and incubated overnight at 4 °C with constant stirring. The supernatants were removed and dialyzed using dialysis membrane tubing (Spectra/Por 1, molecular weight cutoff (MWCO) of 6000–8000, Spectrum Medical Ind., Los Angeles, CA) against 3 L of distilled deionized water at 4 °C, with three exchanges of distilled deionized water. The retentates were freeze dried and stored at −20 °C until experimental use. The purified bovine serum albumin (BSA) was procured from Sigma Chemical Corporation. Purification of recombinant BSA and cellular retinaldehyde-binding protein (CRALBP) produced in *E. coli* pLysS using pET19b construct as His-tagged proteins were carried out using Ni-NTA columns according to the protocol provided by the manufacturer (Qiagen Corporation, Valencia, CA). The pET19b CRALBP construct was a research gift from Dr. John W. Crabb. The BSA construct was prepared from a parent clone obtained from Open Biosystems using standard cloning protocols.

15.2.2 Citrullination and Electrophoretic Separation of Modified Proteins

We subjected 10 μg of purified MBP to solution phase (in-solution) treatment with 2,3-butanedione and antipyrine in a strong acidic (H_3PO_4, H_2SO_4, and acetic acid in presence of $FeCl_3$) environment (monoxime treatment that leads to adduct formation on citrulline moiety) for varying time intervals as indicated in individual experiments. These treatment periods lasted from 30 s to 24 h. Throughout this manuscript, by "treatment" or "monoxime treatment," we refer to the acidic environment conferred by H_3PO_4, H_2SO_4 and $FeCl_3$ (reagent A) reacting with 2,3-butanedione and antipyrine in the presence of acetic acid (reagent B). Reagent A (0.025% $FeCl_3$, 98% H_2SO_4, 85% H_3PO_4; cat#20–255) was mixed in a 1:1 ratio with reagent B (0.5% 2,3-butanedione monoxime, 0.25% antipyrine, and 0.5 M acetic acid;

cat#20–253; Millipore Corporation, Billerica, MA) and used on peptide, proteins, or protein mixtures, usually 2 μl of A + B per 10 μg of protein (1:5) unless stated otherwise. The treatment was conducted at room temperature and in the dark. This modification generates an adduct as shown in Fig. 15.2. Originally an adduct was generated using *N*-butyl urea (Fig. 15.2a) that bears a ureido group (Fig. 15.1b), and a similar adduct acting as an antigen epitope forms within the protein with citrulline as well (Fig. 15.2b). It has been established that in the absence of the ureido group, this adduct formation does not take place (Archibald 1944). Modified proteins were acetone precipitated using 4 volumes of acetone per unit volume of aqueous phase, incubated for 15 min at room temperature, and centrifuged at 10,000 × *g* for 10 min. The precipitate was suspended in 125 mM Tris-Cl, pH 7.5, 100 mM NaCl, and 0.1%SDS, and the modified (adducted) as well as B19control (untreated MBP) proteins were fractionated over 4–20% SDS-PAGE.

For the "in-gel" treatment, 2 μg of purified intact MBP was used in each individual experiment unless stated otherwise. Usually each experimental sample was electrophoresed in separate individual lanes, and each lane was separately incubated with 2,3-butanedione and antipyrine in an acidic environment (H_3PO_4, H_2SO_4, and $FeCl_3$) for the duration comparable to "in-solution" experiments and as indicated for individual experiments. The bands were excised and subjected to in-gel digestion with sequencing grade chymotrypsin (0.1 μg/μl (Cat#11418467001, Roche, Indianapolis, IN) after staining with Coomassie blue for detection and destaining using standard protocols (Patel et al. 2008). The digested peptides were subjected to mass spectrometric analysis on a Q Exactive™ Hybrid Quadrupole-Orbitrap Mass Spectrometer device (Thermo Fisher Scientific).

For the "in-membrane" treatment, intact MBP (10 μg) was transferred to a PVDF membrane (pre-separated by SDS-PAGE using a 4–20% gradient gel) and subjected to the monoxime treatment in a manner similar to "in-solution" and/or "in-gel" treatments. In-gel and in-solution treatments of 10 μg each of BSA, CRALBP, and MBP were also conducted in an identical manner.

15.2.3 Western Blot Analyses

The Western analyses were performed after fractionation of proteins (whether monoxime treated or not) over 4–20% SDS-PAGE. The proteins or modified proteins with appropriate controls were partially transferred to a PVDF membrane and subjected to Western blot analysis using a final concentration of 0.75 μg/μl rabbit polyclonal anti-citrulline antibody (Millipore Corporation, Billerica, MA) and horseradish peroxidase (HRP)-coupled secondary antibody (Millipore Corporation, Billerica, CA). Following the application of ECL reagent (Cat# 32106, Thermo Fisher Scientific, Rockford, IL), the image was developed using an autoradiography film (Cat#1651579, Kodak, Rochester NY).

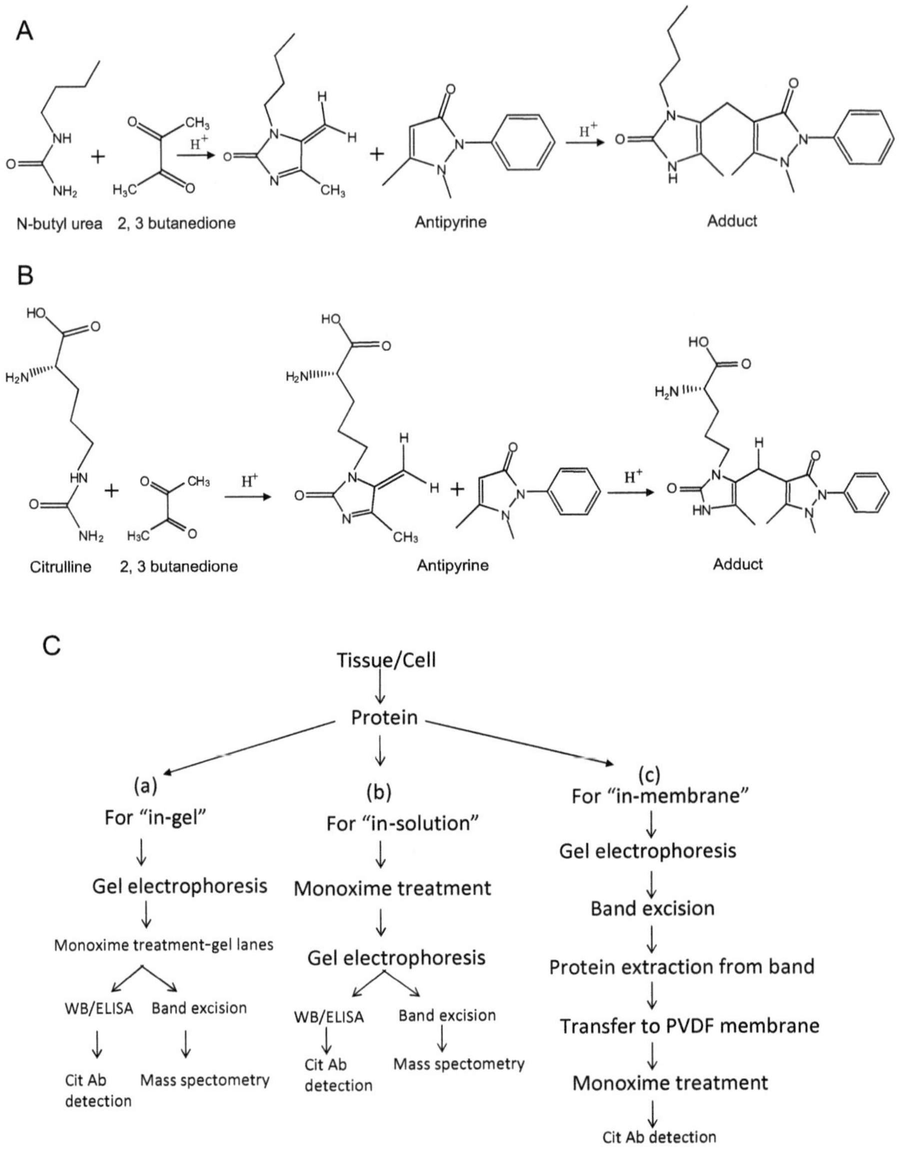

Fig. 15.2 Schematic representation of reaction of monoxime with ureido group. (**a**) Reaction of 2,3-butanedione with *N*-butyl urea and subsequently with antipyrine (second step) to generate adduct. (**b**) Two-step reaction of 2,3-butanedione and antipyrine with citrulline resulting in generation of an adduct. (**c**) Schematic diagram of experimental paradigm. MBP protein was extracted from the human white matter of MS patients and treated with 2,3-butanedione and antipyrine either after gel electrophoretic separation (in-gel), before gel electrophoretic separation (in-solution), or after transfer to PVDF membrane (in-membrane). Efficiency of adduct detection was evaluated after probing with modified anti-citrulline antibody or by mass spectrometric analysis

15.2.4 Extraction of Proteins from Polyacrylamide Gels and Transfer to PVDF Membranes

Bovine serum albumin (5 μg per lane) was electrophoretically separated and treated either "in-gel" or "in-membrane" with 2,3-butanedione and antipyrine in the same manner as soluble MBP.

15.2.5 In-Gel Treatment Prior to PVDF Transfer

Following SDS-PAGE and application of 2,3-butanedione/antipyrine, the band corresponding to BSA was excised, minced, and subjected to acetonitrile extraction by using different concentrations of acetonitrile and water. Different combinations were used for extraction from the gel: 50% acetonitrile; 50% acetonitrile followed by 80% acetonitrile; 50% followed by 80%, followed by 100% acetonitrile; and only 100% acetonitrile. The recovered protein was dried in a SpeedVac and suspended in 10 μl of deionized water; 2 μl (out of 10 μl) were then spotted as a dot on the PVDF membrane and left to dry for 10 min. The blots were blocked with nonfat milk and probed with rabbit polyclonal antibody against modified citrulline (Millipore). Recombinant CRALBP (rCRALBP; 5 μg) was used as a negative control (subjected to similar in-gel monoxime treatment and transferred to PVDF membrane) to demonstrate lack of generalized/nonspecific immunoreactivity of modified anti-citrulline antibody against recombinant, bacterially produced proteins.

15.2.6 In-Membrane Treatment Following Electrophoretic Separation and Transfer

An identical procedure was followed to extract BSA (initial material 10 μg), which was subsequently transferred to PVDF membrane (*a* > 50% transfer was confirmed using pre-stained protein marker) and subjected to monoxime treatment on the membrane only.

15.2.7 Thin-Layer Chromatography (TLC) and Western Blot Analysis of Synthetic Peptides

The synthetic peptides (5 μg each) SWGAEGQRPGFGYGG (Peptide 1) and SWGAEGQ(CIT)PGFGYGG (Peptide 2) were modified with monoxime treatment for 90 min. They were subjected to TLC separation using 30% acetonitrile water as the mobile phase and silica plates (cat#52011, Analtech, Newark, DE). Monoxime-treated

MBP, untreated MBP, MBP that was digested prior to monoxime treatment, and MBP digested after monoxime treatment were also subjected to TLC separation. The proteins and peptides separated on TLC plates were imaged with a Bio-Rad UV transilluminator (ChemiDoc XRS) and subsequently subjected to in situ immunodetection using anti-citrulline antibody after blocking the plates with 2% BSA. In a different experiment designed to confirm results from this approach, TLC-separated peptides were first transferred onto a PVDF membrane and then probed with rabbit polyclonal antibodies against modified citrulline.

The peptide bands from TLC plates subjected to monoxime treatment alone were scratched and extracted using 50% acetonitrile, followed by 80% and 100% acetonitrile. The extracted peptides were subjected to mass spectrometric analysis.

In a different experiment, the MBP (10 μg) or retinal extract (10 μg) were either predigested with sequencing grade chymotrypsin (0.2 μg per μg of substrate protein) and subjected to monoxime treatment or monoxime treated and then subjected to chymotrypsin digestion prior to loading onto TLC and separated as described above.

15.2.8 Enzyme-Linked Immunosorbent Assay (ELISA)

Indicated amounts of purified MBP (1 or 5 μg) were incubated at room temperature with 1:1 reagents A and B (from a citrulline kit, Millipore Corporation, Billerica, MA catalog# 17–347) for 90 min in the dark in ratios as indicated for individual experiments in a 10 μl final reaction. Recombinant, purified BSA was used as a control. The acetone-precipitated proteins were diluted 1:10 or 1:50 (for 1 and 5 μg of MBP, respectively), adjusted to a volume of 100 μl for each sample using phosphate-buffered saline (PBS), and incubated in individual wells in a 96-well plate. After incubation for 1 h, the supernatant was discarded, and the wells in the plate were washed with PBS. The wells were blocked with 2% recombinant purified BSA in PBS for 1 h, washed with PBS, and incubated for 1 h with rabbit polyclonal antibody against modified citrulline. Subsequently a secondary antibody coupled with HRP was added, incubated for 1 h, and washed with PBS. A substrate solution containing O-phenylenediamine in citric acid was then added and incubated for 1 min. The reaction was stopped by addition of 1 M H_2SO_4, and the absorbance measured at 490 nm on a plate reader (Synergy HT, BioTek, Winooski, VT).

15.2.9 Mass Spectrometry of Synthetic and In-Gel Digested Peptides

For identification of MBP digested by chymotrypsin, the peptides were subjected to chromatography with the Thermo EASY-nLC ultra-high pressure HPLC system (Thermo Fisher Scientific) coupled to a Q Exactive™ Hybrid Quadrupole-Orbitrap

Mass Spectrometer device (Thermo Fisher Scientific) and searched for arginine with mass addition of +238. Control peptides without 2,3-butanedione and antipyrine treatment were also subjected to orbitrap to determine the presence of arginines with +1 mass addition.

15.2.10 Immunoprecipitation

Antibody-coupled protein A or G beads were used for all immunoprecipitations (IPs). The coupling was performed in a batchwise fashion, and in each batch, 50 μg of protein A or G Sepharose CL-4B beads (Amersham Pharmacia Biotech, CA) were coupled with 50 μg modified anti-citrulline antibody using dimethylpimelimidate (DMP) or disuccinimidyl suberate (DSS).

For DMP-mediated coupling, the antibody-bead suspension was pre-incubated for 1 h, subjected to the addition of 25 mg of DMP and incubated at room temperature in 50 mM sodium borate buffer pH 8.3 for 2 h. The addition of 25 mg DMP to the suspension was repeated four times. Antibody-conjugated beads were washed and incubated for 1 h with 200 mM ethanolamine, pH 8.0.

For DSS coupling, 50 μg protein A or G Sepharose CL-4B beads (Amersham Pharmacia Biotech, CA) were incubated with 50 μg modified anti-citrulline antibody in PBS in a volume of 1 ml for 1 h at room temperature. A 25 mM DSS (Cat# 21555; Thermo Fisher Scientific) solution was prepared in ethanol and added to the bead-anti-citrulline suspension in aliquots of 30 μl three times at every 6-h intervals at room temperature. The DSS solution obtained was cloudy but did not result in antibody denaturation. The coupling was stopped by adding 1 M Tris-Cl, pH 7.5.

Antibody beads were finally washed with phosphate-buffered saline pH 7.5 and incubated with protein/retinal extracts (~1000 μg per 100 μg of beads coupled anti-citrulline) prepared in 100 mM Tris-Cl buffer pH 7.5, 50 mM NaCl and 0.01% Genapol for DMP-coupled beads or in PBS for DSS-coupled beads. Prior to IP with anti-citrulline coupled beads, 100 μl (10–12 μg/μl) of the protein/protein extract was treated for 90 min with 200 μl of 1:1 mixture of reagents A and B provided in the Millipore citrulline kit. The time period of 90 min was found as optimal and prevented formation of insoluble materials. Following this incubation, the volume was raised to 500 μl using 100 mM Tris-Cl buffer pH 8.0, 50 mM NaCl, and 0.01% Genapol (for DMP) or PBS (for DSS) and incubated with anti-citrulline coupled beads for 1 h at room temperature. After incubation, the beads were recovered by centrifugation at 2500 × *g* for 5 min and washed with 3 × 500 μl of 100 mM Tris-Cl buffer pH 7.5, 100 mM NaCl, and 0.02% Genapol for DMP or with 3 × 500 μl of PBS for DSS-coupled beads. The beads were boiled with 30 μl of 1% SDS for 2 min, and proteins eluted at each stage were measured using the BCA method or a PhstGgel system (GE Healthcare System, CA) to quantify recovered proteins using densitometry. The Coomassie-stained gels were also subjected to a densitometric scan on an Alpha Innotech Imaging system (San Leander, CA), and relative quantification of densitograms were performed using AlphaEase® FC software. Quantification of relative band areas (for IP products or total protein extracts) was determined through comparison to the band area from a known amount of purified BSA (usually 1–4 μg) run in the same gel. We have used two antibodies for immunoprecipitation of citrullinated peptides, monoclonal antibody F95 (Nicholas and Whitaker 2002), and modified anti-citrulline antibody described above. We used extensive mass spectrometry of IP products to validate these antibodies. Our analyses validated that both antibodies precipitate citrullinated peptides. However, F95 possibly precipitate additional epitopes as well depending upon peptide context. The modified anti-citrulline peptide always identified peptides with at least one citrulline in the sequence.

15.3 Results of Deiminated Protein, Mode of Monoxime Treatment, and Cross-Linking

Purified MBP derived from multiple sclerosis patients underwent aggregation when subjected to overnight treatment with 2,3-butanedione and antipyrine in soluble phase, likely due to cross-linking (Fig. 15.3a). A time course analysis of treatment from 1.5 to 24 h showed increased cross-linking as a function of time (Fig. 15.3b). Probing the soluble proteins subjected to treatment for citrullination showed that most cross-linking occurred by 1.5–3 h of incubation (Fig. 15.3b, c). A period of incubation shorter than 90 min resulted in less cross-linking but also less efficient detection of modified protein (Fig. 15.3d, e). It appeared that optimal treatment without significant cross-linking was achieved within 5–10 min for MBP. Coomassie blue staining after transfer of pretreated (2,3-butanedione + antipyrine) protein to PVDF membrane showed less residual protein on lanes treated for a short period (Fig. 15.3d). However, a commensurate transfer of intact band was noticed when PVDF membrane was stained with Ponceau S just after the transfer (Fig. 15.3f). The Ponceau S staining was weak (it bound to proteins weakly and more reversibly) and effectively stained the more compact and intact bands. Therefore, it had a decreased propensity to stain the less compact spread out-modified proteins. In this regard, Coomassie blue and Ponceau S differed, so the point of utilizing these two protein stains was to show that with less chemical treatment, the more compact intact bands transferred most readily to PVDF (stained with Ponceau S, Fig. 15.3f). On the other hand, more modified and cross-linked proteins were stained by Coomassie blue (Fig. 15.3d). The intact bands were not prominent in Coomassie blue staining (Fig. 15.3d), but nevertheless they were present as evidenced by Ponceau S staining (Fig. 15.3f).

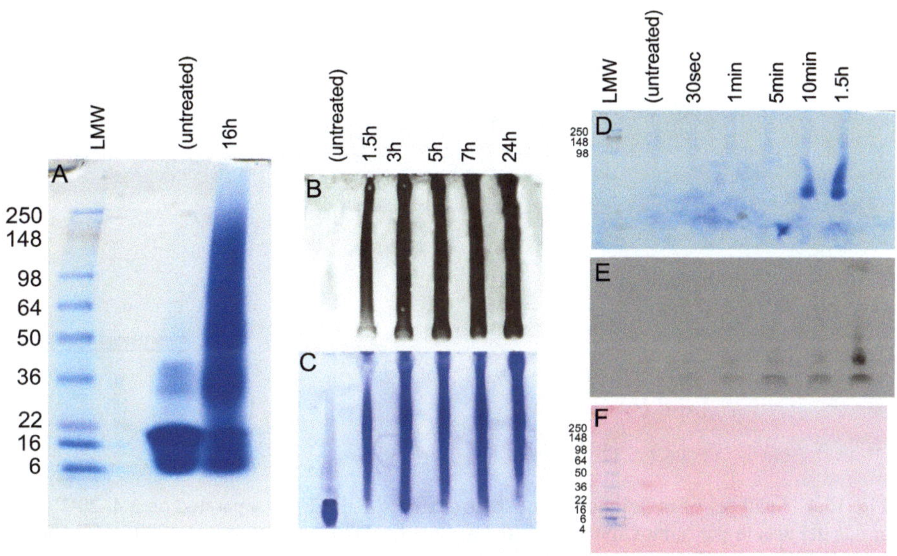

Fig. 15.3 Analyses of duration of monoxime treatment of MBP in solution phase. MBP after monoxime treatment was acetone precipitated and resuspended in 10 μl phosphate-buffered saline (PBS; pH 7.5) and subjected to electrophoretic separation and Western analysis using rabbit polyclonal antibody to modified citrulline. (**a**–**c**) MBP (10 μg) was monoxime treated (2 μl reagents A and B) and, for indicated duration of time in a final reaction volume of 100 μl, acetone precipitated and separated on 4–20% SDS-PAGE. (**a**) and (**c**) are Coomassie-stained gel and (**b**) is Western analysis. (**d**–**f**) MBP (5 μg) reacted with 1 μl of reagents A and B in a final reaction volume of 100 μl and was acetone precipitated after incubation for indicated duration and suspended in 10 μl PBS. (**d**) Coomassie-stained SDS-PAGE. (**e**) Western analyses. (**f**) PVDF membrane in (**e**) stained with Ponceau S just after transfer

15.3.1 In-Gel Monoxime Treatment

This finding of cross-linking prompted us to ask the question of whether immobilization of the protein in polyacrylamide gel matrix would minimize cross-linking but, yet, still allow conversion of protein-bound citrulline into an adduct. Following SDS-PAGE, proteins trapped within the polyacrylamide gel were subjected to monoxime treatment (Fig. 15.4a). Western analysis of in-gel monoxime-treated MBP (treated for a short duration) revealed lack of immunoreactivity with anti-modified citrulline antibody (Fig. 15.4b), consistent with Ponceau S staining of the PVDF membrane just after transfer, which revealed transfer of smaller, intact MBP, likely to be non-deiminated MBP (Fig. 15.4c). Therefore, although transfer of proteins occurs (Fig. 15.4c), lack of detection of modified citrulline suggested insufficient transfer of modified citrulline-containing protein, which, in turn, is likely to be retained within the gel due to cross-linking with gel matrix. A control solution phase monoxime-treated MBP spot introduced just after transfer showed immunoreactivity on the same blot (Fig. 15.4b), suggesting that detection of immunoreactivity is not a limitation. However, the short exposure to treatment (Fig. 15.4a, b) likely

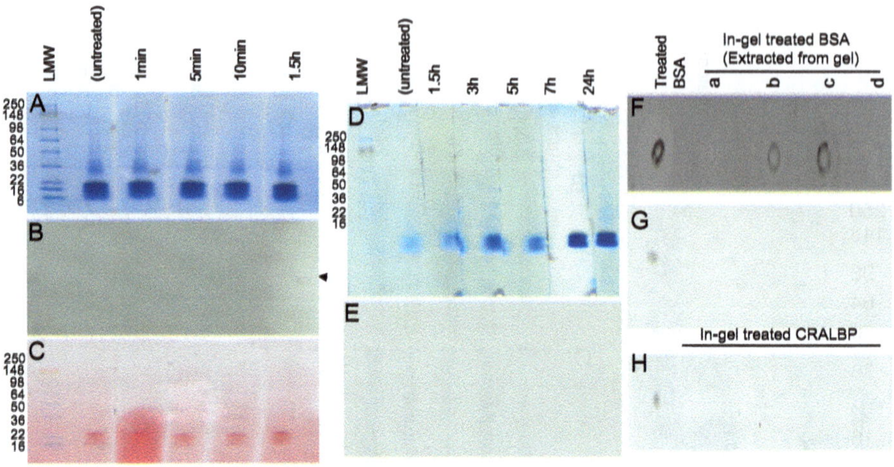

Fig. 15.4 Analyses of in-gel monoxime-treated MBP. MBP (2 μg) separated on a 4–20% acrylamide gel was in-gel monoxime treated (0.4 μl) in a final reaction volume of 100 μl for indicated period of time. (**a**) Coomassie-stained SDS-PAGE. (**b**) The proteins were transferred on a PVDF membrane and subjected to Western analysis using a rabbit polyclonal antibody to modified citrulline. *Arrowhead* indicates solution-treated MBP (~200 ng; monoxime treated for 1.5 h) spotted as a *dot* (positive control). (**c**) Ponceau S stained PVDF membrane just after transfer. (**d**) Coomassie-stained gel and (**e**) Western analysis for modified citrulline of in-gel-treated MBP for indicated durations, all other conditions were identical as in **a**, **f**, and **g**. Recovery of bovine serum albumin (BSA): (*a*) extracted from the gel using 50% acetonitrile; (*b*) extracted using 50% acetonitrile followed by 80% acetonitrile; (*c*) extracted using 50% followed by 80%, followed by 100% acetonitrile; (*d*) extracted using 100% acetonitrile. Blots were probed with a rabbit polyclonal antibody against modified citrulline. (**g**) In solution phase, modified purified recombinant BSA serves as a negative control. (**h**) Recombinant CRALBP was used as a negative control in the same blot as in **d** to demonstrate lack of detection of immunoreactivity using modified anti-citrulline antibody for a recombinant bacterially produced protein. BSA and CRALBP (5 μg each) were reacted with 1 μl reagents A and B (in a final reaction volume of 10 μl made up with water)

generated only low amounts of modified citrulline-containing protein, resulting in suboptimal transfer to PVDF membrane of modified proteins that may be below the threshold limit of Western detection. To address this issue, longer exposures were made but still failed to show immunoreactivity when subjected to Western blot analysis.

In order to determine whether cross-linking occurred with polyacrylamide upon monoxime treatment, we used purified bovine serum albumin (BSA) that was naturally partially citrullinated. In-gel treated, BSA at the end of separation was subjected to mincing with scalpel and extraction with a different combination of solvents (Fig. 15.4f, a–d). The crosslinking was evident by the fact that only a fraction of the modified protein can be recovered (Fig. 15.4f, a–d). The first lane (treated BSA control) was purified BSA modified in solution phase (for all panels of Fig. 15.4f–h) and spotted directly onto the PVDF membrane. Purified, recombinant BSA (non-citrullinated; rBSA) was immobilized on PVDF and subjected to 90 min monoxime treatment, and identical extraction as with the in-gel treated condition (as in Fig. 15.4f) showed no immunoreactivity as expected (Fig. 15.4g). Providing an additional negative control, mammalian cellular retinal aldehyde-binding protein (rCRALBP) that was recombinantly produced in *E. coli,* purified, subjected to in-gel treatment under identical conditions, and extracted using a similar protocol also did not show immunoreactivity (Fig. 15.4h).

15.3.2 In-Membrane Monoxime Treatment

Another method for restricting aggregate formation was treatment of protein after transfer/immobilization onto PVDF membrane. The modified MBP could be detected after 90 min with this approach (Fig. 15.5a, b) in contrast to shorter time periods with soluble proteins (Fig. 15.3e). The lack of detection of modified proteins with shorter time exposures (10 min or less) may be due to steric hindrance or lack of access and thus shielding of citrulline residues in PVDF-immobilized protein (Fig. 15.5b). The fact that the same amount of protein transferred over to the PVDF showed different degrees of intensity for citrullinated MBP suggested that the reaction was incomplete at earlier time points. Taken together with other results (Fig. 15.3b–f), this was consistent with steric hindrance of reactive species in the setting of 2,3-butanedione and antipyrine treatment of transferred proteins, which were less accessible three-dimensionally after they had been transferred to the PVDF membrane, compared to those caged in the gel. Note that in 1.5 h of incubation, the results were very different (Fig. 15.3e versus Fig. 15.5b), which is consistent with steric hindrance or diminished accessibility of chemical modifiers to PVDF-immobilized MBP.

15.3.3 Relationship Between Citrullination and Monoxime Treatment-Induced Aggregation

The aggregation of proteins upon monoxime treatment appeared to be due to citrulline residues. In contrast to deiminated MBP and BSA, recombinant bacterially produced and purified mammalian, non-deiminated proteins such as rBSA and rCRALBP did not show aggregation upon monoxime treatment and failed to show any immunoreactivity when transferred onto a PVDF membrane (where transfer was confirmed using Ponceau S staining, Fig. 15.6).

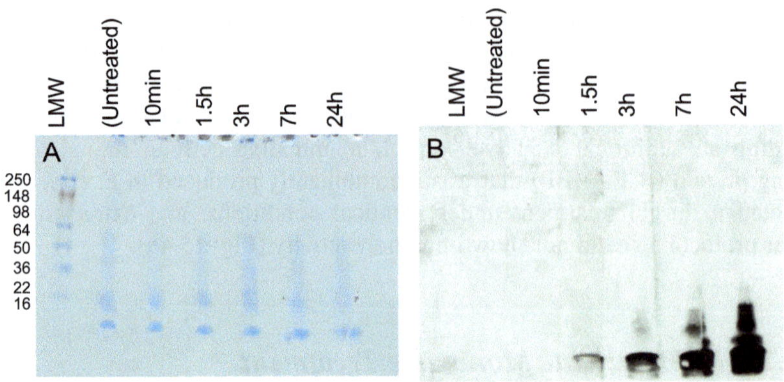

Fig. 15.5 Analysis for deimination of posttransfer MBP monoxime treated on the membrane. (**a**) Coomassie-stained gel after transfer. MBP (10 μg) was separated on SDS-PAGE, transferred to PVDF membrane, and monoxime treated (2 μl spread into the MBP transferred region using 8 μl water rendering final reaction volume on each strip 10 μl). (**b**) PVDF membrane after treatment for indicated time period was subjected to washing with PBS and Western analysis with a rabbit polyclonal antibody against modified citrulline

15.3.4 Monoxime–Protein Ratio and Adduct Formation

Aggregation of citrullinated proteins as a consequence of monoxime treatment could be related to ratio of monoxime reagents to the amount of proteins. To determine the impact of monoxime–protein substrate ratio, we had also used varying amounts of monoxime reagent with a fixed amount of MBP (10 μg) and a fixed time of incubation (10 min) prior to assessing aggregation and amount of ELISA-detectable adduct with and without enzymatic digestion (Fig. 15.7a). In addition, we used a higher amount of MBP (50 μg) while keeping the ratio of reagent constant to determine whether adduct formation showed linear scalability (Fig. 15.7b). Our controls consisted of unmodified rBSA protein, and triplicate wells without protein, the latter serving as a negative control for baseline correction to determine background fluorescence.

15.3.5 Inert Matrix of Thin-Layer Chromatography for Reduction of Cross-Linking Due to Monoxime Treatment

Because MBP was found to be aggregated upon monoxime treatment in solution and cross-linked when subjected to in-gel treatment, we examined the impact of using an inert matrix for separation as well as for treatment. We used thin-layer chromatography matrix (cat#52011, Analtech, Newark, DE) for protein separation (Fig. 15.8a, b). When using TLC separation to assess MBP, with and without

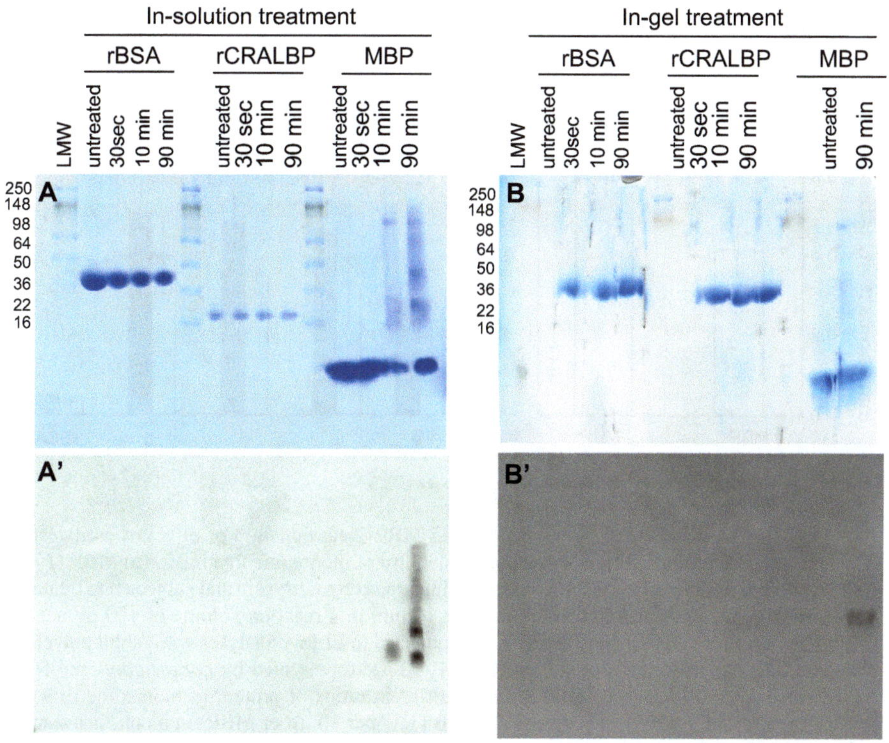

Fig. 15.6 Monoxime treatment of different proteins with 2, 3 butandione and antipyrene. Recombinant purified BSA (10 μg), recombinant purified CRALBP(2 μg) and MBP (10 μg) each was subjected to monoxime treatment carried out using 1 μl monoxime reagent A and B per 5 μg of protein for indicated period or untreated (control) and subjected to electrophoretic separation (4–20% SDS-PAGE) and Coomassie blue staining. (**a**) In-solution treatment and (**b**) In-gel treatment of BSA, CRALBP and MBP as indicated. **a'** and **b'** represents Western analysis of PVDF transferred proteins as in **a** and **b**.

monoxime treatment, both forms were detected using UV fluorescence, but only monoxime-treated protein was detected on a TLC plate subjected to blocking and subsequent probing with a modified anti-citrulline antibody. We also analyzed two synthetic peptides (SWGAEGQRPGFGYGG and SWGAEGQ(CIT)PGFGYGG) where both uncitrullinated and citrullinated versions of the peptides were detected using UV fluorescence, but only citrulline-containing peptide was detected using modified anti-citrulline antibody.

Complementing these results, MBP (or another citrullinated protein; data not shown) showed greater peptide separation when digested first using enzymes such as chymotrypsin, followed by monoxime treatment. Again the digested peptides were detected by UV light, but only citrulline-containing peptides were detected by anti-modified citrulline antibodies. In contrast, MBP (or another citrulline-containing protein) subjected to monoxime treatment followed by chymotrypsin digestion was often poorly separated (data not shown). This impairment of TLC

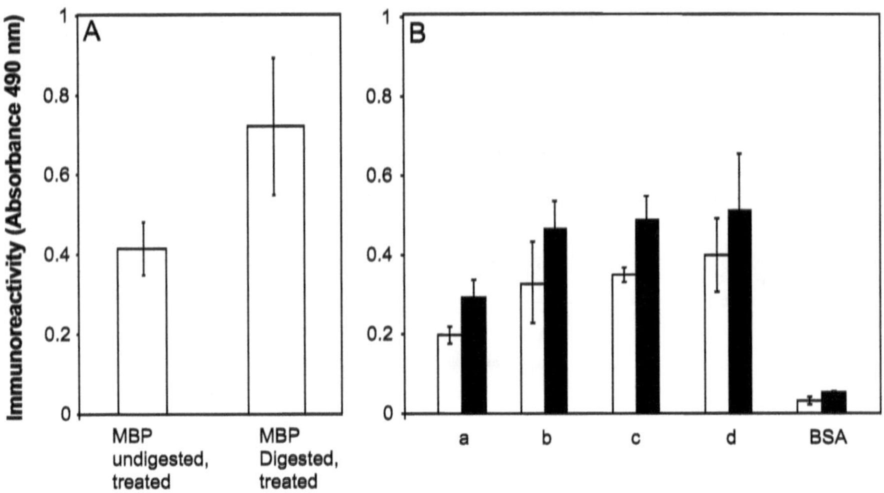

Fig. 15.7 ELISA analyses of monoxime-treated MBP. Determination of effect of predigestion, different ratios of protein to monoxime, and scalability of monoxime treatment. (**a**) MBP (10 μg) was undigested or digested with 2 μg of sequencing grade chymotrypsin and subjected to treatment with 2 μl of reagents A and B (monoxime) for 10 min in a reaction volume of 100 μl, acetone precipitated, resuspended in 10 μl of PBS, and subjected to ELISA analyses with rabbit polyclonal anti-citrulline. (**b**) In-solution treatment of MBP, 10 μg (represented by *hollow bars*) and 50 μg (represented by *solid bars*) were treated with different ratios of protein to monoxime ($a = 0.1$, $b = 0.2$, $c = 2$, and $d = 4$ μl of reagents A and B (1:1) per 10 μg of MBP) in a constant reaction volume of 100 μl; from each reaction mixture in **a** and **b**, 1 μg equivalent MBP protein was subjected to analysis in a volume of 100 μl per ELISA well. Untreated BSA was used as a control

separation of proteins first subjected to monoxime treatment followed by chymotrypsin digestion was further influenced by degree to which proteins were citrullinated. As stated above, the citrullination (especially after monoxime treatment) had an effect on efficient separation by TLC, likely due to aggregation.

15.3.6 Immunoprecipitation of Modified Citrullinated Proteins

Immunoprocipitation was often used as a way of enrichment of proteins, including modified proteins, using antibody against modification. Coupling of rabbit polyclonal antibody with protein A or protein G using dimethylpimelimidate (DMP) or disuccinimidyl suberate (DSS) was optimized. In the DMP coupling method, protein A was able to offer a slightly better coupling with the antibody than protein G (Table 15.1). With DMP coupling, 4–6 rounds of coupling provided better coupling than less incremental treatment (data not shown). Low hydrogen ion concentration elution with glycine was more efficient than elution with Laemmli buffer or elution using a strong detergent such as SDS. The DMP-coupled protein A beads also provided a slightly better capture of proteins from retinal extracts after three PBS

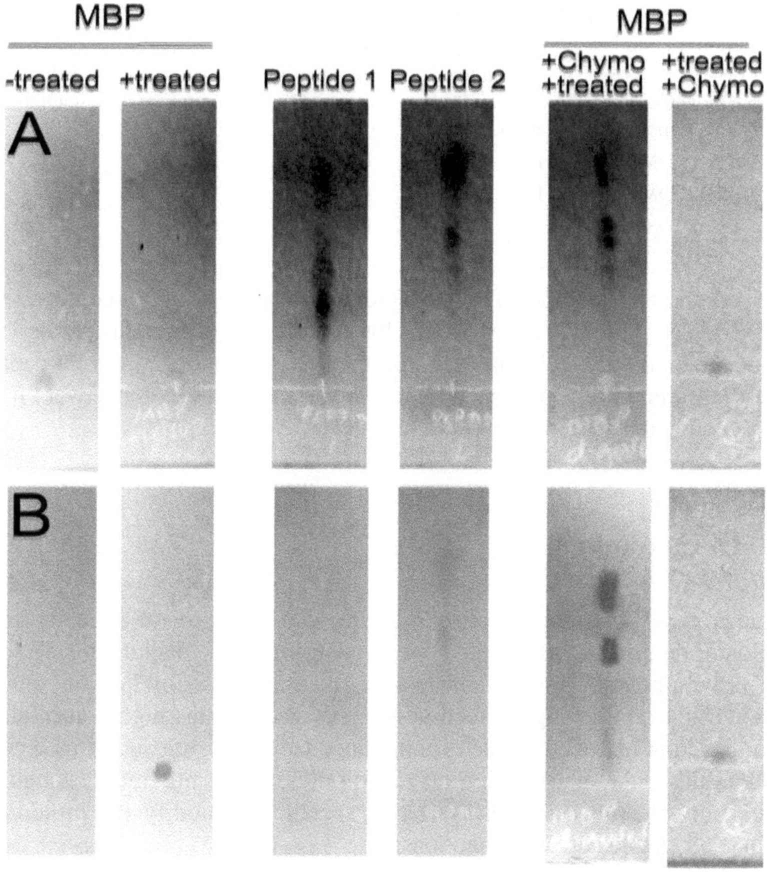

Fig. 15.8 Thin-layer chromatography (TLC) separation of proteins and peptides on a flourophore-bearing TLC plate (Cat#52011 Analtech, Newark, DE). (**a**) TLC separation of MBP (5 μg), peptide 1 (SWGAEGQRPGFGYGG; 5 μg), and peptide 2 (SWGAEGQ(CIT)PGFGYGG; 5 μg) or digested MBP (5 μg digested with 1 μg sequencing grade chymotrypsin) after monoxime treatment as indicated. Sequence of monoxime treatment and chymotrypsin digestion (Chymo) for MBP is as indicated. The detection was performed using fluorescence measured on a Bio-Rad UV transilluminator (Chemidoc XRS). (**b**) In-situ TLC Western analysis using modified anti-citrulline antibody. Chemiluminescence using ECL was detected on an autoradiography film

Table 15.1 Comparison of antibody load parameters on immobilized matrix

Type of beads	Amount of packed beads (μl)	Amount of initial antibody load (T) μg)	Cross-linking reagent	Cross-linking time (h)	Flow through (F) (μg)	Efficiency of crosslinking ($T-F$)/T
Sepharose-A	50	50	DMP	16	14	0.72
Sepharose-G	50	50	DMP	16	14	0.72
Sepharose-A	50	50	DSS	18	21	0.58

DMP dimethylpimelimidate, *DSS* disuccinimidyl suberate

Table 15.2 Immunoprecipitation using different approaches

Type of beads	Coupling method	Amount of packed beads (~100 μg antibody)	Amount of initial retinal extract (*T*) (μg)	Amount of captured protein (*C*) (μg)	Efficiency of capture (*C*/*T*) ×100
Sepharose-A	DMP	138	1000	148	1.48
Sepharose-G	DMP	138	1000	137	1.37
Sepharose-A	DSS	172	1000	110	1.1

washes compared to either DMP-coupled protein G or DSS-coupled protein A. DSS-coupled protein A generated the lowest recovery of proteins (Table 15.2). It is not understood why different coupling methods resulted in different capture of proteins in immunoprecipitation. With minor variations, these trends were found to be reproducible during repetition.

15.4 Discussion

The detection of protein-bound citrulline converted from protein-bound arginine by the action of deiminases using mass spectrometry remains challenging. The formation of protein-bound citrulline results in a change in mass of 1 atomic mass unit (1 amu). The 1 amu change detection not only necessitates highly accurate mass change detection but needs to be differentiated from changes due to incorporation of C-13 in the amino acid, which may also provide the same mass change. Thus, chemical modifications are desirable that specifically modify the protein-bound citrulline and enable suitable detections.

Our results show that the monoxime reaction using 2,3-butanedione and antipyrine formed an adduct with citrulline moieties, adding a monoisotopic mass of +238 on each citrulline, which can be detected with a electrospray ionization-quadrupole ion trap (ESI-QIT) mass spectrometer (Holm et al. 2006). This method had been used with some variation in a number of studies (De Ceuleneer et al. 2012). However, in our hands this method was found not successful in identification of deiminated peptides from MBP and BSA (data not shown). Another method utilized O18 that resulted in mass addition of +1.5 (Kubota et al. 2005), which was not explored in our study. A recent study had used missed trypsin cleavage for detection of citrulline modification using mass spectrometry (Bennike et al. 2013). Our exploration of this method resulted in only limited success. A few other approaches have also been tested; however, monoxime treatment (2,3-butanedione and antipyrine) remained most extensively utilized for detection of citrullines (De Ceuleneer et al. 2012).

Monoxime treatment of synthetic peptides with acidified $FeCl_3$, containing 2,3-butanedione and antipyrine, often did not lead to a complete reaction. The lack of completion was more pronounced with increasing polypeptide complexity. Intermittent stirring or elevation of temperature to 55 °C increased the reaction rate within a 90-min time frame for some peptides. In part, this was attributable to formation of aggregates, specifically with deiminated proteins. These limitations occurred at two levels: (1) by limiting the capture of proteins undergoing deimination and (2) by limiting the identification of peptides with deiminated sites, thus resulting in a tendency to not capture the deiminated sites. The aggregation of deiminated proteins as a consequence of monoxime treatment was shown in Fig. 15.3b. The aggregated citrullinated proteins were transferred on PVDF membranes compared to intact proteins (Fig. 15.3d–f). However, the citrullinated entities were often better identified on aggregated proteins using immunoblot detection (Fig. 15.3d), although their transfer was often inefficient (Fig. 15.4). It should be noted that only a little transfer may result in detection of signal on the immunoblot (Fig. 15.3e), but it is also conceivable that PVDF membrane restricted access to reagents (Fig. 15.5a, b). Steric hindrance of reagent access on PVDF was also corroborated from results presented in Fig. 15.4b, e. To prevent aggregation, we used the mild detergent Genapol C-100. At low concentrations, it was able to keep several proteins in solution phase without denaturing them. Owing to this property, we used 0.02% Genapol in binding and wash buffers. We had not performed a systematic comparison of this detergent with others as yet, but in general, this detergent was helpful in preventing protein aggregation. However, its presence did not hinder aggregation mediated by monoxime treatment. Our investigation presented in Fig. 15.4f–h supported that the negative control for immunoblot citrullination detection upon monoxime treatment results were as intended. Overall, these results were in consonance with recombinant in vitro modified or control proteins presented in Fig. 15.6. These results also corroborated ELISA estimates using monoxime treatment and anti-modified citrulline antibody. The TLC analyses were consistent with complexity of peptide-promoting aggregate formation (Fig. 15.8a, b). However, TLC analyses also suggested high digestibility of citrullinated peptide by chymotrypsin, unlike trypsin (Bennike et al. 2013).

Overall, these results demonstrated the detection ability of protein-bound citrulline using monoxime treatment and its limitations when the detection strategies utilized immunoblot or ELISA. The similar limitations also applied to immunohistochemical or immunocytochemical analysis. We have explored two antibodies for immunoprecipitation: the F95 monoclonal antibody (Nicholas and Whitaker 2002) and a modified anti-citrulline antibody that identified protein-bound citrulline residues after monoxime treatment. To validate if these antibodies indeed identified citrullinated proteins, we performed extensive mass spectrometry of IP products revealing that both antibodies precipitated citrullinated peptides. The F95 precipitated citrullinated proteins, however, it also precipitated >25% peptides that lacked citrulline, thus it possibly precipitated additional epitopes. The modified anti-citrulline peptide always identified peptides with at least one citrulline in the sequence. Coupling with a reagent such as Biotin-CDM (Ganesan et al. 2015) seems to impact the efficiency of immunoprecipitation, possibly by promoting aggregation of citrullinated proteins prior to co-incubation with antibody-bound Sepharose beads. However, immunohistochemical or immunocytochemical analyses usually provided stronger signals with anti-citrulline antibodies when compared to Western blot or mass spectrometric analyses of the same tissue samples. In the event of modified citrulline antibody (after monoxime treatment), the signal intensity showed an increase with length of monoxime treatment; thus, aggregation and other phenomena must have been affecting immunohisto–/cytochemical analyses. These results obviously were approximations for localization and not quantitative. Though the monoxime treatment facilitated detection of protein-bound citrullines, our objective here was to bring attention to the unintended consequences of solution phase aggregation and immobilized phase steric hindrance to the readership.

References

Algeciras, M. E., & Bhattacharya, S. K. (2007). Targeting optic nerve citrullination in glaucoma: A role for protein-arginine deiminase 2 (PAD2) inhibitors. *Drugs of the Future, 32*, 999–1006.

Archibald, R. M. (1944). Determination of citrulline and allantoin and demonstration of citrulline in blood plasma. *The Journal of Biological Chemistry, 156*, 121–142.

Arnold, J. N., Saldova, R., Hamid, U. M., & Rudd, P. M. (2008). Evaluation of the serum N-linked glycome for the diagnosis of cancer and chronic inflammation. *Proteomics, 8*, 3284–3293.

Bennike, T., Lauridsen, K. B., Olesen, M. K., Andersen, V., Birkelund, S., & Stenballe, A. (2013). Optimizing the identification of citrullinated peptides by mass spectrometry: Utilizing the inability of trypsin to cleave after citrullinated amino acids. *Journal of Proteomics & Bioinformatics, 6*, 288–295.

Bhattacharya, S. K., Sinicrope, B., Rayborn, M. E., Hollyfield, J. G., & Bonilha, V. L. (2008). Age-related reduction in retinal deimination levels in the F344BN rat. *Aging Cell, 7*, 441–444.

Boyde, T. R., & Rahmatullah, M. (1980). Optimization of conditions for the colorimetric determination of citrulline, using diacetyl monoxime. *Analytical Biochemistry, 107*, 424–431.

Bradbury, E. M. (1992). Reversible histone modifications and the chromosome cell cycle. *BioEssays, 14*, 9–16.

Chavanas, S., Mechin, M. C., Nachat, R., Adoue, V., Coudane, F., Serre, G., & Simon, M. (2006). Peptidylarginine deiminases and deimination in biology and pathology: Relevance to skin homeostasis. *Journal of Dermatological Science, 44*, 63–72.

De Ceuleneer, M., Van Steendam, K., Dhaenens, M., & Deforce, D. (2012). In vivo relevance of citrullinated proteins and the challenges in their detection. *Proteomics, 12*, 752–760.

Ding, D., Enriquez-Algeciras, M., Dave, K. R., Perez-Pinzon, M., & Bhattacharya, S. K. (2012). The role of deimination in ATP5b mRNA transport in a transgenic mouse model of multiple sclerosis. *EMBO Reports, 13*, 230–236.

Enriquez-Algeciras, M., Ding, D., Chou, T. H., Wang, J., Padgett, K. R., Porciatti, V., & Bhattacharya, S. K. (2011). Evaluation of a transgenic mouse model of multiple sclerosis with noninvasive methods. *Investigative Ophthalmology & Visual Science, 52*, 2405–2411.

Enriquez-Algeciras, M., Ding, D., Mastronardi, F. G., Marc, R. E., Porciatti, V., & Bhattacharya, S. K. (2013). Deimination restores inner retinal visual function in murine demyelinating disease. *The Journal of Clinical Investigation, 123*, 646–656.

Ganesan, V., Schmidt, B., Avula, R., Cooke, D., Maggiacomo, T., Tellin, L., Ascherman, D. P., Bruchez, M. P., & Minden, J. (2015). Immuno-proteomics: Development of a novel reagent for separating antibodies from their target proteins. *Biochimica et Biophysica Acta, 1854*, 592–600.

Gyorgy, B., Toth, E., Tarcsa, E., Falus, A., & Buzas, E. I. (2006). Citrullination: A posttranslational modification in health and disease. *The International Journal of Biochemistry & Cell Biology, 38*, 1662–1677.

Harauz, G., & Musse, A. A. (2007). A tale of two citrullines--structural and functional aspects of myelin basic protein deimination in health and disease. *Neurochemical Research, 32*, 137–158.

Holm, A., Rise, F., Sessler, N., Sollid, L. M., Undheim, K., & Fleckenstein, B. (2006). Specific modification of peptide-bound citrulline residues. *Analytical Biochemistry, 352*, 68–76.

Kubota, K., Yoneyama-Takazawa, T., & Ichikawa, K. (2005). Determination of sites citrullinated by peptidylarginine deiminase using 18O stable isotope labeling and mass spectrometry. *Rapid Communications in Mass Spectrometry, 19*, 683–688.

Macek, B., Mann, M., & Olsen, J. V. (2008). Global and site-specific quantitative phosphoproteomics: Principles and applications. *Annual Review of Pharmacology and Toxicology, 49*, 199–221.

Marletta, M. A., Hurshman, A. R., & Rusche, K. M. (1998). Catalysis by nitric oxide synthase. *Current Opinion in Chemical Biology, 2*, 656–663.

Mechin, M. C., Sebbag, M., Arnaud, J., Nachat, R., Foulquier, C., Adoue, V., Coudane, F., Duplan, H., Schmitt, A. M., Chavanas, S., Guerrin, M., Serre, G., & Simon, M. (2007). Update on peptidylarginine deiminases and deimination in skin physiology and severe human diseases. *International Journal of Cosmetic Science, 29*, 147–168.

Nicholas, A. P., & Whitaker, J. N. (2002). Preparation of a monoclonal antibody to citrullinated epitopes: Its characterization and some applications to immunohistochemistry in human brain. *Glia, 37*, 328–336.

Nita-Lazar, A., Saito-Benz, H., & White, F. M. (2008). Quantitative phosphoproteomics by mass spectrometry: Past, present, and future. *Proteomics, 8*, 4433–4443.

Patel, N., Solanki, E., Picciani, R., Cavett, V., Caldwell-Busby, J. A., & Bhattacharya, S. K. (2008). Strategies to recover proteins from ocular tissues for proteomics. *Proteomics, 8*, 1055–1070.

Ryan, K., & Bauer, D. L. (2008). Finishing touches: Post-translational modification of protein factors involved in mammalian pre-mRNA 3′ end formation. *The International Journal of Biochemistry & Cell Biology, 40*, 2384–2396.

Senshu, T., Sato, T., Inoue, T., Akiyama, K., & Asaga, H. (1992). Detection of citrulline residues in deiminated proteins on polyvinylidene difluoride membrane. *Analytical Biochemistry, 203*, 94–100.

Sims, R. J., III, & Reinberg, D. (2008). Is there a code embedded in proteins that is based on post-translational modifications? *Nature Reviews. Molecular Cell Biology, 9*, 815–820.

Stuehr, D. J. (2004). Enzymes of the L-arginine to nitric oxide pathway. *The Journal of Nutrition, 134*, 2748S–2751S. discussion 2765S–2767S.

Vossenaar, E. R., Zendman, A. J., van Venrooij, W. J., & Pruijn, G. J. (2003). PAD, a growing family of citrullinating enzymes: Genes, features and involvement in disease. *BioEssays, 25*, 1106–1118.

Watanabe, K., Akiyama, K., Hikichi, K., Ohtsuka, R., Okuyama, A., & Senshu, T. (1988). Combined biochemical and immunochemical comparison of peptidylarginine deiminases present in various tissues. *Biochimica et Biophysica Acta, 966*, 375–383.

Wood, D. D., & Moscarello, M. A. (1989). The isolation, characterization, and lipid-aggregating properties of a citrulline containing myelin basic protein. *The Journal of Biological Chemistry, 264*, 5121–5127.

Zarabian, B., Koushesh, F., & Vassef, A. (1987). Modified methods for measuring citrulline and carbamoyl-beta-alanine with reduced light sensitivity and sucrose interference. *Analytical Biochemistry, 166*, 113–119.

Zhang, Q., & Wang, Y. (2008). High mobility group proteins and their post-translational modifications. *Biochimica et Biophysica Acta, 1784*, 1159–1166.

Chapter 16
Citrullination Following Traumatic Brain Injury: A Mechanism for Ongoing Pathology Through Protein Modification

Rachel C. Lazarus, John E. Buonora, Alaa Kamnaksh, Michael N. Flora, James G. Freedy, Gay R. Holstein, Giorgio P. Martinelli, David M. Jacobowitz, Denes Agoston, and Gregory P. Mueller

Abbreviations

ATP	Adenosine triphosphate
C	Control
Ca^{2+}	Calcium
CCI	Controlled cortical impact

R.C. Lazarus
Program in Neuroscience, Uniformed Services University of the Health Sciences, Bethesda, MD, USA

J.E. Buonora
US Army Graduate Program in Anesthesia Nursing, Fort Sam, Houston, TX, USA

A. Kamnaksh • D.M. Jacobowitz • D. Agoston
Department of Anatomy, Physiology, and Genetics, Uniformed Services University of the Health Sciences, C2117, 4301 Jones Bridge Road, Bethesda, MD 20814-4799, USA

Program in Neuroscience, Uniformed Services University of the Health Sciences, Bethesda, MD, USA

M.N. Flora • J.G. Freedy
Department of Anatomy, Physiology, and Genetics, Uniformed Services University of the Health Sciences, C2117, 4301 Jones Bridge Road, Bethesda, MD 20814-4799, USA

G.R. Holstein • G.P. Martinelli
Department of Neurology, Icahn School of Medicine at Mount Sinai, New York, NY, USA

G.P. Mueller (✉)
Center for Neuroscience and Regenerative Medicine, Uniformed Services University of the Health Sciences, Bethesda, MD, USA

Program in Neuroscience, Uniformed Services University of the Health Sciences, Bethesda, MD, USA

Department of Anatomy, Physiology, and Genetics, Uniformed Services University of the Health Sciences, C2117, 4301 Jones Bridge Road, Bethesda, MD 20814-4799, USA
e-mail: gregory.mueller@usuhs.edu

A.P. Nicholas et al. (eds.), *Protein Deimination in Human Health and Disease*,
DOI 10.1007/978-3-319-58244-3_16

GFAP	Glial fibrillary acidic protein
kDa	Kilodalton
mAb	Monoclonal anti-citrulline antibody
MBP	Myelin basic protein
MS	Multiple sclerosis
NMDA	*N*-methyl-D-aspartate
PAD	Peptidylarginine deiminase
TBI	Traumatic brain injury

16.1 Introduction

Traumatic brain injury (TBI) is a major public health issue in the USA, with over 1.7 million cases per year. At least 5.3 million Americans currently live with ongoing disability due to TBI (Walker and Tesco 2013). The long-term consequences of TBI can be complex and progressive: 10–15% of individuals diagnosed with mild TBI continue to suffer from persistent symptoms, while as many as 50% of patients with moderate TBI experience long-term dysfunction (Bales et al. 2009; Walker and Tesco 2013). In addition to the post-injury development of cognitive deficits in attention, memory, and executive function (Bales et al. 2009), a chronic inflammatory state can persist in the brain for months, and even years, following TBI (Opii et al. 2007; Piao et al. 2013). Despite the prevalence of these chronic dysfunctions following TBI, elucidating the mechanisms that underlie these symptoms has proven challenging. It is understood that the post-injury processes of TBI involve a primary and a secondary phase of injury. Primary injury occurs during the initial insult, resulting from the displacement of the physical structures within the brain. Secondary injury occurs over time and involves an interdependent series of cellular dysfunctions that persist long after injury (Giza and Hovda 2001; Nilsson et al. 1996; Park et al. 2008; Walker and Tesco 2013; Werner and Engelhard 2007). Currently, it is not well understood how the acute mechanical injury of TBI, and subsequent secondary injury processes, can result in serious long-term dysfunctions that can persist for years.

16.2 Mechanisms of Secondary Injury: Glutamate Excitotoxicity and Calcium Overload

16.2.1 TBI-Induced Ionic Imbalance and Calcium Excitotoxicity

Primary TBI injury can lead to profound imbalances in ion homeostasis in the brain, largely due to the effects of glutamate excitotoxicity. Physical disruption of neural cellular membranes and the impairment of energy-dependent glutamate uptake mechanisms both lead to a dramatic increase in the extracellular concentration of glutamate, an excitatory amino acid (Walker and Tesco 2013). This rise in extracellular glutamate activates a number of glutamate receptors, including *N*-methyl-D-aspartate (NMDA) receptors (Giza and Hovda 2001). Over-activation of these receptors affects ionic balance by concurrently increasing the flux of potassium out of cells and sodium, chloride, and calcium into cells. These ionic shifts establish conditions that cause neurons and glial cells to swell to a pathological degree (Ray et al. 2002). Furthermore, intracellular magnesium levels are reduced. This unblocks NMDA receptors, leading to an even greater influx of calcium across the cell membrane (Walker and Tesco 2013). This disturbance in ionic balance is harmful, activating damaging intracellular cascades including lipid peroxidation, protein breakdown, and free radical generation (Werner and Engelhard 2007).

Energy metabolism is also significantly impacted by this post-TBI ionic imbalance. As the sodium-potassium pump attempts to restore membrane potential, increasing amounts of ATP are required. However, ATP availability is limited due to decreased oxygen and glucose, stemming from impaired cerebral blood flow and ischemia (Giza and Hovda 2001). The failure of energy resources to correct ionic shifts initiates the release of calcium from intracellular stores (Raijmakers et al. 2005; Szydlowskaa and Tymianskia 2010), further exacerbating the high calcium levels resulting from extracellular influx. This dramatic rise in intracellular calcium is regarded as a hallmark consequence of TBI (Fineman et al. 1993; Floyd et al. 2005; McIntosh et al. 1997; Sun et al. 2008; Weber 2012).

Calcium accumulation is seen within hours of brain injury and persists for several days (Giza and Hovda 2001). This overload can lead to several pathological processes, including the activation of calcium-dependent enzymes (Szydlowskaa and Tymianskia 2010; Weber 2012), the disruption of essential neural structural elements such as neurofilaments and microtubules, as well as the activation of apoptotic genetic signals (Giza and Hovda 2001). Disordered calcium homeostasis also results in an abnormal profile of protein citrullination, generating the formation of potentially antigenic epitopes.

16.2.2 Calcium Overload and PAD Activation

Elevated intracellular calcium activates peptidylarginine deiminases (PADs), enzymes that catalyze the conversion of intra-peptide arginine residues to intra-peptide citrulline residues (Ishigami and Maruyama 2010). Under normal conditions, protein citrullination is essential for a number of basic physiological functions, including epidermal hydration (Rorhbach et al. 2002), epigenetic regulation of gene expression (Spengler and Schell-Toellner 2014), hair growth (Rorhbach et al. 2002), and neural plasticity in the stages of early brain development (Hensvold et al. 2002). However, in pathological states, abnormal hyper-citrullination of proteins can occur. The role of abnormal protein citrullination in pathogenesis is best understood in the case of rheumatoid arthritis, an autoimmune disorder involving progressive inflammation of synovial joints (György et al. 2006). In this disorder, the citrullination of selected proteins (e.g., filaggrin and vimentin) prompts an autoimmune response due to the formation of antigenic epitopes. Protein citrullination is also implicated in several neurodegenerative disorders with altered calcium homeostasis and immune basis, including multiple sclerosis (MS) (Anzilotti et al. 2010), Alzheimer's disease (Ishigami et al. 2005), temporal lobe epilepsy (Asaga and Ishigami 2001), and glaucoma (Bhattacharya et al. 2006).

PAD enzymes exist as five isoforms, PAD1, PAD2, PAD3, PAD4, and PAD6 (Wang and Wang 2013), which are expressed in a tissue-specific manner. PAD1, for example, serves to regulate cornification in the epidermis and is also expressed in the uterus. PAD3 is also expressed in the epidermis, as well as hair follicles, while PAD6 is associated with the ovary and testes, serving to regulate embryonic development (Wang and Wang 2013). PAD2 and PAD4, however, are the only PAD enzymes found in neural tissue. PAD2 is localized to astrocytes (Curis et al. 2005; Jang et al. 2008; Raijmakers et al. 2005) and oligodendrocytes (Gould et al. 2000), while PAD4 is exclusively expressed in neurons (Acharya et al. 2012). PAD2 and PAD4 also exhibit the ability to regulate gene activity through localization to the nucleus, where they mediate histone citrullination (Jang et al. 2011; Lange et al. 2014; Wang and Wang 2013).

While elevated intracellular calcium concentration is a hallmark of brain injury, as well as a necessary prerequisite for PAD activation, little attention has been given to the role of PAD and citrullinated proteins in TBI pathology. Citrullination alters the structure and function of proteins. The conversion of arginine to citrulline results in the loss of a positive charge, which alters tertiary structure, proteolytic susceptibility, and protein-protein interactions (Jang et al. 2008; Lam 2006). In MS, citrullination of myelin basic protein (MBP) limits the ability of this protein to appropriately associate with lipids (György et al. 2006), contributing to demyelination by destabilizing sheath structure (Musse et al. 2008). It has been proposed that the dysfunctional effects of citrullination on myelin sheath structure play a major role in the development of MS (György et al. 2006). Furthermore, citrullinated proteins are also observed within the extracellular plaques seen in postmortem brains affected by Alzheimer's disease, suggesting a functional role for this modification in neurodegenerative pathology (Nicholas 2010).

Perhaps most importantly, in addition to altering both protein structure and function, citrullination has the potential to create "altered-self" epitopes that may be antigenic, prompting the adaptive immune system to mount an autoimmune response against native proteins (Chirivi et al. 2013; Zhao et al. 2010). In this regard, the citrullination of MBP in MS leads not only to myelin degradation but also results in the generation of autoantigenic MBP isomers, which are consequently targeted by T-cell lymphocytes (Tranquill et al. 2000). As noted above, the effects of citrullination are most significantly studied in the context of rheumatoid arthritis, where anti-citrullinated protein antibodies are utilized as diagnostic biomarkers due to the dramatic antigenicity of this modification in inflamed synovial joint spaces (György et al. 2006). Thus, the ability of abnormal citrullination to prompt an adaptive immune response may play a pivotal role in the long-term pathogenesis of TBI. Our work has investigated the expression of citrullination in an adult rodent model of TBI, controlled cortical impact (CCI), 5 days following injury, identifying specific proteins modified following injury, examining what regions and neural cell types were most susceptible to each modification, and studying the impact of gender on the expression of citrullination. The effects of CCI on protein citrullination were determined using a mouse monoclonal anti-citrulline antibody (mAb 6B3, IgG2b).

16.3 Region- and Cell-Specific Protein Citrullination Following Traumatic Brain Injury

The regional effects of CCI on protein citrullination are summarized in Fig. 16.1. CCI produced a marked increase in protein citrullination throughout the injured cortex, extending from the lateral to the lesion site to regions of the cortex not directly impacted by CCI (Fig. 16.1b). Similarly, immunolabeling of the injured ipsilateral hippocampal formation (Fig. 16.1e) and external capsule (Fig. 16.1h) revealed a significant degree of protein citrullination in these regions. Other brain regions, including the amygdala and caudatoputamen, were completely negative for protein citrullination in these CCI animals. Furthermore, no gender difference was observed in the location or degree of protein citrullination following injury. In general, CCI appeared to have little effect on the status of protein citrullination in contralateral brain structures, with the exception of the dorsal hippocampus, where approximately 20% of injured animals (4 of 21) displayed an intense labeling of unusually large and rounded cells (Fig. 16.1f). 6B3 immunolabeling was uniformly low across all regions studied in uninjured, control male and female animals (Fig. 16.1a, d, g).

Dual immunofluorescence revealed astrocytes to be the principal cell type in which protein citrullination was affected by CCI. Figure 16.2 shows that anti-citrulline labeling in the cortex and external capsule was predominantly co-localized with GFAP, a cell-type marker for astrocytes. Similar observations were made in other affected brain regions.

The findings presented in Fig. 16.3 further confirmed that CCI-induced protein citrullination was not significantly associated with neurons (NeuN), microglia/macrophages (Iba1), or oligodendrocytes (MBP) in the cortex. Citrullination was also not significantly associated with these cell types in any other brain regions investigated (not shown).

The finding that TBI-induced protein citrullination is selectively localized to astrocytes is consistent with this glial cell type's role in buffering large shifts in calcium homeostasis following TBI. While influxes in intracellular calcium are also found in neurons following mechanical injury, this rise is far less profound than that observed in astrocytes (Rzigalinski et al. 1998). It has been proposed that the intracellular rise of calcium within astrocytes is directly linked to their neuroprotective effects on neurons (Duffy and MacVicar 1996), perhaps suggesting why the death of astrocytes is observed to precede that of neurons in rodent models of TBI (Floyd et al. 2005).

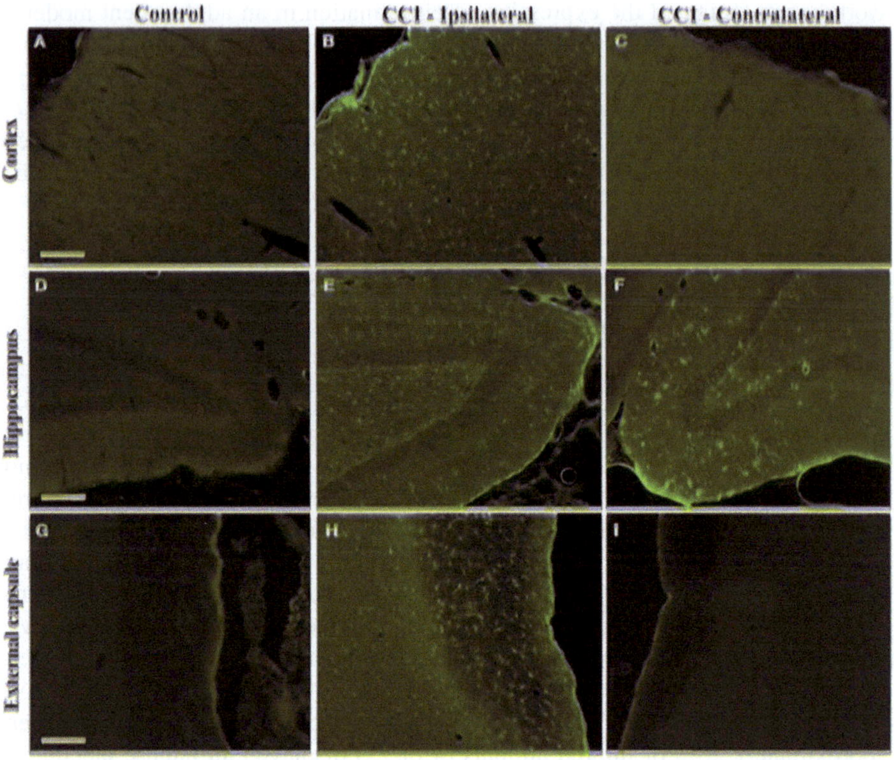

Fig. 16.1 Increased protein citrullination in the cerebral cortex, hippocampus, and external capsule following CCI. Anti-protein citrulline immunolabeling by mAb 6B3 is shown for the control brain regions (**a**, **d**, and **g**), regions ipsilateral to the lesion (**b**, **e**, and **h**), and regions contralateral to the lesion (**c**, **f**, and **g**). Structures represented are the cerebral cortex (**a**–**c**), hippocampus (**d**–**f**), and external capsule (**g**–**i**). Data are representative of 15 control animals (8 males and 7 females) and 21 CCI animals (11 males and 10 females). No gender-based differences were observed. Scale bar = 200 μm

The upregulation of protein citrullination induced by TBI was specific to astrocytes of the cerebral cortex, external capsule, and hippocampus. In contrast, the presence of protein citrullination in other cell types and brain regions was not appreciably changed following injury. The mechanistic basis for this observation may relate to the upregulation of voltage-gated, class C L-type Ca^{2+} channels that are selectively expressed in astrocytes. These channels are particularly sensitive to activation by injury in the regions reported here (Chung et al. 2001). The finding that nimodipine, an L-type calcium channel blocker, protects against excitotoxic damage in cultured astrocytes supports this conclusion (Haun et al. 1992).

Furthermore, this investigation showed that patterns of protein citrullination following TBI were similar between male and female rats. This finding is in contrast to our previous observations concerning the effects of TBI on protein carbonylation (Lazarus et al. 2015), a protein modification associated with free radical generation

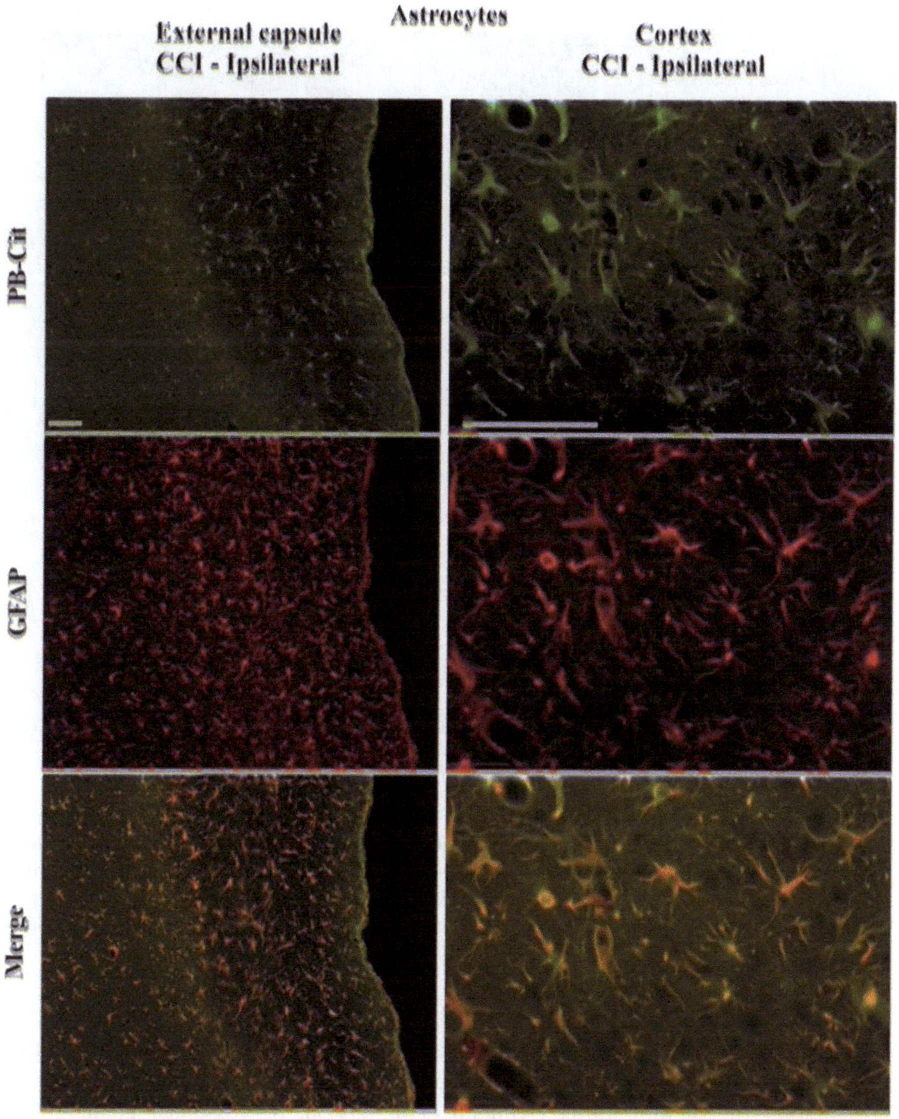

Fig. 16.2 Localization of CCI-induced protein citrullination to astrocytes. Panels on the *left* show the co-localization of mAb 6B3 labeling with anti-GFAP labeling in the external capsule. Panels on the *right* show the co-localization of mAb 6B3 labeling with anti-GFAP labeling in the cerebral cortex. Data are representative of 15 control animals (8 males and 7 females) and 21 CCI animals (11 males and 10 females). No gender-based differences were observed. Scale bar = 200 μm. *PB-cit* protein-bound citrulline

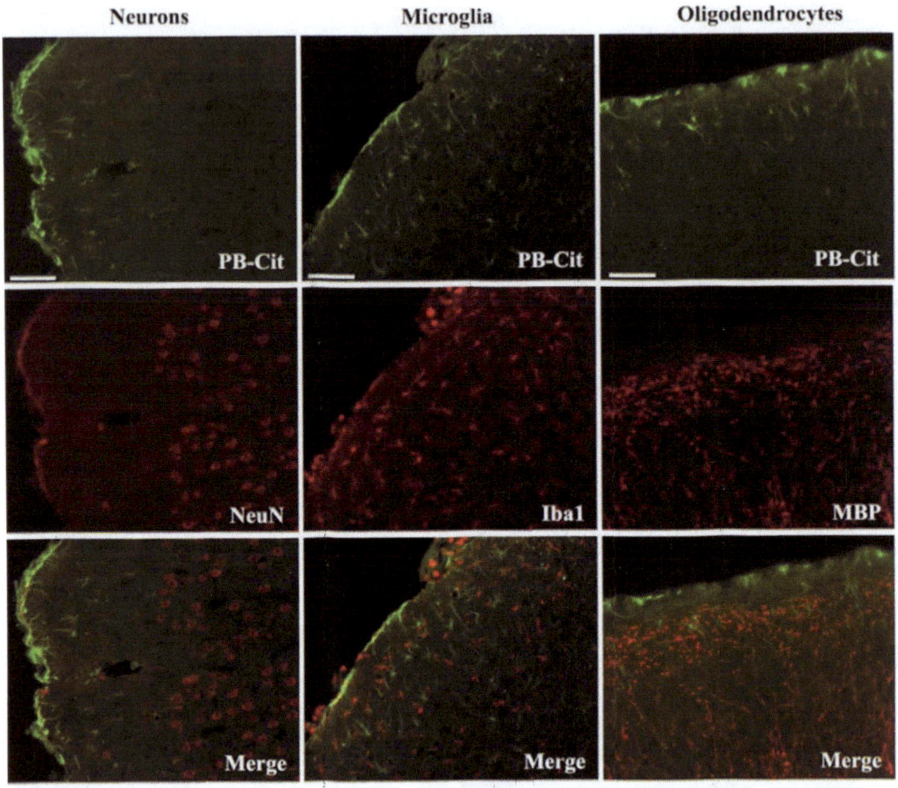

Fig. 16.3 CCI did not affect the status of protein citrullination in neurons, microglia, or oligodendrocytes. Sections of cerebral cortex ipsilateral to CCI were probed with mAb 6B3 to label protein-bound citrulline (PB-cit; *upper panels*) and either anti-NeuN, Iba1, or MBP to label neurons, microglia, or oligodendrocytes, respectively (*middle panels*). The merge of the two signals is presented in the *lower panels*. Data are representative of eight separate experiments. Scale bar = 200 μm

and oxidative stress. Following injury, male rats showed far greater response in protein carbonylation as compared to female rats (Lazarus et al. 2015). Protein carbonylation is a reflection of oxidative stress (Dalle-Donne et al. 2006), whereas citrullination is a marker of calcium influx (Kinloch et al. 2005a, b; Vossenaar et al. 2004). Accordingly, the gender difference observed in carbonylation may be due to the protective antioxidant effects of ovarian steroids (Roof et al. 1997), whereas a similar protective, gender-mediated mechanism does not appear to exist in the case of citrullination. The extent to which these protein modifications contribute to gender differences in TBI complications and mortality (Berry et al. 2009; Goswasser et al. 1998) remains to be elucidated. It should be noted that the present findings focus on only a single time point post-TBI (5 days). In consideration of this, it is possible that gender differences in post-injury citrullination could be evident at other time points.

16.4 Identification of Proteins Citrullinated in Response to TBI

Proteomic analysis revealed that the effects of CCI on protein citrullination were specific to a discrete subset of proteins within the brain proteome. The proteins identified are primarily associated with cytoskeletal structure and metabolic processes (Fig. 16.4). Shown in the upper left panel are the proteomes of control and injured cerebral cortex fractionated by fluid-phase isoelectric focusing. Each pair of lanes, control (C) and CCI (I), shows the proteins present in the four different pI partitions. As visualized by Coomassie staining, CCI did not significantly affect the general pattern of protein staining across the four pI fractions. In contrast, the pattern of protein citrullination was dramatically altered by CCI (upper right panel). Consistent with our immunohistochemistry findings, little protein citrullination was observed in control cortex (C), whereas CCI (I) resulted in the intense labeling of a distinctive subset of the fractionated proteins. The immunoreactive signals of the Western blot were mapped to corresponding features on a Coomassie-stained gel, and proteins were identified by peptide mass fingerprinting and tandem mass spectrometry. The proteins identified (lower panel of Fig. 16.4) may be functionally grouped as cytoskeletal components (including dynamin-1, GFAP, and several forms of tubulin), those involved in metabolic processes (including peroxiredoxin-1, dihydropyrimidinase-related protein 2, and creatine kinase B type), and proteins involved in cell-cell signaling and synaptic transmission (synapsin-2, syntaxin-binding protein 1, and amphiphysin).

The findings presented in Fig. 16.4 are representative of multiple independent analyses. These analyses consistently demonstrated that only a very small subset of the entire brain proteome is affected by TBI-induced citrullination. This consistent selectivity indicates that the mechanism involved is highly specific. A large proportion of the 37 proteins identified here are recognized as being citrullinated in other pathologies, including MS, Alzheimer's disease, prion disease, and rheumatoid arthritis (CM. Bradford et al. 2014; György et al. 2006; Ishigami et al. 2005; Jang et al. 2008; Kinloch et al. 2005b; Nicholas et al. 2005; Tranquill et al. 2000). Specifically, citrullinated GFAP is a characteristic feature in MS and Alzheimer's disease (CM. Bradford et al. 2014; György et al. 2006; Ishigami et al. 2005; Nicholas et al. 2004), and myelin basic protein (MBP), a major component of myelin sheath structure, is profoundly over-citrullinated in MS (CM. Bradford et al. 2014; Tranquill et al. 2000). Similarly, GFAP, tubulin, peroxiredoxin-1, cofilin-1, and alpha/gamma enolase are each selectively citrullinated in prion disease (Jang et al. 2008).

As seen in CCI, blast injury also results in changes in the profile of citrullinated proteins in the brain. As shown in Fig. 16.5, the response to blast injury in porcine is selective, affecting only a relatively small subset of the entire brain proteome, as observed in rodent CCI. In this experiment, porcine were exposed to a single moderate explosive blast (SB; psi = 43–50) or two blasts (RB) 1 week apart. Samples were collected at 2 and 19 weeks post-blast for the SB and RB conditions, respectively.

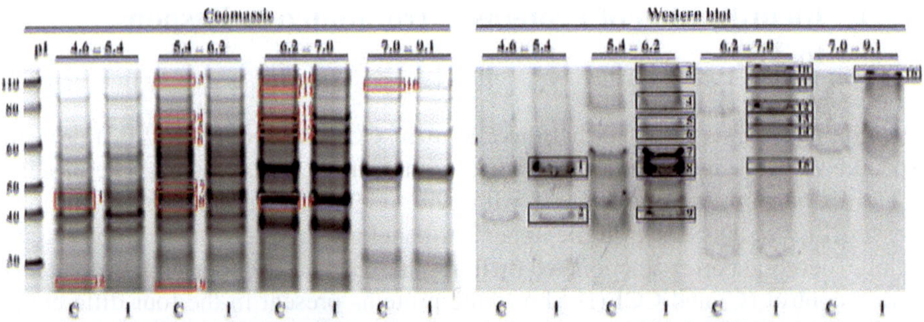

Sample	Score	Expectation	Protein ID	Protein name	MW	% Cov	Accession #	Cit Pep	Prev Report
1	3387	0	ALDOA_RAT	Fructose-bisphosphate aldolase A	39327	68.4	P05065	Y	Y
2	1223	3.70E-119	PRDX1_RAT	Peroxiredoxin-1	22095	60.3	Q63716	N	Y
	1083	1.80E-105	COF1_RAT	Cofilin-1	18521	83.7	P45592	N	Y
	1059	1.10E-102	GSTP1_RAT	Glutathione S-transferase P	23424	52.9	P04906	N	N
	946	2.10E-91	TAGL_RAT	Transgelin	22588	65.7	P31232	N	N
3	7314	0	DYN1_RAT	Dynamin-1	97333	72.9	P21575	Y	N
4	7847	0	SYN2_RAT	Synapsin-2	63417	65	Q63537	Y	N
	1348	0	MOES_RAT	Moesin	67697	51.9	O35763	Y	N
	1241	0	SEPT9_RAT	Septin-9	63152	45.9	Q9QZR6	Y	N
	735	0	ANXA6_RAT	Annexin A6	75706	44.9	P48037	N	N
5	1317	0	STIP1_RAT	Stress-induced-phosphoprotein 1	62530	43.3	O35814	N	N
	1178	8.00E-114	STXB1_RAT	Syntaxin-binding protein 1	67526	46.6	P61765	N	N
6	3206	0	DPYL2_RAT	Dihydropyrimidinase-related protein 2	62239	71.7	P47942	N	N
7	4618	0	ENOA_RAT	Alpha-enolase	47098	77.6	P04764	N	Y
	886	2.00E-85	ENOG_RAT	Gamma-enolase	47111	36.4	P07323	N	Y
	701	6.00E-67	GFAP_RAT	Glial fibrillary acidic protein	49927	35.8	P47819	N	Y
8	1776	0	ALDOC_RAT	Fructose-bisphosphate aldolase C	39259	59.2	P09117	N	Y
	964	8.30E-93	SEPT5_RAT	Neuronal-specific septin-3	40572	32.7	Q9WU34	N	N
9	1516	0	TAGL3_RAT	Transgelin-3	22486	85.4	P37805	N	N
	698	1.20E-66	MBP_RAT	Myelin basic protein	21499	53.9	P02688	Y	Y
10	2309	0	HSP74_RAT	Heat shock 70 kDa protein 4	93997	52.1	O88600	N	N
	1830	0	GRAP1_RAT	GRIP1-associated protein 1	96018	39.7	Q9JHZ4	N	N
11	3751	0	TERA_RAT	Transitional endoplasmic reticulum ATPase	89293	55.5	P46462	Y	N
12	3017	0	NDUS1_RAT	NADH-ubiquinone oxidoreductase 75 kDa subunit, mitochondrial	79363	49	Q66HF1	Y	N
	1883	0	NSF_RAT	Vesicle-fusing ATPase	82600	43.5	Q9QUL6	N	N
	1837	0	GRP78_RAT	78 kDa glucose-regulated protein	72302	43.1	P06761	N	Y
13	7555	0	HSP7C_RAT	Heat shock cognate 71 kDa protein	70827	82.3	P63018	Y	N
	1953	0	ANXA6_RAT	Annexin A6	75706	57.9	P48037	N	N
14	1212	3.10E-118	STXB1_RAT	Syntaxin-binding protein 1	67526	42.1	P61765	N	N
15	3440	0	ACTB_RAT	Actin, cytoplasmic 1	41710	61.1	P60711	N	N
	1943	0	KCRB_RAT	Creatine kinase B-type	42698	54.9	P07335	Y	N
	1488	0	TBB2B_RAT	Tubulin beta-2B chain	49921	47.4	Q3KRE8	N	Y
	1433	0	TBB2A_RAT	Tubulin beta-2A chain	49875	47.4	P85108	N	Y
	1237	1.60E-120	TBB4B_RAT	Tubulin beta-4B chain	49769	46.8	Q6P9T8	N	Y
	1073	4.10E-105	TBB5_RAT	Tubulin beta-5 chain	49639	42.8	P69897	N	Y
	932	5.20E-90	TBB3_RAT	Tubulin beta-3 chain	50386	35.3	Q4QRB4	N	Y
16	2677	0	AMPH_RAT	Amphiphysin	74852	73.5	O08838	Y	N
	1801	0	DREB_RAT	Drebrin	77484	43.3	Q07266	Y	N

Fig. 16.4 Identification of proteins that are citrullinated in response to CCI. Extracts of control (C) and injured (I) cerebral cortex were fractionated by fluid-phase isoelectric focusing into defined pH ranges (shown at top) and then further resolved according to molecular weight using one-dimensional gel electrophoresis. Proteins were then transferred to nitrocellulose membranes and probed for protein-bound citrulline (*right panel*, "Western blot"). Gels run in parallel were visualized with Coomassie (*left panel*). Sixteen features showing increased citrullination in response to CCI (*black numbered boxes, right panel*) were mapped to corresponding Coomassie features (*red numbered boxes, left panel*) and identified by peptide mass finger printing and tandem mass spectrometry. Proteins identified are listed in the *lower panel*. Images are representative of six independent experiments. A total of *n* = 4 CCI and *n* = 4 control animals were examined

These representative data show that there is a pronounced upregulation of protein citrullination following blast injury. Intriguingly, this response is evident for at least 4 months following injury. Accordingly, the profile of protein citrullination in the brain may constitute a long-term biosignature for blast TBI.

In this regard, it is proposed that the link between abnormal TBI-induced protein citrullination and the expression of long-term pathology involves the adaptive immune system. A significant proportion of the proteins identified here (Fig. 16.4) are also recognized as autoantigenic in both neurological and autoimmune-related disorders. Autoantibodies targeting amphiphysin are associated with several neurological disorders, including sensory neuronopathy and encephalopathy (Murinson and Guarnaccia 2008). Additionally, dihydropyrimidinase-related protein 2 (CRMP2) is autoantigenic in autoimmune retinopathy (Adamus et al. 2013). Finally, alpha enolase has been identified as a citrullinated autoantigen in rheumatoid arthritis (Kinloch et al. 2005b), while citrullinated 78 kDa glucose-regulated protein is an autoantigen within the pancreatic beta cells in type 1 diabetes (Rondas et al. 2014). Collectively, these findings support the proposal that injury-induced protein citrullination may generate immunological epitopes that become targets of the adaptive immune system. This process may serve as an underlying basis for chronic and progressive neurological pathology following TBI.

16.5 An In Vitro Model for the Investigation of TBI-Induced Protein Citrullination

An in vitro model of simulated TBI has been developed using normal human astrocytes treated with the calcium ionophore, ionomycin, to induce calcium excitotoxicity. Figure 16.6 shows the results of three separate experiments investigating the effects of ionomycin treatment on the proteolytic processing of GFAP (left panel) and the generation of citrullinated proteins (right panel). The data show that treatment with ionomycin consistently activated the proteolytic processing of intact GFAP (left panel; blue arrows) to produce a distinctive pattern of breakdown products. Probing with mAb 6B3 indicated that one of the GFAP breakdown products is preferentially citrullinated (right panel; orange arrow) in response to simulated TBI. These are consistent with those reported by others on injury-induced proteolytic processing of GFAP using in vitro models of TBI (Okonkwo et al. 2013; Zoltewicz et al. 2012). Importantly, our investigation reveals that one of the breakdown products is heavily citrullinated and thus may serve as the antigen for the development of the anti-GFAP autoantibodies recently reported in TBI (Zhang et al. 2014). Accordingly, the application to this and other in vitro models for simulated TBI may provide novel insights into consequences and mechanisms of TBI and also identify informative biomarkers for assessing brain injury.

16.6 Summary of Findings

This research demonstrates that TBI dramatically upregulates protein citrullination within astrocytes in specific brain regions. This response is specific to a small subset of the neural proteome and long-lasting. A substantial number of the proteins involved have been identified as being citrullinated in other pathologies, including MS, Alzheimer's disease, and rheumatoid arthritis. This indicates a potential role

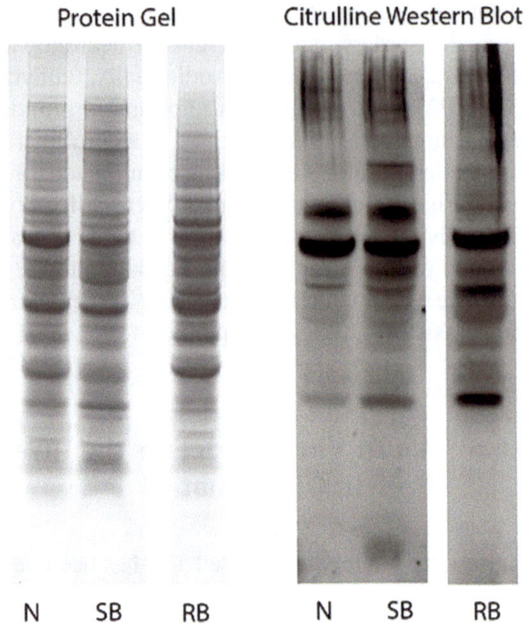

Fig. 16.5 The effects of blast injury on brain protein citrullination are sustained for long periods of time. Anion exchange FPLC fractions of cerebral cortex from single (SB) and repeat blast (RB) porcine were prepared and further fractionated by SDS-PAGE. Gels were stained for total protein (Coomassie; *left panel*) or transferred to PVDF and probed with anti-protein citrulline antibody (6B3)

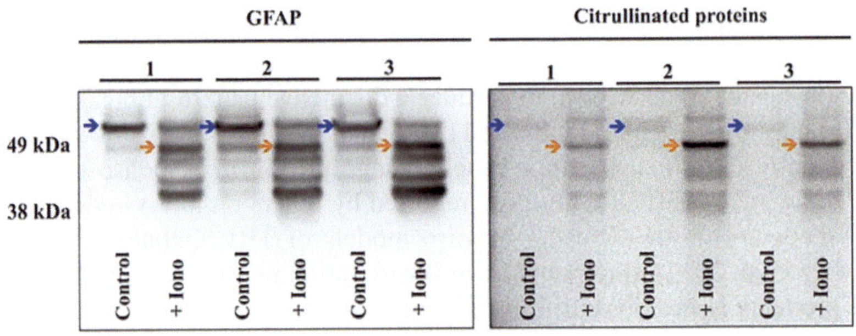

Fig. 16.6 Simulated brain injury in normal human astrocytes reveals a spectrum of GFAP breakdown products and the hyper-citrullination of one GFAP species. Normal human astrocytes were treated with ionomycin (10 μM; 4 h) and analyzed for GFAP (*left panel*) and protein-bound citrulline immunoreactivity (*right panel*) by Western blot. The results of three independent experiments (**a**, **b**, and **c**) are presented, showing the immunoreactivity in extracts prepared from untreated control cells (Control) and cells treated with ionomycin (+ Iono). The *blue arrows* indicate intact GFAP, while the *orange arrows* indicate the hyper-citrullinated GFAP breakdown product

for post-TBI citrulline modification in ongoing neural pathological processes. Interestingly, gender does not affect the degree or distribution of this modification in neural tissue, in distinction to observations involving TBI-induced protein carbonylation (Lazarus et al. 2015). Accordingly, gender differences in the CNS response to TBI may not involve differential responses in protein citrullination.

There are several potential mechanisms through which abnormal protein citrullination may contribute to the pathogenesis of TBI. In addition to changes in protein structure (and thus function), citrullination can have profound effects on protein antigenicity through the creation of epitopes that are potentially neo-antigenic (Vossenaar et al. 2004). This phenomenon is observed in MS, where the citrullination of MBP prompts myelin degeneration due to a loosened aggregation of lipids in the sheath structure (Musse et al. 2008). Citrullination of MBP also increases its susceptibility to degradation by proteinases, including cathepsin D. This degradation leads to the formation and release of antigenic, citrulline-containing peptides, which are presented to peripheral T cells (C. Bradford et al. 2002). In healthy neural tissue, a small fraction of total myelin basic protein contains citrullinated residues; however, this amount increases threefold in multiple sclerosis and presumably occurs at novel sites within the protein, inducing a targeted T-cell response (Anderton 2004; Tranquill et al. 2000). As noted, citrullinated epitopes are the hallmark of the adaptive immune response in rheumatoid arthritis, such that anti-citrullinated protein antibodies serve as a diagnostic biomarker for this disease (Hensvold et al. 2002).

Two other proteins identified here, glucose-regulated protein 78 and amphiphysin, have also been identified as autoantigenic. Citrullinated glucose-regulated protein 78 is recognized as an autoantigen in type 1 diabetes, where it is secreted by pancreatic beta cells, prompting the production of autoantibodies (Rondas et al. 2014). Anti-amphiphysin autoantibodies have been identified in several neurological disorders, including encephalopathy and myelopathy. It is possible that the immune response against amphiphysin is triggered by antigenic citrullinated epitopes. While this type of immune response targeted to neural tissue can be harmful, the brain may also benefit from immune responses following injury. This benefit could involve regulation of localized inflammation and rebuilding of the blood-brain barrier by the immune-modulating "glial scar," a post-injury collection of astrocytes, microglia, macrophages, and extracellular matrix molecules (Rolls et al. 2009). Future research should address the possible detriments and benefits of a citrullination-mediated immune response in neural tissue following injury, including aspects of localized inflammation and immune clearance.

It is not currently understood how the acute, mechanical impact of TBI leads to chronic pathology that can last anywhere from months to years following injury. Protein modifications shared between TBI and neurodegenerative disorders may point to a mechanistic link between physical injury and sustained dysfunction. Specific changes in protein structure and function, seen as early as 5 days following TBI, could lead to the pathological symptomatology mirrored in diseases such as Alzheimer's and multiple sclerosis. Revealing the specific proteomic effects of citrullination could provide mechanistic insight into how acute injury translates into chronic neurodegeneration. We propose that protein citrullination represents a critical mechanistic link between acute physical injury and long-term pathology. Further investigations of TBI-induced citrullination will lend insight into the biological mechanisms of this injury and may also point to proactive measures to counteract long-term pathogenesis.

Acknowledgments Support for this work included a grant from the Defense Medical Research and Development Program (D61_I_10_J6_152) and Department of Defense in the Center for Neuroscience and Regenerative Medicine (G1703D) as well as funding from the Uniformed Services University of the Health Sciences (T0702554 and R07028414).

Disclosure

Some of the findings presented here have appeared in publication (Lazarus Rachel et al. 2015) and in a PhD dissertation (RCL) which is available by public access at http://cdm16005.contentdm.oclc.org/cdm/ref/collection/p15459coll1/id/117618 (Lazarus 2015).

References

Acharya, N. K., Nagele, E. P., Han, M., Coretti, N. J., DeMarshall, C., Kosciuk, M. C., et al. (2012). Neuronal PAD4 expression and protein citrullination: Possible role in production of autoantibodies associated with neurodegenerative disease. *Journal of Autoimmunity, 38*, 369–380.

Adamus, G., Bonnah, R., Brown, L., & David, L. (2013). Detection of autoantibodies against heat shock proteins and collapsin response mediator proteins in autoimmune retinopathy. *BMC Ophthalmology, 13*, 48.

Anderton, S. M. (2004). Post-translational modifications of self antigens: Implications for autoimmunity. *Current Opinion in Immunology, 16*(6), 753–758.

Anzilotti, C., Pratesi, F., Tommasi, C., & Migliorini, P. (2010). Peptidylarginine deiminase 4 and citrullination in health and disease. *Autoimmunity Reviews, 9*(3), 158–160.

Asaga, H., & Ishigami, A. (2001). Protein deimination in the rat brain after kainate administration: Citrulline-containing proteins as a novel marker of neurodegeneration. *Neuroscience Letters, 299*, 5–8.

Bales, J. W., Wagner, A. K., Kline, A. E., & Dixon, C. E. (2009). Persistent cognitive dysfunction after traumatic brain injury: A dopamine hypothesis. *Neuroscience and Biobehavioral Reviews, 33*(7), 981–1003.

Berry, C., Ley, E. J., Tillou, A., Cryer, G., Marguilies, D. R., & Salim, A. (2009). The effect of gender on patients with moderate to severe head injuries. *Journal of Trauma – Injury, Infection &. Critical Care, 67*(5), 950–953.

Bhattacharya, S. K., Crabb, J. S., Bonilha, V. L., Gu, X., Takahara, H., & Crabb, J. W. (2006). Proteomics implicates peptidyl arginine deiminase 2 and optic nerve citrullination in glaucoma pathogenesis. *Investigative Ophthalmology & Visual Science, 47*(6), 2508–2514.

Bradford, C., Nicholas, A. P., Woodroofe, N., & Cross, A. K. (2002). Chapter 10: Deimination in multiple sclerosis and experimental autoimmune encephalomyelitis. In A. Nicholas & S. Bhattacharya (Eds.), *Protein deimination in human health and disease* (pp. 165–185). New York: Springer.

Bradford, C. M., Ramos, I., Cross, A. K., Haddock, A. K., McQuaid, S., Nicholas, A. P., & Woodroofe, M. N. (2014). Localisation of citrullinated proteins in normal appearing white matter and lesions in the central nervous system in multiple sclerosis. *Journal of Neuroimmunology, 273*(1–2), 85–95.

Chirivi, R. G. S., van Rosmalen, J. W. G., Jenniskens, G. J., Pruijn, G. J., & Raats, J. M. H. (2013). Citrullination: A target for disease intervention in multiple sclerosis and other inflammatory diseases? *Journal of Clinical and Cellular Immunology, 4*, 146. doi:10.4172/2155-9899.1000146.

Chung, Y. H., Shin, C. M., Kim, M. J., & Cha, C. I. (2001). Enhanced expression of L-type Ca2+ channels in reactive astrocytes after ischemic injury in rats. *Neuroscience Letters, 302*, 93–96.

Curis, E., Nicolis, I., Moinard, C., Osowska, S., Zerrouk, N., Bénazeth, S., & Cynober, L. (2005). Almost all about citrulline in mammals. *Amino Acids, 29*(3), 177–205.

Dalle-Donne, I., Aldini, G., Carini, M., Colombo, R., Rossi, R., & Milzani, A. (2006). Protein carbonylation, cellular dysfunction, and disease progression. *Journal of Cellular and Molecular Medicine, 10*(2), 389–406.

Duffy, S., & MacVicar, B. A. (1996). In vitro ischemia promotes calcium influx and intracellular calcium release in hippocampal astrocytes. *The Journal of Neuroscience, 16*(1), 71–81.

Fineman, I., Hovda, D. A., Smith, M., Yoshino, A., & Becker, D. P. (1993). Concussive brain injury is associated with a prolonged accumulation of calcium: A 45Ca autoradiographic study. *Brain Research, 624*, 94–102.

Floyd, C. L., Gorin, F. A., & Lyeth, B. G. (2005). Mechanical strain injury increases intracellular sodium and reverses Na+/Ca2+ exchange in cortical astrocytes. *Glia, 51*(1), 35–46.

Giza, C. C., & Hovda, D. A. (2001). The neurometabolic cascade of concussion. *Journal of Athletic Training, 36*(3), 228–235.

Goswasser, Z., Cohen, M., & Keren, O. (1998). Female TBI patients recover better than males. *Brain Injury, 12*, 805–808.

Gould, R. M., Freund, C. M., Palmer, F., & Feinstein, D. L. (2000). Messenger RNAs located in myelin sheath assembly sites. *Journal of Neurochemistry, 75*(5), 1834–1844.

György, B., Tóth, E., Taresa, E., Falus, A., & Buzás, E. I. (2006). Citrullination: A posttranslational modification in health and disease. *The International Journal of Biochemistry & Cell Biology, 38*, 1662–1677.

Haun, S. E., Murphy, E. J., Bates, C. M., & Horrocks, L. A. (1992). Extracellular calcium is a mediator of astroglial injury during combined glucose-oxygen deprivation. *Brain Research, 593*, 45–50.

Hensvold, A. H., Reynisdottir, G., & Catrin, A. I. (2002). Chapter 2: From citrullination to specific immunity and disease in rheumatoid arthritis. In A. Nicholas & S. Bhattacharya (Eds.), *Protein deimination in human health and disease* (pp. 25–40). New York: Springer.

Ishigami, A., & Maruyama, N. (2010). Importance of research on peptidylarginine deiminase and citrullinated proteins in age-related disease. *Geriatrics & Gerontology International, 10*(Suppl 1), S53–S58.

Ishigami, A., Ohsawa, T., Hiratsuka, M., Taguchi, H., Kobayashi, S., Saito, Y., et al. (2005). Abnormal accumulation of citrullinated proteins catalyzed by peptidylarginine deiminase in hippocampal extracts from patients with Alzheimer's disease. *Journal of Neuroscience Research, 80*(1), 120–128.

Jang, B., Kim, E., Choi, J. K., Jin, J. K., Kim, J. I., Ishigami, A., et al. (2008). Accumulation of citrullinated proteins by up-regulated peptidylarginine deiminase 2 in brains of scrapie-infected mice. *The American Journal of Pathology, 173*(4), 1129–1142.

Jang, B., Shin, H. Y., Choi, J. K., Nguyen du, P. T., Jeong, B. H., Ishigami, A., et al. (2011). Subcellular localization of peptidylarginine deiminase 2 and citrullinated proteins in brains of scrapie-infected mice: Nuclear localization of PAD2 and membrane fraction-enriched citrullinated proteins. *Journal of Neuropathology and Experimental Neurology, 70*(2), 116–124.

Kinloch, A., Tatzer, V., Wait, R., Peston, D., Lundberg, K., Donatien, P., et al. (2005a). Identification of citrullinated α-enolase as a candidate autoantigen in rheumatoid arthritis. *Arthritis Research & Therapy, 7*, R1421–R1429.

Kinloch, A., Tatzer, V., Wait, R., Peston, D., Lundberg, K., Donatien, P., et al. (2005b). Identification of citrullinated α-enolase as a candidate autoantigen in rheumatoid arthritis. *Arthritis Research & Therapy, 7*(6), R1421–R1429.

Lam, G. (2006). *Round 4: Citrullinated proteins, peptidylarginine deiminase (PAD), and rheumatoid arthritis*. Baltimore, MD: Johns Hopkins, Arthritis Center.

Lange, S., Rocha-Ferreira, E., Thei, L., Mawjee, P., Bennett, K., Thompson, P. R., et al. (2014). Peptidylarginine deiminases: Novel drug targets for prevention of neuronal damage following hypoxic ischemic insult (HI) in neonates. *Journal of Neurochemistry, 130*(4), 555–562.

Lazarus, R. C. (2015). *Protein modification: A proposed mechanism for the long-term pathogenesis of traumatic brain injury*. Uniformed Services University of the Health Sciences, Bethesda, Maryland 20814. Retrieved from http://cdm16005.contentdm.oclc.org/cdm/ref/collection/p15459coll1/id/117618

Lazarus Rachel, C., Buonora John, E., Flora Michael, N., Freedy James, G., Holstein Gay, R., Martinelli Giorgio, P., Jacobowitz David, M., & Mueller Gregory, P. (2015). Protein citrullination:

A proposed mechanism for pathology in traumatic brain injury. *Frontiers in Neurology, 6*, 204. doi:10.3389/fneur.2015.00204.
Lazarus, R. C., Buonora, J. E., Jacobowitz, D. M., & Mueller, G. P. (2015). Protein carbonylation after traumatic brain injury: Cell specificity, regional susceptibility, and gender differences. *Free Radical Biology and Medicine, 78*, 89–100.
McIntosh, T. K., Saatman, K. E., & Raghupathi, R. (1997). Calcium and the pathogenesis of traumatic CNS injury: Cellular and molecular mechanisms. *The Neuroscientist, 3*(3), 169–175.
Murinson, B. B., & Guarnaccia, J. B. (2008). Stiff-person syndrome with amphiphysin antibodies: Distinctive features of a rare disease. *Neurology, 71*(24), 1955–1958.
Musse, A. A., Li, Z., Ackerly, C. A., Bienzle, D., Lei, H., Poma, R., et al. (2008). Peptidylarginine deiminase 2 (PAD2) overexpression in transgenic mice leads to myelin loss in the central nervous system. *Disease Models & Mechanisms, 1*(4), 229–240.
Nicholas, A. P. (2010). Dual immunofluorescence study of citrullinated proteins in Parkinson diseased substantia nigra. *Neuroscience Letters, 495*, 26–29.
Nicholas, A. P., Sambandam, T., Echols, J. D., & Tourtellotte, W. W. (2004). Increased citrullinated glial fibrillary acidic protein in secondary progressive multiple sclerosis. *The Journal of Comparative Neurology, 473*(1), 128–136.
Nicholas, A. P., Sambandam, T., Echols, J. D., & Barnum, S. R. (2005). Expression of Citrullinated proteins in murine experimental autoimmune encephalomyelitis. *The Journal of Comparative Neurology, 486*, 254–266.
Nilsson, P., Laursen, H., Hillered, L., & Hansen, A. J. (1996). Calcium movements in traumatic brain injury: The role of glutamate receptor-operated ion channels. *Journal of Cerebral Blood Flow & Metabolism, 16*, 262–270.
Okonkwo, D. O., Yue, J. K., Puccio, A. M., Panczykowski, D. M., Inoue, T., McMahon, P. J., et al. (2013). GFAP-BDP as an acute diagnostic marker in traumatic brain injury: Results from the prospective transforming research and clinical knowledge in traumatic brain injury study. *Journal of Neurotrauma, 30*(17), 1490–1497.
Opii, W. O., Nukal, V. N., Sultana, R., Pandya, J. D., Day, K. M., Marchant, M. L., et al. (2007). Proteomic identification of oxidized mitochondrial proteins following experimental traumatic brain injury. *Journal of Neurotrauma, 24*(5), 772–789.
Park, E., Bell, J. D., & Baker, A. J. (2008). Traumatic brain injury: Can the consequences be stopped? *CMAJ, 178*(9), 1163–1170.
Piao, C. S., Stoica, B. A., Wu, J., Sabirzhanov, B., Zhao, Z., Cabatbat, R., et al. (2013). Combined inhibition of cell death induced by apoptosis inducing factor and caspases provides additive neuroprotection in experimental traumatic brain injury. *Neurobiology of Disease, 46*(3), 745–758.
Raijmakers, R., Vogelzangs, J., Croxford, J. L., Wesseling, P., van Venrooij, W. J., & Pruijn, G. J. (2005). Citrullination of central nervous system proteins during the development of experimental autoimmune encephalomyelitis. *The Journal of Comparative Neurology, 486*(3), 243–253.
Ray, S. K., Dixon, C. E., & Banik, N. L. (2002). Molecular mechanisms in the pathogenesis of traumatic brain injury. *Histology and Histopathology, 17*, 1137–1152.
Rolls, A., Shechter, R., & Schwartz, M. (2009). The bright side of the glial scar in CNS repair. *Nature Reviews Neuroscience, 10*, 235–241.
Rondas, D., Crèvecoeur, I., D'Hertog, W., Ferreira, G. B., Staes, A., Garg, A. D., et al. (2014). Citrullinated glucose-regulated protein 78 is an autoantigen in type 1 diabetes. *Diabetes, 64*(2), 573–586.
Roof, R. L., Hoffamn, S. W., & Stein, D. G. (1997). Progesterone protects against lipid peroxidation following traumatic brain injury in rats. *Molecular and Clinical Neuropathology, 31*(1), 1–11.
Rorhbach, A. S., Arandjelovic, S., & Mowen, K. A. (2002). Physiological pathways of PAD activation and citrullinated epitope generation. In A. Nicholas & S. Bhattacharya (Eds.), *Protein deimination in human health and disease* (pp. 1–24). New York: Springer.
Rzigalinski, B. A., Weber, J. T., Willoughby, K. A., & Ellis, E. F. (1998). Intracellular free calcium dynamics in stretch-injured astrocytes. *Journal of Neurochemistry, 70*(6), 2377–2385.

Spengler, J., & Schell-Toellner, D. (2014). Neutrophils and their contributions to autoimmunity in rheumatoid arthritis. In A. Nicholas & S. Bhattacharya (Eds.), *Protein deimination in human health and disease* (pp. 97–111). New York: Springer.

Sun, D. A., Deshpande, L. S., Sombati, S., Baranova, A., Wilson, M. S., Hamm, R. J., & DeLorenzo, R. J. (2008). Traumatic brain injury causes a long-lasting calcium (Ca2+)-plateau of elevated intracellular Ca levels and altered Ca2+ homeostatic mechanisms in hippocampal neurons surviving brain injury. *The European Journal of Neuroscience, 27*(7), 1659–1672.

Szydlowskaa, K., & Tymianskia, M. (2010). Calcium, ischemia and excitotoxicity. *Cell Calcium, 47*(2), 122–129.

Tranquill, L. R., Cao, L., Ling, N., Kalbacher, H., Martin, R. M., & Whitaker, J. N. (2000). Enhanced T cell responsiveness to citrulline-containing myelin basic protein in multiple sclerosis patients. *Multiple Sclerosis, 6*, 220–225.

Vossenaar, E. R., Radstake, T. R. D., van der Heijden, A., van Mansum, M. A. M., Dieteren, C., de Rooij, D. J., et al. (2004). Expression and activity of citrullinating peptidylarginine deiminase enzymes in monocytes and macrophages. *Annals of the Rheumatic Diseases, 63*, 373–381.

Walker, K. R., & Tesco, G. (2013). Molecular mechanisms of cognitive dysfunction following traumatic brain injury. *Frontiers in Aging Neuroscience, 5*, 29.

Wang, S., & Wang, Y. (2013). Peptidylarginine deiminases in citrullination, gene regulation, health and pathogenesis. *Biochimica et Biophysica Acta, 1829*, 1126–1135.

Weber, J. T. (2012). Altered calcium signaling following traumatic brain injury. *Frontiers in Pharmacology, 3*(60), 1–16.

Werner, C., & Engelhard, K. (2007). Pathophysiology of traumatic brain injury. *British Journal of Anaesthesia, 99*(1), 4–9.

Zhang, Z., Zoltewicz, J. S., Mondello, S., Newsom, K. J., Yang, Z., Yang, B., et al. (2014). Human traumatic brain injury induces autoantibody response against glial fibrillary acidic protein and its breakdown products. *PloS One, 9*(3), e92698.

Zhao, Y., Tian, X., & Li, Z. G. (2010). Impact of citrullination upon antigenicity of fibrinogen. *Zhonghua Yi Xue Za Zhi, 90*(9), 628–632.

Zoltewicz, J. S., Scharf, D., Yang, B., Chawla, A., Newsom, K. J., & Fang, L. (2012). Characterization of antibodies that detect human GFAP after traumatic brain injury. *Biomarker Insights, 7*, 71–79.

Chapter 17
Update on Deimination in Alzheimer's Disease

Yoshitaka Kondo, Eun-Kyoung Choi, Yong-Sun Kim, and Akihito Ishigami

Abbreviations

AD	Alzheimer's disease
Anti-CCP	Anti-cyclic citrullinated peptide
Anti-MVC	Anti-mutated citrullinated vimentin
BAEE	Benzoyl-L-arginine ethyl ester
Bz-L-Arg	Benzoyl-L-arginine
Calbindin	Calbindin-D-28K
cAMP	Cyclic adenosine monophosphate
Cit-GFAP	Citrullinated glial fibrillary acidic protein
Cit-hGFAP	Citrullinated human glial fibrillary acidic protein
Cit-rhGFAP	Citrullinated recombinant human glial fibrillary acidic protein
CNS	Central nervous system
dbcAMP	Dibutyryl cyclic adenosine monophosphate
GAPDH	Glyceraldehyde-3-phosphate dehydrogenase
GFAP	Glial fibrillary acidic protein
MALDI-TOF	Matrix-associated laser desorption ionization-time of flight
MAP 2	Microtubule-associated protein 2
MBP	Myelin basic protein
Nef3	Neurofilament 3
NFT	Neurofibrillary tangles
PAD	Peptidylarginine deiminase
PKA	Protein kinase A
qPCR	Quantitative real-time polymerase chain reaction
RA	Rheumatoid arthritis

Y. Kondo (✉) • A. Ishigami
Molecular Regulation of Aging, Tokyo Metropolitan Institute of Gerontology (TMIG), Tokyo, Japan
e-mail: kondo@tmig.or.jp

E.-K. Choi • Y.-S. Kim
Department of Biomedical Gerontology, Graduate School of Hallym University, Anyang, Republic of Korea

A.P. Nicholas et al. (eds.), *Protein Deimination in Human Health and Disease*, DOI 10.1007/978-3-319-58244-3_17

rhGFAP	Recombinant human glial fibrillary acidic protein
rhPAD	Recombinant human peptidylarginine deiminase
RT-PCR	Reverse transcription polymerase chain reaction
SP	Senile plaque
β2-MG	β2-microglobulin

17.1 Introduction

Peptidylarginine deiminases (PADs) are a group of posttranslational modification enzymes that citrullinate (deiminate) protein arginine residues in a calcium ion-dependent manner. Enzymatic citrullination abolishes the positive charges of native protein molecules, inevitably causing significant alterations in their structure and functions. Protein citrullination is important for the formation of the cornified layer of the skin that covers the human body. Despite this beneficial function, protein citrullination can also be detrimental as its accumulation in the brain is a possible cause of Alzheimer's disease (AD). In this chapter, we introduce PADs and their protein citrullination function now considered critical for advancing research on aging and neurodegenerative disorders, especially AD.

Numerous posttranslational modification enzymes participate in age-associated diseases. However, little attention has been paid to one of these groups, the peptidylarginine deiminases (PADs, EC 3.5.3.15) (Kubilus et al. 1980; Rogers and Simmonds 1958; Kubilus and Baden 1983; Ishigami et al. 1996). The function of PADs is to citrullinate (deiminate) protein arginine residues in a calcium ion-dependent manner, yielding citrulline residues. Enzymatic citrullination abolishes the positive charges of native protein molecules, inevitably causing significant alterations in their structure and function (Imparl et al. 1995; Lamensa and Moscarello 1993; Tarcsa et al. 1996). Protein citrullination plays an important physiological role in the cornification, which thickens the protective layer of the skin that covers the human body (Senshu et al. 1995, 1999). However, this modification also has detrimental functions in which its accumulation in the brain constitutes a possible cause of Alzheimer's disease (AD) (Ishigami et al. 2005), prion diseases (Jang et al. 2008, 2010, 2012), Parkinson's disease (Lange et al. 2011), and multiple sclerosis (Moscarello et al. 2007). Research on PADs and citrullinated proteins is devoted to untying the threads of this pathway precisely and usefully in the expectation of contributing to humanity's capacity for healthful longevity. This chapter describes the current state of studies on the expression of PADs and protein citrullination, understanding of which is critical for advancing research on neurodegenerative disorders, especially AD.

17.2 Peptidylarginine Deiminases (PADs)

The group of enzymes collectively called PADs convert protein arginine residues into citrulline residues in the presence of calcium ions (Kubilus et al. 1980; Rogers and Simmonds 1958; Kubilus and Baden 1983; Ishigami et al. 1996). Early reports described three types of PADs termed "PAD I" or "epidermal type," "PAD II" or "muscle type," and "PAD III" or "hair follicle type," each of which differs in their relative activity toward synthetic substrates such as benzoyl-L-arginine ethyl ester (BAEE) or benzoyl-L-arginine (Bz-L-Arg), antigenic properties, and distribution in mammalian tissues (Watanabe et al. 1988; Terakawa et al. 1991). Subsequently, cDNA cloning analyses revealed the existence of five isoforms of PADs (PAD1, PAD2, PAD3, PAD4, and PAD6) in rodents (Ishigami et al. 1998). These isoforms displayed nearly identical amino acid sequences, which are conserved with approximately 59–71% homology (Tsuchida et al. 1993; Ishigami et al. 1998; Watanabe and Senshu 1989; Nishijyo et al. 1997; Rus'd et al. 1999), but appeared to have different tissue-specific expression, as shown by reverse transcription polymerase chain reaction (RT-PCR) or Northern blot analysis (Ishigami et al. 2001). Rat PAD1 mRNA was detected only in the epidermis and stomach; that of rat PAD3 appeared mainly in the epidermis, ovary, and hair follicles, whereas rat PAD2 and PAD4 were more widely expressed, for example, in the epidermis, lung, spleen, stomach, kidney, ovary, and uterus (Fig. 17.1). The epidermis was the only tissue in which four PAD mRNAs were identified, indicating that PADs play functionally important roles during terminal differentiation of epidermal keratinocytes.

All five types of PADs noted above have been cloned in humans to date, i.e., PAD1 (Guerrin et al. 2003), PAD2 (Ishigami et al. 2002), PAD3 (Kanno et al. 2000), PAD4 (Nakashima et al. 1999), and PAD6 (Chavanas et al. 2004). However, the tissue specificity of these human PADs is poorly delineated. PAD3 was found in both the inner and outer root sheaths of the hair follicles, where citrullination of trichohyalin occurs in the process of keratinization (Kanno et al. 2000). PAD4 was present in human myeloid leukemia HL-60 cells induced to differentiate into granulocytes by retinoic acid and later found in peripheral blood granulocytes (Nakashima et al. 1999; Asaga et al. 2001). Recently, identification of the PAD4 gene and citrullinated proteins in tissue samples from patients with rheumatoid arthritis (RA) strongly suggested that these are major factors in the pathogenesis of RA (Suzuki et al. 2003; Liu et al. 2011). In fact, PAD4 was named an autoantigen in some RA patients based on recognition of the conformation-dependent epitopes of PAD4 (Takizawa et al. 2005; Zhao et al. 2008; Halvorsen et al. 2008). Moreover, anti-citrullinated protein/peptide antibodies, such as anti-cyclic citrullinated peptide (anti-CCP) antibodies (Serdaroglu et al. 2008) and anti-mutated citrullinated vimentin (anti-MVC) antibodies (Wagner et al. 2009), displayed a strong sensitivity and specificity that contributed to the diagnosis of RA.

17.3 Detection of All Citrullinated Proteins

To promote a study of PAD and citrullinated proteins, it was essential to establish a method that would detect all citrullinated proteins. However, because PAD replaces the imino group, which is double bonded to the guanidino carbon atom of arginine residues, with an oxygen atom, radioisotopic techniques are not really applicable

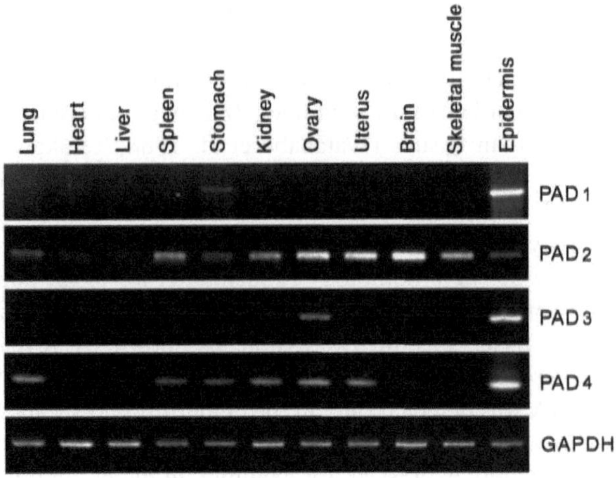

Fig. 17.1 Expression of PAD1, PAD2, PAD3, and PAD4 transcripts in various rat tissue samples analyzed by RT-PCR. Expected sizes were 631 bp for rat PAD1, 428 bp for rat PAD2, 648 bp for rat PAD3, 205 bp for rat PAD4, and 788 bp for glyceraldehyde-3-phosphate dehydrogenase (GAPDH). Reproduced from Ishigami et al. (2001) with permission from Biomedical Research Press

for our purposes here despite their usefulness for the detection of kinases, phosphatases, acetylases, etc. Therefore, we have developed a sensitive method for locating citrulline residues in proteins by generating an antibody that binds specifically to chemically modified citrulline residues. For that purpose, we modified citrulline residues in enzymatically deiminated histones by incubation with diacetyl monoxime and antipyrine in a diluted H_2SO_4 and H_3PO_4 mixture and used that substance as an immunogen in rabbits (Senshu et al. 1992). The resulting rabbit polyclonal antibody was further affinity purified using a modified citrulline column. Chemical modification of citrullinated residues is based on one of the known color reactions of citrulline (Zarabian et al. 1987; Boyde and Rahmatullah 1980). However, the detailed chemical structure of the product has not been elucidated. Therefore, all citrullinated proteins were detected by Western blot analysis (Fig. 17.2) and immunohistochemical methods using this antibody. However, it was essential to modify citrulline residues by incubation of the membrane with diacetyl monoxime and antipyrine in a strong acid mixture after transferring the electrophoresed proteins and specimen onto a slide glass. As little as 3–10 fmol of citrulline residues dotted onto the membrane were detectable regardless of the backbone protein molecules (Senshu et al. 1992). The establishment of this reliable methodology has led to the identification of multiple citrullinated proteins (Senshu et al. 1995; Ishigami et al. 2005; Jang et al. 2008, 2010, 2011) and enabled the following experiments.

17.4 PAD Expression in the Central Nervous System (CNS)

Numerous proteases and posttranslational modification enzymes participate in neurodegeneration such as that observed in patients with AD and Parkinson's disease (Keller et al. 2000; Maccioni et al. 2001). Among the five isoforms of PAD, PAD2 and PAD4 are known to occupy the CNS, although PAD2 is the main member of this

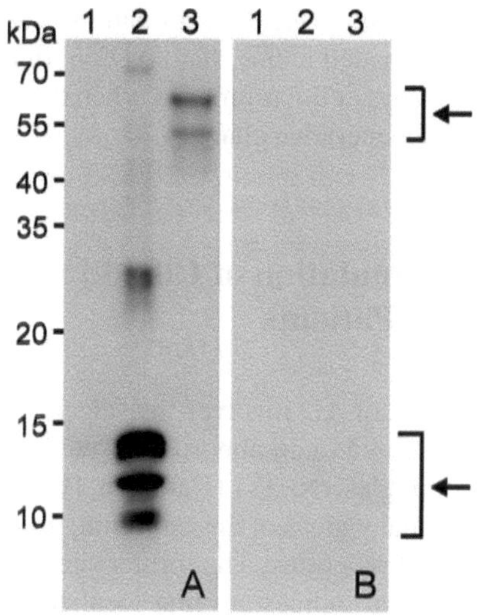

Fig. 17.2 Detection of citrullinated proteins by Western blot analysis using the anti-modified citrulline antibody. For the detection of citrullinated proteins, it is essential to modify citrulline residues by incubation of the membrane with diacetyl monoxime and antipyrine in a diluted H_2SO_4 and H_3PO_4 mixture after transferring the electrophoresed proteins. Signals were detected in (**a**) the chemically modified membrane but not detected in (**b**) a membrane without chemical modification. *Lane 1*, histones from calf thymus; *lane 2*, citrullinated histones; *lane 3*, human skin cornified cell lysate. *Arrows* indicate modified citrullinated histones (*lane 2*, 10–15 kDa) and citrullinated keratins (*lane 3*, 55–65 kDa)

enzyme group that is expressed in the CNS (Kubilus and Baden 1983; Terakawa et al. 1991; Watanabe et al. 1988). Both of these isoforms are present in the myelin sheath, and hypercitrullination of myelin basic protein (MBP) has been shown to result in a loss of myelin sheath integrity in patients with multiple sclerosis (Moscarello et al. 1994; Wood et al. 2008; Musse et al. 2008). Moreover, PAD4, the nuclear isoform of this family of enzymes, is involved in histone citrullination in brain tissue of multiple sclerosis victims (Wakoh et al. 2009). Immunocytochemical studies have localized PAD2 in glial cells, especially astrocytes (Asaga and Ishigami 2000; Vincent et al. 1992; Asaga and Ishigami 2001), microglial cells (Asaga et al. 2002; Vincent et al. 1992), and oligodendrocytes (Akiyama et al. 1999). Additionally, PAD2 expression was later detected in cultured Schwann cells (Keilhoff et al. 2008). Because citrullinated proteins were rarely located in the enzyme-positive glial cells examined by using the anti-modified citrulline antibody method described above (Senshu et al. 1992), we assumed that PAD2 is normally inactive (Senshu et al. 1992; Boyde and Rahmatullah 1980; Jang et al. 2011). However, glial fibrillary acidic protein (GFAP) was highly susceptible to modification by PAD2 in excised rat brains deliberately left at room temperature (Asaga and Senshu 1993). Moreover, under hypoxic conditions (Asaga and Ishigami 2000) and during kainic acid-evoked neurodegeneration (Asaga and Ishigami 2001; Asaga et al. 2002), PAD2 became activated in regions undergoing neurodegeneration and functioned to citrullinate various cerebral proteins, suggesting that protein citrullination occurs in neurodegenerative processes. These findings provided a clue that PAD2 normally remains inactive but becomes active and citrullinates cellular proteins when the intracellular calcium balance is upset during neurodegenerative changes.

17.5 Abnormal Accumulation of Citrullinated Proteins in Brains of AD Patients

The pathological presentation of AD involves the selective death of pyramidal neurons and an accumulation of two main abnormal protein aggregates, senile plaques (SPs) and neurofibrillary tangles (NFTs) (Katzman 1986; Smith 1998). Although NFTs and SPs are found in many areas of the cerebrum, they are concentrated mainly in the hippocampus and cerebral cortex. The former site actually appears to be more important because pathological indices are first localized in that region (Maccioni et al. 2001). Our report indicates that levels of PAD2 are more than threefold higher in the hippocampus than the cortex of rat brains (Asaga and Ishigami 2000).

To elucidate the involvement of protein citrullination in the progress of AD, we examined whether citrullinated proteins are produced in the brains of patients with AD (Ishigami et al. 2005). By Western blot analysis using the anti-modified citrulline antibody, we detected citrullinated proteins of varied molecular weights in hippocampal tissue from patients with AD but not normal humans (Fig. 17.3). Two of the citrullinated proteins were identified as vimentin and GFAP by using two-dimensional gel electrophoresis and MALDI-TOF mass spectrometry (Ishigami et al. 2005). Although citrullination of vimentin and GFAP seems to be much more specific than that of other intracellular proteins, whether this citrullination has physiologically important functions in the brains of AD patients is still unclear. However, vimentin and GFAP were highly susceptible to the attack of PAD2 in vitro; for example, citrullination of vimentin induced disassembly of intermediate filaments (Inagaki et al. 1989).

We also identified citrullinated MBP, which is an authentic marker of oligodendrocytes, in the AD-afflicted hippocampus (Ishigami et al. 2005). Moreover, we found that PAD2 was localized to stage-specific, immature oligodendrocytes in the rat cerebral hemisphere in vitro (Akiyama et al. 1999). Elsewhere, PAD2 cDNA was highly expressed in myelin sheath assembly sites, as others reported after using a combination of subcellular fractionation and suppression subtractive hybridization (Gould et al. 2000). Moreover, the PAD enzyme and citrullinated MBP proved to be relatively enriched in immature myelin, and MBP citrullination played an important role in myelin development and the human demyelinating disease, multiple sclerosis (Moscarello et al. 1994, 2002). Many investigators have suggested that myelin breakdown may be a contributing factor to the pathological effects of AD (Bartzokis 2004; Tian et al. 2004), possibly suggesting that MBP citrullination participates in myelin breakdown.

Interestingly, PAD2 was detected in hippocampal extracts from AD-positive and normal brains, but the amount of PAD2 was markedly greater in the AD tissue.

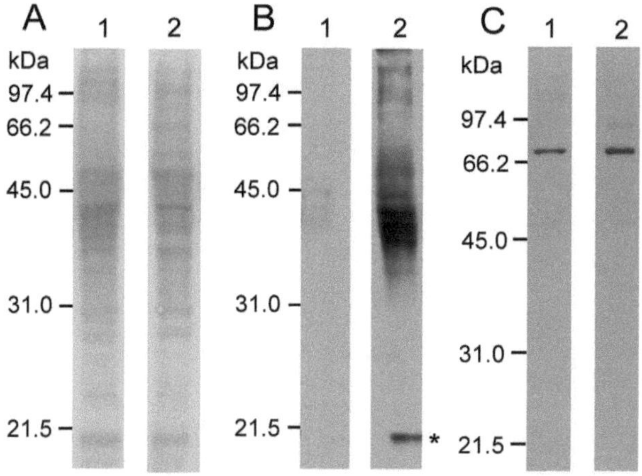

Fig. 17.3 Western blot analysis of citrullinated proteins and PAD2 in hippocampi from the brains of AD patients and normal controls. (**a**) Typical protein profiles detected by amido black staining. (**b**) Citrullinated protein profiles. (**c**) Immunoreactive PAD2 profiles. *Lane 1*, age-matched control; *lane 2*, brain section from an AD patient. *Asterisk* indicates the citrullinated myelin basic protein (MBP). Reproduced from Ishigami et al. (2005) with permission from John Wiley & Sons

Histochemical analysis revealed citrullinated proteins throughout those hippocampal samples, especially in the dentate gyrus and stratum radiatum of the CA1 and CA2 areas (Fig. 17.4). However, the hippocampus from normal brains did not contain measurable citrullinated proteins. Nevertheless, PAD2 immunoreactivity was ubiquitous throughout both the AD-affected and normal hippocampal areas. Still, PAD2 enrichment coincided well with citrullinated protein positivity. Double immunofluorescence staining revealed that citrullinated protein- and PAD2-positive cells also coincided with GFAP-positive cells, but not all GFAP-positive cells were positive for PAD2 (Fig. 17.5). Like GFAP, PAD2 is distributed mainly in astrocytes. These collective results, the abnormal accumulation of citrullinated proteins and abnormal activation of PAD2 in hippocampi of patients with AD, strongly suggest that PAD has an important role in the onset and progression of AD and that citrullinated proteins may become a useful marker for human neurodegenerative diseases.

The mechanism(s) by which citrullinated proteins occur in the hippocampus during AD remains unclear. PAD2 may only become activated, abundant, and functional in the presence of AD because PAD2 content was shown to be notably higher in the hippocampi of persons with AD than in that of normal subjects. Although PAD2 was also present in hippocampal extracts from normal subjects, it remained in a steady state during which no enzyme activation occurred. For enzyme activation, the intracellular calcium concentration must become elevated. To the best of our knowledge, no other factors can regulate PAD activity in vivo or in vitro. A loss of neuronal calcium homeostasis leading to increases in the intracellular calcium concentration has been proposed to play a major role in hypoxic and ischemic brain injury (Hossmann 1999; Choi 1988). In fact, experimental results with simulated ischemia in a primary culture of astrocytes indicated that an influx of extracellular

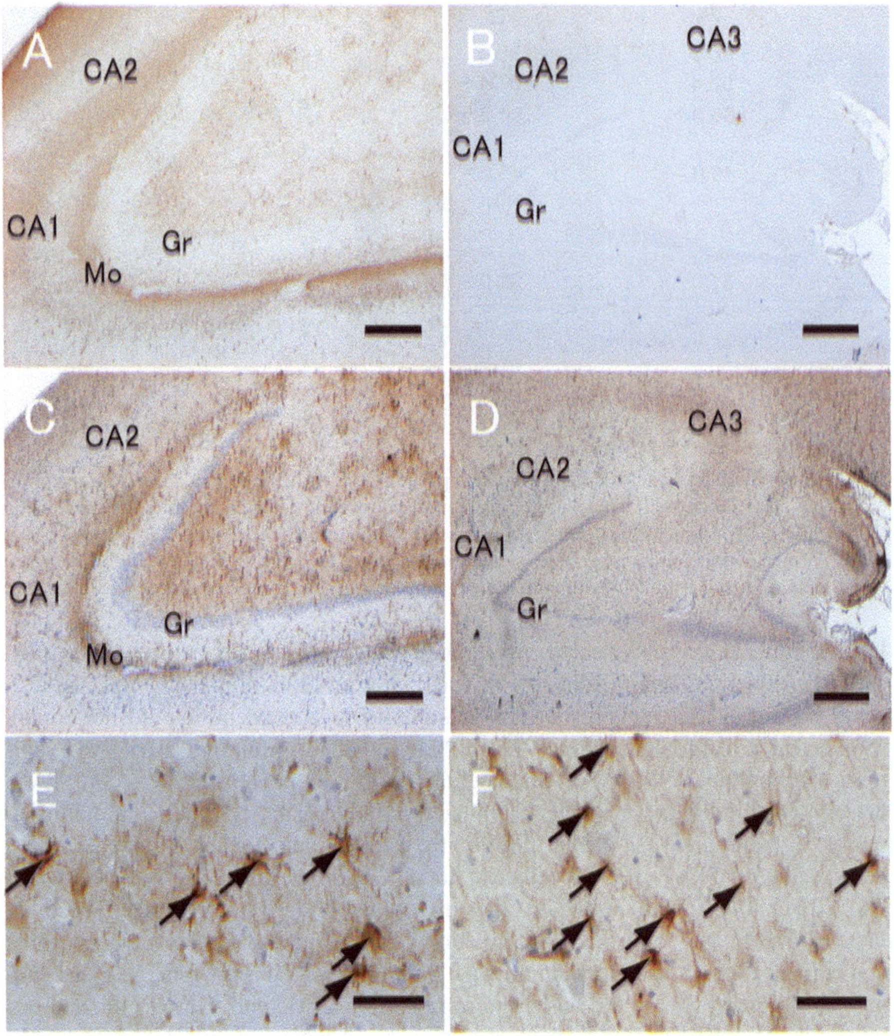

Fig. 17.4 Immunohistochemical staining of citrullinated proteins and PAD2. Hippocampal sections from brains of AD patients (**a**, **c**) and controls (**b**, **d**) were stained for citrullinated proteins (**a**, **b**) and PAD2 (**c**, **d**). (**e**) Higher magnification of **a**. *Arrows* indicate the citrullinated protein-positive cells. (**f**) Higher magnification of **c**. *Arrows* indicate the PAD2-positive cells. *Gr* granule cell layer, *Mo* molecular cell layer. Scale bars: **a**, **b**, **c**, and **d** = 500 μm; **e** and **f** = 50 μm. Reproduced from Ishigami et al. (2005) with permission from John Wiley & Sons

calcium contributes to astroglial injury during ischemia (Haun et al. 1992). Our former report showed that PAD2 activated and citrullinated various cerebral proteins under hypoxic conditions (Asaga and Ishigami 2000) and during kainic acid-evoked neurodegeneration (Asaga and Ishigami 2001; Asaga et al. 2002). Abnormal PAD activation resulted in random protein citrullination, which could then trigger the

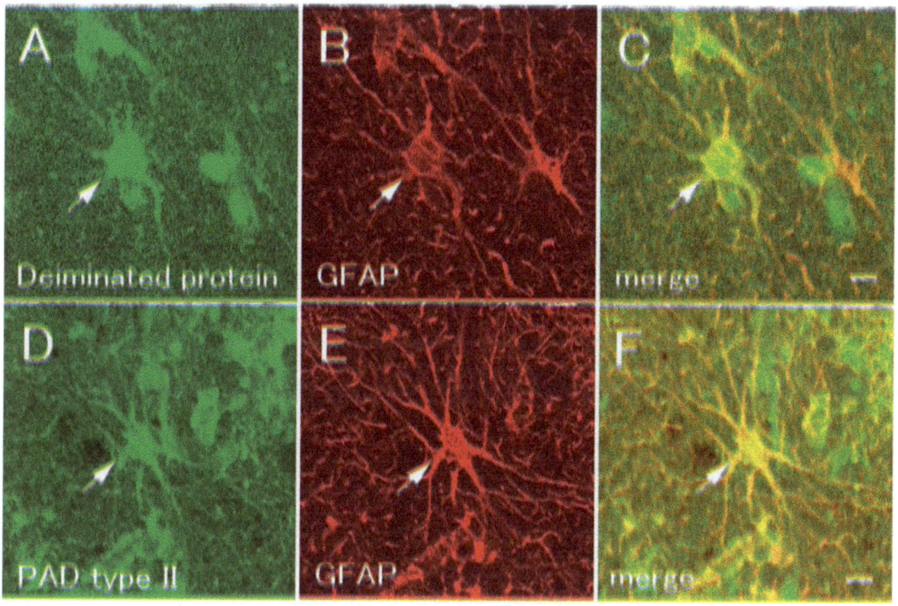

Fig. 17.5 Identification of citrullinated protein-positive and PAD2-positive cells by double immunofluorescence staining. Sections of AD hippocampus were double immunostained with a monoclonal anti-GFAP antibody and polyclonal anti-modified citrulline IgG or polyclonal anti-human PAD2 antibody. The primary antibodies were visualized with anti-rabbit Alexa Fluor 488™ (*green*) and anti-mouse Alexa Fluor 594™ (*red*). (**a** and **d**) Alexa 488 (*green*) for citrullinated protein (**a**) or PAD2 (**d**). (**b** and **e**) Alexa 568 (*red*) for GFAP. (**c** and **f**) Merged views for (**a/b**) and (**d/e**). *Arrows* indicate coincident positions. Scale bar = 5 μm. Reproduced from Ishigami et al. (2005) with permission from John Wiley & Sons

onset of neurodegenerative disease. Therefore, the development of an inhibitory drug specific for PAD could conceivably prevent the onset and progression of neurodegeneration.

17.6 Citrullination Sites of GFAP and Detection of Citrullinated GFAP in AD Brains

Citrullinated (cit-) GFAP was identified in hippocampi from AD patients by using 2-DE analysis and MALDI-TOF mass spectrometry (Ishigami et al. 2005). Although human GFAP contains many arginine residues, specific citrullination site(s) of GFAP in AD brains have been uncovered. At first, to reveal the substrate specificity of PADs, we prepared recombinant human GFAP (rhGFAP) and recombinant human PAD1 (rhPAD1), rhPAD2, rhPAD3, and rhPAD4. rhGFAP was incubated with rhPAD1, rhPAD2, rhPAD3, and rhPAD4, and cit-rhGFAP was detected by Western blot analysis using the anti-modified citrulline antibody (Ishigami et al. 2015).

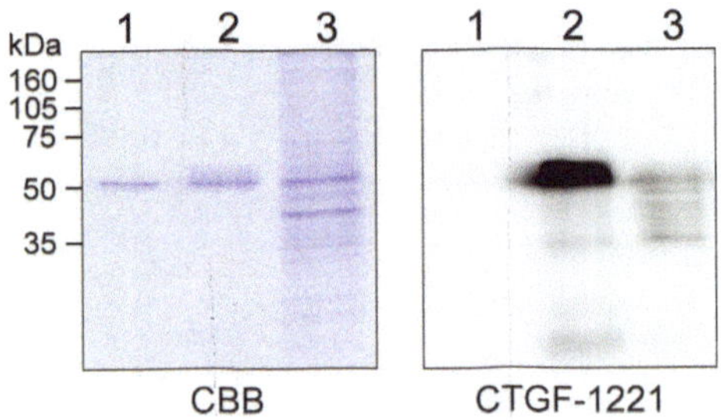

Fig. 17.6 Characterization of anti-cit-rhGFAP monoclonal antibodies. The specificities of the cit-rhGFAP monoclonal antibody CTGF-1221 were determined by Western blot analysis. *Lane 1*, rhGFAP; *lane 2*, cit-rhGFAP; *lane 3*, brain extract from AD patients. Reproduced from Ishigami et al. (2015) with permission from John Wiley & Sons

```
  1 MERRRITSAA RRSYVSSGEM MVGGLAPGRR LGPGTRLSLA RMPPPLPTRV DFSLAGALNA

 61 GFKETRASER AEMMELNDRF ASYIEKVRFL EQQNKALAAE LNQLRAKEPT KLADVYQAEL

121 RELRLRLDQL TANSARLEVE RDNLAQDLAT VRQKLQDETN LRLEAENNLA AYRQEADEAT

181 LARLDLERKI ESLEEEIRFL RKIHEEEVRE LQEQLARQQV HVELDVAKPD LTAALKEIRT

241 QYEAMASSNM HEAEEWYRSK FADLTDAAAR NAELLRQAKH EANDYRRQLQ SLTCDLESLR

301 GTNESLERQM REQEERHVRE AASYQEALAR LEEEGQSLKD EMARHLQEYQ DLLNVKLALD

361 IEIATYRKLL EGEENRITIP VQTFSNLQIR ETSLDTKSVS EGHLKRNIVV KTVEMRDGEV

421 IKESKQEHKD VM
```

Fig. 17.7 Citrullination sites in cit-rhGFAP identified by MS/MS analysis. Underscored sequences were assigned to GFAP-derived peptides by a Mascot MS/MS ion search. Arginine (R) with a *red circle* (R270 and R416) indicates a citrullination site and recognized site with the anti-cit-rhGFAP monoclonal antibody CTGF-1221. Reproduced from Ishigami et al. (2015) with permission from John Wiley & Sons

While rhPAD1 scarcely citrullinated rhGFAP, rhPAD2 specifically citrullinated rhGFAP. rhPAD3 and rhPAD4 exerted no effect on citrullination of rhGFAP. Thus, PAD2 must be a unique player in citrullination of GFAP in the astrocytes of AD brains.

Next, eight independent anti-cit-rhGFAP monoclonal antibodies were developed. Among them, CTGF-1221 reacts specifically with cit-rhGFAP and, notably, with brain extracts from AD patients but not with rhGFAP (Fig. 17.6) (Ishigami et al. 2015). To determine the citrullination site of cit-rhGFAP, the cit-rhGFAP was digested with trypsin, and the citrullinated peptides were affinity purified with the anti-cit-rhGFAP monoclonal antibody CTGF-1221. By MS/MS analysis, two arginine residues (R270

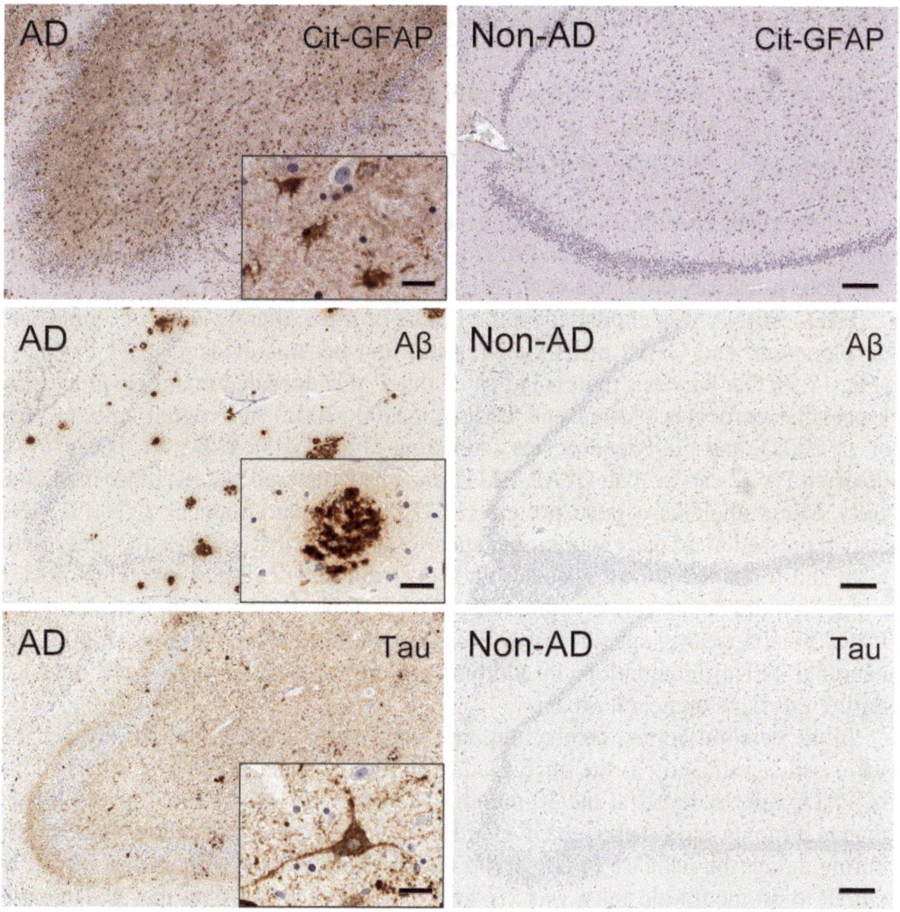

Fig. 17.8 Immunohistochemical staining of cit-GFAP, Aβ, and phosphorylated tau. Hippocampal sections from AD and non-AD brains were stained with the anti-cit-rhGFAP monoclonal antibody CTGF-1221, anti-human Aβ (11–28; 12B2) monoclonal antibody, and anti-human phospho/PHF-tau (AT8) monoclonal antibody, respectively. Scale bars = 200 μm; 20 μm in *insets*. Reproduced from Ishigami et al. (2015) with permission from John Wiley & Sons

and R416) were shown to be citrullinated in cit-rhGFAP and recognized with the anti-cit-rhGFAP monoclonal antibody CTGF-1221 (Fig. 17.7).

We further investigated whether CTGF-1221 is applicable for immunohistochemical staining, clarifying that cit-GFAP was detected in the hippocampus of AD brain tissue and that the cit-GFAP-positive cells could be morphologically identified as astrocytes (Fig. 17.8). Taken together, in the hippocampi of AD brains, PAD2 is solely responsible for the citrullination of GFAP in astrocytes. The anti-cit-rhGFAP monoclonal antibody CTGF-1221, which reacts with R270Cit and R416Cit of cit-hGFAP, could be useful for immunohistochemical research and diagnosis of AD as well as other neurodegenerative diseases.

17.7 Age-Related Changes of PAD2 in the Mouse Brain

Understandably, abnormal protein citrullination by PAD2 has been closely associated with the pathogenesis of neurodegenerative disorders such as AD. Protein citrullination in these patients is thought to develop during the initiation and/or progression of the disease. However, the contribution of the changes in PAD2 levels and consequent citrullination during developmental and aging processes remained unclear. Therefore, we measured PAD2 expression and localization in the brain during those processes (Shimada et al. 2010).

PAD2 mRNA was expressed in the brains of mice after 15 days of embryonic development, and GFAP mRNA expression first became evident just 1 day later (Fig. 17.9). Our previous reports indicated that PAD2 appeared mainly in glial cells, especially astrocytes (Asaga and Ishigami 2000, 2001), microglial cells (Asaga et al. 2002), and oligodendrocytes (Akiyama et al. 1999). However, because we detected PAD2 earlier than GFAP, PAD2 must be expressed in cells other than glial cells. Microtubule-associated protein (MAP 2) and neurofilament 3 (Nef3) were also expressed at an early embryonic stage in amounts that increased slightly until birth and remained almost constant until postnatal day 7 (Fig. 17.9). Thus, PAD2 expression did not correlate with GFAP, MAP2, or Nef3 expression, indicating that PAD2 must appear at specific but still unknown stages and conditions of glial and neuronal cell differentiation. In addition, no citrullinated proteins were detected during this developmental process.

In the cerebral cortex, cerebellum, and hippocampus, PAD2 mRNA expression increased significantly as the mice aged from 3 to 30 months old (Fig. 17.10). That is, PAD2 mRNA levels at the 30-month-old mark were 1.5-fold to 1.6-fold higher than in 3-month-olds. Although GFAP mRNA expression also increased significantly during aging, the increase in GFAP did not correlate closely with that of PAD2, as GFAP in 30-month-old mice was 2.7-fold to 4.7-fold higher than that in 3-month-olds, far exceeding the increase in PAD2. Moreover, Nef3 mRNA expression did not change during aging. Because the change of PAD2 expression levels during aging did not correlate with those of GFAP or Nef3, PAD2 must be expressed only at certain times and under appropriate conditions by neuronal cells and glial cells, including astrocytes (Asaga and Ishigami 2000, 2001), activated microglial cells (Asaga et al. 2002), and stage-specific immature oligodendrocytes (Akiyama et al. 1999). Again, no citrullinated proteins were detected during this aging process.

17.8 Localization of PAD2 in the Cerebral Cortex, Cerebellum, and Hippocampus

Immunohistochemical staining of PAD2, GFAP, and MAP 2 provided new insight into the characteristics of PAD2-positive cells in the cerebral cortex, hippocampus, and cerebellum of 3-month-old mice (Fig. 17.11). In the cerebral cortex and hip-

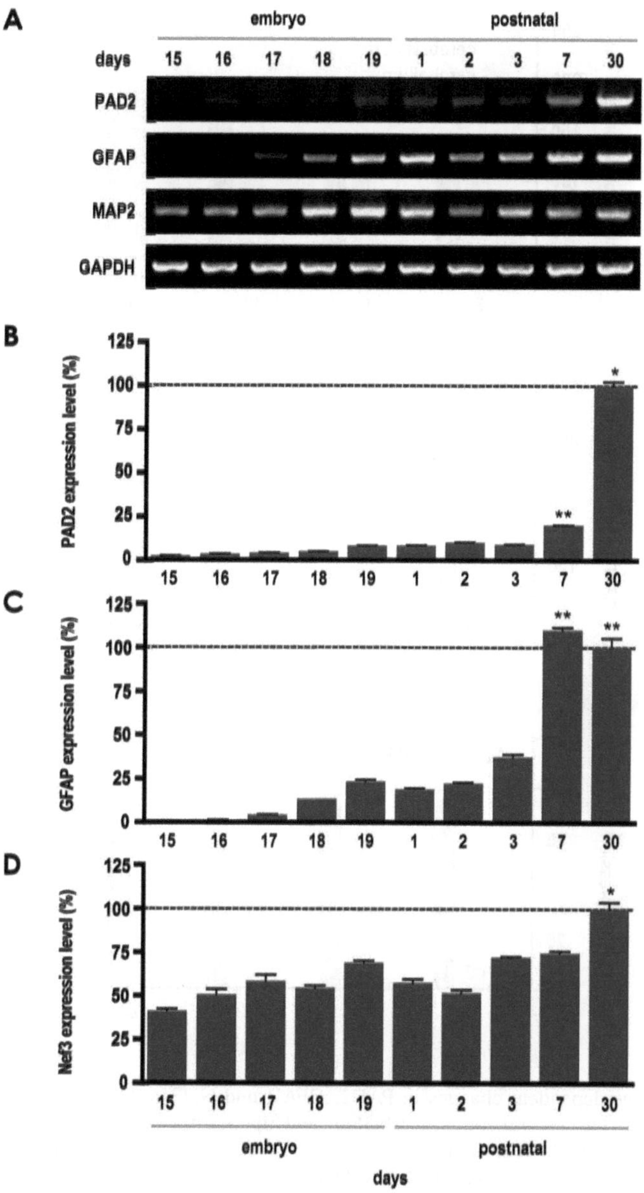

Fig. 17.9 Developmental changes in PAD2, GFAP, MAP 2, and Nef3 mRNA expression levels in the whole brains of mice. (**a**) RT-PCR of PAD2, GFAP, MAP 2, and GAPDH. (**b–d**) Quantitative RT-PCR analysis of PAD2 (**b**), GFAP (**c**), and Nef3 (**d**). Data from quantitative RT-PCR are shown as the percentage of each value with postnatal day 30 taken as 100% and representing a mean ± SEM of five animals. *$p < 0.05$ compared to 15–19 days of embryonic life and 1–3 or 7 days after birth. **$p < 0.05$ compared to 15–19 days of embryonic life and 1–3 days after birth. Reproduced from Shimada et al. (2010) with permission from John Wiley & Sons

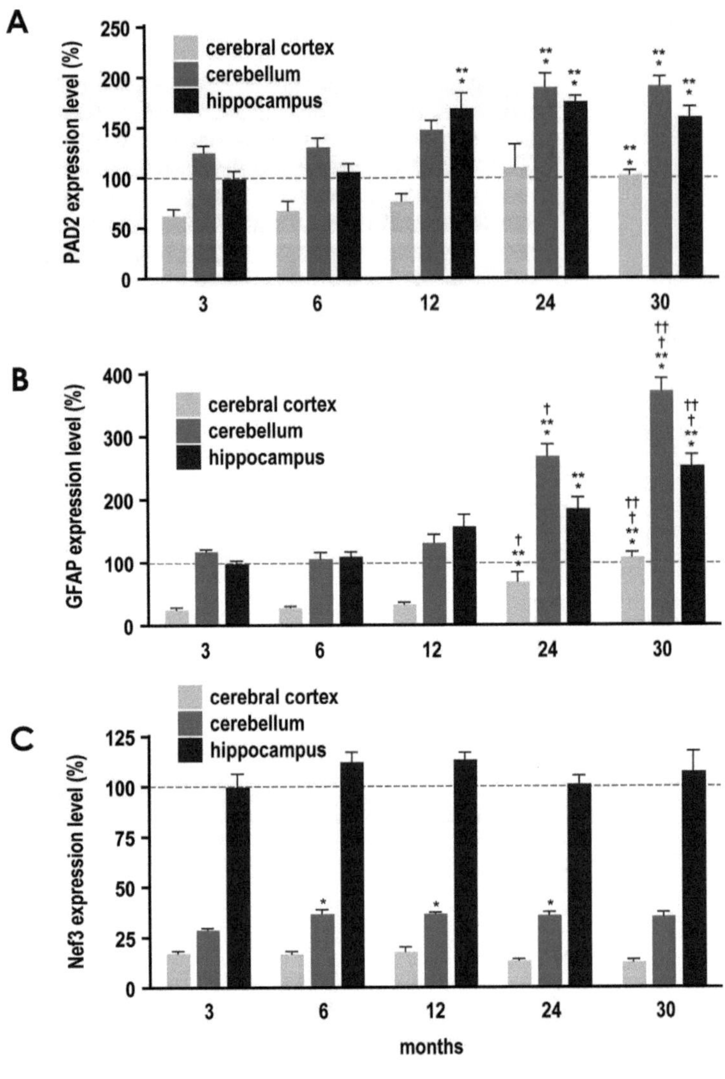

Fig. 17.10 Age-dependent changes in PAD2, GFAP, and Nef3 mRNA expression in the brain. Total RNA from the cerebral cortex, cerebellum, and hippocampi of 3-, 6-, 12-, 24-, and 30-month-old mice was prepared. Quantitative RT-PCR analysis of PAD2 (**a**), GFAP (**b**), and Nef3 (**c**) was carried out. As the endogenous control, GAPDH was quantified simultaneously to normalize each raw data set. Data are expressed in percentages as values in the hippocampi of 3-month-old mice as 100% and represent a mean ± SEM of five animals. $^{*}p < 0.05$ compared to 3-month-old mice. $^{**}p < 0.05$ compared to 6-month-old mice. $^{\dagger}p < 0.05$ compared to 12-month-old mice. $^{\dagger\dagger}p < 0.05$ compared to 24-month-old mice. Reproduced from Shimada et al. (2010) with permission from John Wiley & Sons

pocampus, PAD2-positive signals were detected in neuronal cell bodies that co-stained with MAP 2 but not in dendrites. Not only was MAP 2 staining positive in both the neuronal cell bodies and dendrites, but cells in the cerebral cortex, hippocampus, and cerebellum co-expressed MAP 2 and PAD2. However, GFAP-positive cells, which are considered to be reactive astrocytes, were PAD2-negative in the cerebral cortex, hippocampus, and cerebellum.

In the cerebellum, PAD2-positive staining was present on morphologically characteristic Purkinje-like cells along dense granule cell layers that were not positive for either MAP 2 or GFAP. To confirm that these were actually Purkinje cells, we performed double immunostaining with PAD2 and calbindin-D-28K (calbindin), a known marker of Purkinje cells that is limited to localization in those cells (Servais et al. 2005; Whitney et al. 2008). Calbindin staining was evident as a light magenta coloration on the alkaline phosphate substrate (Fig. 17.12). These Purkinje cells appeared as huge round-shaped cell bodies located between the bottom of the molecule layer and surface of the granule cell layer of cerebellar tissue. PAD2 was stained brown by 3,3′-diaminobenzidine (DAB) used as a chromogenic substrate (Fig. 17.12). Double immunostaining enabled the detection of both calbindin and PAD2 in the same Purkinje cells from the cerebellum, thus assuring the existence of PAD2 in clearly identified Purkinje cells of the cerebellum. Additionally, the characteristic localization of PAD2, GFAP, and MAP 2 in the cerebral cortex, hippocampus, and cerebellum did not change during aging from 3 to 30 months. Thus, PAD2 was localized in neuronal cells of the cerebral cortex and Purkinje cells of the cerebellum (Shimada et al. 2010).

17.9 PAD2 in Purkinje Cells of the Cerebellum

PAD2 has been shown to be expressed in Purkinje cells of the cerebellum (Shimada et al. 2010). The cerebellum functions as the center of learning and control over motion, sensory input, and cognition. Purkinje cells of the cerebellum are the sole output neurons and are important as the integrators and fine-tuners of diverse input signals (Cheron et al. 2008). Accumulated evidence indicates that the dynamic movement of Ca^{2+} plays a key role in the function of Purkinje cells (Matsushita et al. 2002; Erickson et al. 2007). Intracellular Ca^{2+} concentrations become elevated via voltage-dependent calcium channels of plasma membranes or inositol-1,4,5-trisphosphate-dependent Ca^{2+} release from intracellular Ca^{2+} storage sites such as the endoplasmic reticulum (Cheron et al. 2008). Elevations in intracellular Ca^{2+} activate intracellular signal cascades, leading to functional events such as neurotransmitter release (Cheron et al. 2008). Importantly, the PAD enzyme requires an intracellular Ca^{2+} level about a 100-fold higher than normal for its activation (Inagaki et al. 1989). Because Purkinje cells store large amounts of Ca^{2+} corresponding to physiological stimuli (Matsushita et al. 2002), those intracellular Ca^{2+} concentrations can conceivably become elevated transiently in specific, limited areas, such as near the endoplasmic reticulum and plasma membrane. When such a condition

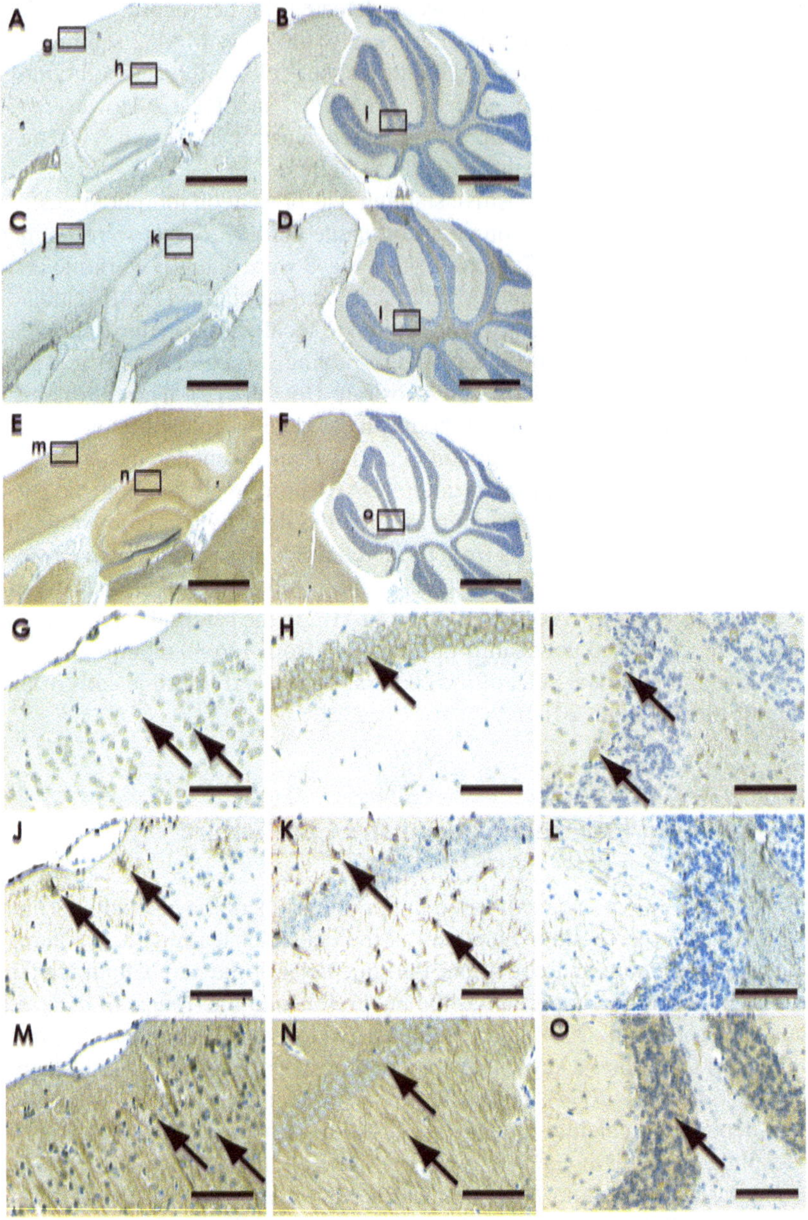

Fig. 17.11 Immunohistochemical staining of PAD2, GFAP, and MAP 2 in the cerebral cortex, hippocampus, and cerebellum of 3-month-old mice. Each brain section was stained with a PAD2 (**a** and **b**), GFAP (**c** and **d**), and MAP 2 (**e** and **f**) antibody. The square area of *g–i* in **a** and **b**, *j–l* in **c** and **d**, and *m–o* in **e** and **f** was magnified for presentation in **g–i**, **j–l**, and **m–o**, respectively. *Arrows* indicate typical stained objects. Scale bars = 1 mm in **a–f** and 100 μm in **g–o**. Reproduced from Shimada et al. (2010) with permission from John Wiley & Sons

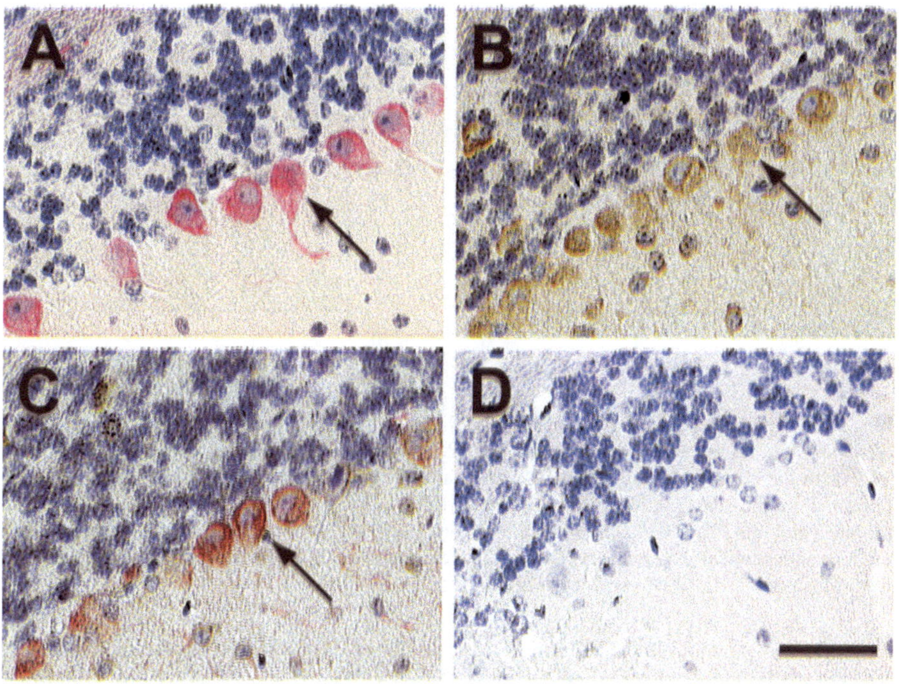

Fig. 17.12 Identification of PAD2-positive cells in the cerebellum by double immunostaining. Sections of the cerebellum from 3-month-old mice were double immunostained with PAD2 and calbindin. (**a**) Alkaline phosphate substrate as a chromogenic substrate (*light magenta*) for calbindin; (**b**) 3,3′-diaminobenzidine (DAB) as a chromogenic substrate (*brown*) for PAD2; (**c**) double immunostaining of PAD2 and calbindin; (**d**) mouse and rabbit IgG were used for control staining. *Arrows* indicate Purkinje cells. Scale bar = 100 μm. Reproduced from Shimada et al. (2010) with permission from John Wiley & Sons

prevails, PAD2 enzymes would become activated and citrullinate various proteins, leading to cell death (Asaga et al. 1998). In fact, quantities of PAD2 and citrullinated proteins have been shown to increase in the brain in vivo during abnormal conditions such as scrapie infection in mice (Jang et al. 2008) and AD in humans (Ishigami et al. 2005).

17.10 Aspects of PAD2 Expression and Protein Citrullination in Neurodegenerative Disorders

The mechanism by which citrullinated proteins occupy the hippocampus during AD remains unclear. PAD2 may only become activated, abundant, and functional in the presence of AD, because the amount of PAD2 was notably higher in the hippocampi of people with AD than in those of normal subjects. Although the mechanism of transcriptional regulation for PAD2 in AD brains remains unclear, dibutyryl cAMP,

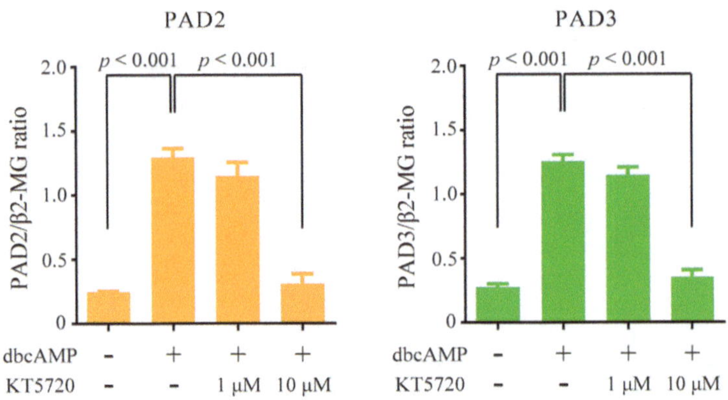

Fig. 17.13 Induction of PAD2 and PAD3 mRNA expression by dibutyryl cAMP (dbcAMP) via cAMP-protein kinase A (PKA) signaling in U-251MG cells. U-251MG cells were incubated for 24 h with the combination of 5 mM dbcAMP and the PKA inhibitor KT5720 (1 or 10 μM) as indicated in the figure. Cells were collected and analyzed by qPCR for PAD2, PAD3, and β2-microglobulin (β2-MG) (as an endogenous control). Values are represented as the means ± SEM of three independent experiments. Each point represents an individual experiment. Reproduced from Masutomi et al. (2016) with permission from John Wiley & Sons

a membrane-permeable cAMP analog, induced gene and protein expression levels of PAD2 and PAD3 in human astrocytoma U-251MG cells via cAMP-PKA signaling, suggesting that cAMP-PKA signaling might be involved in the induction of PAD2 in AD brains (Fig. 17.13) (Masutomi et al. 2016). Although PAD2 was also present in hippocampal extracts from normal subjects, it remained in a steady state during which no enzyme activation occurred. For enzyme activation, the intracellular calcium concentration must become elevated. To the best of our knowledge, no other factors can regulate PAD activity in vivo or in vitro. A loss of neuronal calcium homeostasis, leading to increases in the intracellular calcium concentration, has been proposed to play a major role in hypoxic and ischemic brain injury (Hossmann 1999; Choi 1988). An influx of extracellular calcium could contribute to astroglial injury during ischemia, as suggested on the basis of experimental results with simulated ischemia in a primary culture of astrocytes (Haun et al. 1992). Our reports showed that PAD2 activated and citrullinated various cerebral proteins under hypoxic conditions (Asaga and Ishigami 2000) and during kainic acid-evoked neurodegeneration (Asaga and Ishigami 2001; Asaga et al. 2002). Clearly, from the amount of evidence now available, abnormal PAD activation, which results in random protein citrullination, has the potential to trigger a neurodegenerative disease such as AD.

References

Akiyama, K., Sakurai, Y., Asou, H., & Senshu, T. (1999). Localization of peptidylarginine deiminase type II in a stage-specific immature oligodendrocyte from rat cerebral hemisphere. *Neuroscience Letters, 274*(1), 53–55.

Asaga, H., & Ishigami, A. (2000). Protein deimination in the rat brain: Generation of citrulline-containing proteins in cerebrum perfused with oxygen-deprived media. *Biomedical Research, 21*(4), 197–205.

Asaga, H., & Ishigami, A. (2001). Protein deimination in the rat brain after kainate administration: Citrulline-containing proteins as a novel marker of neurodegeneration. *Neuroscience Letters, 299*(1–2), 5–8.

Asaga, H., & Senshu, T. (1993). Combined biochemical and immunocytochemical analyses of postmortem protein deimination in the rat spinal cord. *Cell Biology International, 17*(5), 525–532.

Asaga, H., Yamada, M., & Senshu, T. (1998). Selective deimination of vimentin in calcium ionophore-induced apoptosis of mouse peritoneal macrophages. *Biochemical and Biophysical Research Communications, 243*(3), 641–646.

Asaga, H., Nakashima, K., Senshu, T., Ishigami, A., & Yamada, M. (2001). Immunocytochemical localization of peptidylarginine deiminase in human eosinophils and neutrophils. *Journal of Leukocyte Biology, 70*(1), 46–51.

Asaga, H., Akiyama, K., Ohsawa, T., & Ishigami, A. (2002). Increased and type II-specific expression of peptidylarginine deiminase in activated microglia but not hyperplastic astrocytes following kainic acid-evoked neurodegeneration in the rat brain. *Neuroscience Letters, 326*(2), 129–132.

Bartzokis, G. (2004). Age-related myelin breakdown: A developmental model of cognitive decline and Alzheimer's disease. *Neurobiology of Aging, 25*(1), 5–18. author reply 49–62.

Boyde, T. R., & Rahmatullah, M. (1980). Optimization of conditions for the colorimetric determination of citrulline, using diacetyl monoxime. *Analytical Biochemistry, 107*(2), 424–431.

Chavanas, S., Mechin, M. C., Takahara, H., Kawada, A., Nachat, R., Serre, G., & Simon, M. (2004). Comparative analysis of the mouse and human peptidylarginine deiminase gene clusters reveals highly conserved non-coding segments and a new human gene, PADI6. *Gene, 330*, 19–27.

Cheron, G., Servais, L., & Dan, B. (2008). Cerebellar network plasticity: From genes to fast oscillation. *Neuroscience, 153*(1), 1–19.

Choi, D. W. (1988). Calcium-mediated neurotoxicity: Relationship to specific channel types and role in ischemic damage. *Trends in Neurosciences, 11*(10), 465–469.

Erickson, M. A., Haburcak, M., Smukler, L., & Dunlap, K. (2007). Altered functional expression of Purkinje cell calcium channels precedes motor dysfunction in tottering mice. *Neuroscience, 150*(3), 547–555.

Gould, R. M., Freund, C. M., Palmer, F., & Feinstein, D. L. (2000). Messenger RNAs located in myelin sheath assembly sites. *Journal of Neurochemistry, 75*(5), 1834–1844.

Guerrin, M., Ishigami, A., Mechin, M. C., Nachat, R., Valmary, S., Sebbag, M., Simon, M., Senshu, T., & Serre, G. (2003). cDNA cloning, gene organization and expression analysis of human peptidylarginine deiminase type I. *The Biochemical Journal, 370*(Pt 1), 167–174.

Halvorsen, E. H., Pollmann, S., Gilboe, I. M., van der Heijde, D., Landewe, R., Odegard, S., Kvien, T. K., & Molberg, O. (2008). Serum IgG antibodies to peptidylarginine deiminase 4 in rheumatoid arthritis and associations with disease severity. *Annals of the Rheumatic Diseases, 67*(3), 414–417.

Haun, S. E., Murphy, E. J., Bates, C. M., & Horrocks, L. A. (1992). Extracellular calcium is a mediator of astroglial injury during combined glucose-oxygen deprivation. *Brain Research, 593*(1), 45–50.

Hossmann, K. A. (1999). The hypoxic brain. Insights from ischemia research. *Adv. Exp. Med. Biol., 474*, 155–169.

Imparl, J. M., Senshu, T., & Graves, D. J. (1995). Studies of calcineurin-calmodulin interaction: Probing the role of arginine residues using peptidylarginine deiminase. *Archives of Biochemistry and Biophysics, 318*(2), 370–377.

Inagaki, M., Takahara, H., Nishi, Y., Sugawara, K., & Sato, C. (1989). Ca2+−dependent deimination-induced disassembly of intermediate filaments involves specific modification of the amino-terminal head domain. *The Journal of Biological Chemistry, 264*(30), 18119–18127.

Ishigami, A., Ohsawa, T., Watanabe, K., & Senshu, T. (1996). All-trans retinoic acid increases peptidylarginine deiminases in a newborn rat keratinocyte cell line. *Biochemical and Biophysical Research Communications, 223*(2), 299–303.

Ishigami, A., Kuramoto, M., Yamada, M., Watanabe, K., & Senshu, T. (1998). Molecular cloning of two novel types of peptidylarginine deiminase cDNAs from retinoic acid-treated culture of a newborn rat keratinocyte cell line. *FEBS Letters, 433*(1–2), 113–118.

Ishigami, A., Asaga, H., Ohsawa, T., Akiyama, K., & Maruyama, N. (2001). Peptidylarginine deiminase type I, type II, type III and type IV are expressed in rat epidermis. *Biomedical Research, 22*(1), 63–65.

Ishigami, A., Ohsawa, T., Asaga, H., Akiyama, K., Kuramoto, M., & Maruyama, N. (2002). Human peptidylarginine deiminase type II: Molecular cloning, gene organization, and expression in human skin. *Archives of Biochemistry and Biophysics, 407*(1), 25–31.

Ishigami, A., Ohsawa, T., Hiratsuka, M., Taguchi, H., Kobayashi, S., Saito, Y., Murayama, S., Asaga, H., Toda, T., Kimura, N., & Maruyama, N. (2005). Abnormal accumulation of citrullinated proteins catalyzed by peptidylarginine deiminase in hippocampal extracts from patients with Alzheimer's disease. *Journal of Neuroscience Research, 80*(1), 120–128.

Ishigami, A., Masutomi, H., Handa, S., Nakamura, M., Nakaya, S., Uchida, Y., Saito, Y., Murayama, S., Jang, B., Jeon, Y. C., Choi, E. K., Kim, Y. S., Kasahara, Y., Maruyama, N., & Toda, T. (2015). Mass spectrometric identification of citrullination sites and immunohistochemical detection of citrullinated glial fibrillary acidic protein in Alzheimer's disease brains. *Journal of Neuroscience Research, 93*(11), 1664–1674.

Jang, B., Kim, E., Choi, J. K., Jin, J. K., Kim, J. I., Ishigami, A., Maruyama, N., Carp, R. I., Kim, Y. S., & Choi, E. K. (2008). Accumulation of citrullinated proteins by up-regulated peptidylarginine deiminase 2 in brains of scrapie-infected mice: A possible role in pathogenesis. *The American Journal of Pathology, 173*(4), 1129–1142.

Jang, B., Jin, J. K., Jeon, Y. C., Cho, H. J., Ishigami, A., Choi, K. C., Carp, R. I., Maruyama, N., Kim, Y. S., & Choi, E. K. (2010). Involvement of peptidylarginine deiminase-mediated post-translational citrullination in pathogenesis of sporadic Creutzfeldt-Jakob disease. *Acta Neuropathologica, 119*(2), 199–210.

Jang, B., Shin, H. Y., Choi, J. K., Nguyen du, P. T., Jeong, B. H., Ishigami, A., Maruyama, N., Carp, R. I., Kim, Y. S., & Choi, E. K. (2011). Subcellular localization of peptidylarginine deiminase 2 and citrullinated proteins in brains of scrapie-infected mice: Nuclear localization of PAD2 and membrane fraction-enriched citrullinated proteins. *Journal of Neuropathology and Experimental Neurology, 70*(2), 116–124.

Jang, B., Jeon, Y. C., Choi, J. K., Park, M., Kim, J. I., Ishigami, A., Maruyama, N., Carp, R. I., Kim, Y. S., & Choi, E. K. (2012). Peptidylarginine deiminase modulates the physiological roles of enolase via citrullination: Links between altered multifunction of enolase and neurodegenerative diseases. *The Biochemical Journal, 445*(2), 183–192.

Kanno, T., Kawada, A., Yamanouchi, J., Yosida-Noro, C., Yoshiki, A., Shiraiwa, M., Kusakabe, M., Manabe, M., Tezuka, T., & Takahara, H. (2000). Human peptidylarginine deiminase type III: Molecular cloning and nucleotide sequence of the cDNA, properties of the recombinant enzyme, and immunohistochemical localization in human skin. *The Journal of Investigative Dermatology, 115*(5), 813–823.

Katzman, R. (1986). Alzheimer's disease. *The New England Journal of Medicine, 314*(15), 964–973.

Keilhoff, G., Prell, T., Langnaese, K., Mawrin, C., Simon, M., Fansa, H., & Nicholas, A. P. (2008). Expression pattern of peptidylarginine deiminase in rat and human Schwann cells. *Developmental Neurobiology, 68*(1), 101–114.

Keller, J. N., Hanni, K. B., & Markesbery, W. R. (2000). Impaired proteasome function in Alzheimer's disease. *Journal of Neurochemistry, 75*(1), 436–439.

Kubilus, J., & Baden, H. P. (1983). Purification and properties of a brain enzyme which deiminates proteins. *Biochimica et Biophysica Acta, 745*(3), 285–291.

Kubilus, J., Waitkus, R. F., & Baden, H. P. (1980). Partial purification and specificity of an arginine-converting enzyme from bovine epidermis. *Biochimica et Biophysica Acta, 615*(1), 246–251.
Lamensa, J. W., & Moscarello, M. A. (1993). Deimination of human myelin basic protein by a peptidylarginine deiminase from bovine brain. *Journal of Neurochemistry, 61*(3), 987–996.
Lange, S., Gogel, S., Leung, K. Y., Vernay, B., Nicholas, A. P., Causey, C. P., Thompson, P. R., Greene, N. D., & Ferretti, P. (2011). Protein deiminases: New players in the developmentally regulated loss of neural regenerative ability. *Developmental Biology, 355*(2), 205–214.
Liu, Y. L., Chiang, Y. H., Liu, G. Y., & Hung, H. C. (2011). Functional role of dimerization of human peptidylarginine deiminase 4 (PAD4). *PloS One, 6*(6), e21314.
Maccioni, R. B., Munoz, J. P., & Barbeito, L. (2001). The molecular bases of Alzheimer's disease and other neurodegenerative disorders. *Archives of Medical Research, 32*(5), 367–381.
Masutomi, H., Kawashima, S., Kondo, Y., Uchida, Y., Jang, B., Choi, E. K., Kim, Y. S., Shimokado, K., & Ishigami, A. (2016). Induction of paptidylarginine deiminase 2 and 3 by dibutyryl cAMP via cAMP-PKA signaling in human astrocytoma U-251MG cells. *Journal of Neuroscience Research, 95*(7), 1503–1512. doi:10.1002/jnr.23959.
Matsushita, K., Wakamori, M., Rhyu, I. J., Arii, T., Oda, S., Mori, Y., & Imoto, K. (2002). Bidirectional alterations in cerebellar synaptic transmission of tottering and rolling Ca2+ channel mutant mice. *The Journal of Neuroscience, 22*(11), 4388–4398.
Moscarello, M. A., Wood, D. D., Ackerley, C., & Boulias, C. (1994). Myelin in multiple sclerosis is developmentally immature. *The Journal of Clinical Investigation, 94*(1), 146–154.
Moscarello, M. A., Pritzker, L., Mastronardi, F. G., & Wood, D. D. (2002). Peptidylarginine deiminase: A candidate factor in demyelinating disease. *Journal of Neurochemistry, 81*(2), 335–343.
Moscarello, M. A., Mastronardi, F. G., & Wood, D. D. (2007). The role of citrullinated proteins suggests a novel mechanism in the pathogenesis of multiple sclerosis. *Neurochemical Research, 32*(2), 251–256.
Musse, A. A., Li, Z., Ackerley, C. A., Bienzle, D., Lei, H., Poma, R., Harauz, G., Moscarello, M. A., & Mastronardi, F. G. (2008). Peptidylarginine deiminase 2 (PAD2) overexpression in transgenic mice leads to myelin loss in the central nervous system. *Disease Models & Mechanisms, 1*(4–5), 229–240.
Nakashima, K., Hagiwara, T., Ishigami, A., Nagata, S., Asaga, H., Kuramoto, M., Senshu, T., & Yamada, M. (1999). Molecular characterization of peptidylarginine deiminase in HL-60 cells induced by retinoic acid and 1alpha,25-dihydroxyvitamin D(3). *The Journal of Biological Chemistry, 274*(39), 27786–27792.
Nishijyo, T., Kawada, A., Kanno, T., Shiraiwa, M., & Takahara, H. (1997). Isolation and molecular cloning of epidermal- and hair follicle-specific peptidylarginine deiminase (type III) from rat. *J. Biochem. (Tokyo), 121*(5), 868–875.
Rogers, G. E., & Simmonds, D. H. (1958). Content of citrulline and other amino acids in a protein of hair follicles. *Nature, 182*, 186–187.
Rus'd, A. A., Ikejiri, Y., Ono, H., Yonekawa, T., Shiraiwa, M., Kawada, A., & Takahara, H. (1999). Molecular cloning of cDNAs of mouse peptidylarginine deiminase type I, type III and type IV, and the expression pattern of type I in mouse. *European Journal of Biochemistry, 259*(3), 660–669.
Senshu, T., Sato, T., Inoue, T., Akiyama, K., & Asaga, H. (1992). Detection of citrulline residues in deiminated proteins on polyvinylidene difluoride membrane. *Analytical Biochemistry, 203*(1), 94–100.
Senshu, T., Akiyama, K., Kan, S., Asaga, H., Ishigami, A., & Manabe, M. (1995). Detection of deiminated proteins in rat skin: Probing with a monospecific antibody after modification of citrulline residues. *The Journal of Investigative Dermatology, 105*(2), 163–169.
Senshu, T., Akiyama, K., Ishigami, A., & Nomura, K. (1999). Studies on specificity of peptidylarginine deiminase reactions using an immunochemical probe that recognizes an enzymatically deiminated partial sequence of mouse keratin K1. *Journal of Dermatological Science, 21*(2), 113–126.

Serdaroglu, M., Cakirbay, H., Deger, O., Cengiz, S., & Kul, S. (2008). The association of anti-CCP antibodies with disease activity in rheumatoid arthritis. *Rheumatology International, 28*(10), 965–970.

Servais, L., Bearzatto, B., Schwaller, B., Dumont, M., De Saedeleer, C., Dan, B., Barski, J. J., Schiffmann, S. N., & Cheron, G. (2005). Mono- and dual-frequency fast cerebellar oscillation in mice lacking parvalbumin and/or calbindin D-28k. *The European Journal of Neuroscience, 22*(4), 861–870.

Shimada, N., Handa, S., Uchida, Y., Fukuda, M., Maruyama, N., Asaga, H., Choi, E. K., Lee, J., & Ishigami, A. (2010). Developmental and age-related changes of peptidylarginine deiminase 2 in the mouse brain. *Journal of Neuroscience Research, 88*(4), 798–806.

Smith, M. A. (1998). Alzheimer disease. *International Review of Neurobiology, 42*, 1–54.

Suzuki, A., Yamada, R., Chang, X., Tokuhiro, S., Sawada, T., Suzuki, M., Nagasaki, M., Nakayama-Hamada, M., Kawaida, R., Ono, M., Ohtsuki, M., Furukawa, H., Yoshino, S., Yukioka, M., Tohma, S., Matsubara, T., Wakitani, S., Teshima, R., Nishioka, Y., Sekine, A., Iida, A., Takahashi, A., Tsunoda, T., Nakamura, Y., & Yamamoto, K. (2003). Functional haplotypes of PADI4, encoding citrullinating enzyme peptidylarginine deiminase 4, are associated with rheumatoid arthritis. *Nature Genetics, 34*(4), 395–402.

Takizawa, Y., Sawada, T., Suzuki, A., Yamada, R., Inoue, T., & Yamamoto, K. (2005). Peptidylarginine deiminase 4 (PADI4) identified as a conformation-dependent autoantigen in rheumatoid arthritis. *Scandinavian Journal of Rheumatology, 34*(3), 212–215.

Tarcsa, E., Marekov, L. N., Mei, G., Melino, G., Lee, S.-C., & Steinert, P. M. (1996). Protein unfolding by peptidylarginine deiminase. Substrate specificity and structural relationships of the natural substrates trichohyalin and filaggrin. *The Journal of Biological Chemistry, 271*(48), 30709–30716.

Terakawa, H., Takahara, H., & Sugawara, K. (1991). Three types of mouse peptidylarginine deiminase: Characterization and tissue distribution. *Journal of Biochemistry (Tokyo), 110*(4), 661–666.

Tian, J., Shi, J., Bailey, K., & Mann, D. M. (2004). Relationships between arteriosclerosis, cerebral amyloid angiopathy and myelin loss from cerebral cortical white matter in Alzheimer's disease. *Neuropathology and Applied Neurobiology, 30*(1), 46–56.

Tsuchida, M., Takahara, H., Minami, N., Arai, T., Kobayashi, Y., Tsujimoto, H., Fukazawa, C., & Sugawara, K. (1993). cDNA nucleotide sequence and primary structure of mouse uterine peptidylarginine deiminase. Detection of a 3′-untranslated nucleotide sequence common to the mRNA of transiently expressed genes and rapid turnover of this enzyme's mRNA in the estrous cycle. *European Journal of Biochemistry, 215*(3), 677–685.

Vincent, S. R., Leung, E., & Watanabe, K. (1992). Immunohistochemical localization of peptidylarginine deiminase in the rat brain. *Journal of Chemical Neuroanatomy, 5*(2), 159–168.

Wagner, E., Skoumal, M., Bayer, P. M., & Klaushofer, K. (2009). Antibody against mutated citrullinated vimentin: A new sensitive marker in the diagnosis of rheumatoid arthritis. *Rheumatology International, 29*(11), 1315–1321.

Wakoh, T., Uekawa, N., Terauchi, K., Sugimoto, M., Ishigami, A., Shimada, J., & Maruyama, M. (2009). Implication of p53-dependent cellular senescence related gene, TARSH in tumor suppression. *Biochemical and Biophysical Research Communications, 380*(4), 807–812.

Watanabe, K., & Senshu, T. (1989). Isolation and characterization of cDNA clones encoding rat skeletal muscle peptidylarginine deiminase. *The Journal of Biological Chemistry, 264*(26), 15255–15260.

Watanabe, K., Akiyama, K., Hikichi, K., Ohtsuka, R., Okuyama, A., & Senshu, T. (1988). Combined biochemical and immunochemical comparison of peptidylarginine deiminases present in various tissues. *Biochimica et Biophysica Acta, 966*(3), 375–383.

Whitney, E. R., Kemper, T. L., Bauman, M. L., Rosene, D. L., & Blatt, G. J. (2008). Cerebellar Purkinje cells are reduced in a subpopulation of autistic brains: A stereological experiment using calbindin-D28k. *Cerebellum, 7*(3), 406–416.

Wood, D. D., Ackerley, C. A., Brand, B., Zhang, L., Raijmakers, R., Mastronardi, F. G., & Moscarello, M. A. (2008). Myelin localization of peptidylarginine deiminases 2 and 4: Comparison of PAD2 and PAD4 activities. *Laboratory Investigation, 88*(4), 354–364.

Zarabian, B., Koushesh, F., & Vassef, A. (1987). Modified methods for measuring citrulline and carbamoyl-beta-alanine with reduced light sensitivity and sucrose interference. *Analytical Biochemistry, 166*(1), 113–119.

Zhao, J., Zhao, Y., He, J., Jia, R., & Li, Z. (2008). Prevalence and significance of anti-peptidylarginine deiminase 4 antibodies in rheumatoid arthritis. *The Journal of Rheumatology, 35*(6), 969–974.

Chapter 18
Deimination in Multiple Sclerosis: The Bad, the Good, and the Ugly

William R. Meador, John R. Rinker, and Anthony P. Nicholas

18.1 Introduction

Multiple sclerosis (MS) is a chronic, autoimmune, demyelinating, and degenerative disease that affects the central nervous system. The majority of patients are diagnosed between the ages of 20 and 40, and it is a leading cause of disability among young adults (Hauser and Goodin 2015; Lavery et al. 2014). MS typically presents with acute attacks of neurologic symptoms such as weakness, imbalance, vision loss, or numbness that are called "relapses," and these occur early in the course of the most common form, historically known as relapsing-remitting MS (RRMS). Over time, increasing numbers of relapses contribute to disability accumulation (Fig. 18.1). At 19 years of disease duration, 75% of RRMS patients will have developed a more progressive course, whereby their worsening occurs in a steady decline and with fewer relapses (Confavreux and Vukusic 2006), a disease process historically known as secondary progressive MS (SPMS). In contrast, about 15% of patients will have progressive neurologic decline from the start, and this has been referred to as primary progressive MS (PPMS) (Lublin and Reingold 1996). In a 2013 effort to align the nomenclature with the underlying pathological changes,

W.R. Meador (✉)
University of Alabama at Birmingham, Birmingham, AL, USA
e-mail: wmeador@uab.edu

J.R. Rinker
University of Alabama at Birmingham, Birmingham, AL, USA

The Birmingham VA Medical Center, Birmingham, AL, USA
e-mail: jrinker@uabmc.edu

A.P. Nicholas
Department of Neurology, University of Alabama at Birmingham
and The Birmingham Veterans Administration Medical Center, Birmingham, AL, USA
e-mail: nicholas@uab.edu

A.P. Nicholas et al. (eds.), *Protein Deimination in Human Health and Disease*,
DOI 10.1007/978-3-319-58244-3_18

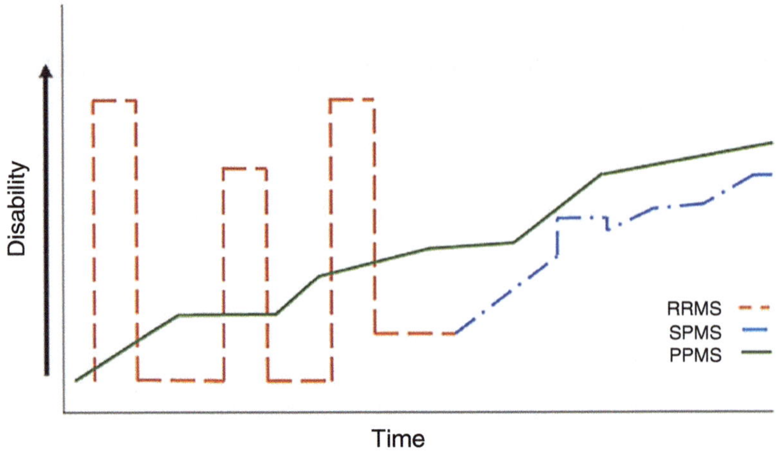

Fig. 18.1 Time course and disability of various MS phenotypes. The three primary clinical patterns of MS disease activity are relapsing-remitting (RRMS), secondary progressive (SPMS), and primary progressive (PPMS). Over time (x-axis), progression of disability (y-axis) differs between the three disease forms. For RRMS patients (*orange dashed line*), the clinical course is highlighted by acute attacks of disability that can return to baseline but sometimes do not. Most of these RRMS patients will progress to SPMS (*blue dashed* and *dotted line*), who worsen over time but no longer have distinct attacks. This pattern can mirror PPMS patients (*green line*), who slowly develop disability over time, without having clear attacks from the onset

these diagnoses were renamed simply relapsing MS (RMS) and progressive MS (PMS) along with other, more subtle changes in diagnostic terminology that are beyond the scope of this chapter (Lublin et al. 2014).

18.2 MS Epidemiology

Since its first, well-organized description in 1865 by Charcot, MS has been increasingly recognized as a significant cause of disability in young people around the globe (Lavery et al. 2014; Compston 2004). MS is currently thought to affect at least 350,000 individuals in the United States and about 2.5 million people throughout the world, with the majority of patients diagnosed between 20 and 49 years of age (Hauser and Goodin 2015). In high prevalence areas, MS affects 2 out of 1000 people with an incidence of 6 per 100,000 (Wynn et al. 1990). The overall prevalence of MS has been increasing since the mid-twentieth century due in part to improved detection and diagnostic capabilities, as well as an increase in incidence. This increase has been found almost exclusively in women living with MS and has resulted in a change in the female-to-male ratio of 1.9 in the mid-twentieth century to 3.2 in the early twenty-first century (Orton et al. 2006). Many reports have referenced this increase in female preponderance as support for an environmental or modifiable factor that may influence who develops MS.

18.3 MS Pathology

MS pathology is driven by an abnormal response of both innate and adaptive immunity within the central nervous system (CNS), with lesions typically occurring within the cerebral hemispheres, posterior fossa structures, and spinal cord. Central in this role, CD4+ T helper lymphocytes drive the inflammatory disease process early after they undergo activation by antigen-presenting cells (APC), with the most evident homeostatic disruption that of a dysregulation of Th1 and Th17 cells (Nuyts et al. 2013). An as-yet unrecognized process activates these autoreactive lymphocytes, so that the immune system targets myelin and causes pathogenic demyelination within the CNS. T helper lymphocytes then activate cytotoxic CD8+ lymphocytes and the innate immune system to cause local destruction of CNS tissue (Serafini et al. 2006). Astrocytes within the acute lesion are damaged and even release pro-inflammatory cytokines to promote further injury. Loss of astrocyte function results in a more permeable blood-brain barrier (BBB), thereby aiding the infiltration of more inflammatory cells and cytokines into the CNS (Brosnan and Raine 2013). In subacute or resolving lesions, astrocytes can take on a reparative role through extending processes to ensheathe vulnerable axons, absorbing harmful components such as glutamate, restoring the extracellular matrix, and even serving as pluripotent cells to promote regeneration of supportive cells. Chronic lesions later show astrogliotic scarring within the lesion and reactive astrocytes along the border of the lesion that are proposed to act as a wall between injured and healthy tissue (Brosnan and Raine 2013).

Over the past decade, increasing awareness has developed on the role of B cell lymphocytes in the pathogenesis and propagation of disease in MS (Blauth et al. 2015). Since the 1940s, it has been well established that unique gamma globulin populations, representing pathologic immunoglobulin populations known as oligoclonal bands (OBs), are present in the spinal fluid of most patients with MS and serve as a useful diagnostic tool (Kabat et al. 1942). The exact antigenic targets of these immunoglobulins are relatively unknown and appear to be inconsistent from patient to patient, but the presence of OBs confers a worse prognosis in patients with MS (Rojas et al. 2012). In a condition known as clinically isolated syndrome (CIS), in which a patient develops a single demyelinating event without complete fulfillment of MS diagnostic criteria, high numbers of OBs indicate that these individuals are more likely to convert to MS (Dalla Costa et al. 2015). These observations clearly support a role for plasma cells, the terminal lineage of B cell lymphocytes, as playing an instrumental role in MS pathogenesis. However, there are likely other roles played by B cells, such as antigen presentation, which contribute to disease activity in MS as well.

18.4 Basic Tenants of MS Treatments

The treatments available for RMS focus on reducing disease activity, thereby reducing disability, improving quality of life, and reducing symptom burden. The first FDA-approved medication for RMS disease modification in the United States was interferon beta-1b in 1993 (The IFNB Multiple Sclerosis Study Group 1993). Since that time, 12 additional, unique medications have been FDA approved in the United States (Table 18.1), with the earlier approved medications tested against placebo and most of the more recent trials using active comparator arms. Many of the medications currently approved have unique mechanisms of action, but they all act by altering the immune system in a manner to reduce immune system-driven disease activity in the CNS. The trials utilized for FDA approval have exclusively used designs that compare active treatment annualized relapse rates to placebo or active comparator annualized relapse rates (Piehl 2014). All of the approved medications have shown some benefit in secondary end points such as reduction of disease burden as measured by MRI, improved quality of life, and disability, although to a lesser degree.

One area that historically had repeatedly failed therapeutic development was that of PMS or progressive disability within RMS patients (Segal and Stuve 2016). Recent trials testing B cell-targeted therapy have shown promise in progressive disease and resulted in approval of the first treatment for PMS in 2017 (Sorensen and Blinkenberg 2016). Other areas of MS treatment that have been poorly addressed include fatigue management, cognitive decline, prevention of the development of

Table 18.1 Disease-modifying therapies approved for treatment of human MS patients in the United States

Agent	Date(s) approved	Mechanism of action
interferon beta-1b	1993, 2009	Impairs migration of lymphocytes and shifts lymphocyte populations to less inflammatory cells
interferon beta-1a	1996, 2002, 2014	Impairs migration of lymphocytes and shifts lymphocyte populations to less inflammatory cells
glatiramer acetate	1996	Amino acid combination aimed at shifting lymphocytes to less inflammatory populations and modulates antibody production
mitoxantrone	2000	Impairs immune cell production
natalizumab	2004	Blocks alpha-4 integrin resulting in reduced lymphocyte entry into CNS
fingolimod	2010	Blocks S1P1 receptor that results in prevention of egress of lymphocytes out of lymphoid tissue
teriflunomide	2012	Dihydroorotate dehydrogenase inhibitor which results in reduced production of lymphocytes
dimethyl fumarate	2013	Activates Nrf2 pathway resulting in impaired lymphocyte activity
alemtuzumab	2014	CD52 monoclonal antibody that results in destruction of T and B lymphocytes
daclizumab	2016	Interleukin-2 receptor antagonist that results in diminished lymphocyte activity
ocrelizumab	2017	Anti-CD20 humanized monoclonal antibody depletes B cells

SPMS, and various other symptomatic features of MS. Although several FDA-approved drugs can reduce relapse rates and residual disability in RMS (Piehl 2014), this success has proven harder to achieve with progressive MS.

18.5 Progressive MS

Progression in MS is a pattern of neurological worsening which is distinct from the more easily recognized relapsing-remitting course (Fig. 18.1). The most distinguishing characteristic of progression is the time course. Unlike relapses, which occur abruptly, worsen over a period of days to weeks, and then spontaneously remit, progression is an insidious and continuous process, in which disability accumulates gradually, over months to years, without sustained regression of symptoms. The worsening in progressive MS is often recognized in retrospect, as the rate of deterioration is such that patients who live with progressive MS may adapt to a worsening symptom, such as leg weakness or gait impairment, without realizing at the time that the process is unrelenting. In acknowledgment of progression being one of the major patterns of disability accumulation in MS, progressive subtypes of MS (PPMS, SPMS, progressive relapsing) were included in the first consensus definitions of MS clinical subtypes in 1996 (Lublin and Reingold 1996). The subtypes were distinguished by (1) the presence or absence of relapses in the overall disease course and (2) the sequence of progression relative to relapses. PPMS was defined as insidious progression from symptom onset, with no relapses during the disease course. SPMS described a progressive course following a period of relapsing-remitting disease. Progressive relapsing disease denoted a primary progressive course in which relapses first occurred after the onset of progression.

Another characteristic that distinguishes progression from relapses is its predilection for certain clinical phenotypes. The most common pattern takes the form of a progressive myelopathy, affecting the spinal cord primarily (Kremenchutzky et al. 2006). As progression becomes evident, patients develop a spastic paraparesis or quadriparesis. Mobility is almost inevitably impaired, and reliance on assistive devices such as a cane, walker, or wheelchair becomes necessary. Symptoms of neurogenic bladder and bowel may also develop. This most common progressive phenotype is well captured in the disability scale most widely used to describe MS disability, the Expanded Disability Status Scale (EDSS), and in its precursor, the Disability Status Scale (DSS) (Kurtzke 1983). While the lower scores on the EDSS are determined by fluctuations in various neurological functional systems (vision, brain stem function, motor systems), which also happen to reflect the symptoms most common to MS relapses, scores in the intermediate range and higher are largely determined by impairments in ambulation. In fact, once a score of 6 is reached (reflecting requirement of unilateral assistance to ambulate), scores are determined entirely by ambulatory impairment and, at scores from 7.5 to 9.5, by loss of functional independence.

Although progression most commonly presents as a progressive myelopathy, other functions may also be involved (Bermel et al. 2010). Cognition is another domain commonly affected by progression, but due at least in part to the difficulty in measuring cognitive impairment characteristic of MS and the insensitivity of the EDSS in capturing cognitive worsening, cognitive progression in MS is less well recognized by clinicians. Progression may affect other domains as well, including vision and cerebellar function, but due to the preponderance of the myelopathic picture in progressive MS and the insensitive instruments for measuring progression in these other functional systems, they are also under-recognized by clinicians, if not by patients.

The division of MS into relapsing and progressive subtypes led in part to the refinement of clinical trial design for investigative therapies in MS. Relapse-related end points such as annualized relapse rate and time to first on-treatment relapse have been the principal metric by which success in disease modification is measured. Having criteria which separated "progressive" from "relapsing," patients enabled investigators to enrich study populations with subjects who were prone to relapse while excluding subjects with slowly worsening (progressive) disease. A small number of trials for relapsing MS have demonstrated sustained change in disability as measured by the EDSS as the primary outcome measure in phase 3 trials (Coles et al. 2012; Jacobs et al. 1996), but demonstrating a benefit in disability accumulation has remained a higher bar for efficacy in such clinical studies.

As phase 3 trials began to be designed around patients with progressive subtypes of disease, change in disability was used as the accepted primary end point. Numerous clinical trials followed, testing drugs that had been successful in cohorts of relapsing MS patients, including interferon preparations, glatiramer acetate, and rituximab (Cohen et al. 2002; Secondary Progressive Efficacy Clinical Trial of Recombinant Interferon-beta-1a in MS (SPECTRIMS) Study Group 2001; Hawker et al. 2009; Leary et al. 2003; The North American Study Group on Interferon beta-1b in Secondary Progressive MS 2004; Wolinsky et al. 2007). Those few studies that have been able to demonstrate a benefit used either a composite end point to demonstrate worsening (Hartung et al. 2002) or enrolled a subject population enriched with markers of inflammatory disease, such as gadolinium-enhancing lesions on brain MRI (European Study Group on Interferon-1b in Secondary Progressive MS 1998). A principal result of these numerous failed trials has been refined strategies in clinical research design for progressive MS, such that progressive patients with characteristics most likely to respond to immune modulation are preferentially selected for inclusion (Ontaneda et al. 2015). The fruits of this attention to trial design have been two recent successful phase 3 trials for progressive MS involving ocrelizumab (Montalban et al. 2016) and siponimod (Kappos et al. 2016). Both drugs are mechanistically similar to drugs known to be effective in relapsing MS but previously unsuccessful in progressive MS (ocrelizumab depletes B cells much the same as rituximab, while siponimod functions similarly to fingolimod). These recent successful trials lead one to at least a couple important conclusions regarding the progressive phase of the disease. First, disability accumulation in progressive MS is at best an inflexible, downstream effect of inflammatory activity occurring behind the blood-brain barrier and at worst a primary degenerative process which is not influenced by immune modulation. Second, patient selection is a key determinant of whether a phase 3 study in progressive MS is likely to be effective.

Perhaps the most important consideration arising from the difficulty in completing a positive trial in progressive MS is whether the mechanisms by which disease-modifying drugs prevent relapses are even capable of having an effect on the progressive disease process. Several observations lend credit to this concern. First, a widely cited observational study of the natural history of untreated MS in France raised the question of whether the progressive phase of MS is "amnestic" to, or independent from, the relapsing phase of the disease (Confavreux et al. 2000). The factor which most accurately predicted the point of transition from relapsing to progressive disease was the age of the patient, regardless of disease duration. Furthermore, when survival curves between relapsing and progressive cohorts were compared, the DSS milestone of 4 (meaning moderate disability with some ambulatory impairment, not yet requiring assistance) appeared to be the point beyond which the clinical trajectory of both relapsing-onset and progressive-onset patients was superimposable. The authors suggest this apparent dissociation of relapses from the progressive phase of the disease implies a fundamentally different substrate to account for worsening due to relapses versus worsening due to progression.

A similar natural history study out of North America also found dissociation between the relapsing phase of MS and the progressive phase (Kremenchutzky et al. 2006). Interestingly, the number of relapses experienced did not appear to influence the trajectory of disability accumulation once subjects reached the progressive phase of the disease, even when the relapsing phase of the disease consisted of only a single relapse.

Yet another pair of studies probed the question whether aggressive immune suppression could influence the natural history of relapsing versus progressive disease (Coles et al. 2006). Investigators used the monoclonal antibody alemtuzumab a potent monoclonal antibody which depletes CD52-positive cells, which include most T cells, B cells, and monocytes. In two separate cohorts, one a group of patients with early RMS and another with established progressive disease, relapsing subjects were found to not only experience a sustained cessation of relapses but also stabilization of the EDSS score. However, in a cohort of PMS subjects, the treatment appeared to have little effect on disability accumulation, even in the absence of relapses. In other words, progression continued, despite near-complete eradication of the adaptive immune system in the peripheral circulation. The implication, thus, is that even highly effective therapy targeting the peripheral immune system has little measurable effect on the PMS disease process.

How, then, can the phenomenon of progression best be explained? The most succinct explanation is that there is a lack of consensus on what the underlying drivers of PMS may be or if it even represents the same process from patient to patient. Theories to explain progression can be broadly divided into inflammation-driven and inflammation-independent processes (Kutzelnigg and Lassmann 2014). It is important to note that while pathological studies of CNS tissue from MS patients reveal a spectrum of inflammation, demyelination, oligodendroglial cytopathy, and neurodegeneration, there are some general distinctions between patients with RMS versus PMS.

Examination of white matter lesions from PMS patients commonly reveals a rim of macrophages and activated microglia with an inactive central area (Prineas et al. 2001). This contrasts with the more active cellular infiltrates and extensive myelin destruction of the acute lesion, more common in early RMS (Brück et al. 1995). Demyelinating plaques in the gray matter (especially in the cerebral cortex) are more abundant in PMS as compared to RMS. The most common gray matter plaques are subpial, extending from the meninges down into the cortex. Despite the lower density of myelin in gray matter, demyelination of gray matter myelin is extensive, and both axonal injury and neuronal death can be observed in gray matter plaques (Peterson et al. 2001). The widespread cortical demyelination seen in primary and secondary progressive MS is also thought to contribute to focal gray matter atrophy in PMS, which has also been shown to correlate with clinical measures of disability.

One potentially important relationship between cortical demyelination and the inflammatory characteristics of PMS is the development of B cell follicles in the meninges of progressive patients (Lovato et al. 2011). These structures may interact extensively with the underlying cortex and may be responsible for much of the cortical demyelination and atrophy in progressive disease (Howell et al. 2011). The proportion of B cells found in MS lesions are known to increase with disease duration; thus, the establishment of what may amount to permanent foci of inflammation within the meninges of MS patients offers one possible explanation for why immune modulators are unable to produce measurable benefit in PMS. These meningeal follicles become established behind the blood-brain barrier and thus are insulated against drugs which act primarily in the peripheral circulation. Another immunological change seen in progressive MS which occurs behind the blood-brain barrier is the development of widespread microglial activation in the radiologically normal-appearing white matter (NAWM). Such widespread changes provide possible bases for diffuse axonal loss, inefficient neural functioning, and resistance to immune-modulating therapy (De Stefano et al. 1999; Kutzelnigg et al. 2005; Seewann et al. 2009; van Horssen et al. 2012).

An alternate or perhaps complementary theory for the pattern of worsening in PMS is the notion that the slow accumulation of disability and the accelerated loss of brain volume are due to a neurodegenerative process which is independent of autoimmunity. One argument holds that due to the loss of trophic support provided by myelin to the axons it supports, the axons eventually succumb to a metabolic deficit that cannot be sustained without the support of myelin. Excitotoxic injury is another theory that may explain axonal loss as a by-product of the inflammatory milieu adjacent to the exposed axon (Pitt et al. 2000). However, not all axonal loss can be accounted for by its proximity to known regions of focal demyelination. Pathological studies of tissue remote from active demyelination have found axonal dropout which cannot be accounted for by active inflammation or from Wallerian-type degeneration. A study of postmortem tissue from 55 MS patients found a disproportionate loss of small-diameter fibers, especially in corticospinal and sensory tracts (DeLuca et al. 2004). One theory that accounts for this axonal destruction that is remotely located from active inflammation attributes the axonal loss to diffuse mitochondrial injury. Dysfunctional mitochondria may lead to an energy deficit, which disproportionately affects small-diameter fibers by inducing oxidative free radical damage (Trapp and Stys 2009).

The process by which relapsing MS, a lesion-centric, inflammatory condition, transitions into PMS, with its widespread involvement of the brain and spinal cord and outwardly less inflammatory phenotype, remains an important gap in the understanding of progression in MS. While theories that bridge these two clinically distinct disease phases continue to evolve, continued investigation into how these two phases bridge to one another will require attention and investigation.

18.6 Deimination in Animal Models and Humans with MS

Deimination, or citrullination, is when arginine amino acids are converted to citrullines by the peptidylarginine deiminase (PAD) family of enzymes (Nicholas and Bhattacharya 2014). In the brain, this process was first discovered to occur with myelin basic protein (MBP), in which a fulminant process of acute demyelination (Marburg type) caused 18 of 19 arginines in MBP to be converted to citrullines (Wood et al. 1996). (For a more detailed discussion on deimination of MBP, see Chap. 19) In fact, deimination was shown to initiate a process of increased myelin unraveling, mirroring an immature isoform of MBP, and ultimately resulted in a widespread autoimmune reaction with eventual brain degeneration (Boggs et al. 1999; Cao et al. 1999; Moscarello et al. 1994; Pritzker et al. 2000).

In our studies of an animal model of MS known as murine experimental autoimmune encephalomyelitis (EAE), we found evidence of increased deimination of both MBP and the astrocyte marker glial fibrillary acidic protein (GFAP) (Nicholas et al. 2005), with previously unforeseen pathology (Fig. 18.2). Using an established monoclonal antibody (F95) against natural and synthetic citrullinated proteins (Nicholas and Whitaker 2002), numerous, small, previously unrecognized "patches" of citrullinated proteins were discovered throughout EAE brains, while EAE spinal cords showed similar but much larger lesions. Using dual-color immunofluorescence, these lesions were found to contain citrullinated MBP and were surrounded by astrocytes immunoreactive for both GFAP and F95. These lesions became evident about the time EAE mice became symptomatic and increased in size and number with increasing disease severity. In some sections of the spinal cord, but not brains of severely debilitated EAE mice, a widespread gliotic response was seen, with astrocytes containing citrullinated GFAP spread throughout the gray and white matter. Western blot analysis of acidic proteins from the brains and spinal cords of EAE mice had higher levels of multiple citrullinated GFAP isoforms as compared to controls, with more F95-positive bands in the EAE brains versus spinal cords.

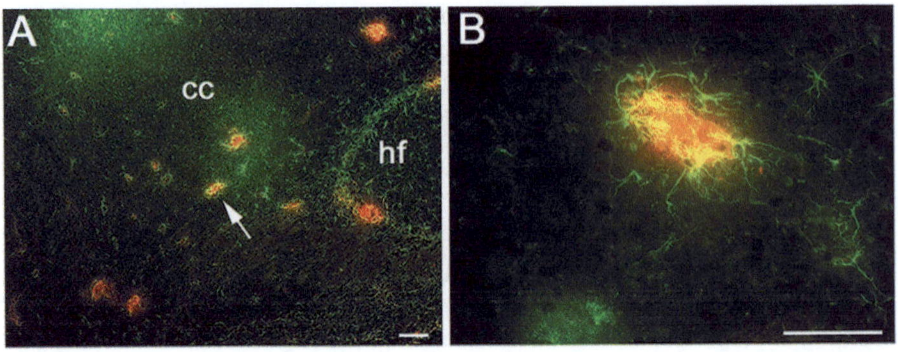

Fig. 18.2 In EAE mice brains, immunoreactive lesions were seen as extracellular patches of citrullinated proteins (*red*) surrounded by GFAP-immunoreactive astrocytes (*green*). The lesion in **a** (*arrow*) is magnified in **b**, showing that a plaque, known to contain deiminated MBP, is sounded by astrocytes, whose processes project into the lesion and demonstrate co-localization (*yellow*) for GFAP and deiminated proteins. cc, cerebral cortex; hf, hippocampal formation. Scale bars = 50 μm (Modified from Nicholas et al. 2005; Bradford et al. 2014a)

(For a more detailed discussion on deimination of GFAP, see Chap. 20) In any case, our results first raised the possibility that citrullination of both GFAP and MBP may contribute to the pathophysiology of EAE and that the brains of EAE mice may contain more pathology than what was previously realized. Using a different citrulline detection method, similar findings were seen in EAE spinal cords in a second report (Raijmakers et al. 2005).

One study that temporarily stagnated enthusiasm for further examining a role for deimination in EAE and MS was when no difference in disease severity was found between control mice with EAE and transgenic mice in which the PAD2 isoform was knocked out (Raijmakers et al. 2006). In this study, PAD2 was preferentially scrutinized primarily because it was known to be the most abundant PAD isoform in the CNS (Akiyama et al. 1990; Watanabe et al. 1988). However, using the F95 antibody, we refuted these findings by showing that deimination persisted in the spinal cords of PAD2 knockout mice with EAE (Bradford et al. 2014a), suggesting that another PAD isoform such as PAD4 may have been responsible for the continued deimination in these animals without PAD2. In support of a role for PAD4 in the pathophysiology of demyelinating disease was a study showing that citrullination of nucleosomal histones by PAD4 was increased in the NAWM from MS patients and animal models of demyelination and that translocation of PAD4 into the nucleus was attributable to elevated tumor necrosis factor-alpha (Mastronardi et al. 2006). In addition, the development of transgenic mice that overexpressed PAD2 showed that these animals spontaneously developed a demyelinating disease with increased citrullination of MBP and clinical worsening over time (Musse et al. 2008). In support of the deleterious effects that protein deimination plays in the CNS, the first generation of PAD inhibitors was shown to reverse or prevent disability in four separate EAE animal models, including autoimmune and transgenic mice that represented both acute and chronic forms of demyelination (Moscarello et al. 2013).

These findings suggest that blocking deimination may also serve as a novel future treatment for both relapsing and progressive MS phenotypes in human patients.

Although we previously demonstrated increased deimination of both MBP and GFAP in EAE mice (Nicholas et al. 2005), the protein with the greatest deimination in our studies of human brains with progressive phenotypes of MS was GFAP, with evidence of pathology extending into normal-looking white and gray matter (Nicholas et al. 2004). In fact, GFAP was first discovered decades ago as the main protein constituent of chronic MS lesions and thus was considered to be the morphologic basis of astrogliosis ever since (Eng 1980). Unfortunately, a study of the temporal changes in brain protein deimination of humans with MS is problematic and primarily limited to autopsy material, so most of these investigations were performed on MS specimens from individuals with more progressive forms of the disease.

Unlike our findings of profound deimination within active demyelinating lesions in EAE brains and spinal cords (Nicholas et al. 2005), our first study of autopsy human brain specimens from American SPMS patients with chronic, advanced disease (Nicholas et al. 2004) showed predominantly "burned-out" plaques, primarily consisting of fibrous scar tissue and devoid of deimination, except within a few astrocytes remaining adjacent to the central blood vessels of old inactive lesions (Fig. 18.3a). However, when examining the NAWM, heavy citrullination was unexpectedly seen in the SPMS specimens as compared to controls. In SPMS brain samples, but not in normal brains, long fibers of co-localized GFAP and citrullinated proteins were seen, extending from the NAWM into the gray matter (Fig. 18.3b), and increased numbers of astrocytes containing citrullinated proteins and GFAP were also present at the junction between the gray and white matter in SPMS brains (Nicholas et al. 2004). In a similar study of SPMS patients from the UK, where some of the postmortem specimens had lesions that were still somewhat inflammatory, evidence for deimination of both GFAP and MBP were demonstrated using multiple techniques, including matrix-assisted laser desorption/ionization-ion mobility separation and tandem mass spectrometry (Bradford et al. 2014b). However, Western blot analysis of brain proteins from the NAWM of more advanced patients primarily showed upregulation of multiple citrullinated GFAP isoforms in SPMS brains as compared to controls (Nicholas et al. 2004; Bradford et al. 2014b). In fact, citrullination was most pronounced in SPMS specimens at the plaque interfaces and was shown to co-localize with GFAP immunoreactivity using dual-color immunofluorescence (Fig. 18.3c). These data raise the possibility that citrullination of GFAP and the role of astrocytes containing it contribute to the pathophysiology of humans with MS, perhaps even more so in the progressive forms of the disease.

In one study of live humans, significantly higher GFAP levels were shown in the cerebrospinal fluid of MS patients compared to control subjects, which was most pronounced in SPMS patients, although GFAP was also significantly increased in RRMS patients during relapses as well (Malmeström et al. 2003). These findings were in contrast with earlier studies that only found elevated GFAP levels in MS patients who were severely disabled (Albrechtsen et al. 1985; Noppe et al. 1986; Petzold et al. 2002). Unfortunately, these studies did not specifically measure the amounts of citrullinated GFAP.

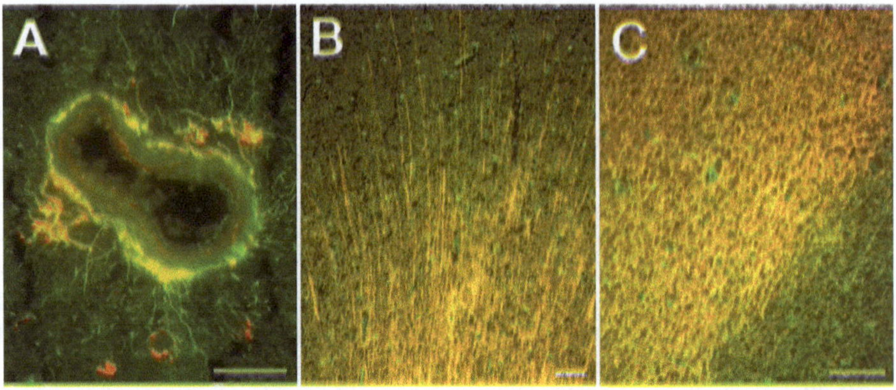

Fig. 18.3 Using dual-color immunofluorescence and the F95 monoclonal antibody against peptidyl-citrulline, examination of chronic SPMS brains and co-localization (*yellow*) of GFAP (*green*) and deiminated proteins (*red*) were often seen (**a**) in astrocytes encircling a central blood vessel within the cores of "burned-out" plaques, suggesting an important role for deiminated proteins in distal astrocyte filaments and foot processes. In NAWM from these same patients (**b**), radiations of GFAP/F95-positive immunoreactive processes extended from the white matter (*bottom*) into the gray matter (*top*). However, the most intense GFAP/F95 co-localization was observed (**c**) in SPMS white matter (*upper left*) at the interface adjacent to old plaques (*lower right*), suggesting that deiminated GFAP in this location may have served to help wall off inflammatory lesions in an earlier, more active phase of this disease. Scale bars = 50 μm (Modified from Nicholas et al. 2004; Bradford et al. 2014a)

One way to measure citrulline in the brains of live humans is with proton magnetic resonance spectroscopy (MRS). In a MRS study of patients with early onset MS, increased citrulline was more often detected in demyelinating lesions and the NAWM of diseased patients, compared to normal controls (Oguz et al. 2009), but this method was unable to determine what specific proteins were deiminated and how many peptidyl-citrullines each one contained. It is also unclear if this method is able to differentiate between free citrulline and peptidyl-citrulline, where only the latter is a product of deimination mediated by PAD enzymes. As a result, the detection of deiminated brain proteins, their locations, identities, and number of peptidyl-citrullines per protein during the time course of live patients with differing phenotypes of MS still remains to be fully studied.

18.7 Deimination in Humans with MS: The Bad, the Good, and the Ugly

As previously noted, properly studying the progression of brain protein deimination throughout the time course of humans with MS is problematic due to the lack of technology to study this process in live patients, as well as the lack of biological specimens from patients with typically nonlethal CIS and RRMS. Although there are a number of various MS phenotypes as already discussed, the majority of patients start probably with CIS and then progress to RMS and later to PMS (Fig. 18.1). Using EAE as a surrogate for active disease (i.e., RMS), we have developed a working hypothesis regarding the progression of brain protein deimination in MS (Fig. 18.4).

In early disease (Fig. 18.4a), we propose that utilizing a leaky BBB, peripheral inflammatory cells enter the brain parenchyma and participate in the demyelinating process, in part mediated by deimination of MBP. A more thorough discussion of the role of deiminated MBP in demyelinating disease is discussed in Chap. 19, but if EAE can be used as a surrogate for early inflammatory demyelinating disease (Fig. 18.2), we have clearly shown that the small extracellular lesions in the white matter of these animals contain deiminated MBP (Nicholas et al. 2005). In order for this process to occur, dormant PAD enzymes must be activated, perhaps due to increased calcium levels resulting from local hypoxia in these pathological areas (Trapp and Stys 2009).

In response to this pathology, PAD in local astrocytes then becomes activated, and these glial cells enlarge, perhaps in part due to deimination of intermediate filaments such as vimentin and GFAP (Fig. 18.4b). A more thorough discussion of the role of astrocytic deiminated GFAP in response to brain injury is discussed in Chap. 20, but again, if EAE can be used as a surrogate for inflammatory MS, we have clearly shown that the small extracellular lesions in the white matter of these animals (Fig. 18.2) are surrounded by astrocytes that express co-localization of GFAP and deiminated proteins. Although astrocytes are critical to the health of neurons and participate in the formation of the BBB, following CNS injury, these cells are also known to become hypertrophic and extend and intertwine their processes to form a mesh-like network that attempts to seal off the injured area (Sofroniew 2009). Since intermediate filaments like vimentin and GFAP have been shown to unwind in vitro when deiminated (Inagaki et al. 1989), perhaps this process contributes to the cytoskeletal changes that take place in astrocytes under these conditions (see Chap. 20, Fig. 20.6).

As a result, it can be theorized that deimination within astrocytes is a normal response to brain injury, not only to help enhance nutrition to and expel waste from struggling neurons but also to phagocytize extracellular debris and shuttle it out of the brain and into the bloodstream and cerebrospinal fluid, due to the anatomical associations of these glial cells with the BBB and ventricular surfaces, respectively. To support this idea in MS, MBP-containing myelin vesicles prepared from rat brains, as a model of degraded myelin, were shown to be phagocytized by rat astrocytes in culture (Gaultier et al. 2009). In addition, highly enriched astrocyte cultures derived from active demyelinated MS lesions of postmortem human adult brains and spinal cords showed an increased proliferating capacity of these cells, as compared to astrocytes derived from similar non-lesioned and normal brain and spinal cord regions (De Groot et al. 1997).

In one case of a patient with rapidly worsening MS, postmortem analysis proved that hypertrophic astrocytes in early and active lesions phagocytize myelin and axonal debris, as well as internalize other glial cells (Morcos et al. 2003). However, in

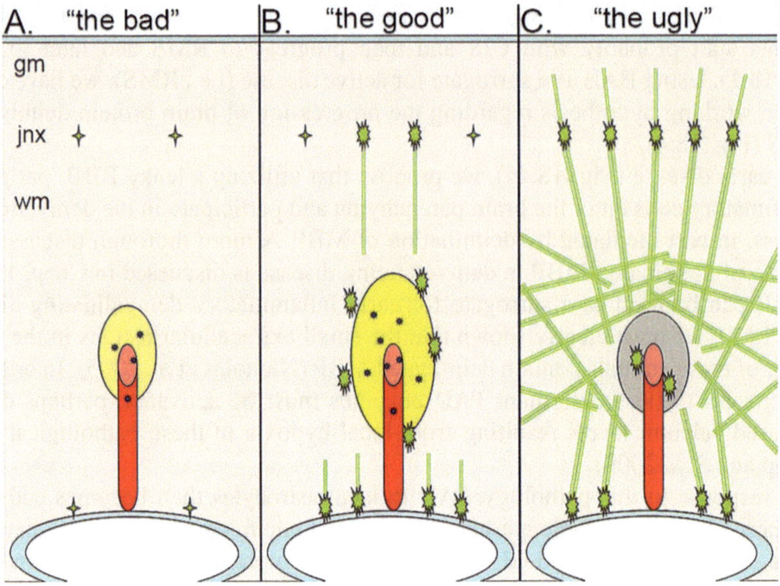

Fig. 18.4 Theoretical progression of deimination in early (**a**), middle (**b**), and late (**c**) stages of multiple sclerosis. As the disease process starts (**a**, "the bad"), peripheral white blood cells (*small blue stars*) enter the white matter (wm) of the brain via "Dawson's fingers," leaky blood vessels (*red tubes*) emerging from the ventricular surfaces (*light blue lines*). Part of the autoimmune reaction may be against hyper-deiminated myelin basic protein (MBP, *yellow*), which may contribute to the extracellular debris found in inflammatory, demyelinating plaques. At this stage, a few non-reactive astrocytes (*thin green stars*) are found at the ventricular surface and the junction (jnx) between the gray matter (gm) and white matter (wm). In response to this invasion (**b**, "the good"), around the plaque (*yellow oval*), astrocytes (*green cells*) become noticeable or "reactive" and enlarge, perhaps in an attempt to phagocytize the extracellular debris, containing deiminated proteins such as MBP (*yellow*), and shuttle it back into the bloodstream. However, if the autoimmune reaction continues unabated, more peripheral white blood cells (*small blue stars*) enter the brain, overwhelming the local astrocytes and the demyelinating plaques enlarge. At this stage, more astrocytes (*green cells*) become apparent at the ventricular surface (*bottom*) and at the gray matter (gm)-white matter (wm) junction (jnx). Many of these cells then send projections (*green lines*) containing deiminated glial fibrillary protein (GFAP) into the normal-appearing white matter (NAWM), perhaps in an attempt to shuttle the debris to the ventricular surfaces and out of the brain into the cerebrospinal fluid. If the disease process progresses further (**c**, "the ugly"), a demyelinating plaque may be replaced by a burned-out scar (*gray oval*), in which a few remaining reactive astrocytes (*green cells*) remain adjacent to its central blood vessel (*red tube*). At this point, other reactive astrocytes (*green cells*) outside of the old plaques have sent extensive networks of projections (*green lines*) containing deiminated GFAP throughout the NAWM, as well as into the gray matter (gm). The highest concentrations of this deiminated GFAP seem to surround the plaques, perhaps in a futile attempt to wall off the spreading demyelinating process

areas that lacked significant inflammation within the adjacent NAWM and within late remyelinating lesions, astrocytes were found to still be active in myelin and axonal debris phagocytosis, with no evidence of cellular internalization. These findings led the authors to conclude that hypertrophic astrocytes not only play an important role in the pathogenesis of MS lesions but also later may exert a continued deleterious effect upon brain tissue in the absence of significant inflammation (Morcos et al. 2003). In a postmortem study of 16 PMS spinal cords, brain regions surrounding demyelinated plaques showed strong evidence that astrocytes in these areas develop chronic alterations in their crucial functions with regard to maintaining neuronal metabolism, oligodendrocyte survival, and synaptic transmission and thus contribute to the unchecked, slow, and extensive expansion of demyelination that occurs in the absence of acute inflammation (Lieury et al. 2014). As a result, it can be suggested that the normal beneficial effect of protein deimination in astrocytes can become destructive if left unchecked (Fig. 18.4c). In order to wall off the spreading demyelination, a meshwork of astrocytic processes surrounds the inflammatory lesion (Sofroniew 2009), and we have shown that this area contains the highest concentration of deiminated GFAP (Fig. 18.3c). As the disease advances to a more progressive stage (Fig. 18.4c), inflammation in these older plaques burn out, leaving behind a fibrous scar. The only evidence of deimination left behind is within a few lingering astrocytes surrounding the central blood vessel (Fig. 18.3a). However, at this point, the NAWM becomes also heavily inundated with fibers containing deiminated GFAP (Fig. 18.3b), perhaps disrupting the normal functioning of this tissue and thus contributing to the pathology of PMS in the absence of active inflammation.

18.8 Summary and Conclusions

The present review provides a possible mechanism for the role of brain protein deimination in the pathophysiology of MS as the disease progresses over time. More research needs to be done in order to clarify and validate or disprove this theory. However, if confirmed, these mechanisms suggest that blocking deimination may provide a novel future treatment for both relapsing and progressive MS phenotypes in human patients.

References

Akiyama, K., Inoue, K., & Senshu, T. (1990). Immunocytochemical demonstration of skeletal muscle type peptidylarginine deiminase in various rat tissues. *Cell Biology International Reports, 14*, 267–273.

Albrechtsen, M., Sorensen, P. S., Gjerris, F., & Bock, E. (1985). High cerebrospinal fluid concentration of glial fibrillary acidic protein (GFAP) in patients with normal pressure hydrocephalus. *Journal of the Neurological Sciences, 70*, 269–274.

Bermel, R. A., Rae-Grant, A. D., & Fox, R. J. (2010). Diagnosing multiple sclerosis at a later age: More than just progressive myelopathy. *Multiple Sclerosis, 16*, 1335–1340.

Blauth, K., Owens, G. P., & Bennett, J. L. (2015). The ins and outs of B cells in multiple sclerosis. *Frontiers in Immunology, 6*, 565.

Boggs, J. M., Rangaraj, G., Koshy, K. M., Ackerley, C., Wood, D. D., & Moscarello, M. A. (1999). Highly deiminated isoform of myelin basic protein from multiple sclerosis brain causes fragmentation of lipid vesicles. *Journal of Neuroscience Research, 57*, 529–535.

Bradford, C., Nicholas, A. P., Woodroofe, N., & Cross, A. K. (2014a). Deimination in multiple sclerosis and experimental autoimmune encephalomyelitis. In A. P. Nicholas & S. K. Bhattacharya (Eds.), *Protein deimination in human health and disease* (pp. 165–185). New York: Springer.

Bradford, C. M., Cross, A. K., Ramos, I., Haddock, G., Nicholas, A. P., McQuaid, S., & Woodroofe, N. (2014b). Localization of citrullinated proteins in normal appearing white matter and lesions in the central nervous system in multiple sclerosis. *Journal of Neuroimmunology, 273*, 85–95.

Brosnan, C. F., & Raine, C. S. (2013). The astrocyte in multiple sclerosis revisited. *Glia, 61*, 453–465.

Brück, W., Porada, P., Poser, S., Rieckmann, P., Hanefeld, F., Kretzschmar, H. A., & Lassmann, H. (1995). Monocyte/macrophage differentiation in early multiple sclerosis lesions. *Annals of Neurology, 38*, 788–796.

Cao, L., Goodin, R., Wood, D., Moscarello, M. A., & Whitaker, J. N. (1999). Rapid release and unusual stability of immunodominant peptide 45-89 from citrullinated myelin basic protein. *Biochemistry, 38*, 6157–6163.

Cohen, J. A., Cutter, G. R., Fischer, J. S., Goodman, A. D., Heidenreich, F. R., Kooijmans, M. F., Sandrock, A. W., Rudick, R. A., Simon, J. H., Simonian, N. A., Tsao, E. C., Whitaker, J. N., & IMPACT Investigators. (2002). Benefit of interferon beta-1a on MSFC progression in secondary progressive MS. *Neurology, 59*, 679–687.

Coles, A. J., Cox, A., Le Page, E., Jones, J., Trip, S. A., Deans, J., Seaman, S., Miller, D. H., Hale, G., Waldmann, H., & Compston, D. A. (2006). The window of therapeutic opportunity in multiple sclerosis: Evidence from monoclonal antibody therapy. *Journal of Neurology, 253*, 98–108.

Coles, A. J., Twyman, C. L., Arnold, D. L., Cohen, J. A., Confavreux, C., Fox, E. J., Hartung, H. P., Havrdova, E., Selmaj, K. W., Weiner, H. L., Miller, T., Fisher, E., Sandbrink, R., Lake, S. L., Margolin, D. H., Oyuela, P., Panzara, M. A., Compston, D. A., & CARE-MS II investigators. (2012). Alemtuzumab for patients with relapsing multiple sclerosis after disease-modifying therapy: A randomized controlled phase 3 trial. *Lancet, 380*, 1829–1839.

Compston, A. (2004). 'The marvellous harmony of the nervous parts': The origins of multiple sclerosis. *Clinical Medicine (London), 4*, 346–354.

Confavreux, C., & Vukusic, S. (2006). Natural history of multiple sclerosis: A unifying concept. *Brain, 129*, 606–616.

Confavreux, C., Vukusic, S., Moreau, T., & Adeleine, P. (2000). Relapses and progression of disability in multiple sclerosis. *NEJM, 343*, 1430–1438.

Dalla Costa, G., Passerini, G., Messina, M. J., Moiola, L., Rodegher, M., Colombo, B., Locatelli, M., Comi, G., Furlan, R., & Martinelli, V. (2015). Clinical significance of the number of oligoclonal bands in patients with clinically isolated syndromes. *Journal of Neuroimmunology, 289*, 62–67.

De Groot, C. J., Langeveld, C. H., Jongenelen, C. A., Montagne, L., Van Der Valk, P., & Dijkstra, C. D. (1997). Establishment of human adult astrocyte cultures derived from postmortem multiple sclerosis and control brain and spinal cord regions: Immunophenotypical and functional characterization. *Journal of Neuroscience Research, 49*, 342–354.

De Stefano, N., Narayanan, S., Matthews, P. M., Francis, G. S., Antel, J. P., & Arnold, D. L. (1999). In vivo evidence for axonal dysfunction remote from focal cerebral demyelination of the type seen in multiple sclerosis. *Brain, 122*, 1933–1939.

DeLuca, G. C., Ebers, G. C., & Esiri, M. M. (2004). Axonal loss in multiple sclerosis: A pathological survey of the corticospinal and sensory tracts. *Brain, 127*, 1009–1018.

Eng, L. F. (1980). The glial fibrillary acidic (GFA) protein. In R. A. Bradshaw & D. M. Schneider (Eds.), *Proteins of the nervous system* (pp. 85–117). New York: Raven Press.

European Study Group on Interferon-1b in Secondary Progressive MS. (1998). Placebo controlled multicentre randomised trial of interferon-1b in treatment of secondary progressive multiple sclerosis. *Lancet, 352*, 1491–1497.

Gaultier, A., Wu, X., Le Moan, N., Takimoto, S., Mukandala, G., Akassoglou, K., Campana, W. M., & Gonias, S. L. (2009). Low-density lipoprotein receptor-related protein 1 is an essential receptor for myelin phagocytosis. *Journal of Cell Science, 122*, 1155–1162.

Hartung, H. P., Gonsette, R., König, N., Kwiecinski, H., Guseo, A., Morrissey, S. P., Krapf, H., Zwingers, T., & M.I.M.S.S. Group. (2002). Mitoxantrone in progressive multiple sclerosis: A placebo-controlled, double-blind, randomized, multicenter trial. *Lancet, 360*, 2018–2025.

Hauser, S., & Goodin, D. S. (2015). Multiple sclerosis and other demyelinating diseases. In D. Kasper, A. Fauci, S. Hauser, D. Longo, J. Jameson, & J. Loscalzo (Eds.), *Harrison's principles of internal medicine*. New York, NY: McGraw-Hill.

Hawker, K., O'Connor, P., Freedman, M. S., Calabresi, P. A., Antel, J., Simon, J., Hauser, S., Waubant, E., Vollmer, T., Panitch, H., Zhang, J., Chin, P., Smith, C. H., & O.T. Group. (2009). Rituximab in patients with primary progressive multiple sclerosis: Results of a randomized double-blind placebo controlled multicenter trial. *Annals of Neurology, 66*, 460–471.

Howell, O. W., Reeves, C. A., Nicholas, R., Carassiti, D., Radotra, B., Gentleman, S. M., Serafini, B., Aloisi, F., Roncaroli, F., Magliozzi, R., & Reynolds, R. (2011). Meningeal inflammation is widespread and linked to cortical pathology in multiple sclerosis. *Brain, 134*, 2755–2771.

Inagaki, M., Takahara, H., Nishi, Y., Sugawara, K., & Sato, C. (1989). Ca^{2+}-dependent deimination-induced disassembly of intermediate filaments involves specific modification of the amino-terminal head domain. *The Journal of Biological Chemistry, 264*, 18119–18127.

Jacobs, L. D., Cookfair, D. L., Rudick, R. A., Herndon, R. M., Richert, J. R., Salazar, A. M., Fischer, J. S., Goodkin, D. E., Granger, C. V., Simon, J. H., Alam, J. J., Bartoszak, D. M., Bourdette, D. N., Braiman, J., Brownscheidle, C. M., Coats, M. E., Cohan, S. L., Dougherty, D. S., Kinkel, R. P., Mass, M. K., Munschauer, F. E., Priore, R. L., Pullicino, P. M., Scherokman, B. J., Whitham, R. H., & T.M.S.C.R. Group. (1996). Intramuscular interferon beta-1a for disease progression in relapsing multiple sclerosis. *Annals of Neurology, 39*, 285–294.

Kabat, E. A., Moore, D. H., & Landow, H. (1942). An electrophoretic study of the protein components in cerebrospinal fluid and their relationship to the serum proteins. *The Journal of Clinical Investigation, 21*, 571–577.

Kappos, L., Bar-Or, A., Cree, B., Fox, R., Giovannoni, G., Gold, R., Vermersch, P., Arnould, S., Sidorenko, T., Wolf, C., Wallstroem, E., & Dahlke, F. (2016). *Efficacy and safety of siponimod in secondary progressive multiple sclerosis—Results of the placebo controlled, double-blind, phase III EXPAND study*. London, UK: ECTRIMS.

Kremenchutzky, M., Rice, G. P. A., Baskervile, J., Wingerchuk, D. M., & Ebers, G. C. (2006). The natural history of multiple sclerosis: A geographically based study 9: Observations on the progressive phase of the disease. *Brain, 129*, 584–594.

Kurtzke, J. F. (1983). Rating neurologic impairment in multiple sclerosis: An expanded disability status scale (EDSS). *Neurology, 33*, 1444–1452.

Kutzelnigg, A., & Lassmann, H. (2014). Pathology of multiple sclerosis and related inflammatory demyelinating diseases. *Handbook of Clinical Neurology, 122*, 15–58.

Kutzelnigg, A., Lucchinetti, C. F., Stadelmann, C., Brück, W., Rauschka, H., Bergmann, M., Schmidbauer, M., Parisi, J. E., & Lassmann, H. (2005). Cortical demyelination and diffuse white matter injury in multiple sclerosis. *Brain, 128*, 2705–2712.

Lavery, A. M., Verhey, L. H., & Waldman, A. T. (2014). Outcome measures in relapsing-remitting multiple sclerosis: Capturing disability and disease progression in clinical trials. *Multiple Sclerosis International, 2014*, 262350.

Leary, S. M., Miller, D. H., Stevenson, V. L., Brex, P. A., Chard, D. T., & Thompson, A. J. (2003). Interferon beta-1a in primary progressive MS: An exploratory, randomized, controlled trial. *Neurology, 60*, 44–51.

Lieury, A., Chanal, M., Androdias, G., Reynolds, R., Cavagna, S., Giraudon, P., Confavreux, C., & Nataf, S. (2014). Tissue remodeling in periplaque regions of multiple sclerosis spinal cord lesions. *Glia, 62*, 1645–1658.

Lovato, L., Willis, S. N., Rodig, S. J., Caron, T., Almendinger, S. E., Howell, O. W., Reynolds, R., O'Connor, K. C., & Hafler, D. A. (2011). Related B cell clones populate the meninges and parenchyma of patients with multiple sclerosis. *Brain, 134*, 534–541.

Lublin, F. D., & Reingold, S. C. (1996). Defining the clinical course of multiple sclerosis: Results of an international survey. National Multiple Sclerosis Society (USA) advisory committee on clinical trials of new agents in multiple sclerosis. *Neurology, 46*, 907–911.

Lublin, F. D., Reingold, S. C., Cohen, J. A., Cutter, G. R., Sørensen, P. S., Thompson, A. J., Wolinsky, J. S., Balcer, L. J., Banwell, B., Barkhof, F., Bebo, B., Calabresi, P. A., Clanet, M., Comi, G., Fox, R. J., Freedman, M. S., Goodman, A. D., Inglese, M., Kappos, L., Kieseier, B. C., Lincoln, J. A., Lubetzki, C., Miller, A. E., Montalban, X., O'Connor, P. W., Petkau, J., Pozzilli, C., Rudick, R. R., Sormani, M. P., Stüve, O., Waubant, E., & Polman, C. H. (2014). Defining the clinical course of multiple sclerosis: The 2013 revisions. *Neurology, 83*, 278–286.

Malmeström, C., Haghighi, S., Rosengren, L., Andersen, O., & Lycke, J. (2003). Neurofilament light protein and glial fibrillary acidic protein as biological markers in MS. *Neurology, 61*, 1720–1725.

Mastronardi, F. G., Wood, D. D., Mei, J., Raijmakers, R., Tseveleki, V., Dosch, H.-M., Probert, L., Casaccia-Bonnefil, P., & Moscarello, M. A. (2006). Increased citrullination of histone H3 in multiple sclerosis brain and animal models of demyelination: A role for tumor necrosis factor-induced peptidylarginine deiminase 4 translocation. *The Journal of Neuroscience, 26*, 11387–11396.

Montalban, X., Hemmer, B., Rammohan, K., Giovannoni, G., De Seze, J., Bar-Or, A., Arnold, D. L., Sauter, A., Masterman, D., Fontoura, P., Garren, H., Chin, P., & Wolinsky, J. (2016). Efficacy and safety of ocrelizumab in primary progressive multiple sclerosis: Results of the phase III double-blind, placebo-controlled ORATORIO study. *Neurology, 86*(Suppl. 16), S49.

Morcos, Y., Lee, S. M., & Levin, M. C. (2003). A role for hypertrophic astrocytes and astrocyte precursors in a case of rapidly progressive multiple sclerosis. *Multiple Sclerosis, 9*, 332–341.

Moscarello, M. A., Wood, D. D., Ackerly, C., & Boulias, C. (1994). Myelin in multiple sclerosis is developmentally immature. *The Journal of Clinical Investigation, 94*, 146–154.

Moscarello, M. A., Lei, H., Mastronardi, F. G., Winer, S., Tsui, H., Li, Z., Ackerley, C., Zhang, L., Raijmakers, R., & Wood, D. D. (2013). Inhibition of peptidyl-arginine deiminases reverses protein-hypercitrullination and disease in mouse models of multiple sclerosis. *Disease Models & Mechanisms, 6*, 467–478.

Musse, A. A., Li, Z., Ackerley, C. A., Bienzle, D., Lei, H., Poma, R., Harauz, G., Moscarello, M. A., & Mastronardi, F. G. (2008). Peptidylarginine deiminase 2 (PAD2) overexpression in transgenic mice leads to myelin loss in the central nervous system. *Disease Models & Mechanisms, 1*, 229–240.

Nicholas, A. P., & Bhattacharya, S. K. (2014). *Deimination in human health and disease*. New York: Springer.

Nicholas, A. P., & Whitaker, J. N. (2002). Preparation of a monoclonal antibody to citrullinated epitopes: Its characterization and some applications to immunohistochemistry in human brain. *Glia, 37*, 328–336.

Nicholas, A. P., Sambandam, T., Echols, J. D., & Tourtellotte, W. W. (2004). Increased citrullinated glial fibrillary acidic protein in secondary progressive multiple sclerosis. *The Journal of Comparative Neurology, 473*, 128–136.

Nicholas, A. P., Sambandam, T., Echols, J. D., & Barnum, S. R. (2005). Expression of citrullinated proteins in murine experimental autoimmune encephalomyelitis. *The Journal of Comparative Neurology, 486*, 254–266.

Noppe, M., Crols, R., Andries, D., & Lowenthal, A. (1986). Determination in human cerebrospinal fluid of glial fibrillary acidic protein, S-100 and myelin basic protein as indices of non-specific or specific central nervous tissue pathology. *Clinica Chimica Acta, 155*, 143–150.

Nuyts, A. H., Lee, W. P., Bashir-Dar, R., Berneman, Z. N., & Cools, N. (2013). Dendritic cells in multiple sclerosis: Key players in the immunopathogenesis, key players for new cellular immunotherapies? *Multiple Sclerosis, 19*, 995–1002.

Oguz, K. K., Kurne, A., Aksu, A. O., Karabulut, E., Serdaroglu, A., Teber, S., Haspolat, S., Senbil, N., Kurul, S., & Anlar, B. (2009). Assessment of citrullinated myelin by 1H-MR spectroscopy in early-onset multiple sclerosis. *AJNR. American Journal of Neuroradiology, 30*, 716–721.

Ontaneda, D., Fox, R. J., & Chataway, J. (2015). Clinical trials in progressive multiple sclerosis: Lessons learned and future perspectives. *Lancet Neurology, 14*, 208–223.

Orton, S. M., Herrera, B. M., Yee, I. M., Valdar, W., Ramagopalan, S. V., Sadovnick, A. D., Ebers, G. C., & C.C.S. Group. (2006). Sex ratio of multiple sclerosis in Canada: A longitudinal study. *Lancet Neurology, 5*, 932–936.

Peterson, J. W., Bö, L., Mörk, S., Chang, A., & Trapp, B. D. (2001). Transected neurites, apoptotic neurons, and reduced inflammation in cortical multiple sclerosis lesions. *Annals of Neurology, 50*, 389–400.

Petzold, A., Eikelenboom, M. J., Gveric, D., Keir, G., Chapman, M., Lazeron, R. H. C., Cuzner, M. L., Polman, C. H., Uitdehaag, B. M. J., Thompson, E. J., & Giovannoni, G. (2002). Markers for different glial cell responses in multiple sclerosis: Clinical and pathological correlations. *Brain, 125*, 1462–1473.

Piehl, F. (2014). A changing treatment landscape for multiple sclerosis: Challenges and opportunities. *Journal of Internal Medicine, 275*, 364–381.

Pitt, D., Werner, P., & Raine, C. S. (2000). Glutamate excitotoxicity in a model of multiple sclerosis. *Nature Medicine, 6*, 67–70.

Prineas, J. W., Kwon, E. E., Cho, E. S., Sharer, L. R., Barnett, M. H., Oleszak, E. L., Hoffman, B., & Morgan, B. P. (2001). Immunopathology of secondary-progressive multiple sclerosis. *Annals of Neurology, 50*, 646–657.

Pritzker, L. B., Joshi, S., Gowan, J. J., Harauz, G., & Moscarello, M. A. (2000). Deimination of myelin basic protein. 1. Effect of deimination of arginyl residues of myelin basic protein on its structure and susceptibility to digestion by cathepsin D. *Biochemistry, 39*, 5374–5381.

Raijmakers, R., Vogelzangs, J., Croxford, J. L., Wesseling, P., van Venrooij, W. J., & Pruijn, G. J. M. (2005). Citrullination of central nervous system proteins during the development of experimental autoimmune encephalomyelitis. *The Journal of Comparative Neurology, 486*, 243–253.

Raijmakers, R., Vogelzangs, J., Raats, J., Panzenbeck, M., Corby, M., Jiang, H., Thibodeau, M., Haynes, N., van Venrooij, W. J., Pruijn, G. J., & Werneburg, B. (2006). Experimental autoimmune encephalomyelitis induction in peptidylarginine deiminase 2 knockout mice. *The Journal of Comparative Neurology, 498*, 217–226.

Rojas, J. I., Tizio, S., Patrucco, L., & Cristiano, E. (2012). Oligoclonal bands in multiple sclerosis patients: Worse prognosis? *Neurological Research, 34*, 889–892.

Secondary Progressive Efficacy Clinical Trial of Recombinant Interferon-beta-1a in MS (SPECTRIMS) Study Group. (2001). Randomized controlled trial of interferon beta-1a in secondary progressive MS: Clinical results. *Neurology, 56*, 1496–1504.

Seewann, A., Vrenken, H., van der Valk, P., Blezer, E. L., Knol, D. L., Castelijns, J. A., Polman, C. H., Pouwels, P. J., Barkhof, F., & Geurts, J. J. (2009). Diffusely abnormal white matter in chronic multiple sclerosis: Imaging and histopathologic analysis. *Archives of Neurology, 66*, 601–609.

Segal, B. M., & Stuve, O. (2016). Primary progressive multiple sclerosis—Why we are failing. *Lancet, 387*, 1032–1034.

Serafini, B., Rosicarelli, B., Magliozzi, R., Stigliano, E., Capello, E., Mancardi, G. L., & Aloisi, F. (2006). Dendritic cells in multiple sclerosis lesions: Maturation stage, myelin uptake, and interaction with proliferating T cells. *Journal of Neuropathology and Experimental Neurology, 65*, 124–141.

Sofroniew, M. V. (2009). Molecular dissection of reactive astrogliosis and glial scar formation. *Trends in Neurosciences, 32*, 638–647.

Sorensen, P. S., & Blinkenberg, M. (2016). The potential role for ocrelizumab in the treatment of multiple sclerosis: Current evidence and future prospects. *Therapeutic Advances in Neurological Disorders, 9*, 44–52.

The IFNB Multiple Sclerosis Study Group. (1993). Interferon beta-1b is effective in relapsing-remitting multiple sclerosis. I. Clinical results of a multicenter, randomized, double-blind, placebo-controlled trial. *Neurology, 43*, 665–661.

The North American Study Group on Interferon beta-1b in Secondary Progressive MS. (2004). Interferon beta-1b in secondary progressive MS: Results from a 3-year controlled study. *Neurology, 63*, 1788–1795.

Trapp, B. D., & Stys, P. K. (2009). Virtual hypoxia and chronic necrosis of demyelinated axons in multiple sclerosis. *Lancet Neurology, 8*, 280–291.

van Horssen, J., Singh, S., van der Pol, S., Kipp, M., Lim, J. L., Peferoen, L., Gerritsen, W., Kooi, E. J., Witte, M. E., Geurts, J. J., de Vries, H. E., Peferoen-Baert, R., van den Elsen, P. J., van der Valk, P., & Amor, S. (2012). Clusters of activated microglia in normal-appearing white matter show signs of innate immune activation. *Neuroinflammation, 9*, 156.

Watanabe, K., Akiyama, K., Hikichi, K., Ohtsuka, R., Okuyama, A., & Senshu, T. (1988). Combined biochemical and immunochemical comparison of peptidylarginine deiminases present in various tissues. *Biochimica et Biophysica Acta, 966*, 375–383.

Wolinsky, J. S., Narayana, P. A., O'Connor, P., Coyle, P. K., Ford, C., Johnson, K., Miller, A., Pardo, L., Kadosh, S., Ladkani, D., & P.T.S. Group. (2007). Glatiramer acetate in primary progressive multiple sclerosis: Results of a multinational, multicenter, double-blind, placebo-controlled trial. *Annals of Neurology, 61*, 14–24.

Wood, D. D., Bilbao, J. M., O'Connors, P., & Moscarello, M. A. (1996). Acute multiple sclerosis (Marburg type) is associated with developmentally immature myelin basic protein. *Annals of Neurology, 40*, 18–24.

Wynn, D. R., Rodriguez, M., O'Fallon, W. M., & Kurland, L. T. (1990). A reappraisal of the epidemiology of multiple sclerosis in Olmsted County, Minnesota. *Neurology, 40*, 780–786.

Chapter 19
Turning White Matter "Inside-Out" by Hyper-deimination of Myelin Basic Protein (MBP)

George Harauz

19.1 Introduction

The myelin sheath of the nervous system provides trophic and metabolic support to underlying axons and facilitates faster action potential transmission by saltatory conduction. In the mammalian central nervous system, the classic isoforms of myelin basic protein (MBP) are essential to the development and maintenance of the sheath. These proteins are multifunctional by virtue of being intrinsically disordered and thus conformationally dynamic and adaptable, with multiple associations with membranes and other proteins that are modulated by combinatorial post-translational modifications. The enzymatic deimination (citrullination) of the predominant adult 18.5-kDa splice isoform of MBP is associated with myelin in both the healthy and developing brain and with myelin from patients afflicted with multiple sclerosis (MS). In this disease, hyper-deimination of MBP can arise initially due to cytodegenerative triggers affecting the oligodendrocyte-neuron complex and can become irreversibly self-propagating. We review here the extensive evidence that an initial aberrant calcium influx leading to up-regulation of peptidylarginine deiminases is probably a central, if not key, molecular mechanism of the pathogenesis of MS and one that precedes the neuroinflammatory response. The loss of net positive charge of MBP caused by this modification changes the networking ability of MBP, and biophysical mechanisms can explain structural destabilization of myelin as an early molecular event in lesion formation. The inhibition of peptidylarginine deiminases could serve as an additional disease-modifying therapeutic measure to ameliorate the demyelination cascade and the rate of disease progression and thus to maintain and prolong "neurological reserve" in MS patients.

G. Harauz, Ph.D. (✉)
Department of Molecular and Cellular Biology, Biophysics Interdepartmental Group, and Collaborative Program in Neuroscience, University of Guelph,
50 Stone Road East, Guelph, ON, Canada, N1G 2W1
e-mail: gharauz@uoguelph.ca

A.P. Nicholas et al. (eds.), *Protein Deimination in Human Health and Disease*,
DOI 10.1007/978-3-319-58244-3_19

19.1.1 Myelin: The Basics and the Basic Protein

Myelin is a complex assembly of lipids and proteins that is an essential component of the central and peripheral nervous systems of higher vertebrates (Morell 1984; Lazzarini et al. 2004; Rasband and Macklin 2012). In the mammalian central nervous system (CNS), myelin arises from oligodendrocytes that extend multiple membrane processes to ensheathe neuronal axons, forming compacted myelin internodes (Fig. 19.1) (Baumann and Pham-Dinh 2001; Trapp and Kidd 2004; Sherman and Brophy 2005). This "white matter" contains a high proportion of lipids (roughly 70% by weight), primarily phospholipids (phosphatidylethanolamine, phosphatidylserine, phosphatidylcholine, the phosphatidylinositides), plasmalogens, galactolipids (galactrocerebrosides and sulphatides), sphingomyelin, and cholesterol. The predominant protein families in adult CNS white matter are myelin basic protein (MBP) and proteolipid protein (PLP), of which there are different splice variants. Other proteins are found in lesser proportions: 2′3′-cyclic nucleotide 3′-phosphodiesterase (CNPase), myelin-associated glycoprotein (MAG), and myelin-oligodendrocyte glycoprotein (MOG), to name a few.

White matter is formed from birth and develops until the early twenties in humans: it is an intricate, heterogeneous network, and it is altered continuously throughout life (Miller et al. 2012; Young et al. 2013; Snaidero et al. 2014; Tomassy et al. 2014; de Hoz and Simons 2015). Despite this plasticity, though, the myelin proteins (including MBP) are by and large long-lived, along with other essential proteins such as lens proteins, nucleoporins, and histones (Toyama et al. 2013; Friedrich et al. 2016). The myelin wrapping forces axonal ion channels to be clustered at periodic nodes of Ranvier, and action potentials proceed more quickly via saltatory conduction (Castelfranco and Hartline 2016). Our focus in this chapter is on MBP because of its centrality in maintaining the CNS myelin of higher-evolved organisms (Nawaz et al. 2013).

In humans, multiple sclerosis (MS) is an enigmatic disease characterized by the active degradation and loss of CNS myelin (Compston and Coles 2002; Hauser and Oksenberg 2006; Trapp and Nave 2008; Lassmann and van Horssen 2011; Hauser et al. 2013; Lassmann 2014; Schaeffer et al. 2015). The causes of the disease are multifactorial, and the early cellular events that lead to the formation of an MS "plaque", including regions of demyelination with neuroinflammation and infiltration of activated lymphocytes (B and T cells) and macrophages, as well as microglia, are unknown. A clinical diagnosis of MS means that MS plaques are already well formed and visible by magnetic resonance imaging because of the compromised blood-brain barrier. The disease has long been considered to be primarily an autoimmune disorder; indeed, MBP comprises several immunodominant epitopes that activate T cells (Wucherpfennig et al. 1997; Hansen et al. 2011). But the autoimmune and neuroinflammatory viewpoint is gradually being modified to one of the disease being initially neurodegenerative (Stys 2010; Stys et al. 2012; Traka et al. 2016). For these reasons, our work is devoted to unravelling the fundamental basis of myelin architecture and how it deteriorates during early MS disease progression (Harauz et al. 2009; Harauz and Boggs 2013; Vassall et al. 2015a). A detailed

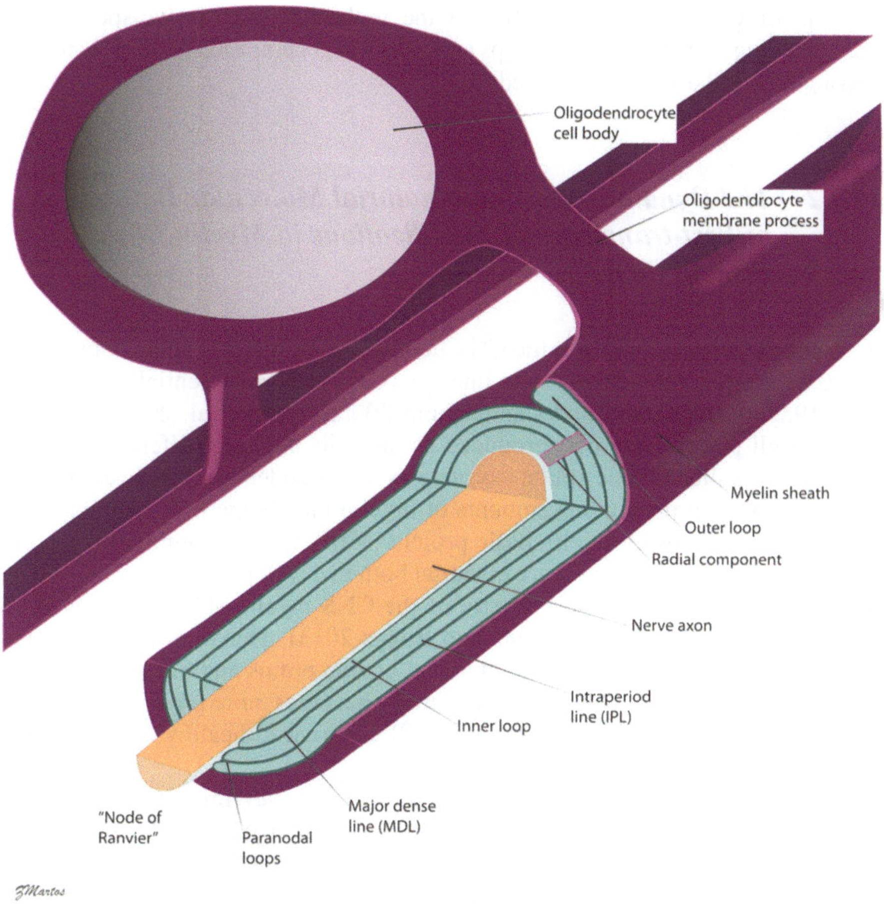

Fig. 19.1 Schematic view of the overall organization of CNS myelin. Oligodendrocytes extend multiple membrane processes that wrap around several axons forming the compacted, internodal, myelin sheath of the central nervous system (CNS) (Baumann and Pham-Dinh 2001). Transmission electron micrographs of stained cross-sections of myelin show the juxtaposition of tens of spiral wrappings of the membrane. The major dense line (MDL) is the cytoplasmic compartment roughly 2–3 nm in thickness and is electron-dense. The intraperiod line (IPL) is the extracellular space and roughly twice as thick as the MDL or slightly more. The major myelin proteins in the adult brain are the cytoplasmic 18.5-kDa MBP (14-kDa MBP in rodents) and the transmembrane proteolipid protein (PLP). Other minor myelin proteins include myelin-associated glycoprotein (MAG), myelin-oligodendrocyte protein (MOG), and myelin 2′,3′-cyclic nucleotide 3′-phosphodiesterase (CNPase). The lipid compositions of the cytoplasmic and extracellular membrane leaflets are different. The paranodal loops are larger, cytoplasmic compartments that maintain continuity with the oligodendrocyte cell body, and the oligodendrocyte and neuronal membranes interact at the periaxonal region. During myelination, additional cytoplasmic channels traverse internodal myelin. The radial component of mature CNS myelin additionally maintains structural integrity. The nodes of Ranvier contain the ion channels that generate action potentials (Na^+/K^+-ATPases, Na^+/Ca^{2+}-exchangers, and voltage-gated Na^+-channels) and are not depicted here for simplicity. This figure has been adapted and redrawn from a previously published version (DeBruin and Harauz 2007). Figure courtesy of Ms. Zoë Martos

understanding of multifaceted MBP at the molecular level is indispensable to understanding how the brain develops and works in general and specifically to deciphering how MS arises and proceeds.

19.1.2 The Dynamic and Combinatorial Molecular Barcode of Post-translational Modifications in Myelin Basic Protein (MBP)

In oligodendrocytes that form the CNS myelin sheath, diverse splice isoforms of the gene in the oligodendrocyte lineage (*Golli*) are differentially expressed (Fig. 19.2) (Campagnoni and Campagnoni 2008; Fulton et al. 2010a). The so-called Golli proteins are first produced as the cells begin to differentiate. Then come the so-called classic splice isoforms as the oligodendrocytes become committed to forming prodigious amounts of myelin membrane lipids and proteins. The classic isoforms of myelin basic protein (MBP) range in nominal molecular mass from 14 to 21.5 kDa and are essential both to developing and to maintaining the structural and metabolic integrity of the CNS myelin sheath (Boggs 2006, 2008a; Harauz et al. 2009; Harauz and Boggs 2013). The 18.5-kDa splice isoform predominates in the adult human and bovine brains and has been the most studied. In rodents, the 14-kDa splice isoform becomes more prominent in adulthood, although the 18.5-kDa protein is still essential to healthy myelination in these species (Palma et al. 1997).

The *shiverer* mouse line has an ablated gene missing several classic MBP exons and consequently sparse and poorly compacted myelin (Fig. 19.2) (Privat et al. 1979; Dupouey et al. 1979; Chernoff 1981). Indeed, even *shiverer* heterozygotes, which still have intact 18.5-kDa MBP but in lesser amounts, have been shown recently to display subtle myelination defects (Poggi et al. 2016). For this reason, it has long been accepted that the protein forms myelin internodes by adhesion of oligodendrocyte cytoplasmic membrane leaflets to each other and that this is its fundamental role. Yet, many in vitro and cell culture studies suggest that the protein participates also in dynamic processes such as cytoskeletal turnover at leading edges of membrane ruffles and processes, phosphoinositide sequestration, Fyn-SH3-mediated signalling pathways during myelin formation and restructuring, and regulation of voltage-operated calcium channels (reviewed in (Boggs 2008a; Harauz et al. 2009; Harauz and Boggs 2013; Vassall et al. 2015a)). The late Dr. Mario Moscarello considered MBP to be the "executive" molecule of the myelin sheath (Moscarello 1997; Schachter et al. 2014), having long before expounded that its loss of function was pivotal in demyelination in MS (Wood et al. 1975). We have referred to MBP tongue-in-cheek as a "MyelStone" by the analogy that it organizes the CNS myelin membrane much like the basic histone proteins organize nuclear chromatin (Vassall et al. 2015a; Wenderski and Maze 2016).

There are two salient points to note about the 18.5-kDa MBP variant, which will henceforth be referred to simply as "MBP". First, it is structurally polymorphic,

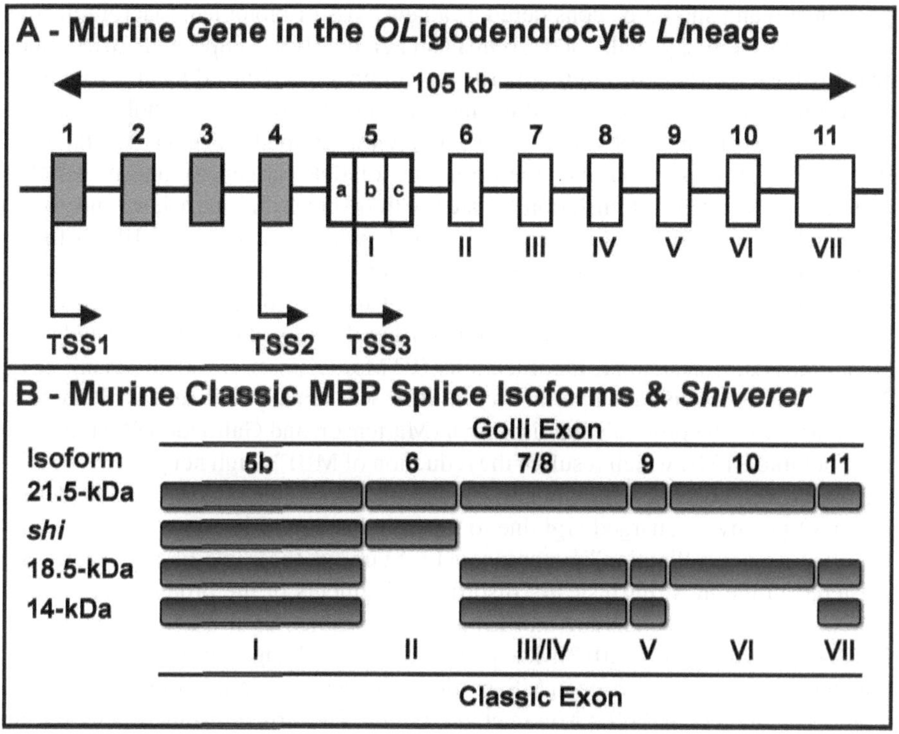

Fig. 19.2 The gene in the oligodendrocyte lineage (*Golli*) and classic MBP isoforms. (**a**) The gene in the oligodendrocyte lineage (*Golli*) consists of 11 exons in both the human and mouse species (Campagnoni et al. 1993; Pribyl et al. 1993; Campagnoni and Campagnoni 2008) with three transcription start sites (TSS) and many developmentally regulated splice isoforms. The early developmental Golli proteins arise from TSS1. The so-called "classic" isoforms arise primarily from TSS3 and are denoted by Roman numerals. (**b**) The full-length 21.5-kDa classic MBP has a nontraditional nuclear localization signal within its exon-II-encoded segment and will not be considered further here (Smith et al. 2012c, 2013). The 18.5-kDa classic MBP predominates in adult human and bovine brain, and in the developing mouse brain, and is synthesized at the outer membrane (Muller et al. 2013). In adult rodents, though, the 14-kDa isoform becomes predominant in compact myelin (it arises from TSS2 and some unusual splicing events) (Rasband and Macklin 2012). Here, we consider mainly the 18.5-kDa isoform that we shall refer to simply as "MBP". The *shiverer* mouse is missing most of its classic MBP and is unable to form tightly compacted myelin. This figure has been adapted and redrawn from previously published versions (Harauz et al. 2004, 2013; Harauz and Libich 2009; Harauz and Boggs 2013)

being an exemplary intrinsically disordered protein (IDP) (Boggs 2008a; Harauz et al. 2009, 2013; Harauz and Libich 2009; Vassall et al. 2015a). For most of the twentieth century, we have thought of all proteins as having a "function" defined by their three-dimensional "structure" which arises directly from their amino acid sequence. This conceptual framework works well for enzymes and other proteins whose arrangements have been determined primarily by X-ray crystallography. Since the start of the twenty-first century, we have come to realize that roughly a third of the eukaryotic proteome does not have such a rigid structure-function relationship and should be considered as a new class of biological macromolecule, i.e. the IDPs (Uversky et al. 2000; Tompa and Fersht 2015; Tompa et al. 2015). Such proteins have a variety of binding partners, can undergo induced local disorder-to-order transitions (e.g. coupled folding and binding), and often act as hubs in signalling and/or structural networks. A recent review entitled "The Dark Matter of Biology" encapsulates many of the properties of IDPs in general and of MBP in particular: diverse post-translational states with diverse (rare, transient, and weak) interactions (Ross 2016). Another appropriately whimsical view of MBP is that it constitutes the "dark matter of CNS white matter".

Second, MBP is extremely basic because of its large proportion of arginyl and lysyl residues, and every splice isoform is further diversified by extensive and combinatorial post-translational modifications (PTMs) (Kim et al. 2003; Zhang et al. 2012a). This extra degree of variability in MBP was already realized in 1969 only a few years after the protein's first isolation (Martenson and Gaitonde 1969a, b). The most notable PTMs, which result in the reduction of MBP's high net positive charge, are reversible phosphorylation of seryl and threonyl residues and irreversible deimination of positively charged arginine to neutral citrulline (Fig. 19.3). For the latter, we also use "citrullination" synonymously. Many of these modification sites are clustered in the most intrinsically disordered segments of the protein, particularly the N- and C-termini (Harauz and Libich 2009; Libich et al. 2010; Harauz et al. 2013; Harauz and Boggs 2013). The pattern of PTMs changes during myelin development and becomes aberrant in MS, particularly by hyper-deimination (Moscarello et al. 1994; Kim et al. 2003; Musse et al. 2008b). The structure-function repertoire of MBP is thus modulated by its PTMs, a feature common with most IDPs (Bah and Forman-Kay 2016). Specifically, charge reduction both locally and globally affects the conformational properties of IDPs significantly (Marsh and Forman-Kay 2010; Mao et al. 2010, 2013).

19.1.3 Whence and Whither MBP Deimination

As discussed in detail elsewhere in the first and in this second volume on *Deimination in Human Health and Disease*, protein deimination is an irreversible post-translational modification of positively charged peptidylarginine to uncharged peptidyl-citrulline (Cit). The modification is catalysed by Ca^{2+}-dependent enzymes called peptidylarginine deiminases (PAD or PADI, EC 3.5.3.15), of which five isoforms (PAD1, PAD2, PAD3, PAD4/PAD5, and PAD6) are known in mammals (Vossenaar et al. 2003; Bicker and Thompson 2013; Gudmann et al. 2015; Tu et al. 2016). The PAD2 isotype encoded by the *Padi2* gene has the broadest tissue distribution—it is the most abundant one found in the brain and glial cells, including oligodendrocytes, and is the one that deiminates MBP (Lamensa and Moscarello 1993; Pritzker et al. 1999; Raijmakers et al. 2006; van Beers et al. 2013). The PAD4 isozyme has also been demonstrated to be active in the brain and deiminates MBP in PAD2-knockout mice (Mastronardi et al. 2006; Wood et al. 2008). In this chapter,

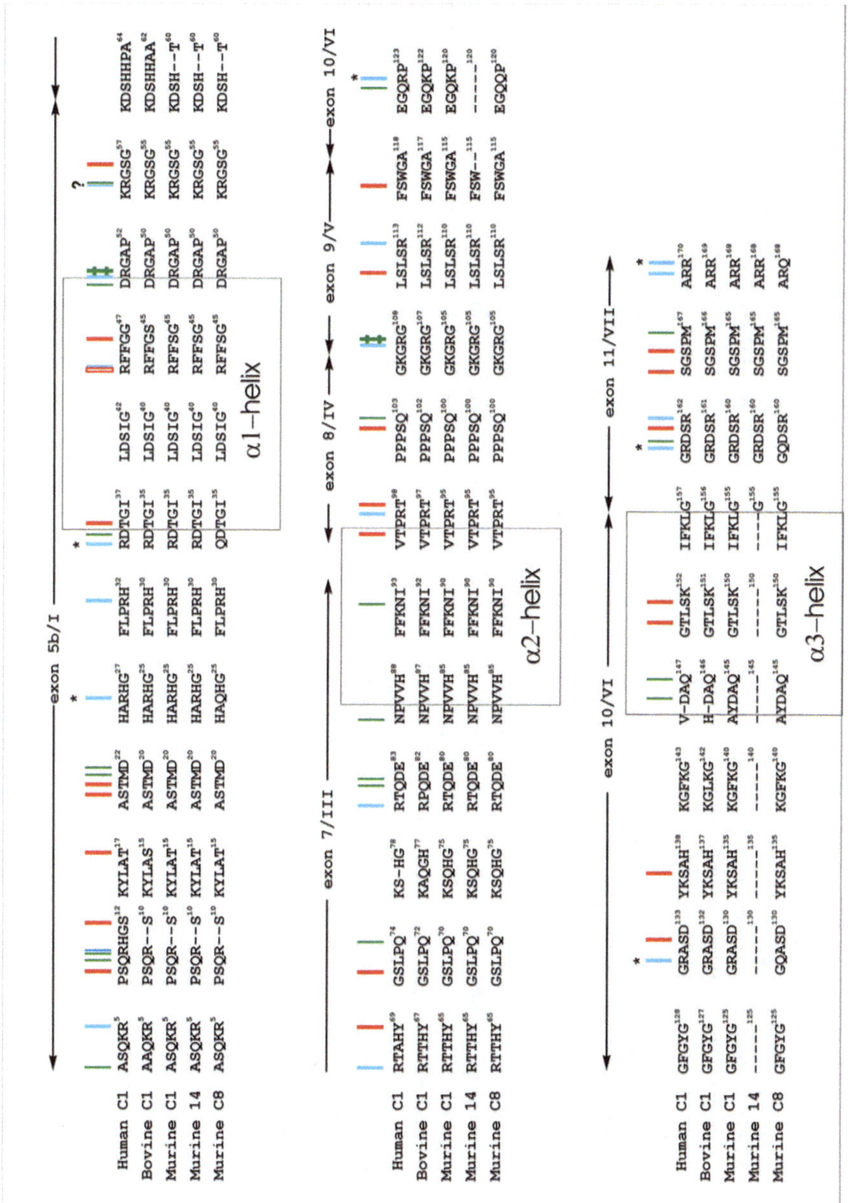
exon 5b/I
α1–helix
exon 7/III
exon 8/IV
exon 9/V
exon 10/VI
α2–helix
exon 10/VI
exon 11/VII
α3–helix
Human C1
Bovine C1
Murine C1
Murine 14
Murine C8

Fig. 19.3 Amino acid sequences of human, bovine, and murine 18.5-kDa MBP isoforms. The 18.5-kDa classic MBP has been studied from human (h), bovine (b), and murine (m) brain, and often specific residues are referred to by their number within that particular context: 170 residues, 169 residues, and 168 residues, respectively. Here, all three sequences are aligned and compared (Harauz and Boggs 2013). The 18.5-kDa MBP contains three major α-helical membrane-anchoring motifs (α-MoRFs, Fig. 19.4) (Vassall et al. 2015a). The "C1" designation means the unmodified C1 component, whereas we use "C8" to designate the enzymatically or pseudo-deiminated protein. The molecular barcode is due to diverse and combinatorial post-translational modifications (Kim et al. 2003; Harauz et al. 2004; Harauz and Boggs 2013): *red bar*, phosphorylation (S, T, Y); *hollow red bar*, an unusual case of reported arginyl phosphorylation (Kim et al. 2003); *ice blue bar*, deimination/citrullination of arginine; *green bar*, diverse PTMs (acylation, deamidation, ADP-ribosylation, sulfoxide oxidation of methionine, racemization); and *crossed green bar*, methylation (either ω-N^{G}-mono-methylation or symmetric ω-N^{G},N'^{G}-di-methylation particularly of hR107/bR106/mR104). The density of PTMs is greatest at both termini, which are also the most intrinsically disordered. Recently, residue hR49 has also been observed to be either mono- or di-methylated in MBP from cerebella of controls and MS patients (Friedrich et al. 2016). The human/murine residues hT95/mT92 and hT98/mT95 are MAP kinase phosphorylation sites, lying in the vicinity of a putative molecular switch comprising the α2-helix and proline-rich SH3 ligand. *Asterisks* denote the six main deimination sites that were first identified in the natural C8 component of the 170-residue human 18.5-kDa protein (hR25, hR33, hR122, hR130, hR159, and hR170) (Wood and Moscarello 1989, 1997; Moscarello 1997, 2014). These primary arginine deimination sites in the human protein correspond to mR23, mR31, mK119, mR127, mR157, and mR168 in the 168-residue murine 18.5-kDa MBP and were used to generate a pseudo-deiminated rmC8 variant by glutamine substitution (Bates et al. 2003). Residue human hR107 (bovine R106 or murine mR104) is not deiminated when it is methylated, but can be otherwise (Pritzker et al. 2000b). A *question mark* denotes the residue human hR54 (murine mR52) which is not deiminated by either PAD2 or PAD4 in vitro for unknown reasons, perhaps due to conformational inaccessibility (Wood et al. 2008), but is deiminated to a measurable extent in normal human brain and can also be ADP-ribosylated (Kim et al. 2003). It was not identified as being deiminated in a recent study, so it remains enigmatic (Friedrich et al. 2016). This figure has been adapted and updated from previously published versions (Harauz et al. 2009; Polverini et al. 2011). Detailed tabulations are found in Harauz et al. (2004) and Harauz and Boggs (2013) we discuss the effects of deimination of MBP on its properties both in vitro and in myelin. The presence in myelin of the non-coded amino acid citrulline was first discovered over 45 years ago in the laboratory of Dr. Moscarello (Finch et al. 1971). The paper concluded with the statement: "Whether or not the presence of citrulline in a membrane protein has any functional significance remains to be determined". Now, over 45 years later, we are still wrestling with the details, although we accept the pivotal significance of this PTM.

The diverse effects of deimination of MBP on its conformation and interactions were reviewed in detail almost a decade ago (Harauz and Musse 2007; Zand 2008; Boggs et al. 2008; Mastronardi and Moscarello 2008; Boggs 2008b), concurrently with appraisals of its involvement in MS pathogenesis (Moscarello et al. 2002b, 2007; Mastronardi and Moscarello 2008). Subsequently, new studies have focussed on understanding the extent of citrullination in MS and on the search for inhibitors of deimination as potential therapeutic measures (Wood et al. 2008; Musse et al. 2008b, c; Moscarello et al. 2013; Wei et al. 2013; Bello et al. 2013; Moscarello 2014; Bradford et al. 2014b; Tu et al. 2016). Many of these later investigations were reviewed comprehensively in several chapters of the first volume of this series on protein deimination and some elsewhere (Nicholas and Bhattacharya 2014; Moscarello 2014; Nicholas et al. 2014; Bradford et al. 2014a; Yang et al. 2016).

Here, this foundation of literature provides the context into which to integrate recent biochemical and biophysical studies on the effects of deimination on MBP conformation and interactions (Ahmed et al. 2010; Bamm et al. 2010; Homchaudhuri et al. 2010; Smith et al. 2010, 2011, 2012b; Suresh et al. 2010; Boggs et al. 2011; Wang et al. 2011; Vassall et al. 2015a, 2016). It is becoming appreciated that biophysical considerations need to be integrated into models of myelination and remyelination (Ozgen et al. 2014, 2016; Chang et al. 2016; Micu et al. 2016). The aim here is to bring an additional reductionist and mechanistic viewpoint into the discussion of MS, as is being done for amyloidogenic disorders such as Alzheimer's and Parkinson's diseases where one can find several thousand biochemical and biophysical papers on prions, amyloid-β peptides, α-synuclein, or the microtubule-associated protein tau (e.g. Mor et al. (2016) and Šimic et al. (2016)).

19.2 Charge Components of MBP in Healthy and Diseased Myelin

19.2.1 Fractionation of White Matter MBP into Charged Components: The Deiminated Fraction or "C8 Component"

Myelin basic protein can be extracted from brain or spinal cord white matter (myelin) by de-lipidation with organic solvents, acid solubilization, and ethanol precipitation. This preparation can be resuspended and further fractionated on a cation-exchange column at alkaline pH (or by urea-alkaline tube gel electrophoresis), yielding a number of elution peaks containing charge components that are denoted as C1, C2, C3, C4, C5, and sometimes C6 (Chou et al. 1976; Cheifetz et al. 1984; Cheifetz and Moscarello 1985; Moscarello et al. 1986; Wood and Moscarello 1989; Wood et al. 1996; Wood and Moscarello 1997; Kim et al. 2003). These components have been referred to as "charge isomers", although the word "isomers" is too easily confused with "isoforms" and is not preferred.

The C1 component is the most highly positively charged, being modified only by N-terminal acylation; additionally, there may be single or double symmetric methylation of a highly conserved central arginyl residue, hR107 in the human 170-residue protein (murine mR104, bovine bR106; Fig. 19.3) (Moscarello et al. 1992; Wood and Moscarello 1997; Pritzker et al. 2000b). Successive components C2–C5 or C6 are reduced by a single unit of positive charge, due to a succession of PTMs. These modifications at different sites are primarily the aforementioned phosphorylation by diverse kinases, and deimination by PAD2, but also could include methylation, deamidation, glucosylation, and ADP-ribosylation (Wood and Moscarello 1997; Kim et al. 2003; Majava et al. 2010; Zhang et al. 2012a; Friedrich et al. 2016). There is no C7 component, and the unbound, flow-through fraction from the cation-exchange column was originally denoted simply as the C8 component. This experimental fraction contains many proteins, not just MBP, and the latter population displays a mixture of assorted PTMs. The MBP from the flow-through fraction needs to be purified further by high-performance liquid chromatography (HPLC) for detailed compositional analyses (Wood and Moscarello 1989; Pritzker et al. 2000a, b).

19.2.2 Changing Proportions of Citrullinated MBP in the Developing and MS-Afflicted Brain

Whereas the 18.5-kDa MBP population isolated from the normal adult human brain is mainly MBP-C1 and roughly 20% MBP-C8, the MBP in early childhood (<5 years) human brain consists almost entirely of MBP-C8 (Moscarello et al. 1994). These quantifications were achieved from the ratios of MBP-C8 to MBP-C1 elution peaks on cation-exchange columns, with some caveats to be mentioned below. The proportion of MBP-C1 increases and that of MBP-C8 gradually decreases during adolescence, concomitant with the timeline of full CNS myelin development over the first two decades of life (Miller et al. 2012). Throughout adulthood, myelin undergoes continuous transformation and plasticity (Young et al. 2013; Wang and Young 2014), but the MBP-C1/MBP-C8 ratios remain constant in the healthy brain until the seventh or so decade of life (Moscarello et al. 1994). With increasing age, both the activity of PAD2 and the amount of deiminated MBP (along with other proteins) also increase (Ishigami and Maruyama 2010; Shimada et al. 2010; Bradford et al. 2014a, b; Friedrich et al. 2016). The point to keep in mind is that the MBP-C8 chromatographic fraction extracted from human or rodent brain is still a heterogeneous population of this protein species and a fully inherent part of *healthy* developing and adult CNS myelin.

In MS, however, it was recognized that the severity of the disease correlates with the degree of deimination of MBP: there is roughly 20% MBP-C8 in un-afflicted brain; 35–45% MBP-C8 in most MS cases, meaning the most common relapsing-remitting pattern; and rising to 90% in an acute case of Marburg MS (Moscarello et al. 1994; Wood et al. 1996). The MBP-C8 from normal human brain has roughly six arginyl residues that are deiminated (Fig. 19.3 and Sect. 19.3.1) (Wood and Moscarello 1989). In our work, we have always assumed that these same sites would be the predominant ones modified in MBP-C8 from MS-afflicted tissue, although it can be expected that more arginyl residues are deiminated as the disease progresses in any one individual (Sect. 19.3.3). In the fulminating Marburg's variant of MS, the proportion of MBP-C8 component was the same as in early childhood, and 18 of the 19 arginyl residues in this human 18.5-kDa protein were deiminated (ascertained by amino acid analysis), leading to the thesis that the protein's PTM pattern is reverting to that found in an early developmental state (Wood et al. 1996).

Recognizing the substantial and deleterious impact of hyper-deimination of MBP, Dr. Moscarello posited early on that "unstable myelin structure due to altered lipid or protein composition may result in demyelinating diseases, such as multiple sclerosis" (Boggs and Moscarello 1978) and that "a primary defect in myelin basic protein is responsible for the initial changes in myelin in MS, and precedes the autoimmune response" (Wood and Moscarello 1997). Indeed, MBP deimination has been shown to precede the actual loss of myelin (Bradford et al. 2014a, b). Like rheumatoid arthritis (Suzuki et al. 2007; Hensvold et al. 2014; van Beers and Pruijn 2014; de Smit et al. 2014; Makrygiannakis 2014; Spengler and Scheel-Toellner 2014), MS could be considered in part a "post-translational disease", because this modification of MBP could explain many aspects of disease pathogenesis. This idea can be considered further by analogy of MBP's physicochemical properties with those of other IDPs such as α-synuclein and the microtubule-associated protein tau that are implicated in other neurodegenerative disorders (Uversky 2014, 2015; Zienowicz et al. 2015). To understand how MBP deimination drives demyelination, we first discuss the fundamental ways by which MBP contributes to myelin architecture at the molecular level.

19.3 Molecular Aspects of MBP Deimination

19.3.1 The Arginyl Residues in 18.5-kDa MBP Can (Almost) All Be Deiminated

By amino acid analysis of the 170-residue human 18.5-kDa MBP isoform extracted from normal brain (cation-exchange chromatography followed by HPLC (high-performance liquid chromatography)), Wood and Moscarello defined primarily six arginyl residues that were citrullinated: hR25, hR31, hR122, hR130, hR159, and hR170 (Fig. 19.3) (Wood and Moscarello 1989). In subsequent reviews, residue hR33 instead of hR31 was reported to be one these six primary citrullination sites and is the one denoted in Fig. 19.3 (Wood and Moscarello 1997; Moscarello 1997, 2014). In an extreme case, 18 of the 19 arginyl residues of human 18.5-kDa MBP were deiminated in protein isolated from the brain of a patient who died of the fulminating Marburg's variant of MS, ascertained by amino acid analysis and quantitation of citrulline (Wood et al. 1996). These naturally deiminated forms of human MBP have been referred to specifically as hMBP-Cit_6 ("C8") and hMBP-Cit_{18} ("MC8"), respectively, in our earlier studies (Beniac et al. 1999, 2000). Here, the "C8" designation has now changed to refer to particular MBP forms per se, not just the heterogeneous flow-through fraction of cation-exchange chromatography. Later mass spectrometric analyses of natural human MBP charge components showed partial deimination at multiple sites throughout the protein sequence (Kim et al. 2003). A newer report confirms this observation and is discussed separately in the Appendix (Friedrich et al. 2016).

Although it appears that, in principle, any free arginyl residue could be found and thus modified in MBP-C8 extracted from the brain, some in vitro experiments show that the protein's environment is important because it affects its local or overall conformation and thus the accessibility of PAD target sites. The in vitro deimination of bovine brain bMBP-C1 component, using purified natural PAD2, yielded products that were then referred to as "bMBP-Cit_n" with "n" being the average number of citrulline residues per mole of MBP, but that were otherwise heterogeneous (in terms of precisely which arginyl residues were modified) (Lamensa and Moscarello 1993; Pritzker et al. 2000a, b). The methylation of hR107 (human 18.5-kDa sequence numbering, corresponding to bovine bR106 and murine hR104, Fig. 19.3) protected it from enzymatic citrullination, a type of antagonistic crosstalk (Pritzker et al. 2000b; Raijmakers et al. 2007; Fuhrmann et al. 2015; Fuhrmann and Thompson 2016).

A more exhaustive analysis using recombinant PAD2 (Raijmakers et al. 2005) to deiminate human component hMBP-C1 in vitro showed that all arginyl residues except for hR54 (human 18.5-kDa sequence numbering, murine mR52, Fig. 19.3) were accessible to the enzyme under aqueous conditions (Wood et al. 2008). Yet this residue is partially deiminated in human brain MBP as shown by mass spectrometry, but can also be ADP-ribosylated (Kim et al. 2003). A later study of bovine 18.5-kDa MBP identified five sites found to be endogenously deiminated in vivo (bR41, bR47, bR63, bR96, bR129) and more sites that could be deiminated by PAD2 in vitro (bR23, bR29, bR31, bR41, bR47, bR52, bR63, bR78, bR96, bR106, bR112, bR129, bR158, bR161, bR168, br169) (Jin et al. 2013). Thus, it appears that almost all arginyl residues in 18.5-kDa MBP *can* be converted to citrulline, but there is a hierarchy of modification and crosstalk with other PTMs.

This putative crosstalk may represent conformational (steric) effects that are also dependent on whether the protein is free or membrane associated, i.e. its environment.

19.3.2 Pseudo-deimination by Site-Directed Mutagenesis Mimics Natural Deimination, but the C8-Moniker Remains

Biochemical and biophysical studies such as fluorescence and magnetic resonance spectroscopy require well-defined and reproducible protein preparations. Our group has relied heavily on recombinant variants of the 168-residue murine 18.5-kDa MBP isoform, starting with the basic form that we now generally call "rmC1" because it is recombinant, murine, and unmodified by the bacterial host used to produce it (Bates et al. 2000). The in vitro deimination of rmC1 by purified PAD2 has also yielded heterogeneous populations that were then referred to as "rmMBP-Cit_n" (Ishiyama et al. 2001).

To ensure homogeneity and reproducibility of our recombinant protein preparations in studying this PTM, we generated a quasi- or pseudo-deiminated variant formed by Arg/Lys → Gln substitutions: mR23Q, mR31Q, mK119Q, mR127Q, mR157Q, and mR168Q (Bates et al. 2002; Musse et al. 2006). Note now the murine 18.5-kDa sequence numbering—sequences are compared explicitly in Fig. 19.3 and in our recent review (Harauz and Boggs 2013). Moreover, we do not mean that lysyl residues are deiminated, since the hR122 in human MBP is replaced conservatively by bK121 and mK119 in the bovine and murine proteins, respectively. Like enzymatic deimination, each pseudo-deimination reduces the net positive charge of the protein by 1 unit. For the recombinant murine ("rm") components, rmC1 has a net charge of +19 at neutral pH compared to +13 for rmC8. These latter constructs have a C-terminal LEH_6-tag to facilitate purification by nickel affinity chromatography (Bates et al. 2000, 2002). We have recently removed this tag to yield new "untagged" preparations called UTC1 and UTC8 more suited for quantitative spectroscopic studies on the protein's conformational transitions and interactions, particularly with divalent cations (Smith et al. 2010; Baran et al. 2010; Zienowicz et al. 2015; Vassall et al. 2015b, 2016). These variants have net positive charges of +20 and +14, respectively, at neutral pH.

Over the course of time, then, the "C8 component" designation has come in the literature to represent either purified naturally deiminated or recombinant pseudo-deiminated MBP per se, primarily for convenience in distinguishing "normal, healthy C1" and "MS disease-associated C8" (Bates et al. 2003; D'Souza et al. 2005; D'Souza and Moscarello 2006; Musse et al. 2006; Homchaudhuri et al. 2010; Ahmed et al. 2010; Suresh et al. 2010; Boggs et al. 2011; Wang et al. 2011; Vassall et al. 2015b, 2016).

19.3.3 Various Methods to Assess the Degree of Brain MBP Deimination

It should be noted again, though, that the nomenclature over the last few decades has not been consistent even when discussing natural protein preparations isolated from brain and that various C1 versus C8 comparisons in different papers have been based on ratios of peaks derived from (1) cation-exchange columns (Moscarello et al. 1994; Wood et al. 1996), (2) alkaline-urea slab or tube gel electrophoresis (Palma et al. 1997), or (3) HPLC (Wood and Moscarello 1989; Cao et al. 1999; Wood et al. 2008).

Immunolocalization studies of MBP-C8 have relied on antibodies raised against tetra-citrulline peptides (McLaurin et al. 1993; Moscarello et al. 1994; Wood et al. 1996) or against deca-citrulline peptides (Nicholas and Whitaker 2002), the latter known as the F95 antibody (Nicholas et al. 2014; Bradford et al. 2014a, b). Both antibodies have been shown to detect deiminated MBP by ELISA (enzyme-linked immunosorbent assay), but can also detect other deiminated proteins in Western blots or tissue sections. Of note is the demonstration that high levels of citrullinated MBP can be detected immunohistochemically in areas of ongoing demyelination in human brain (Bradford et al. 2014b).

Although the 18.5-kDa MBP isoform has been the most studied, the smaller 14-kDa isoform is also a major component of rodent CNS myelin and confounds quantification of PTMs by ratios of chromatographic elution peaks because the two protein populations are mixed in most brain extracts from experimental animals. Nevertheless, it has been concluded that both splice variants are similarly post-translationally modified, including an increase in proportion of less-cationic components in an animal model exhibiting spontaneous demyelination, called the ND4 mouse line (Mastronardi et al. 1996a, b; Palma et al. 1997). Referring to Fig. 19.3, it is interesting that the 41 amino acids encoded by exon-10 of *Golli* and that are missing in 14-kDa MBP compared to 18.5-kDa MBP contain only one arginyl residue that could be deiminated (mR127). Thus, it could be expected that 14-kDa MBP would be as likely to contain citrulline as the 18.5-kDa splice variant.

We do not know to what extent any of the early developmental *Golli* proteins are deiminated. Studies of ageing and MS brain tissues by Western blotting and immunolocalization using the anti-deca-citrulline F95 antibody show a generalized increase in relative amounts of deiminated proteins in general, and glial fibrillary acidic protein (GFAP) has specifically been identified as being thus modified (Nicholas et al. 2004, 2005, 2014; Ishigami et al. 2005; Bradford et al. 2014a, b). We mention in passing the full-length classic 21.5-kDa MBP isoform, which has a nuclear localization signal, yet unknown interaction partners in the nucleus and yet unknown PTM modification patterns (Smith et al. 2012c, 2013; Harauz and Boggs 2013). A top-down proteomic analysis has shown that deimination of classic 14- and 18.5-kDa MBP isoforms occurs in species as divergent as rattlesnake and cattle (Zhang et al. 2012a). It is thus conceivable that the early developmental Golli and other classic MBP isoforms (like 21.5-kDa and 14-kDa) are also deiminated to some extent, but this information is still lacking.

19.3.4 The "Executive" Molecule Has Protein-Protein Interactions Throughout Myelin

In the compacted, multilayered internode regions of CNS myelin, MBP adheres the cytoplasmic leaflets of the oligodendrocyte membrane to each other to form the major dense line observed in transmission electron micrographs of thinly sectioned tissue (Fig. 19.1) (Trapp and Kidd 2004; Inouye and Kirschner 2015). This line is only about 3 nm thick, and the protein forms a two-dimensional "molecular sieve" that restricts diffusion of some membrane proteins from paranodal loops into compact myelin (Aggarwal et al. 2011; Kattnig et al. 2012; Bakhti et al. 2014). It is thus hard to envisage how protein-modifying enzymes can access membrane-associated MBP to generate the dynamic molecular barcode or even how it can act as a hub in protein-protein networks. The key is first that developing myelin is relatively open and second that mature myelin also has many noncompacted domains. Thus, there are many other places in myelin where MBP could still be accessible to protein-modifying enzymes, such as in paranodal loops, outer loops, or around inner cytoplasmic channels that run through the structure (Fig. 19.1) (Velumian et al. 2011; Snaidero et al. 2014; Tomassy et al. 2014; Rinholm et al. 2016; Velte 2016). These regions are the commuter routes and suburbs of the myelin sheath, so to speak.

The mRNA for 18.5-kDa MBP has a 3′ untranslated region that directs it to the distal cell processes where the protein is synthesized and then quickly adsorbs to single leaflets of the oligodendrocyte membrane (Muller et al. 2013). The protein attaches primarily by strong electrostatic interactions and specifically via three amphipathic α-helical membrane-anchoring motifs (Fig. 19.4) (Harauz et al. 2009; Vassall et al. 2015a). Here in this environment, before getting sandwiched by another membrane leaflet, the protein could potentially interact with cytoskeletal proteins such as actin and tubulin, signalling proteins such as Fyn-SH3, voltage-operated calcium channels, and calcium-activated calmodulin. All of these protein-protein associations have been studied in vitro, and co-localization with cytoskeletal and SH3-domain-containing proteins has been demonstrated in membrane ruffles of immortalized oligodendroglial cells in culture (Smith et al. 2011, 2012a, b; De Avila et al. 2014). Presumably, the protein could now be modified by kinases, for example, as it would still be exposed to the cytosol when just adsorbed to a single membrane leaflet.

Eventually though, the charge neutralization of the protein by adsorption to the phospholipid bilayer would facilitate condensation of two membrane leaflets to form a major dense line only about 3 nm thick (Trapp and Kidd 2004; Min et al. 2009; Aggarwal et al. 2013; Lee et al. 2014, 2015; Ozgen et al. 2014; Snaidero and Simons 2014; Snaidero et al. 2014). We are still developing molecular models of how MBP can adsorb to the membrane, self-assemble with itself and with proteolipid protein, and paste two membrane leaflets together to form compact myelin, and many details remain unknown (Fig. 19.4) (Ahmed et al. 2010; Kattnig et al. 2012; Vassall et al. 2015a, 2016). The point here, though, is that MBP experiences different environments during myelinogenesis and even within fully formed, mature myelin, which is by no means a static entity.

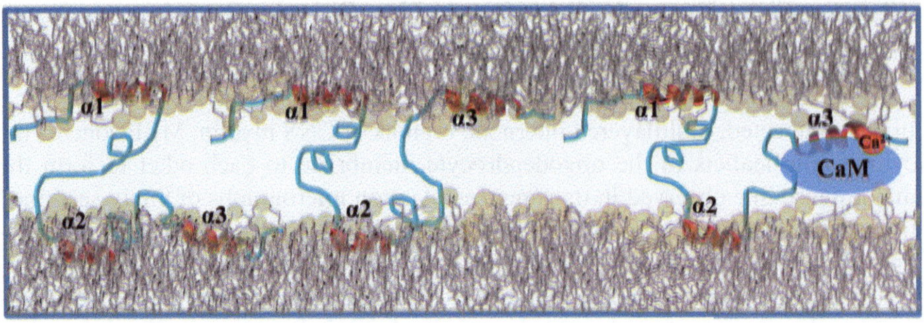

Fig. 19.4 Models of adhesion of inner oligodendrocyte leaflets by 18.5-kDa MBP. We envisage that MBP adopts a minimal hairpin conformation which would allow different α-helical segments to interact simultaneously with apposing membrane leaflets, possibly with an unspecific global topology. A hairpin-like structure could also allow any one of the α-helices to detach from the membrane and interact with other partners, whilst the other two α-helices remain tethered to apposing membrane bilayers. The α2-helix is an immunodominant epitope and is part of a putative molecular switch that contains an SH3 ligand (Bates et al. 2004; Musse et al. 2006; Vassall et al. 2013; De Avila et al. 2014). The C-terminal domain containing the α3-helix "moonlights" as a target for calcium-activated calmodulin (Libich et al. 2003; Majava et al. 2008; Libich and Harauz 2008a; Homchaudhuri et al. 2010; Majava et al. 2010; Bamm et al. 2011; Wang et al. 2011; Vassall et al. 2016). The N-terminal domain containing the α1-helix is sufficient to polymerize actin and tubulin in vitro and is a secondary target for calcium-activated calmodulin (Bamm et al. 2010, 2011; Zienowicz et al. 2015). Such protein-protein interactions would presumably happen in places where myelin is not fully condensed, e.g. at paranodal loops or in oligodendrocyte membrane processes before compaction has occurred. For the 14-kDa isoform that lacks the α3-helix, the α1- and α2-helices suffice for the protein to compact myelin membranes. This isoform predominates in adult rodent myelin. The *shiverer* mouse MBP would only have the α1-helix (see Fig. 19.2). Deimination (at the six primary sites distributed across the 18.5-kDa protein sequence, Fig. 19.3) affects the disposition of the central α2-helix of MBP, representing an immunodominant epitope in MS, rising closer to the membrane surface and being less tilted (Musse et al. 2006; Musse and Harauz 2007; Ahmed et al. 2010). This figure is adapted with permission from reference (Vassall et al. 2015a)

19.3.5 When and Where in CNS Myelin Does MBP Get Deiminated?

We do not know exactly when, or where, or why MBP is deiminated in differentiating oligodendrocytes and in mature myelin. Myelin proteomic analyses are overwhelmed by the sheer overall quantities of MBP and proteolipid protein (Werner and Jahn 2010; de Monasterio-Schrader et al. 2012; Panfoli et al. 2014), and subpopulations of citrullinated MBP would be difficult enough to detect by mass spectrometry because the mass change is only 1 Da (Hensen and Pruijn 2014; Clancy et al. 2016). Spatial mapping of myelin by emerging techniques of mass spectrometric imaging (Hanrieder et al. 2015; Lahiri et al. 2016) would particularly face this problem. At present, it perhaps would be the most insightful to combine approaches such as (1) immunohistochemistry of thinly sectioned myelin tissue using F95 (or even new anti-MBP-Cit$_n$ antibodies yet to be created), (2) subcellular fractionation of myelin into microdomains by differential detergent extractions prior to proteomics (Gielen et al. 2006; DeBruin and Harauz 2007; Sonnino et al. 2015; Aureli et al. 2015), (3) indirect mass spectrometric detection by fragmentation of citrullinated peptides in brain extracts (Jin et al. 2013), or (4) detection of general PAD activity (Hensen and Pruijn 2014). However, such "-omics" studies are exceptionally tedious, time-consuming, and expensive. Despite this current uncertainty, it remains that MBP citrullination occurs to a significant extent in both the healthy developing and in the adult CNS (Kim et al. 2003; Friedrich et al. 2016).

We have hypothesized that the specific pattern of PTMs of MBP reflects its localization in distinct membrane microdomains in healthy myelin and that citrullinated MBP may be involved in regular myelin remodelling in the normal adult (DeBruin et al. 2005, 2006; DeBruin and Harauz 2007; Harauz and Musse 2007). This idea is consistent with the ongoing realization that CNS myelin is plastic and dynamic in the healthy adult (Miller et al. 2012; Young et al. 2013; Snaidero et al. 2014; Tomassy et al. 2014; de Hoz and Simons 2015; Silbereis et al. 2016). In developing CNS myelin, the high degree of MBP deimination probably facilitates developmental plasticity (Wang and Young 2014; Pajevic et al. 2014). (The deimination of linker histones, analogously, decondenses chromatin (Christophorou et al. 2014).) In adult CNS myelin, the degree of MBP deimination must be in equilibrium with other PTMs (DeBruin and Harauz 2007; Harauz and Musse 2007). Although the foremost populations of MBP are stable and long-lived (Toyama et al. 2013; Friedrich et al. 2016), MBP half-lives of a few weeks have been reported (Sabri et al. 1974; Quarles et al. 2006). The hierarchy, interplay, and crosstalk of deimination, phosphorylation, methylation, and racemization affect not only the protein conformation and degree of membrane adhesion on and between oligodendrocyte membranes in compact myelin but also MBP's interactions with cytoskeletal and signalling proteins in vitro and in cultured cells (Harauz and Musse 2007; Homchaudhuri et al. 2010; Smith et al. 2010; Boggs et al. 2011; Vassall et al. 2013, 2016; Friedrich et al. 2016). Again, MBP appears to organize CNS myelin in much the same way that basic histones organize chromatin (Vassall et al. 2015a; Wenderski and Maze 2016).

Specifically, the PAD2 isotype (the one that deiminates MBP) has been observed by immunoelectron microscopy to be found at myelin peri-axonal structures (Wood et al. 2008), consistent with this notion. Presumably, other protein-modifying enzymes such as kinases could also modify MBP here, e.g. the phosphorylation of murine mT95 (human hT98, bovine bT97) within the segment mT92-mP93-mR94-mT95-mP96 (a putative SH3 ligand and part of a putative molecular switch; Figs. 19.3 and 19.4) by mitogen-activated protein kinases (MAP kinases) is regulated by action potential generation in axons (Murray and Steck 1984; Atkins et al. 1997; Atkins et al. 1999). Indeed, the recent demonstrations of cytoplasmic diffusion and even of mitochondrial movement within myelin, although slow, demonstrates that the structure is not an unassailable fortress and that where there is a cellular will, there is a cytoplasmic way for these enzymes to go to work (Velumian et al. 2011; Snaidero et al. 2014; Tomassy et al. 2014; Rinholm et al. 2016; Velte 2016).

Even wherever and whenever PAD2 can gain access to MBP, the enzyme is activated by calcium levels that are 100-fold higher than normal intracellular concentrations, which must be as tightly regulated in developing and mature oligodendrocyte cell bodies as in other cells. Indeed, classic MBP isoforms serve to inhibit voltage-operated calcium channels (VOCCs) in cultured and primary oligodendrocytes in cell culture (Smith et al. 2011). Interestingly, pseudo-deimination of the 18.5-kDa MBP did not have as strong a diminutive effect as the unmodified form, indicating that membrane association of this MBP isoform was essential per se and that the strength of attachment to the lipid bilayer was also important. These observations with the classic MBP isoforms stand in contrast to the early developmental Golli proteins that enhance VOCC activity (Paez et al. 2009a, b, c; Fulton et al. 2010b).

We have not further investigated the mechanism of VOCC inhibition by classic MBP, but surmise that it may involve competition of binding with the β2E subunit of the channel with anionic phospholipids, including phosphoinositides (Kim et al. 2015a, b; Kim and Suh 2016). The overall calcium concentrations in the intracellular compartments of mature myelin might differ from those of normal cells, but still must be controlled. In particular, calcium spikes must be highly localized in healthy oligodendrocytes. Indeed, it has recently been demonstrated that axonal action potentials can transiently raise Ca^{2+} levels in myelin via *N*-methyl-D-aspartate receptors (NMDARs) (Micu et al. 2006, 2016). Such influxes of intracellular Ca^{2+} from the peri-axonal space must generally be temporary and highly localized and might be part of a signalling process during myelination (Friess et al. 2016).

So how does an imbalance of intracellular calcium levels arise in the first place? In any one individual, there might be many independent and minor traumatic factors that, over a prolonged period, precipitate local cytodegenerative events that must include an abnormal and prolonged myelinic Ca^{2+}-influx (through either VOCCs or NMDARs), which then activates the PAD enzymes (Stirling and Stys 2010; Ferretti et al. 2014; Lazarus et al. 2015; Tu et al. 2016). The incipient molecular events may include oxidative damage to myelin components due to reactive molecules such as extracellular haemoglobin (Bamm and Harauz 2014), "myelin mis-structuring" or "micro-amyloidogenesis" during development (Harauz and Boggs 2013; Vassall et al. 2015a), or "abnormal regulation of the fine nanostructure of myelin" (Micu et al. 2016). Proton-gated Ca^{2+}-channels such as the TRP (transient receptor potential) family can allow Ca^{2+} entry into the cell under ischaemic conditions and occasionally even spontaneously (Hamilton et al. 2016; Saab and Nave 2016). Oligodendrocyte apoptosis then becomes a seed for myelin destruction and lesion formation (Lucchinetti et al. 2000; Barnett and Prineas 2004; Lassmann 2004; Lassmann and van Horssen 2011; Moscarello et al. 2013; Lassmann 2014). Such scenarios have been presented more recently as an "inside-out" idea of MS pathogenesis (Stys et al. 2012; Stys 2013; Micu et al. 2016; Weil et al. 2016; Traka et al. 2016).

We might never be able to predict that the disease will arise in any individual, but here we can propose a mechanistic explanation of how demyelination can propagate when the normal levels of MBP deimination become disturbed. In order to link Dr. Moscarello's "primary defect in MBP" to myelin destruction, we now change tack to discuss events first from a structural protein-centric perspective and then from a cell biological stance.

19.4 Effects of MBP Deimination on Intra- and Intermolecular Associations

19.4.1 Deimination of MBP Affects Its Intrinsic "Unstructure"

The eukaryotic proteome comprises roughly a third of proteins whose conformations are in whole, or in significant part, "intrinsically unstructured or disordered". The now-accepted term is "intrinsically disordered proteins" (IDPs) (e.g. (van der Lee et al. 2014; Yan et al. 2016; Uversky 2016)). These IDPs have dynamic conformations and associations, being highly post-translationally modified, promiscuous (weak and transient interactions with diverse surfaces, protein partners, and small-molecule ligands, not always simultaneously), and multifunctional (Tompa and Fersht 2015; Tompa et al. 2015; Bah and Forman-Kay 2016; Ross 2016). Different interaction partners might interact at different times with the same target segment of an IDP, a phenomenon called "moonlighting". These targets are referred to as molecular recognition fragments (MoRF) and potentially undergo disorder-to-order transitions upon binding.

A seminal paper by Uversky and colleagues identified MBP as an IDP on the basis of its charge-hydrophobicity profile (Uversky et al. 2000; see also Hill et al. 2002; Harauz et al. 2004). The "structure" of MBP has thus been probed in this light in a variety of environments, aqueous, membrane mimetic, and membrane associated, and in complex with numerous other proteins: to define local and global disorder-to-order transitions and interaction motifs (Harauz et al. 2004; Zhong et al. 2007; Libich and Harauz 2008a, b; Ahmed et al. 2009; Libich et al. 2010; Ahmed et al. 2010; Wang et al. 2011; Vassall et al. 2013, 2015b, 2016; De Avila et al. 2014; Zienowicz et al. 2015). This effort has solidified MBP as being a "poster protein" for this class (Harauz et al. 2009, 2013; Harauz and Libich 2009; Vassall et al. 2015a). To understand myelin architecture in health and disease, we must think of MBP in this conceptual framework, especially examining how its myriad PTMs affect its networking properties.

Briefly, MBP has three amphipathic α-helical segments that associate with the surface of the phospholipid bilayer (examples of α-MoRFs (α-helical molecular recognition fragments)) and a loose tertiary conformation that is minimally a hairpin or paperclip shape (Fig. 19.4). Like other IDPs, its "tertiary structure" is really an ensemble of rapidly interconverting conformations that depend on environment, i.e. whatever its interaction partners are at the time. As the protein is synthesized on the ribosome, it will attach quickly to the inner face of the oligodendrocyte membrane. Thus, it experiences an environment of decreasing dielectric constant, which causes a shift in the conformational ensemble with small but measurable energy barriers (Vassall et al. 2015b, 2016). At some stage here, the protein is accessible to protein-modifying enzymes, such as protein kinases and deiminases, as discussed above.

The combinatorial PTMs of MBP reduce its overall net charge and local charge distribution (e.g. at the central molecular switch) and will modulate the ensemble of conformations sampled by the protein and the interaction motifs that are exposed

(Mao et al. 2010; Marsh and Forman-Kay 2010; Babu et al. 2012). The environment of the protein is the most influential determinant of its secondary and tertiary structure composition, and the differences between unmodified and pseudo-deiminated recombinant forms appear to be more evident in a membrane environment, as probed by diverse spectroscopic approaches (Bates et al. 2002, 2003; Musse et al. 2006; Ahmed et al. 2010; Boggs et al. 2011; Wang et al. 2011).

Recently, we have applied a number of complementary fluorescence-based techniques to perform thermodynamic analyses of the conformational transitions that the protein undergoes as its environment changes (from aqueous to membrane associated) and how pseudo-deimination and pseudo-phosphorylation affect the energy of these structural interconversions (Vassall et al. 2015b, 2016). The experimental design was based on untagged recombinant forms introduced above, the unmodified UTC1 and the pseudo-deiminated UTC8 (Smith et al. 2010). First, in increasing concentrations of trifluoroethanol to mimic the decreasing dielectric-constant environment as the protein approaches the membrane, we showed that the unmodified UTC1 variant underwent transitions from a disordered extended ↔ intermediate compact ↔ α-helical extended form, with small but distinct energy barriers (Vassall et al. 2015b). This study was important because it demonstrated that MBP has an underlying, definable, morphological state even though it is so flexible.

In contrast, fluorescence-based thermodynamic analyses of the conformational transitions of the pseudo-deiminated UTC8 showed only slight differences in Gibbs free energy. Nevertheless, it was confirmed that there *are* distinct tertiary conformational differences by evaluating the kinetics of digestion with the protease cathepsin D (CatD, EC 3.4.23.5) (Vassall et al. 2016). This enzyme probes the accessibility of the two Phe-Phe pairs that are essential for membrane association of MBP in myelin (murine mF42-mF43 and mF86-mF87, Fig. 19.3), and the UTC8 variant was observed to be digested more quickly than UTC1 was. This result was consistent with previous protease digestion studies using CatD, and other proteolytic enzymes, on natural or enzymatically deiminated MBP-Cit_n preparations that were degraded faster (Cao et al. 1999; Pritzker et al. 2000a, b; D'Souza and Moscarello 2006). Protein-protein interaction studies with the hexa-histidine-tagged rmC1 and rmC8 constructs also showed changes in the binding of protein ligands such as calcium-activated calmodulin, for example (Libich et al. 2003; Wang et al. 2011; Vassall et al. 2016).

The apparently slight energy changes between the conformational ensembles of UTC1 and UTC8 could, in retrospect, be explained by the presence of only six pseudo-deiminated residues in the recombinant protein construct that was used. (These sites corresponded to those defined originally in hMBP-C8 from normal human brain (Wood and Moscarello 1989).) A much earlier study than our 2016 one had used enzymatically deiminated natural bovine bMBP-C1; with these protein preparations, it was found that there was a significant conformational switch between seven and nine (on *average*) citrullinations (Pritzker et al. 2000a) and that methylation of bR106 (bovine 18.5-kDa sequence numbering, human hR107, murine mR104) decreased the rate of enzymatic citrullination (Pritzker et al. 2000b). Thus, it is clear first that deimination of MBP alters the conformational ensemble distribution of 18.5-kDa MBP, especially as the number of modifications increases. Second, it is seen that other PTMs such as methylation act antagonistically in a form of crosstalk (Raijmakers et al. 2007), as is being observed in the basic histone proteins that compact DNA (Venne et al. 2014; Zhang et al. 2014; Fuhrmann et al. 2015; Fuhrmann and Thompson 2016). All in all, we suspect, but have yet to prove, that hyper-deimination of 18.5-kDa MBP beyond seven to nine sites pushes the protein beyond a major structural boundary (Vassall et al. 2016).

The deimination-induced conformational shifts in 18.5-kDa MBP have implications for MS pathogenesis in an intriguing way. Although seeming to act primarily as a membrane adhesive, and as a hub in protein-membrane and protein-protein interaction networks (Harauz et al. 2009; Vassall et al. 2015a), D'Souza and Moscarello have observed an unusual and unexpected property of MBP: "self-proteolysis" (D'Souza et al. 2005)! They demonstrated that hMBP-C1 (sic.) from MS tissue underwent autocatalytic cleavage at slightly alkaline pH and that it showed a faster time course of cleavage than hMBP-C1 isolated from normal individuals. The same phenomenon was observed with recombinant murine rmC1 (unmodified) and rmC8 (pseudo-deiminated), preparations that could not be considered contaminated by other proteases. The mechanism of this unusual autocatalysis has not been investigated further, to our knowledge, but we can hypothesize that naturally deiminated MBP-Cit$_n$ would undergo progressive conformational conversions that would enhance its self-destruction (cf., (Di Salvo et al. 2013)).

One of the self-proteolytic fragments observed by D'Souza and Moscarello consisted of residues hN84-hF89 (human 18.5-kDa sequence numbering, corresponding to murine mN81-mF86; Fig. 19.3), which is important because it is an immunodominant epitope that activates T cells in MS (Wucherpfennig et al. 1997; Hansen et al. 2011). But in CNS myelin, the protein is generally associated with membranes that can presumably prevent its self-autolysis. Moreover, the immunodominant epitope is one of the amphipathic α-helical membrane-anchoring segments of 18.5-kDa MBP (Bates et al. 2004; Libich and Harauz 2008a; Ahmed et al. 2010; Vassall et al. 2013). Specifically, the two Phe-Phe pairs of 18.5-kDa MBP are embedded in the membrane (but not equivalently) and contribute to compaction and self-assembly (Bates et al. 2003; Aggarwal et al. 2013). How do these residues become exposed to proteases such as cathepsin D?

19.4.2 Deimination of MBP Diminishes Its Interactions with Myelin Membranes and Renders It More Susceptible to Proteolytic Degradation and Immune Recognition

The highly basic nature of MBP means that it readily associates with negatively charged phospholipid membranes and, by electrostatic and some hydrophobic associations, maintains two inner leaflets of the oligodendrocyte membrane together in close apposition to form the major dense line of the myelin sheath (Figs. 19.1 and 19.4) (Boggs et al. 1982, 1997; Jo and Boggs 1995; Bates et al. 2003; Min et al. 2009; Lee et al. 2014, 2015; Ozgen et al. 2014). One simple assay for this primary "function" of MBP is to observe the degree of aggregation of lipid vesicles (mimicking the composition of the cytoplasmic leaflet of oligodendrocytes) as MBP is added. As expected, net charge reduction due to deimination reduces the protein's ability to do so (Moscarello et al. 1986; Wood and Moscarello 1989; Jo and Boggs 1995; Boggs et al. 1997; Bates et al. 2002). The highly deiminated MBP fraction from the Marburg's variant of MS (MC8 or hMBP-Cit_{18}) even fragments the lipid vesicles (Boggs et al. 1999). Thus, hyper-deimination of MBP results in a loosening of the compact myelin multilayers in MS tissue. This overall structural destabilization was dramatically illustrated (by immunoelectron microscopy) to delocalize MBP from the major dense line of myelin (McLaurin et al. 1993) and by atomic force microscopy to diminish the ability of the protein to stack lipid bilayers (Suresh et al. 2010).

By site-directed spin labelling (SDSL) and electron paramagnetic resonance (EPR) spectroscopy of membrane-reconstituted MBP, we have shown that the immunodominant epitope is within the highly conserved, central amphipathic α2-helical segment of MBP-C1 that lies on the surface of the bilayer, with the mF86-mF87 pair being most deeply embedded, penetrating up to 12 Å into the bilayer (refer to Figs. 19.3 and 19.4) (Bates et al. 2004). In the pseudo-deiminated rmC8 variant, this primary immunodominant epitope formed a more highly surface-exposed and shorter amphipathic α-helix (Musse et al. 2006; Musse and Harauz 2007). Moreover, cathepsin D digested lipid-associated rmC8 ~ 3 times faster than it did rmC1. Molecular details of the membrane association of this central immunodominant epitope have been refined by solid-state NMR spectroscopy of membrane-reconstituted MBP (Zhong et al. 2007; Ahmed et al. 2010).

Subsequently, we evaluated the myelin membrane association of the 3rd amphipathic α3-helical segment of MBP, again by SDSL-EPR spectroscopy (Homchaudhuri et al. 2010). Pseudo-deimination across the protein sequence also rendered this segment exposed to the aqueous phase and thus more easily detached by calcium-activated calmodulin, thereby potentially affecting signalling pathways in myelin turnover (Fig. 19.4) (ibid.; see also Majava et al. 2010; Wang et al. 2011).

All in all, these combined biochemical and biophysical studies have provided atomic-level detail of membrane-anchoring segments of MBP and direct evidence of decreased MBP-membrane or altered MBP-protein interaction for the deiminated form. These findings suggest a mechanism for initial loss of myelin stability: de-compaction of the multilayered sheath by decreased electrostatic protein-membrane interactions and increased exposure of the two Phe-Phe pairs to digestion by enzymes such as cathepsin D, with ensuing digestion fragments that would no longer be able to adhere membrane leaflets together.

We can begin to envisage several consequences of this structural destabilization, starting a flowchart in Fig. 19.5. For example, de-compacted myelin can be more easily taken up by macrophages that would digest MBP, generating fragments to be presented on MHC class II molecules (Sospedra and Martin 2005; Lassmann and

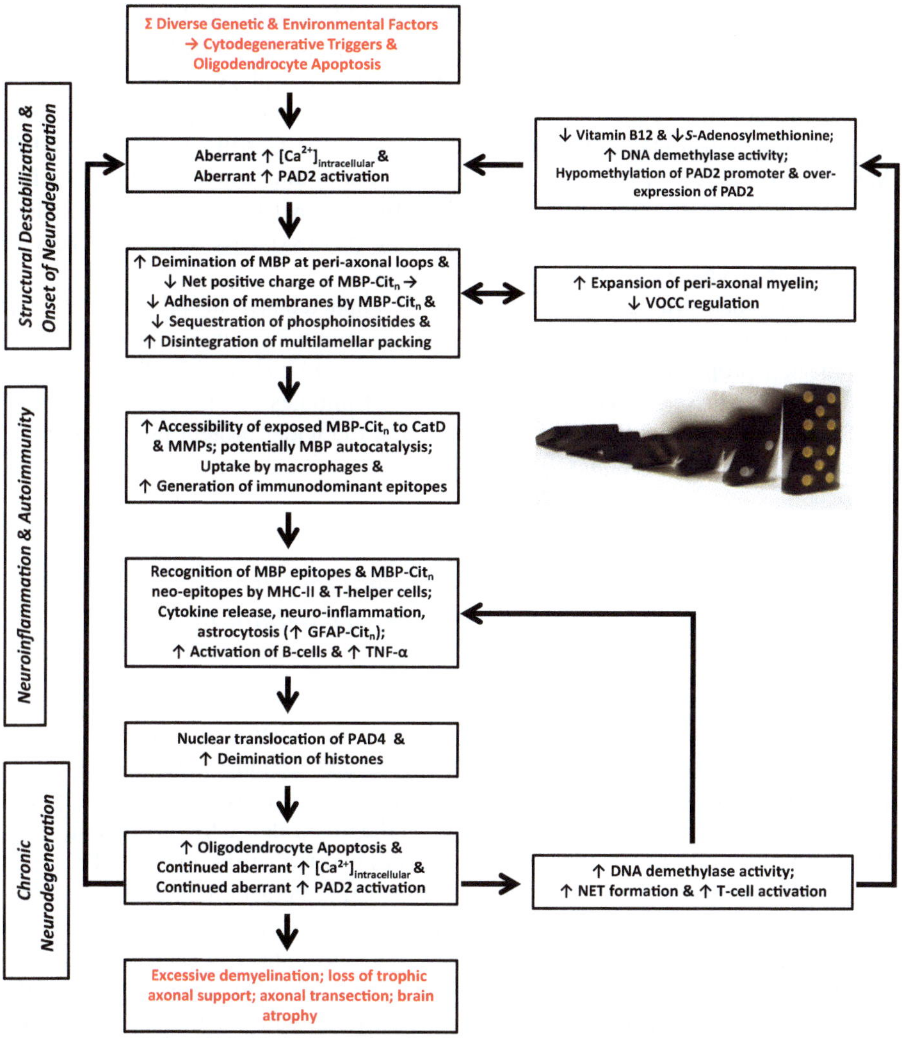

Fig. 19.5 Domino-demyelination by hyper-deimination. An aberrant calcium influx that increases PAD2/PAD4 expression and activates the enzyme in oligodendrocytes leads to hyper-deimination of MBP and other proteins, with diverse deleterious effects that can be propagated via positive feedback to form a cascade of myelin destruction depicted here. We posit that perturbations of myelin structure interaction networks play a central role in demyelination, according to the thesis that "a primary defect in myelin basic protein is responsible for initial changes in myelin in MS, and precedes the autoimmune response" (Wood and Moscarello 1997). This figure is founded on schemes from references (Moscarello et al. 2007; Mastronardi and Moscarello 2008; Musse et al. 2008b; Moscarello 2014). The domino image was obtained from Wikipedia

van Horssen 2011; Lassmann 2014). A deiminated MBP that is detaching from the bilayer surface would be more susceptible to self-proteolysis (D'Souza et al. 2005) and to matrix metalloproteases such as stromelysin (MMP-3) that generate a 17-residue peptide that can be easily accommodated by MHC class II molecules

(D'Souza and Moscarello 2006; Mastronardi and Moscarello 2008). There are also catalytic antibodies called abzymes that circulate in the serum of MS patients and recognize the immunodominant epitope (Ponomarenko et al. 2006; Belogurov et al. 2008). Moreover, T cells can also recognize neo-epitopes on protein fragments derived from citrullinated MBP (Deraos et al. 2008; Carrillo-Vico et al. 2010; Moscarello et al. 2013; Yang et al. 2016). These autoimmune attacks contribute further to myelin destruction. Finally, modifications to MBP by lipid-derived aldehydes (derived during the neuroinflammatory process) have similar effects to deimination (Cygan et al. 2011).

The continued structural destabilization of compacted myelin and provocation of autoimmune response in MS by exposure of an immunodominant epitope can thus become self-propagating (Fig. 19.5) (Matsuo et al. 1997; Husted 2006; Harauz and Musse 2007; Musse and Harauz 2007). So what initiates this process and can the CNS put a brake on it? We need to discuss first the wider protein-protein associations of MBP within myelin.

19.4.3 Equilibrium Disruption of Modified Variants and the Myelin Network by Hyper-deimination of MBP in MS

It has been proposed that MBP self-assembles on the inner leaflet of the oligodendrocyte membrane to form a "molecular sieve" that excludes integral membrane proteins from adjacent regions that "zipper" together to extrude excess cytoplasm and then seal the two leaflets together to form the major dense line (Aggarwal et al. 2013; Bakhti et al. 2014). One intriguing finding was that the two Phe-Phe pairs within MBP appeared to mediate the protein-protein interaction, in addition to penetrating into the lipid bilayer to anchor the protein as we had shown earlier (Bates et al. 2003) (Figs. 19.3 and 19.4). Further biophysical study is needed to see how these phenomena are linked, such as if decreased adhesion of the protein to the membrane leaflet will, in turn, affect its interactions with neighbouring proteins. There is already transmission electron microscopical evidence for this proposition, as we discuss next (Ishiyama et al. 2001; Hill et al. 2002; Harauz et al. 2004).

In our early electron microscopy studies of two-dimensional arrays of MBP variants formed on lipid monolayers of different compositions, pseudo-deimination at six sites (the rmC8 variant) and enzymatic deimination at an average of nine sites (rmMBP-Cit$_9$) resulted in a different kind of ordering, a switch from closely packed fibrils to what appeared to be hexagonal close packing (Ishiyama et al. 2001; Hill et al. 2002; Harauz et al. 2004). Using pulse EPR spectroscopy (DEER, double electron-electron resonance), the group of Dr. Dariush Hinderberger in Mainz, Germany, and we have demonstrated that the distribution of rmC1 networks is neither entirely regular nor entirely random (Kattnig et al. 2012) and is similar for the less-charged rmC8 variant (Vassall et al. 2016). This result stands in contrast to the results obtained by atomic force microscopy that showed a dramatically reduced ability of rmC8 to stack lipid bilayers compared to rmC1 (Suresh et al. 2010). Considering these diverse data together, we argue that MBP deimination disrupts the arrangement of the molecular sieve and that this effect would be exacerbated by further citrullination beyond some boundary of seven to nine residues (Figs. 19.3 and 19.5). More recent high-resolution microscopical analyses emphasize the importance of the lipid environment on the lateral organization of MBP and PLP in the myelin membrane (Ozgen et al. 2014, 2016).

In its putative cytoskeletal turnover roles, deiminated MBP has both a reduced ability to polymerize and bundle actin and to tether actin microfilaments to membranes (Hill and Harauz 2005; Boggs et al. 2005, 2011; Ahmed et al. 2009; Bamm et al. 2010; Smith et al. 2012b). Associated with these functions is MBP's role as a PIPmodulin—a protein that sequesters phosphoinositides such as $PI(4,5)P_2$ and $PI(3,4,5)P_3$—the interaction of deiminated MBP with these lipids is significantly decreased (Musse et al. 2008a). It has been suggested that phosphoinositide sequestration could release actin-severing proteins at the outer membrane processes (Boggs et al. 2012; Nawaz et al. 2015; Zuchero et al. 2015). Presumably, deimination of MBP could also modulate this function by affecting the number of phosphatidylinositol lipid molecules bound to each protein. There is also a structural consideration of the membrane per se. We have shown that 18.5-kDa MBP, incubated with high proportions of phosphoinositides, can form tubular vesicles in vitro (Ishiyama et al. 2001, 2002). These curious structures could reflect a membrane *modelling* function of MBP, an idea that has not further been explored. But changes in MBP's barcode of PTMs affect this role as well.

Finally, it should be noted that the overall lipid composition of the myelin membrane appears to become altered during demyelination, thereby affecting further the ability of all MBP isomers to maintain the sheath synergistically with other components (Ohler et al. 2004; Min et al. 2011; Boggs et al. 2012; Lee et al. 2014, 2015; Ozgen et al. 2014; Frid et al. 2015). A recent biophysical study of membranes with altered lipid compositions demonstrated structural transitions from lamellar to inverted hexagonal phases (Shaharabani et al. 2016). It was proposed over a decade ago, on the basis of surface force measurements, that vesiculation of myelin membranes occurred during the demyelination process (Ohler et al. 2004). Moreover, the lipid compositions of the inner and outer membrane leaflets are different. In particular, MBP interacts strongly with the outer leaflet monosialoganglioside G_{M1} forming monomeric and oligomeric protein-lipid particles (Beniac et al. 1997, 1999, 2000; Ridsdale et al. 1997). The exposure of extracellular leaflet lipids such as G_{M1} would thus accelerate further the direct, physical fragmentation and disintegration of the planar bilayer structure myelin membrane (vesiculation). Again, the deiminated protein interacts differently than the unmodified one, forming less-condensed assemblies (Beniac et al. 2000). Whereas the addition of MBP counteracts the spontaneous vesiculation of myelin-lipid vesicles in vitro (Fraser et al. 1986), the altered PTMs of the protein and the altered lipid composition of the myelin membrane in diseased myelin work synergistically to destabilize it.

Deiminated MBP has been found in high levels specifically in active lesions in the brains of MS patients (Bradford et al. 2014b), not just in whole brain preparations (Kim et al. 2003; Friedrich et al. 2016). Although this hyper-deimination of MBP in MS might be indicative of normal remyelination attempts (as in the myelination process of the developing CNS), it seems more likely in adult myelin to contribute directly to physical disintegration of the sheath and thus create a catastrophic feedback loop in the absence of other balancing developmental signals (Franklin 2002; Sherman and Brophy 2005; Franklin and Gallo 2014; Kremer et al. 2016). For these reasons, we have projected a centrality of citrullinated MBP in the destructive cascade in Fig. 19.5. In many ways, MBP's roles are like those of other IDPs involved in other neurodegenerative diseases in which protein-protein interactions are disrupted (Uversky 2014; Uversky 2015) or of IDPs exhibiting "conformational noise" in cancer (Mahmoudabadi et al. 2013). To understand how potentially to attenuate this process, we now review the role of the PAD2 and PAD4 isozymes in MS pathogenesis.

19.5 Peptidylarginine Deiminases in Demyelinating Disease

19.5.1 Up-regulation of Peptidylarginine Deiminase Isozymes 2 and 4 in MS and Animal Models of MS

The PAD2 isotype is the most abundant one found in the cytosol of oligodendrocytes, but its membrane-bound fraction in these cells is preponderant compared to the levels of soluble enzyme (Lamensa and Moscarello 1993; Pritzker et al. 1999; Moscarello et al. 2002b). Moreover, phospholipid association reduces this isozyme's Ca^{2+} dependence by almost twofold (Musse et al. 2008c). The three-dimensional structure of PAD2 has only recently been solved by X-ray crystallography, and it has been revealed that it binds up to six Ca^{2+} ions in a sequential manner (Slade et al. 2015). Continuing our earlier discussion of accessibility, PAD2 will act first on the protein that is found at myelin peri-axonal structures where it has, indeed, been immunolocalized (Wood et al. 2008). Moreover, the membrane association of PAD2 would mean that it would encounter membrane-associated MBP quite readily. This same study has immunolocalized the enzyme to clusters in myelin, which was considered by the authors at that time to be suggestive of vesicular localization (ibid.).

Newer approaches of tissue preparation involve high-pressure freezing, automatic freeze-substitution, and cryosectioning to preserve better the integrity of the myelin tissue (Möbius et al. 2010; Möbius et al. 2016; Weil et al. 2016; Velte 2016). Indeed, the vesiculation phenomenon has recently been observed in situ by cryo-electron microscopy (Weil et al. 2016). Other developing microscopic imaging modalities may prove valuable in exploring further the relationship between PAD2 and regions of the myelin membrane in the process of fragmentation (Velumian et al. 2011; Christensen et al. 2014; D'Este et al. 2016; Cicerone 2016). Nevertheless, it can be appreciated from our discussion of myelin architecture (Fig. 19.1) that enhanced levels of PAD2 would modify MBP where accessible and gradually nibble away at the molecular sieve (Figs. 19.4 and 19.5).

The isozyme PAD4 is also found in brain tissue and is nuclear localized in oligodendrocytes (Wood et al. 2008). Here, as in other cells, it functions in chromatin remodelling by deiminating histones (Arita et al. 2006). Interestingly, the PAD2 isozyme has also been found to operate in the nucleus, where it citrullinates histone H3 just as PAD4 does (Zhang et al. 2012b). This phenomenon has been reviewed in Fuhrmann et al. (2015) and Fuhrmann and Thompson (2016). Moreover, PAD2 has been found to be up-regulated in inflamed joints in rheumatoid arthritis, meaning that PAD4 is not exclusively operative in that disease either (Kinloch et al. 2008). The two isozymes are both implicated in the pathogenesis of MS with diverse effects that we review next.

The human disease MS is characterized by patches of demyelination in the brain and spinal cord (Compston and Coles 2002; Hauser and Oksenberg 2006; Popescu and Lucchinetti 2012). These MS white matter plaques consist of myelin sheath, macrophages containing undigested myelin fragments, microglia with phagocytosis of myelin, and axonal debris. Magnetic resonance imaging demonstrates these areas of neuroinflammation and demyelination as bright lesions (Neema et al. 2007). We have come to appreciate that normal-appearing white matter (NAWM) is already altered in the MS brain (Lassmann and van Horssen 2011; de Groot et al. 2013; Lassmann 2014), including by elevated PAD expression, activity, and citrullination of glial proteins such as MBP, GFAP, and histones (Nicholas et al. 2004; Mastronardi et al. 2006; Moscarello et al. 2013; Bradford et al. 2014a, b). In such tissue, the promoter for the gene that expresses PAD2 is *hypo*-methylated—hyper-methylation would silence this gene's expression (Mastronardi et al. 2007a; Calabrese et al. 2012). This phenomenon may reflect an epigenetic factor in susceptibility to MS (Zhou et al. 2014; van den Elsen et al. 2014).

The mechanism of recruitment of PAD4 into the deimination cascade may involve tumour-necrosis factor-α (TNFα), a pro-inflammatory cytokine (Mastronardi et al. 2006). Interestingly, the exposure of B cells to the immunodominant epitope of MBP (residues 85-99) induces production of TNF-α as well as of interleukin 6, as shown in cells derived from patients with relapsing-remitting MS (Nielsen et al. 2016). The hyper-deimination of histones consequent to PAD4 nuclear translocation leads to irreversible changes in, and apoptosis of, oligodendrocytes in MS (Mastronardi et al. 2006; Shanshiashvili et al. 2012). Further effects of PAD4 include the increased production of NET ("neutrophil extracellular trap") that can reduce the activation threshold of T cells (Li et al. 2010; Moscarello et al. 2013). These cellular facets of myelin degeneration are linked to our structural considerations in the flowchart in Fig. 19.5, which we collectively entitle "domino demyelination". One key locus that we link to MBP hyper-deimination is oligodendrocyte apoptosis that has been argued as being incipient in MS lesion formation (Barnett and Prineas 2004; Locatelli et al. 2012; Moscarello et al. 2013; Traka et al. 2016).

Clearly, then, the control of the degree of deimination of MBP by inhibitors of PAD expression or enzymatic activity is an attractive alternative to the predominant immunosuppressive strategies and may represent a means of slowing the degradation of myelin and ameliorating the neurological consequences of the disease. The search for such potential therapeutic measures requires appropriate animal models in which PAD expression is demonstrably associated with the demyelinating phenotype.

19.5.2 *Animal Models for Multiple Sclerosis Demonstrate Only Specific Aspects of the Disease, but Still Demonstrate the Importance of Deimination*

Multiple sclerosis is a heterogeneous disease with high variability between individuals, although roughly four major patterns can be discerned: two comprise neuroinflammatory demyelination with macrophage or T cell infiltration; two pertain to loss of oligodendrocytes by apoptosis or other causes (Lucchinetti et al. 2000; Lassmann et al. 2001). There are several animal models, but none of these truly recapitulates all aspects of the disease (Miller and Fyffe-Maricich 2010; Procaccini et al. 2015).

The autoimmune and neuroinflammatory facets of MS are generally probed in rodent and primate models of experimental allergic/autoimmune encephalomyelitis (EAE), in which administration of self-antigens, such as MBP or MOG fragments, precipitates the pathology (Terry et al. 2016). In such models, MBP deimination appears at first glance to be an epiphenomenon, meaning that it is observed, but is not the central mechanism underlying the demyelinating phenotype and may simply reflect increased apoptosis of glial cells (Nicholas et al. 2005; Raijmakers et al. 2005; Raijmakers et al. 2006; Artemiadis and Anagnostouli 2010; Caprariello et al. 2012). A different animal model exhibiting spontaneous demyelination is the ND4 mouse line overexpressing the DM20 isoform of proteolipid protein (PLP) (Mastronardi et al. 1993; DeBruin et al. 2006). Increases in overall PAD activity and PAD mRNA levels have been observed in ND4 mice as demyelination progresses (Mastronardi et al. 1996a, b; Moscarello et al. 2002b). In another PAD2-knockout mouse strain, PAD2 was proved to be the particular isozyme responsible for MBP deimination in the brain (Raijmakers et al. 2006; van Beers et al. 2013). The co-involvement of PAD4 with PAD2 is suggested by the observation that histone H3 is also deiminated in both human MS brain and in the ND4 murine model for demyelinating disease (Mastronardi et al. 2006; Wood et al. 2008; Mastronardi and Moscarello 2008).

The animal model most pertinent to this discussion at present is a transgenic mouse line (PD2), containing multiple copies of PAD2 cDNA under the control of the MBP promoter (Musse et al. 2008b). Using previously established criteria, clinical scores of neurological deficits were more severe in heterozygous PD2 mice than in their normal littermates. The increase in PAD2 expression and activity in white matter was demonstrated by immunohistochemistry, reverse-transcriptase PCR (polymerase chain reaction), enzyme activity assays, and by the increased deimination of MBP. Light and electron microscopy revealed more severe focal demyelination and thinner myelin in the homozygous compared with heterozygous PD2 mice. Quantitation of the disease-associated molecules GFAP and the glycoprotein marker CD68 ("cluster of differentiation 68"), as measured by immunoslot blots, were indicative of astrocytosis and macrophage activation. However, there was no apparent increase in lymphocytic infiltration, showing that the PD2 model was primarily one of neurodegenerations. Concurrently, the elevated levels of pro-inflammatory cytokine TNF-α and nuclear histone deimination supported the contribution of increased PAD4 activity to the domino-demyelination cascade depicted in Fig. 19.5. Although a *Padi2*-knockout mouse line had previously been described in which the amount of citrullination in the CNS was diminished but in which there was no demyelination (Raijmakers et al. 2006), the PD2 overexpressing line supports the idea of a causal link.

The final animal model to note is one for toxic yet reversible demyelination, induced chemically by the copper-chelating compound cuprizone (*bis*-cyclohexanone oxaldihydrazone) (Torkildsen et al. 2008; Gudi et al. 2014; Praet et al. 2014). When cuprizone is added to the diet of rodents, apoptosis of oligodendrocytes is induced (potentially by affecting copper-dependent mitochondrial enzymes), and demyelination ensues. Restoration of a normal diet results in remyelination by resident oligodendrocyte progenitor cells. Although this model is not considered to reflect MS per se, it can serve to examine early molecular and cellular events that lead to oligodendrocyte apoptosis and potential mechanisms of remyelination (Miller and Fyffe-Maricich 2010). At time of writing, the presence of citrullinated proteins has not yet been reported in this animal model but could confidently be anticipated, and this matter should be explored by combined histochemical, immuno-microscopical, mass spectrometric imaging and magnetic resonance imaging approaches to probe the time course and regional heterogeneity of demyelination patterns (Tagge et al. 2016). The recent demonstration of an "axo-myelinic synapse" and transient increases in intracellular calcium concentrations via NMDA receptors could provide further insight into where and how PAD2 is activated in vivo and on its immediate protein targets (Micu et al. 2006, 2016). Such studies would provide further evidence for MBP deimination being an early molecular event that precipitates demyelination, as opposed to being a bystander effect.

19.5.3 Myelin Membrane Vesiculation Proceeds Radially During the Demyelination Cascade

The domino-demyelination scheme depicted in Fig. 19.5 incorporates phenomena such as myelin vesiculation due to the loss of the net positive charge of MBP by hyper-deimination, leading to decreased membrane adhesion and a positive feedback loop of increasing PAD2 activity. The vesiculation of myelin has been suggested several times in the past on the basis of quite disparate observations, as indicated above (Ohler et al. 2004; Wood et al. 2008). It has been demonstrated in a recent morphological study of brain tissue from EAE and cuprizone-treated mouse models and from rats that had been injected with anti-MOG (myelin oligodendrocyte glycoprotein) or anti-AQP4 (aquaporin 4), in addition to cuprizone-treated animals (Weil et al. 2016). The anti-MOG and anti-AQP4 treatments induce features of neuromyelitis optica (NMO), which is a proven autoimmune disease targeting glia, although not an appropriate model for MS (Popescu and Lucchinetti 2012, 2016). Nevertheless, Weil et al. used high-pressure freezing for optimal preservation of brain tissue for sectioning before transmission electron microscopy and observed in all models that demyelination proceeded by fragmentation that started at the inner tongue and moved outwards. They correlated the demyelination with loss of MBP (and thus of MBP's "adhesive function") as observed by antibody labelling and with increases in intracellular calcium levels using the ionophore ionomycin to restore partially MBP's membrane association. The molecular mechanism that they proffered was that the calcium ions competed with MBP for $PI(4,5)P_2$ and displaced the protein from the membrane.

In light of the extensive literature already reviewed here, it can reasonably and confidently be predicted that the increased intracellular calcium levels would activate PAD2 that would then deiminate MBP where it is most accessible, at inner loops, and that this would be the mechanism making the protein lose its adhesive function and disrupting its extensive networking. We argue here that the simple mechanism proposed by Weil et al. of MBP displacement by Ca^{2+} does not reflect the full range of molecular events (Fig. 19.5). There are several reasons to say so. First, the inner leaflet of the myelin membrane comprises other negatively charged phospholipids in far greater proportion than phosphoinositides with which MBP still interacts, as would any other peripheral membrane protein (Inouye and Kirschner 1988; Bates et al. 2003; Lee et al. 2014; Whited and Johs 2015; Shaharabani et al. 2016). Any diminished interaction with $PI(4,5)P_2$ would not displace the protein from the membrane (Boggs et al. 2012). Second, zinc (Zn^{2+}) is another divalent cation that is essential in the brain and found at high concentrations (at 50 μM in myelin) and stabilizes myelin by binding to MBP rather than dislodging the protein from the membrane (Earl et al. 1988; Tsang et al. 1997; Smith et al. 2010; Baran et al. 2010). Even monovalent cations enhance vesicular membrane adhesion by MBP in vitro by shielding the charges of the phospholipid headgroups (Jo and Boggs 1995). So Ca^{2+} ions would not be expected simply to displace MBP, but do affect membrane structure. Third, deimination of MBP and GFAP has been demonstrated in EAE as noted above, in addition to the PD2 overexpressing mouse line that established a causal link with demyelination (Nicholas et al. 2005; Raijmakers et al. 2005, 2006; Musse et al. 2008b; Artemiadis and Anagnostouli 2010; Caprariello et al. 2012; Wei et al. 2013; Moscarello et al. 2013; Bradford et al. 2014a). Fourth, deimination has been implicated in optical neurodegeneration in both humans and in the spontaneously demyelinating ND4 mouse line (Bhattacharya 2009; Enriquez-Algeciras et al. 2013; Ding et al. 2014). Finally, the effects of MBP deimination on one of its many functions (i.e. membrane adhesion) have already been studied extensively and have led to detailed molecular mechanisms of physical destabilization and priming of an autoimmune response (Fig. 19.5) (Wood and Moscarello 1989; Whitaker and Mitchell 1996; Boggs et al. 1999, 2005, 2011; Cao et al. 1999; Bates et al. 2003; D'Souza and Moscarello 2006; Musse et al. 2006; Harauz and Musse 2007; Musse and Harauz 2007; Ahmed et al. 2010; Vassall et al. 2016). These biophysical considerations already begin to explain how vesiculation occurs.

Again, it would be worthwhile to investigate and confirm PAD2 up-regulation and MBP deimination directly in other demyelination models such as cuprizone treatment and various PAD-knockout lines (Lewis and Nacht 2016), as well as in clinical situations of *all* demyelinating syndromes, including MS (Banwell et al. 2007; Oguz et al. 2009; Popescu and Lucchinetti 2012; Hardy et al. 2016). It cannot be overemphasized that there is overwhelming evidence that "demyelinating disease results from aberrant or excessive post-translational processing [of MBP]" (Wood and Moscarello 1997) and that the "post-translational modifications [of MBP] have an important role in both maintaining [myelin] membrane integrity and regulating metabolic reactions" (Moscarello 2014) (see also Harauz and Musse 2007; Harauz et al. 2009; Harauz and Boggs 2013). One cannot discuss tauopathies without referring to tau hype-phosphorylation (Šimic et al. 2016) or epigenetics without discussing histone modifications (van den Elsen et al. 2014; Wenderski and Maze 2016). Hyper-deimination of MBP caused by aberrant Ca^{2+} influx and up-regulation of PAD2 and PAD4 is equally a significant event in demyelinating diseases.

19.5.4 Potential Therapeutic Measures Targeting PAD Expression and Activity

One of the first anti-PAD compounds to be proposed was paclitaxel (a drug commonly used in cancer chemotherapy and clinically known as Taxol®), whose methyl ester sidechain (TSME) shares structural features with a synthetic PAD substrate, viz. the benzoyl-L-arginine ethyl ester (BAEE) (Pritzker and Moscarello 1998; Musse et al. 2008c). This compound has been shown to alleviate clinical symptoms in spontaneously demyelinating ND4 mice, potentially by competitive inhibition, but keeping in mind that the compound has many side-effects that might have been responsible for the amelioration of phenotype (Moscarello et al. 2002a). Taxol's beneficial effect in these mice was significantly enhanced in combination with vitamin B12 as an exogenous methyl donor to promote re-methylation of the gene promoter for PAD2, suggesting that PAD2 indeed was being the affected target (Mastronardi et al. 2007b). A similar synergy had previously been observed with interferon-β and vitamin B12 (Mastronardi et al. 2004).

More recently, the small molecule 2-chloroacetamidine (a PAD active-site inhibitor) dramatically attenuated disease in a number of animal models that were each reminiscent of the four major patterns of human MS: autoimmune, ND4, and the most recent PD2 mouse line (Moscarello et al. 2013). Because this compound covalently modifies the enzyme, it is not suitable for clinical use, but several designer probes were then identified with potential for future evaluation (Wei et al. 2013; Bello et al. 2013). These compounds continue to be examined not only for MS but also other chronic neurodegenerative syndromes such as glaucoma and Parkinson's, Alzheimer's, and Creutzfeldt-Jakob diseases (Witalison et al. 2015; Subramanian et al. 2015; Lewis and Nacht 2016; Tu et al. 2016; Kuzina et al. 2016). For MS, alternative disease-modifying therapies that are not immunomodulatory or anti-inflammatory, especially in the early stages of the disease after initial diagnosis, could contribute significantly to the health and well-being of individual patients in the context of personalized medicine and maintenance of overall brain health (Giovannoni et al. 2016; Simone and Chitnis 2016; Winkelmann et al. 2016).

19.6 Deimination of MBP in Ageing and MS Brain

This chapter had been essentially completed upon the publication of a new proteomic (mass spectrometric) analysis of the PTMs of MBP from autopsied brain material of 8 MS patients and 21 controls (Friedrich et al. 2016). Although these authors did not fractionate the extracted MBP into either splice isoforms or charge components, in contrast to the first comprehensive proteomic work performed 13 years earlier (Kim et al. 2003), they extended their analysis to include a hitherto little-considered PTM of the protein—racemization of aspartic acid (L-Asp) in particular to alternate forms such as D-Asp, L-isoAsp, and D-isoAsp. Their data were also presented according to the subject's age as well as disease state. The racemization of residues such as Asp and Asn due to spontaneous cyclization reactions is especially prevalent in long-lived proteins such as lens proteins (Reissner and Aswad 2003; Truscott and Friedrich 2016). Myelin-associated proteins, including MBP, are considered to be long-lived (Toyama et al. 2013).

Both racemization and deimination per se were observed to be generally higher in MBP from older subjects. Twelve arginyl residues were converted to citrulline (hR25, hR31, hR33, hR49, hR65, hR97, hR107, hR130, hR159, hR162, hR169, and hR170, human 18.5-kDa sequence numbering as in Fig. 19.3) in control MBP. Most of these sites (but not all of them) were more likely to be deiminated in MS-MBP. Because the protein extracts were not fractionated by chromatography, we do not know the average number of citrulline residues per protein and how this number varied with age. Additionally, residues hR43, hR79, and hR122 were deiminated in MBP from brains of MS patients. Residue hR54 was not identified as being deiminated in these samples, unlike in other studies of MS brain tissue (Kim et al. 2003), but we note that this site is also ADP-ribosylated (ibid.) and recalcitrant to enzymatic modification in vitro (Wood et al. 2008). Specifically, residue hR49 is noteworthy because its degree of deimination increased with age and moreover was accompanied by isomerization of hD48 to isoAsp48 in MS patients. It was suggested that this correlated pair of PTMs could potentially represent a novel antigenic form, in addition to representing an instance of crosstalk of PTMs, as we have mentioned in the main text above.

It is intriguing that the residues in 18.5-kDa MBP most likely to be modified in MS brain (hD34 converted from L-Asp34 to D-Asp34, D-isoAsp34, or L-isoAsp34 and the aforementioned pair hD48-hR49 to isoAsp48-Cit49) flank or lie just within the first of the protein's three membrane-anchoring α-helices (the α1-helix in Figs. 19.3 and 19.4) and are sequestered within the lipid bilayer as previously demonstrated spectroscopically (Bates et al. 2003; Harauz et al. 2009, 2013; Harauz and Libich 2009). Similarly, residue hD82 is significantly racemized in greater proportion in aged control MBP and MS-MBP and lies within the α2-helix, a major immunodominant epitope (Bates et al. 2004; Musse et al. 2006; Musse and Harauz 2007; Ahmed et al. 2010). The residue hQ147 that is more likely to be deamidated in MS-MBP lies within the α3-helix, another potential immunodominant epitope and the primary Ca^{2+}-calmodulin binding site (Libich et al. 2003, 2010; Majava et al. 2008; Libich and Harauz 2008a; Homchaudhuri et al. 2010). Friedrich et al. state that "Sites of racemization are generated by spontaneous processes which occur more rapidly in unstructured regions of a protein" (Friedrich et al. 2016). Their observation of elevated degrees of racemization in the normally most structured regions of 18.5-kDa MBP, in or proximal to its membrane-anchoring α-helical MoRFs, perhaps reflects the aberrant extent of dissociation of the protein from the membrane in MS and the altered environment that it experiences (Figs. 19.4 and 19.5) (cf., (Reissner and Aswad 2003)).

Continued and detailed proteomic analyses are needed to catalogue comprehensively the PTMs of all MBP splice isoforms in healthy and diseased brains as a function of age. The study of Friedrich et al. is thus a welcome addition to our body of knowledge (Friedrich et al. 2016), but further information is needed on the number of individual MBP species and the pattern of PTMs that define each one. Particular attention must be given to normal-appearing white matter, not just lesions per se. Continued biochemical and biophysical investigations of the effects of these PTMs on MBP's conformational states, and on its interactions with membranes and proteins, are also needed to define more fully this protein's versatile and multifunctional nature (Vassall et al. 2015a, 2016).

19.7 Conclusions

"Whether or not the presence of citrulline in [myelin basic protein] has any functional significance remains to be determined" (Finch et al. 1971). This statement still holds true, to some extent. Deimination is an impactful modification of any protein and it cannot be without purpose. The deimination of MBP in normal, healthy CNS myelin has been suggested to reflect normal events of plasticity and restoration, but this conjecture remains to be proven. The hyper-deimination of MBP in the CNS represents an important early molecular change that precipitates the destabilization of myelin in MS brain white matter and contributes to feedback loops that accelerate myelin destruction both by structural destabilization and by continued up-regulation of the peptidylarginine deiminases (Fig. 19.5).

The hyper-deimination of MBP in MS might reflect, in part, attempts at remyelination. These events ultimately fail because the molecular and cellular backdrop of the developing brain differs from that of the adult brain and because of the overwhelming cascade of consequences arising from the hyper-deimination: the PADs modify a host of target proteins in the CNS, with eventual apoptosis of oligodendrocytes.

We cannot prevent MS because we do not know the root causes. Nevertheless, the development of tissue-specific inhibitors of PAD expression and activity represents a promising avenue for treating the disease by ameliorating the symptoms caused by demyelination and neurodegeneration and enhancing remyelination and repair. We stress that a more detailed developmental mapping of myelin deimination and modification of crosstalk in situ, although technically difficult, is essential to our understanding of both the normal CNS white matter and the myriad neurodegenerative diseases that affect it, not just MS.

Acknowledgements Our research on MBP deimination in MS has been funded largely by the Multiple Sclerosis Society of Canada (especially Studentships and Fellowships) and by the Canadian Institutes of Health Research (operating grants to GH and colleagues). Overall program support has been by the Natural Sciences and Engineering Research Council of Canada (Discovery Grant RG121541 to GH) and the Canada Research Chairs Program (GH is a Tier 1 CRC in Myelin Biology). The author is grateful to Drs. Vladimir Bamm (Molecular and Cellular Biology, Guelph), Joan Boggs (Research Institute, Hospital for Sick Children, Toronto), Lillian DeBruin (Wilfrid Laurier University, Waterloo), Lakshmi Kotra (Pharmaceutical Sciences, Toronto), Scott Ryan (Molecular and Cellular Biology, Guelph), and Peter Stys (Hotchkiss Brain Institute, Calgary) for helpful comments on the manuscript and especially to Ms. Zoë Martos for her rendition of myelin in Fig. 19.1. There are no conflicts of interests to declare.

References

Aggarwal, S., Yurlova, L., Snaidero, N., Reetz, C., Frey, S., Zimmermann, J., Pähler, G., Janshoff, A., Friedrichs, J., Muller, D. J., Goebel, C., & Simons, M. (2011). A size barrier limits protein diffusion at the cell surface to generate lipid-rich myelin-membrane sheets. *Developmental Cell, 21*(3), 445–456. doi:10.1016/j.devcel.2011.08.001.

Aggarwal, S., Snaidero, N., Pahler, G., Frey, S., Sanchez, P., Zweckstetter, M., Janshoff, A., Schneider, A., Weil, M. T., Schaap, I. A., Gorlich, D., & Simons, M. (2013). Myelin membrane assembly is driven by a phase transition of myelin basic proteins into a cohesive protein meshwork. *PLoS Biology, 11*(6), e1001577. doi:10.1371/journal.pbio.1001577.

Ahmed, M. A. M., Bamm, V. V., Shi, L., Steiner-Mosonyi, M., Dawson, J. F., Brown, L., Harauz, G., & Ladizhansky, V. (2009). Induced secondary structure and polymorphism in an intrinsically disordered structural linker of the CNS: Solid-state NMR and FTIR spectroscopy of myelin basic protein bound to actin. *Biophysical Journal, 96*(1), 180–191. doi:10.1016/j.bpj.2008.10.003.

Ahmed, M. A. M., Bamm, V. V., Harauz, G., & Ladizhansky, V. (2010). Solid-state NMR spectroscopy of membrane-associated myelin basic protein—Conformation and dynamics of an immunodominant epitope. *Biophysical Journal, 99*(4), 1247–1255. doi:10.1016/j.bpj.2010.06.022.

Arita, K., Shimizu, T., Hashimoto, H., Hidaka, Y., Yamada, M., & Sato, M. (2006). Structural basis for histone N-terminal recognition by human peptidylarginine deiminase 4. *Proceedings of the National Academy of Sciences of the United States of America, 103*(14), 5291–5296.

Artemiadis, A. K., & Anagnostouli, M. C. (2010). Apoptosis of oligodendrocytes and post-translational modifications of myelin basic protein in multiple sclerosis: Possible role for the early stages of multiple sclerosis. *European Neurology, 63*(2), 65–72. doi:10.1159/000272940.

Atkins, C. M., Chen, S. J., Klann, E., & Sweatt, J. D. (1997). Increased phosphorylation of myelin basic protein during hippocampal long-term potentiation. *Journal of Neurochemistry, 68*(5), 1960–1967.

Atkins, C. M., Yon, M., Groome, N. P., & Sweatt, J. D. (1999). Regulation of myelin basic protein phosphorylation by mitogen-activated protein kinase during increased action potential firing in the hippocampus. *Journal of Neurochemistry, 73*(3), 1090–1097.

Aureli, M., Grassi, S., Prioni, S., Sonnino, S., & Prinetti, A. (2015). Lipid membrane domains in the brain. *Biochimica et Biophysica Acta, 1851*(8), 1006–1016. doi:10.1016/j.bbalip.2015.02.001.

Babu, M. M., Kriwacki, R. W., & Pappu, R. V. (2012). Structural biology. Versatility from protein disorder. *Science, 337*(6101), 1460–1461. doi:10.1126/science.1228775.

Bah, A., & Forman-Kay, J. D. (2016). Modulation of intrinsically disordered protein function by post-translational modifications. *The Journal of Biological Chemistry, 291*(13), 6696–6705. doi:10.1074/jbc.R115.695056.

Bakhti, M., Aggarwal, S., & Simons, M. (2014). Myelin architecture: Zippering membranes tightly together. *Cellular and Molecular Life Sciences, 71*(7), 1265–1277. doi:10.1007/s00018-013-1492-0.

Bamm, V. V., & Harauz, G. (2014). Hemoglobin as a source of iron overload in multiple sclerosis: Does multiple sclerosis share risk factors with vascular disorders? *Cellular and Molecular Life Sciences, 71*(10), 1789–1798. doi:10.1007/s00018-014-1570-y.

Bamm, V. V., Ahmed, M. A., & Harauz, G. (2010). Interaction of myelin basic protein with actin in the presence of dodecylphosphocholine micelles. *Biochemistry, 49*(32), 6903–6915. doi:10.1021/bi100308d.

Bamm, V. V., De Avila, M., Smith, G. S. T., Ahmed, M. A., & Harauz, G. (2011). Structured functional domains of myelin basic protein: Cross talk between actin polymerization and Ca(2+)-dependent calmodulin interaction. *Biophysical Journal, 101*(5), 1248–1256. doi:10.1016/j.bpj.2011.07.035.

Banwell, B., Ghezzi, A., Bar-Or, A., Mikaeloff, Y., & Tardieu, M. (2007). Multiple sclerosis in children: Clinical diagnosis, therapeutic strategies, and future directions. *Lancet Neurology, 6*(10), 887–902. doi:10.1016/S1474-4422(07)70242-9.

Baran, C., Smith, G. S. T., Bamm, V. V., Harauz, G., & Lee, J. S. (2010). Divalent cations induce a compaction of intrinsically disordered myelin basic protein. *Biochemical and Biophysical Research Communications, 391*(1), 224–229. doi:10.1016/j.bbrc.2009.11.036.

Barnett, M. H., & Prineas, J. W. (2004). Relapsing and remitting multiple sclerosis: Pathology of the newly forming lesion. *Annals of Neurology, 55*(4), 458–468.

Bates, I. R., Matharu, P., Ishiyama, N., Rochon, D., Wood, D. D., Polverini, E., Moscarello, M. A., Viner, N. J., & Harauz, G. (2000). Characterization of a recombinant murine 18.5-kDa myelin basic protein. *Protein Expression and Purification, 20*(2), 285–299.

Bates, I. R., Libich, D. S., Wood, D. D., Moscarello, M. A., & Harauz, G. (2002). An Arg/Lys-->Gln mutant of recombinant murine myelin basic protein as a mimic of the deiminated form implicated in multiple sclerosis. *Protein Expression and Purification, 25*(2), 330–341.

Bates, I. R., Boggs, J. M., Feix, J. B., & Harauz, G. (2003). Membrane-anchoring and charge effects in the interaction of myelin basic protein with lipid bilayers studied by site-directed spin labeling. *The Journal of Biological Chemistry, 278*(31), 29041–29047.

Bates, I. R., Feix, J. B., Boggs, J. M., & Harauz, G. (2004). An immunodominant epitope of myelin basic protein is an amphipathic alpha-helix. *The Journal of Biological Chemistry, 279*(7), 5757–5764.

Baumann, N., & Pham-Dinh, D. (2001). Biology of oligodendrocyte and myelin in the mammalian central nervous system. *Physiological Reviews, 81*(2), 871–927.

Bello, A. M., Wasilewski, E., Wei, L., Moscarello, M. A., & Kotra, L. P. (2013). Interrogation of the active sites of protein arginine deiminases (PAD1, -2, and -4) using designer probes. *ACS Medicinal Chemistry Letters, 4*(2), 249–253. doi:10.1021/ml300377d.

Belogurov, A. A., Jr., Kurkova, I. N., Friboulet, A., Thomas, D., Misikov, V. K., Zakharova, M. Y., Suchkov, S. V., Kotov, S. V., Alehin, A. I., Avalle, B., Souslova, E. A., Morse, H. C., III, Gabibov, A. G., & Ponomarenko, N. A. (2008). Recognition and degradation of myelin basic protein peptides by serum autoantibodies: Novel biomarker for multiple sclerosis. *Journal of Immunology, 180*(2), 1258–1267.

Beniac, D. R., Luckevich, M. D., Czarnota, G. J., Tompkins, T. A., Ridsdale, R. A., Ottensmeyer, F. P., Moscarello, M. A., & Harauz, G. (1997). Three-dimensional structure of myelin basic protein. I. Reconstruction via angular reconstitution of randomly oriented single particles. *The Journal of Biological Chemistry, 272*(7), 4261–4268.

Beniac, D. R., Wood, D. D., Palaniyar, N., Ottensmeyer, F. P., Moscarello, M. A., & Harauz, G. (1999). Marburg's variant of multiple sclerosis correlates with a less compact structure of myelin basic protein. *Molecular Cell Biology Research Communications, 1*(1), 48–51.

Beniac, D. R., Wood, D. D., Palaniyar, N., Ottensmeyer, F. P., Moscarello, M. A., & Harauz, G. (2000). Cryoelectron microscopy of protein-lipid complexes of human myelin basic protein charge isomers differing in degree of citrullination. *Journal of Structural Biology, 129*(1), 80–95.

Bhattacharya, S. K. (2009). Retinal deimination in aging and disease. *IUBMB Life, 61*(5), 504–509. doi:10.1002/iub.184.

Bicker, K. L., & Thompson, P. R. (2013). The protein arginine deiminases: Structure, function, inhibition, and disease. *Biopolymers, 99*(2), 155–163. doi:10.1002/bip.22127.

Boggs, J. M. (2006). Myelin basic protein: A multifunctional protein. *Cellular and Molecular Life Sciences, 63*(17), 1945–1961.

Boggs, J. M. (2008a). *Myelin basic protein*. Hauppauge, NY: Nova Science Publishers.

Boggs, J. M. (2008b). Myelin basic protein interactions with actin and tubulin *in vitro*: Binding, assembly, and regulation. In J. M. Boggs (Ed.), *Myelin basic protein* (pp. 149–167). New York: Nova Science Publishers.

Boggs, J. M., & Moscarello, M. A. (1978). Structural organization of the human myelin membrane. *Biochimica et Biophysica Acta, 515*(1), 1–21.

Boggs, J. M., Moscarello, M. A., & Papahadjopoulos, D. (1982). Structural organization of myelin: Role of lipid-protein interactions determined in model systems. In P. C. Jost & O. H. Griffith (Eds.), *Lipid-protein interactions* (pp. 1–51). New York: Wiley-Interscience.

Boggs, J. M., Yip, P. M., Rangaraj, G., & Jo, E. (1997). Effect of posttranslational modifications to myelin basic protein on its ability to aggregate acidic lipid vesicles. *Biochemistry, 36*(16), 5065–5071.

Boggs, J. M., Rangaraj, G., Koshy, K. M., Ackerley, C., Wood, D. D., & Moscarello, M. A. (1999). Highly deiminated isoform of myelin basic protein from multiple sclerosis brain causes fragmentation of lipid vesicles. *Journal of Neuroscience Research, 57*(4), 529–535.

Boggs, J. M., Rangaraj, G., Hill, C. M. D., Bates, I. R., Heng, Y. M., & Harauz, G. (2005). Effect of arginine loss in myelin basic protein, as occurs in its deiminated charge isoform, on mediation of actin polymerization and actin binding to a lipid membrane *in vitro*. *Biochemistry, 44*(9), 3524–3534.

Boggs, J. M., Bates, I. R., Musse, A. A., & Harauz, G. (2008). Interactions of the 18.5 kDa myelin basic protein with lipid bilayers: Studies by electron paramagnetic resonance spectroscopy and implications for generation of autoimmunity in multiple sclerosis. In J. M. Boggs (Ed.), *Myelin basic protein* (pp. 105–125). New York: Nova Science Publishers.

Boggs, J. M., Rangaraj, G., Heng, Y. M., Liu, Y., & Harauz, G. (2011). Myelin basic protein binds microtubules to a membrane surface and to actin filaments *in vitro*: Effect of phosphorylation and deimination. *Biochimica et Biophysica Acta-Biomembranes, 1808*(3), 761–773. doi:10.1016/j.bbamem.2010.12.016.

Boggs, J. M., Rangaraj, G., & Dicko, A. (2012). Effect of phosphorylation of phosphatidylinositol on myelin basic protein-mediated binding of actin filaments to lipid bilayers *in vitro*. *Biochimica et Biophysica Acta-Biomembranes, 1818*(9), 2217–2227. doi:10.1016/j.bbamem.2012.04.006.

Bradford, C. M., Nicholas, A. P., Woodroofe, N., & Cross, A. K. (2014a). Deimination in multiple sclerosis and experimental autoimmune encephalomyelitis. In A. P. Nicholas & S. K. Bhattacharya (Eds.), *Protein Deimination in human health and disease* (pp. 165–185). New York: Springer.

Bradford, C. M., Ramos, I., Cross, A. K., Haddock, G., McQuaid, S., Nicholas, A. P., & Woodroofe, M. N. (2014b). Localisation of citrullinated proteins in normal appearing white matter and lesions in the central nervous system in multiple sclerosis. *Journal of Neuroimmunology, 273*(1-2), 85–95. doi:10.1016/j.jneuroim.2014.05.007.

Calabrese, R., Zampieri, M., Mechelli, R., Annibali, V., Guastafierro, T., Ciccarone, F., Coarelli, G., Umeton, R., Salvetti, M., & Caiafa, P. (2012). Methylation-dependent PAD2 upregulation in multiple sclerosis peripheral blood. *Multiple Sclerosis, 18*(3), 299–304. doi:10.1177/1352458511421055.

Campagnoni, A. T., & Campagnoni, C. W. (2008). The properties and functions of the GOLLI myelin basic proteins. In J. M. Boggs (Ed.), *Myelin basic protein* (pp. 1–17). New York: Nova Science Publishers.

Campagnoni, A. T., Pribyl, T. M., Campagnoni, C. W., Kampf, K., Amur-Umarjee, S., Landry, C. F., Handley, V. W., Newman, S. L., Garbay, B., & Kitamura, K. (1993). Structure and developmental regulation of Golli-mbp, a 105-kilobase gene that encompasses the myelin basic protein gene and is expressed in cells in the oligodendrocyte lineage in the brain. *The Journal of Biological Chemistry, 268*(7), 4930–4938.

Cao, L., Goodin, R., Wood, D., Moscarello, M. A., & Whitaker, J. N. (1999). Rapid release and unusual stability of immunodominant peptide 45-89 from citrullinated myelin basic protein. *Biochemistry, 38*(19), 6157–6163. doi:10.1021/bi982960s.

Caprariello, A. V., Mangla, S., Miller, R. H., & Selkirk, S. M. (2012). Apoptosis of oligodendrocytes in the central nervous system results in rapid focal demyelination. *Annals of Neurology, 72*(3), 395–405. doi:10.1002/ana.23606.

Carrillo-Vico, A., Leech, M. D., & Anderton, S. M. (2010). Contribution of myelin autoantigen citrullination to T cell autoaggression in the central nervous system. *Journal of Immunology, 184*(6), 2839–2846. doi:10.4049/jimmunol.0903639.

Castelfranco, A. M., & Hartline, D. K. (2016). Evolution of rapid nerve conduction. *Brain Research*. doi:10.1016/j.brainres.2016.02.015.

Chang, K. J., Redmond, S. A., & Chan, J. R. (2016). Remodeling myelination: Implications for mechanisms of neural plasticity. *Nature Neuroscience, 19*(2), 190–197. doi:10.1038/nn.4200l.

Cheifetz, S., & Moscarello, M. A. (1985). Effect of bovine basic protein charge microheterogeneity on protein-induced aggregation of unilamellar vesicles containing a mixture of acidic and neutral phospholipids. *Biochemistry, 24*(8), 1909–1914.

Cheifetz, S., Moscarello, M. A., & Deber, C. M. (1984). NMR investigation of the charge isomers of bovine myelin basic protein. *Archives of Biochemistry and Biophysics, 233*(1), 151–160.

Chernoff, G. F. (1981). *Shiverer*: An autosomal recessive mutant mouse with myelin deficiency. *The Journal of Heredity, 72*(2), 128.

Chou, F. C., Chou, C. H., Shapira, R., & Kibler, R. F. (1976). Basis of microheterogeneity of myelin basic protein. *The Journal of Biological Chemistry, 251*(9), 2671–2679.

Christensen, P. C., Brideau, C., Poon, K. W., Doring, A., Yong, V. W., & Stys, P. K. (2014). High-resolution fluorescence microscopy of myelin without exogenous probes. *NeuroImage, 87*, 42–54. doi:10.1016/j.neuroimage.2013.10.050.

Christophorou, M. A., Castelo-Branco, G., Halley-Stott, R. P., Oliveira, C. S., Loos, R., Radzisheuskaya, A., Mowen, K. A., Bertone, P., Silva, J. C., Zernicka-Goetz, M., Nielsen, M. L., Gurdon, J. B., & Kouzarides, T. (2014). Citrullination regulates pluripotency and histone H1 binding to chromatin. *Nature, 507*(7490), 104–108. doi:10.1038/nature12942.

Cicerone, M. (2016). Molecular imaging with CARS micro-spectroscopy. *Current Opinion in Chemical Biology, 33*, 179–185. doi:10.1016/j.cbpa.2016.05.010.

Clancy, K. W., Weerapana, E., & Thompson, P. R. (2016). Detection and identification of protein citrullination in complex biological systems. *Current Opinion in Chemical Biology, 30*(1), 1–6. doi:10.1016/j.cbpa.2015.10.014.
Compston, A., & Coles, A. (2002). Multiple sclerosis. *Lancet, 359*(9313), 1221–1231. doi:10.1016/S0140-6736(02)08220-X.
Cygan, N. K., Scheinost, J. C., Butters, T. D., & Wentworth, P., Jr. (2011). Adduction of cholesterol 5,6-secosterol aldehyde to membrane-bound myelin basic protein exposes an immunodominant epitope. *Biochemistry, 50*(12), 2092–2100. doi:10.1021/bi200109q.
D'Este, E., Kamin, D., Velte, C. J., Gottfert, F., Simons, M., & Hell, S. W. (2016). Subcortical cytoskeleton periodicity throughout the nervous system. *Scientific Reports, 6*, 22741. doi:10.1038/srep22741.
D'Souza, C. A., & Moscarello, M. A. (2006). Differences in susceptibility of MBP charge isomers to digestion by stromelysin-1 (MMP-3) and release of an immunodominant epitope. *Neurochemical Research, 31*(8), 1045–1054.
D'Souza, C. A., Wood, D. D., She, Y. M., & Moscarello, M. A. (2005). Autocatalytic cleavage of myelin basic protein: An alternative to molecular mimicry. *Biochemistry, 44*(38), 12905–12913.
De Avila, M., Vassall, K. A., Smith, G. S. T., Bamm, V. V., & Harauz, G. (2014). The proline-rich region of 18.5-kDa myelin basic protein binds to the SH3-domain of Fyn tyrosine kinase with the aid of an upstream segment to form a dynamic complex *in vitro*. *The Biochemical Journal, 34*(6), e00157. doi:10.1042/BSR20140149.
de Groot, M., Verhaaren, B. F., de Boer, R., Klein, S., Hofman, A., van der Lugt, A., Ikram, M. A., Niessen, W. J., & Vernooij, M. W. (2013). Changes in normal-appearing white matter precede development of white matter lesions. *Stroke, 44*(4), 1037–1042. doi:10.1161/STROKEAHA.112.680223.
de Hoz, L., & Simons, M. (2015). The emerging functions of oligodendrocytes in regulating neuronal network behaviour. *BioEssays, 37*(1), 60–69. doi:10.1002/bies.201400127.
de Monasterio-Schrader, P., Jahn, O., Tenzer, S., Wichert, S. P., Patzig, J., & Werner, H. B. (2012). Systematic approaches to central nervous system myelin. *Cellular and Molecular Life Sciences, 69*(17), 2879–2894. doi:10.1007/s00018-012-0958-9.
de Smit, M. J., Westra, J., Nesse, W., Vissink, A., van WInkelhoff, A. J., & Brouwer, E. (2014). Protein citullination: The link between rheumatoid arthritis and periodontitis? In A. P. Nicholas & S. K. Bhattacharya (Eds.), *Protein deimination in human health and disease* (pp. 69–82). New York: Springer.
DeBruin, L. S., & Harauz, G. (2007). White matter rafting—Membrane microdomains in myelin. *Neurochemical Research, 32*(2), 213–228.
DeBruin, L. S., Haines, J. D., Wellhauser, L. A., Radeva, G., Schonmann, V., Bienzle, D., & Harauz, G. (2005). Developmental partitioning of myelin basic protein into membrane microdomains. *Journal of Neuroscience Research, 80*(2), 211–225.
DeBruin, L. S., Haines, J. D., Bienzle, D., & Harauz, G. (2006). Partitioning of myelin basic protein into membrane microdomains in a spontaneously demyelinating mouse model for multiple sclerosis. *Biochemistry and Cell Biology, 84*(6), 993–1005.
Deraos, G., Chatzantoni, K., Matsoukas, M. T., Tselios, T., Deraos, S., Katsara, M., Papathanasopoulos, P., Vynios, D., Apostolopoulos, V., Mouzaki, A., & Matsoukas, J. (2008). Citrullination of linear and cyclic altered peptide ligands from myelin basic protein (MBP(87-99)) epitope elicits a Th1 polarized response by T cells isolated from multiple sclerosis patients: Implications in triggering disease. *Journal of Medicinal Chemistry, 51*(24), 7834–7842.
Di Salvo, C., Barreca, D., Lagana, G., di Bella, M., Tellone, E., Ficarra, S., & Bellocco, E. (2013). Myelin basic protein: Structural characterization of spherulites formation and preventive action of trehalose. *International Journal of Biological Macromolecules, 57*, 63–68. doi:10.1016/j.ijbiomac.2013.02.022.
Ding, D., Enriquez-Algericas, M., Bonilha, V. L., & Bhattacharya, S. K. (2014). Deimination in ocular tissues: Present and future. In A. P. Nicholas & S. K. Bhattacharya (Eds.), *Protein deimination in human health and disease* (pp. 295–304). New York: Springer.

Dupouey, P., Jacque, C., Bourre, J. M., Cesselin, F., Privat, A., & Baumann, N. (1979). Immunochemical studies of myelin basic protein in *shiverer* mouse devoid of major dense line of myelin. *Neuroscience Letters, 12*(1), 113–118.

Earl, C., Chantry, A., Mohammad, N., & Glynn, P. (1988). Zinc ions stabilise the association of basic protein with brain myelin membranes. *Journal of Neurochemistry, 51*(3), 718–724.

Enriquez-Algeciras, M., Ding, D., Mastronardi, F. G., Marc, R. E., Porciatti, V., & Bhattacharya, S. K. (2013). Deimination restores inner retinal visual function in murine demyelinating disease. *The Journal of Clinical Investigation, 123*(2), 646–656. doi:10.1172/JCI64811.

Ferretti, P., Langun, S., Pong, U. K., & Raivich, G. (2014). The role of deimination as a response to trauma and hypoxic injury in the developing CNS. In A. P. Nicholas & S. K. Bhattacharya (Eds.), *Protein deimination in human health and disease* (pp. 281–294). New York: Springer.

Finch, P. R., Wood, D. D., & Moscarello, M. A. (1971). The presence of citrulline in a myelin protein fraction. *FEBS Letters, 15*(2), 145–148.

Franklin, R. J. (2002). Why does remyelination fail in multiple sclerosis? *Nature Reviews. Neuroscience, 3*(9), 705–714.

Franklin, R. J., & Gallo, V. (2014). The translational biology of remyelination: Past, present, and future. *Glia, 62*(11), 1905–1915. doi:10.1002/glia.22622.

Fraser, P. E., Moscarello, M. A., Rand, R. P., & Deber, C. M. (1986). Spontaneous vesicularization of myelin lipids is counteracted by myelin basic protein. *Biochimica et Biophysica Acta, 863*(2), 282–288.

Frid, K., Einstein, O., Friedman-Levi, Y., Binyamin, O., Ben-Hur, T., & Gabizon, R. (2015). Aggregation of MBP in chronic demyelination. *Annals of Clinical Translational Neurology, 2*(7), 711–721. doi:10.1002/acn3.207.

Friedrich, M. G., Hancock, S. E., Raftery, M. J., & Truscott, R. J. (2016). Isoaspartic acid is present at specific sites in myelin basic protein from multiple sclerosis patients: Could this represent a trigger for disease onset? *Acta Neuropathologica Communications, 4*(1), 83. doi:10.1186/s40478-016-0348-x.

Friess, M., Hammann, J., Unichenko, P., Luhmann, H. J., White, R., & Kirischuk, S. (2016). Intracellular ion signaling influences myelin basic protein synthesis in oligodendrocyte precursor cells. *Cell Calcium, 60*(5), 322–330. doi:10.1016/j.ceca.2016.06.009.

Fuhrmann, J., & Thompson, P. R. (2016). Protein arginine methylation and citrullination in epigenetic regulation. *ACS Chemical Biology, 11*(3), 654–668. doi:10.1021/acschembio.5b00942.

Fuhrmann, J., Clancy, K. W., & Thompson, P. R. (2015). Chemical biology of protein arginine modifications in epigenetic regulation. *Chemical Reviews, 115*(11), 5413–5461. doi:10.1021/acs.chemrev.5b00003.

Fulton, D., Paez, P. M., & Campagnoni, A. T. (2010a). The multiple roles of myelin protein genes during the development of the oligodendrocyte. *ASN Neuro, 2*(1), e00027. doi:10.1042/AN20090051.

Fulton, D., Paez, P. M., Fisher, R., Handley, V., Colwell, C. S., & Campagnoni, A. T. (2010b). Regulation of L-type Ca(++) currents and process morphology in white matter oligodendrocyte precursor cells by golli-myelin proteins. *Glia, 58*(11), 1292–1303. doi:10.1002/glia.21008.

Gielen, E., Baron, W., Vandeven, M., Steels, P., Hoekstra, D., & Ameloot, M. (2006). Rafts in oligodendrocytes: Evidence and structure-function relationship. *Glia, 54*(6), 499–512. doi:10.1002/glia.20406.

Giovannoni, G., Butzkueven, H., Dhib-Jalbut, S., Hobart, J., Kobelt, G., Pepper, G., Sormani, M. P., Thalheim, C., Traboulsee, A., & Vollmer, T. (2016). Brain health: Time matters in multiple sclerosis. *Multiple Sclerosis and Related Disorders, 9*(Supplement 1), S5–S48. doi:10.1016/j.msard.2016.07.003.

Gudi, V., Gingele, S., Skripuletz, T., & Stangel, M. (2014). Glial response during cuprizone-induced de- and remyelination in the CNS: Lessons learned. *Frontiers in Cellular Neuroscience, 8*, 73. doi:10.3389/fncel.2014.00073.

Gudmann, N. S., Hansen, N. U., Jensen, A. C., Karsdal, M. A., & Siebuhr, A. S. (2015). Biological relevance of citrullinations: Diagnostic, prognostic and therapeutic options. *Autoimmunity, 48*(2), 73–79. doi:10.3109/08916934.2014.962024.

Hamilton, N. B., Kolodziejczyk, K., Kougioumtzidou, E., & Attwell, D. (2016). Proton-gated Ca(2+)-permeable TRP channels damage myelin in conditions mimicking ischaemia. *Nature, 529*(7587), 523–527. doi:10.1038/nature16519.

Hanrieder, J., Malmberg, P., & Ewing, A. G. (2015). Spatial neuroproteomics using imaging mass spectrometry. *Biochimica et Biophysica Acta, 1854*(7), 718–731. doi:10.1016/j.bbapap.2014.12.026.

Hansen, B. E., Nielsen, C. H., Madsen, H. O., Ryder, L. P., Jakobsen, B. K., & Svejgaard, A. (2011). The HLA-DP2 protein binds the immunodominant epitope from myelin basic protein, MBP85-99, with high affinity. *Tissue Antigens, 77*(3), 229–234. doi:10.1111/j.1399-0039.2010.01614.x.

Harauz, G., & Boggs, J. M. (2013). Myelin management by the 18.5-kDa and 21.5-kDa classic myelin basic protein isoforms. *Journal of Neurochemistry, 125*(3), 334–361. doi:10.1111/jnc.12195.

Harauz, G., & Libich, D. S. (2009). The classic basic protein of myelin—Conserved structural motifs and the dynamic molecular barcode involved in membrane adhesion and protein-protein interactions. *Current Protein and Peptide Science, 10*(3), 196–215.

Harauz, G., & Musse, A. A. (2007). A tale of two citrullines—Structural and functional aspects of myelin basic protein deimination in health and disease. *Neurochemical Research, 32*(2), 137–158.

Harauz, G., Ishiyama, N., Hill, C. M. D., Bates, I. R., Libich, D. S., & Farès, C. (2004). Myelin basic protein—Diverse conformational states of an intrinsically unstructured protein and its roles in myelin assembly and multiple sclerosis. *Micron, 35*(7), 503–542.

Harauz, G., Ladizhansky, V., & Boggs, J. M. (2009). Structural polymorphism and multifunctionality of myelin basic protein. *Biochemistry, 48*(34), 8094–8104. doi:10.1021/bi901005f.

Harauz, G., Libich, D. S., Polverini, E., & Vassall, K. A. (2013). The classic protein of myelin—Conserved structural motifs and the dynamic molecular barcode involved in membrane adhesion, protein-protein interactions, and pathogenesis in multiple sclerosis. In B. M. Dunn (Ed.), *Advances in protein and peptide science* (pp. 1–53). Bentham Science Publishers. (e-book; http://benthamscience.com/ebooks/9781608054879/index.htm).

Hardy, T. A., Tobin, W. O., & Lucchinetti, C. F. (2016). Exploring the overlap between multiple sclerosis, tumefactive demyelination and Balo's concentric sclerosis. *Multiple Sclerosis, 22*(8), 986–992. doi:10.1177/1352458516641776.

Hauser, S. L., & Oksenberg, J. R. (2006). The neurobiology of multiple sclerosis: Genes, inflammation, and neurodegeneration. *Neuron, 52*(1), 61–76. doi:10.1016/j.neuron.2006.09.011.

Hauser, S. L., Chan, J. R., & Oksenberg, J. R. (2013). Multiple sclerosis: Prospects and promise. *Annals of Neurology, 74*(3), 317–327. doi:10.1002/ana.24009.

Hensen, S. M., & Pruijn, G. J. (2014). Methods for the detection of peptidylarginine deiminase (PAD) activity and protein citrullination. *Molecular & Cellular Proteomics, 13*(2), 388–396. doi:10.1074/mcp.R113.033746.

Hensvold, A. H., Reynisdottir, G., & Catrina, A. I. (2014). From citrullination to specific imunity and disease in rheumatoid arthritis. In A. P. Nicholas & S. K. Bhattacharya (Eds.), *Protein deimination in human health and disease* (pp. 25–40). New York: Springer.

Hill, C. M. D., & Harauz, G. (2005). Charge effects modulate actin assembly by classic myelin basic protein isoforms. *Biochemical and Biophysical Research Communications, 329*(1), 362–369.

Hill, C. M. D., Bates, I. R., White, G. F., Hallett, F. R., & Harauz, G. (2002). Effects of the osmolyte trimethylamine-*N*-oxide on conformation, self-association, and two-dimensional crystallization of myelin basic protein. *Journal of Structural Biology, 139*(1), 13–26.

Homchaudhuri, L., De Avila, M., Nilsson, S. B., Bessonov, K., Smith, G. S. T., Bamm, V. V., Musse, A. A., Harauz, G., & Boggs, J. M. (2010). Secondary structure and solvent accessibility of a calmodulin-binding C-terminal segment of membrane-associated myelin basic protein. *Biochemistry, 49*(41), 8955–8966. doi:10.1021/bi100988p.

Husted, C. (2006). Structural insight into the role of myelin basic protein in multiple sclerosis. *Proceedings of the National Academy of Sciences of the United States of America, 103*(12), 4339–4340.

Inouye, H., & Kirschner, D. A. (1988). Membrane interactions in nerve myelin: II. Determination of surface charge from biochemical data. *Biophysical Journal, 53*(2), 247–260.

Inouye, H., & Kirschner, D. A. (2015). Evolution of myelin ultrastructure and the major structural myelin proteins. *Brain Research*. doi:10.1016/j.brainres.2015.10.037.

Ishigami, A., & Maruyama, N. (2010). Importance of research on peptidylarginine deiminase and citrullinated proteins in age-related disease. *Geriatrics & Gerontology International, 10*(Supplement 1), S53–S58. doi:10.1111/j.1447-0594.2010.00593.x.

Ishigami, A., Ohsawa, T., Hiratsuka, M., Taguchi, H., Kobayashi, S., Saito, Y., Murayama, S., Asaga, H., Toda, T., Kimura, N., & Maruyama, N. (2005). Abnormal accumulation of citrullinated proteins catalyzed by peptidylarginine deiminase in hippocampal extracts from patients with Alzheimer's disease. *Journal of Neuroscience Research, 80*(1), 120–128.

Ishiyama, N., Bates, I. R., Hill, C. M. D., Wood, D. D., Matharu, P., Viner, N. J., Moscarello, M. A., & Harauz, G. (2001). The effects of deimination of myelin basic protein on structures formed by its interaction with phosphoinositide-containing lipid monolayers. *Journal of Structural Biology, 136*(1), 30–45.

Ishiyama, N., Hill, C. M. D., Bates, I. R., & Harauz, G. (2002). The formation of helical tubular vesicles by binary monolayers containing a nickel-chelating lipid and phosphoinositides in the presence of basic polypeptides. *Chemistry and Physics of Lipids, 114*(1), 103–111.

Jin, Z., Fu, Z., Yang, J., Troncosco, J., Everett, A. D., & Van Eyk, J. E. (2013). Identification and characterization of citrulline-modified brain proteins by combining HCD and CID fragmentation. *Proteomics, 13*(17), 2682–2691. doi:10.1002/pmic.201300064.

Jo, E., & Boggs, J. M. (1995). Aggregation of acidic lipid vesicles by myelin basic protein: Dependence on potassium concentration. *Biochemistry, 34*(41), 13705–13716.

Kattnig, D. R., Bund, T., Boggs, J. M., Harauz, G., & Hinderberger, D. (2012). Lateral self-assembly of 18.5-kDa myelin basic protein (MBP) charge component-C1 on membranes. *Biochimica et Biophysica Acta-Biomembranes, 1818*(11), 2636–2647. doi:10.1016/j.bbamem.2012.06.010.

Kim, D. I., & Suh, B. C. (2016). Differential interaction of beta2e with phosphoinositides: A comparative study between beta2e and MARCKS. *Channels (Austin, Tex.), 10*(3), 238–246. doi:10.1080/19336950.2015.1124311.

Kim, J. K., Mastronardi, F. G., Wood, D. D., Lubman, D. M., Zand, R., & Moscarello, M. A. (2003). Multiple sclerosis: An important role for post-translational modifications of myelin basic protein in pathogenesis. *Molecular & Cellular Proteomics, 2*(7), 453–462.

Kim, D. I., Kang, M., Kim, S., Lee, J., Park, Y., Chang, I., & Suh, B. C. (2015a). Molecular basis of the membrane interaction of the beta2e subunit of voltage-gated Ca(2+) channels. *Biophysical Journal, 109*(5), 922–935. doi:10.1016/j.bpj.2015.07.040.

Kim, D. I., Park, Y., Jang, D. J., & Suh, B. C. (2015b). Dynamic phospholipid interaction of beta2e subunit regulates the gating of voltage-gated Ca2+ channels. *The Journal of General Physiology, 145*(6), 529–541. doi:10.1085/jgp.201411349.

Kinloch, A., Lundberg, K., Wait, R., Wegner, N., Lim, N. H., Zendman, A. J., Saxne, T., Malmstrom, V., & Venables, P. J. (2008). Synovial fluid is a site of citrullination of autoantigens in inflammatory arthritis. *Arthritis and Rheumatism, 58*(8), 2287–2295. doi:10.1002/art.23618.

Kremer, D., Gottle, P., Hartung, H. P., & Kury, P. (2016). Pushing forward: Remyelination as the new frontier in CNS diseases. *Trends in Neurosciences, 39*(4), 246–263. doi:10.1016/j.tins.2016.02.004.

Kuzina, E. S., Kudriaeva, A. A., Glagoleva, I. S., Knorre, V. D., Gabibov, A. G., & Belogurov, A. A. (2016). Deimination of the myelin basic protein decelerates its proteasome-mediated metabolism. *Doklady. Biochemistry and Biophysics, 469*(1), 277–280. doi:10.1134/S1607672916040116.

Lahiri, S., Sun, N., Solis-Mezarino, V., Fedisch, A., Ninkovic, J., Feuchtinger, A., Götz, M., Walch, A., & Imhof, A. (2016). *In situ* detection of histone variants and modifications in mouse brain using imaging mass spectrometry. *Proteomics, 16*(3), 437–447. doi:10.1002/pmic.201500345.

Lamensa, J. W., & Moscarello, M. A. (1993). Deimination of human myelin basic protein by a peptidylarginine deiminase from bovine brain. *Journal of Neurochemistry, 61*(3), 987–996.

Lassmann, H. (2004). Cellular damage and repair in multiple sclerosis. In R. A. Lazzarini, J. W. Griffin, H. Lassman, K.-A. Nave, R. H. Miller, & B. D. Trapp (Eds.), *Myelin biology and disorders* (pp. 733–762). San Diego: Elsevier Academic Press.
Lassmann, H. (2014). Mechanisms of white matter damage in multiple sclerosis. *Glia, 62*(11), 1816–1830. doi:10.1002/glia.22597.
Lassmann, H., & van Horssen, J. (2011). The molecular basis of neurodegeneration in multiple sclerosis. *FEBS Letters, 585*(23), 3715–3723. doi:10.1016/j.febslet.2011.08.004.
Lassmann, H., Bruck, W., & Lucchinetti, C. (2001). Heterogeneity of multiple sclerosis pathogenesis: Implications for diagnosis and therapy. *Trends in Molecular Medicine, 7*(3), 115–121.
Lazarus, R. C., Buonora, J. E., Flora, M. N., Freedy, J. G., Holstein, G. R., Martinelli, G. P., Jacobowitz, D. M., & Mueller, G. P. (2015). Protein citrullination: A proposed mechanism for pathology in traumatic brain injury. *Frontiers in Neurology, 6*(1), 204-1–204-14. doi:10.3389/fneur.2015.00204.
Lazzarini, R. A., Griffin, J. W., Lassman, H., Nave, K.-A., Miller, R. H., & Trapp, B. D. (2004). *Myelin biology and disorders*. San Diego: Elsevier Academic Press.
Lee, D. W., Banquy, X., Kristiansen, K., Kaufman, Y., Boggs, J. M., & Israelachvili, J. N. (2014). Lipid domains control myelin basic protein adsorption and membrane interactions between model myelin lipid bilayers. *Proceedings of the National Academy of Sciences of the United States of America, 111*(8), E768–E775. doi:10.1073/pnas.1401165111.
Lee, D. W., Banquy, X., Kristiansen, K., Min, Y., Ramachandran, A., Boggs, J. M., & Israelachvili, J. N. (2015). Adsorption mechanism of myelin basic protein on model substrates and its bridging interaction between the two surfaces. *Langmuir, 31*(10), 3159–3166. doi:10.1021/acs.langmuir.5b00145.
Lewis, H. D., & Nacht, M. (2016). iPAD or PADi-'tablets' with therapeutic disease potential? *Current Opinion in Chemical Biology, 33*(1), 169–178. doi:10.1016/j.cbpa.2016.06.020.
Li, P., Li, M., Lindberg, M. R., Kennett, M. J., Xiong, N., & Wang, Y. (2010). PAD4 is essential for antibacterial innate immunity mediated by neutrophil extracellular traps. *The Journal of Experimental Medicine, 207*(9), 1853–1862. doi:10.1084/jem.20100239.
Libich, D. S., & Harauz, G. (2008a). Backbone dynamics of the 18.5-kDa isoform of myelin basic protein reveals transient alpha-helices and a calmodulin-binding site. *Biophysical Journal, 94*(12), 4847–4866.
Libich, D. S., & Harauz, G. (2008b). Solution NMR and CD spectroscopy of an intrinsically disordered, peripheral membrane protein: Evaluation of aqueous and membrane-mimetic solvent conditions for studying the conformational adaptability of the 18.5 kDa isoform of myelin basic protein (MBP). *European Biophysics Journal, 37*(6), 1015–1029.
Libich, D. S., Hill, C. M. D., Bates, I. R., Hallett, F. R., Armstrong, S., Siemiarczuk, A., & Harauz, G. (2003). Interaction of the 18.5-kDa isoform of myelin basic protein with Ca2+-calmodulin: Effects of deimination assessed by intrinsic Trp fluorescence spectroscopy, dynamic light scattering, and circular dichroism. *Protein Science, 12*(7), 1507–1521.
Libich, D. S., Ahmed, M. A. M., Zhong, L., Bamm, V. V., Ladizhansky, V., & Harauz, G. (2010). Fuzzy complexes of myelin basic protein: NMR spectroscopic investigations of a polymorphic organizational linker of the central nervous system. *Biochemistry and Cell Biology, 88*(2), 143–155. doi:10.1139/o09-123.
Locatelli, G., Wortge, S., Buch, T., Ingold, B., Frommer, F., Sobottka, B., Kruger, M., Karram, K., Buhlmann, C., Bechmann, I., Heppner, F. L., Waisman, A., & Becher, B. (2012). Primary oligodendrocyte death does not elicit anti-CNS immunity. *Nature Neuroscience, 15*(4), 543–550. doi:10.1038/nn.3062.
Lucchinetti, C., Brück, W., Parisi, J., Scheithauer, B., Rodriguez, M., & Lassmann, H. (2000). Heterogeneity of multiple sclerosis lesions: Implications for the pathogenesis of demyelination. *Annals of Neurology, 47*(6), 707–717.
Mahmoudabadi, G., Rajagopalan, K., Getzenberg, R. H., Hannenhalli, S., Rangarajan, G., & Kulkarni, P. (2013). Intrinsically disordered proteins and conformational noise: Implications in cancer. *Cell Cycle, 12*(1), 26–31. doi:10.4161/cc.23178.

Majava, V., Petoukhov, M. V., Hayashi, N., Pirila, P., Svergun, D. I., & Kursula, P. (2008). Interaction between the C-terminal region of human myelin basic protein and calmodulin: Analysis of complex formation and solution structure. *BMC Structural Biology, 8*(1), 10.

Majava, V., Wang, C., Myllykoski, M., Kangas, S. M., Kang, S. U., Hayashi, N., Baumgartel, P., Heape, A. M., Lubec, G., & Kursula, P. (2010). Structural analysis of the complex between calmodulin and full-length myelin basic protein, an intrinsically disordered molecule. *Amino Acids, 39*(1), 59–71. doi:10.1007/s00726-009-0364-2.

Makrygiannakis, D. (2014). From genes and environment to anti-citrulline immunity in rheumatoid arthritis: The role of the lungs. In A. P. Nicholas & S. K. Bhattacharya (Eds.), *Protein deimination in human health and disease* (pp. 83–96). New York: Springer.

Mao, A. H., Crick, S. L., Vitalis, A., Chicoine, C. L., & Pappu, R. V. (2010). Net charge per residue modulates conformational ensembles of intrinsically disordered proteins. *Proceedings of the National Academy of Sciences of the United States of America, 107*(18), 8183–8188. doi:10.1073/pnas.0911107107.

Mao, A. H., Lyle, N., & Pappu, R. V. (2013). Describing sequence-ensemble relationships for intrinsically disordered proteins. *The Biochemical Journal, 449*(2), 307–318. doi:10.1042/BJ20121346.

Marsh, J. A., & Forman-Kay, J. D. (2010). Sequence determinants of compaction in intrinsically disordered proteins. *Biophysical Journal, 98*(10), 2383–2390. doi:10.1016/j.bpj.2010.02.006.

Martenson, R. E., & Gaitonde, M. (1969a). Comparative studies of highly basic proteins of ox brain and rat brain. Microheterogeneity of basic encephalitogenic (myelin) protein. *Journal of Neurochemistry, 16*(3), 889–898.

Martenson, R. E., & Gaitonde, M. K. (1969b). Electrophoretic analysis of the highly basic proteins of the rat brain fraction which induces experimental allergic encephalomyelitis. *Journal of Neurochemistry, 16*(3), 333–347.

Mastronardi, F. G., & Moscarello, M. A. (2008). Deimination of myelin basic protein by PAD enzymes, and their role in multiple sclerosis. In J. M. Boggs (Ed.), *Myelin basic protein* (pp. 31–49). New York: Nova Science Publishers.

Mastronardi, F. G., Ackerley, C. A., Arsenault, L., Roots, B. I., & Moscarello, M. A. (1993). Demyelination in a transgenic mouse: A model for multiple sclerosis. *Journal of Neuroscience Research, 36*(3), 315–324. doi:10.1002/jnr.490360309.

Mastronardi, F. G., Ackerley, C. A., Roots, B. I., & Moscarello, M. A. (1996a). Loss of myelin basic protein cationicity in DM20 transgenic mice is dosage dependent. *Journal of Neuroscience Research, 44*(4), 301–307.

Mastronardi, F. G., Mak, B., Ackerley, C. A., Roots, B. I., & Moscarello, M. A. (1996b). Modifications of myelin basic protein in DM20 transgenic mice are similar to those in myelin basic protein from multiple sclerosis. *The Journal of Clinical Investigation, 97*(2), 349–358.

Mastronardi, F. G., Min, W., Wang, H., Winer, S., Dosch, M., Boggs, J. M., & Moscarello, M. A. (2004). Attenuation of experimental autoimmune encephalomyelitis and nonimmune demyelination by IFN-beta plus vitamin B12: Treatment to modify notch-1/sonic hedgehog balance. *Journal of Immunology, 172*(10), 6418–6426.

Mastronardi, F. G., Wood, D. D., Mei, J., Raijmakers, R., Tseveleki, V., Dosch, H. M., Probert, L., Casaccia-Bonnefil, P., & Moscarello, M. A. (2006). Increased citrullination of histone H3 in multiple sclerosis brain and animal models of demyelination: A role for tumor necrosis factor-induced peptidylarginine deiminase 4 translocation. *The Journal of Neuroscience, 26*(44), 11387–11396.

Mastronardi, F. G., Noor, A., Wood, D. D., Paton, T., & Moscarello, M. A. (2007a). Peptidylarginine deiminase 2 CpG island in multiple sclerosis white matter is hypomethylated. *Journal of Neuroscience Research, 85*(9), 2006–2016.

Mastronardi, F. G., Tsui, H., Winer, S., Wood, D. D., Selvanantham, T., Galligan, C., Fish, E. N., Dosch, H. M., & Moscarello, M. A. (2007b). Synergy between paclitaxel plus an exogenous methyl donor in the suppression of murine demyelinating diseases. *Multiple Sclerosis, 13*(5), 596–609.

Matsuo, A., Lee, G. C., Terai, K., Takami, K., Hickey, W. F., McGeer, E. G., & McGeer, P. L. (1997). Unmasking of an unusual myelin basic protein epitope during the process of myelin degeneration in humans: A potential mechanism for the generation of autoantigens. *The American Journal of Pathology, 150*(4), 1253–1266.

McLaurin, J., Ackerley, C. A., & Moscarello, M. A. (1993). Localization of basic proteins in human myelin. *Journal of Neuroscience Research, 35*(6), 618–628.

Micu, I., Jiang, Q., Coderre, E., Ridsdale, A., Zhang, L., Woulfe, J., Yin, X., Trapp, B. D., McRory, J. E., Rehak, R., Zamponi, G. W., Wang, W., & Stys, P. K. (2006). NMDA receptors mediate calcium accumulation in myelin during chemical ischaemia. *Nature, 439*(7079), 988–992. doi:10.1038/nature04474.

Micu, I., Plemel, J. R., Lachance, C., Proft, J., Jansen, A. J., Cummins, K., van Minnen, J., & Stys, P. K. (2016). The molecular physiology of the axo-myelinic synapse. *Experimental Neurology, 276*(1), 41–50. doi:10.1016/j.expneurol.2015.10.006.

Miller, R. H., & Fyffe-Maricich, S. L. (2010). Restoring the balance between disease and repair in multiple sclerosis: Insights from mouse models. *Disease Models & Mechanisms, 3*(9-10), 535–539. doi:10.1242/dmm.001958.

Miller, D. J., Duka, T., Stimpson, C. D., Schapiro, S. J., Baze, W. B., McArthur, M. J., Fobbs, A. J., Sousa, A. M., Sestan, N., Wildman, D. E., Lipovich, L., Kuzawa, C. W., Hof, P. R., & Sherwood, C. C. (2012). Prolonged myelination in human neocortical evolution. *Proceedings of the National Academy of Sciences of the United States of America, 109*(41), 16480–16485. doi:10.1073/pnas.1117943109.

Min, Y., Kristiansen, K., Boggs, J. M., Husted, C., Zasadzinski, J. A., & Israelachvili, J. (2009). Interaction forces and adhesion of supported myelin lipid bilayers modulated by myelin basic protein. *Proceedings of the National Academy of Sciences of the United States of America, 106*(9), 3154–3159. doi:10.1073/pnas.0813110106.

Min, Y., Alig, T. F., Lee, D. W., Boggs, J. M., Israelachvili, J. N., & Zasadzinski, J. A. (2011). Critical and off-critical miscibility transitions in model extracellular and cytoplasmic myelin lipid monolayers. *Biophysical Journal, 100*(6), 1490–1498. doi:10.1016/j.bpj.2011.02.009.

Möbius, W., Cooper, B., Kaufmann, W. A., Imig, C., Ruhwedel, T., Snaidero, N., Saab, A. S., & Varoqueaux, F. (2010). Electron microscopy of the mouse central nervous system. *Methods in Cell Biology, 96*(1), 475–512. doi:10.1016/S0091-679X(10)96020-2.

Möbius, W., Nave, K. A., & Werner, H. B. (2016). Electron microscopy of myelin: Structure preservation by high-pressure freezing. *Brain Research, 1641*(PtA), 92–100. doi:10.1016/j.brainres.2016.02.027.

Mor, D. E., Ugras, S. E., Daniels, M. J., & Ischiropoulos, H. (2016). Dynamic structural flexibility of alpha-synuclein. *Neurobiology of Disease, 88*, 66–74. doi:10.1016/j.nbd.2015.12.018.

Morell, P. (1984). *Myelin* (2nd ed.pp. 1–545). New York: Plenum Press.

Moscarello, M. A. (1997). Myelin basic protein, the "executive" molecule of the myelin membrane. In B. H. J. Juurlink, R. M. Devon, J. R. Doucette, A. J. Nazarali, D. J. Schreyer, & V. M. K. Verge (Eds.), *Cell biology and pathology of myelin: Evolving biological concepts and therapeutic approaches* (pp. 13–25). New York: Plenum Press.

Moscarello, M. A. (2014). Protein hypercitrullination in CNS demyelinating disease reversed by PAD inhibition. In A. P. Nicholas & S. K. Bhattacharya (Eds.), *Protein deimination in human health and disease* (pp. 187–218). New York: Springer.

Moscarello, M. A., Brady, G. W., Fein, D. B., Wood, D. D., & Cruz, T. F. (1986). The role of charge microheterogeneity of basic protein in the formation and maintenance of the multilayered structure of myelin: A possible role in multiple sclerosis. *Journal of Neuroscience Research, 15*(1), 87–99.

Moscarello, M. A., Pang, H., Pace-Asciak, C. R., & Wood, D. D. (1992). The N terminus of human myelin basic protein consists of C2, C4, C6, and C8 alkyl carboxylic acids. *The Journal of Biological Chemistry, 267*(14), 9779–9782.

Moscarello, M. A., Wood, D. D., Ackerley, C., & Boulias, C. (1994). Myelin in multiple sclerosis is developmentally immature. *The Journal of Clinical Investigation, 94*(1), 146–154.

Moscarello, M. A., Mak, B., Nguyen, T. A., Wood, D. D., Mastronardi, F., & Ludwin, S. K. (2002a). Paclitaxel (Taxol) attenuates clinical disease in a spontaneously demyelinating transgenic mouse and induces remyelination. *Multiple Sclerosis, 8*(2), 130–138.

Moscarello, M. A., Pritzker, L., Mastronardi, F. G., & Wood, D. D. (2002b). Peptidylarginine deiminase: A candidate factor in demyelinating disease. *Journal of Neurochemistry, 81*(2), 335–343.

Moscarello, M. A., Mastronardi, F. G., & Wood, D. D. (2007). The role of citrullinated proteins suggests a novel mechanism in the pathogenesis of multiple sclerosis. *Neurochemical Research, 32*(2), 251–256.

Moscarello, M. A., Lei, H., Mastronardi, F. G., Winer, S., Tsui, H., Li, Z., Ackerley, C., Zhang, L., Raijmakers, R., & Wood, D. D. (2013). Inhibition of peptidyl-arginine deiminases reverses protein-hypercitrullination and disease in mouse models of multiple sclerosis. *Disease Models & Mechanisms, 6*(2), 467–478. doi:10.1242/dmm.010520.

Muller, C., Bauer, N. M., Schafer, I., & White, R. (2013). Making myelin basic protein—From mRNA transport to localized translation. *Frontiers in Cellular Neuroscience, 7*, 169. doi:10.3389/fncel.2013.00169.

Murray, N., & Steck, A. J. (1984). Impulse conduction regulates myelin basic protein phosphorylation in rat optic nerve. *Journal of Neurochemistry, 43*(1), 243–248.

Musse, A. A., & Harauz, G. (2007). Molecular "negativity" may underlie multiple sclerosis: Role of the myelin basic protein family in the pathogenesis of MS. *International Review of Neurobiology, 79*(1), 149–172.

Musse, A. A., Boggs, J. M., & Harauz, G. (2006). Deimination of membrane-bound myelin basic protein in multiple sclerosis exposes an immunodominant epitope. *Proceedings of the National Academy of Sciences of the United States of America, 103*(12), 4422–4427.

Musse, A. A., Gao, W., Homchaudhuri, L., Boggs, J. M., & Harauz, G. (2008a). Myelin basic protein as a "PI(4,5)P2-modulin": A new biological function for a major central nervous system protein. *Biochemistry, 47*(39), 10372–10382.

Musse, A. A., Li, Z., Ackerley, C. A., Bienzle, D., Lei, H., Poma, R., Harauz, G., Moscarello, M. A., & Mastronardi, F. G. (2008b). Peptidylarginine deiminase 2 (PAD2) expression in a transgenic mouse leads to specific central nervous system (CNS) myelin instability. *Disease Models & Mechanisms, 1*(4/5), 229–240.

Musse, A. A., Polverini, E., Raijmakers, R., & Harauz, G. (2008c). Kinetics of human peptidylarginine deiminase 2 (hPAD2)—Reduction of Ca^{2+}-dependence by phospholipids and assessment of proposed inhibition by paclitaxel side chains. *Biochemistry and Cell Biology, 86*(5), 437–447.

Nawaz, S., Schweitzer, J., Jahn, O., & Werner, H. B. (2013). Molecular evolution of myelin basic protein, an abundant structural myelin component. *Glia, 61*(8), 1364–1377.

Nawaz, S., Sanchez, P., Schmitt, S., Snaidero, N., Mitkovski, M., Velte, C. J., Bruckner, B. R., Alexopoulos, I., Czopka, T., Jung, S. Y., Rhee, J. S., Janshoff, A., Witke, W., Schaap, I. A., Lyons, D. A., & Simons, M. (2015). Actin filament turnover drives leading edge growth during myelin sheath formation in the central nervous system. *Developmental Cell, 34*(2), 139–151. doi:10.1016/j.devcel.2015.05.013.

Neema, M., Stankiewicz, J., Arora, A., Guss, Z. D., & Bakshi, R. (2007). MRI in multiple sclerosis: What's inside the toolbox? *Neurotherapeutics, 4*(4), 602–617. doi:10.1016/j.nurt.2007.08.001.

Nicholas, A. P., & Bhattacharya, S. K. (2014). *Protein deimination in human health and disease* (1st ed.). New York: Springer.

Nicholas, A. P., & Whitaker, J. N. (2002). Preparation of a monoclonal antibody to citrullinated epitopes: Its characterization and some applications to immunohistochemistry in human brain. *Glia, 37*(4), 328–336.

Nicholas, A. P., Sambandam, T., Echols, J. D., & Tourtellotte, W. W. (2004). Increased citrullinated glial fibrillary acidic protein in secondary progressive multiple sclerosis. *The Journal of Comparative Neurology, 473*(1), 128–136.

Nicholas, A. P., Sambandam, T., Echols, J. D., & Barnum, S. R. (2005). Expression of citrullinated proteins in murine experimental autoimmune encephalomyelitis. *The Journal of Comparative Neurology, 486*(3), 254–266.

Nicholas, A. P., Lu, L., Heaven, M., Kadish, I., van Groen, T., Accaviti-Loper, M. A., Wewering, S., Kofskey, D., Gambetti, P., & Brenner, M. (2014). Ongoing studies of demination in neurodegenerative diseases using the F95 antibody. In A. P. Nicholas & S. K. Bhattacharya (Eds.), *Protein deimination in human health and disease* (pp. 257–280). New York: Springer.

Nielsen, C. H., Bornsen, L., Sellebjerg, F., & Brimnes, M. K. (2016). Myelin basic protein-induced production of tumor necrosis factor-alpha and interleukin-6, and presentation of the immunodominant peptide MBP85-99 by B cells from patients with relapsing-remitting multiple sclerosis. *PloS One, 11*(1), e0146971. doi:10.1371/journal.pone.0146971.

Oguz, K. K., Kurne, A., Aksu, A. O., Karabulut, E., Serdaroglu, A., Teber, S., Haspolat, S., Senbil, N., Kurul, S., & Anlar, B. (2009). Assessment of citrullinated myelin by 1H-MR spectroscopy in early-onset multiple sclerosis. *AJNR. American Journal of Neuroradiology, 30*(4), 716–721. doi:10.3174/ajnr.A1425.

Ohler, B., Graf, K., Bragg, R., Lemons, T., Coe, R., Genain, C., Israelachvili, J., & Husted, C. (2004). Role of lipid interactions in autoimmune demyelination. *Biochimica et Biophysica Acta, 1688*(1), 10–17.

Ozgen, H., Schrimpf, W., Hendrix, J., de Jonge, J. C., Lamb, D. C., Hoekstra, D., Kahya, N., & Baron, W. (2014). The lateral membrane organization and dynamics of myelin proteins PLP and MBP are dictated by distinct galactolipids and the extracellular matrix. *PloS One, 9*(7), e101834. doi:10.1371/journal.pone.0101834.

Ozgen, H., Baron, W., Hoekstra, D., & Kahya, N. (2016). Oligodendroglial membrane dynamics in relation to myelin biogenesis. *Cellular and Molecular Life Sciences, 73*(17), 3291–3310. doi:10.1007/s00018-016-2228-8.

Paez, P. M., Fulton, D., Colwell, C. S., & Campagnoni, A. T. (2009a). Voltage-operated Ca(2+) and Na(+) channels in the oligodendrocyte lineage. *Journal of Neuroscience Research, 87*(15), 3259–3266.

Paez, P. M., Fulton, D. J., Spreuer, V., Handley, V., Campagnoni, C. W., & Campagnoni, A. T. (2009b). Regulation of store-operated and voltage-operated Ca++ channels in the proliferation and death of oligodendrocyte precursor cells by Golli proteins. *ASN Neuro, 1*(1), e00003.

Paez, P. M., Fulton, D. J., Spreuer, V., Handley, V., Campagnoni, C. W., Macklin, W. B., Colwell, C., & Campagnoni, A. T. (2009c). Golli myelin basic proteins regulate oligodendroglial progenitor cell migration through voltage-gated Ca2+ influx. *The Journal of Neuroscience, 29*(20), 6663–6676. doi:10.1523/JNEUROSCI.5806-08.2009.

Pajevic, S., Basser, P. J., & Fields, R. D. (2014). Role of myelin plasticity in oscillations and synchrony of neuronal activity. *Neuroscience, 276*, 135–147. doi:10.1016/j.neuroscience.2013.11.007.

Palma, A. E., Owh, P., Fredric, C., Readhead, C., & Moscarello, M. A. (1997). Characterization of myelin basic protein charge microheterogeneity in developing mouse brain and in the transgenic shiverer mutant. *Journal of Neurochemistry, 69*(4), 1753–1762.

Panfoli, I., Bruschi, M., Santucci, L., Calzia, D., Ravera, S., Petretto, A., & Candiano, G. (2014). Myelin proteomics: The past, the unexpected and the future. *Expert Review of Proteomics, 11*(3), 345–354. doi:10.1586/14789450.2014.900444.

Poggi, G., Boretius, S., Mobius, W., Moschny, N., Baudewig, J., Ruhwedel, T., Hassouna, I., Wieser, G. L., Werner, H. B., Goebbels, S., Nave, K. A., & Ehrenreich, H. (2016). Cortical network dysfunction caused by a subtle defect of myelination. *Glia, 64*(11), 2025–2040. doi:10.1002/glia.23039.

Polverini, E., Coll, E. P., Tieleman, D. P., & Harauz, G. (2011). Conformational choreography of a molecular switch region in myelin basic protein—Molecular dynamics shows induced folding and secondary structure type conversion upon threonyl phosphorylation in both aqueous and membrane-associated environments. *Biochimica et Biophysica Acta-Biomembranes, 1808*(3), 674–683. doi:10.1016/j.bbamem.2010.11.030.

Ponomarenko, N. A., Durova, O. M., Vorobiev, I. I., Belogurov, A. A., Jr., Kurkova, I. N., Petrenko, A. G., Telegin, G. B., Suchkov, S. V., Kiselev, S. L., Lagarkova, M. A., Govorun, V. M., Serebryakova, M. V., Avalle, B., Tornatore, P., Karavanov, A., Morse, H. C., III, Thomas, D., Friboulet, A., & Gabibov, A. G. (2006). Autoantibodies to myelin basic protein catalyze site-specific degradation of their antigen. *Proceedings of the National Academy of Sciences of the United States of America, 103*(2), 281–286. doi:10.1073/pnas.0509849103.

Popescu, B. F., & Lucchinetti, C. F. (2012). Pathology of demyelinating diseases. *Annual Review of Pathology, 7*, 185–217. doi:10.1146/annurev-pathol-011811-132443.

Popescu, B. F., & Lucchinetti, C. F. (2016). Immunopathology: Autoimmune glial diseases and differentiation from multiple sclerosis. *Handbook of Clinical Neurology, 133*(1), 95–106. doi:10.1016/B978-0-444-63432-0.00006-2.

Praet, J., Guglielmetti, C., Berneman, Z., Van der Linden, A., & Ponsaerts, P. (2014). Cellular and molecular neuropathology of the cuprizone mouse model: Clinical relevance for multiple sclerosis. *Neuroscience and Biobehavioral Reviews, 47*, 485–505. doi:10.1016/j.neubiorev.2014.10.004.

Pribyl, T. M., Campagnoni, C. W., Kampf, K., Kashima, T., Handley, V. W., McMahon, J., & Campagnoni, A. T. (1993). The human myelin basic protein gene is included within a 179-kilobase transcription unit: Expression in the immune and central nervous systems. *Proceedings of the National Academy of Sciences of the United States of America, 90*(22), 10695–10699.

Pritzker, L. B., & Moscarello, M. A. (1998). A novel microtubule independent effect of paclitaxel: The inhibition of peptidylarginine deiminase from bovine brain. *Biochimica et Biophysica Acta, 1388*(1), 154–160.

Pritzker, L. B., Nguyen, T. A., & Moscarello, M. A. (1999). The developmental expression and activity of peptidylarginine deiminase in the mouse. *Neuroscience Letters, 266*(3), 161–164.

Pritzker, L. B., Joshi, S., Gowan, J. J., Harauz, G., & Moscarello, M. A. (2000a). Deimination of myelin basic protein. 1. Effect of deimination of arginyl residues of myelin basic protein on its structure and susceptibility to digestion by cathepsin D. *Biochemistry, 39*(18), 5374–5381.

Pritzker, L. B., Joshi, S., Harauz, G., & Moscarello, M. A. (2000b). Deimination of myelin basic protein. 2. Effect of methylation of MBP on its deimination by peptidylarginine deiminase. *Biochemistry, 39*(18), 5382–5388.

Privat, A., Jacque, C., Bourre, J. M., Dupouey, P., & Baumann, N. (1979). Absence of the major dense line in myelin of the mutant mouse "*shiverer*". *Neuroscience Letters, 12*(1), 107–112.

Procaccini, C., De Rosa, V., Pucino, V., Formisano, L., & Matarese, G. (2015). Animal models of multiple sclerosis. *European Journal of Pharmacology, 759*(1), 182–191. doi:10.1016/j.ejphar.2015.03.042.

Quarles, R. H., Macklin, W. B., & Morell, P. (2006). Myelin formation, structure, and biochemistry. In G. J. Siegel, R. W. Albers, S. T. Brady, & D. L. Price (Eds.), *Basic neurochemistry—Molecular, cellular, and medical aspects* (pp. 51–71). San Diego: Elsevier Academic Press.

Raijmakers, R., Vogelzangs, J., Croxford, J. L., Wesseling, P., Van Venrooij, W. J., & Pruijn, G. J. (2005). Citrullination of central nervous system proteins during the development of experimental autoimmune encephalomyelitis. *The Journal of Comparative Neurology, 486*(3), 243–253.

Raijmakers, R., Vogelzangs, J., Raats, J., Panzenbeck, M., Corby, M., Jiang, H., Thibodeau, M., Haynes, N., Van Venrooij, W. J., Pruijn, G. J., & Werneburg, B. (2006). Experimental autoimmune encephalomyelitis induction in peptidylarginine deiminase 2 knockout mice. *The Journal of Comparative Neurology, 498*(2), 217–226.

Raijmakers, R., Zendman, A. J., Egberts, W. V., Vossenaar, E. R., Raats, J., Soede-Huijbregts, C., Rutjes, F. P., van Veelen, P. A., Drijfhout, J. W., & Pruijn, G. J. (2007). Methylation of arginine residues interferes with citrullination by peptidylarginine deiminases *in vitro*. *Journal of Molecular Biology, 367*(4), 1118–1129.

Rasband, M. N., & Macklin, W. B. (2012). Myelin structure and biochemistry. In S. T. Brady, G. J. Siegel, R. W. Albers, & D. L. Price (Eds.), *Basic neurochemistry: Principles of molecular, cellular, and medical neurobiology* (pp. 180–199). San Diego: Academic Press/Elsevier.

Reissner, K. J., & Aswad, D. W. (2003). Deamidation and isoaspartate formation in proteins: Unwanted alterations or surreptitious signals? *Cellular and Molecular Life Sciences, 60*(7), 1281–1295. doi:10.1007/s00018-003-2287-5.

Ridsdale, R. A., Beniac, D. R., Tompkins, T. A., Moscarello, M. A., & Harauz, G. (1997). Three-dimensional structure of myelin basic protein. II. Molecular modeling and considerations of predicted structures in multiple sclerosis. *The Journal of Biological Chemistry, 272*(7), 4269–4275.

Rinholm, J. E., Vervaeke, K., Tadross, M. R., Tkachuk, A. N., Kopek, B. G., Brown, T. A., Bergersen, L. H., & Clayton, D. A. (2016). Movement and structure of mitochondria in oligodendrocytes and their myelin sheaths. *Glia*. doi:10.1002/glia.22965.

Ross, J. L. (2016). The dark matter of biology. *Biophysical Journal, 111*(5), 909–916. doi:10.1016/j.bpj.2016.07.037.

Saab, A. S., & Nave, K. A. (2016). Neuroscience: A mechanism for myelin injury. *Nature, 529*(7587), 474–475. doi:10.1038/nature16865.

Sabri, M. I., Bone, A. H., & Davison, A. N. (1974). Turnover of myelin and other structural proteins in the developing rat brain. *The Biochemical Journal, 142*(3), 499–507.

Schachter, H., McLaurin, J., & Harauz, G. (2014). A tribute to Dr. Mario Moscarello. *Multiple Sclerosis Journal, 20*(7), 48–49.

Schaeffer, J., Cossetti, C., Mallucci, G., & Pluchino, S. (2015). Multiple sclerosis. In M. J. Zigmond, J. T. Coyle, & L. P. Rowland (Eds.), *Neurobiology of brain disorders* (pp. 497–520). Amsterdam: Elsevier/Academic Press.

Shaharabani, R., Ram-On, M., Avinery, R., Aharoni, R., Arnon, R., Talmon, Y., & Beck, R. (2016). Structural transition in myelin membrane as initiator of multiple sclerosis. *Journal of the American Chemical Society, 138*(37), 12159–12165. doi:10.1021/jacs.6b04826.

Shanshiashvili, L. V., Kalandadze, I. V., Ramsden, J. J., & Mikeladze, D. G. (2012). Adhesive properties and inflammatory potential of citrullinated myelin basic protein peptide 45-89. *Neurochemical Research, 37*(9), 1959–1966. doi:10.1007/s11064-012-0816-z.

Sherman, D. L., & Brophy, P. J. (2005). Mechanisms of axon ensheathment and myelin growth. *Nature Reviews. Neuroscience, 6*(9), 683–690.

Shimada, N., Handa, S., Uchida, Y., Fukuda, M., Maruyama, N., Asaga, H., Choi, E. K., Lee, J., & Ishigami, A. (2010). Developmental and age-related changes of peptidylarginine deiminase 2 in the mouse brain. *Journal of Neuroscience Research, 88*(4), 798–806. doi:10.1002/jnr.22255.

Silbereis, J. C., Pochareddy, S., Zhu, Y., Li, M., & Sestan, N. (2016). The cellular and molecular landscapes of the developing human central nervous system. *Neuron, 89*(2), 248–268. doi:10.1016/j.neuron.2015.12.008.

Šimic, G., Babic-Leko, M., Wray, S., Harrington, C., Delalle, I., Jovanov-Miloševic, N., Bažadona, D., Buée, L., de Silva, R., Di Giovanni, G., Wischik, C., & Hof, P. R. (2016). Tau protein hyperphosphorylation and aggregation in Alzheimer's disease and other tauopathies, and possible neuroprotective strategies. *Biomolecules, 6*(1), 1–28. doi:10.3390/biom6010006.

Simone, M., & Chitnis, T. (2016). Use of disease-modifying therapies in pediatric MS. *Current Treatment Options in Neurology, 18*(8), 36. doi:10.1007/s11940-016-0420-7.

Slade, D. J., Fang, P., Dreyton, C. J., Zhang, Y., Fuhrmann, J., Rempel, D., Bax, B. D., Coonrod, S. A., Lewis, H. D., Guo, M., Gross, M. L., & Thompson, P. R. (2015). Protein arginine deiminase 2 binds calcium in an ordered fashion: Implications for inhibitor design. *ACS Chemical Biology, 10*(4), 1043–1053. doi:10.1021/cb500933j.

Smith, G. S. T., Chen, L., Bamm, V. V., Dutcher, J. R., & Harauz, G. (2010). The interaction of zinc with membrane-associated 18.5 kDa myelin basic protein: An attenuated total reflectance-Fourier transform infrared spectroscopic study. *Amino Acids, 39*(3), 739–750. doi:10.1007/s00726-010-0513-7.

Smith, G. S. T., Paez, P. M., Spreuer, V., Campagnoni, C. W., Boggs, J. M., Campagnoni, A. T., & Harauz, G. (2011). Classical 18.5-and 21.5-kDa isoforms of myelin basic protein inhibit calcium influx into oligodendroglial cells, in contrast to golli isoforms. *Journal of Neuroscience Research, 89*(4), 467–480. doi:10.1002/jnr.22570.

Smith, G. S. T., De Avila, M., Paez, P. M., Spreuer, V., Wills, M. K. B., Jones, N., Boggs, J. M., & Harauz, G. (2012a). Proline substitutions and threonine pseudophosphorylation of the SH3 ligand of 18.5-kDa myelin basic protein decrease its affinity for the Fyn-SH3 domain and alter process development and protein localization in oligodendrocytes. *Journal of Neuroscience Research, 90*(1), 28–47. doi:10.1002/jnr.22733.

Smith, G. S. T., Homchaudhuri, L., Boggs, J. M., & Harauz, G. (2012b). Classic 18.5- and 21.5-kDa myelin basic protein isoforms associate with cytoskeletal and SH3-domain proteins in the immortalized N19-oligodendroglial cell line stimulated by phorbol ester and IGF-1. *Neurochemical Research, 37*(6), 1277–1295. doi:10.1007/s11064-011-0700-2.

Smith, G. S. T., Seymour, L. V., Boggs, J. M., & Harauz, G. (2012c). The 21.5-kDa isoform of myelin basic protein has a non-traditional PY-nuclear-localization signal. *Biochemical and Biophysical Research Communications, 422*(4), 670–675. doi:10.1016/j.bbrc.2012.05.051.

Smith, G. S. T., Samborska, B., Hawley, S. P., Klaiman, J. M., Gillis, T. E., Jones, N., Boggs, J. M., & Harauz, G. (2013). Nucleus-localized 21.5-kDa myelin basic protein promotes oligodendrocyte proliferation and enhances neurite outgrowth in coculture, unlike the plasma membrane-associated 18.5-kDa isoform. *Journal of Neuroscience Research, 91*(3), 349–362. doi:10.1002/jnr.23166.

Snaidero, N., & Simons, M. (2014). Myelination at a glance. *Journal of Cell Science, 127*(Pt 14), 2999–3004. doi:10.1242/jcs.151043.

Snaidero, N., Mobius, W., Czopka, T., Hekking, L. H., Mathisen, C., Verkleij, D., Goebbels, S., Edgar, J., Merkler, D., Lyons, D. A., Nave, K. A., & Simons, M. (2014). Myelin membrane wrapping of CNS axons by PI(3,4,5)P3-dependent polarized growth at the inner tongue. *Cell, 156*(1-2), 277–290. doi:10.1016/j.cell.2013.11.044.

Sonnino, S., Aureli, M., Mauri, L., Ciampa, M. G., & Prinetti, A. (2015). Membrane lipid domains in the nervous system. *Frontiers in Bioscience, 20*(1), 280–302.

Sospedra, M., & Martin, R. (2005). Immunology of multiple sclerosis. *Annual Review of Immunology, 23*(1), 683–747.

Spengler, J., & Scheel-Toellner, D. (2014). Neutrophils and their contribution to autoimmunity in rheumatoid arthritis. In A. P. Nicholas & S. K. Bhattacharya (Eds.), *Protein deimination in human health and disease* (pp. 97–112). New York: Springer.

Stirling, D. P., & Stys, P. K. (2010). Mechanisms of axonal injury: Internodal nanocomplexes and calcium deregulation. *Trends in Molecular Medicine, 16*(4), 160–170. doi:10.1016/j.molmed.2010.02.002.

Stys, P. K. (2010). Multiple sclerosis: Autoimmune disease or autoimmune reaction? *The Canadian Journal of Neurological Sciences, 37*(Supplement 2), S16–S23.

Stys, P. K. (2013). Pathoetiology of multiple sclerosis: Are we barking up the wrong tree? *F1000Prime Reports, 5*(1), 20. doi:10.12703/P5-20.

Stys, P. K., Zamponi, G. W., van Minnen, J., & Geurts, J. J. G. (2012). Will the real multiple sclerosis please stand up? *Nature Reviews. Neuroscience, 13*(7), 507–514. doi:10.1038/nrn3275.

Subramanian, V., Knight, J. S., Parelkar, S., Anguish, L., Coonrod, S. A., Kaplan, M. J., & Thompson, P. R. (2015). Design, synthesis, and biological evaluation of tetrazole analogs of Cl-amidine as protein arginine deiminase inhibitors. *Journal of Medicinal Chemistry, 58*(3), 1337–1344. doi:10.1021/jm501636x.

Suresh, S., Wang, C., Nanekar, R., Kursula, P., & Edwardson, J. M. (2010). Myelin basic protein and myelin protein 2 act synergistically to cause stacking of lipid bilayers. *Biochemistry, 49*(16), 3456–3463. doi:10.1021/bi100128h.

Suzuki, A., Yamada, R., & Yamamoto, K. (2007). Citrullination by peptidylarginine deiminase in rheumatoid arthritis. *Annals of the New York Academy of Sciences, 1108*(1), 323–339.

Tagge, I., O'Connor, A., Chaudhary, P., Pollaro, J., Berlow, Y., Chalupsky, M., Bourdette, D., Woltjer, R., Johnson, M., & Rooney, W. (2016). Spatio-temporal patterns of demyelination and remyelination in the cuprizone mouse model. *PloS One, 11*(4), e0152480-1–e0152480-24. doi:10.1371/journal.pone.0152480.

Terry, R. L., Ifergan, I., & Miller, S. D. (2016). Experimental autoimmune encephalomyelitis in mice. *Methods in Molecular Biology, 1304*(1), 145–160. doi:10.1007/7651_2014_88.

Tomassy, G. S., Berger, D. R., Chen, H. H., Kasthuri, N., Hayworth, K. J., Vercelli, A., Seung, H. S., Lichtman, J. W., & Arlotta, P. (2014). Distinct profiles of myelin distribution along single axons of pyramidal neurons in the neocortex. *Science, 344*(6181), 319–324. doi:10.1126/science.1249766.

Tompa, P., & Fersht, A. (2015). *Structure and function of intrinsically disordered proteins* (1st ed.pp. 1–359). London: Chapman and Hall/CRC Press.

Tompa, P., Schad, E., Tantos, A., & Kalmar, L. (2015). Intrinsically disordered proteins: Emerging interaction specialists. *Current Opinion in Structural Biology, 35*, 49–59. doi:10.1016/j.sbi.2015.08.009.

Torkildsen, O., Brunborg, L. A., Myhr, K. M., & Bo, L. (2008). The cuprizone model for demyelination. *Acta Neurologica Scandinavica. Supplementum, 188*, 72–76. doi:10.1111/j.1600-0404.2008.01036.x.

Toyama, B. H., Savas, J. N., Park, S. K., Harris, M. S., Ingolia, N. T., Yates, J. R., III, & Hetzer, M. W. (2013). Identification of long-lived proteins reveals exceptional stability of essential cellular structures. *Cell, 154*(5), 971–982. doi:10.1016/j.cell.2013.07.037.

Traka, M., Podojil, J. R., McCarthy, D. P., Miller, S. D., & Popko, B. (2016). Oligodendrocyte death results in immune-mediated CNS demyelination. *Nature Neuroscience, 19*(1), 65–74. doi:10.1038/nn.4193.

Trapp, B. D., & Kidd, G. J. (2004). Structure of the myelinated axon. In R. A. Lazzarini, J. W. Griffin, H. Lassman, K.-A. Nave, R. H. Miller, & B. D. Trapp (Eds.), *Myelin biology and disorders* (pp. 3–27). San Diego: Elsevier Academic Press.

Trapp, B. D., & Nave, K. A. (2008). Multiple sclerosis: an immune or neurodegenerative disorder? *Annual Review of Neuroscience, 31*(1), 247–269.

Truscott, R. J., & Friedrich, M. G. (2016). The etiology of human age-related cataract. Proteins don't last forever. *Biochimica et Biophysica Acta, 1860*(1 Pt B), 192–198. doi:10.1016/j.bbagen.2015.08.016.

Tsang, D., Tsang, Y. S., Ho, W. K., & Wong, R. N. (1997). Myelin basic protein is a zinc-binding protein in brain: Possible role in myelin compaction. *Neurochemical Research, 22*(7), 811–819.

Tu, R., Grover, H. M., & Kotra, L. P. (2016). Peptidyl arginine deiminases and neurodegenerative diseases. *Current Medicinal Chemistry, 23*(2), 104–114. doi:10.2174/0929867323666151118120710.

Uversky, V. N. (2014). The triple power of D(3): Protein intrinsic disorder in degenerative diseases. *Frontiers in Bioscience, 19*, 181–258.

Uversky, V. N. (2015). Intrinsically disordered proteins and their (disordered) proteomes in neurodegenerative disorders. *Frontiers in Aging Neuroscience, 7*(1), 18-1–18-8. doi:10.3389/fnagi.2015.00018.

Uversky, V. N. (2016). Dancing protein clouds: The strange biology and chaotic physics of intrinsically disordered proteins. *The Journal of Biological Chemistry, 291*(13), 6681–6688. doi:10.1074/jbc.R115.685859.

Uversky, V. N., Gillespie, J. R., & Fink, A. L. (2000). Why are "natively unfolded" proteins unstructured under physiologic conditions? *Proteins, 41*(3), 415–427.

van Beers, J. J. B. C., & Pruijn, G. J. M. (2014). The role of synovial citrullinated proteins in the pathophysiology of rheumatoid arthritis. In A. P. Nicholas & S. K. Bhattacharya (Eds.), *Protein deimination in human health and disease* (pp. 41–68). New York: Springer.

van Beers, J. J., Zendman, A. J., Raijmakers, R., Stammen-Vogelzangs, J., & Pruijn, G. J. (2013). Peptidylarginine deiminase expression and activity in PAD2 knock-out and PAD4-low mice. *Biochimie, 95*(2), 299–308. doi:10.1016/j.biochi.2012.09.029.

van den Elsen, P. J., van Eggermond, M. C., Puentes, F., van der Valk, P., Baker, D., & Amor, S. (2014). The epigenetics of multiple sclerosis and other related disorders. *Multiple Sclerosis and Related Disorders, 3*(2), 163–175. doi:10.1016/j.msard.2013.08.007.

van der Lee, R., Buljan, M., Lang, B., Weatheritt, R. J., Daughdrill, G. W., Dunker, A. K., Fuxreiter, M., Gough, J., Gsponer, J., Jones, D. T., Kim, P. M., Kriwacki, R. W., Oldfield, C. J., Pappu, R. V., Tompa, P., Uversky, V. N., Wright, P. E., & Babu, M. M. (2014). Classification of intrinsically disordered regions and proteins. *Chemical Reviews, 114*(13), 6589–6631. doi:10.1021/cr400525m.

Vassall, K. A., Bessonov, K., De Avila, M., Polverini, E., & Harauz, G. (2013). The effects of threonine phosphorylation on the stability and dynamics of the central molecular switch region of 18.5-kDa myelin basic protein. *PloS One, 8*(7), e68175. doi:10.1371/journal.pone.0068175.

Vassall, K. A., Bamm, V. V., & Harauz, G. (2015a). MyelStones: The executive roles of myelin basic protein in myelin assembly and destabilization in multiple sclerosis. *The Biochemical Journal, 472*(1), 17–32. doi:10.1042/BJ20150710.

Vassall, K. A., Jenkins, A. D., Bamm, V. V., & Harauz, G. (2015b). Thermodynamic analysis of the disorder-to-alpha-helical transition of 18.5-kDa myelin basic protein reveals an equilibrium intermediate representing the most compact conformation. *Journal of Molecular Biology, 427*(10), 1977–1992. doi:10.1016/j.jmb.2015.03.011.

Vassall, K. A., Bamm, V. V., Jenkins, A. D., Velte, C. J., Kattnig, D. R., Boggs, J. M., Hinderberger, D., & Harauz, G. (2016). Substitutions mimicking deimination and phosphorylation of 18.5-kDa myelin basic protein exert local structural effects that subtly influence its global folding. *Biochimica et Biophysica Acta-Biomembranes, 1858*(6), 1262–1277. doi:10.1016/j.bbamem.2016.02.024.

Velte, C. J. (2016, June 27). Biogenesis and maintenance of cytoplasmic domains in myelin of the cental nevous system. Doctor rerum naturalium dissertation, Georg-August University School of Science (GAUSS), Göttingen, Germany.

Velumian, A. A., Samoilova, M., & Fehlings, M. G. (2011). Visualization of cytoplasmic diffusion within living myelin sheaths of CNS white matter axons using microinjection of the fluorescent dye Lucifer Yellow. *NeuroImage, 56*(1), 27–34. doi:10.1016/j.neuroimage.2010.11.022.

Venne, A. S., Kollipara, L., & Zahedi, R. P. (2014). The next level of complexity: Crosstalk of posttranslational modifications. *Proteomics, 14*(4–5), 513–524. doi:10.1002/pmic.201300344.

Vossenaar, E. R., Zendman, A. J., Van Venrooij, W. J., & Pruijn, G. J. (2003). PAD, a growing family of citrullinating enzymes: Genes, features and involvement in disease. *BioEssays, 25*(11), 1106–1118.

Wang, S., & Young, K. M. (2014). White matter plasticity in adulthood. *Neuroscience, 276*(1), 148–160. doi:10.1016/j.neuroscience.2013.10.018.

Wang, C., Neugebauer, U., Burck, J., Myllykoski, M., Baumgartel, P., Popp, J., & Kursula, P. (2011). Charge isomers of myelin basic protein: Structure and interactions with membranes, nucleotide analogues, and calmodulin. *PloS One, 6*(5), e19915. doi:10.1371/journal.pone.0019915.

Wei, L., Wasilewski, E., Chakka, S. K., Bello, A. M., Moscarello, M. A., & Kotra, L. P. (2013). Novel inhibitors of protein arginine deiminase with potential activity in multiple sclerosis animal model. *Journal of Medicinal Chemistry, 56*(4), 1715–1722. doi:10.1021/jm301755q.

Weil, M. T., Mobius, W., Winkler, A., Ruhwedel, T., Wrzos, C., Romanelli, E., Bennett, J. L., Enz, L., Goebels, N., Nave, K. A., Kerschensteiner, M., Schaeren-Wiemers, N., Stadelmann, C., & Simons, M. (2016). Loss of myelin basic protein function triggers myelin breakdown in models of demyelinating diseases. *Cell Reports, 16*(2), 314–322. doi:10.1016/j.celrep.2016.06.008;.

Wenderski, W., & Maze, I. (2016). Histone turnover and chromatin accessibility: Critical mediators of neurological development, plasticity, and disease. *BioEssays, 38*(5), 410–419. doi:10.1002/bies.201500171.

Werner, H. B., & Jahn, O. (2010). Myelin matters: Proteomic insights into white matter disorders. *Expert Review of Proteomics, 7*(2), 159–164. doi:10.1586/epr.09.105.

Whitaker, J. N., & Mitchell, G. W. (1996). A possible role for altered myelin basic protein in multiple sclerosis. *Annals of Neurology, 40*(1), 3–4.

Whited, A. M., & Johs, A. (2015). The interactions of peripheral membrane proteins with biological membranes. *Chemistry and Physics of Lipids, 192*(1), 51–59. doi:10.1016/j.chemphyslip.2015.07.015.

Winkelmann, A., Loebermann, M., Reisinger, E. C., Hartung, H. P., & Zettl, U. K. (2016). Disease-modifying therapies and infectious risks in multiple sclerosis. *Nature Reviews. Neurology, 12*(4), 217–233. doi:10.1038/nrneurol.2016.21.

Witalison, E. E., Thompson, P. R., & Hofseth, L. J. (2015). Protein arginine deiminases and associated citrullination: Physiological functions and diseases associated with dysregulation. *Current Drug Targets, 16*(7), 700–710.

Wood, D. D., & Moscarello, M. A. (1989). The isolation, characterization, and lipid-aggregating properties of a citrulline containing myelin basic protein. *The Journal of Biological Chemistry, 264*(9), 5121–5127.

Wood, D. D., & Moscarello, M. A. (1997). Molecular biology of the Glia: Components of myelin—Myelin basic protein—The implication of post-translational changes for demyelinating disease. In W. C. Russell (Ed.), *Molecular biology of multiple sclerosis* (pp. 37–54). Chichester: Wiley.

Wood, D. D., Vail, W. J., & Moscarello, M. A. (1975). The localization of the basic protein and N-2 in diseased myelin. *Brain Research, 93*(3), 463–471.

Wood, D. D., Bilbao, J. M., O'Connors, P., & Moscarello, M. A. (1996). Acute multiple sclerosis (Marburg type) is associated with developmentally immature myelin basic protein. *Annals of Neurology, 40*(1), 18–24.

Wood, D. D., Ackerley, C. A., Brand, B., Zhang, L., Raijmakers, R., Mastronardi, F. G., & Moscarello, M. A. (2008). Myelin localization of peptidylarginine deiminases 2 and 4: Comparison of PAD2 and PAD4 activities. *Laboratory Investigation, 88*(4), 354–364. doi:10.1038/labinvest.3700748.

Wucherpfennig, K. W., Catz, I., Hausmann, S., Strominger, J. L., Steinman, L., & Warren, K. G. (1997). Recognition of the immunodominant myelin basic protein peptide by autoantibodies and HLA-DR2-restricted T cell clones from multiple sclerosis patients. Identity of key contact residues in the B-cell and T-cell epitopes. *The Journal of Clinical Investigation, 100*(5), 1114–1122.

Yan, J., Dunker, A. K., Uversky, V. N., & Kurgan, L. (2016). Molecular recognition features (MoRFs) in three domains of life. *Molecular BioSystems, 12*(3), 697–710. doi:10.1039/c5mb00640f.

Yang, L., Tan, D., & Piao, H. (2016). Myelin basic protein citrullination in multiple sclerosis: A potential therapeutic target for the pathology. *Neurochemical Research, 41*(8), 1845–1856. doi:10.1007/s11064-016-1920-2.

Young, K. M., Psachoulia, K., Tripathi, R. B., Dunn, S. J., Cossell, L., Attwell, D., Tohyama, K., & Richardson, W. D. (2013). Oligodendrocyte dynamics in the healthy adult CNS: Evidence for myelin remodeling. *Neuron, 77*(5), 873–885. doi:10.1016/j.neuron.2013.01.006.

Zand, R. (2008). Posttranslational modifications of myelin basic proteins. In J. M. Boggs (Ed.), *Myelin basic protein* (pp. 19–29). New York: Nova Science Publishers.

Zhang, C., Walker, A. K., Zand, R., Moscarello, M. A., Yan, J. M., & Andrews, P. C. (2012a). Myelin basic protein undergoes a broader range of modifications in mammals than in lower vertebrates. *Journal of Proteome Research, 11*(10), 4791–4802. doi:10.1021/pr201196e.

Zhang, X., Bolt, M., Guertin, M. J., Chen, W., Zhang, S., Cherrington, B. D., Slade, D. J., Dreyton, C. J., Subramanian, V., Bicker, K. L., Thompson, P. R., Mancini, M. A., Lis, J. T., & Coonrod, S. A. (2012b). Peptidylarginine deiminase 2-catalyzed histone H3 arginine 26 citrullination facilitates estrogen receptor alpha target gene activation. *Proceedings of the National Academy of Sciences of the United States of America, 109*(33), 13331–13336. doi:10.1073/pnas.1203280109.

Zhang, C., Gao, S., Molascon, A. J., Wang, Z., Gorovsky, M. A., Liu, Y., & Andrews, P. C. (2014). Bioinformatic and proteomic analysis of bulk histones reveals PTM crosstalk and chromatin features. *Journal of Proteome Research, 13*(7), 3330–3337. doi:10.1021/pr5001829.

Zhong, L., Bamm, V. V., Ahmed, M. A., Harauz, G., & Ladizhansky, V. (2007). Solid-state NMR spectroscopy of 18.5 kDa myelin basic protein reconstituted with lipid vesicles: Spectroscopic characterisation and spectral assignments of solvent-exposed protein fragments. *Biochimica et Biophysica Acta-Biomembranes, 1768*(12), 3193–3205.

Zhou, Y., Simpson, S., Jr., Holloway, A. F., Charlesworth, J., van der Mei, I., & Taylor, B. V. (2014). The potential role of epigenetic modifications in the heritability of multiple sclerosis. *Multiple Sclerosis Journal, 20*(2), 135–140. doi:10.1177/1352458514520911.

Zienowicz, A., Bamm, V. V., Vassall, K. A., & Harauz, G. (2015). Myelin basic protein is a glial microtubule-associated protein—Characterization of binding domains, kinetics of polymerization, and regulation by phosphorylation and a lipidic environment. *Biochemical and Biophysical Research Communications, 461*(1), 136–141. doi:10.1016/j.bbrc.2015.03.181.

Zuchero, J. B., Fu, M. M., Sloan, S. A., Ibrahim, A., Olson, A., Zaremba, A., Dugas, J. C., Wienbar, S., Caprariello, A. V., Kantor, C., Leonoudakus, D., Lariosa-Willingham, K., Kronenberg, G., Gertz, K., Soderling, S. H., Miller, R. H., & Barres, B. A. (2015). CNS myelin wrapping is driven by actin disassembly. *Developmental Cell, 34*(2), 152–167. doi:10.1016/j.devcel.2015.06.011.

Chapter 20
The Significance of Deiminated GFAP in Neurodegenerative Diseases with Special Emphasis on Alexander Disease

Michael Brenner and Anthony P. Nicholas

Abbreviations

AlzD	Alzheimer's disease
AxD	Alexander disease
EAE	Experimental autoimmune encephalomyelitis
GFAP	Glial fibrillary acidic protein
IF	Intermediate filament
KO	Gene knockout
LC-MS/MS	Liquid chromatography-tandem mass spectrometry
mass spec	Mass spectroscopy
MBP	Myelin basic protein
MS	Multiple sclerosis
MW	Molecular weight
PAD	Peptidylarginine deiminase
rh	Recombinant human

20.1 Introduction

A primary impetus for interest in a role for deiminated proteins in neurodegenerative diseases has been suggested that deimination of myelin basic protein (MBP) is a primary step in the development of multiple sclerosis (MS) (reviewed in Chap. 19).

M. Brenner (✉)
Department of Neurobiology, University of Alabama at Birmingham, Birmingham, AL, USA
e-mail: michaelb@uab.edu

A.P. Nicholas
Department of Neurology, University of Alabama at Birmingham and
The Birmingham Veterans Administration Medical Center, Birmingham, AL, USA
e-mail: nicholas@uab.edu

A.P. Nicholas et al. (eds.), *Protein Deimination in Human Health and Disease*,
DOI 10.1007/978-3-319-58244-3_20

In brief, it is proposed that an unknown precipitating event causes increased peptidylarginine deiminase (PAD) activity, leading to deimination of MBP. This in turn results in disruption of the myelin structure, increased susceptibility of MBP to proteolysis, and release of MBP peptides that precipitate the autoimmune response that results in clinical symptoms. It was surprising, therefore, that a survey of proteins deiminated in MS, as well as in other CNS disorders, suggested that glial fibrillary acidic protein (GFAP) is the more predominant target. This finding raises the possibility that deimination of GFAP could have a general role in neurodegenerative disorders. This chapter explores this premise, with special emphasis on Alexander disease (AxD) as a likely candidate for an effect, since this disease is caused by dominant mutations in the GFAP gene. We begin with a description of GFAP and AxD, and then review evidence that astrocytes are a major repository for PAD in the CNS, and finally that GFAP is a primary target of its activity in both normal conditions and in several neurodegenerative disorders. This is followed by a discussion of the possible physiological effects of GFAP deimination and how these might influence disease states.

20.2 The GFAP Protein

GFAP is a member of the intermediate filament (IF) protein family, so called because the approximately 10 nm diameter of the polymers that they form is intermediate between those of actin and tubulin (reviewed in Snider and Omary 2014; Koster et al. 2015). They are a large protein family, encoded by over 70 different genes. All have the same basic structure of a central alpha helical rod domain flanked by random coil head and tail domains (Fig. 20.1). The central rod domain is highly conserved among IFs in both length and sequence, whereas the head and tail domains have only limited conservation within IF subclasses. These subclasses include the acidic and basic keratins (types I and II); GFAP, vimentin, desmin, and synemin (type III); neurofilaments (type IV); nuclear lamins (type V); and the eye lens proteins CP49 and filensin (type VI). Except for the almost ubiquitously expressed nuclear lamins, presence of IF proteins is typically restricted to a particular cell type and may be developmentally regulated. GFAP is predominantly expressed in astrocytes, but is present at lower levels in other cell types, including liver stellate cells, enteric glia, Müller cells in the retina, and fibroblasts. The synthesis of GFAP commences as astrocytes differentiate, and it is upregulated as part of the reactive response of astrocytes to almost any CNS perturbation, including injury or disease. It is a common misconception that cytoplasmic IFs like GFAP are responsible for maintaining cell shape; but instead, a primary function is to render cells resistant to shearing. Defects in IFs can produce extreme cell fragility, as seen in several skin blistering disorders, due to keratin mutations, or myopathies, due to desmin mutations. Other functions of IFs will be discussed later in the context of specific neurodegenerative diseases.

Polymerization of IF proteins into filaments follows different pathways depending on IF type (reviewed in Snider and Omary 2014; Koster et al. 2015). For example,

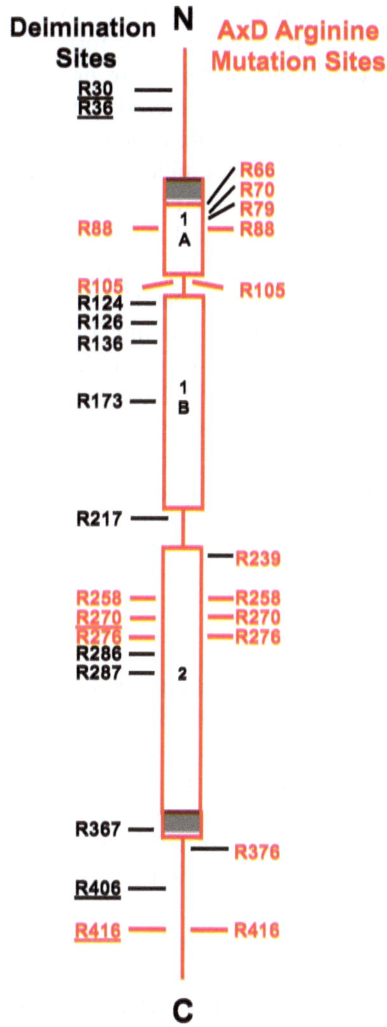

Fig. 20.1 Structure of GFAP with locations of arginines that have been found deiminated or mutated in AxD. The three open rectangular boxes (1A, 1B, 2) represent the helical rod domains of GFAP; these structural motifs are highly conserved among most intermediate filament proteins. The *solid lines* joining these segments are non-helical linker regions, and the *solid lines* at either end are the nonconserved, random coil, *N*-terminal, and *C*-terminal regions. The *gray box* before segment 1A is a nonconserved pre-helical sequence important for initiation of rod formation; the *gray box* at the end of segment 2 represents a highly conserved sequence that is critical for polymerization. Deiminated sites experimentally identified for GFAP are shown on the left. All were present in GFAP digested with PAD2 in vitro, while underlined sites were also identified in GFAP isolated from Alzheimer's brains (all five underlined sites) or normal controls (all five except the R270 site). Deimination sites whose mutation produces AxD are displayed in *red font*. The 11 arginine sites of the 47 present in human GFAP that are identified as mutated in AxD are shown on the right (the 59 other amino acid sites also identified as mutated in AxD have not been shown). The deimination sites were identified as reported by Jin et al. (2013), and R270 and R416 were also identified in vitro by Ishigami et al. (2015)

the keratins initially form heterodimers containing an acidic and a basic member, whereas type III IFs can either homodimerize or form a heterodimer with another type III IF. In particular, GFAP and vimentin are present in astrocytes to varying extents and may form homodimers or heterodimers (reviewed in Pekny 2001; Brenner 2014). The dimers form by the alpha helical domains coiling around each other in parallel orientation, and then combine side by side in an antiparallel alignment, forming a tetramer lacking end-to-end polarity. Tetramer formation is believed to be extremely rapid under physiological conditions. The 10 nm filaments are formed by tetramers polymerizing both end to end and side by side. For analytical studies, fully formed filaments may be separated from smaller oligomers such as tetramers by centrifugation in a buffer containing Triton X-100 (Tang et al. 2010). Type III IFs are in dynamic equilibrium with soluble IF proteins, but the exchange rate has been estimated to be only about 1% per hour. The rapid disassembly and reassembly of IF networks that occur during cell division are believed to be controlled by reversible phosphorylation of serine and threonine residues in the head region by cell cycle-dependent protein kinases.

20.3 Alexander Disease

AxD is a prime candidate for a disorder that might be affected by GFAP deimination. This astrogliopathy is a rare but usually fatal disease caused by dominant, gain-of-function mutations in GFAP (reviewed in Brenner et al. 2009; Prust et al. 2011). Although the causative mutations are usually present from the beginning of life (Li et al. 2006a), clinical signs may appear at any age. These signs can differ markedly among individuals, but two general classes of AxD have been recognized (Prust et al. 2011). Type I AxD most commonly presents before 4 years of age with impaired physical and mental development, seizures, ataxia, and speech and swallowing difficulties and is often accompanied by megalencephaly. Type II AxD can occur at any age but most often after 4 years and as late as old age. Unlike the type I class, developmental delay, seizures, and megalencephaly are not features of type II patients, but the two types share the common occurrence of ataxia and speech and swallowing difficulties. A very wide range of additional symptoms may also be present, including headache, sleep disturbances, vision problems, autonomic dysfunction and palatal myoclonus. Unifying the two forms, and the hallmark of the disease, is the presence of protein aggregates in astrocytes called Rosenthal fibers. These aggregates are composed of multiple different proteins, but their primary component is ubiquitinated GFAP (Heaven et al. 2016). Several animal models have been made to facilitate the study of AxD (Messing et al. 1998; Hagemann et al. 2006; Tanaka et al. 2007; Wang et al. 2011). The two most widely used, and which will be mentioned later, are a mouse carrying a transgene producing elevated expression of wild-type human GFAP (Messing et al. 1998) and one carrying a knock-in mutation of the endogenous mouse GFAP gene that is homologous to a common human AxD mutation (Hagemann et al. 2006).

20.4 PADs in Astrocytes

A prerequisite for GFAP deimination is the presence of PAD in astrocytes. Multiple studies not only support this location but also indicate that astrocytes are the primary repository of PAD in the CNS of rodents and man (Vincent et al. 1992; Asaga and Senshu 1993; Asaga and Ishigami 2000, 2001; Asaga et al. 2002; Ishigami et al. 2005; Acharya et al. 2012; Bradford et al. 2014).

When the PAD isotypes present in the CNS were separately analyzed, multiple studies detected only PAD2 protein by immunohistochemistry or immunoblotting or PAD2 mRNA by RT-PCR (Asaga and Ishigami 2001; Asaga et al. 2002), whereas others (particularly more recently) have also observed the presence of PAD4 (Acharya et al. 2012). When detected, PAD4 is found in neurons rather than astrocytes (Acharya et al. 2012). The inconsistent findings for PAD4 could be explained by a sensitivity issue, since Bradford et al. (2014) detected both PAD2 and PAD4 mRNA in control human brain, but the level of PAD4 mRNA was about 100 times lower than that for PAD2. The predominance of PAD2 for protein deimination in the CNS, at least for the mouse, is supported by a failure to detect any deiminated protein in the brain or spinal cord of control mice or mice with experimental autoimmune encephalomyelitis (EAE) in which the PAD2 gene had been knocked out (Raijmakers et al. 2006). Subsequently, however, deiminated myelin basic protein has been observed in these mice, presumably due to PAD4 activity (Wood et al. 2008). In striking contrast to the findings cited above, Shimada et al. (2010) found that PAD2 is present in neurons of the cerebral cortex and in cerebellar Purkinje cells, but not in astrocytes. As a note of caution, although the monoclonal antibody developed for this study was shown to yield a clean signal of the size expected for PAD2 in immunoblots of human and mouse brain extracts, it was not stated whether it was tested against extracts from the PAD2 gene knockout (KO) mouse or against PAD4. Finally, there has been a single report of the possible presence of PAD3 in chick astrocytes (Lange et al. 2011).

20.5 Deimination of GFAP

As might be expected from the predominant presence of PAD2 in astrocytes, GFAP has consistently been reported to be a major citrullinated protein in both normal (Fig. 20.2) and diseased human brains. Techniques used to document this identification have included double-label immunofluorescence, one- and two-dimensional (1D and 2D) immunoblotting, electron microscopy of immunogold labeling, and mass spectroscopy (mass spec) of samples isolated by 1D or 2D gel electrophoresis or by immunoprecipitation for peptidyl-citrulline (Senshu et al. 1992; Nicholas and Whitaker 2002; Nicholas et al. 2003, 2004, 2005, 2014; Sambandam et al. 2004; Ishigami et al. 2005; Raijmakers et al. 2005; Bhattacharya et al. 2006; Grant et al. 2007; Jang et al. 2008, 2010, 2011; Nicholas 2011; Jin et al. 2013; Bradford et al. 2014; Ishigami et al. 2015). Of relevance to AxD, Nicholas and Whitaker (2002) observed that many of the astrocytes staining for peptidyl-citrulline in normal human brain were subpial or perivascular, which are the same sites at which Rosenthal fibers preferentially accumulate in this disorder.

An unusual feature of the immunoblots of the Nicholas laboratory (Nicholas et al. 2003, 2004, 2014; Sambandam et al. 2004) is the observation of multiple species staining for both GFAP and peptidyl-citrulline at apparent molecular weights (MWs) of about 55, 65, and 70 kDa, in addition to the expected band at 50 kDa (human GFAP MW is 49.9 kDa) (Fig. 20.3). Interestingly, the 50 kDa band stains most strongly for GFAP, but one of the more slowly migrating species stains strongest for peptidyl-citrulline. Ishigami et al. (2015) have observed that some forms of citrullinated GFAP may migrate somewhat more slowly than the unmodified protein in an SDS-polyacrylamide gel, but this has not been reported in studies from multiple other laboratories. Possibly the more slowly migrating GFAP forms observed by the Nicholas laboratory were created during the 6-day, multistep procedure used in their studies to isolate the proteins for immunoblotting. If so, the relative peptidyl-citrulline staining intensities indicate that GFAP is much more prone to this alteration when deiminated.

The level of deiminated GFAP, and more generally of peptidyl-citrulline, is elevated in several neurodegenerative conditions. These include rodent EAE models (Nicholas et al. 2005; Raijmakers et al. 2005; Grant et al. 2007), MS (Nicholas et al. 2004; Bradford et al. 2014), Alzheimer's disease (AlzD) (Ishigami et al. 2005; Jin et al. 2013; Nicholas 2013; Nicholas et al. 2014; Ishigami et al. 2015), scrapie (Jang et al. 2008), Creutzfeldt-Jakob disease (Jang et al. 2010), and glaucoma (Bhattacharya et al. 2006). In the latter case, suggestive of a direct role for astrocytes was the finding that primary human cultures responded to increased pressure by increasing their level of PAD2 and deiminated proteins. Cultured human primary astrocytes were also used by Sambandam et al. (2004) in their investigation of the effect of hypoxia on deimination. Immunoblotting indicated a large increase in both PAD2 and deiminated GFAP in response to reduced oxygen. In contrast, although both PAD and deiminated proteins were observed primarily in astrocytes when rats were subjected to hypoxia by reducing the oxygenation of a perfluorochemical perfusate, there was no change in PAD levels in response to hypoxia (Asaga and Ishigami 2000). In addition, the increased deimination was in a broad group of proteins migrating more slowly than GFAP on an immunoblot. An acute increase in GFAP deimination in response to seizures is indicated by elevated immunostaining of deiminated proteins in rat cerebrum astrocytes 2 h after kainic acid treatment (Asaga and Ishigami 2001). On the other hand, a decrease compared to controls in both PAD2 and deiminated proteins in astrocytes was observed 7 days after kainic acid treatment (Asaga et al. 2002). Together, these observations suggest that PAD2 is rapidly activated in astrocytes following seizure activity, but then lost. This loss could be due to changes in intracellular calcium levels; or alternatively, the acute activation could be occurring in severely injured cells that leak calcium to activate PAD2 and then die. Deiminated GFAP has also been reported present in Parkinson's disease, but it was not stated whether the level was increased over controls (Nicholas 2011). In AxD, large

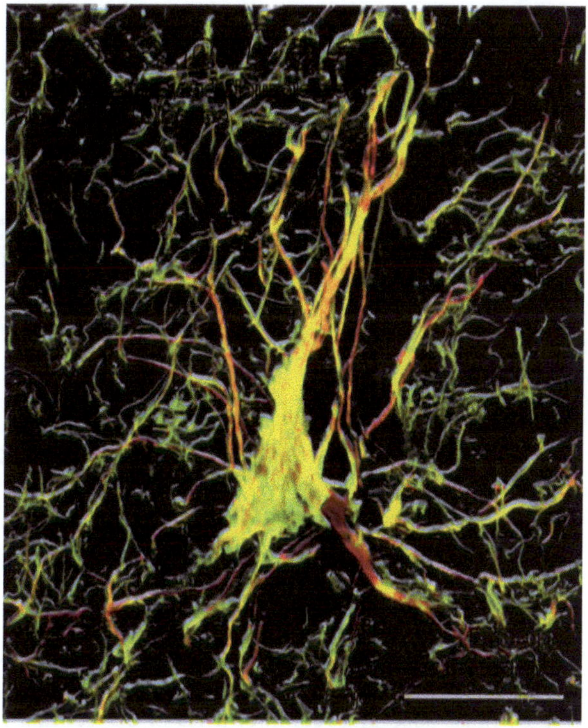

Fig. 20.2 Confocal microscopy of co-localized (*yellow/orange*) immunofluorescent staining for deiminated proteins with the F95 monoclonal antibody against peptidyl-citrulline (*red*) and GFAP (*green*) in the cell body and processes of a protoplasmic astrocyte from the granular cell layer of a human cerebellum without known neurological disease. Scale bar = 50 μm [Modified from Nicholas and Whitaker (2002)]

increases in GFAP deimination have been indicated in both human patients (Fig. 20.3) and a mouse model (Figs. 20.4 and 20.5) (Nicholas et al. 2014).

The increased amount of citrullinated GFAP in several neurodegenerative diseases could simply reflect the increased levels of total GFAP produced by the reactive astrocytes present in the disorder or could be due to greater deimination per GFAP molecule. This question was addressed by Raijmakers et al. (2005) in the case of the EAE mouse model by comparing the relative staining intensities for GFAP and comigrating peptidyl-citrulline on immunoblots of extracts from control and EAE mouse brain. Although not quantified, visually the signal intensity for peptidyl-citrulline increased markedly more than that for GFAP between the control and EAE samples, indicating a greater degree of citrullination per protein.

For most of these reports, citrullination of GFAP is inferred from correlation of the signal locations for GFAP and peptidyl-citrulline, rather than being directly demonstrated. Images obtained by immunostaining generally lack the resolution to distinguish between peptidyl-citrulline staining of GFAP filaments and more general staining of astrocytes. This caveat of correlation persists even for the immunogold staining for peptidyl-citrulline associated with astrocytic intermediate filaments (Jang et al. 2011), since it is possible that the staining was of a protein associated with GFAP rather than GFAP itself. For example, the same study observed localization of PAD2, itself a substrate for deimination, with the GFAP filaments.

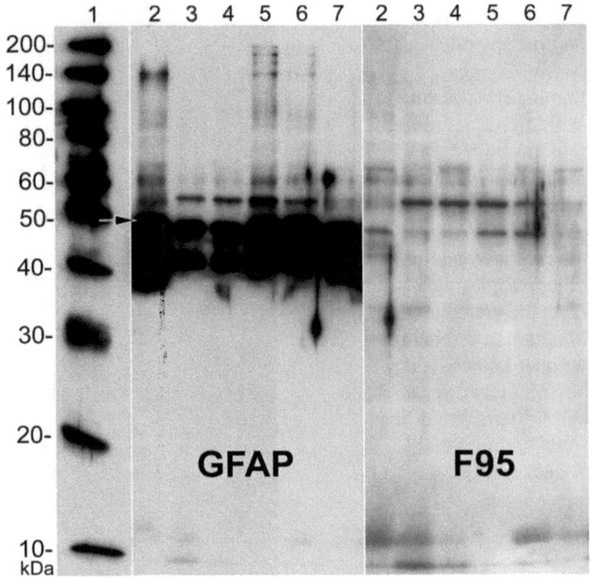

Fig. 20.3 Immunoblot for GFAP and citrullinated proteins (F95) from 1-year-old (lanes 2–4) and 2-year-old (lanes 5–7) AxD patients (lanes 3–6) and age-matched controls (lanes 2 and 7), using extracts from the cerebrum (lanes 2, 3, and 5) and cerebellum (lanes 4, 6, and 7). Lane 1 contains molecular weight markers. The arrow depicts the expected position of human GFAP (50 kDa) [Modified from Nicholas et al. (2014)]

The same reservation of possible co-positioning applies to many of the experiments in which GFAP was presumed to be deiminated based on its identification by mass spec in a region of a 1D or 2D gel that produced a peptidyl-citrulline signal, but the GFAP was not directly demonstrated to contain peptidyl-citrulline (Ishigami et al. 2005; Jang et al. 2008, 2010; Bradford et al. 2014). More direct evidence for the presence of citrullinated GFAP was obtained by Bhattacharya et al. (2006) by identification of GFAP by liquid chromatography-tandem mass spectrometry (LC-MS/MS) among proteins immunoprecipitated by an anti-citrulline antibody. However, in that study, controls were absent to address the possibility that GFAP was present in the immunoprecipitate as a nonspecific contaminant.

Although there are caveats concerning the conclusions in many of the individual studies demonstrating GFAP deimination, taken as a group, they provide compelling evidence, given the multiple properties shared by the peptidyl-citrulline reacting protein and GFAP. These include general morphology in astrocytes, associated with astrocytic intermediate filaments, molecular weight, and isoelectric point. Most convincing, however, is the direct detection by several groups of deiminated arginines within GFAP by mass spec. Grant et al. (2007) searched for citrullinated proteins in spinal cord extracts of control and EAE model rats using the isobaric tags for relative and absolute quantification (iTRAQ) proteomic method (Ross et al. 2004). They observed a deimination in the head domain of rat GFAP at R27. This arginine is conserved in human GFAP as R29, but human R29 has not been observed to be deiminated (see below), perhaps due to its different context (PTRHL in rat, PGRRL in human). The deiminated peptide was 1.3 times more abundant in the EAE extracts than in the controls, but it was not determined if this was due to an increase in citrullination per molecule or to an EAE-induced increase in total GFAP.

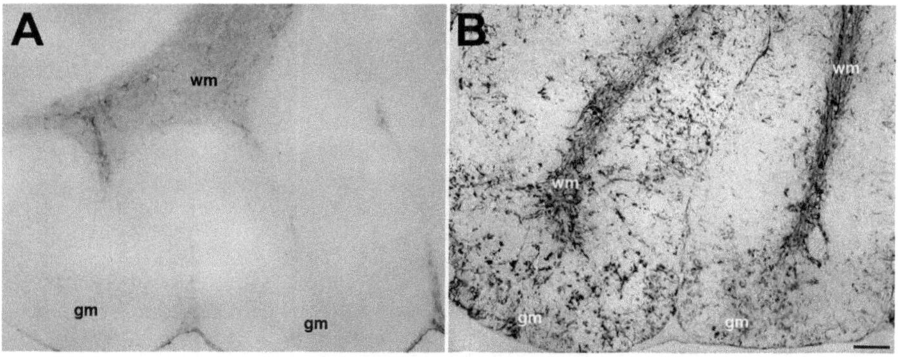

Fig. 20.4 Immunostaining for citrullinated proteins in the cerebellum of (**a**) control mice versus (**b**) AxD model mice that overexpress GFAP. wm, white matter; gm, gray matter; ventricular surface, bottom [Modified from Nicholas et al. (2014)]

Deimination of human GFAP in controls and AlzD brains was analyzed by Jin et al. (2013) using an LC-MS/MS method modified for detection of citrullinated proteins. The AlzD extracts revealed deimination at five sites, R30, R36, R270, R406, and R416. The control extracts also displayed deimination at each of these sites except for R270, but R270 was found deiminated in analysis of commercially obtained purified GFAP, suggesting its absence in the control extracts was a sensitivity issue. Of these five sites, R270 is in coil 2B, whereas the others are in the head or tail domains (Fig. 20.1). No comparison was made between the level of citrullination at the several sites in GFAP from the AlzD and control brains, nor was the extent of deimination determined for any of the sites. The in vitro deimination of the commercial GFAP with purified PAD2 resulted in citrullination of these same five sites, plus 12 additional ones: R88, R105, R124, R126, R136, R173, R217, R258, R276, R286, R287, and R367 (Fig. 20.1). Unfortunately, the properties of this deiminated GFAP were not investigated. As noted in Fig. 20.1, mutation of some of the GFAP arginines subject to deimination causes AxD, raising the possibility that their chronic deimination could have AxD-like consequences.

Ishigami et al. (2015) used a different approach to identify citrullinated peptides in GFAP. Recombinant human (rh) GFAP was deiminated in vitro with rhPAD2 and then used to generate monoclonal antibodies that recognized citrullinated GFAP, but not unmodified GFAP. Surprisingly, when one of these monoclonal antibodies was used to isolate peptides from a tryptic digest of deiminated rhGFAP, two different peptidyl-citrulline peptides were identified that have no apparent homology: FADLTDAAA<u>R</u>NAELLR, which was citrullinated at R270, and TVEM<u>R</u>DGEVIKESKQEHKDVM, which was citrullinated at R416 (citrullination sites underlined; note that the R416 peptide had three missed cleavages). Both of these citrullination sites had previously been observed in vivo by Jin et al. (2013) in control and AlzD subjects. Although not remarked upon by the authors, the isolation of these two different peptides suggests two possibilities. One is that the monoclonal antibody they used recognizes multiple peptidyl-citrulline sites rather than being

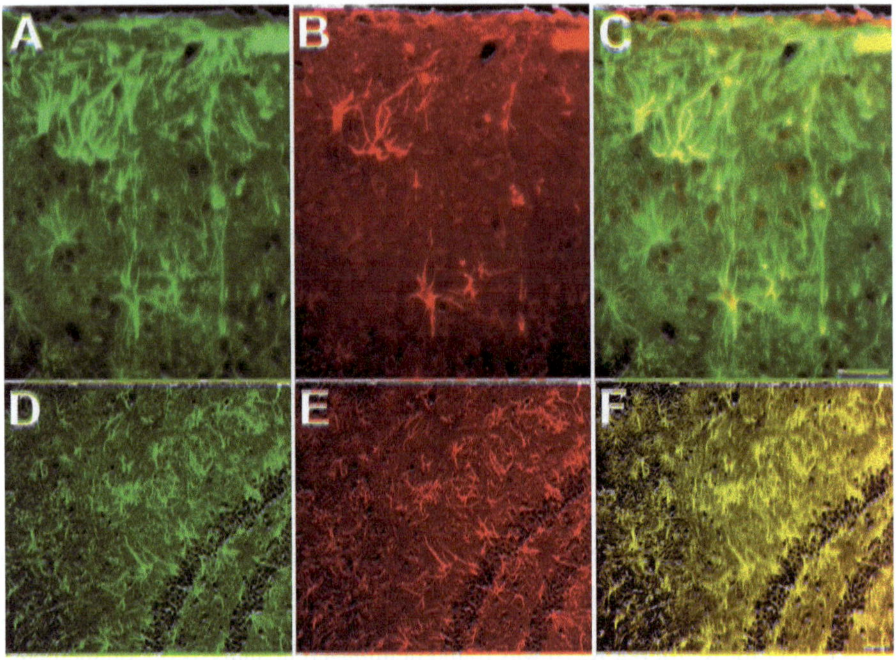

Fig. 20.5 Co-localization (*yellow*) of GFAP (*green*) and peptidyl-citrulline (F95 antibody, *red*) immunofluorescent staining in the telencephalon (**a**–**c**) near the cortical surface (*top*) and in the hippocampus (**d**–**f**) of GFAP-overexpressing transgenic mice demonstrate that the most reactive astrocytes were also immunopositive for deiminated proteins. Scale bars = 50 μm [Modified from Nicholas et al. (2014)]

specific for GFAP citrullination; the other is that the antibody is specific for one of the peptide sites, and the other was present as a contaminant, since a negative control for their peptide isolation in which the antibody was omitted was not described. Accordingly, citrullinated GFAP cannot rigorously be concluded to be the target when this same monoclonal antibody stained AlzD brain but not control brain, although this is suggested by the staining being strongest in astrocytes.

20.6 Vimentin Deimination

Studies of GFAP KO mice have indicated that some of the functions of GFAP are subtle, whereas knockout of both GFAP and vimentin produces more profound effects (reviewed in Brenner 2014; Pekny et al. 2014). Thus, the functional consequences of GFAP deimination could depend on whether vimentin is also deiminated. Since these two proteins are homologous type III IF proteins present in astrocytes, it is not surprising that deiminated vimentin has also been identified. Inagaki et al. (1989) obtained extensive deimination in vitro using purified vimentin and PAD2.

In vivo, immunoblotting has indicated the presence of citrullinated vimentin in rat uterus (Senshu et al. 1992), as an autoantigen in rheumatoid arthritis (Vossenaar et al. 2004), in a human leukemia cell line treated with the calcium ionophore A23187 to activate PAD (Hojo-Nakashima et al. 2009), and in cultured differentiating osteoclast precursors (Harre et al. 2012). The latter study identified a single deimination site at R113, which is in coil 1A (Fig. 20.1); and Wang et al. (2016) identified deiminations in vimentin in synovial fluid of patients with rheumatoid arthritis at R270 and R273 (both near the beginning of coil 2A) and R401 (near the end of coil 2A). Thus, none of the vimentin deimination sites identified in vivo is in the head domain, as was found in the in vitro study of Inagaki et al. (1989). Thus, selection of sites for deimination is influenced by the local environment. In the CNS, citrullinated vimentin has been detected in the brains of AlzD patients (Ishigami et al. 2005) and Creutzfeldt-Jakob disease patients (Jang et al. 2010) by mass spec analysis of spots on a 2D gel comigrating with a peptidyl-citrulline immunostaining signal.

20.7 Biological Effects of Vimentin and GFAP Deimination

The first investigation of possible functional effects of deimination was the in vitro study of Inagaki et al. (1989). Data were primarily presented for vimentin, but similar observations were stated to have been made for GFAP. Purified, soluble mouse vimentin was deiminated to varying levels using purified PAD2 and then tested for its ability to polymerize. The extent of deimination was evaluated by amino acid analysis, and polymerization was assayed either by centrifugation or by electron microscopic visualization of filaments. These assays revealed that deimination to a level of about eight moles of citrulline per mole of vimentin resulted in complete inhibition of filament formation. Lower levels of deimination were examined only by the solubility assay, with the lowest amount, about one mole citrulline per mole vimentin, producing only a marginal effect. The amino acid composition of peptides produced from the highly deiminated vimentin revealed that a head domain peptide containing 12 arginines averaged about eight modified residues and a tail domain peptide had a deimination at R450, which is homologous to R416 in human GFAP (Fig. 20.1). No deimination was observed in the central rod domain. The inability of the deiminated vimentin to polymerize is consistent with this function being dependent on arginine-rich regions in the N-terminal head domain (Shoeman et al. 2002), a requirement shared by GFAP (Chen and Liem 1994). The tail region deimination could also contribute to defective polymerization, since it occurs in an RGD sequence that is conserved among type III intermediate filament proteins and has also been found to have a role in filament formation (McCormick et al. 1993; Chen and Liem 1994). Consistent with these in vitro results, in their study treating a human leukemia cell line with the calcium ionophore A23187, Hojo-Nakashima et al. (2009) observed that the filamentous vimentin network in affected cells collapsed around the nucleus. Although this collapse could be an indirect effect of the calcium ionophore, it is consistent with vimentin being a primary deiminated protein in these cells.

In addition to its effect on polymerization, in vitro deimination of the head region of vimentin inhibited its phosphorylation, presumably due to dependence of the kinases on arginine residues for site recognition (Inagaki et al. 1989). Like deiminated vimentin, phosphorylated vimentin is assembly incompetent. Several signal-regulated kinases phosphorylate serine and threonine residues in the head region of vimentin during mitosis, leading to the reversible dissolution of the filaments, a process that appears important for successful mitosis (Matsuyama et al. 2013). Whether the inhibition of phosphorylation by deimination would compromise mitosis is unclear, since deimination itself may produce the disassembly of IFs, albeit irreversibly. On the other hand, phosphorylation of type III IFs facilitates binding of 14-3-3 proteins (Li et al. 2006b), which in turn stimulates cell growth, migration, and malignancy (reviewed in Snider and Omary 2014), and it is questionable whether this effect of phosphorylation would be mimicked by deimination.

There are several caveats to drawing conclusions for the biological effects of GFAP deimination from the Inagaki et al. study. Although it was reported that similar effects on polymerization to those obtained for vimentin were also observed for GFAP, centrifugation data for GFAP were presented for only a single time point in the deimination reaction, and there was no electron microscopic visualization of polymerization or determination of the deimination sites. These sites could differ between vimentin and GFAP, since the sequences of the head domains are not conserved. In addition, both the deimination and polymerization reactions were performed in vitro, which may not be predictive of conditions in vivo. For example, some altered vimentin and GFAP proteins that do not polymerize in vitro nevertheless do so in vivo and vice versa (McCormick et al. 1993; Chen and Liem 1994). Another concern is that the deimination was performed using soluble IF protein rather than filaments, whereas over 90% of the total GFAP in the brain is filamentous (Heaven et al. 2016). The authors do state that deimination of intact GFAP filaments leads to their dissociation, but no relevant data or experimental protocols were presented.

Perhaps the greatest concern for drawing biological consequences of GFAP deimination from the Inagaki et al. study is that the level of GFAP deimination in vivo is not known. This question was not addressed by any of the studies demonstrating GFAP deimination, including those using mass spec to directly identify peptidyl-citrulline in GFAP. If the level approached the eight moles of citrulline per mole of protein monomer that was primarily investigated by Inagaki et al., it might be expected that more extensive deimination would have been detected by the mass spec analyses. Also not addressed in the studies using in vivo tissues is the effect of GFAP deimination on its polymerization properties, but pertinent information can be extracted from some of the data presented. Several of the dual fluorescence immunostaining studies cited earlier show essentially identical morphologies for GFAP and peptidyl-citrulline signals (as examples, see Fig. 20.2 above and other images in Nicholas and Whitaker 2002; Nicholas et al. 2003, 2005; Nicholas 2011). In addition, the electron microscopic visualization of peptidyl-citrulline immunogold staining indicated labeling of astrocyte filaments (Jang et al. 2011). Although caveats concerning interpretation have been mentioned earlier, these observations suggest that deiminated GFAP can be incorporated into normal-appearing filaments.

Furthermore, immunoblot data of Raijmakers et al. (2005) suggest that deiminated GFAP polymerizes with a facility similar to that of unmodified GFAP. In their study of GFAP deimination in a mouse EAE model, the distribution of deiminated GFAP was monitored between Triton X-100 soluble and insoluble fractions. Solubility in Triton X-100 is commonly used to separate polymerized IFs from monomers and small oligomers. Although the amount of each of these fractions loaded on the gels was not specified, the use of equal volumes for each fraction and the observation that generally about 10% of GFAP is Triton X-100 soluble (Heaven et al. 2016) suggest that relatively similar percentages of each fraction were analyzed. If this was indeed the case, their data show that a large proportion of the deiminated GFAP is present in the insoluble fraction and thus retains its ability to polymerize.

Seemingly contrary data for the solubility of citrullinated GFAP was subsequently presented by this same group (Raijmakers et al. 2006), again using a mouse EAE model. Immunoblots for peptidyl-citrulline showed a much stronger signal corresponding to the size expected for GFAP in a Triton X-100 soluble fraction than in the insoluble fraction. However, in this instance, sonication was used to prepare the extracts, and the previous publication showed that this could result in the solubilization of a significant fraction of otherwise insoluble GFAP. Thus, in the absence of immunoblotting for GFAP, which was not performed in this later study, no definitive conclusion can be drawn from this result regarding the effect of citrullination on GFAP solubility.

There are several possible explanations for why these in vivo observations of polymerized deiminated GFAP differ from the absence of polymerization observed by Inagaki et al. (1989) described earlier. One possibility is that only a small fraction of GFAP molecules may be deiminated. Although GFAP lacking its head domain cannot self-polymerize when expressed in cultured cells lacking cytoplasmic intermediate filaments, it can incorporate into filaments when expressed in vimentin-containing cells (Chen and Liem 1994), such as astrocytes that contain both of these IFs. Therefore, if only a minor fraction of GFAP is deiminated, it might similarly assemble into filaments with unmodified GFAP in vivo. Another possibility is that the extent of deimination per molecule is low. As mentioned in the earlier discussion, the failure to polymerize was noted for vimentin having about eight sites deiminated per molecule, whereas GFAP, with an average of only a single site deiminated, had almost normal polymerization. Finally, as also suggested earlier, in vitro polymerization properties may not adequately reflect those in vivo.

20.8 Possible Functional Roles for GFAP Deimination in Brain Trauma and Neurodegenerative Diseases

Observations summarized above suggest that GFAP deimination has an important role in neurological diseases. These findings include (1) PAD2 appears to be the predominant PAD in the CNS and is primarily (if not entirely) localized to astrocytes,

(2) identification of GFAP as a major deiminated protein, and (3) increased levels of both PAD2 and GFAP deimination are seen in several neurodegenerative diseases. However, it is not known what role GFAP deimination has in neurodegeneration or even if it is positive or negative. Studies of GFAP KO mice have attributed multiple functions to GFAP, any one of which might be affected by its deimination (reviewed in Brenner 2014). These include contributing to structural support of the vasculature, regulation of blood flow, facilitating blood-brain barrier formation and maintenance, hippocampal long-term potentiation, cerebellar motor learning and long-term depression, sequestration of damaged tissue, glutamate transporter trafficking, and suppression of neurite extension and neuron proliferation in the adult brain. In addition, a recently discovered nonstructural role for GFAP is facilitating inhibition of chaperone-mediated autophagy by GTP (Bandyopadhyay et al. 2010). Since a phosphorylated form of GFAP has been implicated in this process, this function may be compromised by deimination. Chaperone-mediated autophagy may contribute to the progression of several neurodegenerative disorders, including Parkinson's disease, AlzD, Huntington's disease, and normal aging (Wang and Mao 2014).

Combined knockout of both GFAP and vimentin reveals additional functions that could be affected by their deimination (reviewed in Brenner 2014; Pekny et al. 2014). These include vesicle trafficking, protection against mechanical stress, volume control, formation of the glial scar following injury, and suppression via Notch signaling of neural stem cell differentiation into astrocytes and neurons. Still other possible roles for GFAP can be inferred from analogy to vimentin and keratin intermediate filaments, which have been discovered to participate in several signaling pathways involving growth, cell division, and apoptosis (reviewed in Snider and Omary 2014). In particular, Hsu et al. (2014) observed that overexpression of vimentin partially alleviates apoptosis in Jurkat cells caused by activation of PAD2.

Whether GFAP deimination has a role in any of these functions has not yet been investigated, but an indication can be discerned from the more general question of whether GFAP itself has any role in neurodegenerative disorders. This has been addressed by using GFAP KO or GFAP/vimentin double KO mice in models of neurodegenerative diseases (Brenner 2014; Pekny et al. 2014). Despite increased levels of citrullinated GFAP in a mouse scrapie model, no effect on the disease was observed in a GFAP KO. Both harmful and helpful effects have been observed in several other mouse injury models. Deleterious effects of a GFAP KO or double KO of GFAP and vimentin have included a larger infarct size following ischemia, more rapid progression and severity of a Batten disease mouse model, and greater acute loss of synapses and increased healing time after traumatic injury. In AlzD model mice, increases in both plaque load and dystrophic neurites have been noted. Multiple studies have reported that deiminated GFAP increases in this disease (Ishigami et al. 2005; Jin et al. 2013; Nicholas 2013; Nicholas et al. 2014; Ishigami et al. 2015), although Acharya et al. (2012) reported that the increase in peptidylcitrulline in AlzD is confined to neurons. Beneficial effects have been observed in several models of traumatic injury (see Chap. 16 for more details). These positive effects have included increased differentiation of astrocytes and neurons from neural stem cells, better long-term behavioral recovery from spinal cord injury, and increased neuronal migration, axonal growth, and neurite extension.

The availability of PAD inhibitors and KO mice makes possible a more direct test of the role of deiminated GFAP. A PAD2 KO mouse was used by Raijmakers et al. (2006) to examine the impact of protein deimination on an EAE model of MS. They observed that the KO had no effect on disease onset, severity, or duration, despite being unable to detect protein deimination in the brain and spinal cord that was readily apparent in wild-type EAE mice. These findings suggest that the GFAP deimination accompanying EAE has no role in the disease. A caveat for this conclusion is the report of Bradford et al. (2014) that spinal cords from EAE control and PAD2 KO mice stain equally for peptidyl-citrulline. This discrepancy may be due to the employment of different detection methods in these two investigations. Raijmakers et al. used an antibody against citrulline residues that are first chemically modified (the method of Senshu et al. 1992), whereas Bradford et al. used an antibody that directly recognizes peptidyl-citrulline (F95 antibody, Nicholas et al. 2014). The conclusion that deiminated GFAP has no role in EAE thus awaits further study, although its presence in the PAD2 KO mice would presumably require discovering that a PAD other than PAD2 is also present in astrocytes. The more general conclusion that deiminated proteins are without effect in EAE was challenged by Moscarello et al. (2013), who found that the general PAD inhibitor 2-chloroacetamidine provided strong protection in several mouse EAE models. This outcome was attributed to the presence of PAD4.

PAD2 KO mice were also used to evaluate the role of GFAP deimination in a mouse model of AxD. The mice used carried a combination of two genetic changes. One is a transgene expressing human GFAP ($GFAP^{TG}$), and the other is an R236H alteration of the endogenous GFAP gene, which is homologous to the highly severe R239H AxD mutation. Each of these changes alone results in some characteristics of AxD, such as greater susceptibility to kainic acid induced seizures and production of Rosenthal fibers, but the mice live a normal life span. However, the $GFAP^{TG}$/R236H double transgenic mouse dies by about 30 days of age, apparently from seizures (Hagemann et al. 2006). Combining the $GFAP^{TG}$/R236H double transgenic mouse with the PAD2 KO significantly increased the survival time ($p < 0.01$) (A.N. and M.B., unpublished observations). In most instances, life span was increased about a week, but one mouse appeared still healthy when euthanized at 6 months of age. These indications that PAD2 has a deleterious effect on AxD progression were contrary to expectations. The in vitro studies of GFAP deimination (Inagaki et al. 1989) and findings for MBP (Chap. 19) suggest that PAD2 activity would decrease GFAP polymerization and increase its degradation. Since elevated GFAP levels contribute to AxD, probably through aberrant polymerization (Li et al. 2006a), GFAP deimination was expected to have a protective effect. A simple explanation of the contrary outcome is that chronic deimination of GFAP mimics the effects of disease-causing mutations in AxD (see Fig. 20.1), but other mechanisms are possible. A practical consequence of this finding is that PAD2 inhibitors may extend survival of AxD patients, providing the first treatment available for this deadly disease.

20.9 Conclusions, Conjectures, and Future Directions

PAD2 accounts for nearly all of the PAD activity in the CNS, and most of this PAD2 is present in astrocytes. Multiple studies have identified GFAP as a primary PAD2 target and have observed that the level of deiminated GFAP is markedly increased in a variety of CNS disorders. Presumably, deimination of GFAP provides some critical response to CNS disturbance, but how deimination affects GFAP function and its biological consequences await further investigation.

GFAP is one of the main IFs in astrocytes, a critical glial cell with anatomical and functional associations with neurons, cerebral blood vessels, and the ventricular system (Fig. 20.6a). A possible role for GFAP (and vimentin) deimination is to contribute to the reorganization of the cytoskeleton that occurs as part of the astrocytic reactive response (Fig. 20.6b). Following CNS injury, astrocytes become hypertrophic and may extend and intertwine their processes to form a mesh-like network that seals off the injured area (Sofroniew 2009). As a result, GFAP deimination could facilitate formation of this glial scar by causing depolymerization of the preexisting intermediate filaments. Two teleological considerations, however, militate against this role. The first is that the concentration of calcium present in reactive astrocytes, likely less than 250 nM (Matsuda et al. 1996; Cheli et al. 2016), is far below that required for half maximal activity of PAD2, which is about 50 μM. Why would PAD2 have evolved such a high calcium requirement if its primary function was in conditions that support less than 0.5% of its maximum activity? Much higher calcium levels, up to about 100 μM, could be transiently experienced by the enzyme, if it were sequestered near a calcium channel (Csordas et al. 1999), but evidence for such a localization is lacking. The second teleological challenge is that if GFAP deimination were used for cytoskeleton rearrangement, the deiminated GFAP would need to be degraded and new GFAP synthesized prior to reformation of filaments. A more resource-sparing means for cytoskeleton rearrangement would be reversible phosphorylation, as occurs during mitosis (Matsuyama et al. 2013).

Another possible function of GFAP deamination that is not subject to these concerns is depolymerization of the filament network in the course of apoptosis (Fig. 20.6c). Although the intracellular calcium level may also be no more than 250 nM when apoptosis is initiated in astrocytes (Matsuda et al. 1996; Takuma et al. 1999), activation of caspases may lead to cleavage of calcium transporter and channel proteins resulting in breakdown of the calcium concentration gradient between the astrocytes and the cerebrospinal fluid (Orrenius et al. 2003), where the calcium concentration is about 1 mM. Dissolution of GFAP filaments has been observed in an astrocyte apoptosis model induced by treatment with 3-nitropropionate (Fukuda et al. 1998), although an effect of the drug itself cannot be ruled out. If excessive deimination of GFAP does result in astrocyte apoptosis, ultimately the same fate may eventually affect their associated neurons that may not be able to survive without the critical assistance of these supporting glial cells (Fig. 20.6c).

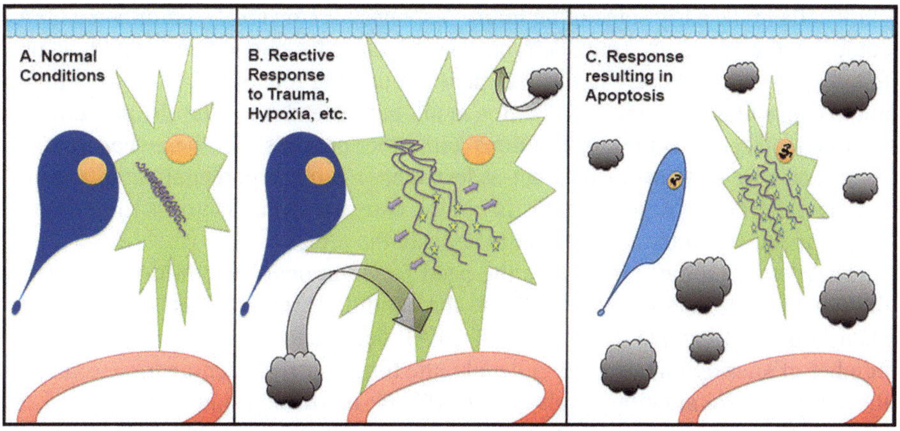

Fig. 20.6 Theoretical functions of IF deimination in astrocytes. Under normal conditions (**a**), astrocytes (*green*) extend processes that interact with ependymal cells (*light blue*) lining the ventricular system (*top*), endothelial cells of central nervous system blood vessels (*red*, *bottom*), and neurons (*dark blue*). Within astrocytes, IFs such as GFAP and vimentin (*purple*) are found in tightly wound fibrils that are critical to the cytoskeletal integrity of these cells. However, in response to injury (**b**), PAD enzymes are activated in astrocytes, resulting in IF deimination, in which peptidyl arginines are transformed to peptidyl-citrullines (*yellow stars*). Due to the loss of the positive charges on these arginines that contribute to salt bridges between the fibrils, IFs may unwind, which may then contribute to expansion of the intracellular IF matrix (*purple arrows*) resulting in astrocyte enlargement that may serve to enhance associations with neurons (to provide these cells nutrition and assist in dispelling waste), as well as with the cerebrospinal fluid and blood systems, possibly in an attempt to clear waste products and extracellular debris (*black clouds*) via phagocytosis (*black arrows*). Alternatively, deimination of IFs may be part of an apoptotic program that includes collapse of the cytoskeleton (**c**) with condensation of nuclear material (*black lines* within *orange* ovals that represent cell nuclei) and blebbing off of extracellular debris for phagocytosis. Loss of astrocytic support could ultimately cause neuronal cell death (shrunken *blue* cell in panel **c**)

Both of these possibilities posit that deimination causes depolymerization of GFAP filaments, followed by rapid degradation of the deiminated protein. However, neither provides an explanation for the observed presence of deiminated GFAP in both diseased and control brains that is associated with normal-appearing intermediate filaments. Inagaki et al. (1989) reported that highly deiminated GFAP was incompetent for self-polymerization in vitro. However, multiple studies have shown that altered forms of GFAP that cannot self-polymerize can nevertheless copolymerize with an excess of wild-type GFAP (Chen and Liem 1994; Der Perng et al. 2006; Flint et al. 2012; Moeton et al. 2016). Thus, the deiminated GFAP visualized in normal-appearing filaments may be present as a low molar fraction of the total GFAP, or each deiminated GFAP monomer may have only one or a few citrullinations. The origin and significance of this filamentous deiminated GFAP, however, are unclear. In control brain, it could be the product of a low background level of PAD activity or could identify cells that had recently experienced some perturbation. In the injured brain, the elevated level of filamentous deiminated GFAP could mark cells in the early stage of an injury response; or conversely, deiminated GFAP could be the remnants from deimination-driven cytoskeletal reorganization or a failed apoptotic attempt. Finally, it is possible that the role of GFAP deimination is not to cause filament depolymerization at all, but instead is to alter the properties of the filaments, such as their interactions with other proteins. For these scenarios, the puzzle of why PAD evolved to act at less than 1% of its potential activity would remain. Alternatively, perhaps this is not the case, and some factor is to be discovered that can substitute for calcium to activate the protein.

These uncertainties underscore the need for additional basic information about the functions of deiminated GFAP and the conditions under which it is produced. Data of interest include the fraction of GFAP deiminated in a given condition, the number of sites deiminated per molecule, the time course of deimination, effects of the observed deimination levels on GFAP polymerization and stability, and the effects of incorporation of deiminated GFAP on filament properties. Determining whether GFAP deimination has a role in glial scar formation or apoptosis, or in any of the multiple other activities attributable to GFAP enumerated above, can now be performed through the use of tools such as the PAD2 KO mice, PAD2-overexpressing mice, and PAD inhibitors. Studies performed to date indicate that GFAP deimination has a questionable role in the EAE mouse model of MS, but has a critical one in a mouse model of AxD. Extending these approaches to other CNS disorders should produce findings relevant to both basic biological processes and for clinical application.

References

Acharya, N. K., Nagele, E. P., Han, M., Coretti, N. J., DeMarshall, C., Kosciuk, M. C., Boulos, P. A., & Nagele, R. G. (2012). Neuronal PAD4 expression and protein citrullination: Possible role in production of autoantibodies associated with neurodegenerative disease. *Journal of Autoimmunity, 38*, 369–380.

Asaga, H., & Ishigami, A. (2000). Protein deimination in the rat brain: Generation of citrulline-containing proteins in cerebrum perfused with oxygen-deprived media. *Biomedical Research, 21*, 197–205.

Asaga, H., & Ishigami, A. (2001). Protein deimination in the rat brain after kainate administration: Citrulline-containing proteins as a novel marker of neurodegeneration. *Neuroscience Letters, 299*, 5–8.

Asaga, H., & Senshu, T. (1993). Combined biochemical and immunocytochemical analyses of postmortem protein deimination in the rat spinal cord. *Cell Biology International, 17*, 525–532.

Asaga, H., Akiyama, K., Ohsawa, T., & Ishigami, A. (2002). Increased and type II-specific expression of peptidylarginine deiminase in activated microglia but not hyperplastic astrocytes following kainic acid-evoked neurodegeneration in the rat brain. *Neuroscience Letters, 326*, 129–132.

Bandyopadhyay, U., Sridhar, S., Kaushik, S., Kiffin, R., & Cuervo, A. M. (2010). Identification of regulators of chaperone-mediated autophagy. *Molecular Cell, 39*, 535–547.

Bhattacharya, S. K., Crabb, J. S., Bonilha, V. L., Gu, X., Takahara, H., & Crabb, J. W. (2006). Proteomics implicates peptidyl arginine deiminase 2 and optic nerve citrullination in glaucoma pathogenesis. *Investigative Ophthalmology & Visual Science, 47*, 2508–2514.

Bradford, C. M., Ramos, I., Cross, A. K., Haddock, G., McQuaid, S., Nicholas, A. P., & Woodroofe, M. N. (2014). Localisation of citrullinated proteins in normal appearing white matter and lesions in the central nervous system in multiple sclerosis. *Journal of Neuroimmunology, 273*, 85–95.

Bradford, C., Nicholas, A. P., Woodroofe, N., & Cross, A. K. (2014). Deimination in multiple sclerosis and experimental autoimmune encephalomyelitis. In A. P. Nicholas & S. K. Bhattacharya (Eds.), *Protein deimination in human health and disease* (pp. 165–185). New York: Springer.

Brenner, M. (2014). Role of GFAP in CNS injuries. *Neuroscience Letters, 565*, 7–13.

Brenner, M., Goldman, J. E., Quinlan, R. A., & Messing, A. (2009). Alexander disease: A genetic disorder of astrocytes. In V. Parpura & P. Haydon (Eds.), *Astrocytes in (Patho)physiology of the nervous system* (pp. 591–648). New York: Springer.

Cheli, V. T., Santiago Gonzalez, D. A., Smith, J., Spreuer, V., Murphy, G. G., & Paez, P. M. (2016). L-type voltage-operated calcium channels contribute to astrocyte activation *in vitro*. *Glia, 64*, 1396–1415.

Chen, W. J., & Liem, R. K. (1994). The endless story of the glial fibrillary acidic protein. *Journal of Cell Science, 107*, 2299–2311.

Csordas, G., Thomas, A. P., & Hajnoczky, G. (1999). Quasi-synaptic calcium signal transmission between endoplasmic reticulum and mitochondria. *The EMBO Journal, 18*, 96–108.

Der Perng, M., Su, M., Wen, S. F., Li, R., Gibbon, T., Prescott, A. R., Brenner, M., & Quinlan, R. A. (2006). The Alexander disease-causing glial fibrillary acidic protein mutant, R416W, accumulates into Rosenthal fibers by a pathway that involves filament aggregation and the association of alpha b-crystallin and HSP27. *American Journal of Human Genetics, 79*, 197–213.

Flint, D., Li, R., Webster, L. S., Naidu, S., Kolodny, E., Percy, A., van der Knaap, M., Powers, J. M., Mantovani, J. F., Ekstein, J., Goldman, J. E., Messing, A., & Brenner, M. (2012). Splice site, frameshift, and chimeric GFAP mutations in Alexander disease. *Human Mutation, 33*, 1141–1148.

Fukuda, A., Deshpande, S. B., Shimano, Y., & Nishino, H. (1998). Astrocytes are more vulnerable than neurons to cellular Ca2+ overload induced by a mitochondrial toxin, 3-nitropropionic acid. *Neuroscience, 87*, 497–507.

Grant, J. E., Hu, J., Liu, T., Jain, M. R., Elkabes, S., & Li, H. (2007). Post-translational modifications in the rat lumbar spinal cord in experimental autoimmune encephalomyelitis. *Journal of Proteome Research, 6*, 2786–2791.

Hagemann, T. L., Connor, J. X., & Messing, A. (2006). Alexander disease-associated glial fibrillary acidic protein mutations in mice induce Rosenthal fiber formation and a white matter stress response. *The Journal of Neuroscience, 26*, 11162–11173.

Harre, U., Georgess, D., Bang, H., Bozec, A., Axmann, R., Ossipova, E., Jakobsson, P. J., Baum, W., Nimmerjahn, F., Szarka, E., Sarmay, G., Krumbholz, G., Neumann, E., Toes, R., Scherer, H. U., Catrina, A. I., Klareskog, L., Jurdic, P., & Schett, G. (2012). Induction of osteoclastogenesis and bone loss by human autoantibodies against citrullinated vimentin. *The Journal of Clinical Investigation, 122*, 1791–1802.

Heaven, M. R., Flint, D., Randall, S. M., Sosunov, A., Wilson, L., Barnes, S., Goldman, J., Muddiman, D. C., & Brenner, M. (2016). Composition of Rosenthal fibers, the protein aggregate hallmark of Alexander disease. *Journal of Proteome Research, 15*, 2265–2282.

Hojo-Nakashima, I., Sato, R., Nakashima, K., Hagiwara, T., & Yamada, M. (2009). Dynamic expression of peptidylarginine deiminase 2 in human monocytic leukaemia THP-1 cells during macrophage differentiation. *Journal of Biochemistry, 146*, 471–479.

Hsu, P. C., Liao, Y. F., Lin, C. L., Lin, W. H., Liu, G. Y., & Hung, H. C. (2014). Vimentin is involved in peptidylarginine deiminase 2-induced apoptosis of activated Jurkat cells. *Molecules and Cells, 37*, 426–434.

Inagaki, M., Takahara, H., Nishi, Y., Sugawara, K., & Sato, C. (1989). Ca2+-dependent deimination-induced disassembly of intermediate filaments involves specific modification of the amino-terminal head domain. *The Journal of Biological Chemistry, 264*, 18119–18127.

Ishigami, A., Ohsawa, T., Hiratsuka, M., Taguchi, H., Kobayashi, S., Saito, Y., Murayama, S., Asaga, H., Toda, T., Kimura, N., & Maruyama, N. (2005). Abnormal accumulation of citrullinated proteins catalyzed by peptidylarginine deiminase in hippocampal extracts from patients with Alzheimer's disease. *Journal of Neuroscience Research, 80*, 120–128.

Ishigami, A., Masutomi, H., Handa, S., Nakamura, M., Nakaya, S., Uchida, Y., Saito, Y., Murayama, S., Jang, B., Jeon, Y. C., Choi, E. K., Kim, Y. S., Kasahara, Y., Maruyama, N., & Toda, T. (2015). Mass spectrometric identification of citrullination sites and immunohistochemical detection of citrullinated glial fibrillary acidic protein in Alzheimer's disease brains. *Journal of Neuroscience Research, 93*, 1664–1674.

Jang, B., Kim, E., Choi, J. K., Jin, J. K., Kim, J. I., Ishigami, A., Maruyama, N., Carp, R. I., Kim, Y. S., & Choi, E. K. (2008). Accumulation of citrullinated proteins by up-regulated peptidylarginine deiminase 2 in brains of scrapie-infected mice—A possible role in pathogenesis. *The American Journal of Pathology, 173*, 1129–1142.

Jang, B., Jin, J. K., Jeon, Y. C., Cho, H. J., Ishigami, A., Choi, K. C., Carp, R. I., Maruyama, N., Kim, Y. S., & Choi, E. K. (2010). Involvement of peptidylarginine deiminase-mediated post-translational citrullination in pathogenesis of sporadic Creutzfeldt–Jakob disease. *Acta Neuropathologica, 119*, 199–210.

Jang, B., Shin, H. Y., Choi, J. K., Nguyen, d. P. T., Jeong, B. H., Ishigami, A., Maruyama, N., Carp, R. I., Kim, Y. S., & Choi, E. K. (2011). Subcellular localization of peptidylarginine deiminase 2 and citrullinated proteins in brains of scrapie-infected mice: Nuclear localization of PAD2 and membrane fraction-enriched citrullinated proteins. *Journal of Neuropathology and Experimental Neurology, 70*, 116–124.

Jin, Z., Fu, Z., Yang, J., Troncosco, J., Everett, A. D., & Van Eyk, J. E. (2013). Identification and characterization of citrulline-modified brain proteins by combining HCD and CID fragmentation. *Proteomics, 13*, 2682–2691.

Koster, S., Weitz, D. A., Goldman, R. D., Aebi, U., & Herrmann, H. (2015). Intermediate filament mechanics *in vitro* and in the cell: From coiled coils to filaments, fibers and networks. *Current Opinion in Cell Biology, 32*, 82–91.

Lange, S., Gogel, S., Leung, K. Y., Vernay, B., Vocholas, A. P., Sausey, C. P., Thompson, P. R., Greene, N. D., & Feretti, P. (2011). Protein deiminases: New players in the developmentally regulated loss of neural regenerative ability. *Developmental Biology, 355*, 205–214.

Li, R., Johnson, A. B., Salomons, G. S., van der Knaap, M. S., Rodriguez, D., Boespflug-Tanguy, O., Gorospe, J. R., Goldman, J. E., Messing, A., & Brenner, M. (2006a). Propensity for paternal inheritance of de novo mutations in Alexander disease. *Human Genetics, 119*, 137–144.

Li, H., Guo, Y., Teng, J., Ding, M., AC, Y., & Chen, J. (2006b). 14-3-3 gamma affects dynamics and integrity of glial filaments by binding to phosphorylated GFAP. *Journal of Cell Science, 119*, 4452–4461.

Matsuda, T., Takuma, K., Nishiguchi, E., Hashimoto, H., Azuma, J., & Baba, A. (1996). Involvement of Na+-Ca2+ exchanger in reperfusion-induced delayed cell death of cultured rat astrocytes. *The European Journal of Neuroscience, 8*, 951–958.

Matsuyama, M., Tanaka, H., Inoko, A., Goto, H., Yonemura, S., Kobori, K., Hayashi, Y., Kondo, E., Itohara, S., Izawa, I., & Inagaki, M. (2013). Defect of mitotic vimentin phosphorylation causes microophthalmia and cataract via aneuploidy and senescence in lens epithelial cells. *The Journal of Biological Chemistry, 288*, 35626–35635.

McCormick, M. B., Kouklis, P., Syder, A., & Fuchs, E. (1993). The roles of the rod end and the tail in vimentin IF assembly and IF network formation. *The Journal of Cell Biology, 122*, 395–407.

Messing, A., Head, M. W., Galles, K., Galbreath, E. J., Goldman, J. E., & Brenner, M. (1998). Fatal encephalopathy with astrocyte inclusions in GFAP transgenic mice. *The American Journal of Pathology, 152*, 391–398.

Moeton, M., Stassen, O. M., Sluijs, J. A., van der Meer, V. W., Kluivers, L. J., van Hoorn, H., Schmidt, T., Reits, E. A., van Strien, M. E., & Hol, E. M. (2016). GFAP isoforms control intermediate filament network dynamics, cell morphology, and focal adhesions. *Cellular and Molecular Life Sciences, 73*, 4101–4120.

Moscarello, M. A., Lei, H., Mastronardi, F. G., Winer, S., Tsui, H., Li, Z., Ackerley, C., Zhang, L., Raijmakers, R., & Wood, D. D. (2013). Inhibition of peptidyl-arginine deiminases reverses protein-hypercitrullination and disease in mouse models of multiple sclerosis. *Disease Models & Mechanisms, 6*, 467–478.

Nicholas, A. P. (2011). Dual immunofluorescence study of citrullinated proteins in Parkinson diseased substantia nigra. *Neuroscience Letters, 495*, 26–29.

Nicholas, A. P. (2013). Dual immunofluorescence study of citrullinated proteins in Alzheimer diseased frontal cortex. *Neuroscience Letters, 545*, 107–111.

Nicholas, A. P., & Whitaker, J. N. (2002). Preparation of a monoclonal antibody to citrullinated epitopes: Its characterization and some applications to immunohistochemistry in human brain. *Glia, 37*, 328–336.

Nicholas, A. P., King, J. L., Sambandam, T., Echols, J. D., Gupta, K. B., McInnis, C., & Whitaker, J. N. (2003). Immunohistochemical localization of citrullinated proteins in adult rat brain. *The Journal of Comparative Neurology, 459*, 251–266.

Nicholas, A. P., Sambandam, T., Echols, J. D., & Tourtellotte, W. W. (2004). Increased citrullinated glial fibrillary acidic protein in secondary progressive multiple sclerosis. *The Journal of Comparative Neurology, 473*, 128–136.

Nicholas, A. P., Sambandam, T., Echols, J. D., & Barnum, S. R. (2005). Expression of citrullinated proteins in murine experimental autoimmune encephalomyelitis. *The Journal of Comparative Neurology, 486*, 254–266.

Nicholas, A. P., Lu, L., Heaven, M., Kadish, I., Van Groen, T., Accaviti-Loper, M. A., Wewering, S., Kofskey, D., Gambetti, P., & Brenner, M. (2014). Ongoing studies of deimination in neurodegenerative diseases using the F95 antibody. In S. Bhattacharya & A. P. Nicholas (Eds.), *Protein deimination in human health and disease* (pp. 257–280). New York: Springer.

Orrenius, S., Zhivotovsky, B., & Nicotera, P. (2003). Regulation of cell death: The calcium-apoptosis link. *Nature Reviews. Molecular Cell Biology, 4*, 552–565.

Pekny, M. (2001). Astrocytic intermediate filaments: Lessons from GFAP and vimentin knock-out mice. *Progress in Brain Research, 132*, 23–30.

Pekny, M., Wilhelmsson, U., & Pekna, M. (2014). The dual role of astrocyte activation and reactive gliosis. *Neuroscience Letters, 565*, 30–38.

Prust, M., Wang, J., Morizono, H., Messing, A., Brenner, M., Gordon, E., Hartka, T., Sokohl, A., Schiffmann, R., Gordish-Dressman, H., Albin, R., Amartino, H., Brockman, K., Dinopoulos, A., Dotti, M. T., Fain, D., Fernandez, R., Ferreira, J., Fleming, J., Gill, D., Griebel, M., Heilstedt, H., Kaplan, P., Lewis, D., Nakagawa, M., Pedersen, R., Reddy, A., Sawaishi, Y., Schneider, M., Sherr, E., Takiyama, Y., Wakabayashi, K., Gorospe, J. R., & Vanderver, A. (2011). GFAP mutations, age at onset, and clinical subtypes in Alexander disease. *Neurology, 77*, 1287–1294.

Raijmakers, R., Vogelzangs, J., Croxford, J. L., Wesseling, P., van Venrooij, W. J., & Pruijn, G. J. (2005). Citrullination of central nervous system proteins during the development of experimental autoimmune encephalomyelitis. *The Journal of Comparative Neurology, 486*, 243–253.

Raijmakers, R., Vogelzangs, J., Raats, J., Panzenbeck, M., Corby, M., Jiang, H., Thibodeau, M., Haynes, N., van Venrooij, W. J., Pruijn, G. J., & Werneburg, B. (2006). Experimental autoimmune encephalomyelitis induction in peptidylarginine deiminase 2 knockout mice. *The Journal of Comparative Neurology, 498*, 217–226.

Ross, P. L., Huang, Y. N., Marchese, J. N., Williamson, B., Parker, K., Hattan, S., Khainovski, N., Pillai, S., Dey, S., Daniels, S., Purkayastha, S., Juhasz, P., Martin, S., Bartlet-Jones, M., He, F., Jacobson, A., & Pappin, D. J. (2004). Multiplexed protein quantitation in *Saccharomyces cerevisiae* using amine-reactive isobaric tagging reagents. *Molecular & Cellular Proteomics, 3*, 1154–1169.

Sambandam, T., Belousova, M., Accaviti-Loper, M. A., Blanquicett, C., Guercello, V., Raijmakers, R., & Nicholas, A. P. (2004). Increased peptidylarginine deiminase type II in hypoxic astrocytes. *Biochemical and Biophysical Research Communications, 325*, 1324–1329.

Senshu, T., Sato, T., Inoue, T., Akiyama, K., & Asaga, H. (1992). Detection of citrulline residues in deiminated proteins on polyvinylidene difluoride membrane. *Analytical Biochemistry, 203*, 94–100.

Shimada, N., Handa, S., Uchida, Y., Fukuda, M., Maruyama, N., Asaga, H., Choi, E. K., Lee, J., & Ishigami, A. (2010). Developmental and age-related changes of peptidylarginine deiminase 2 in the mouse brain. *Journal of Neuroscience Research, 88*, 798–806.

Shoeman, R. L., Hartig, R., Berthel, M., & Traub, P. (2002). Deletion mutagenesis of the amino-terminal head domain of vimentin reveals dispensability of large internal regions for intermediate filament assembly and stability. *Experimental Cell Research, 279*, 344–353.

Snider, N. T., & Omary, M. B. (2014). Post-translational modifications of intermediate filament proteins: Mechanisms and functions. *Nature Reviews. Molecular Cell Biology, 15*, 163–177.

Sofroniew, M. V. (2009). Molecular dissection of reactive astrogliosis and glial scar formation. *Trends in Neurosciences, 32*, 638–647.

Takuma, K., Lee, E., Kidawara, M., Mori, K., Kimura, Y., Baba, A., & Matsuda, T. (1999). Apoptosis in Ca2 + reperfusion injury of cultured astrocytes: Roles of reactive oxygen species and NF-kappaB activation. *The European Journal of Neuroscience, 11*, 4204–4212.

Tanaka, K. F., Takebayashi, H., Yamazaki, Y., Ono, K., Naruse, M., Iwasato, T., Itohara, S., Kato, H., & Ikenaka, K. (2007). Murine model of Alexander disease: Analysis of GFAP aggregate formation and its pathological significance. *Glia, 55*, 617–631.

Tang, G., Perng, M. D., Wilk, S., Quinlan, R., & Goldman, J. E. (2010). Oligomers of mutant glial fibrillary acidic protein (GFAP) inhibit the proteasome system in Alexander disease astrocytes, and the small heat shock protein alpha B-crystallin reverses the inhibition. *The Journal of Biological Chemistry, 285*, 10527–10537.

Vincent, S. R., Leung, E., & Watanabe, K. (1992). Immunohistochemical localization of peptidylarginine deiminase in the rat brain. *Journal of Chemical Neuroanatomy, 5*, 159–168.

Vossenaar, E. R., Despres, N., Lapointe, E., van der Heijden, A., Lora, M., Senshu, T., van Venrooij, W. J., & Menard, H. A. (2004). Rheumatoid arthritis specific anti-Sa antibodies target citrullinated vimentin. *Arthritis Research & Therapy, 6*, R142–R150.

Wang, G., & Mao, Z. (2014). Chaperone-mediated autophagy: Roles in neurodegeneration. *Translational Neurodegeneration, 3*, 20. doi:10.1186/2047-9158-3-20.

Wang, L., Colodner, K. J., & Feany, M. B. (2011). Protein misfolding and oxidative stress promote glial-mediated neurodegeneration in an Alexander disease model. *The Journal of Neuroscience, 31*, 2868–2877.

Wang, F., Chen, F. F., Gao, W. B., Wang, H. Y., Zhao, N. W., Xu, M., Gao, D. Y., Yu, W., Yan, X. L., Zhao, J. N., & Li, X. J. (2016). Identification of citrullinated peptides in the synovial fluid of patients with rheumatoid arthritis using LC-MALDI-TOF/TOF. *Clinical Rheumatology, 35*, 2185–2194.

Wood, D. D., Ackerley, C. A., Brand, B., Zhang, L., Raijmakers, R., Mastronardi, F. G., & Moscarello, M. A. (2008). Myelin localization of peptidylarginine deiminases 2 and 4: Comparison of PAD2 and PAD4 activities. *Laboratory Investigation, 88*, 354–364.

Chapter 21
Treatment of Prostate Cancer Using Deimination Antagonists and Microvesicle Technology

Sigrun Lange, Sharad Kholia, Uchini S. Kosgodage, and Jameel M. Inal

21.1 Background

Cellular MV release and PAD-mediated protein deimination are calcium-dependent events that are elevated in a number of human diseases, including autoimmunity and cancer (Berckmans et al. 2005; Migliorini et al. 2005; Nomura et al. 2008; Antwi-Baffour et al. 2010; Anzilotti et al. 2010; Castellana et al. 2010; Mohanan et al. 2012; McElwee et al. 2012; Nicholas et al. 2014). To address a putative interplay between these two mechanisms, we demonstrated in a recent prostate cancer study (Kholia et al. 2015) that PAD enzyme activation affects MV biogenesis through deimination of cytoskeletal actins and nuclear histones. Therefore, PAD-mediated MV biogenesis reveals novel synergistic roles in cancer progression.

21.2 Cellular Microvesicles

Microvesicles (MVs) are phospholipid-rich vesicles that are ubiquitously released from the cell membrane of diverse cell types upon stimulation and/or apoptosis (Piccin et al. 2007; Raposo and Stoorvogel 2013; Fig. 21.1). MV release depends on calcium ion influx, which occurs either through pores created by sublytic complement or calcium released by the endoplasmic reticulum through various

S. Lange (✉)
Department of Biomedical Sciences, University of Westminster, London W1W 6 UW, UK

University College London School of Pharmacy, London WC1N 1AX, UK
e-mail: S.Lange@westminster.ac.uk

S. Kholia • U.S. Kosgodage • J.M. Inal
Cellular and Molecular Immunology Research Centre, School of Human Sciences, London Metropolitan University, London N7 8DB, UK

A.P. Nicholas et al. (eds.), *Protein Deimination in Human Health and Disease*,
DOI 10.1007/978-3-319-58244-3_21

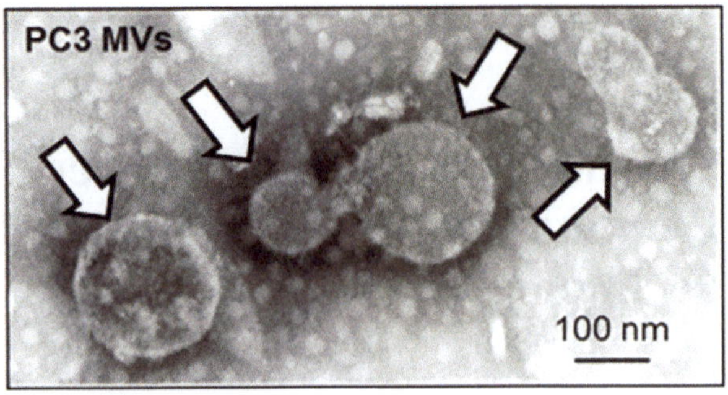

Fig. 21.1 Electron micrograph of microvesicles (MVs). Transmission electron microscopy of purified MVs (*arrows*) released from metastatic PC3 prostate cancer cells and isolated by differential centrifugation at 25,000 g for 90 min

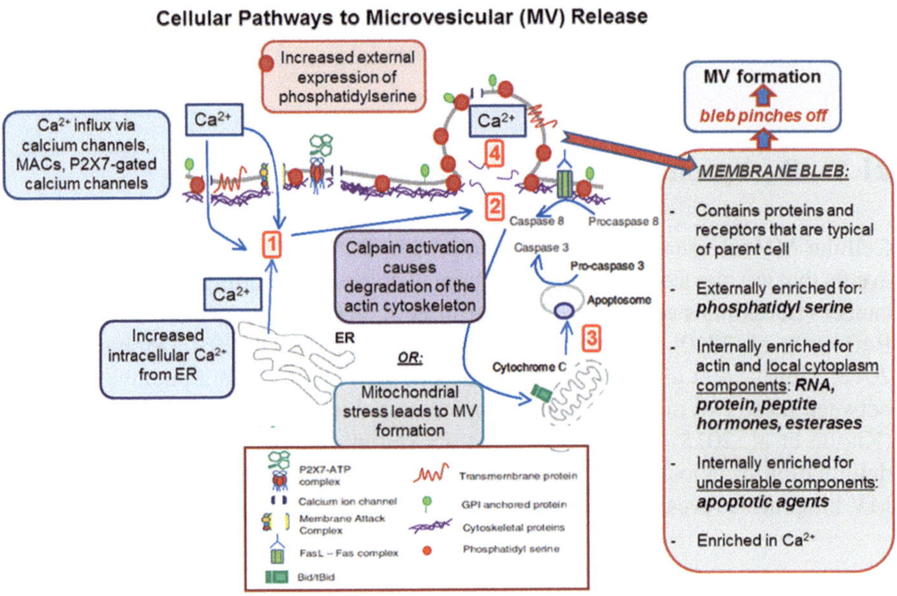

Fig. 21.2 Cellular pathways to microvesicular release. Ca^{2+} influx via calcium channels, MACs or P2X7-gated calcium channels leads to a microenvironment with increased $[Ca2+]_i$, resulting in the deactivation of cytoskeletal-associated enzymes, membrane instabilities and cytoskeletal protein degradation. Intracellular Ca^{2+} may also originate from organelles, in particular the endoplasmic reticulum (ER). (*1*) Flippase, floppase and aminophospholipid translocase are unable to maintain membrane asymmetry leading to increased external expression of phosphatidylserine. (*2*) The simultaneous activation of calpain by free Ca_i^{2+} causes the degradation of the actin cytoskeleton, leading to the formation of the membrane bleb. (*3*) Mitochondrial stress can lead to MV formation via a different pathway to Ca^{2+} activation. Proapoptotic Bcl-2, Bax and Bak are inserted into the mitochondrial outer membrane causing an increase in membrane permeability. This lead to leakage of reactive oxygen species, cytochrome C and apoptosis-inducing factor into the cytoplasm and stimulation of the caspase cascade, resulting in cytoskeletal degradation and the formation of the apoptosome that in turn leads to apoptosis. If the 'spill' is relatively small and the damage is minimal, the cell can use the apoptosome to form a MV and export the hazardous agents (pseudoapoptosis). Stimulation of FasL leads to the activation of caspase 8 and the subsequent activation of caspase 3, resulting in the translocation of tBid to the mitochondrial membrane and the leaking of cytochrome C, which leads to the formation of the apoptosome. The membrane bleb expresses receptors and proteins typical of the parent cell, enriched externally for phosphatidyl serine. (*4*) Internally, the bleb is enriched for actin and contains a sample of the 'local' cytoplasm components such as RNA, protein, peptide hormones and esterases. The bleb also contains undesirable components such as apoptotic agents, notably $Ca2^{+}_{i}$ which is enriched in MVs as a part of the cell's calcium homeostatic mechanisms. The bleb finally pinches off, forming the MV (Modified from Inal et al. 2012)

calcium channels on activated cells. This increase in cytosolic calcium results in cytoskeletal reorganisation, loss of membrane asymmetry, membrane blebbing and subsequent MV formation and release (Fig. 21.2). The cytoskeletal rearrangement is known to be facilitated by the activation of various enzymes, including calpain, gelsolin, scramblase and protein kinase, and the simultaneous inhibition of translocase and phosphatases (Inal et al. 2012). MVs play physiological roles as mediators of intracellular communication, such as transferring growth factors, microRNAs and enzymes between cells, and play roles in diverse processes such as differentiation, migration and angiogenesis (Ansa-Addo et al. 2010; Muralidharan-Chari et al. 2010; Roos et al. 2010).

21.3 MVs in Cancer

Accumulative evidence implicates MVs in the pathogenesis of various diseases, including cancer, either directly or indirectly. The presence of MVs has been noticed in cancer patients since the 1970s (Friend et al. 1978), and elevated MV levels in the blood from cancer patients compared to healthy individuals has been demonstrated by various investigators (Ginestra et al. 1998; Kim et al. 2003; Zwicker et al. 2009). MV release has also been shown to aid tumour spread and survival as MVs transport various pathological growth factor receptors, soluble proteins and microRNAs (Muralidharan-Chari et al. 2010; Roos et al. 2010; Inal et al. 2012). Interestingly, MV shedding from cancer cells also aids increased active drug efflux and thus contributes to their resistance to chemotherapeutic agents (Bebawy et al. 2009; Jorfi and Inal 2013). In addition, inhibition of microvesiculation has been shown to render cancer cells more susceptible to anticancer drug treatment (Jorfi and Inal 2013). Inhibition of microvesiculation has also been shown to reduce the dose of docetaxel required to limit tumour growth in vivo (Jorfi et al. 2015).

21.4 PADs in Cancer

In patients suffering from malignant tumours, the overexpression of the PAD2 and PAD4 isozymes has been reported in blood and tissues (Nakashima et al. 1999; Wang et al. 2010; Cherrington et al. 2012). PAD4, which is commonly found in the nucleus (Asaga et al. 2001; Nakashima et al. 2002), is the only isozyme that contains a classic nuclear localisation signal and has been shown to translocate to the nucleus in response to tumour necrosis factor alpha (TNFα) upregulation (Mastronardi et al. 2006). PAD4 acts as a transcriptional co-regulator for various factors including p53, p300, p21 and ELK1, and this regulatory function may be mediated via deimination of the N-terminal tails of various histone proteins (Li et al. 2008; Tanikawa et al. 2009; Zhang et al. 2011). PAD4 plays a role in apoptosis as it regulates p53 gene activity during DNA damage by acting as a co-mediator of gene transcription and epigenetic cross talk with histone deacetylase 2 (HDAC2) (Li et al. 2010). PAD4 is also co-localised with cytokeratin (CK), which is an established tumour marker and occurs in various isoforms, some of which are deiminated (Chang & Han 2006). Deiminated CK isoforms (Vossenaar et al. 2003; Ellsworth et al. 2008; Zwicker et al. 2009) become resistant to caspase-mediated cleavage, which contributes to the disruption of apoptosis in cancer tumours (Chang & Han 2006). In addition, PAD4 acts as a cofactor in epidermal growth factor-mediated target gene activity, activating the expression of proto-oncogene *c-fos* (Zhang et al. 2011), interacting with p53 and influencing the expression of its target genes (Yao et al. 2008; Li et al. 2008; Tanikawa et al. 2009; Guo and Fast 2011). PAD4 is also linked with oestrogen receptor target gene activity via histone tail deimination (Wang et al. 2004). In spite of lacking a classic nuclear translocation site such as that found in PAD4, both PAD2 and PAD3 have also been localised and detected in the nucleus (Lange et al. 2011; Cherrington et al. 2012; Lange et al. 2014; Kin Pong et al. 2014). In addition, PAD2, which is the most widely expressed isozyme in the body (Watanabe et al. 1988), has been shown to deiminate histone 3 and play a role in gene regulation in cancer cells (Cuthbert et al. 2004; Cherrington et al. 2012; McElwee et al. 2012; Zhang et al. 2012).

21.5 The Interplay of MVs and PADs in Prostate Cancer

Our recent study on PADs in prostate cancer (Kholia et al. 2015) showed that PAD2 and 4 were present at significantly elevated levels in metastatic prostate PC3 cancer cells compared to a control benign (PNT2) prostate cell line (Kholia et al. 2015). Upon stimulating cells to microvesiculate, the nuclear translocation of PAD4 and that of PAD2, albeit at lower levels, was observed in the metastatic PC3 cancer cells (Fig. 21.3; Kholia et al. 2015). This nuclear translocation was prevented when the cells were pretreated with the PAD inhibitor Cl-Am prior to stimulating the cells for MV release (Fig. 21.3). Cl-Am also had a dose-dependent inhibitory effect on MV release in both PC3 (Fig. 21.4) and normal prostate control cell lines (PNT2) (Kholia et al. 2015). It still needs to be elucidated whether nuclear translocation of specific PAD isozymes is directly involved in MV release, as increased translocation of PAD4 was particularly observed. Another possibility is that the MV release cascade causing the translocation of PADs to the nucleus may cause upregulation in TNFα as part of the inflammatory response. The finding that PAD inhibition with Cl-Am was also observed to affect MV release from control prostate cells (PNT2) indicates a significant role for PAD activity in MV release in both cancerous and non-cancerous cells and impacts cell communication via this pathway in a variety of scenarios. As known, Cl-Am is a pan-PAD inhibitor, and thus it remains to be elucidated which particular PAD isozyme plays the main role in the MV release mechanism or whether they are collectively involved with equally important, albeit slightly different, roles.

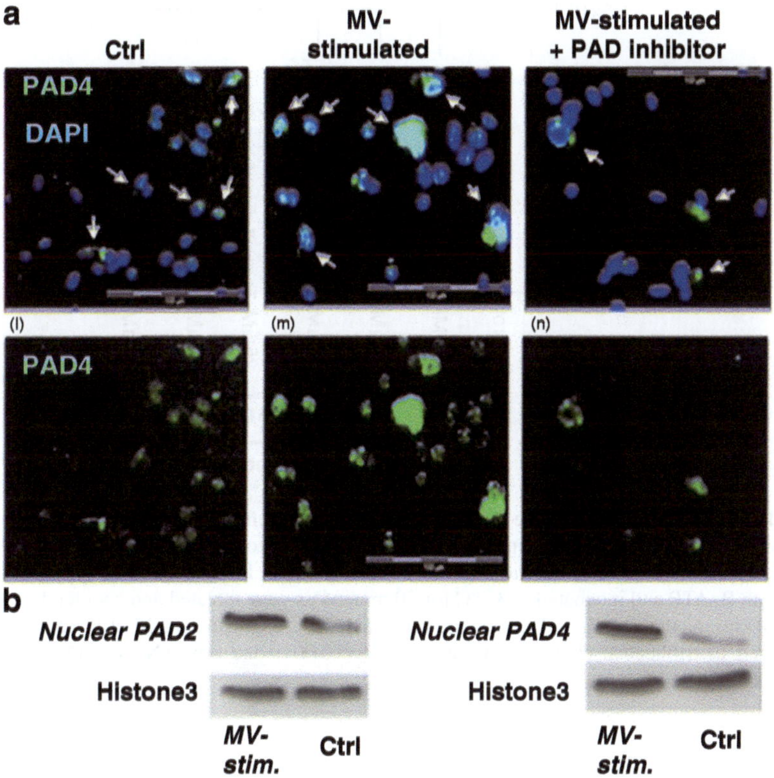

Fig. 21.3 Nuclear PAD2 and PAD4 translocation is increased in PC3 cancer cells during microvesiculation and reduced upon PAD inhibition. (**a**) PAD4 nuclear translocation is increased upon MV stimulation and reduced with PAD inhibitor (Cl-Am). (**b**) Nuclear preparations indicate increased expression of PAD2 and PAD4 by Western blot during BzATP-induced MV release in PC3 cells. Scale bars: 100 mm, Histone3: internal nuclear control (Modified from Kholia et al. 2015)

21.6 Deiminated Target Proteins Involved in MV Biogenesis

So far, the mechanism of MV release is known to be regulated by either influx of $Ca2^{+}$ that is released by the ER or through $Ca2^{+}$ that enters the cell through pores generated by sublytic complement (Inal et al. 2012) or through the various calcium channels (Salzer et al. 2002; Pizzirani et al. 2007) on activated cells (Fig. 21.2). This leads to calpain-mediated cleavage of a variety of cytoskeletal actin filaments, causing reorganisation of the cytoskeleton that in turn facilitates MV release (Lemoinne et al. 2014). During microvesiculation, both β- and F-actin stress fibres play important roles in the redistribution of the actin cytoskeleton during the formation process of the MVs. This happens through the activation of Rho/Rho-associated kinase

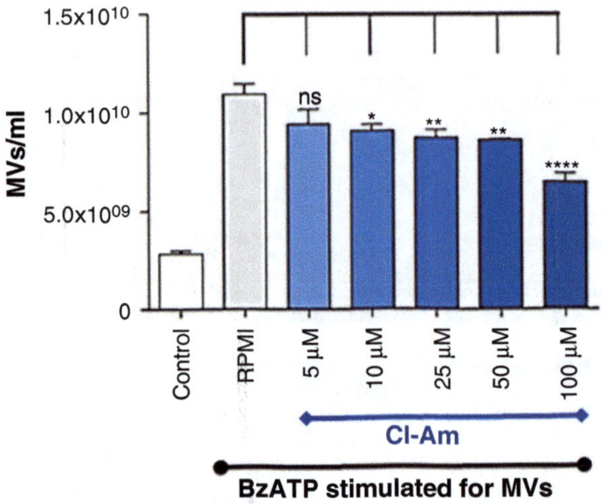

Fig. 21.4 In PC3 cancer cells, the PAD inhibitor Cl-Am significantly reduces BzATP-stimulated cellular MV release in a dose-dependent manner, as compared to non-stimulated PC3 control cells. Cells pretreated with or without various concentrations of the PAD inhibitor Cl-Am were stimulated with BzATP and incubated at 37 °C for 30 min. MVs were isolated and counted by nanoparticle tracking analysis (NTA). The data are represented as the mean ± SEM of three experiments performed in triplicates. (**$P < 0.005$ and ***$P < 0.001$; one-way ANOVA) (Modified from Kholia et al. 2015)

(ROCK) pathways during apoptosis and thrombin stimulation (Coleman et al. 2001). Interestingly, in sera and synovial fluid from RA patients, deiminated β- and γ-actins have been detected (van Beers et al. 2013; van Beers and Pruijn 2014). β-actin is one of six different human actin isoforms and is a cytoskeletal protein involved in cell structure and integrity, cell migration and movement (Bunnell et al. 2011). β-actin has also been identified as a substrate for PAD2 in ionomycin-activated neutrophils (Darrah et al. 2012). Our prostate cancer study showed that the presence of deiminated β-actin increased in cells that were stimulated for MV release and that pretreatment with Cl-Am markedly decreased β-actin deimination (Fig. 21.5; Kholia et al. 2015). So far, our work implies that PAD-mediated deimination of target proteins that are involved in cytoskeletal rearrangement, such as β-actin, actin alpha 1 and glyceraldehyde-3-phosphate dehydrogenase, is an essential step for successful MV biogenesis. Upon PAD upregulation, as observed in cancer cells and in cells stimulated for MV release, increased deimination of cytoskeletal actin filaments probably aids the cytoskeletal reorganisation necessary for successful MV release. PAD inhibition using Cl-Am inhibits this deimination and renders the cytoskeleton less flexible, which significantly impacts the capability of effective MV formation and shedding (Fig. 21.6).

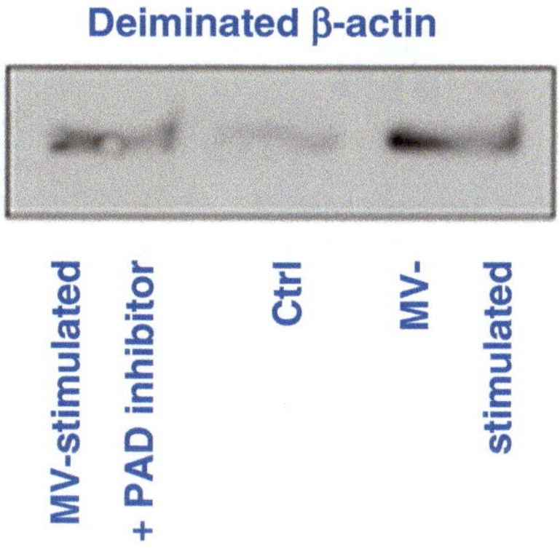

Fig. 21.5 MV stimulation of PC3 cancer cells leads to increased β-actin deimination that is reduced upon PAD inhibition. PC3 cells were stimulated with BzATP to microvesiculate, and total deiminated proteins were immunoprecipitated using the F95 pan-deimination protein antibody (Nicholas and Whitaker 2002). Increased deiminated β-actin was observed in PC3 cells stimulated to microvesiculate with BzATP (Lane 3) compared to untreated, control PC3 cells (Lane 2). Upon treatment with 10 mM Cl-Am prior to MV stimulation, a significant decrease was observed in the levels of deiminated β-actin (Lane 1) compared to MV-stimulated cells (Lane 3). Equal amounts of protein extract immunoprecipitated with F95 were loaded as verified by Coomassie *blue* staining (not shown) (Modified from Kholia et al. 2015)

21.7 PAD Inhibitors in Synergy with Chemotherapeutic Drugs

Cytotoxic effects of the PAD inhibitors Cl-Am and F-Am on several other cancer cell lines (HL60, MCF-7 and HT-29) have been shown, while normal cells were not affected (Luo et al. 2006; Knuckley et al. 2008). In addition, PAD inhibitors have been shown to affect differentiation of HL60 and HT29 cells (Slack et al. 2011). PAD inhibitors in synergy with chemotherapeutic agents such as doxorubicin show increased cell cytotoxicity in cancer cells (Slack et al. 2011), and paclitaxel, which is used to treat neoplasms such as breast and ovarian cancers, can inhibit PAD2 enzymatic activity (Pritzker and Moscarello 1998). The role PAD-mediated MVs release in this context had previously been overlooked until our prostate cancer study (Kholia et al. 2015). MV release may act through both PAD2 and PAD4 depending on target protein preference and thus act through both cytoskeletal and epigenetic pathways. Distinct substrate preferences have indeed been demonstrated for the different PAD isozymes (Tarcsa et al. 1996; Knuckley et al. 2010; Darrah et al. 2012; Assohou-Luty et al. 2014). In general, PAD4 shows more restrictive substrate specificity compared to PAD2 (Assohou-Luty et al. 2014). PAD4 seems to prefer sequences with highly disordered conformation, while PAD2 has a broader sequence specificity (Arita et al. 2004), which might partly be reflected by the broader tissue expression of PAD2. In a study by Darrah et al. (2012), PAD4 was the only isozyme able to deiminate histone H3, while PAD2 was able to deiminate

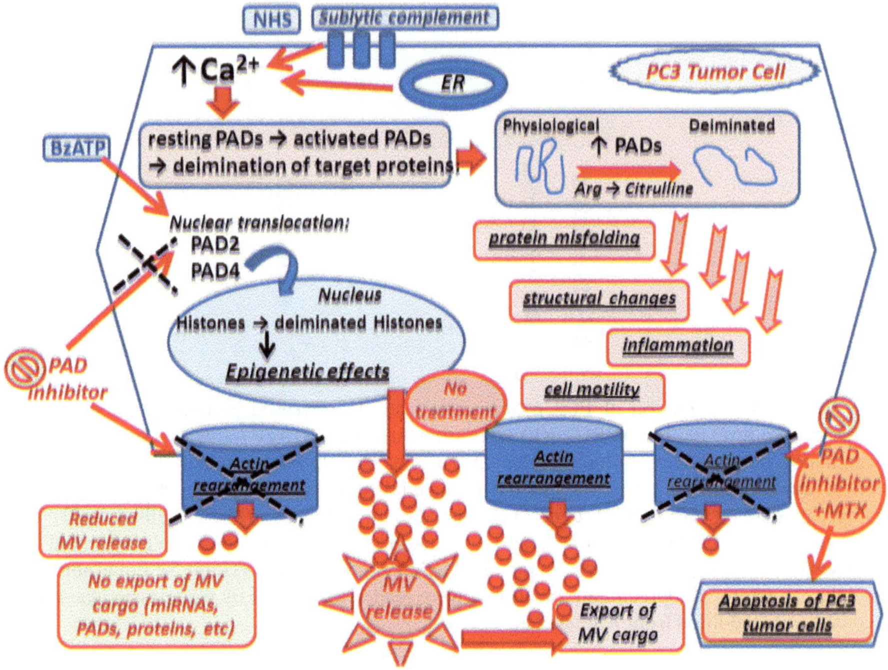

Fig. 21.6 The proposed role of PADs in microvesicular biogenesis and the potential therapeutic application of PAD inhibitors in anticancer therapy. PAD2 and PAD4 are cytosolic enzymes, which, during the course of cells being stimulated to microvesiculate (e.g. using BzATP stimulation of P2X channels or NHS as a source of sublytic complement), by raised intracellular calcium levels, results in PAD enzymes (PAD2 and PAD4) translocating to the nucleus or remaining partially in the cytosol (PAD2). These PAD isozymes play a potential novel role in the biogenesis of MV release. This may happen by influencing actin-cytoskeleton cleavage and actin rearrangement and/or by nuclear histone deimination affecting gene regulation through still unknown mechanisms involved in microvesicle biogenesis. The pharmacological inhibition of PADs abrogates the release of MVs, and when combined with the anticancer drug methotrexate (MTX), works synergistically to induce increased cytotoxic effects and apoptosis of PC3 tumour cells (Modified from Kholia et al. 2015)

β- and γ-actins. The selectivity of PAD inhibitors and chemotherapeutic agents is thus of great interest. We showed that Cl-Am and MTX exerted synergistic cytotoxic effects on PC3 cells compared to MTX alone and significantly reduced MV biogenesis and release (Kholia et al. 2015). This change in MV shedding will prevent the cancer cells from carrying out active drug efflux through MV shedding and thus sensitize them to chemotherapeutic drugs (Fig. 21.6). Accordingly, other studies in our group show that neoplastic cells are rendered more sensitive to cancer drugs when MV biogenesis is inhibited (Jorfi and Inal 2013). Most recently we have also shown that MV inhibition via a calpeptin inhibitor or siRNA renders PC3 cells more susceptible to chemotherapy with docetaxel and limits tumour growth in vivo (Jorfi et al. 2015). This further points out the important role of MVs generation for tumour growth.

21.8 MVs and PADs in the Spread of Tumorigenesis

MVs play various roles in cellular cross talk and act as carriers of active molecules that can have an effect at sites distal from the origin of MV release (Piccin et al. 2007; Inal et al. 2013). Both PAD2 and PAD4 isozymes are expressed in patients with malignant tumours, and PAD expression is elevated in cancers, which may cause increased MV release and contribute to cancer progression (Kholia et al. 2015). Interestingly, the presence of PAD4 has been reported in the plasma of cancer patients without an explanation as yet of how it reaches that location (Chang and Han 2006). It is tempting to speculate how tumour cells might increase MV release in a PAD4-dependent mechanism through which the PAD enzymes may be packaged in MVs and carried into plasma where they could then deiminate target proteins such as antithrombin and aid in the spread of tumorigenesis indirectly. In the same vein, PAD enzymes may be packaged in MVs from non-cancerous cells in response to inflammation and elevated TNFα levels. For example, MVs may play a similar role in diseases such as RA, possibly explaining the unknown mechanism by which PAD molecules gain access to the extracellular space to deiminate extracellular proteins such as collagen and fibrinogen (Anzilotti et al. 2006; Yoshida et al. 2006; Zhao et al. 2008;). However, NETosis has also been suggested to be a method by which PAD4 can be released from cells to generate deiminated antigens (Jones et al. 2009; Dwivedi et al. 2012). One could postulate that PAD isozymes carried in MVs from different cells differ between physiological and disease states, vary between diseases and thus play important roles in both autoimmune diseases and cancer (Kholia et al. 2015).

21.9 Conclusion

Our recent studies have shown novel roles for PAD2 and PAD4 in the biogenesis of cellular microvesiculation (Kholia et al. 2015). MV release was significantly reduced in metastatic prostate PC3 cancer cells with the pan-PAD inhibitor Cl-Am, also making them more susceptible to chemotherapeutic drugs (MTX). Significant reduction of deiminated actin and changes in PAD translocation were observed upon PAD inhibition with Cl-Am. PADs may thus affect MV biogenesis directly through deimination of cellular actins and other cytoskeletal components, facilitating cytoskeletal rearrangement for effective MV biogenesis or indirectly, through histone deimination, as supported by observed changes in nuclear PAD translocation upon MV release and in response to pharmacological PAD inhibition. Epigenetic regulation via PAD-mediated histone deimination needs further elucidation in MV biogenesis. Increased MV shedding from cancer cells, facilitated in part by upregulated PAD expression, appears to be a protective mechanism of cancer cells, as this acts as a repellent for chemotherapeutic drugs such as MTX (Fig. 21.6). Ongoing work aims at further understanding the role of histone deimination in MV biogenesis, identifying further deiminated target proteins and dissecting the roles of individual PAD isozymes. The use of targeted isozyme-selective PAD inhibitors to control MV release poses as promising combinatory therapies for various forms of cancers.

References

Ansa-Addo, E. A., Lange, S., Stratton, D., et al. (2010). Human plasma membrane-derived vesicles halt proliferation and induce differentiation of THP-1 acute monocytic leukemia cells. *Journal of Immunology, 185*, 5236–5246.

Antwi-Baffour, S., Kholia, S., Aryee, Y. K., et al. (2010). Human plasma membrane-derived vesicles inhibit the phagocytosis of apoptotic cells—Possible role in SLE. *Biochemical and Biophysical Research Communications, 398*, 278–283.

Anzilotti, C., Merlini, G., Pratesi, F., et al. (2006). Antibodies to viral citrullinated peptide in rheumatoid arthritis. *The Journal of Rheumatology, 33*, 647–651.

Anzilotti, C., Pratesi, F., Tommasi, C., & Migliorini, P. (2010). Peptidylarginine deiminase 4 and citrullination in health and disease. *Autoimmunity Reviews, 9*, 158–160.

Arita, K., Hashimoto, H., Shimizu, T., et al. (2004). Structural basis for Ca(2+)-induced activation of human PAD4. *Nature Structural & Molecular Biology, 11*(8), 777–783.

Asaga, H., Nakashima, K., Senshu, T., et al. (2001). Immunocytochemical localization of peptidylarginine deiminase in human eosinophils and neutrophils. *Journal of Leukocyte Biology, 70*, 46–51.

Assohou-Luty, C., Raijmakers, R., Benckhuijsen, W. E., et al. (2014). The human peptidylarginine deiminases type 2 and type 4 have distinct substrate specificities. *Biochimica et Biophysica Acta, 1844*(4), 829–836.

Bebawy, M., Combes, V., Lee, E., et al. (2009). Membrane microparticles mediate transfer of P-glycoprotein to drug sensitive cancer cells. *Leukemia, 23*(9), 1643.

Berckmans, R. J., Nieuwland, R., Kraan, M. C., et al. (2005). Synovial microparticles from arthritic patients modulate chemokine and cytokine release by synoviocytes. *Arthritis Research & Therapy, 7*, R536–R544.

Bunnell, T. M., Burbach, B. J., Shimizu, Y., & Ervasti, J. M. (2011). beta-Actin specifically controls cell growth, migration, and the G-actin pool. *Molecular Biology of the Cell, 22*, 4047–4058.

Castellana, D., Toti, F., & Freyssinet, J. M. (2010). Membrane microvesicles: Macromessengers in cancer disease and progression. *Thrombosis Research, 125*, S84–S88.

Chang, X., & Han, J. (2006). Expression of peptidylarginine deiminase type 4 (PAD4) in various tumors. *Molecular Carcinogenesis, 45*, 183–196.

Cherrington, B. D., Zhang, X., McElwee, J. L., et al. (2012). Potential role for PAD2 in gene regulation in breast cancer cells. *PloS One, 7*, e41242.

Coleman, M. L., Sahai, E. A., Yeo, M., et al. (2001). Membrane blebbing during apoptosis results from caspase-mediated activation of ROCK I. *Nature Cell Biology, 3*, 339–345.

Cuthbert, G. L., Daujat, S., Snowden, A. W., et al. (2004). Histone deimination antagonizes arginine methylation. *Cell, 118*, 545–553.

Darrah, E., Rosen, A., Giles, J. T., & Andrade, F. (2012). Peptidylarginine deiminase 2, 3 and 4 have distinct specificities against cellular substrates: Novel insights into autoantigen selection in rheumatoid arthritis. *Annals of the Rheumatic Diseases, 71*(1), 92–98.

Dwivedi, N., Upadhyay, J., Neeli, I., et al. (2012). Felty's syndrome autoantibodies bind to deiminated histones and neutrophil extracellular traps. *Arthritis and Rheumatism, 64*(4), 982–992.

Ellsworth, R. E., Vertrees, A., Love, B., Hooke, J. A., Ellsworth, D. L., & Shriver, C. D. (2008). Chromosomal alterations associated with the transition from in situ to invasive breast cancer. *Annals of Surgical Oncology, 15*, 2519–2525.

Friend, C., Marovitz, W., Henie, G., Henie, W., Tsuei, D., Hirschhorn, K., et al. (1978). Observations on cell lines derived from a patient with Hodgkin's disease. *Cancer Research, 38*, 2581–2591.

Ginestra, A., La, P., Saladino, F., et al. (1998). The amount and proteolytic content of vesicles shed by human cancer cell lines correlates with their in vitro invasiveness. *Anticancer Research, 18*, 3433–3437.
Guo, Q., & Fast, W. (2011). Citrullination of inhibitor of growth 4 (ING4) by peptidylarginine deminase 4 (PAD4) disrupts the interaction between ING4 and p53. *The Journal of Biological Chemistry, 286*, 17069–17078.
Inal, J. M., Ansa-Addo, E. A., Stratton, D., et al. (2012). Microvesicles in health and disease. *Archivum Immunologiae et Therapiae Experimentalis, 60*, 107–121.
Inal, J. M., Kosgodage, U., Azam, S., et al. (2013). Blood/plasma secretome and microvesicles. *Biochimica et Biophysica Acta, 1834*, 2317–2325.
Jones, J. E., Causey, C. P., Knuckley, B., et al. (2009). Protein arginine deiminase 4 (PAD4): Current understanding and future therapeutic potential. *Current Opinion in Drug Discovery & Development, 12*(5), 616–627.
Jorfi, S., & Inal, J. M. (2013). The role of microvesicles in cancer progression and drug resistance. *Biochemical Society Transactions, 41*, 293–298.
Jorfi, S., Ansa-Addo, E. A., Kholia, S., et al. (2015). Inhibition of microvesiculation sensitizes prostate cancer cells to chemotherapy and reduces docetaxel dose required to limit tumor growth in vivo. *Scientific Reports, 5*, 13006.
Kholia, S., Jorfi, S., Thompson, P. R., et al. (2015). A novel role for peptidylarginine deiminases in microvesicle release reveals therapeutic potential of PAD inhibition in sensitizing prostate cancer cells to chemotherapy. *Journal of Extracellular Vesicles, 4*, 26192.
Kim, H. K., Song, K. S., Park, Y. S., et al. (2003). Elevated levels of circulating platelet microparticles, VEGF, IL-6 and RANTES in patients with gastric cancer: Possible role of a metastasis predictor. *European Journal of Cancer, 39*, 184–191.
Kin Pong, U., Subramanian, V., Nicholas, A. P., et al. (2014). Modulation of calcium-induced cell death in human neural stem cells by the novel peptidylarginine deiminase-AIF pathway. *Biochimica et Biophysica Acta, 1843*, 1162–1171.
Knuckley, B., Luo, Y., & Thompson, P. R. (2008). Profiling protein arginine deiminase 4 (PAD4): A novel screen to identify PAD4 inhibitors. *Bioorganic & Medicinal Chemistry, 16*, 739–745.
Knuckley, B., Causey, C. P., Jones, J. E., et al. (2010). Substrate specificity and kinetic studies of PADs 1, 3, and 4 identify potent and selective inhibitors of protein arginine deiminase 3. *Biochemistry, 49*, 4852–4863.
Lange, S., Gögel, S., Leung, K. Y., et al. (2011). Protein deiminases: New players in the developmentally regulated loss of neural regenerative ability. *Developmental Biology, 355*, 205–214.
Lange, S., Rocha-Ferreira, E., Thei, L., et al. (2014). Peptidylarginine deiminases: Novel drug targets for prevention of neuronal damage following hypoxic ischemic insult (HI) in neonates. *Journal of Neurochemistry, 130*, 555–562.
Lemoinne, S., Thabut, D., Housset, C., Moreau, R., Valla, D., Boulanger, C. M., et al. (2014). The emerging roles of microvesicles in liver diseases. Nature Reviews. *Gastroenterology & Hepatology, 11*, 350–361.
Li, P., Yao, H., Zhang, Z., et al. (2008). Regulation of p53 target gene expression by peptidylarginine deiminase 4. *Molecular and Cellular Biology, 28*, 4745–4758.
Li, P., Wang, D., Yao, H., et al. (2010). Coordination of PAD4 and HDAC2 in the regulation of p53-target gene expression. *Oncogene, 29*, 3153–3162.
Luo, Y., Knuckley, B., Lee, Y. H., et al. (2006). A fluoroacetamidine-based inactivator of protein arginine deiminase 4: Design, synthesis, and in vitro and in vivo evaluation. *Journal of the American Chemical Society, 128*, 1092–1093.
Mastronardi, F. G., Wood, D. D., Mei, J., et al. (2006). Increased citrullination of histone H3 in multiple sclerosis brain and animal models of demyelination: A role for tumor necrosis factor-induced peptidylarginine deiminase 4 translocation. *The Journal of Neuroscience, 26*, 11387–11396.
McElwee, J. L., Mohanan, S., Griffith, O. L., et al. (2012). Identification of PADI2 as a potential breast cancer biomarker and therapeutic target. *BMC Cancer, 12*, 500.

Migliorini, P., Pratesi, F., Tommasi, C., & Anzilotti, C. (2005). The immune response to citrullinated antigens in autoimmune diseases. *Autoimmunity Reviews, 4*, 561–564.

Mohanan, S., Cherrington, B. D., Horibata, S., et al. (2012). Potential role of peptidylarginine deiminase enzymes and protein citrullination in cancer pathogenesis. *Biochemistry Research International, 2012*, 895343.

Muralidharan-Chari, V., Clancy, J. W., Sedgwick, A., & D'Souza-Schorey, C. (2010). Microvesicles: Mediators of extracellular communication during cancer progression. *Journal of Cell Science, 123*, 1603–1611.

Nakashima, K., Hagiwara, T., Ishigami, A., et al. (1999). Molecular characterization of peptidylarginine deiminase in HL-60 cells induced by retinoic acid and 1alpha,25-dihydroxyvitamin D3. *The Journal of Biological Chemistry, 274*, 27786–27792.

Nakashima, K., Hagiwara, T., & Yamada, M. (2002). Nuclear localization of peptidylarginine deiminase V and histone deimination in granulocytes. *The Journal of Biological Chemistry, 277*, 49562–49568.

Nicholas, A. P., & Whitaker, J. N. (2002). Preparation of a monoclonal antibody to citrullinated epitopes: Its characterization and some applications to immunohistochemistry in human brain. *Glia, 37*, 328–336.

Nicholas, A. P., Lu, L., Heaven, M., et al. (2014). Ongoing studies of deimination in neurodegenerative diseases using the F95 antibody. In A. P. Nicholas & S. K. Bhattacharya (Eds.), *Protein deimination in human health and disease* (pp. 257–280). New York: Springer.

Nomura, S., Ozaki, Y., & Ikeda, Y. (2008). Function and role of microparticles in various clinical settings. *Thrombosis Research, 123*, 8–23.

Piccin, A., Murphy, W. G., & Smith, O. P. (2007). Circulating microparticles: Pathophysiology and clinical implications. *Blood Reviews, 21*, 157–171.

Pizzirani, C., Ferrari, D., Chiozzi, P., et al. (2007). Stimulation of P2 receptors causes release of IL-1beta-loaded microvesicles from human dendritic cells. *Blood, 109*, 3856–3864.

Pritzker, L. B., & Moscarello, M. A. (1998). A novel microtubule independent effect of paclitaxel: The inhibition of peptidylarginine deiminase from bovine brain. *Biochimica et Biophysica Acta, 1388*(1), 154–160.

Raposo, G., & Stoorvogel, W. (2013). Extracellular vesicles: Exosomes, microvesicles and friends. *The Journal of Cell Biology, 200*, 373–383.

Roos, M. A., Gennero, L., Denysenko, T., et al. (2010). Microparticles in physiological and in pathological conditions. *Cell Biochemistry and Function, 28*, 539–548.

Salzer, U., Hinterdorfer, P., Hunger, U., et al. (2002). Ca(++)-dependent vesicle release from erythrocytes involves stomatin-specific lipid rafts, synexin (annexin VII), and sorcin. *Blood, 99*, 2569–2577.

Slack, J. L., Causey, C. P., & Thompson, P. R. (2011). Protein arginine deiminase 4: A target for an epigenetic cancer therapy. *Cellular and Molecular Life Sciences, 68*, 709–720.

Tanikawa, C., Ueda, K., Nakagawa, H., et al. (2009). Regulation of protein Citrullination through p53/PADI4 network in DNA damage response. *Cancer Research, 69*, 8761–8769.

Tarcsa, E., Marekov, L. N., Mei, G., et al. (1996). Protein unfolding by peptidylarginine deiminase. Substrate specificity and structural relationships of the natural substrates trichohyalin and filaggrin. *The Journal of Biological Chemistry, 271*, 30709–30716.

van Beers, J. J. B. V., & Pruijn, G. J. M. (2014). Chapter 3: The role of synovial citrullinated proteins in the pathophysiology of rheumatoid arthritis. In S. Bhattacharya & A. Nicholas (Eds.), *Protein deimination in human health and disease*. New York: Springer; ISBN 978-1-4614-8316-8.

van Beers, J. J., Schwarte, C. M., Stammen-Vogelzangs, J., et al. (2013). The rheumatoid arthritis synovial fluid citrullinome reveals novel citrullinated epitopes in apolipoprotein E, myeloid nuclear differentiation antigen, and beta-actin. *Arthritis and Rheumatism, 65*, 69–80.

Vossenaar, E. R., Nijenhuis, S., Helsen, M. M., et al. (2003). Citrullination of synovial proteins in murine models of rheumatoid arthritis. *Arthritis and Rheumatism, 48*, 2489–2500.

Wang, Y., Wysocka, J., Sayegh, J., et al. (2004). Human PAD4 regulates histone arginine methylation levels via demethylimination. *Science, 306*, 279–283.

Wang, L., Chang, X., Yuan, G., et al. (2010). Expression of peptidylarginine deiminase type 4 in ovarian tumors. *International Journal of Biological Sciences, 6*, 454–464.

Watanabe, K., Akiyama, K., Hikichi, K., et al. (1988). Combined biochemical and immunochemical comparison of peptidylarginine deiminases present in various tissues. *Biochimica et Biophysica Acta, 966*, 375–383.

Yao, H., Li, P., Venters, B. J., et al. (2008). Histone Arg modifications and p53 regulate the expression of OKL38, a mediator of apoptosis. *The Journal of Biological Chemistry, 283*, 20060–20068.

Yoshida, M., Tsuji, M., Kurosaka, D., et al. (2006). Autoimmunity to citrullinated type II collagen in rheumatoid arthritis. *Modern Rheumatology, 16*(5), 276–281.

Zhang, X., Gamble, M. J., Stadler, S., et al. (2011). Genome-wide analysis reveals PADI4 cooperates with Elk-1 to activate c-Fos expression in breast cancer cells. *PLoS Genetics, 7*, e1002112.

Zhang, X., Bolt, M., Guertin, M. J., et al. (2012). Peptidylarginine deiminase 2-catalyzed histone H3 arginine 26 citrullination facilitates estrogen receptor alpha target gene activation. *Proceedings of the National Academy of Sciences of the United States of America, 109*(33), 13331–13336.

Zhao, X., Okeke, N. L., Sharpe, O., et al. (2008). Circulating immune complexes contain citrullinated fibrinogen in rheumatoid arthritis. *Arthritis Research & Therapy, 10*(4), R94.

Zwicker, J. I., Liebman, H. A., Neuberg, D., et al. (2009). Tumor-derived tissue factor-bearing microparticles are associated with venous thromboembolic events in malignancy. *Clinical Cancer Research, 15*, 6830–6840.

Chapter 22
Citrullination in Inflammatory-Driven Carcinogenesis of the Colon

Erin E. Witalison and Lorne J. Hofseth

22.1 Inflammatory-Driven Carcinogenesis

When a cell is damaged or a foreign pathogen is detected, an acute inflammatory response is triggered to isolate the stimulus and prevent any further harm to the surrounding cells. Inflammation is associated with increased blood flow to the inflamed tissues, allowing for the recruitment of pro-inflammatory cells that release cytokines to aid in the destruction of pathogens and irreparably damaged cells (Green et al. 1990; Vane et al. 1994; Nathan 2002). Once the stimuli have been neutralized, typically inflammation will subside. In cases of excessive stimuli or where the inflammatory response does not abate, chronic inflammation can occur. As in acute inflammation, chronic inflammation causes the production of pro-inflammatory cytokines that release free radicals with the intent to target and kill invading pathogens (Wink et al. 1991; Seibert and Masferrer 1994). However, due to the prolonged exposure of cells to an environment with high levels of DNA-damaging free radicals, many chronic inflammatory conditions are associated with a heightened risk for the development of cancer in the inflamed areas (Eaden et al. 2001; Hussain et al. 2003). In this section, several chronic inflammatory diseases (caused by a variety of stimuli) and their associated cancers will be discussed (Fig. 22.1).

E.E. Witalison • L.J. Hofseth (✉)
Department of Drug Discovery and Biomedical Sciences, South Carolina College of Pharmacy, University of South Carolina, Columbia, SC, USA
e-mail: Hofseth@cop.sc.edu

A.P. Nicholas et al. (eds.), *Protein Deimination in Human Health and Disease*,
DOI 10.1007/978-3-319-58244-3_22

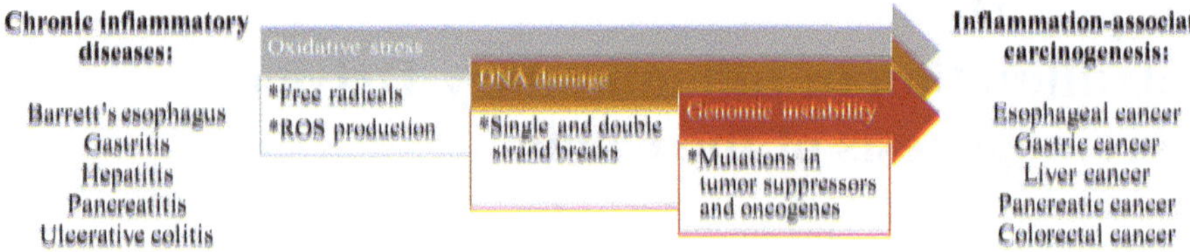

Fig. 22.1 Chronic inflammatory diseases progress to inflammation-associated carcinogenesis through three main steps. Chronic inflammation produces oxidative stress, which can damage DNA. The buildup of unrepaired DNA damage contributes to genomic instability and an imbalance of tumor suppressor genes and oncogenes, giving rise to the inflammation-associated cancers

22.1.1 Barrett's Esophagus and Esophageal Cancer

To begin with, a condition called Barrett's esophagus occurs after repeated exposure to stomach acid, which causes dysplasia in the epithelial cells lining the lower esophagus. The repeated exposure to acidic gastric fluids is often caused by a hiatal hernia where there is insufficient closing of the cardiac sphincter at the junction of the esophagus and the stomach, allowing for frequent acid reflux and heartburn (Shaheen and Richter 2009). The persistent injury to the cells results in an inflammatory response. Then, the squamous epithelial cells lining the esophagus respond to the acidic environment in an attempt to create a more acid-resistant esophageal lining (similar to the lining of the stomach). Unfortunately, chronic inflammation will eventually lead to varying degrees of dysplasia as the cells are exposed to DNA-damaging reactive oxygen and nitrogen species and change to a more columnar morphology (Shaheen and Richter 2009).

The duration and severity of the disease can significantly impact the risk of developing esophageal adenocarcinoma, which has a devastating 5-year survival rate of less than 25% (Shaheen and Richter 2009; Weston et al. 2000; Koppert et al. 2005). In fact, patients with Barrett's esophagus are about ten times more likely to develop esophageal adenocarcinoma than the general population without Barrett's esophagus (Hvid-Jensen et al. 2011). Once Barrett's esophagus has been diagnosed, the standard protocol is to repair the hiatal hernia to stop the reoccurring acid reflux, then monitor the affected areas of the esophagus with routine surveillance endoscopies with the collection of tissue biopsies to determine histological changes. The main treatment options for high-grade dysplasia currently are either the surgical removal of the esophagus or a series of ablation treatments to methodically destroy the dysplastic cells (Shaheen and Richter 2009). Yearly endoscopies are recommended after the irregular cells have been successfully eliminated, in order to avoid the reoccurrence of Barrett's esophagus and further prevent progression to esophageal cancer (Shaheen and Richter 2009). Overall, Barrett's esophagus is an example of a disease that is characterized by chronic inflammation, resulting in genetic and cellular alterations that contribute to carcinogenesis.

22.1.2 Gastritis and Gastric Cancer

Another example of inflammation-driven carcinogenesis is chronic gastritis developing into gastric cancer. Chronic gastritis is a well-studied, precancerous condition that is commonly caused by *Helicobacter pylori* (*H. pylori*) infection (Chen et al. 2016). Up to 85% of patients with chronic gastritis have *H. pylori* infection and are, consequently, at an increased risk of developing gastric cancer because *H. pylori* infection is strongly associated with gastric cancer (Chen et al. 2016; Dooley et al. 1989; Humans IWGotEoCRt 1994). The progression of gastritis to gastric cancer is initiated with an infection of *H. pylori*, which triggers an immune response that becomes pro-inflammatory to destroy the bacteria. Then, as the *H. pylori* infection persists, chronic inflammation leads to a buildup of free radicals that damage surrounding cells. Also, within the cells that are damaged, imbalances of oncogenes/tumor suppressors and cell proliferation/apoptosis often occur (Sipponen 2002). Finally, these cells become abnormal and form cancerous lesions. The *H. pylori* infection is not responsible for the progression from gastritis to gastric cancer, but it is responsible for the original immune response that causes gastritis (Chen et al. 2016).

The use of antibiotics to treat the *H. pylori* infection is administered during the early stages of gastritis and can prevent the development of gastric cancer (Humans IWGotEoCRt 1994; Sipponen 2002). However, when gastritis, which is characterized by continuous *H. pylori* infection, is left untreated, as in many patients from developing countries where antibiotics are not readily available, gastric cancer can develop over time (Sipponen 2002). The severity of chronic gastritis also contributes to the likelihood of tumorigenesis (Humans IWGotEoCRt 1994). The prompt treatment of the immune-stimulating *H. pylori* infiltration is ultimately crucial to suppress the pro-cancerous events that can lead to the initiation of gastric tumors.

22.1.3 Hepatitis and Liver Cancer

Hepatitis is inflammation of the liver that can be instigated by a number of stimuli, including viruses, parasites, drug metabolism, bacteria, and alcohol consumption. With viral infection being the most common source of hepatitis, viral hepatitis is caused by infection with any of the five known hepatitis viruses (hepatitis A-E) (World Health Organization 2016). Hepatitis B and C viruses are both known to cause chronic hepatitis, but there is a large difference in the percentage of patients that develop chronic hepatitis. Since hepatitis C is more virulent (approximately 80% of patients developing chronic hepatitis) than hepatitis B (only about 5% of patients develop chronic hepatitis), treatment for hepatitis C is critical (Chen and Morgan 2006). Unfortunately, patients with viral hepatitis infection are often asymptomatic in the early stages of the disease, with initial diagnoses often occurring not until later stages when symptoms finally manifest themselves (McCaughan et al. 1992).

Chronic hepatitis, as seen upon prolonged hepatitis C infection, causes repeated damage to the cells of the liver, which leads to cirrhosis (Chen and Morgan 2006). Cirrhosis causes scarring and irreparable damage to the liver, making liver transplants an often necessary route, although the infection nearly always reappears if the virus has not been clinically cleared (Charlton 2001). About 10–15% of patients who have developed chronic hepatitis, as a result of at least 6 months of hepatitis C viral infection, are found to develop cirrhosis within 20 years of the initial infection (Chen and Morgan 2006; Tong et al. 1995). In turn, cirrhosis is shown to progress to hepatocellular carcinoma after about 30 years since the initial hepatitis C infection (Chen and Morgan 2006; Tong et al. 1995). Based on these clinical outcomes, the main goal of hepatitis C treatments is to stop this chronic inflammatory disease from progressing to cirrhosis and hepatocellular carcinoma. Combinations of antiviral treatments are the primary options for patients with persistent hepatitis C infection, yet the effects of therapeutics may not be enough to sustain clearance of the virus for at least 12 weeks to be considered clinically free of the virus or cause excessive treatment-related toxicity (Thompson and Holmes 2015). Even with the success of antiviral drugs, the damage from chronic hepatitis and cirrhosis may still be too much, contributing to the overall morbidity of the disease.

22.1.4 Pancreatitis and Pancreatic Cancer

Chronic pancreatitis differs from the previously mentioned chronic inflammatory diseases because it is most commonly caused by a lifestyle factor—alcohol consumption. Chronic pancreatitis is characterized by severe abdominal pain and loss of pancreatic function, alongside numerous other symptoms (Fasanella et al. 2007). About 50% of chronic pancreatitis cases are directly related to heavy alcohol consumption, and other factors, like genetics, can also increase susceptibility for the disease (Yadav et al. 2011; Pandol et al. 2011). The exact mechanism of how alcohol induces chronic pancreatitis remains unclear; nevertheless, it is postulated that alcohol metabolism may be an important contributor to the onset of the disease, and in general, alcohol may sensitize the pancreas to inflammation (Pandol et al. 2007, 2011). Indeed, it has been shown that inflammatory signaling is triggered and leads to the advancement of acute pancreatitis to chronic pancreatitis (Pandol et al. 2007, 2011). The increased activation of inflammatory mediators, like TNF-α and NF-κB, then causes the release of pro-inflammatory cytokines to attempt to combat the cellular injury caused by heavy, prolonged alcohol consumption (Pandol et al. 2007; Gukovsky et al. 2008). Even with the moderation or cessation of alcohol intake, chronic pancreatitis and the associated pain may not subside due to irreparable damage (Pandol et al. 2011).

As in other chronic inflammatory diseases, pancreatic cells that are exposed to oxidative stress from free radical interaction for an extended length of time (months to years) are prone to genetic aberrations that give rise to pancreatic cancer.

Pancreatic cancer has a high mortality rate with a 5-year survival rate of only 7% (National Cancer Institute, National Institutes of Health 2016). Therefore, the need for treatments for this chronic condition is very high to impede the progression to pancreatic cancer. However, currently, most chronic pancreatitis treatments are focused on managing the excruciating abdominal pain, while anti-inflammatory and antioxidant treatments are secondary (Fasanella et al. 2007).

22.1.5 Ulcerative Colitis and Colorectal Cancer

A final example of a chronic inflammatory disease that precedes inflammation-associated cancer is ulcerative colitis (UC). UC is an inflammatory bowel disease that affects the mucosal layer of the colon and rectum with clinical symptoms of rectal bleeding, pain, and diarrhea (Ford et al. 2013). Environmental factors, diet, and genetics play multifaceted roles in the onset of UC with an abnormal immune response and chronic inflammation (Ford et al. 2013; Ordás et al. 2012). Inducible nitric oxide synthase (iNOS) and cyclooxygenase-2 (Cox-2) are pro-inflammatory cytokines that serve as valuable, long-standing inflammatory markers for the diagnosis of UC (Singer et al. 1996, 1998; Godkin et al. 1996; Kimura et al. 1998; Hendel and Nielsen 1997). In chronic UC, the production of reactive oxygen and nitrogen species increases, while antioxidant enzymes, like superoxide dismutase 1 (SOD1), glutathione peroxidase 1 (GPx1), and catalase, are downregulated (Lih-Brody et al. 1996; Ren et al. 2014; Cetinkaya et al. 2006; Sakthivel and Guruvayoorappan 2013). The imbalance between free radical production and antioxidant enzymes creates an oxidative stress overload, which further contributes to the buildup of unrepaired DNA damage that is fundamentally involved in the progression to UC-associated colorectal cancer (CRC) (McKenzie et al. 1996; Marnett 2000; Itzkowitz and Yio 2004).

As seen in other chronic inflammatory diseases, the risk of developing UC-associated CRC increases yearly for patients who have been suffering from the chronic form of the disease for more than a decade, with about two out of ten patients developing UC-associated CRC in their lifetime (Eaden et al. 2001). Likewise, CRC is ranked highly in the United States as the second leading cause of cancer-related deaths in men and women (American Cancer Society 2015). The risk associated with the development of CRC is also dependent on the severity of the disease. Therefore, patients with moderate to severe UC have routine colonoscopies to determine that cancerous lesions have not appeared and are put on more effective therapeutics to manage the flare-ups of the disease, although many of these are accompanied with potentially severe side effects (Kavanagh et al. 2014; Hata et al. 2003). Several new UC treatments and cancer preventatives are being tested that target protein arginine deiminases (PADs; discussed in Sect. 22.2), which are shown to be upregulated in animal and human samples of UC and CRC.

22.2 Protein Arginine Deiminases and Citrullination

Protein arginine deiminases (PADs) are members of a family of isozymes that catalyze citrullination. This posttranslational modification converts peptidyl-arginine to peptidyl-citrulline through a hydrolytic reaction (Arita et al. 2004; Knuckley et al. 2010). The conversion of the cationic arginine to the neutral citrulline can affect the folding of individual proteins and can even disturb interactions between proteins (Knuckley et al. 2010; Tarcsa et al. 1996; Vossenaar et al. 2003). PAD sequence homology, tissue distribution, substrates, and the consequences of citrullination will be discussed throughout this section.

22.2.1 Organisms with PADs

In humans, there are five PAD isozymes (PAD1–PAD4, PAD6) with PAD6 being the only one without documented enzymatic activity (Rus'd et al. 1999; Ishigami et al. 2002; Chavanas et al. 2004; Raijmakers et al. 2007). Humans (*Homo sapiens*), mice (*Mus musculus*), and rats (*Rattus norvegicus*) all are known to share the five isozymes of PADs (Ying et al. 2009). PAD isozymes are also predicted to be found in other organisms, such as PAD2 in zebra fish (*Danio rerio*), frogs (*Xenopus laevis*), chickens (*Gallus gallus*), and turkeys (*Meleagris gallopavo*) (Ying et al. 2009; Shimizu et al. 2014). PAD isozymes have not been detected in some lower organisms like fruit flies (*Drosophila melanogaster*) and nematodes (*Caenorhabditis elegans*); however, PADs were discovered in the bacteria *Porphyromonas gingivalis* (see Chap. 7 for more details) (Shimizu et al. 2014; McGraw et al. 1999). Furthermore, there is a strong percentage of sequence homology (approximately 70–95%) between organisms that are found to express PADs (Vossenaar et al. 2003). The sequence homology shared between different organisms has several benefits, including similar tissue distribution and targeted substrates, as well as similar activity between organisms (e.g., humans and mice).

22.2.2 Tissue Distribution in Humans

The different PAD isozymes have individual yet overlapping distributions across an assortment of human tissues. PAD1 is found in the uterus, hair follicles, and the epidermis, while PAD3 is also expressed in hair follicles and the epidermis (Rus'd et al. 1999; Chavanas et al. 2006; Guerrin et al. 2003). Since PAD2 and PAD4 are found in inflammatory cells (i.e., monocytes, macrophages, neutrophils), they are found in numerous locations throughout the human body (Vossenaar et al. 2004). These sites of PAD2 and PAD4 expressions include the central nervous system (CNS), spleen, skeletal muscle, mammary glands, colon, and various tumor types

(Lamensa and Moscarello 1993; Watanabe et al. 1988; Watanabe and Senshu 1989; Chang et al. 2009). PAD6 is sometimes referred to as "ePAD" because of its expression in the early embryo, eggs, and ovaries (Chavanas et al. 2004; Kan et al. 2011).

22.2.3 Substrates

In addition to the multitude of tissues that can express the PAD isozymes, there is an assortment of substrates that PADs can target. PADs do not typically citrullinate total arginines found in the amino acid sequence of a protein, but certain arginines are more susceptible to citrullination. The amino acid sequence surrounding arginine residues and the protein structure, as well as the overall number of arginines present, affects the probability of an arginine being citrullinated by PADs (Tarcsa et al. 1996; Nomura 1992). As mirrored in their tissue distribution, PAD1/3 and PAD2/4 have individual and overlapping known substrates. For instance, PAD1 and PAD3 can citrullinate filaggrin, while PAD2 and PAD4 both target histones for citrullination (Senshu et al. 1995; Zhang et al. 2012; Kan et al. 2012). Then, separately, PAD1 targets keratin, and PAD3 targets trichohyalin and vimentin (Senshu et al. 1995). Likewise, PAD2 also citrullinates actin and myelin basic protein in the CNS, while PAD4 can citrullinate substrates like nucleophosmin, nuclear lamin C, and p21 (Lamensa and Moscarello 1993; Hagiwara et al. 2002; Tanikawa et al. 2012; Li et al. 2008). Furthermore, PAD6 does not have any known substrates, since its activity has not yet been discovered in mammals.

22.2.4 Consequences of Citrullinated Proteins

The seemingly minute posttranslational modification in the primary structure of proteins that happens as a result of citrullination can have a myriad of effects. Due to the varied tissue distributions and the diversity of substrates targeted by each PAD isozyme, these effects can be seen at the cellular and organismal level. For instance, citrullinated proteins can spawn an immune response because the improperly folded protein structures are not recognized as "self," which, in turn, causes the production of anti-citrullinated protein antibodies. Then, inflammation occurs in response to the presence of these antibodies (Khandpur et al. 2013). Anti-citrullinated protein antibodies are found in the blood from patients with rheumatoid arthritis (RA), and their presence serves as a valuable marker for diagnosing RA (Kolfenbach et al. 2010). Correspondingly, in UC patients, blood levels of anti-citrullinated protein antibodies are elevated to levels similar to what are seen in RA patients (Chumanevich et al. 2011). As discussed in detail later in this chapter, increased PAD expression is also implicated in the onset of UC and further concomitant with the progression to UC-associated CRC.

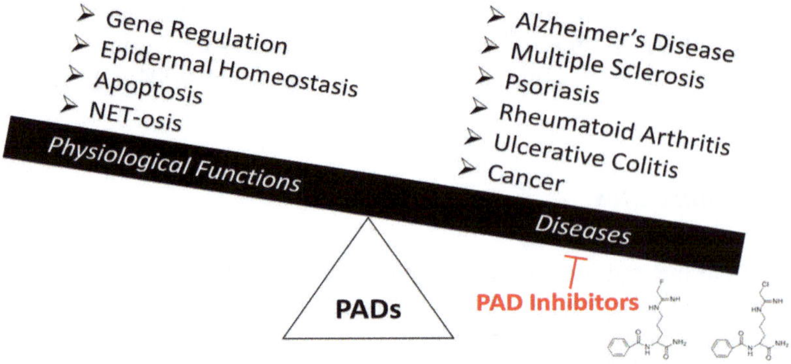

Fig. 22.2 PAD antagonists, like Cl-amidine, help tip the balance toward physiological PAD functions by suppressing the diseases that are associated with aberrant PAD activity (Witalison et al. 2015a)

Since PAD2 is found in tissues of the CNS, there is a distinct association with PAD2 and neurodegenerative diseases like Alzheimer's disease (see Chap. 17) and multiple sclerosis (MS; see Chap. 18). In both diseases, the levels of PAD activity and citrullinated proteins increase significantly (Lamensa and Moscarello 1993; Ishigami et al. 2005; Méchin et al. 2007). As found in the brains of patients suffering from Alzheimer's disease, there is a colocalization of activated PADs and citrullinated proteins at sites of inflammation and neurodegeneration (Acharya et al. 2012). Moreover, the myelin basic protein (MBP) is the main target of PADs in MS patients, causing improper signal transduction due to the inversely proportional lack of cationic arginines and rise in neutral citrullines (Cao et al. 1999). The percentage of citrullinated MBP is indicative of the severity of MS in patients, further demonstrating the pathological consequences of citrullination catalyzed by PADs (Méchin et al. 2007; Wood et al. 1996). As such, PAD inhibitors are critical in attenuating the burden of PAD dysregulation and restoring physiological PAD functions (Fig. 22.2) (Witalison et al. 2015a).

22.3 Citrullination and Colitis

As a consequence of recruited pro-inflammatory cells carrying PADs (i.e., PAD2 and PAD4) to sites of inflammation, PADs are implicated in many inflammatory diseases, including Alzheimer's disease (Ishigami et al. 2005; Acharya et al. 2012), MS (Moscarello et al. 2013), RA (Jones et al. 2009; Bicker and Thompson 2013), and UC (Chumanevich et al. 2011; Chen et al. 2008). In fact, a haplotype of the PADI4 gene has been discovered to be linked to RA and UC genetic predisposition. This susceptibility locus is found at human chromosome 1p36, the location of the PADI4 gene (Cho et al. 1998; Suzuki et al. 2003). Yet, the onset of RA and UC is thought to be attributed to a combination of genetic, environmental, and immunity

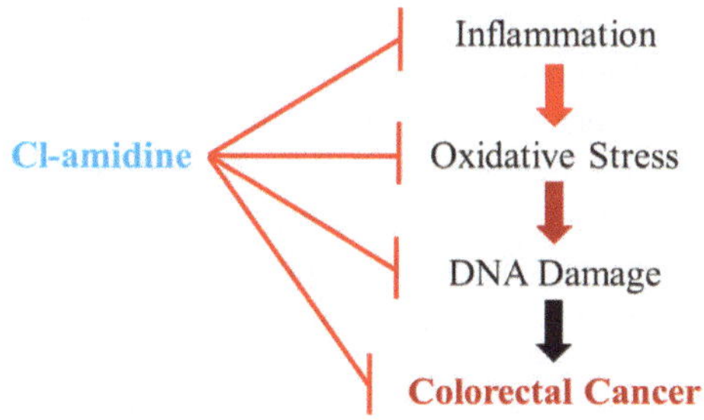

Fig. 22.3 PAD inhibition, by Cl-amidine, successfully prevents the progression from ulcerative colitis to colitis-associated colorectal cancer. Cl-amidine inhibits chronic inflammatory markers (i.e., iNOS and Cox-2), oxidative stress, and DNA damage

factors. In regard to PADs in RA, the citrullination of proteins and the production of anti-citrullinated protein antibodies have been very well-studied in cellular and animal models, as well as in patient-derived samples (Foulquier et al. 2007; Willis et al. 2011; Chang et al. 2013; Damgaard et al. 2014). However, very few research groups have focused on exploring the role of PADs in UC.

Currently it has been established that the PADs are upregulated in cellular models of inflammation, murine models of colitis, and the tissues of UC patients (Chumanevich et al. 2011; Witalison et al. 2015b; Makrygiannakis et al. 2006). Witalison et al. have shown that using a pan-PAD inhibitor (Cl-amidine) caused suppression of inflammatory cell activation and protected epithelial cells from DNA damage in cellular and animal models of UC (Fig. 22.3) (Witalison et al. 2015b). Cl-amidine works by covalent modification of the active site of calcium-bound/calcium-activated PADs (Luo et al. 2006) and, to date, has not demonstrated any off-target effects. It is recognized that Cl-amidine can work by inducing apoptosis in inflammatory cells to decrease the release of pro-inflammatory cytokines like iNOS (Chumanevich et al. 2011; Witalison et al. 2015b). Cl-amidine has also demonstrated toxicity targeting pro-cancerous cell lines (HL-60, MCF-7, and HT-29), over differentiated or noncancerous cells lines (HL-60 granulocytes and NIH 3T3) (Slack et al. 2011). This selectivity is thought to occur, since pro-cancerous and pro-inflammatory cells often have higher PAD levels (Chang et al. 2009; Chang and Han 2006), therefore making the effects of Cl-amidine more pronounced than in benign cells not exhibiting PAD activity.

In addition, key regulatory and anti-inflammatory molecules are frequently downregulated in UC, including antioxidant enzymes such as catalase, GPx1, and SOD1 (Ren et al. 2014; Cetinkaya et al. 2006; Sakthivel and Guruvayoorappan 2013). Cl-amidine treatment rescued these antioxidant enzyme levels in a stimulated mouse macrophage cell line and in mice with chemically induced colitis (Witalison et al. 2015b). Correspondingly, Cl-amidine is capable of upregulating p53 expression in inflammatory and epithelial cell populations, further indicating individual mechanisms of how Cl-amidine can suppress iNOS and DNA damage in inflammatory and epithelial cells, respectively (Chumanevich et al. 2011; Witalison et al. 2015b;

Cui et al. 2013). By using Cl-amidine (administered at an effective dose as low as 5 mg/kg) as a treatment for chemically induced UC in mice, significant reductions in PAD levels and activity were also seen across multiple studies (Chumanevich et al. 2011; Witalison et al. 2015b). In parallel with PAD inhibition, overall disease activity and inflammatory measures were subsequently reduced by Cl-amidine treatments. Taken altogether, experimental outcomes indicate that PAD inhibition is a promising target for UC drug development. Studies are currently underway to develop more potent, selective, and bioavailable compounds with the goals of better understanding the physiological and pathological roles of PADs and for future therapeutic testing on PAD-associated diseases (Bicker et al. 2012; Subramanian et al. 2015; Knight et al. 2015).

22.4 Citrullination and Colitis-Associated Colorectal Cancer

Many chronic inflammatory and fibrotic diseases of the gastrointestinal tract that have high PADs levels are at an increased risk for the initiation and progression to cancer. For example, patients with pancreatitis/pancreatic cancer and hepatitis/hepatocellular carcinoma are reported to have elevated plasma levels of PAD4 as compared to healthy control patients (Chang et al. 2009). Increased protein citrullination was also seen in a preclinical murine model of liver fibrosis that was attenuated by treatment with the PAD inhibitor Cl-amidine (Vassiliadis et al. 2013). Similarly, levels of PAD2 and PAD4 are higher in UC and associated CRC samples as compared to normal tissues (Chang et al. 2009; Chumanevich et al. 2011; Chang and Han 2006). UC patients are indeed at an increased risk of developing CRC (Eaden et al. 2001). CRC is the second leading cause of cancer-related deaths in both men and women in the United States and is characterized by several hallmarks of many forms of cancer, including mutations or deletions in p53 and tumor suppressor miRNAs that regulate the cell cycle and apoptosis (American Cancer Society 2015; Brentnall et al. 1994; Burmer et al. 1992; Rogler 2014; Necela et al. 2011; Young et al. 2012).

Based on a microarray analysis, Cui et al. demonstrated that the PAD antagonist Cl-amidine increased the putative tumor suppressor, miR-16 (Cui et al. 2013). MiR-16 is often downregulated in CRC and targets cell cycle regulators like Cyclin D1 and Cyclin E1 (Young et al. 2012). Cl-amidine treatment was also found to significantly upregulate miR-16 expression in human colon cancer cell lines. PAD inhibition by Cl-amidine and by PAD4 siRNA also caused cell cycle arrest at G1 in human colon cancer cell lines (Cui et al. 2013). To verify that Cl-amidine caused the G1 arrest, miR-16 knockdown and overexpression were examined. The cells that had low levels of miR-16 did not produce G1 arrest after Cl-amidine treatment; however, miR-16 overexpression caused G1 arrest even without Cl-amidine treatment (Cui et al. 2013). Interestingly, the effects of Cl-amidine were found to be dependent on the p53 status of the tested cell lines, and furthermore, Cl-amidine increased p53 expression in cells with a wild-type p53 status (Cui et al. 2013). Taken altogether, the cellular results further indicate that PADs suppress miR-16 to possibly prompt the progression to tumor initiation.

In a mouse model of UC-associated CRC, where prolonged duration of dextran sulfate sodium (DSS)-induced UC and a carcinogenic agent, azoxymethane (AOM), leads to tumorigenesis within 10 weeks, Cl-amidine-treated mice showed significantly reduced tumor incidence and multiplicity. Since Cl-amidine was successful at stopping the development of tumors, the events leading up to carcinogenesis were investigated, including miR-16 expression. As corroborated by the cellular study that showed Cl-amidine rescued miR-16 expression, Cl-amidine increased miR-16 expression and subsequently downregulated miR-16 cell cycle targets (Cyclin D1 and Cyclin E1) at 5 weeks in the AOM/DSS mouse model of UC-associated CRC (Witalison et al. 2015c). In these mice, at 5 weeks after the start of the experiment, inflammation occurred in mice that were not treated with Cl-amidine, but tumors had not developed in any of the mice (Witalison et al. 2015c). Since tumors had not occurred yet at 5 weeks but were present at 10 weeks, this establishes the AOM/DSS mouse model of UC-associated CRC as a viable model to study pro-inflammatory/premalignant and malignant environments and the events contributing to/resulting from tumorigenesis.

References

Acharya, N. K., Nagele, E. P., Han, M., Coretti, N. J., DeMarshall, C., Kosciuk, M. C., et al. (2012). Neuronal PAD4 expression and protein citrullination: Possible role in production of autoantibodies associated with neurodegenerative disease. *Journal of Autoimmunity, 38*(4), 369–380.

American Cancer Society. (2015). What are the key statistics about colorectal cancer? [updated 2015 Aug 13; cited 2016 Jan 4]. Retrieved from http://www.cancer.org/cancer/colonandrectumcancer/detailedguide/colorectal-cancer-key-statistics.

Arita, K., Hashimoto, H., Shimizu, T., Nakashima, K., Yamada, M., & Sato, M. (2004). Structural basis for Ca(2+)-induced activation of human PAD4. *Nature Structural & Molecular Biology, 11*(8), 777–783.

Bicker, K. L., & Thompson, P. R. (2013). The protein arginine deiminases: Structure, function, inhibition, and disease. *Biopolymers, 99*(2), 155–163.

Bicker, K. L., Anguish, L., Chumanevich, A. A., Cameron, M. D., Cui, X., Witalison, E., et al. (2012). D-amino acid based protein arginine deiminase inhibitors: Synthesis, pharmacokinetics, and in cellulo efficacy. *ACS Medicinal Chemistry Letters, 3*(12), 1081–1085.

Brentnall, T. A., Crispin, D. A., Rabinovitch, P. S., Haggitt, R. C., Rubin, C. E., Stevens, A. C., et al. (1994). Mutations in the p53 gene: An early marker of neoplastic progression in ulcerative colitis. *Gastroenterology, 107*(2), 369–378.

Burmer, G. C., Rabinovitch, P. S., Haggitt, R. C., Crispin, D. A., Brentnall, T. A., Kolli, V. R., et al. (1992). Neoplastic progression in ulcerative colitis: Histology, DNA content, and loss of a p53 allele. *Gastroenterology, 103*(5), 1602–1610.

Cao, L., Goodin, R., Wood, D., Moscarello, M. A., & Whitaker, J. N. (1999). Rapid release and unusual stability of immunodominant peptide 45-89 from citrullinated myelin basic protein. *Biochemistry, 38*(19), 6157–6163.

Cetinkaya, A., Bulbuloglu, E., Kantarceken, B., Ciralik, H., Kurutas, E. B., Buyukbese, M. A., et al. (2006). Effects of L-carnitine on oxidant/antioxidant status in acetic acid-induced colitis. *Digestive Diseases and Sciences, 51*(3), 488–494.

Chang, X., & Han, J. (2006). Expression of peptidylarginine deiminase type 4 (PAD4) in various tumors. *Molecular Carcinogenesis, 45*(3), 183–196.

Chang, X., Han, J., Pang, L., Zhao, Y., Yang, Y., & Shen, Z. (2009). Increased PADI4 expression in blood and tissues of patients with malignant tumors. *BMC Cancer, 9*, 40.

Chang, X., Xia, Y., Pan, J., Meng, Q., Zhao, Y., & Yan, X. (2013). PADI2 is significantly associated with rheumatoid arthritis. *PloS One, 8*(12), e81259.

Charlton, M. (2001). Hepatitis C infection in liver transplantation. *American Journal of Transplantation, 1*(3), 197–203.

Chavanas, S., Méchin, M. C., Takahara, H., Kawada, A., Nachat, R., Serre, G., et al. (2004). Comparative analysis of the mouse and human peptidylarginine deiminase gene clusters reveals highly conserved non-coding segments and a new human gene, PADI6. *Gene, 330*, 19–27.

Chavanas, S., Méchin, M. C., Nachat, R., Adoue, V., Coudane, F., Serre, G., et al. (2006). Peptidylarginine deiminases and deimination in biology and pathology: Relevance to skin homeostasis. *Journal of Dermatological Science, 44*(2), 63–72.

Chen, S. L., & Morgan, T. R. (2006). The natural history of hepatitis C virus (HCV) infection. *International Journal of Medical Sciences, 3*(2), 47–52.

Chen, C. C., Isomoto, H., Narumi, Y., Sato, K., Oishi, Y., Kobayashi, T., et al. (2008). Haplotypes of PADI4 susceptible to rheumatoid arthritis are also associated with ulcerative colitis in the Japanese population. *Clinical Immunology, 126*(2), 165–171.

Chen, X. Z., Schöttker, B., Castro, F. A., Chen, H., Zhang, Y., Holleczek, B., et al. (2016). Association of helicobacter pylori infection and chronic atrophic gastritis with risk of colonic, pancreatic and gastric cancer: A ten-year follow-up of the ESTHER cohort study. *Oncotarget*. Epub ahead of print.

Cho, J. H., Nicolae, D. L., Gold, L. H., Fields, C. T., LaBuda, M. C., Rohal, P. M., et al. (1998). Identification of novel susceptibility loci for inflammatory bowel disease on chromosomes 1p, 3q, and 4q: Evidence for epistasis between 1p and IBD1. *Proceedings of the National Academy of Sciences of the United States of America, 95*(13), 7502–7507.

Chumanevich, A. A., Causey, C. P., Knuckley, B. A., Jones, J. E., Poudyal, D., Chumanevich, A. P., et al. (2011). Suppression of colitis in mice by Cl-amidine: A novel peptidylarginine deiminase inhibitor. *American Journal of Physiology. Gastrointestinal and Liver Physiology, 300*(6), G929–G938.

Cui, X., Witalison, E. E., Chumanevich, A. P., Chumanevich, A. A., Poudyal, D., Subramanian, V., et al. (2013). The induction of microRNA-16 in colon cancer cells by protein arginine deiminase inhibition causes a p53-dependent cell cycle arrest. *PloS One, 8*(1), e53791.

Damgaard, D., Senolt, L., Nielsen, M. F., Pruijn, G. J., & Nielsen, C. H. (2014). Demonstration of extracellular peptidylarginine deiminase (PAD) activity in synovial fluid of patients with rheumatoid arthritis using a novel assay for citrullination of fibrinogen. *Arthritis Research & Therapy, 16*(6), 498.

Dooley, C. P., Cohen, H., Fitzgibbons, P. L., Bauer, M., Appleman, M. D., Perez-Perez, G. I., et al. (1989). Prevalence of helicobacter pylori infection and histologic gastritis in asymptomatic persons. *The New England Journal of Medicine, 321*(23), 1562–1566.

Eaden, J. A., Abrams, K. R., & Mayberry, J. F. (2001). The risk of colorectal cancer in ulcerative colitis: A meta-analysis. *Gut, 48*(4), 526–535.

Fasanella, K. E., Davis, B., Lyons, J., Chen, Z., Lee, K. K., Slivka, A., et al. (2007). Pain in chronic pancreatitis and pancreatic cancer. *Gastroenterology Clinics of North America, 36*(2), 335–364.

Ford, A. C., Moayyedi, P., & Hanauer, S. B. (2013). Ulcerative colitis. *BMJ, 346*, f432.

Foulquier, C., Sebbag, M., Clavel, C., Chapuy-Regaud, S., Al Badine, R., Méchin, M. C., et al. (2007). Peptidyl arginine deiminase type 2 (PAD-2) and PAD-4 but not PAD-1, PAD-3, and PAD-6 are expressed in rheumatoid arthritis synovium in close association with tissue inflammation. *Arthritis and Rheumatism, 56*(11), 3541–3553.

Godkin, A. J., De Belder, A. J., Villa, L., Wong, A., Beesley, J. E., Kane, S. P., et al. (1996). Expression of nitric oxide synthase in ulcerative colitis. *European Journal of Clinical Investigation, 26*(10), 867–872.

Green, S. J., Mellouk, S., Hoffman, S. L., Meltzer, M. S., & Nacy, C. A. (1990). Cellular mechanisms of nonspecific immunity to intracellular infection: Cytokine-induced synthesis of toxic nitrogen oxides from L-arginine by macrophages and hepatocytes. *Immunology Letters, 25*(1–3), 15–19.

Guerrin, M., Ishigami, A., Méchin, M. C., Nachat, R., Valmary, S., Sebbag, M., et al. (2003). cDNA cloning, gene organization and expression analysis of human peptidylarginine deiminase type I. *The Biochemical Journal, 370*(Pt 1), 167–174.

Gukovsky, I., Lugea, A., Shahsahebi, M., Cheng, J. H., Hong, P. P., Jung, Y. J., et al. (2008). A rat model reproducing key pathological responses of alcoholic chronic pancreatitis. *American Journal of Physiology: Gastrointestinal and Liver Physiology, 294*(1), G68–G79.

Hagiwara, T., Nakashima, K., Hirano, H., Senshu, T., & Yamada, M. (2002). Deimination of arginine residues in nucleophosmin/B23 and histones in HL-60 granulocytes. *Biochemical and Biophysical Research Communications, 290*(3), 979–983.

Hata, K., Watanabe, T., Kazama, S., Suzuki, K., Shinozaki, M., Yokoyama, T., et al. (2003). Earlier surveillance colonoscopy programme improves survival in patients with ulcerative colitis associated colorectal cancer: Results of a 23-year surveillance programme in the Japanese population. *British Journal of Cancer, 89*(7), 1232–1236.

Hendel, J., & Nielsen, O. H. (1997). Expression of cyclooxygenase-2 mRNA in active inflammatory bowel disease. *The American Journal of Gastroenterology, 92*(7), 1170–1173.

Humans IWGotEoCRt. (1994). *Schistosomes, liver flukes and Helicobacter pylori*. Lyon, France: World Health Organization, International Agency for Research on Cancer.

Hussain, S. P., Hofseth, L. J., & Harris, C. C. (2003). Radical causes of cancer. *Nature Reviews. Cancer, 3*(4), 276–285.

Hvid-Jensen, F., Pedersen, L., Drewes, A. M., Sørensen, H. T., & Funch-Jensen, P. (2011). Incidence of adenocarcinoma among patients with Barrett's esophagus. *The New England Journal of Medicine, 365*(15), 1375–1383.

Ishigami, A., Ohsawa, T., Asaga, H., Akiyama, K., Kuramoto, M., & Maruyama, N. (2002). Human peptidylarginine deiminase type II: Molecular cloning, gene organization, and expression in human skin. *Archives of Biochemistry and Biophysics, 407*(1), 25–31.

Ishigami, A., Ohsawa, T., Hiratsuka, M., Taguchi, H., Kobayashi, S., Saito, Y., et al. (2005). Abnormal accumulation of citrullinated proteins catalyzed by peptidylarginine deiminase in hippocampal extracts from patients with Alzheimer's disease. *Journal of Neuroscience Research, 80*(1), 120–128.

Itzkowitz, S. H., & Yio, X. (2004). Inflammation and cancer IV. Colorectal cancer in inflammatory bowel disease: The role of inflammation. *American Journal of Physiology. Gastrointestinal and Liver Physiology, 287*(1), G7–17.

Jones, J. E., Causey, C. P., Knuckley, B., Slack-Noyes, J. L., & Thompson, P. R. (2009). Protein arginine deiminase 4 (PAD4): Current understanding and future therapeutic potential. *Current Opinion in Drug Discovery & Development, 12*(5), 616–627.

Kan, R., Yurttas, P., Kim, B., Jin, M., Wo, L., Lee, B., et al. (2011). Regulation of mouse oocyte microtubule and organelle dynamics by PADI6 and the cytoplasmic lattices. *Developmental Biology, 350*(2), 311–322.

Kan, R., Jin, M., Subramanian, V., Causey, C. P., Thompson, P. R., & Coonrod, S. A. (2012). Potential role for PADI-mediated histone citrullination in preimplantation development. *BMC Developmental Biology, 12*, 19.

Kavanagh, D. O., Carter, M. C., Keegan, D., Doherty, G., Smith, M. J., Hyland, J. M., et al. (2014). Management of colorectal cancer in patients with inflammatory bowel disease. *Techniques in Coloproctology, 18*(1), 23–28.

Khandpur, R., Carmona-Rivera, C., Vivekanandan-Giri, A., Gizinski, A., Yalavarthi, S., Knight, J. S., et al. (2013). NETs are a source of citrullinated autoantigens and stimulate inflammatory responses in rheumatoid arthritis. *Science Translational Medicine, 5*(178), 178ra40.

Kimura, H., Hokari, R., Miura, S., Shigematsu, T., Hirokawa, M., Akiba, Y., et al. (1998). Increased expression of an inducible isoform of nitric oxide synthase and the formation of peroxynitrite in colonic mucosa of patients with active ulcerative colitis. *Gut, 42*(2), 180–187.

Knight, J. S., Subramanian, V., O'Dell, A. A., Yalavarthi, S., Zhao, W., Smith, C. K., et al. (2015). Peptidylarginine deiminase inhibition disrupts NET formation and protects against kidney, skin and vascular disease in lupus-prone MRL/lpr mice. *Annals of the Rheumatic Diseases, 74*(12), 2199–2206.

Knuckley, B., Causey, C. P., Jones, J. E., Bhatia, M., Dreyton, C. J., Osborne, T. C., et al. (2010). Substrate specificity and kinetic studies of PADs 1, 3, and 4 identify potent and selective inhibitors of protein arginine deiminase 3. *Biochemistry, 49*(23), 4852–4863.

Kolfenbach, J. R., Deane, K. D., Derber, L. A., O'Donnell, C. I., Gilliland, W. R., Edison, J. D., et al. (2010). Autoimmunity to peptidyl arginine deiminase type 4 precedes clinical onset of rheumatoid arthritis. *Arthritis and Rheumatism, 62*(9), 2633–2639.

Koppert, L. B., Wijnhoven, B. P., van Dekken, H., Tilanus, H. W., & Dinjens, W. N. (2005). The molecular biology of esophageal adenocarcinoma. *Journal of Surgical Oncology, 92*(3), 169–190.

Lamensa, J., & Moscarello, M. A. (1993). Deimination of human myelin basic protein by a peptidylarginine deiminase from bovine brain. *Journal of Neurochemistry, 61*(3), 987–996.

Li, P., Yao, H., Zhang, Z., Li, M., Luo, Y., Thompson, P. R., et al. (2008). Regulation of p53 target gene expression by peptidylarginine deiminase 4. *Molecular and Cellular Biology, 28*(15), 4745–4758.

Lih-Brody, L., Powell, S. R., Collier, K. P., Reddy, G. M., Cerchia, R., Kahn, E., et al. (1996). Increased oxidative stress and decreased antioxidant defenses in mucosa of inflammatory bowel disease. *Digestive Diseases and Sciences, 41*(10), 2078–2086.

Luo, Y., Arita, K., Bhatia, M., Knuckley, B., Lee, Y. H., Stallcup, M. R., et al. (2006). Inhibitors and inactivators of protein arginine deiminase 4: Functional and structural characterization. *Biochemistry, 45*(39), 11727–11736.

Makrygiannakis, D., af Klint, E., Lundberg, I. E., Löfberg, R., Ulfgren, A. K., Klareskog, L., et al. (2006). Citrullination is an inflammation-dependent process. *Annals of the Rheumatic Diseases, 65*(9), 1219–1222.

Marnett, L. J. (2000). Oxyradicals and DNA damage. *Carcinogenesis, 21*(3), 361–370.

McCaughan, G. W., McGuinness, P. H., Bishop, G. A., Painter, D. M., Lien, A. S., Tulloch, R., et al. (1992). Clinical assessment and incidence of hepatitis C RNA in 50 consecutive RIBA-positive volunteer blood donors. *The Medical Journal of Australia, 157*(4), 231–233.

McGraw, W. T., Potempa, J., Farley, D., & Travis, J. (1999). Purification, characterization, and sequence analysis of a potential virulence factor from Porphyromonas gingivalis, peptidylarginine deiminase. *Infection and Immunity, 67*(7), 3248–3256.

McKenzie, S. J., Baker, M. S., Buffinton, G. D., & Doe, W. F. (1996). Evidence of oxidant-induced injury to epithelial cells during inflammatory bowel disease. *The Journal of Clinical Investigation, 98*(1), 136–141.

Méchin, M. C., Sebbag, M., Arnaud, J., Nachat, R., Foulquier, C., Adoue, V., et al. (2007). Update on peptidylarginine deiminases and deimination in skin physiology and severe human diseases. *International Journal of Cosmetic Science, 29*(3), 147–168.

Moscarello, M. A., Lei, H., Mastronardi, F. G., Winer, S., Tsui, H., Li, Z., et al. (2013). Inhibition of peptidyl-arginine deiminases reverses protein-hypercitrullination and disease in mouse models of multiple sclerosis. *Disease Models & Mechanisms, 6*(2), 467–478.

Nathan, C. (2002). Points of control in inflammation. *Nature, 420*(6917), 846–852.

National Cancer Institute, National Institutes of Health. (2016). SEER Stat Fact Sheets: Pancreas Cancer. [cited 2016 Mar 17]. Retrieved from http://seer.cancer.gov/statfacts/html/pancreas.html.

Necela, B. M., Carr, J. M., Asmann, Y. W., & Thompson, E. A. (2011). Differential expression of microRNAs in tumors from chronically inflamed or genetic (APC(Min/+)) models of colon cancer. *PloS One, 6*(4), e18501.

Nomura, K. (1992). Specificity and mode of action of the muscle-type protein-arginine deiminase. *Archives of Biochemistry and Biophysics, 293*(2), 362–369.

Ordás, I., Eckmann, L., Talamini, M., Baumgart, D. C., & Sandborn, W. J. (2012). Ulcerative colitis. *Lancet, 380*(9853), 1606–1619.

Pandol, S. J., Saluja, A. K., Imrie, C. W., & Banks, P. A. (2007). Acute pancreatitis: Bench to the bedside. *Gastroenterology, 132*(3), 1127–1151.

Pandol, S. J., Lugea, A., Mareninova, O. A., Smoot, D., Gorelick, F. S., Gukovskaya, A. S., et al. (2011). Investigating the pathobiology of alcoholic pancreatitis. *Alcoholism, Clinical and Experimental Research, 35*(5), 830–837.

Raijmakers, R., Zendman, A. J., Egberts, W. V., Vossenaar, E. R., Raats, J., Soede-Huijbregts, C., et al. (2007). Methylation of arginine residues interferes with citrullination by peptidylarginine deiminases in vitro. *Journal of Molecular Biology, 367*(4), 1118–1129.

Ren, W., Yin, J., Wu, M., Liu, G., Yang, G., Xion, Y., et al. (2014). Serum amino acids profile and the beneficial effects of L-arginine or L-glutamine supplementation in dextran sulfate sodium colitis. *PloS One, 9*(2), e88335.

Rogler, G. (2014). Chronic ulcerative colitis and colorectal cancer. *Cancer Letters, 345*(2), 235–241.

Rus'd, A. A., Ikejiri, Y., Ono, H., Yonekawa, T., Shiraiwa, M., Kawada, A., et al. (1999). Molecular cloning of cDNAs of mouse peptidylarginine deiminase type I, type III and type IV, and the expression pattern of type I in mouse. *European Journal of Biochemistry, 259*(3), 660–669.

Sakthivel, K. M., & Guruvayoorappan, C. (2013). Amentoflavone inhibits iNOS, COX-2 expression and modulates cytokine profile, NF-κB signal transduction pathways in rats with ulcerative colitis. *International Immunopharmacology, 17*(3), 907–916.

Seibert, K., & Masferrer, J. L. (1994). Role of inducible cyclooxygenase (COX-2) in inflammation. *Receptor, 4*(1), 17–23.

Senshu, T., Akiyama, K., Kan, S., Asaga, H., Ishigami, A., & Manabe, M. (1995). Detection of deiminated proteins in rat skin: Probing with a monospecific antibody after modification of citrulline residues. *The Journal of Investigative Dermatology, 105*(2), 163–169.

Shaheen, N. J., & Richter, J. E. (2009). Barrett's oesphagus. *Lancet, 373*(9666), 850–861.

Shimizu, A., Handa, K., Honda, T., Abe, N., Kojima, T., & Takahara, H. (2014). Three isozymes of peptidylarginine deiminase in the chicken: Molecular cloning, characterization, and tissue distribution. *Comparative Biochemistry and Physiology. Part B, Biochemistry & Molecular Biology, 167*, 65–73.

Singer, I. I., Kawka, D. W., Scott, S., Weidner, J. R., Mumford, R. A., Riehl, T. E., et al. (1996). Expression of inducible nitric oxide synthase and nitrotyrosine in colonic epithelium in inflammatory bowel disease. *Gastroenterology, 111*(4), 871–885.

Singer, I. I., Kawka, D. W., Schloemann, S., Tessner, T., Riehl, T., & Stenson, W. F. (1998). Cyclooxygenase 2 is induced in colonic epithelial cells in inflammatory bowel disease. *Gastroenterology, 115*(2), 297–306.

Sipponen, P. (2002). Gastric cancer: Pathogenesis, risks, and prevention. *Journal of Gastroenterology, 37*(Suppl 13), 39–44.

Slack, J. L., Causey, C. P., & Thompson, P. R. (2011). Protein arginine deiminase 4: A target for an epigenetic cancer therapy. *Cellular and Molecular Life Sciences, 68*(4), 709–720.

Subramanian, V., Knight, J. S., Parelkar, S., Anguish, L., Coonrod, S. A., Kaplan, M. J., et al. (2015). Design, synthesis, and biological evaluation of tetrazole analogs of Cl-amidine as protein arginine deiminase inhibitors. *Journal of Medicinal Chemistry, 58*(3), 1337–1344.

Suzuki, A., Yamada, R., Chang, X., Tokuhiro, S., Sawada, T., Suzuki, M., et al. (2003). Functional haplotypes of PADI4, encoding citrullinating enzyme peptidylarginine deiminase 4, are associated with rheumatoid arthritis. *Nature Genetics, 34*(4), 395–402.

Tanikawa, C., Espinosa, M., Suzuki, A., Masuda, K., Yamamoto, K., Tsuchiya, E., et al. (2012). Regulation of histone modification and chromatin structure by the p53-PADI4 pathway. *Nature Communications, 3*, 676.

Tarcsa, E., Marekov, L. N., Mei, G., Melino, G., Lee, S. C., & Steinert, P. M. (1996). Protein unfolding by peptidylarginine deiminase. Substrate specificity and structural relationships of the natural substrates trichohyalin and filaggrin. *The Journal of Biological Chemistry, 271*(48), 30709–30716.

Thompson, A. J., & Holmes, J. A. (2015). Treating hepatitis C—What's new? *Australian Prescriber, 38*(6), 191–197.

Tong, M. J., el-Farra, N. S., Reikes, A. R., & Co, R. L. (1995). Clinical outcomes after transfusion-associated hepatitis C. *The New England Journal of Medicine, 332*(22), 1463–1466.

Vane, J. R., Mitchell, J. A., Appleton, I., Tomlinson, A., Bishop-Bailey, D., Croxtall, J., et al. (1994). Inducible isoforms of cyclooxygenase and nitric-oxide synthase in inflammation. *Proceedings of the National Academy of Sciences of the United States of America, 91*(6), 2046–2050.

Vassiliadis, E., Veidal, S. S., Kristiansen, M. N., Hansen, C., Jorgensen, M., Leeming, D. J., et al. (2013). Peptidyl arginine deiminase inhibitor effect on hepatic fibrogenesis in a CCl4 pre-clinical model of liver fibrosis. *American Journal of Translational Research, 5*(4), 465–469.

Vossenaar, E. R., Zendman, A. J., van Venrooij, W. J., & Pruijn, G. J. (2003). PAD, a growing family of citrullinating enzymes: Genes, features and involvement in disease. *BioEssays, 25*(11), 1106–1118.

Vossenaar, E. R., Radstake, T. R., van der Heijden, A., van Mansum, M. A., Dieteren, C., de Rooij, D. J., et al. (2004). Expression and activity of citrullinating peptidylarginine deiminase enzymes in monocytes and macrophages. *Annals of the Rheumatic Diseases, 63*(4), 373–381.

Watanabe, K., & Senshu, T. (1989). Isolation and characterization of cDNA clones encoding rat skeletal muscle peptidylarginine deiminase. *The Journal of Biological Chemistry, 264*(26), 15255–15260.

Watanabe, K., Akiyama, K., Hikichi, K., Ohtsuka, R., Okuyama, A., & Senshu, T. (1988). Combined biochemical and immunochemical comparison of peptidylarginine deiminases present in various tissues. *Biochimica et Biophysica Acta, 966*(3), 375–383.

Weston, A. P., Sharma, P., Topalovski, M., Richards, R., Cherian, R., & Dixon, A. (2000). Long-term follow-up of Barrett's high-grade dysplasia. *The American Journal of Gastroenterology, 95*(8), 1888–1893.

Willis, V. C., Gizinski, A. M., Banda, N. K., Causey, C. P., Knuckley, B., Cordova, K. N., et al. (2011). N-α-benzoyl-N5-(2-chloro-1-iminoethyl)-L-ornithine amide, a protein arginine deiminase inhibitor, reduces the severity of murine collagen-induced arthritis. *Journal of Immunology, 186*(7), 4396–4404.

Wink, D. A., Kasprzak, K. S., Maragos, C. M., Elespuru, R. K., Misra, M., Dunams, T. M., et al. (1991). DNA deaminating ability and genotoxicity of nitric oxide and its progenitors. *Science, 254*(5034), 1001–1003.

Witalison, E. E., Thompson, P. R., & Hofseth, L. J. (2015a). Protein arginine deiminases and associated citrullination: Physiological functions and diseases associated with dysregulation. *Current Drug Targets, 16*(7), 700–710.

Witalison, E. E., Cui, X., Hofseth, A. B., Subramanian, V., Causey, C. P., Thompson, P. R., et al. (2015b). Inhibiting protein arginine deiminases has antioxidant consequences. *The Journal of Pharmacology and Experimental Therapeutics, 353*(1), 64–70.

Witalison, E. E., Cui, X., Causey, C. P., Thompson, P. R., & Hofseth, L. J. (2015c). Molecular targeting of protein arginine deiminases to suppress colitis and prevent colon cancer. *Oncotarget, 6*(34), 36053–36062.

Wood, D. D., Bilbao, J. M., O'Connors, P., & Moscarello, M. A. (1996). Acute multiple sclerosis (Marburg type) is associated with developmentally immature myelin basic protein. *Annals of Neurology, 40*(1), 18–24.

World Health Organization. (2016). Health topics: Hepatitis. [cited 2016 Mar 16]. Retrieved from http://www.who.int/topics/hepatitis/en/.

Yadav, D., Timmons, L., Benson, J. T., Dierkhising, R. A., & Chari, S. T. (2011). Incidence, prevalence, and survival of chronic pancreatitis: A population-based study. *The American Journal of Gastroenterology, 106*(12), 2192–2199.

Ying, S., Dong, S., Kawada, A., Kojima, T., Chavanas, S., Méchin, M. C., et al. (2009). Transcriptional regulation of peptidylarginine deiminase expression in human keratinocytes. *Journal of Dermatological Science, 53*(1), 2–9.

Young, L. E., Moore, A. E., Sokol, L., Meisner-Kober, N., & Dixon, D. A. (2012). The mRNA stability factor HuR inhibits microRNA-16 targeting of COX-2. *Molecular Cancer Research, 10*(1), 167–180.

Zhang, X., Bolt, M., Guertin, M. J., Chen, W., Zhang, S., Cherrington, B. D., et al. (2012). Peptidylarginine deiminase 2-catalyzed histone H3 arginine 26 citrullination facilitates estrogen receptor α target gene activation. *Proceedings of the National Academy of Sciences of the United States of America, 109*(33), 13331–13336.

Chapter 23
Development of the Protein Arginine Deiminase (PAD) Inhibitors

Aaron Muth and Paul R. Thompson

23.1 Introduction

The protein arginine deiminases (PADs) catalyze the hydrolysis of positively charged arginine residues to generate neutral citrulline (Fig. 23.1) (Jones et al. 2009; Fuhrmann and Thompson 2016; Fuhrmann et al. 2015). This reaction is a calcium-dependent process wherein calcium binding is necessary for the formation of the active site and results in a >10,000-fold increase in activity (Liu et al. 2011; Kearney et al. 2005). There are five known PAD isozymes (PAD1–PAD4 and PAD6), of which only PAD1–PAD4 are catalytically active (Raijmakers et al. 2007). These enzymes contain a high degree of sequence similarity; however, they are uniquely distributed throughout the body. PAD1 is principally expressed in the uterus and epidermis, PAD2 is expressed in most tissue and cell types, PAD3 is found in the skin and hair follicles, PAD4 is primarily expressed in neutrophils and other myeloid derived cells, and PAD6 is found in oocytes and embryos (Jones et al. 2009; Fuhrmann and Thompson 2016; Fuhrmann et al. 2015).

Aberrant PAD activity and subsequent hypercitrullination has been extensively investigated primarily due to its role in disease development and progression. Specifically, aberrant PAD activity has been associated with a variety of autoimmune diseases, including rheumatoid arthritis (RA) (Luban and Li 2010), systemic lupus erythematosus (SLE) (Hakkim et al. 2010; Villanueva et al. 2011), ulcerative colitis (Chumanevich et al. 2011a), and inflammatory bowel disease (IBD) (Makrygiannakis et al. 2006); a variety of neurodegenerative diseases (i.e., multiple sclerosis (MS) (Moscarello et al. 2007; Musse et al. 2008), Parkinson's disease (Nicholas 2011), Alzheimer's diseases (Nicholas 2013)); as well as certain types of cancer (Mohanan et al. 2012).

A. Muth • P.R. Thompson (✉)
University of Massachusetts Medical School,
LRB 826, 364 Plantation Street, Worcester, MA 01605, USA
e-mail: Paul.Thompson@umassmed.edu

A.P. Nicholas et al. (eds.), *Protein Deimination in Human Health and Disease*,
DOI 10.1007/978-3-319-58244-3_23

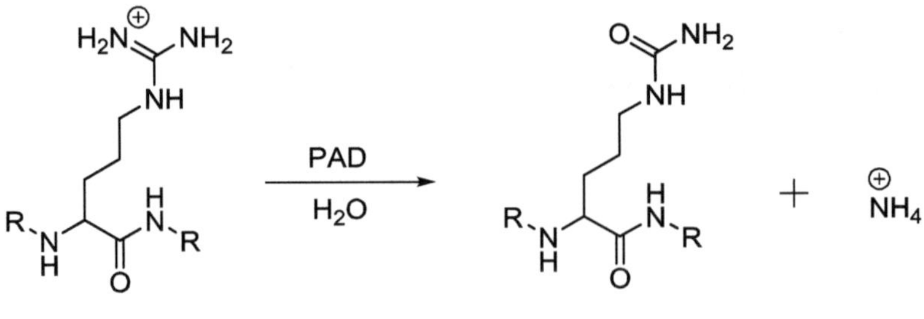

Fig. 23.1 Hydrolysis of peptidyl-arginine (*left*) to peptidyl-citrulline (*right*), which is catalyzed by the protein arginine deiminases (PADs)

With regard to specific isozymes, PAD4 represses the expression of p53 target genes, and its inhibition or knockdown triggers apoptosis in a variety of tumor-derived cell lines (Li et al. 2008). Notably, pan-PAD inhibitors decrease tumor burden in models of breast and colon cancer (McElwee et al. 2012a; Witalison et al. 2015a). PAD4 also contributes to NETosis, a pro-inflammatory form of cell death that is aberrantly upregulated in numerous autoimmune disorders including RA (Khandpur et al. 2013), atherosclerosis (Knight et al. 2014a), and lupus (Knight et al. 2014b), and it is noteworthy that PAD inhibitors show efficacy in these and other inflammatory diseases (McElwee et al. 2012a; Knight et al. 2014a, b; Willis et al. 2011; Willis 2012; Chumanevich et al. 2011b; Witalison et al. 2015b; Moscarello et al. 2013; Knight et al. 2013; Lange et al. 2011, 2014; Wang et al. 2012a).

Upregulated PAD2 activity is also linked to the development of certain cancers and inflammatory diseases (Jones et al. 2009). Specifically, PAD2 is highly expressed in luminal breast cancers, and its expression positively correlates with the levels of the HER2 proto-oncogene (McElwee et al. 2012a; Guertin et al. 2014; Zhang et al. 2012). When PAD2 activity is inhibited in MCF10-DCIS cells, a breast cancer cell line, proliferation is decreased and apoptosis is increased (McElwee et al. 2012a). PAD2 activity is also upregulated in MS where it citrullinates myelin basic protein (MBP) leading to demyelination (Lamensa and Moscarello 1993).

PAD3 levels are elevated in spinal cord injury models, and pan-PAD inhibitors reduce apoptosis and tissue loss in these models (Lange et al. 2011). Recently, PAD3 was shown to enter the nucleus and citrullinate histones to regulate gene expression in response to gonadotropin (Li et al. 2016); however, the broader role of this isozyme in gene regulation is only beginning to be explored.

PAD1 activity appears to be dysregulated in psoriasis; (Ishida-Yamamoto et al. 2000) however, the actual function of this isozyme has not been deeply examined. The extent to which the PADs are involved in a variety of diseases is more fully summarized in several recent reviews and other chapters of this book (Jones et al. 2009; Fuhrmann and Thompson 2016; Fuhrmann et al. 2015; Lewis et al. 2015; Tu et al. 2016).

Given the distinct roles of these isozymes in human disease and cell signaling, there is a pressing need to develop potent and isozyme-specific inhibitors for use as therapeutics and/or probes to decipher the full complement of processes regulated by these enzymes.

23.2 PAD Enzyme Structures

One of the keys in developing PAD inhibitors has been determining the crystal structures of this family of enzymes. The structures of PAD1 (Saijo et al. 2016), PAD2 (Slade et al. 2015), and PAD4 (Arita et al. 2004) are currently known; however, structures of both the inactive (apo) and active (holo) states have only been determined for PAD2 and PAD4. It is important to again note that calcium is required for the formation of the holo state (preferentially targeted by the irreversible inhibitors detailed later), while a new class of inhibitors has targeted the apo state. The most important difference between these two states is that the catalytic cysteine (C647 for PAD2 and C645 for PAD4) moves into the active site when enough calcium is present and the enzyme is in the holo state. These structures have contributed to the development of certain inhibitors and will certainly aid in future inhibitor development. The PAD1 crystal structure was found to have a very similar catalytic domain to PAD2 and PAD4 (Fig. 23.2), which was expected, since it possesses 50.2% amino acid similarity to PAD2 and 54.8% similarity to PAD4 (Saijo et al. 2016). However, even at high concentrations of $CaCl_2$ (200 mM), only four Ca^{2+}ions were found in the PAD1 monomer (Saijo et al. 2016). Additionally, the *N*-terminus of the *N*-terminal immunoglobulin-like domain (IgG-1) adopts a slightly different orientation in the structure of PAD1, relative to both PAD2 and PAD4, that precludes dimer formation (Saijo et al. 2016). As all of the structures are determined, this information will be invaluable in not only improving potency but also isozyme selectivity, ultimately allowing us to better understand the biology associated with each PAD isozyme.

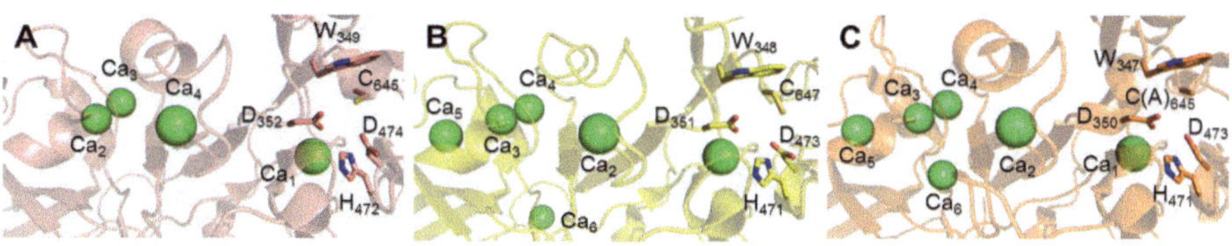

Fig. 23.2 (**a**) HoloPAD1 crystal structure with Ca^{2+} (*purple*, PDB ID: 5HP5). (**b**) HoloPAD2 F221/222A mutant crystal structure with Ca^{2+} (*yellow*, PDB ID: 4N2C). (**c**) HoloPAD4 crystal structure with Ca^{2+} (*orange*, PDB ID: 1WD9)

23.3 PAD Inhibitor Development

23.3.1 *First-Generation Irreversible PAD Inhibitors*

PAD inhibitor development began with the identification of 2-chloroacetamidine as a modest inhibitor of PAD4 (Stone et al. 2005). Specifically, it was identified as a covalent and irreversible inhibitor, which decreases enzyme activity in a time-dependent manner (Stone et al. 2005). In parallel, the Thompson group identified the PAD substrate benzoyl-L-arginine amide (BAA, Fig. 23.3a **1**, PAD substrate) as a core scaffold (Kearney et al. 2005) and showed that the terminal guanidinium could be replaced by either a chloroacetamidine or fluoroacetamidine warhead to generate the first-generation inhibitors F-amidine (Fig. 23.3a, **2**) and Cl-amidine (Fig. 23.3a, **3**) (Luo et al. 2006a, b). These two inhibitors were then found to covalently modify the catalytic cysteine (C645 in PAD4, Fig. 23.3) conserved in all PAD isozymes. It is important to note that the chloroacetamidine warhead demonstrated higher reactivity with the cysteine thiol than the fluoroacetamidine warhead, which was expected (Luo et al. 2006a, b). These two warheads also found great utility due to their high stability at neutral pH for prolonged periods of time (>24 h) (Luo et al. 2006a, b). These early inhibitors demonstrated that the chloroacetamidine warhead was more potent than the fluoroacetamidine warhead when evaluated against PAD1, PAD2, PAD3, and PAD4. Another key lesson learned from Cl-amidine (Fig. 23.3a, **3**) was in its efficacy in a variety of animal models. Specifically, Cl-amidine (Fig. 23.3a, **3**) has shown utility in treating animal models of RA, lupus, ulcerative colitis, atherosclerosis, and breast cancer (Chumanevich et al. 2011a; Khandpur et al. 2013; Knight et al. 2013, 2014c; Lange et al. 2011; Wang et al. 2012b; Causey et al. 2011; Knight and Kaplan 2012; Smith et al. 2014; McElwee et al. 2012b). Since PAD activity is dysregulated in all of these diseases, Cl-amidine (Fig. 23.3a, **3**) proved invaluable in demonstrating the disease relevance of the PADs and ultimately the importance of developing inhibitors of these enzymes.

A

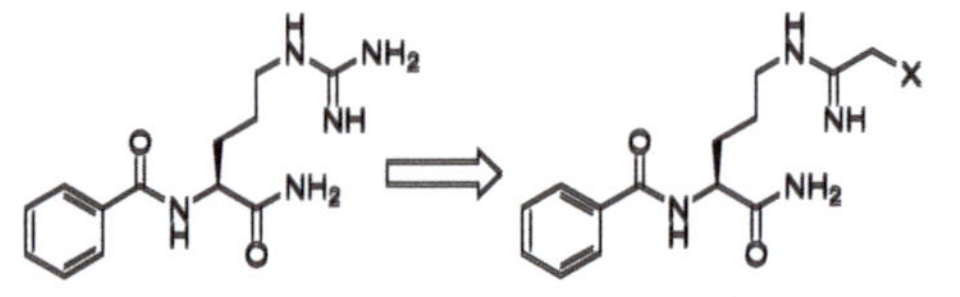

B

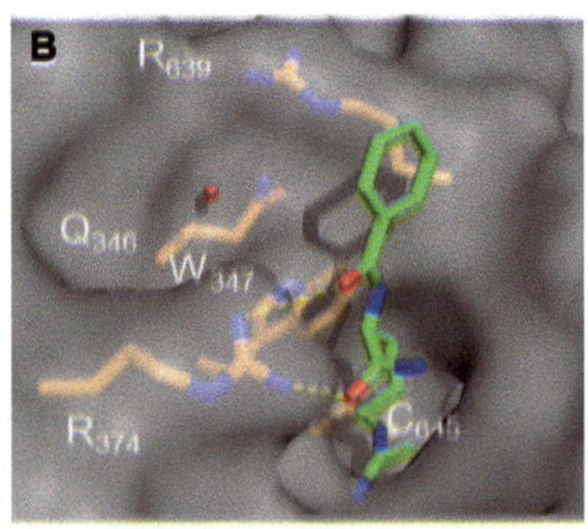

Entry	PAD1 k_{inact}/K_I ($M^{-1}min^{-1}$)	PAD1 k_{inact}/K_I ($M^{-1}min^{-1}$)	PAD1 k_{inact}/K_I ($M^{-1}min^{-1}$)	PAD1 k_{inact}/K_I ($M^{-1}min^{-1}$)
2	2800	380	170	3000
3	37000	1200	2000	13000

Fig. 23.3 (**a**) Development of PAD inhibitors F-amidine and Cl-amidine from benzoyl-L-arginine amide (BAA, **1**). (**b**) PAD inhibition data for F-amidine (**2**) and Cl-amidine (**3**). (**c**) Co-crystal structure of F-amidine (**2**) bound to PAD4 (PDB ID: 2DW5)

23.3.2 *Preliminary Structure Activity Relationships (SARs) for F-Amidine and Cl-Amidine*

After the initial discovery of F-amidine (Fig. 23.3a, **2**) and Cl-amidine (Fig. 23.3a, **3**), a preliminary SAR was explored around the length of the side chain linker distance to the haloacetamidine warhead (Luo et al. 2006b). Interestingly, the Thompson group found that a one carbon homologation (F4-amidine, Fig. 23.4, **4**, and Cl4-amidine, Fig. 23.4, **5**) resulted in the first PAD3-selective inhibitors (Luo et al. 2006b; Knuckley et al. 2010a). Specifically, these compounds demonstrated 10- to 50-fold selective inhibition of PAD3 over PAD1 and PAD4 (Knuckley et al. 2010a).

23.3.3 *Enhanced Inhibition of PADs*

After exploring the side chain linker length, the Thompson group then generated a SAR for the benzoyl moiety of F-amidine (Fig. 23.3a, **2**) and Cl-amidine (Fig. 23.3a, **3**). This extensive SAR found that a carboxylic acid functional group at the *ortho*-position of the benzoyl moiety dramatically improved inhibition of PAD1–PAD4 (*o*-F-amidine, Fig. 23.5a, **6**, and *o*-Cl-amidine, Fig. 23.5a, **7**) (Causey et al. 2011). The key to the enhancement in PAD4 inhibition appeared to be through enhanced interactions with the indole nitrogen of W347 and guanidinium group R374 (Fig. 23.5b, which depicts the co-crystal structure of PAD4 bound to *o*-F-amidine, Fig. 23.5a, **6**) (Causey et al. 2011). In particular, the gained interaction with W347 (conserved in all of the PADs) appears to be key for the enhanced inhibition (Causey et al. 2011).

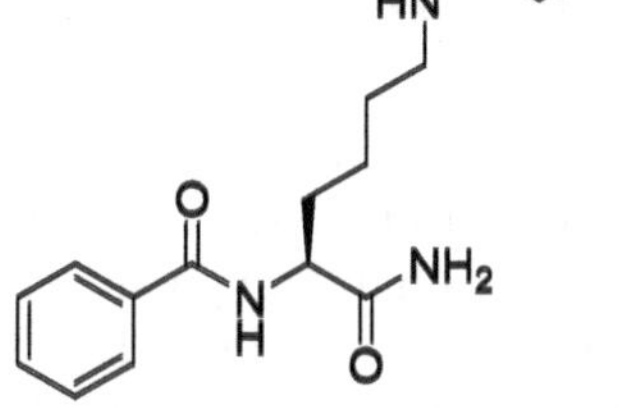

4: X=F; F4-amidine

IC_{50} (μM)	Isozyme Selectivity
PAD1=7965	1.0
PAD3=64.4	124
PAD4=648	12

5: X=Cl; Cl4-amidine

IC_{50} (μM)	Isozyme Selectivity
PAD1=256	2.5
PAD3=12.8	50
PAD4=640	1.0

Fig. 23.4 Structures, IC_{50} values, and isozyme selectivity of the PAD3-selective inhibitors F4-amidine (**4**) and Cl4-amidine (**5**)

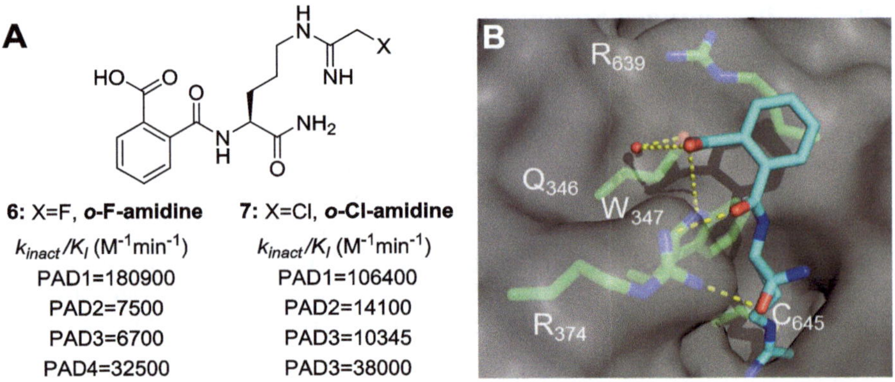

Fig. 23.5 (**a**) Structures and corresponding inhibition values for *o*-F-amidine (**6**) and *o*-Cl-amidine (**7**). (**b**) *o*-F-amidine (**6**) bound to PAD4 (PDB ID: 3B1T)

23.3.4 Identification of Potent and Selective PAD4 Inhibitors

In an effort to identify potent PAD4 inhibitors, in 2011, the Thompson group screened a tripeptide library of haloacetamidine-based compounds (Jones et al. 2012). This proved fruitful as two hits were identified, TDFA (Fig. 23.6a, **8**) and TDCA (Fig. 23.6a, **9**), as potent and selective PAD4 inhibitors (Jones et al. 2012). Interestingly, this threonine/aspartic acid/ornithine haloacetamidine also appears to take advantage of a gained interaction with W347 in PAD4 to enhance inhibition as seen in Fig. 23.6b. It was also seen that this improved inhibition was gained through interactions with Q346, R374, and R639. In particular, the interaction with R639 was noted as being of particular importance for PAD4-selective inhibition, as this residue is unique to PAD4 (Jones et al. 2012).

23.3.5 Discovery of the PAD4-Selective Inhibitor Streptonigrin

In 2010, the Thompson group utilized a derivative of F-amidine (**2**), rhodamine-conjugated F-amidine (RFA (**26**) discussed later), to conduct a fluopol-activity-based protein profiling (ABPP) high-throughput screen (HTS) assay (Knuckley et al. 2010b). Through this screen, streptonigrin (Fig. 23.7, **10**) was discovered to be a potent, selective, and irreversible inhibitor of PAD4 (Knuckley et al. 2010b; Dreyton et al. 2014). Currently, the binding location and residue of modification is unknown; however, SAR and mass spectrometric results suggest that covalent modification of its quinoline-5,8-dione Michael acceptor is necessary for enzyme inactivation (Dreyton et al. 2014). Streptonigrin is known to inhibit a number of other cellular processes (i.e., inhibition of DNA and RNA synthesis) (Bolzan and Bianchi 2001) and is highly cytotoxic across multiple cell lines (Dreyton et al. 2014). As such, we do not recommend its use as a probe of PAD4 activity. Nevertheless, since

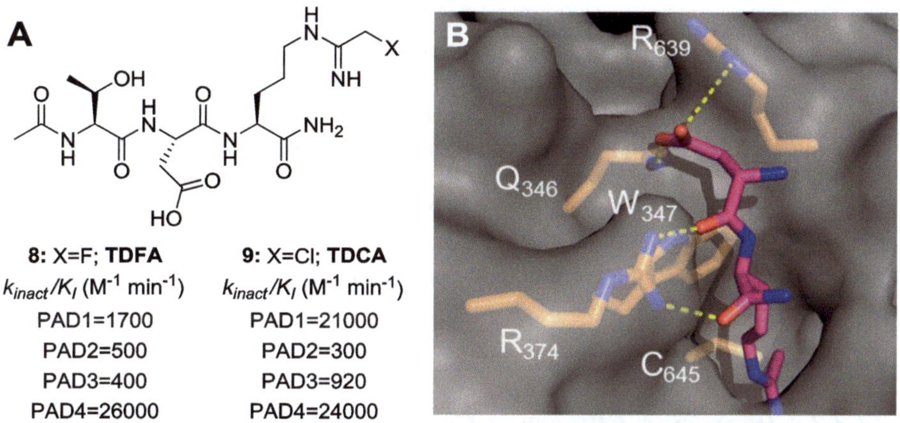

Fig. 23.6 (**a**) Structures and corresponding inhibition values for the PAD4-selective inhibitors TDFA (**8**) and TDCA (**9**). (**b**) TDFA (**8**) bound to PAD4 (PDB ID: 4DKT)

Fig. 23.7 Structures, k_{inact}/K_I values, and isozyme selectivity of the PAD4-selective inhibitor streptonigrin (**10**)

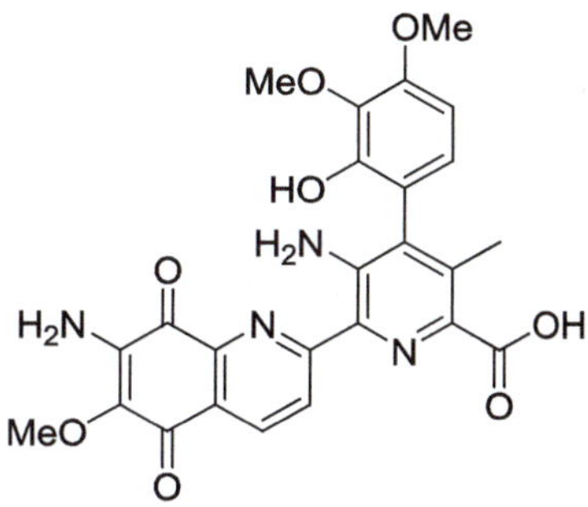

10: Streptonigrin

k_{inact}/K_I (M^{-1} min^{-1})	Isozyme Selectivity
PAD1=3700	1.1
PAD2=12000	3.4
PAD3=3500	1.0
PAD4=440000	126

its structure is significantly different from previous haloacetamidine-based inhibitors, it presents an interesting scaffold for future modifications in an effort to improve selectivity.

23.3.6 Identification of PAD1-Selective Inhibitors

While many of the haloacetamidine-based inhibitors proved to be potent PAD inhibitors in vitro, the intracellular potencies of many of these compounds were relatively poor. One potential reason for this was thought to be the metabolic instability of peptide-based inhibitors. In 2011, the Thompson group attempted to address this issue by utilizing D-ornithine or D-lysine in place of the typically used L-ornithine or L-lysine (Bicker et al. 2012). While the D-ornithine derivatives (D-F-amidine, Fig. 23.8, **11**, and D-Cl-amidine, Fig. 23.8, **12**) did demonstrate improved metabolic stability when administered to mice, they also proved to be fairly selective for PAD1 (8–400-fold selective) (Bicker et al. 2012). Note that Cl-amidine (Fig. 23.3a, **3**) is completely degraded in 2–4 h, whereas D-Cl-amidine (Fig. 23.8, **12**) was still present. With the elucidation of the PAD1 crystal structure, some light can now be shed on to the binding mode of these inhibitors, as well as aid in future inhibitor development.

23.3.7 Improving Efficacy in Cell-Based Assays

As mentioned above, many of these haloacetamidine-based inhibitors, which are predicated on a peptide core, have shown deficiencies in cell-based assays. This is primarily due to metabolic instability and/or cell permeability. In order to overcome the cell permeability issues that these intrinsically hydrophilic inhibitors present, the Wang group developed YW3-56 (Fig. 23.9, **13**), which increased the hydrophobicity of the Cl-amidine scaffold (Wang et al. 2012b). This was accomplished through replacement of the *N*-terminal benzoyl group with a naphthoyl moiety, as well as functionalizing the C-terminal carboxamide to the benzamide moiety. This resulted in improved cellular efficacy (U2OS cells EC_{50} ~ 2.5 μM) while maintaining in vitro efficacy (PAD4 IC_{50} = 1–5 μM). These results indicated that increased hydrophobicity could further enhance intracellular efficacy and potentially translate to improved efficacy in certain animal models.

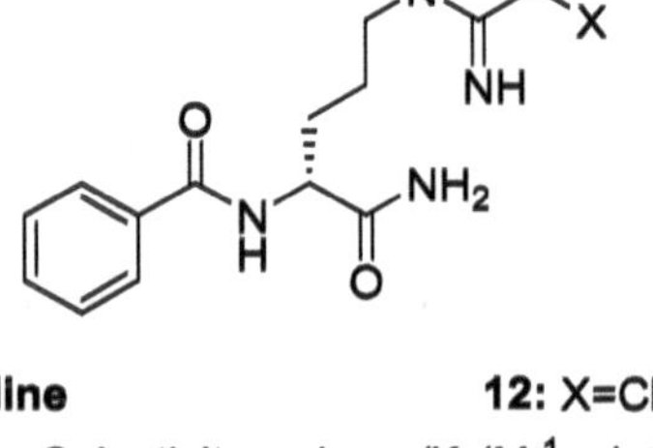

11: X=F; D-F-amidine k_{inact}/K_I (M^{-1} min^{-1})	Isozyme Selectivity	**12: X=Cl; D-Cl-amidine** k_{inact}/K_I (M^{-1} min^{-1})	Isozyme Selectivity
PAD1=1220	407	PAD1=13500	208
PAD2=160	53	PAD2=270	4.2
PAD3=3	1.0	PAD3=65	1.0
PAD4=130	43	PAD4=1400	22

Fig. 23.8 Structures, k_{inact}/K_I values, and isozyme selectivity of the PAD1-selective inhibitors D-F-amidine (**11**) and D-Cl-amidine (**12**)

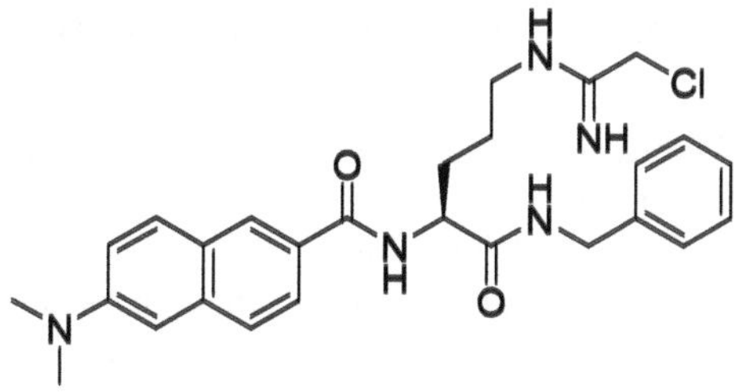

13: YW3-56

IC_{50} (μM)	CLogP
PAD4=1-5	3.40

Fig. 23.9 Structure, PAD4 inhibition, and CLogP value for YW3-56 (**13**)

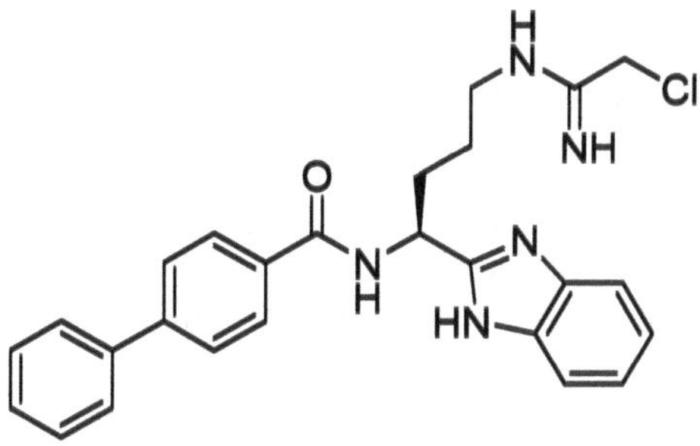

14: BB-Cl-Amidine

k_{inact}/K_I (M^{-1} min^{-1})	CLogP
PAD1=16100	4.17
PAD2=4100	
PAD3=6800	
PAD4=13300	

Fig. 23.10 Structure, PAD inhibition, and CLogP value for BB-Cl-amidine (**14**)

This approach was further optimized through isosteric replacement of certain moieties found in peptide-based inhibitors. This led to the Thompson group replacing the C-terminal carboxamide of Cl-amidine with a benzimidazole moiety yielding the next-generation inhibitor BB-Cl-amidine (Fig. 23.10, **14**) (Knight et al. 2015). They also replaced the *N*-terminal phenyl ring with a biphenyl moiety. Introduction of both these replacements causes a dramatic increase in overall hydrophobicity (Cl-amidine, Fig. 23.3a, **3**, CLogP = −0.23 and BB-Cl-amidine, Fig. 23.10, **14**, CLogP = 4.17) in hopes of improving cell permeability through passive diffusion. While demonstrating similar in vitro efficacy to Cl-amidine (Fig. 23.3a, **3**), BB-Cl-amidine (Fig. 23.10, **14**) proved superior when evaluated in a variety of cell-based assays as well as different animal models (Knight et al. 2015; Ghari et al. 2016; Horibata et al. 2015).

23.3.8 Tetrazole-Based PAD Inhibitors

The isosteric replacement approach was also utilized to replace the C-terminal carboxamide of Cl-amidine with a tetrazole moiety (Fig. 23.11, **15–18**) (Subramanian et al. 2015). While this class was not designed to improve cell permeability, it instead displayed a marked improvement in metabolic stability. This class of compounds was also interesting due to a marked increase in PAD inhibition across all isozymes for unsubstituted tetrazoles (Fig. 23.11, **15** and **16**) (Subramanian et al. 2015). Interestingly, when the tetrazole was substituted with a *tert*-butyl group (Fig. 23.11, **17** and **18**), a dramatic increase in PAD2 inhibition was observed (Subramanian et al. 2015). While these inhibitors did not demonstrate improved efficacy in cell-based assays, they do provide insight into enhanced PAD inhibition, specifically PAD2 inhibition.

23.3.9 Development of Reversible PAD Inhibitors

In 2015, a novel class of reversible PAD4 inhibitors was discovered. A screen of DNA-encoded libraries with PAD4 with and without calcium unveiled GSK121 (Fig. 23.12, **19**) as a potent inhibitor that bound to the apo state (inactive form) of

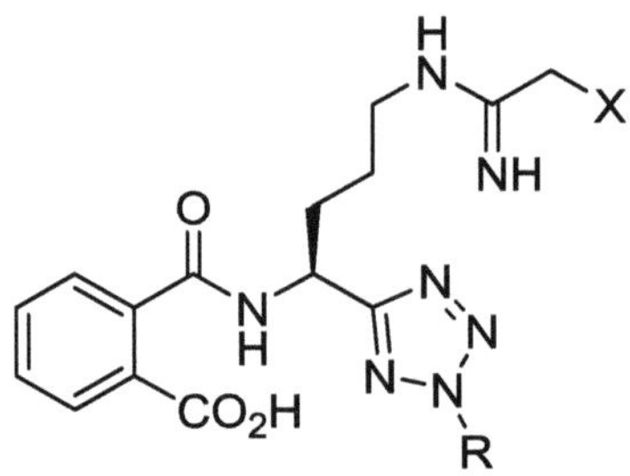

15-18: Tetrazole-Based

15: R=H, X=F
k_{inact}/K_I (M^{-1} min^{-1})
PAD1=81720
PAD2=2550
PAD3=3800
PAD4=46800

16: R=H, X=Cl
k_{inact}/K_I (M^{-1} min^{-1})
PAD1=82110
PAD2=9700
PAD3=12400
PAD4=159330

17: R=^{t}Bu, X=F
k_{inact}/K_I (M^{-1} min^{-1})
PAD1=100120
PAD2=120900
PAD3=15680
PAD4=25950

18: R=^{t}Bu, X=Cl
k_{inact}/K_I (M^{-1} min^{-1})
PAD1=162200
PAD2=60000
PAD3=17500
PAD4=25400

Fig. 23.11 Structure and PAD inhibition of tetrazole-based PAD inhibitors (**15–18**)

PAD4 (Lewis et al. 2015). A detailed SAR led to the optimized compounds GSK199 (Fig. 23.12, **20**) and GSK484 (Fig. 23.12, **21**), which demonstrated potent inhibition of PAD4 when low concentrations of calcium were used (0.2 mM) (Lewis et al. 2015). Notably, when higher levels of calcium were used, potencies for both compounds were reduced by ≥5-fold. These inhibitors were also found to selectively inhibit PAD4 by more than 35-fold as compared to other PAD isozymes. Kinetic analysis and co-crystal structures indicated that these inhibitors bound to the apo form of PAD4, which prevents the calcium-induced movement of Cys645 into the active site, yielding the active holo form of the enzyme.

The co-crystal structure of GSK199 (Fig. 23.12, **20**) bound to PAD4 (Fig. 23.13) illustrates the unique binding of these inhibitors (Lewis et al. 2015). Specifically in the apo state (Fig. 23.13a), a unique binding region exists where D473 and H471 can be accessed. While the aminopiperidine moiety of GSK199 (Fig. 23.12, **20**) overlaps with haloacetamidine binding region, the two classes of inhibitors approach this binding region from different directions. In the presence of calcium (Fig 23.13b), this unique binding region for GSK199 (Fig. 23.12, **20**) is closed off and not accessible. The discovery of these inhibitors has demonstrated the importance of pursuing compounds that bind to the apo state of all PAD isozymes.

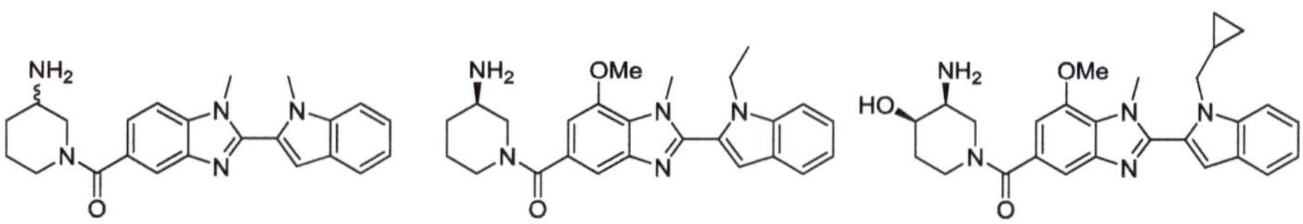

19: GSK121
IC_{50} with 0 mM Ca^{2+}=3.2 μM
IC_{50} with 2 mM Ca^{2+}=ND

20: GSK199
IC_{50} with 0 mM Ca^{2+}=200 nM
IC_{50} with 2 mM Ca^{2+}=1.0 μM

21: GSK484
IC_{50} with 0 mM Ca^{2+}=50 nM
IC_{50} with 2 mM Ca^{2+}=250 nM

Fig. 23.12 Structures and IC_{50} values of reversible PAD inhibitors GSK121 (**19**), GSK199 (**20**), and GSK484 (**21**) in the presence and absence of calcium

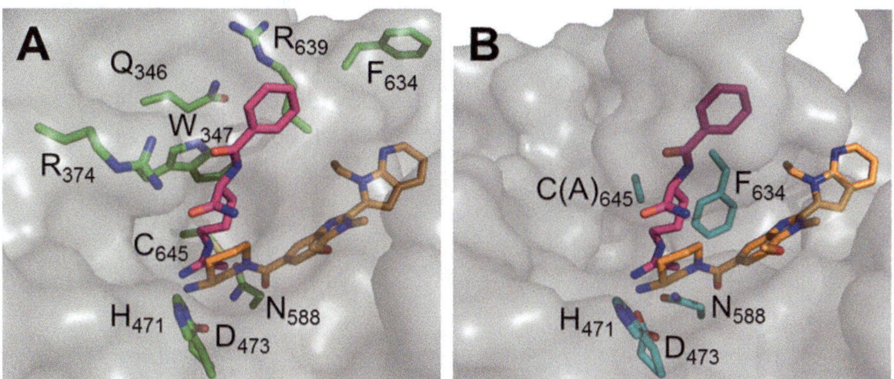

Fig. 23.13 (**a**) F-amidine (**2**, *magenta*) bound to holoPAD4 (PDB ID: 2DW5) and GSK199 (**20**, *orange*) superimposed into holoPAD4. (**b**) GSK199 (**20**, *orange*) bound to apoPAD4 (PDB ID: 4X8G) and F-amidine (**2**, *magenta*) superimposed into apoPAD4

23.3.10 Next-Generation PAD3 Inhibitors

While the first class of PAD3-selective inhibitors was discovered in 2010, the Ellman group identified a novel scaffold for potent and selective PAD3 inhibition in 2015 (Jamali et al. 2015). Through a substrate-based fragment method, a hydantoin-containing, chloroacetamidine-based series of compounds were found to inhibit PAD3 selectively by >10-fold (Fig. 23.14, **22**) (Jamali et al. 2015). Through an extensive SAR, these inhibitors also demonstrated an increase in PAD3 potency as compared to Cl-amidine and F-amidine (Jamali et al. 2015).

Additional work by the Ellman group generated further improvements on PAD3 potency and selectivity in 2016 (Fig. 23.15, **23–25**) (Jamali et al. 2016). These inhibitors explored the introduction of conformational constraints and amide replacement with different isosteres to enhance PAD3 inhibition. These inhibitors also demonstrated efficacy in a PAD3 cell-based assay, illustrating future utility in cell-based systems (Jamali et al. 2016). This work along with the discovery of BB-Cl-amidine (Fig. 23.10, **14**) has demonstrated that haloacetamidine-based inhibitors are viable in cell-based assays and aided in understanding the rules for their cell permeability.

Fig. 23.14 Structure, k_{inact}/K_I values, and isozyme selectivity of a hydantoin-based PAD3-selective inhibitor (**22**)

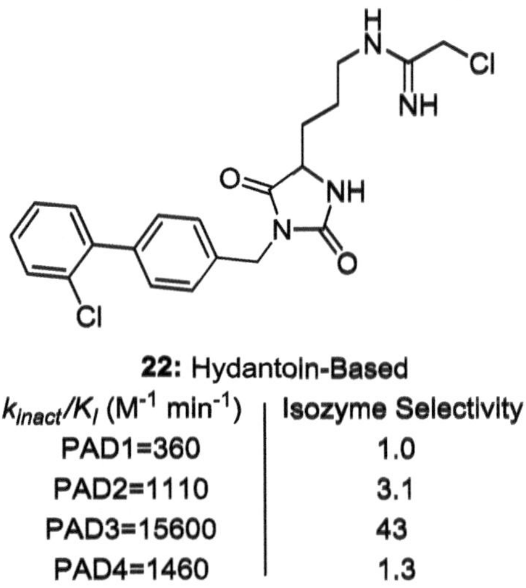

22: Hydantoin-Based

k_{inact}/K_I (M^{-1} min^{-1})	Isozyme Selectivity
PAD1=360	1.0
PAD2=1110	3.1
PAD3=15600	43
PAD4=1460	1.3

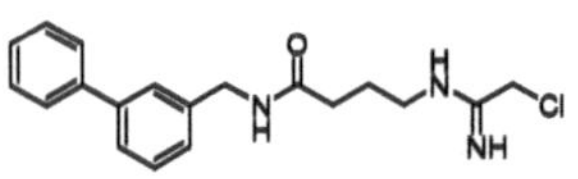

23

IC_{50} (μM)	Isozyme Selectivity
PAD1=80	1.0
PAD2=46	1.7
PAD3=5.8	14
PAD4=39	2.1

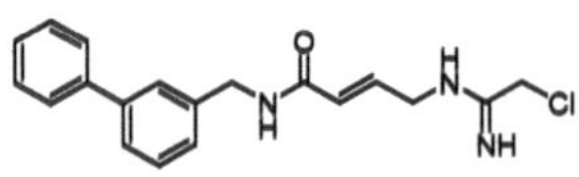

24

IC_{50} (μM)	Isozyme Selectivity
PAD1=125	1.0
PAD2=18	6.9
PAD3=5.6	22
PAD4=53	2.4

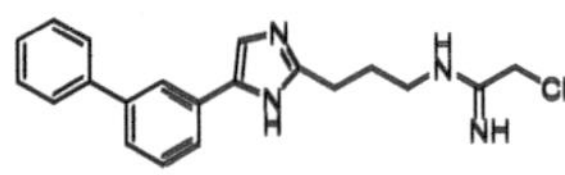

25

IC_{50} (μM)	Isozyme Selectivity
PAD1=34	1.0
PAD2=22	1.5
PAD3=5.4	6.3
PAD4=28	1.2

Fig. 23.15 Structure, IC_{50} values, and isozyme selectivity of PAD3-selective inhibitors (**23–25**)

23.4 Animal Models of PAD-Related Diseases

An integral part of PAD inhibitor and probe development has been the development and evaluation of certain animal models of PAD-related diseases. To date, animal models have been developed for a variety of different PAD-related diseases; however, models of RA, SLE, atherosclerosis, thrombosis, and IBD have been extensively investigated (Lewis and Nacht 2016).

23.4.1 Rheumatoid Arthritis (RA)

One of the most well-investigated deimination-related animal models is that of RA. It is known that PAD2 and PAD4 are released into the joint (i.e., synovial fluid) leading to the development of anti-citrullinated protein antibodies (ACPA) (Van Steendam et al. 2011; Puszczewicz and Iwaszkiewicz 2011; van Boekel et al. 2002; Masson-Bessiere et al. 2001; Burska et al. 2014; Berlyne 1998; Swaminathan and Shah 2011). In 2011, a collagen-induced arthritis (CIA) model was treated with Cl-amidine (Fig 23.3a, **3**) and demonstrated an ability to ameliorate arthritis (Willis et al. 2011). This was key to the development of irreversible PAD inhibitors, such as Cl-amidine (Fig 23.3a, **3**), as decreased levels of citrullinated proteins were observed in both serum and synovial samples. This marked a significant find in the field, as a peptide-based irreversible PAD inhibitor demonstrated efficacy, paving the way for future advancement. Specifically, BB-Cl-amidine (Fig. 23.10, **14**) has demonstrated improved efficacy in the CIA model, potentially due to an improvement in cell permeability and metabolic stability (Ghari et al. 2016). BB-Cl-amidine (Fig. 23.10, **14**) also showed an ability to reduce expression of pro-inflammatory cytokines when coadministered with JQ1 (a bromodomain and extraterminal (BET) domain inhibitor) and was even able to reduce the effects of arthritis after onset (Kawalkowska et al. 2016).

23.4.2 Systemic Lupus Erythematosus (SLE)

The development of an animal model for SLE has also proved to be invaluable for both improving our understanding of PAD biology and also in the generation of PAD inhibitors and probes. SLE is of importance due to its role as an autoimmune disease also defined by the production of autoantibodies, which is also potentially due to abnormal NET formation (Hakkim et al. 2010; Brinkmann et al. 2004). While the exact role that NETs play (if any) is not completely understood (see Chap. 8), in a MRL/lpr lupus mouse model, both Cl-amidine (Fig. 23.3a, **3**, 10 mg/kg) (Knight et al. 2013) and BB-Cl-amidine (Fig. 23.10, **14**, 1 mg/kg) (Knight et al. 2015) treatment led to a reduction in NETosis and disease severity, with BB-Cl-amidine (Fig. 23.10, **14**) proving to the more potent of the two inhibitors. This data suggested that PAD inhibition could reduce symptoms of lupus in mice; however, the exact mechanism and particular isozyme necessary are not completely understood.

23.4.3 Atherosclerosis

The development of animal models for atherosclerosis has garnered much interest in the PAD field due to its role in the mortality of patients with RA or SLE (Salmon and Roman 2008). It has also been identified that PAD4, citrullinated proteins, and NETs are found in atherosclerotic plaques (Sokolove et al. 2013). The role of the PADs in atherosclerosis was then further investigated utilizing the apolipoprotein-E-deficient ($ApoE^{-/-}$) mouse model of atherosclerosis (Meir and Leitersdorf 2004). When treated with Cl-amidine (Fig. 23.3a, **3**, 10 mg/kg), a decrease in NETosis was observed and ultimately a reduction in disease severity (Knight et al. 2014a). However, in contrast to SLE treatment, if atheroma formation had already begun in the $ApoE^{-/-}$ mouse model, the effects of Cl-amidine (Fig. 23.3a, **3**) treatment were drastically reduced (Knight et al. 2014a).

23.4.4 Thrombosis

Thrombosis is the formation of a blood clot inside a blood vessel, and NETs have been identified as playing a role in stimulating this process. This occurs by facilitating the deposition of fibrin, as well as promoting the activation of platelets (resulting in increased blood clots), as well as feeding back into this loop and promoting NET formation (Fuchs et al. 2010). Treatment of New Zealand Mixed (NZM) mice with Cl-amidine (Fig. 23.3a, **3**) then led to a significant delay in thrombosis while inhibiting NET formation, again suggesting a strong link between NETosis and thrombosis (Knight et al. 2013). In addition, $PAD4^{-/-}$ mice lacked extracellular citrullinated proteins and provided protection from thrombosis while also demonstrating an ability to be rescued by the introduction of neutrophils, again demonstrating the link between aberrant PAD activity and thrombosis (Martinod et al. 2013).

23.4.5 Irritable Bowel Disease (IBD)

Similar to the diseases discussed above, PAD4 has been implicated in the onset and progression of IBD as well. Specifically, the presence of NETs has been identified in certain biopsy samples; (Savchenko et al. 2011) however, a study using mouse model lacking the PADs has not been reported, and so studying this link has proven challenging. This problem has been addressed partially through the treatment of the DSS-mediated model of colitis with Cl-amidine (Fig. 23.3a, **3**, as low as 5 mg/kg) which led to a decrease in PAD activity and inflammation leading to a decrease in IBD severity (Chumanevich et al. 2011b).

23.4.6 Other PAD-Related Conditions

As indicated above, these five diseases are not an exhaustive list of animal models developed for PAD-related diseases. While Cl-amidine (Fig. 23.3a, **3**) has been used extensively in these five disease animal models, it has also shown efficacy in animal models of breast cancer (McElwee et al. 2012b), hypoxic ischemia (Lange et al. 2014), diabetic wound healing (Fadini et al. 2016), and spinal cord injury and regeneration (Lange et al. 2011). Other PAD inhibitors such as YW3-56 (Fig. 23.9, **13**) and 2-chloroacetamidine have demonstrated efficacy in animal models of hemorrhagic shock (He et al. 2016) and multiple sclerosis (MS) (Moscarello et al. 2013), respectively. The discovery of new PAD inhibitors and probes will continue to aid in the development of animal models of PAD-related diseases, while the current animal models have proven to be a key component in inhibitor development.

23.5 PAD Probe Development

23.5.1 First-Generation PAD Probes

The development of PAD probes soon followed the discovery of F-amidine (Fig. 23.3a, **2**) and Cl-amidine (Fig. 23.3a, **3**). Specifically, these inhibitors were conjugated with a fluorophore to give rhodamine-conjugated F-amidine (RFA, Fig. 23.16, **26**) and rhodamine-conjugated Cl-amidine (RCA, Fig. 23.16, **27**) (Knuckley et al. 2008). These probes were then found to efficiently label PAD4 and demonstrated their utility in the development of different assays which identified novel PAD2 and PAD4 inhibitors (Jones et al. 2012; Lewallen et al. 2014). These early probes also helped set the stage for building future PAD probes.

23.5.2 "Clickable" PAD Probes

The first-generation PAD probes RFA (Fig. 23.16, **26**) and RCA (Fig. 23.16, **27**) proved to be excellent starting points for developing a series of highly versatile "clickable" PAD probes that could be utilized both in vitro and in cellulo (Fig. 23.17, **28–31**) (Slack et al. 2011). These probes again utilized the F-amidine (Fig. 23.3a, **2**) and Cl-amidine (Fig. 23.3a, **3**) scaffolds and installed either an azide or alkyne functionality at the 4-position of the phenylamide. This allowed these probes to then undergo a "click" reaction with either a fluorophore or biotin. The power of this versatility was observed in their ability to effectively label cellular PADs in cell lysates (Slack et al. 2011). This scaffold demonstrated the ability to more fully investigate the biological roles and impact of the PADs. These probes also highlighted the differences between the chloroacetamidine and fluoroacetamidine

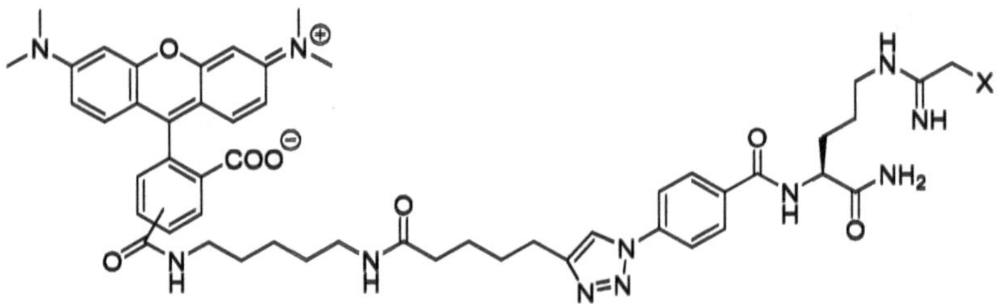

26: X=F; **RFA**
27: X=Cl; **RCA**

Fig. 23.16 Structures of the PAD probes RFA (**26**) and RCA (**27**)

Fig. 23.17 Structures of "clickable" PAD probes F-amidine-azide (**28**), Cl-amidine-azide (**29**), F-amidine-YNE (**30**), and Cl-amidine-YNE (**31**)

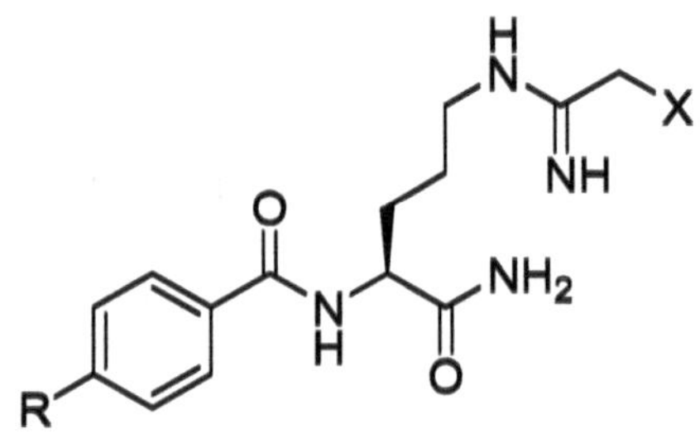

28: R=N_3, X=F; **F-amidine-azide**
29: R=N_3, X=Cl; **Cl-amidine-azide**
30: R=CCH, X=F; **F-amidine-YNE**
31: R=CCH, X=Cl; **Cl-amidine-YNE**

warheads. As discussed earlier, the chloroacetamidine warhead is more potent and produced more efficient labeling at lower concentrations; however, it also proved to be less selective when labeling cell lysates. These two lessons will greatly aid the development of potent and selective PAD probes and inhibitors going forward.

23.6 Conclusions

The development of PAD inhibitors was initially inspired by a known substrate that was modified to generate a class of irreversible inhibitors. These haloacetamidine-based inhibitors have gone through a variety of iterations and will continue to see improved potency and selectivity as the structures of each isozyme are uncovered. The use of HTS capabilities has proved invaluable, not only for the haloacetamidine-based inhibitors but also for identifying novel scaffolds which bind to allosteric sites and both apo and holo states of these enzymes. These discoveries have also ushered in novel probes that have greatly aided in inhibitor development and in understanding the physiological roles of the PADs in cell-based systems. These discoveries have demonstrated that treating diseases related to PAD dysregulation can be achievable through a variety of approaches.

References

Arita, K., Hashimoto, H., Shimizu, T., Nakashima, K., Yamada, M., & Sato, M. (2004). Structural basis for Ca(2+)-induced activation of human PAD4. *Nature Structural & Molecular Biology, 11*, 777–783.

Berlyne, G. M. (1998). Carbamylated proteins and peptides in health and in uremia. *Nephron, 79*, 125–130.

Bicker, K. L., Anguish, L., Chumanevich, A. A., Cameron, M. D., Cui, X., Witalison, E., Subramanian, V., Zhang, X., Chumanevich, A. P., Hofseth, L. J., Coonrod, S. A., & Thompson, P. R. (2012). D-amino acid based protein arginine deiminase inhibitors: Synthesis, pharmacokinetics, and in cellulo efficacy. *ACS Medicinal Chemistry Letters, 3*, 1081–1085.

Bolzan, A. D., & Bianchi, M. S. (2001). Genotoxicity of streptonigrin: A review. *Mutation Research, 488*, 25–37.

Brinkmann, V., Reichard, U., Goosmann, C., Fauler, B., Uhlemann, Y., Weiss, D. S., Weinrauch, Y., & Zychlinsky, A. (2004). Neutrophil extracellular traps kill bacteria. *Science, 303*, 1532–1535.

Burska, A. N., Hunt, L., Boissinot, M., Strollo, R., Ryan, B. J., Vital, E., Nissim, A., Winyard, P. G., Emery, P., & Ponchel, F. (2014). Autoantibodies to posttranslational modifications in rheumatoid arthritis. *Mediators of Inflammation, 2014*, 492873.

Causey, C. P., Jones, J. E., Slack, J. L., Kamei, D., Jones, L. E., Subramanian, V., Knuckley, B., Ebrahimi, P., Chumanevich, A. A., Luo, Y., Hashimoto, H., Sato, M., Hofseth, L. J., & Thompson, P. R. (2011). The development of N-alpha-(2-carboxyl)benzoyl-N(5)-(2-fluoro-1-iminoethyl)-l-ornithine amide (o-F-amidine) and N-alpha-(2-carboxyl)benzoyl-N(5)-(2-chloro-1-iminoethyl)-l-ornithine amide (o-Cl-amidine) as second generation protein arginine deiminase (PAD) inhibitors. *Journal of Medicinal Chemistry, 54*, 6919–6935.

Chumanevich, A. A., Causey, C. P., Knuckley, B. A., Jones, J. E., Poudyal, D., Chumanevich, A. P., Davis, T., Matesic, L. E., Thompson, P. R., & Hofseth, L. J. (2011a). Suppression of colitis in mice by Cl-amidine: A novel peptidylarginine deiminase inhibitor. *American Journal of Physiology. Gastrointestinal and Liver Physiology, 300*, G929–G938.

Chumanevich, A. A., Causey, C. P., Knuckley, B. A., Jones, J. E., Poudyal, D., Chumanevich, A. P., Davis, T., Matesic, L. E., Thompson, P. R., & Hofseth, L. J. (2011b). Suppression of colitis in mice by Cl-amidine: A novel peptidylarginine deiminase (PAD) inhibitor. *American Journal of Physiology. Gastrointestinal and Liver Physiology, 300*, G929–G938.

Dreyton, C. J., Anderson, E. D., Subramanian, V., Boger, D. L., & Thompson, P. R. (2014). Insights into the mechanism of streptonigrin-induced protein arginine deiminase inactivation. *Bioorganic & Medicinal Chemistry, 22*, 1362–1369.

Fadini, G. P., Menegazzo, L., Rigato, M., Scattolini, V., Poncina, N., Bruttocao, A., Ciciliot, S., Mammano, F., Ciubotaru, C. D., Brocco, E., Marescotti, M. C., Cappellari, R., Arrigoni, G., Millioni, R., Vigili de Kreutzenberg, S., Albiero, M., & Avogaro, A. (2016). NETosis delays diabetic wound healing in mice and humans. *Diabetes, 65*, 1061–1071.

Fuchs, T. A., Brill, A., Duerschmied, D., Schatzberg, D., Monestier, M., Myers, D. D., Jr., Wrobleski, S. K., Wakefield, T. W., Hartwig, J. H., & Wagner, D. D. (2010). Extracellular DNA traps promote thrombosis. *Proceedings of the National Academy of Sciences of the United States of America, 107*, 15880–15885.

Fuhrmann, J., & Thompson, P. R. (2016). Protein arginine methylation and citrullination in epigenetic regulation. *ACS Chemical Biology, 11*, 654–668.

Fuhrmann, J., Clancy, K. W., & Thompson, P. R. (2015). Chemical biology of protein arginine modifications in epigenetic regulation. *Chemical Reviews, 115*, 5413–5461.

Ghari, F., Quirke, A. M., Munro, S., Kawalkowska, J., Picaud, S., McGouran, J., Subramanian, V., Muth, A., Williams, R., Kessler, B., Thompson, P. R., Fillipakopoulos, P., Knapp, S., Venables, P. J., & La Thangue, N. B. (2016). Citrullination-acetylation interplay guides E2F-1 activity during the inflammatory response. *Science Advances, 2*, e1501257.

Guertin, M. J., Zhang, X., Anguish, L., Kim, S., Varticovski, L., Lis, J. T., Hager, G. L., & Coonrod, S. A. (2014). Targeted H3R26 deimination specifically facilitates estrogen receptor binding by modifying nucleosome structure. *PLoS Genetics, 10*, e1004613.

Hakkim, A., Furnrohr, B. G., Amann, K., Laube, B., Abed, U. A., Brinkmann, V., Herrmann, M., Voll, R. E., & Zychlinsky, A. (2010). Impairment of neutrophil extracellular trap degradation is associated with lupus nephritis. *Proceedings of the National Academy of Sciences of the United States of America, 107*, 9813–9818.

He, W., Zhou, P., Chang, Z., Liu, B., Liu, X., Wang, Y., Li, Y., & Alam, H. B. (2016). Inhibition of peptidylarginine deiminase attenuates inflammation and improves survival in a rat model of hemorrhagic shock. *The Journal of Surgical Research, 200*, 610–618.

Horibata, S., Vo, T. V., Subramanian, V., Thompson, P. R., & Coonrod, S. A. (2015). Utilization of the soft agar colony formation assay to identify inhibitors of tumorigenicity in breast cancer cells. *Journal of Visualized Experiments*, (99), e52727.

Ishida-Yamamoto, A., Senshu, T., Takahashi, H., Akiyama, K., Nomura, K., & Iizuka, H. (2000). Decreased deiminated keratin K1 in psoriatic hyperproliferative epidermis. *The Journal of Investigative Dermatology, 114*, 701–705.

Jamali, H., Khan, H. A., Stringer, J. R., Chowdhury, S., & Ellman, J. A. (2015). Identification of multiple structurally distinct, nonpeptidic small molecule inhibitors of protein arginine deiminase 3 using a substrate-based fragment method. *Journal of the American Chemical Society, 137*, 3616–3621.

Jamali, H., Khan, H. A., Tjin, C. C., & Ellman, J. A. (2016). Cellular activity of new small molecule protein arginine deiminase 3 (PAD3) inhibitors. *ACS Medicinal Chemistry Letters, 7*, 847–851.

Jones, J. E., Causey, C. P., Knuckley, B., Slack-Noyes, J. L., & Thompson, P. R. (2009). Protein arginine deiminase 4 (PAD4): Current understanding and future therapeutic potential. *Current Opinion in Drug Discovery & Development, 12*, 616–627.

Jones, J. E., Slack, J. L., Fang, P., Zhang, X., Subramanian, V., Causey, C. P., Coonrod, S. A., Guo, M., & Thompson, P. R. (2012). Synthesis and screening of a haloacetamidine containing library to identify PAD4 selective inhibitors. *ACS Chemical Biology, 7*, 160–165.

Kawalkowska, J., Quirke, A. M., Ghari, F., Davis, S., Subramanian, V., Thompson, P. R., Williams, R. O., Fischer, R., La Thangue, N. B., & Venables, P. J. (2016). Abrogation of collagen-induced arthritis by a peptidyl arginine deiminase inhibitor is associated with modulation of T cell-mediated immune responses. *Scientific Reports, 6*, 26430.

Kearney, P. L., Bhatia, M., Jones, N. G., Yuan, L., Glascock, M. C., Catchings, K. L., Yamada, M., & Thompson, P. R. (2005). Kinetic characterization of protein arginine deiminase 4: A transcriptional corepressor implicated in the onset and progression of rheumatoid arthritis. *Biochemistry, 44*, 10570–10582.

Khandpur, R., Carmona-Rivera, C., Vivekanandan-Giri, A., Gizinski, A., Yalavarthi, S., Knight, J. S., Friday, S., Li, S., Patel, R. M., Subramanian, V., Thompson, P., Chen, P., Fox, D. A., Pennathur, S., & Kaplan, M. J. (2013). NETs are a source of citrullinated autoantigens and stimulate inflammatory responses in rheumatoid arthritis. *Science Translational Medicine, 5*, 178ra140.

Knight, J. S., & Kaplan, M. J. (2012). Lupus neutrophils: 'NET' gain in understanding lupus pathogenesis. *Current Opinion in Rheumatology, 24*, 441–450.

Knight, J. S., Zhao, W., Luo, W., Subramanian, V., O'Dell, A. A., Yalavarthi, S., Hodgin, J. B., Eitzman, D. T., Thompson, P. R., & Kaplan, M. J. (2013). Peptidylarginine deiminase inhibition is immunomodulatory and vasculoprotective in murine lupus. *The Journal of Clinical Investigation, 123*, 2981–2993.

Knight, J. S., Luo, W., O'Dell, A. A., Yalavarthi, S., Zhao, W., Subramanian, V., Guo, C., Grenn, R. C., Thompson, P. R., Eitzman, D. T., & Kaplan, M. J. (2014a). Peptidylarginine deiminase inhibition reduces vascular damage and modulates innate immune responses in murine models of atherosclerosis. *Circulation Research, 114*, 947–956.

Knight, J. S., Subramanian, V., O'Dell, A. A., Yalavarthi, S., Zhao, W., Smith, C. K., Hodgin, J. B., Thompson, P. R., & Kaplan, M. J. (2014b). Peptidylarginine deiminase inhibition disrupts NET formation and protects against kidney, skin and vascular disease in lupus-prone MRL/lpr mice. *Annals of the Rheumatic Diseases, 74*, 2199–2206.

Knight, J. S., Luo, W., O'Dell, A. A., Yalavarthi, S., Zhao, W., Subramanian, V., Guo, C., Grenn, R. C., Thompson, P. R., Eitzman, D. T., & Kaplan, M. J. (2014c). Peptidylarginine deiminase inhibition reduces vascular damage and modulates innate immune responses in murine models of atherosclerosis. *Circulation Research, 114*, 947–956.

Knight, J. S., Subramanian, V., O'Dell, A. A., Yalavarthi, S., Zhao, W., Smith, C. K., Hodgin, J. B., Thompson, P. R., & Kaplan, M. J. (2015). Peptidylarginine deiminase inhibition disrupts NET formation and protects against kidney, skin and vascular disease in lupus-prone MRL/lpr mice. *Annals of the Rheumatic Diseases, 74*, 2199–2206.

Knuckley, B., Luo, Y., & Thompson, P. R. (2008). Profiling protein arginine deiminase 4 (PAD4): A novel screen to identify PAD4 inhibitors. *Bioorganic & Medicinal Chemistry, 16*, 739–745.

Knuckley, B., Causey, C. P., Jones, J. E., Bhatia, M., Dreyton, C. J., Osborne, T. C., Takahara, H., & Thompson, P. R. (2010a). Substrate specificity and kinetic studies of PADs 1, 3, and 4 identify potent and selective inhibitors of protein arginine deiminase 3. *Biochemistry, 49*, 4852–4863.

Knuckley, B., Jones, J. E., Bachovchin, D. A., Slack, J., Causey, C. P., Brown, S. J., Rosen, H., Cravatt, B. F., & Thompson, P. R. (2010b). A fluopol-ABPP HTS assay to identify PAD inhibitors. *Chemical Communications, 46*, 7175–7177.

Lamensa, J. W., & Moscarello, M. A. (1993). Deimination of human myelin basic protein by a peptidylarginine deiminase from bovine brain. *Journal of Neurochemistry, 61*, 987–996.

Lange, S., Gogel, S., Leung, K. Y., Vernay, B., Nicholas, A. P., Causey, C. P., Thompson, P. R., Greene, N. D., & Ferretti, P. (2011). Protein deiminases: New players in the developmentally regulated loss of neural regenerative ability. *Developmental Biology, 355*, 205–214.

Lange, S., Rocha-Ferreira, E., Thei, L., Mawjee, P., Bennett, K., Thompson, P. R., Subramanian, V., Nicholas, A. P., Peebles, D., Hristova, M., & Raivich, G. (2014). Peptidylarginine deiminases: Novel drug targets for prevention of neuronal damage following hypoxic ischemic insult (HI) in neonates. *Journal of Neurochemistry, 130*, 555–562.

Lewallen, D. M., Bicker, K. L., Madoux, F., Chase, P., Anguish, L., Coonrod, S., Hodder, P., & Thompson, P. R. (2014). A FluoPol-ABPP PAD2 high-throughput screen identifies the first calcium site inhibitor targeting the PADs. *ACS Chemical Biology, 9*, 913–921.

Lewis, H. D., & Nacht, M. (2016). iPAD or PADi-'tablets' with therapeutic disease potential? *Current Opinion in Chemical Biology, 33*, 169–178.

Lewis, H. D., Liddle, J., Coote, J. E., Atkinson, S. J., Barker, M. D., Bax, B. D., Bicker, K. L., Bingham, R. P., Campbell, M., Chen, Y. H., Chung, C. W., Craggs, P. D., Davis, R. P., Eberhard, D., Joberty, G., Lind, K. E., Locke, K., Maller, C., Martinod, K., Patten, C., Polyakova, O., Rise, C. E., Rudiger, M., Sheppard, R. J., Slade, D. J., Thomas, P., Thorpe, J., Yao, G., Drewes, G., Wagner, D. D., Thompson, P. R., Prinjha, R. K., & Wilson, D. M. (2015). Inhibition of PAD4 activity is sufficient to disrupt mouse and human NET formation. *Nature Chemical Biology, 11*, 189–191.

Li, P., Yao, H., Zhang, Z., Li, M., Luo, Y., Thompson, P. R., Gilmour, D. S., & Wang, Y. (2008). Regulation of p53 target gene expression by peptidylarginine deiminase 4. *Molecular and Cellular Biology, 28*, 4745–4758.

Li, G., Hayward, I. N., Jenkins, B. R., Rothfuss, H. M., Young, C. H., Nevalainen, M. T., Muth, A., Thompson, P. R., Navratil, A. M., & Cherrington, B. D. (2016). Peptidylarginine deiminase 3 (PAD3) is upregulated by prolactin stimulation of CID-9 cells and expressed in the lactating mouse mammary gland. *PloS One, 11*, e0147503.

Liu, Y. L., Chiang, Y. H., Liu, G. Y., & Hung, H. C. (2011). Functional role of dimerization of human peptidylarginine deiminase 4 (PAD4). *PloS One, 6*, e21314.

Luban, S., & Li, Z. G. (2010). Citrullinated peptide and its relevance to rheumatoid arthritis: An update. *International Journal of Rheumatic Diseases, 13*, 284–287.

Luo, Y., Knuckley, B., Lee, Y. H., Stallcup, M. R., & Thompson, P. R. (2006a). A fluoroacetamidine-based inactivator of protein arginine deiminase 4: Design, synthesis, and in vitro and in vivo evaluation. *Journal of the American Chemical Society, 128*, 1092–1093.

Luo, Y., Arita, K., Bhatia, M., Knuckley, B., Lee, Y. H., Stallcup, M. R., Sato, M., & Thompson, P. R. (2006b). Inhibitors and inactivators of protein arginine deiminase 4: Functional and structural characterization. *Biochemistry, 45*, 11727–11736.

Makrygiannakis, D., af Klint, E., Lundberg, I. E., Lofberg, R., Ulfgren, A. K., Klareskog, L., & Catrina, A. I. (2006). Citrullination is an inflammation-dependent process. *Annals of the Rheumatic Diseases, 65*, 1219–1222.

Martinod, K., Demers, M., Fuchs, T. A., Wong, S. L., Brill, A., Gallant, M., Hu, J., Wang, Y., & Wagner, D. D. (2013). Neutrophil histone modification by peptidylarginine deiminase 4 is critical for deep vein thrombosis in mice. *Proceedings of the National Academy of Sciences of the United States of America, 110*, 8674–8679.

Masson-Bessiere, C., Sebbag, M., Girbal-Neuhauser, E., Nogueira, L., Vincent, C., Senshu, T., & Serre, G. (2001). The major synovial targets of the rheumatoid arthritis-specific antifilaggrin autoantibodies are deiminated forms of the alpha- and beta-chains of fibrin. *Journal of Immunology, 166*, 4177–4184.

McElwee, J. L., Mohanan, S., Griffith, O. L., Breuer, H. C., Anguish, L. J., Cherrington, B. D., Palmer, A. M., Howe, L. R., Subramanian, V., Causey, C. P., Thompson, P. R., Gray, J. W., & Coonrod, S. A. (2012a). Identification of PADI2 as a potential breast cancer biomarker and therapeutic target. *BMC Cancer, 12*, 500.

McElwee, J. L., Mohanan, S., Griffith, O. L., Breuer, H. C., Anguish, L. J., Cherrington, B. D., Palmer, A. M., Howe, L. R., Subramanian, V., Causey, C. P., Thompson, P. R., Gray, J. W., & Coonrod, S. A. (2012b). Identification of PADI2 as a potential breast cancer biomarker and therapeutic target. *BMC Cancer, 12*, 500.

Meir, K. S., & Leitersdorf, E. (2004). Atherosclerosis in the apolipoprotein-E-deficient mouse: A decade of progress. *Arteriosclerosis, Thrombosis, and Vascular Biology, 24*, 1006–1014.

Mohanan, S., Cherrington, B. D., Horibata, S., McElwee, J. L., Thompson, P. R., & Coonrod, S. A. (2012). Potential role of peptidylarginine deiminase enzymes and protein citrullination in cancer pathogenesis. *Biochemistry Research International, 2012*, 895343.

Moscarello, M. A., Mastronardi, F. G., & Wood, D. D. (2007). The role of citrullinated proteins suggests a novel mechanism in the pathogenesis of multiple sclerosis. *Neurochemical Research, 32*, 251–256.

Moscarello, M. A., Lei, H., Mastronardi, F. G., Winer, S., Tsui, H., Li, Z., Ackerley, C., Zhang, L., Raijmakers, R., & Wood, D. D. (2013). Inhibition of peptidyl-arginine deiminases reverses protein-hypercitrullination and disease in mouse models of multiple sclerosis. *Disease Models & Mechanisms, 6*, 467–478.

Musse, A. A., Li, Z., Ackerley, C. A., Bienzle, D., Lei, H., Poma, R., Harauz, G., Moscarello, M. A., & Mastronardi, F. G. (2008). Peptidylarginine deiminase 2 (PAD2) overexpression in transgenic mice leads to myelin loss in the central nervous system. *Disease Models & Mechanisms, 1*, 229–240.

Nicholas, A. P. (2011). Dual immunofluorescence study of citrullinated proteins in Parkinson diseased substantia nigra. *Neuroscience Letters, 495*, 26–29.

Nicholas, A. P. (2013). Dual immunofluorescence study of citrullinated proteins in Alzheimer diseased frontal cortex. *Neuroscience Letters, 545*, 107–111.

Puszczewicz, M., & Iwaszkiewicz, C. (2011). Role of anti-citrullinated protein antibodies in diagnosis and prognosis of rheumatoid arthritis. *Archives of Medical Science, 7*, 189–194.

Raijmakers, R., Zendman, A. J., Egberts, W. V., Vossenaar, E. R., Raats, J., Soede-Huijbregts, C., Rutjes, F. P., van Veelen, P. A., Drijfhout, J. W., & Pruijn, G. J. (2007). Methylation of arginine residues interferes with citrullination by peptidylarginine deiminases in vitro. *Journal of Molecular Biology, 367*, 1118–1129.

Saijo, S., Nagai, A., Kinjo, S., Mashimo, R., Akimoto, M., Kizawa, K., Yabe-Wada, T., Shimizu, N., Takahara, H., & Unno, M. (2016). Monomeric form of peptidylarginine deiminase type I revealed by X-ray crystallography and small-angle X-ray scattering. *Journal of Molecular Biology, 428*, 3058–3073.

Salmon, J. E., & Roman, M. J. (2008). Subclinical atherosclerosis in rheumatoid arthritis and systemic lupus erythematosus. *The American Journal of Medicine, 121*, S3–S8.

Savchenko, A. S., Inoue, A., Ohashi, R., Jiang, S., Hasegawa, G., Tanaka, T., Hamakubo, T., Kodama, T., Aoyagi, Y., Ushiki, T., & Naito, M. (2011). Long pentraxin 3 (PTX3) expression and release by neutrophils in vitro and in ulcerative colitis. *Pathology International, 61*, 290–297.

Slack, J. L., Causey, C. P., Luo, Y., & Thompson, P. R. (2011). Development and use of clickable activity based protein profiling agents for protein arginine deiminase 4. *ACS Chemical Biology, 6*, 466–476.

Slade, D. J., Fang, P., Dreyton, C. J., Zhang, Y., Fuhrmann, J., Rempel, D., Bax, B. D., Coonrod, S. A., Lewis, H. D., Guo, M., Gross, M. L., & Thompson, P. R. (2015). Protein arginine deiminase 2 binds calcium in an ordered fashion: Implications for inhibitor design. *ACS Chemical Biology, 10*, 1043–1053.

Smith, C. K., Vivekanandan-Giri, A., Tang, C., Knight, J. S., Mathew, A., Padilla, R. L., Gillespie, B. W., Carmona-Rivera, C., Liu, X., Subramanian, V., Hasni, S., Thompson, P. R., Heinecke, J. W., Saran, R., Pennathur, S., & Kaplan, M. J. (2014). Neutrophil extracellular trap-derived enzymes oxidize high-density lipoprotein: An additional proatherogenic mechanism in systemic lupus erythematosus. *Arthritis & Rhematology, 66*, 2532–2544.

Sokolove, J., Brennan, M. J., Sharpe, O., Lahey, L. J., Kao, A. H., Krishnan, E., Edmundowicz, D., Lepus, C. M., Wasko, M. C., & Robinson, W. H. (2013). Brief report: Citrullination within the atherosclerotic plaque: A potential target for the anti-citrullinated protein antibody response in rheumatoid arthritis. *Arthritis and Rheumatism, 65*, 1719–1724.

Stone, E. M., Schaller, T. H., Bianchi, H., Person, M. D., & Fast, W. (2005). Inactivation of two diverse enzymes in the amidinotransferase superfamily by 2-chloroacetamidine: Dimethylargininase and peptidylarginine deiminase. *Biochemistry, 44*, 13744–13752.

Subramanian, V., Knight, J. S., Parelkar, S., Anguish, L., Coonrod, S. A., Kaplan, M. J., & Thompson, P. R. (2015). Design, synthesis, and biological evaluation of tetrazole analogs of Cl-amidine as protein arginine deiminase inhibitors. *Journal of Medicinal Chemistry, 58*, 1337–1344.

Swaminathan, S., & Shah, S. V. (2011). Novel inflammatory mechanisms of accelerated atherosclerosis in kidney disease. *Kidney International, 80*, 453–463.

Tu, R., Grover, H. M., & Kotra, L. P. (2016). Peptidyl arginine deiminases and neurodegenerative diseases. *Current Medicinal Chemistry, 23*, 104–114.

van Boekel, M. A., Vossenaar, E. R., van den Hoogen, F. H., & van Venrooij, W. J. (2002). Autoantibody systems in rheumatoid arthritis: Specificity, sensitivity and diagnostic value. *Arthritis Research, 4*, 87–93.

Van Steendam, K., Tilleman, K., & Deforce, D. (2011). The relevance of citrullinated vimentin in the production of antibodies against citrullinated proteins and the pathogenesis of rheumatoid arthritis. *Rheumatology (Oxford), 50*, 830–837.

Villanueva, E., Yalavarthi, S., Berthier, C. C., Hodgin, J. B., Khandpur, R., Lin, A. M., Rubin, C. J., Zhao, W., Olsen, S. H., Klinker, M., Shealy, D., Denny, M. F., Plumas, J., Chaperot, L., Kretzler, M., Bruce, A. T., & Kaplan, M. J. (2011). Netting neutrophils induce endothelial damage, infiltrate tissues, and expose immunostimulatory molecules in systemic lupus erythematosus. *Journal of Immunology, 187*, 538–552.

Wang, Y., Li, P., Wang, S., Hu, J., Chen, X. A., Wu, J., Fisher, M., Oshaben, K., Zhao, N., Gu, Y., Wang, D., & Chen, G. (2012a). Anticancer PAD inhibitors regulate the autophagy flux and the mammalian target of rapamycin complex 1 activity. *The Journal of Biological Chemistry, 287*, 25941–25953.

Wang, Y., Li, P., Wang, S., Hu, J., Chen, X. A., Wu, J., Fisher, M., Oshaben, K., Zhao, N., Gu, Y., Wang, D., Chen, G., & Wang, Y. (2012b). Anticancer peptidylarginine deiminase (PAD) inhibitors regulate the autophagy flux and the mammalian target of rapamycin complex 1 activity. *The Journal of Biological Chemistry, 287*, 25941–25953.

Willis, V. (2012). The role of citrullination in the development of mouse and human inflammatory arthritis. In *Molecular biology* (p. 146). Denver, CO: University of Colorado.

Willis, V. C., Gizinski, A. M., Banda, N. K., Causey, C. P., Knuckley, B., Cordova, K. N., Luo, Y., Levitt, B., Glogowska, M., Chandra, P., Kulik, L., Robinson, W. H., Arend, W. P., Thompson, P. R., & Holers, V. M. (2011). N-alpha-benzoyl-N5-(2-Chloro-1-Iminoethyl)-L-ornithine amide, a protein arginine deiminase inhibitor, reduces the severity of murine collagen-induced arthritis. *Journal of Immunology, 186*, 4396–4404.

Witalison, E. E., Cui, X., Causey, C. P., Thompson, P. R., & Hofseth, L. J. (2015a). Molecular targeting of protein arginine deiminases to suppress colitis and prevent colon cancer. *Oncotarget, 6*, 36053–36062.

Witalison, E. E., Cui, X., Hofseth, A. B., Subramanian, V., Causey, C. P., Thompson, P. R., & Hofseth, L. J. (2015b). Inhibiting protein arginine deiminases has antioxidant consequences. *The Journal of Pharmacology and Experimental Therapeutics, 353*, 64–70.

Zhang, X., Bolt, M., Guertin, M. J., Chen, W., Zhang, S., Cherrington, B. D., Slade, D. J., Dreyton, C. J., Subramanian, V., Bicker, K. L., Thompson, P. R., Mancini, M. A., Lis, J. T., & Coonrod, S. A. (2012). Peptidylarginine deiminase 2-catalyzed histone H3 arginine 26 citrullination facilitates estrogen receptor alpha target gene activation. *Proceedings of the National Academy of Sciences of the United States of America, 109*, 13331–13336.

Index

A.P. Nicholas et al. (eds.), *Protein Deimination in Human Health and Disease*,
DOI 10.1007/978-3-319-58244-3

MIX
Papier aus verantwortungsvollen Quellen
Paper from responsible sources
FSC® C105338
FSC
www.fsc.org

If you have any concerns about our products,
you can contact us on
ProductSafety@springernature.com

In case Publisher is established outside the EU,
the EU authorized representative is:
Springer Nature Customer Service Center GmbH
Europaplatz 3, 69115 Heidelberg, Germany

Printed by Libri Plureos GmbH
in Hamburg, Germany